PRINCIPLES OF PHYSIOLOGY

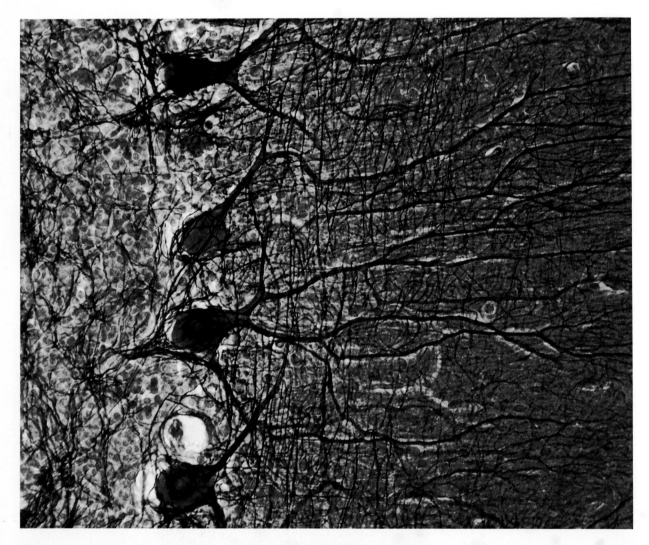

Photomicrograph of human brain-nerve cells, by Manfred Kage.

PRINCIPLES OF
PHYSIOLOGY

Edited by

ROBERT M. BERNE, M.D., D.Sc. (Hon.)

Alumni Professor of Physiology
Department of Physiology
University of Virginia School of Medicine
Charlottesville, Virginia

MATTHEW N. LEVY, M.D.

Chief of Investigative Medicine
Mount Sinai Medical Center;
Professor of Physiology and Biophysics and of
    Biomedical Engineering
Case Western Reserve University
Cleveland, Ohio

*With **620** illustrations*

The C. V. Mosby Company

ST. LOUIS • BALTIMORE • PHILADELPHIA • TORONTO 1990

Editor Stephanie Bircher Manning
Developmental Editor Elaine Steinborn
Assistant Editors Anne Gunter and Jo Salway
Project Manager Patricia Gayle May
Book Design John Rokusek and Candace Conner
Cover Design Candace Conner

Cover Photo © Manfred Kage-Peter Arnold Inc.

Printed in the United States of America

The C. V. Mosby Company
11830 Westline Industrial Drive, St. Louis, Missouri 63146

Library of Congress Cataloging-in-Publication Data

Principles of physiology / edited by Robert M. Berne,
    Matthew N. Levy.   p.   cm.
  Includes bibliographical references.
  ISBN 0-8016-0548-2
  1. Human physiology. I. Berne, Robert M., - . II. Levy,
Matthew N.,
  [DNLM: 1. Physiology.   QT 104 P957]
QP34.5.P744 1990
612—dc20
DNLM/DLC
for Library of Congress   89-14598
                                                      CIP

TSI/VH/VH   9 8 7 6 5 4 3 2 1

Contributors

ROBERT M. BERNE, M.D., D.Sc. (Hon.)
Alumni Professor of Physiology
Department of Physiology
University of Virginia School of Medicine
Charlottesville, Virginia

SAUL M. GENUTH, M.D.
Professor of Medicine
Case Western Reserve University School of Medicine
Mount Sinai Medical Center
Cleveland, Ohio

BRUCE M. KOEPPEN, M.D., Ph.D.
Associate Professor of Medicine and Physiology
Department of Medicine
Division of Nephrology
University of Connecticut Health Center
Farmington, Connecticut

HOWARD C. KUTCHAI, Ph.D.
Professor
Department of Physiology
University of Virginia School of Medicine
Charlottesville, Virginia

MATTHEW N. LEVY, M.D.
Chief of Investigative Medicine
Mount Sinai Medical Center;
Professor of Physiology and Biophysics and of Biomedical
  Engineering
Case Western Reserve University
Cleveland, Ohio

RICHARD A. MURPHY, Ph.D.
Professor
Department of Physiology
University of Virginia School of Medicine
Charlottesville, Virginia

LORING B. ROWELL, Ph.D.
Professor
Department of Physiology and Biophysics
University of Washington School of Medicine
Seattle, Washington

BRUCE A. STANTON, Ph.D.
Associate Professor
Department of Physiology
Dartmouth Medical School
Hanover, New Hampshire

WILLIAM D. WILLIS, JR., M.D., Ph.D.
Professor and Director
The Marine Biomedical Institute
Department of Anatomy and Neurosciences
The University of Texas Medical Branch at Galveston
Galveston, Texas

Preface

Principles of Physiology has been carefully designed to present the important features of mammalian physiology in a clear and concise manner. General principles and underlying mechanisms are emphasized, and relatively nonvital details are minimized. Considerable attention is directed to cell physiology, which serves as the basis for body functions. Not only is the first section of the text devoted to this topic, but it is also provided as foundational information in each succeeding section. We have tried to show that the processes that take place in individual cells are usually applicable to each organ system as a whole.

The major emphasis in *Principles of Physiology* is on *regulation*. The mechanisms that regulate the functions of the individual organ systems are thoroughly described. They are then applied in the complex interactions between the systems as they maintain the constant internal environment so important for optimal function of the constituent cells. All systems are then tied together in the final chapter describing exercise physiology. We hope that this will explain how the body successfully integrates its varied functions to perform a specific task.

Because the intent of this text is to offer, in a clear and concise presentation, *all* information needed to master a complete course in physiology, the use of mathematics has been minimized, and succinct, lucid descriptions substituted wherever feasible. Furthermore, controversial issues in physiology have been purposely omitted to allow ample room for the explanation of important physiological mechanisms.

To contribute to our goal of clarity, many color illustrations are used to portray concepts as simply as possible. When sequential mechanisms are involved, multipaneled diagrams have been designed to illustrate each step clearly. Block diagrams are used to depict the interrelationships among the various factors that may affect a specific function. Finally, figures are included that reiterate some of the concepts that appear in the text and serve to inform the reader about important investigative techniques.

For clarity and simplicity, the sources of statements or assertions in the text are not cited. Brief bibliographies are included at the end of each chapter to direct the student to more detailed information. The references listed in these bibliographies are mainly review articles or recent scientific papers that can be of interest to the reader.

Robert M. Berne
Matthew N. Levy

Contents

CELL PHYSIOLOGY

HOWARD C. KUTCHAI

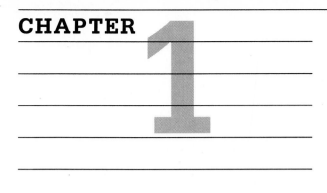

PART I

CHAPTER 1

Cellular Membranes and Transmembrane Transport of Solutes and Water

CELLULAR MEMBRANES

Each cell is surrounded by a plasma membrane that separates it from the extracellular milieu. The plasma membrane serves as a permeability barrier that allows the cell to maintain a cytoplasmic composition far different from the composition of the extracellular fluid. The plasma membrane contains enzymes, receptors, and antigens that play central roles in the interaction of the cell with other cells and with hormones and other regulatory agents in the extracellular fluid.

The membranes that enclose the various organelles divide the cell into discrete compartments and allow the localization of particular biochemical processes in specific organelles. Many vital cellular processes take place in or on the membranes of the organelles. Striking examples are the processes of electron transport and oxidative phosphorylation, which occur on, within, and across the mitochondrial inner membrane.

Most biologic membranes have certain features in common. However, in keeping with the diversity of membrane functions, the composition and structure of the membranes differ from one cell to another and among the membranes of a single cell.

Membrane Structure

Proteins and phospholipids are the most abundant constituents of cellular membranes. A phospholipid molecule has a polar head group and two very nonpolar, hydrophobic fatty acyl chains (Figure 1-1, *A*). In an aqueous environment phospholipids tend to form structures that allow the fatty acyl chains to be kept from contact with water. One such structure is the lipid bilayer (Figure 1-1, *B*). Many phospholipids, when dispersed in water, spontaneously form lipid bilayers. Most of the phospholipid molecules in biologic membranes have a lipid bilayer structure.

Figure 1-2 depicts the *fluid mosaic model* of membrane structure. This model is consistent with many of the properties of biological membranes. Note the bilayer structure of most of the membrane phospholipids. The membrane proteins are of two major classes: (1) *integral* or *intrinsic* membrane proteins that are embedded in the phospholipid bilayer and (2) *peripheral* or *extrinsic* membrane proteins that are associated with the surface of the membrane. The peripheral membrane proteins interact with the membrane predominantly by charge interactions with integral membrane proteins. Thus peripheral proteins may often be removed from the membrane by altering the ionic composition of the medium. **Integral membrane proteins have important hydrophobic interactions with the interior of the membrane.** These hydrophobic interactions can be disrupted only by detergents that make the integral proteins soluble by interacting hydrophobically with nonpolar amino acid side chains.

Cellular membranes are fluid structures in which many of the constituent molecules are free to diffuse in the plane of the membrane. Most lipid and proteins

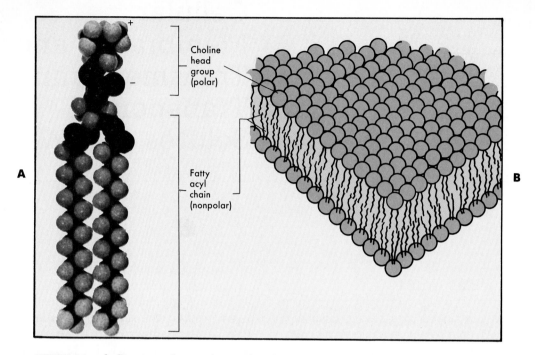

FIGURE 1-1   A, Structure of a membrane phospholipid molecule, in this case a phosphatidyl-choline. **B,** Structure of a phospholipid bilayer. The open circles represent the polar head groups of the phospholipid molecules. The wavy lines represent the fatty acyl chains of the phospholipids.

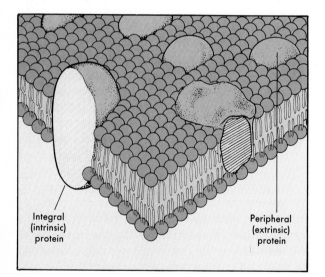

FIGURE 1-2   Schematic representation of the fluid mosaic model of membrane structure. The integral proteins are embedded in the lipid bilayer matrix of the membrane, and the peripheral proteins are associated with the external surfaces of integral membrane proteins.

can move freely in the bilayer plane, but they "flip-flop" from one phospholipid monolayer to the other at much slower rates. A large hydrophilic moiety is unlikely to flip-flop if it must be dragged through the nonpolar interior of the lipid bilayer.

In some cases membrane components are not free to diffuse in the plane of the membrane. Examples of this motional constraint are the sequestration of acetylcholine receptors (integral membrane proteins) at the motor endplate of skeletal muscle and the presence of different membrane proteins in the apical and basolateral plasma membranes of epithelial cells. At present little is known about the ways in which membrane constituents are restrained from lateral diffusion. The cytoskeleton appears to tether certain membrane proteins.

MEMBRANE COMPOSITION

Lipid Composition

Major Phospholipids In animal cell membranes the most abundant phospholipids are often the choline-containing phospholipids: the lecithins (phosphatidylcholines) and the sphingomyelins. Next in abundance are frequently the amino phospholipids: phosphatidylserine and phosphatidylethanolamine. Other important phospholipids that are present in smaller amounts are phosphatidylglycerol, phosphatidylinositol, and cardiolipin.

Cholesterol Cholesterol is a major constituent of plasma membranes, and its steroid nucleus lies parallel to the fatty acyl chains of membrane phospholipids.

Glycolipids Glycolipids are not abundant, but they have important functions. Glycolipids are found mostly in plasma membranes, where their carbohydrate moieties protrude from the external surface of the membrane.

Asymmetry of Lipid Distribution In many membranes the lipid components are not distributed uniformly across the bilayer. The glycolipids of the plasma membrane are located almost exclusively in the outer monolayer. Asymmetry of phospholipids also occurs. In the red blood cell membrane, for example, the outer monolayer contains most of the choline-containing phospholipids, whereas the inner monolayer contains most of the amino phospholipids.

Membrane Proteins

The protein composition of membranes may be simple or complex. The highly specialized membranes of the sarcoplasmic reticulum of skeletal muscle and the disks of the rod outer segment of the retina contain only a few different proteins. Plasma membranes, by contrast, perform many functions and may have more than 100 different protein constituents. Membrane proteins include enzymes, transport proteins, and receptors for hormones and neurotransmitters.

Glycoproteins Some membrane proteins are glycoproteins with covalently bound carbohydrate side chains. As with glycolipids, the carbohydrate chains of glycoproteins are located almost exclusively on the external surfaces of plasma membranes. Cell surface carbohydrate has important functions. **The negative surface charge of cells is ascribable to the negatively charged sialic acid of glycolipids and glycoproteins.**

Asymmetry of Membrane Proteins The Na^+, K^+-ATPase of the plasma membrane and the Ca^{++} pump protein (Ca^{++}-ATPase) of the sarcoplasmic reticulum membrane are examples of the asymmetric functions of membrane proteins. In both cases ATP is split on the cytoplasmic face of the membrane, and some of the energy liberated is used to pump ions in specific directions across the membrane. **In the case of the Na^+, K^+-ATPase, K^+ is pumped into the cell and Na^+ is pumped out, whereas the Ca^{++}-ATPase actively pumps Ca^{++} into the sarcoplasmic reticulum.**

MEMBRANES AS PERMEABILITY BARRIERS

Biological membranes serve as permeability barriers. Most of the molecules present in living systems are highly soluble in water and poorly soluble in nonpolar solvents. Thus such molecules are poorly soluble in the nonpolar environment in the interior of the lipid bilayer of biological membranes. As a consequence, biological membranes pose a formidable barrier to most water-soluble molecules. The plasma membrane is a permeability barrier between the cytoplasm and the extracellular fluid. This barrier allows the maintenance of large concentration differences of many substances between the cytoplasm and the extracellular fluid.

The localization of various cellular processes in certain organelles depends on the barrier properties of cellular membranes. For example, the inner mitochondrial membrane is impermeable to the enzymes and substrates of the tricarboxylic acid cycle, allowing the localization of the tricarboxylic cycle in the mitochondrial matrix. The spatial organization of chemical and physical processes in the cell depends on the barrier functions of cellular membranes.

The passage of important molecules across membranes at controlled rates is central to the life of the cell. Examples are the uptake of nutrient molecules, the discharge of waste products, and the release of secreted molecules. In some cases molecules move from one side of a membrane to another without actually moving through the membrane itself. In other cases molecules cross a particular membrane by passing through or between the molecules that make up the membrane.

TRANSPORT ACROSS, BUT NOT THROUGH, MEMBRANES

Endocytosis Endocytosis is the process that allows material to enter the cell without passing

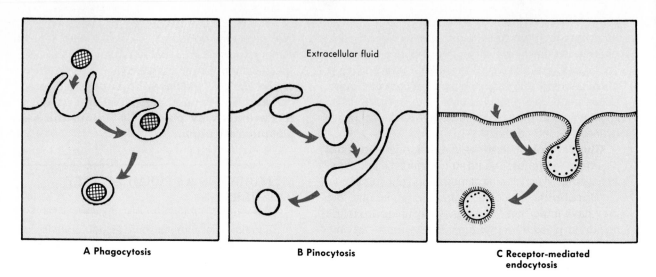

FIGURE 1-3   Schematic depiction of endocytotic processes. **A,** Phagocytosis of a solid particle. **B,** Pinocytosis of extracellular fluid. **C,** Receptor-mediated endocytosis by coated pits.

through the membrane (Figure 1-3); it includes phagocytosis and pinocytosis. The uptake of particulate material is termed *phagocytosis* (Figure 1-3, *A*). The uptake of soluble molecules is called *pinocytosis* (Figure 1-3, *B*). Sometimes special regions of the plasma membrane are involved in endocytosis. In these regions the cytoplasmic surface is covered with bristles made primarily of a protein called *clathrin.* These clathrin-covered regions are called *coated pits,* and their endocytosis gives rise to *coated vesicles* (Figure 1-3, *C*). The coated pits are involved in receptor-mediated endocytosis. Proteins to be taken up are recognized and bound by specific membrane receptor proteins in the coated pits. The binding often leads to aggregation of receptor-ligand complexes, and the aggregation triggers endocytosis in ways that are not yet understood. Endocytosis is an active process that requires metabolic energy. Endocytosis also can occur in regions of the plasma membrane that do not contain coated pits.

Exocytosis Molecules can be ejected from cells by exocytosis, a process that resembles endocytosis in reverse. The release of neurotransmitters, which is considered in more detail in Chapter 4, takes place by exocytosis. Exocytosis is responsible for the release of secretory proteins by many cells; the release of pancreatic proenzymes from the acinar cells of the pancreas is a well-studied example. In such cases the proteins to be secreted are stored in secretory vesicles in the cytoplasm. A stimulus to secrete causes the secretory vesicles to fuse with the plasma membrane and to release the vesicle contents by exocytosis.

Fusion of Membrane Vesicles The contents of one type of organelle can be transferred to another organelle by fusion of the membranes of the organelles. In some cells secretory products are transferred from the endoplasmic reticulum to the Golgi apparatus by fusion of endoplasmic reticulum vesicles with membranous sacs of the Golgi apparatus. Fusion of phagocytic vesicles with lysosomes allows the phagocytosed material to be digested.

TRANSPORT OF MOLECULES THROUGH MEMBRANES

The traffic of molecules through biological membranes is vital for most cellular processes. Some molecules move through biological membranes simply by diffusing among the molecules that make up the membrane, whereas the passage of other molecules involves the mediation of specific transport proteins in the membrane.

Oxygen, for example, is a small molecule that is fairly soluble in nonpolar solvents. It crosses biological membranes by diffusing among membrane lipid molecules. Glucose, on the other hand, is a much larger molecule that is not very soluble in the membrane lipids. Glucose enters cells via a specific glucose transport protein in the plasma membrane.

Diffusion

Diffusion is the process whereby atoms or molecules intermingle because of their random thermal (Brownian) motion. Imagine a container divided into two compartments by a removable partition. A much larger number of molecules of a compound is placed on side A than on side B, and then the partition is removed. Every molecule is in random thermal motion. It is equally probable that a molecule that begins on side A will move to side B in a given time as it is that a molecule beginning on side B will end up on side A. Because many more molecules are present on side A, the total number of molecules moving from side A to side B will be greater than the number moving from side B to side A. In this way the number of molecules on side A will decrease, whereas the number of molecules on side B will increase. This process of net diffusion of molecules will continue until the concentration of molecules on side A equals that on side B. Thereafter the rate of diffusion of molecules from A to B will equal that from B to A, and no further net movement will occur; a dynamic equilibrium exists.

Range of Diffusion Diffusion is a rapid process when the distance over which it takes place is small. This can be appreciated from a relation derived by Einstein. He considered the random movements of molecules that are originally located at $x = 0$. Because a given molecule is equally likely to diffuse in one direction as in the other, the average displacement of all the molecules that begin at $x = 0$ will be zero. The average displacement squared, $(\Delta x)^2$, which is a positive quantity, is represented by

$$\overline{(\Delta x)^2} = 2\ Dt \qquad (1)$$

where t is the time elapsed since the molecules started diffusing and D is a constant of proportionality called the *diffusion coefficient*. The Einstein relation (equation 1) tells us how far the average molecule will diffuse in a given time (t). It is a useful but rough estimate of the time scale of a particular diffusion process.

Einstein's relation reveals that the time required for diffusion increases with the square of the distance over which diffusion occurs. Thus a tenfold increase in the diffusion distance means that the diffusion process will require about 100 times longer to reach a given degree of completion. Table 1-1 shows the results of calculations using Einstein's relation for a typical, small, water-soluble solute. It can be seen

Table 1-1 Time Required for Diffusion To Occur over Various Diffusion Distances*

Diffusion Distance (μm)	Time Required for Diffusion
1	0.5 msec
10	50 msec
100	5 seconds
1000 (1 mm)	8.3 minutes
10,000 (1 cm)	14 hours

*The time required for the "average" molecule (with diffusion coefficient taken to be 1×10^{-5} cm·sec^{-1}) to diffuse the required distance was computed from the Einstein relation (equation 1).

that diffusion is extremely rapid on a microscopic scale of distance. For macroscopic distances diffusion is rather slow. A cell that is 100 μm away from the nearest capillary can receive nutrients from the blood by diffusion with a time lag of only 5 seconds or so. This is sufficiently fast to satisfy the metabolic demands of many cells. However, a nerve axon that is 1 cm long cannot rely on diffusion for the intracellular transport of vital metabolites, since the 14 hours required for diffusion over the 1 cm distance is too long on the time scale of cellular metabolism. Some nerve fibers are longer than 1 m. Therefore it is no wonder that intracellular axonal transport systems are involved in transporting important molecules along nerve fibers. Because of the slowness of diffusion over macroscopic distances, it is not surprising that even small multicellular organisms have evolved circulatory systems to bring the individual cells of the organisms within a reasonable diffusion range of nutrients.

Diffusion Coefficient The diffusion coefficient (D) is proportional to the speed with which the diffusing molecule can move in the surrounding medium. The larger the molecule and the more viscous the medium, the smaller is D.

Einstein obtained the following equation for the diffusion coefficient of a spherical solute molecule that is much larger than the surrounding solvent molecules:

$$D = kT/ (6\pi r\eta) \qquad (2)$$

where

k = Boltzmann's constant
T = Absolute temperature (kT is proportional to the average kinetic energy of a solute molecule)
r = Molecular radius
η = Viscosity of the medium

The equation is called the *Stokes-Einstein equation,* and the molecular radius defined by this equation is known as the Stokes-Einstein radius.

For large molecules equation 2 predicts that D will be inversely proportional to the radius of the diffusing molecule. Because the molecular weight (MW) is approximately proportional to r^3, D should be inversely proportional to $\sqrt[3]{MW}$. Thus a molecule that has ⅛ the mass of another molecule will have a diffusion coefficient only twice as large as the other molecule. For smaller solutes, with a molecular weight less than about 300, D is inversely proportional to $\sqrt{MW}$ rather than $\sqrt[3]{MW}$.

Diffusion Across a Membrane Diffusion leads to a state in which the concentration of the diffusing species is constant in space and time. Diffusion across cellular membranes tends to equalize the concentrations on the two sides of the membrane. The diffusion rate across a membrane is proportional to the area of the membrane and to the difference in concentration of the diffusing substance on the two sides of the membrane. *Fick's first law of diffusion* states that

$$J = -DA \frac{\Delta c}{\Delta x} \qquad (3)$$

where

  J = Net rate of diffusion in moles or grams per unit time
  D = Diffusion coefficient of the diffusing solute in the membrane
  A = Area of the membrane
  Δc = Concentration difference across the membrane
  Δx = Thickness of the membrane

Diffusive Permeability of Cellular Membranes

Permeability to Lipid-Soluble Molecules The plasma membrane serves as a diffusion barrier that enables the cell to maintain cytoplasmic concentrations of many substances that differ greatly from their extracellular concentrations. As early as the turn of the century, the relative impermeability of the plasma membrane to most water-soluble substances was attributed to its "lipoid nature."

The hypothesis that the plasma membrane has a lipoid character is supported by experiments showing that compounds that are soluble in nonpolar solvents (e.g., benzene or olive oil) enter cells more readily than do water-soluble substances of similar molecular weight. Figure 1-4 shows the relationship between membrane permeability and solubility in a nonpolar solvent for a number of different solutes. The ratio of the solubility of the solute in olive oil to its solubility in water is used as a measure of solubility in nonpolar solvents. This ratio is called the *olive oil/water partition coefficient.* The permeability of the plasma membrane to a particular substance increases with the "lipid solubility" of the substance. For compounds with the same olive oil/water partition coefficient, permeability decreases with increasing molecular weight. As described previously, the fluid mosaic model of membrane structure envisions the plasma membrane as a lipid bilayer with proteins embedded in it (Figure 1-2). The data of Figure 1-4 support the idea that the lipid bilayer is the principal barrier to substances that permeate the membrane by simple diffusion.

Permeability to Water-Soluble Molecules Very small, uncharged, water-soluble molecules pass through cell membranes much more rapidly than predicted by their lipid solubility. For example, water permeates cell membranes much more rapidly than is predicted from its molecular radius and its olive oil/water partition coefficient. The reason for the unusually high permeability to water is controversial. Some evidence suggests that very small water-soluble molecules can pass between adjacent phospholipid molecules without actually dissolving in the region occupied by the fatty acid side chains. Other evidence suggests that membrane proteins are responsible for the high membrane permeability to water.

As the size of uncharged, water-soluble molecules increases, their membrane permeativity decreases. Most plasma membranes are essentially impermeable to water-soluble molecules whose molecular weights are greater than about 200.

Because of their charge, ions are relatively insoluble in lipid solvents, and thus membranes are not very permeable to most ions. Ionic diffusion across membranes occurs mainly through protein "channels" that span the membrane. Some channels are highly specific with respect to the ions allowed to pass, whereas others allow all ions below a certain size to pass. Some ion channels are controlled by the voltage difference across the membrane, and others are controlled by neurotransmitters or certain other regulatory molecules.

Although certain water-soluble molecules such as sugars and amino acids are essential for cellular survival, they do not cross plasma membranes appreciably by simple diffusion. **Plasma membranes have specific proteins that allow the transfer of vital**

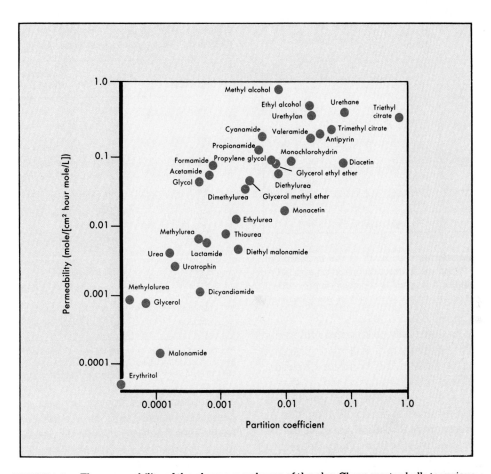

FIGURE 1-4   The permeability of the plasma membrane of the alga *Chara ceratophylla* to various nonelectrolytes as a function of the lipid solubility of the solutes. Lipid solubility is represented on the abscissa by the olive oil/water partition coefficient. (Redrawn from Christensen HN: Biological transport, ed 2, Menlo Park, Calif, 1975, WA Benjamin. Data from Collander, R: Trans Faraday Soc 33:985, 1937.)

metabolites into or out of the cell. The characteristics of membrane protein-mediated transport are discussed later.

Osmosis

Osmosis is defined as the flow of water across a semipermeable membrane from a compartment in which the solute concentration is lower to one in which the solute concentration is greater. A *semipermeable membrane* is defined as a membrane permeable to water but impermeable to solutes. Osmosis takes place because the presence of solute decreases the chemical potential of water. Water tends to flow from where its chemical potential is higher to where its chemical potential is lower. Other effects caused

by the decrease of the chemical potential of water (because of the presence of solute) include reduced vapor pressure, lower freezing point, and higher boiling point of the solution as compared with pure water. Because these properties, and osmotic pressure as well, depend on the **concentration of the solute present** rather than on its chemical properties, they are called *colligative properties*.

Osmotic Pressure In Figure 1-5 a semipermeable membrane separates a solution from pure water. Water flows from side B to side A by osmosis because the presence of solute on side A reduces the chemical potential of water in the solution. Pushing on the piston will increase the chemical potential of the water in the solution of side A and slow down the net rate of osmosis. If the force on the piston is increased grad-

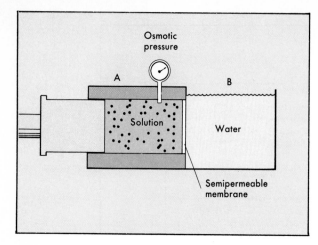

FIGURE 1-5   Schematic representation of the definition of osmotic pressure. When the hydrostatic pressure applied to the solution in chamber A is equal to the osmotic pressure of that solution, there will be no net water flow across the membrane.

Substance	i	Molecular Weight	φ
NaCl	2	58.5	0.93
KCl	2	74.6	0.92
HCl	2	36.6	0.95
NH_4Cl	2	53.5	0.92
$NaHCO_3$	2	84.0	0.96
$NaNO_3$	2	85.0	0.90
KSCN	2	97.2	0.91
KH_2PO_4	2	136.0	0.87
$CaCl_2$	3	111.0	0.86
$MgCl_2$	3	95.2	0.89
Na_2SO_4	3	142.0	0.74
K_2SO_4	3	174.0	0.74
$MgSO_4$	2	120.0	0.58
Glucose	1	180.0	1.01
Sucrose	1	342.0	1.02
Maltose	1	342.0	1.01
Lactose	1	342.0	1.01

Table 1-2   Osmotic Coefficients (φ) of Certain Solutes of Physiological Interest

Reproduced with permission from Lifson N and Visscher MB: Osmosis in living systems. In Glasser O, editor: Medical physics, vol 1. Copyright © 1944 by Year Book Medical Publishers, Inc, Chicago.

ually, a pressure is eventually reached at which net water flow stops. Application of still more pressure will cause water to flow in the opposite direction, from side A to side B. The pressure on side A that is just sufficient to keep pure water from entering is called the *osmotic pressure* of the solution on side A.

The osmotic pressure of a solution depends on the **number of particles in solution.** Thus the degree of ionization of the solute must be taken into account. A 1 M solution of glucose, a 0.5 M solution of NaCl, and 0.333 M solution of $CaCl_2$ theoretically should have the same osmotic pressure. (Actually their osmotic pressures will differ somewhat because of the deviations of real solutions from ideal solution theory.) Important equations that pertain to osmotic pressure and the other colligative properties were derived by van't Hoff. One form of *van't Hoff's law* for calculation of osmotic pressure is

$$\pi = RTic \qquad (4)$$

where

π = Osmotic pressure
R = Ideal gas constant
T = Absolute temperature
 i = Number of ions formed by dissociation of a solute molecule
c = Molar concentration of solute (moles of solute per liter of solution)

This equation applies more exactly as the solution becomes more dilute.

Equation 4 does not predict precisely the osmotic pressures of real solutions. At the concentrations of many substances in cytoplasm and extracellular fluids, the deviations from ideality may be substantial. For example, sodium is the principal cation of the extracellular fluids, and chloride is the main anion. Na^+ is present at about 150 mEq/L and Cl^- at about 120 mEq/L. NaCl solutions in this concentration range differ considerably in their osmotic pressures from the predictions of van't Hoff's law.

One way of correcting for the deviations of real solutions from the predictions of van't Hoff's law is to use a correction factor called the *osmotic coefficient* (φ). Including the osmotic coefficient, equation 4 becomes

$$\pi = RT\phi ic \qquad (5)$$

The osmotic coefficient may be greater or less than one. It is less than one for electrolytes of physiologic importance, and for all solutes it approaches one as the solution becomes more and more dilute. The term φic can be regarded as the osmotically effective concentration, and φic often is referred to as the *osmolar concentration*, with units in osmoles per liter.

Values of the osmotic coefficient depend on the concentration of the solute and on its chemical properties. Table 1-2 lists osmotic coefficients for several solutes. These values apply fairly well at the concentrations of these solutes in the extracellular fluids of mammals. The value of ϕ may vary with concentration. More precise values of ϕ can be obtained from tables in handbooks that list values of ϕ for different substances as functions of concentration. Solutions of proteins deviate greatly from van't Hoff's law, and different proteins may deviate to different extents. The deviations from ideality are frequently greater for proteins than for simpler solutes.

Measurement of Osmotic Pressure The osmotic pressure of a solution can be obtained by determining the pressure required to prevent water from entering the solution across a semipermeable membrane (Figure 1-5).

However, this method is time consuming and technically difficult. More often, therefore, the osmotic pressure is estimated from another colligative property, such as depression of the freezing point. The relation that describes the depression of the freezing point of water by a solute is

$$\Delta T_f = 1.86\ \phi ic \qquad \textbf{(6)}$$

where ΔT_f is the freezing point depression in degrees centigrade. Thus the effective osmotic concentration (in osmoles per liter) is

$$\phi ic = \Delta T_f/1.86 \qquad \textbf{(7)}$$

When the freezing point depression of a multicomponent solution is determined, the effective osmolar concentration (in osmoles per liter) of the solution as a whole can be obtained.

If the total osmotic pressures of two solutions (as measured by freezing point depression or by the osmotic pressure developed across a true semipermeable membrane) are equal, the solutions are said to be *isoosmotic* (or *isosmotic*). If solution A has greater osmotic pressure than solution B, A is said to be *hyperosmotic* with respect to B. If solution A has less total osmotic pressure than solution B, A is said to be *hypoosmotic* to B.

Osmotic Swelling and Shrinking of Cells

The plasma membranes of most of the body's cells are relatively impermeable to many of the solutes of the interstitial fluid but are highly permeable to water. Therefore, when the osmotic pressure of the interstitial fluid is increased, water leaves the cells by osmo-

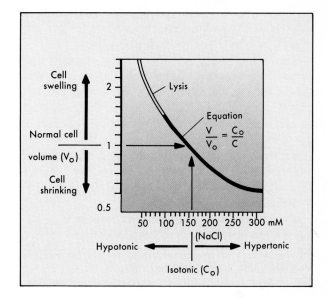

FIGURE 1-6   The osmotic behavior of human red blood cells in NaCl solutions. At 154 mM NaCl (isotonic), the red cell has its normal volume. It shrinks in more concentrated (hypertonic) solutions and swells in more dilute (hypotonic) solutions.

sis and the cells shrink. Thus the cellular solutes become more concentrated until the effective osmotic pressure of the cytoplasm is again equal to that of the interstitial fluid. Conversely, if the osmotic pressure of the extracellular fluid is decreased, water enters the cells. The cells will continue to swell until the intracellular and extracellular osmotic pressures are equal.

Red blood cells are often used to illustrate the osmotic properties of cells because they are readily obtained and are easily studied. Within a certain range of external solute concentrations, the red cell behaves as an *osmometer*, since its volume is inversely related to the solute concentration in the extracellular medium. In Figure 1-6 the red cell volume, as a fraction of its normal volume in plasma, is shown as a function of the concentration of NaCl solution in which the red cells are suspended. At a NaCl concentration of 154 mM (308 mM particles), the volume of the cells is the same as their volume in plasma; this concentration of NaCl is said to be *isotonic* to the red cell. A concentration of NaCl greater than 154 mM is called *hypertonic* (causes cells to shrink), and a solution less concentrated than 154 mM a termed *hypotonic* (cells swell). When red cells have swollen to about 1.4 times their original volume, some cells *lyse* (burst). At this volume the properties of the red cell membrane abruptly change; hemoglobin leaks out of the cell, and the membrane becomes transiently permeable to other large molecules as well.

The intracellular substances that produce an osmotic pressure that just balances the osmotic pressure of the extracellular fluid include hemoglobin, K^+, organic phosphates (e.g., ATP and 2,3-diphosphoglycerate), and glycolytic intermediates. Regardless of the chemical nature of its contents, the red cell behaves as though it were filled with a solution of impermeant molecules with an osmotically effective concentration of 286 milliosmolar ($0.93 \times 2 \times 0.154$ M = 0.286 M).

Osmotic Effects of Permeant Solutes Permeating solutes eventually equilibrate across the plasma membrane. For this reason permeating solutes exert only a transient effect on cell volume.

Consider a red blood cell placed in a large volume of 0.154 M NaCl, containing 0.050 M glycerol. Initially, because of the extracellular NaCl and glycerol, the osmotic pressure of the extracellular fluid will exceed that of the cell interior, and the cell will shrink. With time, however, glycerol will equilibrate across the plasma membrane of the red cell, and the cell will swell back toward its original volume. **The steady-state volume of the cell will be determined only by the impermeant solutes in the extracellular fluid.** In this case the impermeant solutes (NaCl) have a total concentration that is isotonic, so the final volume of the cell will be equal to the normal red cell volume. Because the red cell ultimately returns to its normal volume, the solution (0.050 M glycerol in 0.154 M NaCl) is isotonic. Because the red cell initially shrinks when put in this solution, the solution is hyperosmotic with respect to the normal red cell. The transient changes in cell volume depend on equilibration of glycerol across the membrane. Had we used urea (a more rapidly permeating substance), the cell would have reached steady-state volume sooner.

The following rules help predict the volume changes a cell will undergo when suspended in solutions of permeant and impermeant solutes:

1. The steady-state volume of the cell is determined only by the concentration of impermeant particles in the extracellular fluid.
2. Permeant particles cause only transient changes in cell volume.
3. The time course of the transient changes is more rapid the greater the permeativity of the permeant molecule.

Magnitudes of Osmotic Flows Caused by Permeating Solutes In the preceding example it was explained that permeants, such as glycerol, exert only a transient osmotic effect. It is sometimes important to determine the rate of the osmotic flow caused by a particular permeant.

When a difference of hydrostatic pressure (ΔP) causes water flow across a membrane, the rate of water flow ($\dot{V}_w$) is

$$\dot{V}_w = L\Delta P \qquad (8)$$

where L is a constant of proportionality called the *hydraulic conductivity*.

Osmotic flow of water across a membrane is directly proportional to the osmotic pressure difference ($\Delta\pi$) of the solutions on the two sides of the membrane; thus

$$\dot{V}_w = L\Delta\pi \qquad (9)$$

Equation 9 holds only for osmosis caused by impermeant solutes. Permeant solutes cause less osmotic flow. The greater the permeability of a solute, the less the osmotic flow it causes. Table 1-3 shows the osmotic water flows induced across a porous membrane by solutes of different molecular size. The solutions have identical freezing points, so the total osmotic pressures are the same. The larger the solute molecule, the more impermeable the membrane is to the solute, and the greater the osmotic water flow it causes.

Equation 9 can be rewritten to take solute permeability into account by including σ, the *reflection coefficient*.

$$\dot{V}_w = \sigma L\Delta\pi \qquad (10)$$

σ is a dimensionless number that ranges from 1 for completely impermeant solutes to 0 for extremely permeant solutes. σ is a property of a particular solute and a particular membrane and represents the osmotic flow induced by the solute as a fraction of the theoretical maximum osmotic flow (Table 1-3).

Protein-Mediated Membrane Transport

Certain substances enter or leave cells by way of specific carriers or channels that are intrinsic proteins of the plasma membrane. Transport via such protein carriers or channels is called *protein-mediated transport* or simply *mediated transport*. Specific ions or molecules may cross the membranes of mitochondria,

Table 1-3   Osmotic Water Flow Across a Porous Dialysis Membrane Caused by Various Solutes*

Gradient Producing the Water Flow	Net Volume Flow (μl/minute)*	Solute Radius (Å)	Reflection Coefficient (σ)
D_2O	0.06	1.9	0.0024
Urea	0.6	2.7	0.024
Glucose	5.1	4.4	0.205
Sucrose	9.2	5.3	0.368
Raffinose	11	6.1	0.440
Inulin	19	12	0.760
Bovine serum albumin	25.5	37	1.02
Hydrostatic pressure	25		

Data from Durbin RP: J Gen Physiol 44:315, 1960. Reproduced from The Journal of General Physiology by copyright permission of The Rockefeller University Press.

*Flow is expressed as microliters per minute caused by a 1 M concentration difference of solute across the membrane. The flows are compared with the flow caused by a theoretically equivalent hydrostatic pressure.

endoplasmic reticulum, and other organelles by mediated transport. Mediated transport systems include active transport and facilitated transport processes, which have several properties in common. The principal distinction between these two processes is that *active transport* is capable of "pumping" a substance against a gradient of concentration (or electrochemical potential), whereas *facilitated transport* tends to equilibrate the substance across the membrane.

Properties of Mediated Transport

1. Transport is more rapid than that of other molecules that have a similar molecular weight and lipid solubility and that cross the membrane by simple diffusion.

2. The transport rate shows *saturation kinetics*: as the concentration of the transported compound is increased, the rate of transport at first increases, but eventually a concentration is reached after which the transport rate increases no further. At this point the transport system is said to be *saturated* with the transported compound.

3. The mediating protein has *chemical specificity*: only molecules with the requisite chemical structure are transported. The specificity of most transport systems is not absolute, and in general it is broader than the specificity of most enzymes.

4. Structurally related molecules may compete for transport. Typically the presence of one transport substrate will decrease the transport rate of a second substrate by competing for the transport protein. The competition is analogous to *competitive inhibition* of an enzyme.

5. Transport may be inhibited by compounds that are not structurally related to transport substrates. An inhibitor may bind to the transport protein in a way that decreases its affinity for the normal transport substrate. The compound phloretin does not resemble a sugar molecule, yet it strongly inhibits red cell sugar transport. Active transport systems, which require some link to metabolism, may be inhibited by metabolic inhibitors. The rate of Na^+ transport out of cells by the Na^+, K^+-ATPase is decreased by substances that interfere with ATP generation.

Facilitated Transport   Sometimes called facilitated diffusion, facilitated transport occurs via a transport protein that is not linked to metabolic energy. Facilitated transport has the properties discussed previously, except that it is not generally depressed by metabolic inhibitors. Because facilitated transport processes are not linked to energy metabolism, they cannot move substances against concentration gradients. Facilitated transport systems act to equalize concentrations of the transported substances in the cytoplasm and extracellular fluid.

Monosaccharides enter muscle cells by facilitated transport. Glucose, galactose, arabinose, and 3-O-methylglucose compete for the same carrier. The rate of transport shows saturation kinetics. The nonphysiologic stereoisomer L-glucose enters the cells very slowly, and nontransported sugars, such as mannitol or sorbose, enter muscle cells very slowly, if at all. Phloretin inhibits sugar uptake, and insulin stimulates it. In the absence of insulin, glucose transport is rate limiting for glucose use, so that insulin is a major

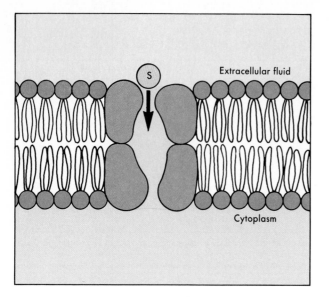

FIGURE 1-7   Hypothetical model of a transport protein. This is a model of the mechanism of monosaccharide transport by the sugar transport protein of the human red blood cell. The protein is postulated to be a tetramer. Binding of sugars is proposed to cause conformational changes that allow sugar molecules to enter and leave the central cavity of the transport protein.

physiological regulator of muscle glucose metabolism.

The molecular details of protein-mediated transport processes are not well understood. Current evidence suggests that most transport proteins span the membrane and are multimeric. Figure 1-7 depicts a hypothetic model that has been proposed for the monosaccharide transport protein of the membrane of the human red blood cell. Conformational changes of the protein, induced by monosaccharide binding, may allow a sugar molecule to enter and leave the central cavity.

Active Transport   These processes have most of the properties of facilitated transport. In addition, active transport systems can concentrate their substrates against concentration or electrochemical potential gradients. This requires energy, so active transport processes must be linked to energy metabolism in some way. Active transport systems may use ATP directly, or they may be linked more indirectly to metabolism. Because of their dependence on metabolism, active transport processes may be inhibited by any substance that interferes with energy metabolism.

Mediated transport is analogous in many respects to an enzyme-catalyzed chemical reaction. When the transport process is in the steady state, the unidirectional flux from cytoplasm to extracellular fluid is exactly equal to the unidirectional flux from extracellular fluid to cytoplasm. Because influx equals efflux, no net flux occurs across the plasma membrane, and the cellular concentration is constant in the steady state. For a facilitated transport process across the plasma membrane, influx and efflux are frequently symmetric processes.

In creating a concentration difference of a transported substrate, an active transport system accomplishes work. The ultimate source of the energy to do transport work is metabolic energy. How is metabolic energy harnessed to do transport work? One way involves the cyclic phosphorylation and dephosphorylation of the transport protein, which causes the protein to alternate between two conformational states. In one state the substrate-binding site of the transport protein faces the extracellular fluid and binds the substrate with a particular affinity. In the alternate state the binding site faces the cytosol and binds the substrate with a different affinity. The expenditure of ATP drives the transport cycle of the protein. Because active transport is directly powered by ATP, this is an example of a *primary active transport* process.

Examples of active transport processes include the following:

Transport powered by the phosphorylation of a transport protein   In the cytoplasm of most animal cells the concentration of Na^+ is much less and the concentration of K^+ is much greater than their extracellular concentrations. These concentration gradients are brought about by the action of a Na^+, K^+ pump in the plasma membrane; Na^+ is pumped out of the cell, and K^+ is pumped into the cell. The Na^+, K^+ pump activity is the result of an integral membrane protein called the *Na^+, K^+-ATPase*. The Na^+, K^+-ATPase transports three sodium ions out of the cell and transports two potassium ions into the cell for each ATP hydrolyzed. The cyclic phosphorylation and dephosphorylation of the protein causes it to alternate between two conformations, E1 and E2. In the E1 conformation the ion-binding sites of the protein have a high affinity for Na^+ and a low affinity for K^+, and the binding sites face the cytoplasm. In the E2 conformation the ion-binding sites face the extracellular fluid, and their affinities favor the binding of K^+ and the dissociation of Na^+. In this way the Na^+, K^+-ATPase alternates between the E1 and the

E2 conformations and transports K^+ into the cell and Na^+ out of the cell by a process resembling *molecular peristalsis*.

Because the Na^+, K^+-ATPase uses the energy in the terminal phosphate bond of ATP to power its transport cycle, it is said to be a *primary active transport system*. A transport process powered by some other high-energy metabolic intermediate or linked directly to a primary metabolic reaction would also be classified as primary active transport.

Transport powered by the gradient of another species: secondary active transport The last section emphasized that energy is required to create a concentration gradient of a transported substance. Once created, a concentration gradient represents a store of chemical potential energy that can be used to perform work. When the substance flows down its gradient, it releases energy that can be harnessed to do work. In many cell types the concentration gradient of Na^+ created by the Na^+, K^+-ATPase, is used to actively transport certain other solutes into the cell. Many cells take up the neutral, hydrophilic amino acids by a membrane transport protein that links the inward transport of Na^+ down its electrochemical potential gradient to the inward transport of amino acids against their gradients of concentration (Figure 1-8). The energy for the transport of the amino acid is not provided directly by ATP or some other high-energy metabolite, but indirectly, from the gradient of Na^+ that is itself actively transported. Thus the amino acid is said to be transported by *secondary active transport*. In the secondary active transport of amino acids, both the rate of amino acid transport and the extent to which the amino acid is accumulated depend on the electrochemical potential gradient of Na^+.

Other Membrane Transport Processes

Calcium Transport Under most circumstances the concentration of Ca^{++} in the cytosol of cells is maintained at low levels, below 10^{-7} M, whereas the concentration of Ca^{++} in extracellular fluids is of the order of 10^{-3} M. The plasma membranes of most cells contain a Ca^{++}-ATPase that helps to maintain the large gradient of Ca^{++} across the plasma membrane. The plasma membrane Ca^{++}-ATPase is different from the Ca^{++}-ATPase that is responsible for sequestering Ca^{++} in the sarcoplasmic reticulum of muscle (see Chapter 12). However, the plasma membrane Ca^{++}-ATPase shares several important properties

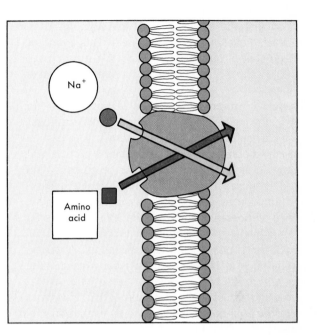

FIGURE 1-8   Many cells take up certain neutral amino acids by secondary active transport. The transport protein binds both Na^+ and the amino acid. Na^+ is transported down its electrochemical potential gradient, and the protein uses the energy released by Na^+ flux to transport the amino acid against a concentration gradient.

with the Ca^{++}-ATPase of the sarcoplasmic reticulum and with the Na^+, K^+-ATPase of plasma membranes. These proteins carry out the primary active transport of ions across membranes, and they use the energy of the terminal phosphate bond of ATP to accomplish this task. The transport cycle of these proteins is driven by the cyclic phosphorylation and dephosphorylation of an aspartate residue of the transport protein. The proteins alternate between two distinct conformational states: one in which binding sites for ions face the cytosol, and the other in which the binding sites have different affinity and face the other side of the membrane.

Many excitable cells, such as those of the heart, have an additional mechanism for controlling the level of intracellular Ca^{++}. A *sodium/calcium exchange protein* in the plasma membrane uses the energy in the Na^+ gradient to extrude Ca^{++} from the cell. In heart cells the rapid, transient changes in intracellular Ca^{++} appear to be mediated by the sodium/calcium exchange protein, whereas the resting level of intracellular Ca^{++} is set mainly by the Ca^{++}-ATPase (see Chapters 13 and 17).

In most cells the rate at which Ca^{++} leaks into the

cell down its electrochemical potential gradient is slow, so that the energy cost of maintaining a low intracellular level of Ca^{++} is not great. This contrasts with the cost of pumping Na^+ and K^+; running the Na^+, K^+ pump is a major item in the energy budget of many cells.

Sugar Transport Glucose is a primary fuel for most of the cells of the body, but glucose diffuses across plasma membranes rather slowly. The plasma membranes of many cell types contain a protein that mediates the facilitated transport of glucose and related monosaccharides. Red blood cells, hepatocytes, adipocytes, and muscle cells (skeletal, cardiac, smooth) all possess facilitated transport systems for uptake of glucose. The uptake of glucose in these cell types depends neither on the electrochemical potential difference of Na^+ across the plasma membrane nor in any direct way on cellular metabolism. In adipocytes and muscle cells, transport of glucose across the plasma membrane is increased by insulin. Insulin increases glucose transport by causing more glucose transport proteins to be inserted into the plasma membrane. The source of the newly inserted protein is a preformed pool of transporters in the membranes of the endoplasmic reticulum within the cell.

Amino Acid Transport Most of the cells in the body synthesize proteins and therefore require amino acids. Several different amino acid transport systems are present in plasma membranes. The amino acid transport systems include three distinct systems: for neutral, for basic, and for acidic amino acids. Amino acid transport systems overlap significantly in specificities, and the distribution of the different systems varies from one cell type to another. Some of these transport systems are secondary active transport processes powered by the concentration gradient of Na^+. Others are facilitated transport systems.

Transport of Nucleosides and Nucleic Acid Bases Nucleotides enter most cells very slowly. Nucleosides and nucleic acid bases enter mammalian cells by facilitated transport. A single facilitated transport system, with broad specificity, may be responsible for the transport of nucleosides. The transport system has higher affinities for purine

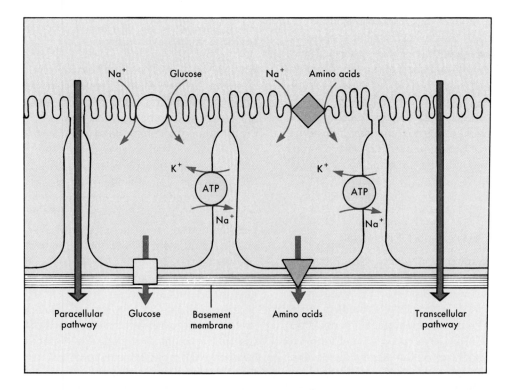

FIGURE 1-9   Epithelial transport processes that occur in the small intestine and renal tubules. Certain epithelia are polarized such that the transport processes on one side of the cell differ from those on the other side.

nucleosides than for pyrimidine nucleosides; cytosine nucleosides have especially low affinity for the transport system.

The transport of the bases of nucleic acids is less well characterized. The number of distinct transport systems for nucleobases is a matter of controversy. Purines are transported much more rapidly than pyrimidines. Cytosine is the most poorly transported base; its transport may not be mediated by a membrane protein at all.

Transport Across Epithelia Frequently epithelial cells are polarized with respect to their transport properties; that is, the transport properties of the plasma membrane facing one side of the epithelial cell layer are different from those of the membrane facing the other side.

The epithelial cells of the small intestine and the proximal tubule of the kidney are good examples of this polarity. The complement of membrane transport proteins in the brush border that faces the lumen of the small bowel or the renal tubule differs from the transport protein composition of the basolateral plasma membrane of the cell. The tight junctions that join the epithelial cells side to side prevent mixing of the transport proteins of the luminal and basolateral plasma membranes. The brush border plasma membranes of these epithelia contain very few Na^+, K^+-ATPase molecules, which reside mainly in the basolateral plasma membrane. Glucose (and galactose) and neutral amino acids enter these epithelial cells at the brush border by secondary active transport systems driven by the Na^+ gradient. However, these substances leave the cells at the basolateral membrane by facilitated transport systems (Figure 1-9).

The tight junctions that join the cells are leaky to water and small water-soluble molecules and ions. There are thus two types of pathways for transport across the epithelia: (1) *transcellular pathways*, through the cells, and (2) *paracellular pathways*, in between the cells (Figure 1-9). The nature and significance of transport across the epithelia of the intestine and the renal tubule are discussed in Chapters 31 and 32, respectively.

BIBLIOGRAPHY

Journal Articles

Carafoli E and Scarpa A, editors: Transport ATPases. In Annals of the New York Academy of Sciences, vol 402, New York, 1982, New York Academy of Sciences.

Books and Monographs

Andreoli TE et al, editors: Physiology of membrane disorders, ed 2, New York, 1986, Plenum Press.

Christensen HN: Biological transport, ed 2, Menlo Park, Calif, 1975, WA Benjamin, Inc.

Davson H: a textbook of general physiology, ed 4, Baltimore, 1970, Williams & Wilkins.

Finean JB and Michell RH, editors: Membrane structure, New York, 1981, Elsevier/North-Holland Biomedical Press.

Finean JB et al: Membranes and their cellular functions, ed 2, New York, 1978, John Wiley & Sons, Inc.

Kotyk A and Janacek K: Cell membrane transport: principles and techniques, ed 2, New York, 1975, Plenum Publishing Corp.

Martonosi AN, editor. Membranes and transport, vol 2, New York, 1982, Plenum Press.

CHAPTER 2

Ionic Equilibria and Resting Membrane Potentials

Most animal cells have an electrical potential difference (voltage) across their plasma membranes. The cytoplasm is usually electrically negative relative to the extracellular fluid. The electrical potential difference across the plasma membrane in resting cells is called a *resting membrane potential*. The resting membrane potential plays a central role in the excitability of nerve and muscle cells and in certain other cellular responses.

The major purpose of this chapter is to discuss the ways that concentration gradients of certain ions across the plasma membrane generate the resting membrane potential. The first part of the chapter deals with some fundamental definitions and concepts that describe the flow of ions across membranes.

IONIC EQUILIBRIA

Electrochemical Potentials of Ions

A membrane separates aqueous solutions in two chambers (A and B). The ion X^+ is at a higher concentration on side A than on side B (Figure 2-1). If no electrical potential difference exists between side A and side B, X^+ will tend to diffuse from side A to side B, just as if it were an uncharged molecule. If, however, side A is electrically negative with respect to side B, the situation is more complex. The tendency of X^+ to diffuse from side A to side B because of the concentration difference remains, but now X^+ also tends to move in the opposite direction (from B to A) because of the electrical potential difference across

the membrane. The direction of net X^+ movement depends on whether the effect of the concentration difference or the effect of the electrical potential difference is larger. By comparing the two tendencies—concentration and electrical—one can predict the direction of net X^+ movement.

The quantity that allows us to compare the relative contributions of ionic concentration and electrical potential is called the *electrochemical potential* (μ) of an ion. The electrochemical potential difference of X^+ across the membrane is defined as

$$\Delta\mu(X) = \mu_A(X) - \mu_B(X) = RT\ln \frac{[X]_A}{[X]_B} + zF(E_A - E_B) \quad \textbf{(1)}$$

where

$$\begin{aligned}
\Delta\mu &= \text{Electrochemical potential difference of} \\
&\quad \text{the ion between sides A and B of the} \\
&\quad \text{membrane} \\
R &= \text{Gas constant} \\
T &= \text{Absolute temperature} \\
\ln\frac{[X]_A}{[X]_B} &= \text{Natural logarithm of concentration ratio of } X^+ \text{ on the two sides of the membrane} \\
z &= \text{Charge number of the ion (} + 2 \text{ for} \\
&\quad Ca^{++}, -1 \text{ for } Cl^-, \text{ etc.)} \\
F &= \text{Faraday's number} \\
E_A - E_B &= \text{Electrical potential difference across the} \\
&\quad \text{membrane}
\end{aligned}$$

The first term [$RT\ln ([X]_A/[X]_B)$] on the right-hand side of equation 1 is the tendency for X^+ ions to move from A to B because of the concentration difference, and the second term [$zF(E_A - E_B)$] is the tendency for the ions to move from A to B because of the electrical potential difference. The first term represents the

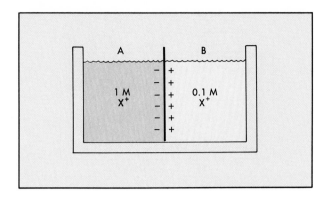

FIGURE 2-1 X^+ is present at 1 M in chamber A and at 0.1 M in chamber B. A concentration force for X^+ tends to cause X^+ to flow from A to B. However, chamber A is electrically negative with respect to chamber B, so an electrical force tends to cause X^+ to flow from B to A.

potential energy difference between a mole of X^+ ions on side A and a mole of X^+ ions on side B as a result of the concentration difference. The second term represents the potential energy difference between a mole of X^+ ions on side A and a mole of X^+ ions on side B caused by the electrical potential difference between A and B. Thus $\Delta\mu$ describes the difference in potential energy between a mole of X^+ ions on side A and a mole of X^+ ions on side B resulting from both concentration and electrical potential differences; hence the name *electrochemical potential*. The unit of electrochemical potential, and of both terms on the right hand side of equation 1, is energy/mole.

The X^+ ions will tend to move from higher to lower electrochemical potential. We defined $\Delta\mu$ as the electrochemical potential of the ion on side A minus that on side B. If $\Delta\mu$ is positive, the ions will tend to move from A to B; if $\Delta\mu$ is zero, there is no net tendency for the ions to move at all; and if $\Delta\mu$ is negative, the ions will tend to move from side B to side A.

If μ_A is greater than μ_B, ions will tend to flow spontaneously from side A to side B. To cause ions to flow from B to A, work must be done. Specifically, $\mu_A - \mu_B$ is the minimal amount of work that must be done to cause 1 mole of ions to flow from B to A. When ions flow from A to B, on the other hand, energy is dissipated. In theory this energy can be harnessed to perform work. The maximal amount of work that can be done by 1 mole of ions flowing from A to B is $\mu_A - \mu_B$. **An electrochemical potential difference of an ion across a membrane thus represents potential energy that can be used to perform work.**

Electrochemical Equilibrium and the Nernst Equation

In equation 1, $\Delta\mu$ may be thought of as the net force on the ion, whereas $RT\ln([X]_A/[X]_B)$ is the force caused by the concentration difference, and $zF(E_A - E_B)$ is the force caused by the electrical potential difference. When the two forces are equal and opposite, $\Delta\mu = 0$, and there is no net force on the ion. When there is no net force on the ion, no net movement of the ion will occur, and the ion is said to be in *electrochemical equilibrium* across the membrane. At equilibrium $\Delta\mu = 0$. From equation 1, therefore, at equilibrium:

$$RT\ln\frac{[X]_A}{[X]_B} + zF(E_A - E_B) = 0 \qquad (2)$$

Solving for $E_A - E_B$, we obtain

$$E_A - E_B = \frac{-RT}{zF}\ln\frac{[X]_A}{[X]_B} = \frac{RT}{zF}\ln\frac{[X]_B}{[X]_A} \qquad (3)$$

Equation 3 is called the *Nernst equation*. The condition of equilibrium was assumed in its derivation, and the Nernst equation is valid only for ions at equilibrium. It allows one to compute the electrical potential difference, $E_A - E_B$, required to produce an electrical force, $zF(E_A - E_B)$, that is equal and opposite to the concentration force, $RT\ln([X]_A/[X]_B)$.

Use of Nernst Equation It is often convenient to convert the Nernst equation to a form involving logarithm to the base 10 (log) rather than natural logarithms ($\ln X = 2.303 \log X$). Because biological potentials are usually expressed in millivolts (mV), the units of R may be selected so RT comes out in millivolts. At 29.2° C the quantity $2.303\, RT/F$ is equal to 60 mV. Because this quantity is proportional to the absolute temperature, it changes by approximately $1/273$ for each centigrade degree. Thus the value of 60 mV for $2.303\, RT/F$ holds approximately for most experimental conditions, and a useful form of the Nernst equation is

$$E_A - E_B = \frac{-60\text{ mV}}{z}\log\frac{[X]_A}{[X]_B} = \frac{60\text{ mV}}{z}\log\frac{[X]_B}{[X]_A} \qquad (4)$$

Examples of Uses of Nernst Equation Example 1. In Figure 2-2, K^+ is 10 times more concentrated in chamber A than in chamber B. Following is a calculation of the electrical potential difference that must exist between the chambers for K^+ to be in equilibrium across the membrane.

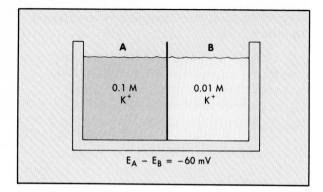

FIGURE 2-2   A membrane separates chambers containing different K^+ concentrations. At an electrical potential difference ($E_A - E_B$) of -60 mV, K^+ is in electrochemical equilibrium across the membrane.

Because we have specified that K^+ should be in equilibrium, the Nernst equation will hold.

$$E_A - E_B = \frac{-60 \text{ mV}}{+1} \log \frac{[K^+]_A}{[K^+]_B}$$
$$= -60 \text{ mV} \log \frac{0.1}{0.01} \qquad \textbf{(5)}$$
$$= -60 \text{ mV} \log(10) = -60 \text{ mV}$$

The Nernst equation tells us that at equilibrium, side A must be 60 mV negative relative to side B. We can see that this polarity is correct, since K^+ will tend to move from B to A due to the electrical force, which will counteract the tendency for it to move from A to B because of the concentration difference.

This example shows that an electrical potential difference of about 60 mV is required to balance a tenfold concentration difference of a univalent ion. This is a useful rule of thumb.

Example 2. In Figure 2-3 the Nernst equation can help decide whether HCO_3^- is in equilibrium. If HCO_3^- is not in equilibrium, the Nernst equation allows us to predict the direction of net flow of HCO_3^-.

The Nernst equation tells us the electrical potential difference, $E_A - E_B$, that will just balance the concentration difference of HCO_3^- across the membrane.

$$E_A - E_B = \frac{-60 \text{ mV}}{z} \log \frac{[HCO_3^-]_A}{[HCO_3^-]_B}$$
$$= \frac{-60 \text{ mV}}{-1} \log \frac{1}{0.1} \qquad \textbf{(6)}$$
$$= 60 \text{ mV} \log(10) = 60 \text{ mV}$$

Thus a potential difference of $+60$ mV between A and B would just balance the tendency of HCO_3^- to move from A to B because of its concentration differ-

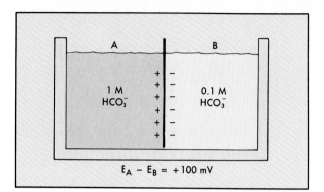

FIGURE 2-3   A membrane separates chambers that contain different HCO_3^- concentrations. $E_A - E_B = +100$ mV. HCO_3^- is not in electrochemical equilibrium. If $E_A - E_B$ were $+60$ mV, HCO_3^- would be in equilibrium. $E_A - E_B$ is stronger than it needs to be to just balance the tendency for HCO_3^- to move from A to B because of its concentration difference. Thus net movement of HCO_3^- from B to A will occur.

ence. However, $E_A - E_B$ is actually $+100$ mV. Therefore the electrical force is in the right direction to balance the concentration force, but it is 40 mV larger than it needs to be to just balance the concentration force. Because the electrical force on HCO_3^- is larger than the concentration force, it will determine the direction of net HCO_3^- movement. Net HCO_3^- flow will occur from B to A.

In brief, the Nernst equation can be used to predict the direction that ions will tend to flow:

1. If the potential difference measured across a membrane is equal to the potential difference calculated from the Nernst equation for a particular ion, then that ion is in electrochemical equilibrium across the membrane, and no net flow of that ion will occur across the membrane.

2. If the measured electrical potential is of the same sign as that calculated from the Nernst equation for a particular ion but is larger in magnitude than the calculated value, then the electrical force is larger than the concentration force, and net movement of that particular ion will tend to occur in the direction determined by the electrical force.

3. When the electrical potential difference is of the same sign but is numerically less than that calculated from the Nernst equation for a particular ion, then the concentration force is larger than the electrical force, and net movement of that ion tends to occur in the direction determined by the concentration difference.

4. If the electrical potential difference measured

across the membrane is of the sign opposite to that predicted by the Nernst equation for a particular ion, then the electrical and concentration forces are in the same direction. Thus that ion cannot be in equilibrium, and it will tend to flow in the direction determined by both electrical and concentration forces.

In the following discussion on electrical activity of nerve and muscle cells, the principles of ionic equilibrium are used to explain the ionic mechanisms of the resting membrane potential and the action potential.

Gibbs-Donnan Equilibrium

Cytoplasm typically contains proteins, organic polyphosphates, and other ionized substances that

Initial concentrations

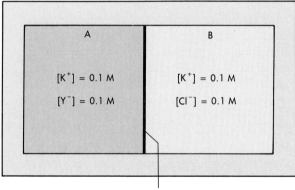

Membrane permeable to H_2O, K^+, and Cl^- but impermeable to Y^-

Equilibrium concentrations

A
[Y^-] = 0.1 M
[K^+] = 0.133...M
[Cl^-] = 0.033...M

B
[K^+] = [Cl^-] = 0.0666...M

FIGURE 2-4 *Top,* Before a Gibbs-Donnan equilibrium is established, a membrane separates two aqueous compartments. The membrane is permeable to water, K^+, and Cl^- but impermeable to Y^-. *Bottom,* Ion concentrations after Gibbs-Donnan equilibrium has been attained.

cannot permeate the plasma membrane. Cytoplasm also contains Na^+, K^+, Cl^-, and other ions to which the plasma membrane is somewhat permeable. The steady-state properties of this mixture of permeant and impermeant ions are described by the *Gibbs-Donnan equilibrium.*

Consider a membrane separating a solution of KCl from a solution of KY, where Y^- is an anion to which the plasma membrane is completely impermeable (Figure 2-4, *top*). The membrane is permeable to water, K^+, and Cl^-. Suppose that initially chamber A contains a 0.1 M solution of KY and that chamber B contains an equal volume of 0.1 M KCl. Because [Cl^-]$_B$ exceeds [Cl^-]$_A$, there will be a net flow of Cl^- from chamber B to chamber A. Negatively charged Cl^- ions flowing from side B to side A will create an electrical potential difference (side A negative) that will cause K^+ also to flow from side B to side A. Essentially the same number of K^+ ions as Cl^- ions will flow from side B to side A to preserve electroneutrality. The *principle of electroneutrality* states that any macroscopic region of a solution must contain an equal number of positive and negative charges. In reality slight separation of charges may occur, but in chemical terms the imbalance between positive and negative charges is very small and impossible to measure by chemical techniques.

If enough time passes, those components of the system that can permeate the membrane—K^+ and Cl^- in this example—will come to equilibrium. At equilibrium both $\Delta\mu_{K+}$ and $\Delta\mu_{Cl-}$ must equal zero. From equation 1

$$\Delta\mu_{K^+} = RT\ln\frac{[K^+]_A}{[K^+]_B} + F(E_A - E_B) = 0$$

$$\Delta\mu_{Cl^-} = RT\ln\frac{[Cl^-]_A}{[Cl^-]_B} - F(E_A - E_B) = 0 \qquad (7)$$

Adding these two equations and dividing the result by RT gives

$$\ln\frac{[K^+]_A}{[K^+]_B} + \ln\frac{[Cl^-]_A}{[Cl^-]_B} = 0 \qquad (8)$$

This gives

$$\ln\frac{[K^+]_A}{[K^+]_B} = -\ln\frac{[Cl^-]_A}{[Cl^-]_B} = \ln\frac{[Cl^-]_B}{[Cl^-]_A} \qquad (9)$$

Thus

$$[K^+]_A/[K^+]_B = [Cl^-]_B[Cl^-]_A \qquad (10)$$

Cross multiplying gives

$$[K^+]_A[Cl^-]_A = [K^+]_B[Cl^-]_B \qquad (11)$$

Equation 11 is called the *Donnan relationship* (or the Gibbs-Donnan equation) and holds for any pair of univalent cation and anion in equilibrium between the two chambers. If other ions that could attain an equilibrium distribution were present, the same reasoning and an equation similar to equation 11 would apply to them as well.

Example of Gibbs-Donnan Equilibrium The Donnan relationship and the principle of electroneutrality makes it possible to determine the equilibrium concentrations of the components in the problem posed at the beginning of this section. The initial ion concentrations are shown in Figure 2-4, *top*. If y represents the change in $[Cl^-]$ when Cl^- moves from chamber B to chamber A, the equilibrium value of $[Cl^-]_B$ can be denoted as $0.1 - y$. By the electroneutrality principle $[K^+]_B = [Cl^-]_B = 0.1 - y$. If the volumes of chambers A and B are the same, at equilibrium, $[Cl^-]_A = y$, and $[K^+]_A = 0.1 + y$. Substituting these concentrations into the Donnan relationship:

$$[K^+]_A\,[Cl^-]_A = [K^+]_B\,[Cl^-]_B$$
$$(0.1 + y)\,(y) = (0.1 - y)\,(0.1 - y) \qquad (12)$$

Solving this equation gives $y = 0.0333$, so that at equilibrium we obtain the concentrations shown in Figure 2-4, *bottom*.

In this Gibbs-Donnan equilibrium both K^+ and Cl^- (but not Y^-) are in electrochemical equilibrium. This means that both K^+ and Cl^- must satisfy the Nernst equation, so that the equilibrium transmembrane potential difference can be computed from the Nernst equation for either K^+ or Cl^-.

$$\begin{aligned}
E_A - E_B &= \frac{-60\ mV}{+1} \log \frac{[K^+]_A}{[K^+]_B} \\
&= -60\ mV \log \frac{0.133}{0.0666} \qquad (13) \\
&= \frac{-60\ mV}{-1} \log \frac{[Cl^-]_A}{[Cl^-]_B} \\
&= 60\ mV \log \frac{0.0333}{0.0666} \\
&= -60\ mV \log 2 = -60\ mV\ (0.3) = -18\ mV
\end{aligned}$$

Note that only the permeant ions attain equilibrium. The impermeant ion, Y^-, cannot reach an equilibrium distribution. It may not be evident that water also will not achieve equilibrium, unless provision is made for that to occur. The sum of the concentrations

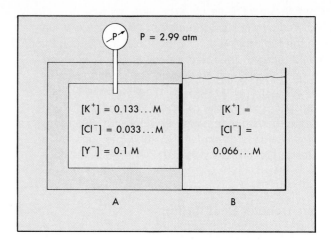

FIGURE 2-5   A hydrostatic pressure of 2.99 atm is required to prevent water from flowing from chamber B to chamber A in the Gibbs-Donnan equilibrium in Figure 2-4.

of K^+ and Cl^- ions on side A in the preceding example exceeds that on side B. This is a general property of Gibbs-Donnan equilibria. Taking the impermeant Y^- into account as well, the total concentration of osmotically active ions is considerably greater on side A than on side B. Water will tend to flow by osmosis from side B to side A until the total osmotic pressure of the two solutions is equal. However, then ions will flow to set up a new Gibbs-Donnan equilibrium, and that requires there be more osmotically active ions on the side with Y^-. All the water from side B will end up on side A unless water is restrained from moving.

This can be done by enclosing the solution on side A in a rigid container (Figure 2-5). Then, as fluid flows from side B to side A, pressure will build up in chamber A and that pressure will oppose further osmotic water flow. The pressure in chamber A at equilibrium will be equal to the difference between the total osmotic pressures of the solutions in chambers A and B. In this example the approximate hydrostatic pressure (P) in chamber A at equilibrium (at 0° C) is

$$\begin{aligned}
P &= \Delta\pi_{K+} + \Delta\pi_{Cl-} + \pi_{Y-} \\
&= RT(\Delta[K^+] + \Delta[Cl^-] + [Y^-]) \\
&= RT(0.06667 - 0.03333 + 0.1) \qquad (14) \\
&= (22.4\ atm)(0.13333) = 2.99\ atm
\end{aligned}$$

Regulation of Cell Volume

K^+ and Cl^- are nearly in equilibrium across many plasma membranes, and their distribution is influenced by the predominantly negatively charged impermeant ions, such as proteins and nucleotides, in the cytoplasm. K^+ and Cl^- approximately satisfy the Donnan relationship. This being the case, why does the osmotic imbalance discussed previously not cause the cells to swell and finally burst? One reason is that cells actively pump Na^+ out of the cytoplasm to the extracellular fluid, decreasing the osmotic pressure of the cytoplasm and increasing that of the extracellular fluid. Much of the pumping of Na^+ is done by the Na^+ pump, the Na^+, K^+-ATPase, in the plasma membrane. The Na^+, K^+-ATPase splits an ATP and uses some of the energy released to extrude 3 Na^+ from the cytoplasm and to pump 2 K^+ into the cell. Whereas K^+ is only slightly removed from an equilibrium distribution, Na^+ is pumped out against a large electrochemical potential difference.

When the ATP production of the cell is compromised (in the presence of metabolic inhibitors or low O_2 levels), or when the Na^+, K^+-ATPase is specifically inhibited, the cells swell.

RESTING MEMBRANE POTENTIALS

Communication between nerve cells depends on an electrical disturbance that is propagated in the plasma membrane and is called an *action potential.* In striated muscle an action potential propagates rapidly over the entire cell surface, allowing the cell to contract synchronously. The action potential and the ionic mechanisms that account for its properties are discussed in Chapter 3. All cells that can produce action potentials have sizable resting membrane potentials across their plasma membranes. Most inexcitable cells also have a resting membrane potential.

The resting membrane potential of many cells can be measured with glass microelectrodes that have tip diameters of about 0.1 μm and that can puncture the plasma membrane of some cells without greatly injuring the cell. The electrical potential difference between the tip of a microelectrode inside a skeletal muscle cell and a reference electrode in the extracellular fluid is about -90 mV. **The resting membrane potential is necessary for the cell to fire an action potential.**

Ions that are actively transported are not in electrochemical equilibrium across the plasma membrane. It is shown later that the flow of ions across the plasma membrane, down their electrochemical potential gradients, is directly responsible for generating most of the resting membrane potential. To understand how the electrochemical potential gradient of an ion can give rise to a transmembrane difference in electrical potential, let us first consider a model system known as a concentration cell.

Concentration Cell

In Figure 2-6 the membrane that separates chambers A and B is permeable to cations but not to anions. Initially no electrical potential difference exists across the membrane. K^+ will flow from A to B because of the concentration force acting on it. Cl^- has the same force on it, but it cannot flow because the membrane is impermeable to anions. The flow of K^+ from A to B will transfer net positive charge to side B and leave a very slight excess of negative charges behind on side A. Side A will thus become electrically negative to side B (Figure 2-6). This electrical force is oppositely directed to the concentration force on K^+

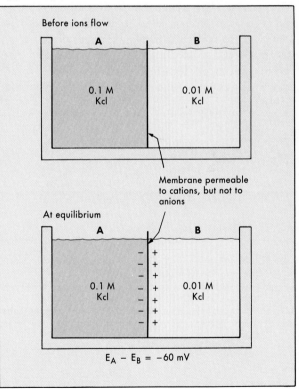

FIGURE 2-6   *Top,* A concentration cell. The membrane, which is permeable to cations but not to anions, separates KCl solutions of different concentrations. *Bottom,* The concentration cell after electrochemical equilibrium has been established. The flow of an infinitesimal amount of K^+ generated an electrical potential difference across the membrane that is equal to the equilibrium potential for K^+.

The more K^+ that flows, the larger is the opposing electrical force. Net K^+ flow will stop when the electrical force just balances the concentration force; that is, when the electrical potential difference is equal to the equilibrium (Nernst) potential for K^+. That is,

$$E_A - E_B = \frac{-60 \text{ mV}}{+1} \log \frac{0.1}{0.01} \quad \textbf{(15)}$$
$$= -60 \text{ mV} \log (10) = -60 \text{ mV}$$

Only a very small amount of K^+ flows from A to B before equilibrium is reached. This is because the separation of positive and negative charges requires a large amount of work. The potential difference that builds up to oppose further K^+ movement is a manifestation of that work.

The K^+ concentration difference in this example acts much like a battery. The natural tendency for any ion that can flow is to seek equilibrium; thus K^+ tends to flow until its equilibrium potential difference is established. As explained later, when more than one type of ion can permeate a membrane, each ion "strives" to make the transmembrane potential difference equal to its equilibrium potential. The more permeant the ion, the greater is its ability to force the electrical potential difference toward its equilibrium potential.

Distribution of Ions Across Plasma Membranes

In most tissues a number of ions are not in equilibrium between the extracellular fluid and the cytoplasm. Table 2-1 gives the concentrations of Na^+, K^+, and Cl^- in the extracellular fluid and in the cytoplasmic water of frog skeletal muscle and squid giant axon. Intracellular ion concentrations for mammalian muscle are similar to those for frog muscle.

Chloride is nearly in equilibrium across the plasma membranes of both frog muscle and squid axon. This is known because chloride's equilibrium potential, as calculated from the Nernst equation, is about equal to the measured transmembrane potential difference. In both tissues K^+ has a concentration force that tends to make it flow out of the cell. The electrical force on K^+ is oppositely directed to the concentration force. If the $E_{in} - E_{out}$ in frog muscle were -105 mV, electrical and concentration forces on K^+ would exactly balance. Because $E_{in} - E_{out}$ is only -90 mV, the concentration force is greater than the electrical force. Therefore K^+ has a net tendency to flow out of the cell. In frog muscle and in squid axon both the concentration and the electrical forces on Na^+ tend to cause it to flow into the cell. Na^+ is the ion farthest from an equilibrium distribution. The larger the difference between the measured membrane potential and the equilibrium potential for an ion, the larger is the net force tending to make that ion flow.

Active Ion Pumping and Resting Potential

The Na^+, K^+-ATPase, located in the plasma membrane, uses the energy of the terminal phosphate ester bond of ATP to extrude Na^+ actively from the cell and to take K^+ actively into the cell. The Na^+, K^+

Table 2-1 Distribution of Na^+, K^+ and Cl^- Across the Plasma Membranes of Frog Muscle and Squid Axon

	Extracellular Fluid (mM)	Cytoplasm (mM)	Approximate Equilibrium Potential (mV)	Actual Resting Potential (mV)
FROG MUSCLE				
[Na$^+$]	120	9.2	+67	
[K$^+$]	2.5	140	−105	
[Cl$^-$]	120	3 to 4	−89 to −96	−90
SQUID AXON				
[Na$^+$]	460	50	+58	
[K$^+$]	10	400	−96	
[Cl$^-$]	540	About 40	About −68	−70

Data from Katz B: Nerve, muscle, and synapse, New York, 1966, McGraw-Hill Book Co. Copyright © 1966 by McGraw-Hill Book Co. Used with permission of McGraw-Hill Book Co.

pump is responsible for the high intracellular K^+ concentration and the low intracellular Na^+ concentration. Because the pump moves a larger number of Na^+ ions out than K^+ ions in (3 Na^+ to 2 K^+), it causes a net transfer of positive charge out of the cell and thus contributes to the resting membrane potential. Because it brings about net movement of charge across the membrane, the pump is termed *electrogenic*.

The size of the pump's electrogenic contribution to the resting potential can be estimated by completely inhibiting the pump with a cardiac glycoside, such as ouabain. Such studies show that in some cells the electrogenic Na^+, K^+ pump is responsible for a large fraction of the resting potential. In most vertebrate nerve and skeletal muscle cells, however, the direct contribution of the pump to the resting potential is usually small—less than 5 mV. The resting membrane potential in nerve and skeletal muscle results mainly from the diffusion of ions down their electrochemical potential gradients. The ionic gradients are maintained by active ion pumping. In other types of excitable cells, electrogenic pumping of ions may contribute more to the resting membrane potential. In certain smooth muscle cells, for example, the electrogenic effect of the Na^+, K^+ pump may be responsible for 20 mV or more of the resting membrane potential.

Generation of Resting Membrane Potential by Ion Gradients

The earlier section on concentration cells shows how an ion gradient can act as a battery. When a number of ions are distributed across a membrane, all being removed from electrochemical equilibrium, each ion will tend to force the transmembrane potential toward its own equilibrium potential, as calculated from the Nernst equation. The more permeable the membrane to a particular ion, the greater strength that ion will have in forcing the membrane potential toward its equilibrium potential. In frog muscle (Table 2-1) the Na^+ concentration difference can be regarded as a battery that tries to make $E_{in} - E_{out}$ equal to +67 mV. The K^+ concentration difference resembles a battery that attempts to make $E_{in} - E_{out}$ equal to −105 mV. The Cl^- concentration difference resembles a battery trying to make $E_{in} - E_{out}$ equal to −90 mV.

Chord Conductance Equation The way in which the interplay of ion gradients creates the resting membrane potential (Em) is also illustrated by a simple mathematic model. If we consider the distribution of K^+, Na^+, and Cl^- across the plasma membrane of a cell, then the following equation predicts the transmembrane potential difference across the membrane:

$$E_m = (g_K/g_t)E_K + (g_{Na}/g_t)E_{Na} + (g_{Cl}/g_t)E_{Cl} \qquad \textbf{(16)}$$

where

$$g_t = g_K + g_{Na} + g_{Cl}$$

and the g represents the conductances of the membrane to the ion indicated by the subscript. Conductance is the reciprocal of resistance (g = 1/R). The more permeable the membrane to a particular ion, the greater is the conductance of the membrane to that ion.

Equation 16 is called the *chord conductance equation*. It states that the membrane potential is a weighted average of the equilibrium potentials of all the ions to which the membrane is permeable, in this case K^+, Na^+, and Cl^-. The weighting factor for each ion is the fraction of the total ionic conductance of the membrane (the sum of the individual ionic conductances) that results from the conductance of the ion in question. Note that the sum of the weighting factors for the ions must equal 1, so that if one weighting factor grows larger, the others must become smaller. The chord conductance equation shows that the greater the conductance of the membrane to a particular ion, the greater the ability of that ion to bring the membrane potential toward the equilibrium potential of that ion.

For the frog muscle fiber discussed earlier, $E_{in} - E_{out} = -90$ mV. The membrane potential is much closer to E_K (−105 mV) than to E_{Na} (+67 mV). The chord conductance equation predicts that in resting muscle g_K is about 10 times larger than g_{Na}. This has been confirmed by ion flux measurements with radioactive tracers. In resting squid axon ($E_m = -70$ mV) the chord conductance equation predicts that g_K is about five times larger than g_{Na}. In other types of excitable cells the relationship between g_K and g_{Na} is somewhat different. Other ions also may play a role in generating the resting membrane potential. Resting membrane potentials vary from about −7 mV or so in human erythrocytes to −30 mV in some types of smooth muscle and up to −90 mV or more in vertebrate skeletal muscle and cardiac ventricular cells.

Roles of Na^+, K^+-ATPase in Establishing Resting Membrane Potential: Direct Versus Indirect Effects

The Na^+, K^+ pump establishes gradients of Na^+ and K^+ across the plasma membranes of the cells. Since the amount of Na^+ pumped out is larger than the amount of K^+ pumped in, the pump transfers net charge across the membrane and contributes *directly* to the resting membrane potential. In vertebrate skeletal and cardiac muscle and in nerve this electrogenic activity of the pump is directly responsible for only a small fraction of the resting membrane potential. The major portion of the resting membrane potential in these tissues is a result of the diffusion of Na^+ and K^+ down their electrochemical potential gradients, with each ion trying to bring the transmembrane potential toward its own equilibrium potential. This contribution to the resting membrane potential is **indirectly** caused by the Na^+, K^+-ATPase. The relative magnitudes of the direct and indirect contributions of the Na^+, K^+-ATPase to the resting membrane potential vary from one cell type to another.

BIBLIOGRAPHY
Books and Monographs

Aidley DJ: The physiology of excitable cells, ed 2, Cambridge, 1978, Cambridge University Press.

Davson H: A textbook of general physiology, ed 4, Baltimore, 1970, Williams & Wilkins.

Junge D: Nerve and muscle excitation, ed 2, Sunderland, Mass, 1981, Sinauer Associates, Inc.

Kandel ER and Schwartz JH: Principles of neural science, ed 2, New York, 1985, Elsevier Science Publishing Co, Inc.

Katz B: Nerve, muscle, and synapse, New York, 1966, McGraw-Hill Book Co.

Kuffler SW et al: From neuron to brain, ed 2, Sunderland, Mass, 1984, Sinauer Associates, Inc.

CHAPTER 3

Generation and Conduction of Action Potentials

An action potential is a rapid change in the membrane potential followed by a return to the resting membrane potential (Figure 3-1). The size and shape of action potentials differ considerably from one excitable tissue to another. An action potential is propagated with the same shape and size along the whole length of a nerve or muscle cell. The action potential is the basis of the signal-carrying ability of nerve cells. An action potential allows all parts of a long muscle cell to contract almost simultaneously. This chapter discusses the ionic currents that generate action potentials and the ways in which action potentials are propagated and conducted.

EXPERIMENTAL OBSERVATION OF MEMBRANE POTENTIALS

Our knowledge of the ionic mechanisms of action potentials was first obtained from experiments on the squid giant axon. The large diameter (up to 0.5 mm) of the squid giant axon makes it a convenient object

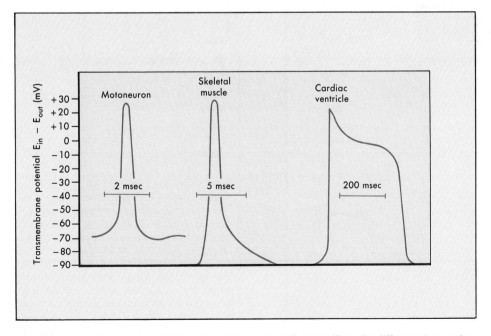

FIGURE 3-1 Action potentials from three vertebrate cell types. Note the different time scales. (Redrawn from Flickinger CJ et al: Medical cell biology, Philadelphia, 1979, WB Saunders Co.)

for electrophysiologic research with intracellular electrodes. The frog sartorius muscle is another useful preparation.

The following experiment illustrates some basic aspects of the resting membrane potential. A sartorius muscle is removed from a frog and put in a dish of fluid with composition simular to the frog's extracellular fluid. Two microelectrodes (tip diameter less than 0.5 μm) are placed in the extracellular fluid. When both microelectrodes are placed in the extracellular fluid, no electrical potential difference between them is observed. One electrode is then moved slowly toward a muscle cell until it penetrates the plasma membrane. At the instant the microelectrode penetrates the membrane, an abrupt change of the potential difference between the two electrodes is observed. The intracellular electrode suddenly becomes about 90 mV negative with respect to the external electrode. This −90 mV potential difference is the resting membrane potential of the muscle fiber.

A third microelectrode placed in the same cell also will register −90 mV relative to the external solution. In the resting muscle cell no net internal current flows, and thus no potential difference exists between these two intracellular electrodes. In the absence of perturbing influences, the resting membrane potential remains at −90 mV.

SUBTHRESHOLD RESPONSES: THE LOCAL RESPONSE

Figure 3-2 illustrates the results of an experiment in which the membrane potential of an axon of a shore crab is perturbed by passing rectangular pulses of current across the plasma membrane. Current pulses are depolarizing or hyperpolarizing, depending on the direction of current flow. The terms *depolarizing* and *hyperpolarizing* may be confusing. A change of the membrane potential from −90 mV to −70 mV is

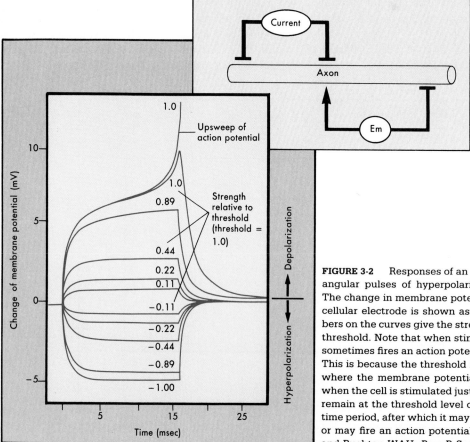

FIGURE 3-2   Responses of an axon of the shore crab to rectangular pulses of hyperpolarizing or depolarizing current. The change in membrane potential as recorded by an extracellular electrode is shown as a function of time. The numbers on the curves give the strength of the current relative to threshold. Note that when stimulated to threshold, the axon sometimes fires an action potential, but sometimes does not. This is because the threshold represents a metastable state where the membrane potential is delicately poised. Hence when the cell is stimulated just to the threshold, the cell may remain at the threshold level of depolarization for a variable time period, after which it may return to the resting potential or may fire an action potential. (Redrawn from Hodgkin AL and Rushton WAH: Proc R Soc B133:97, 1946.)

a *depolarization* because it is a decrease of the potential difference, or polarization, across the cell membrane. If the membrane potential changes from -90 mV to -100 mV, the polarization of the membrane has increased; this is *hyperpolarization*.

The larger the current passed, the larger is the pertubation of the membrane potential. As shown in Figure 3-2, in response to depolarizing current pulses above a certain *threshold* strength, the cell fires an action potential.

When subthreshold current pulses are passed, the size of the potential change observed depends on the distance of the recording electrode from the point of current passage (Figure 3-3, *A*). The closer the recording electrode to the site of current passage, the larger is the potential change observed. The size of the potential change is found to decrease exponentially with distance from the site of current passage (Figure 3-3, *B*). The distance over which the potential change decreases to 1/e (37%) of its maximal value is called the *length constant* (or space constant). (e is the base of natural logarithms and is equal to 2.7182. . . .) A length constant of 1 to 3 mm is typical for mammalian nerve or muscle cells. Because these potential

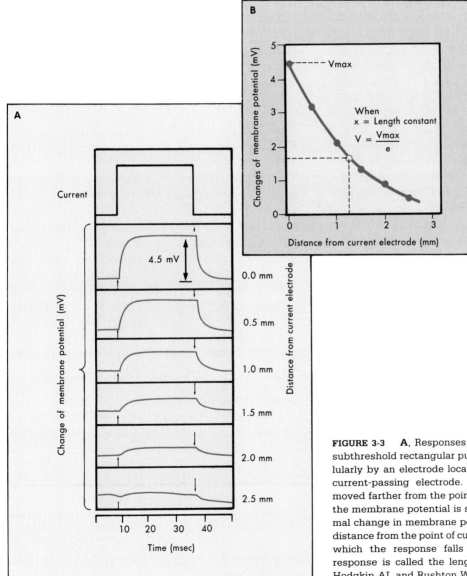

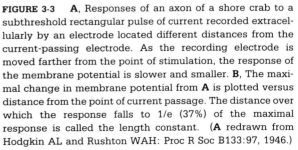

FIGURE 3-3   A, Responses of an axon of a shore crab to a subthreshold rectangular pulse of current recorded extracellularly by an electrode located different distances from the current-passing electrode. As the recording electrode is moved farther from the point of stimulation, the response of the membrane potential is slower and smaller. **B,** The maximal change in membrane potential from **A** is plotted versus distance from the point of current passage. The distance over which the response falls to 1/e (37%) of the maximal response is called the length constant. (**A** redrawn from Hodgkin AL and Rushton WAH: Proc R Soc B133:97, 1946.)

changes are observed primarily near the site of current passage and the changes are not propagated along the length of the cell (as are action potentials), they are called *local responses*.

ACTION POTENTIALS

If progressivley larger depolarizing current pulses are applied, a point is reached at which a different sort of response, the *action potential*, occurs (Figures 3-2 and 3-4). **An action potential is triggered when the depolarization is sufficient for the membrane potential to reach a threshold value,** which is near −55 mV for squid giant axon. The action potential differs from the local depolarizing response in two important ways: (1) it is a much larger response, with the polarity of the membrane potential actually reversing (the cell interior becoming positive with respect to the exterior); and (2) the action potential is propagated without decrement down the entire length of the nerve or muscle fiber. The size and shape of an action potential remain the same as it travels along the fiber; it does not decrease in size with distance, contrary to the local response. When a stimulus larger than the threshold stimulus is applied, the size and shape of the action potential do not change; the size of the action potential does not increase with increased stimulus strength. A stimulus either fails to elicit an action potential (a sub-

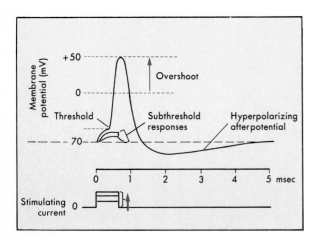

FIGURE 3-4 Responses of the membrane potential of a squid giant axon to increasing pulses of depolarizing current. When the cell is depolarized to threshold, it fires an action potential.

threshold stimulus), or it produces a full-sized action potential. For this reason the action potential is an *all-or-none response*.

Shape of Action Potential

The form of an action potential of a squid giant axon is shown in Figure 3-4. Once the membrane is depolarized to the threshold, an explosive depolarization occurs, which completely depolarizes the membrane and even overshoots, so that the membrane becomes polarized in the reverse direction. The peak of the action potential reaches about +50 mV. The membrane potential then returns toward the resting membrane potential almost as rapidly as it was depolarized. After repolarization, a transient hyperpolarization occurs that is known as the *hyperpolarizing afterpotential*. It persists for about 4 msec. The following section discusses the ionic currents that cause the various phases of the action potential.

IONIC MECHANISMS OF ACTION POTENTIAL

In Chapter 2 the resting membrane potential was seen to be a weighted sum of the equilibrium potentials for Na^+, K^+, Cl^-, etc. The weighting factor for each ion is the fraction that its conductance contributes to the total ionic conductance of the membrane (the chord conductance equation). In squid giant axon the resting membrane potential (E_m) is about −70 mV. E_K is about −100 mV in squid axon, so an increase in g_K would hyperpolarize the membrane. E_{Cl} is about −70 mV, so an increase in g_{Cl} would stabilize E_m at −70 mV. An increase in g_{Na} of sufficient magnitude would cause depolarization and reversal of the membrane polarity, since E_{Na} is about +60 mV in squid axon. A decrease in g_K also tends to depolarize the membrane.

In the 1950s Hodgkin and Huxley showed that the action potential of squid giant axon is caused by successive conductance increases to sodium and potassium ions. They found that the conductance to Na^+, g_{Na}, increases very rapidly during the early part of the action potential (Figure 3-5). The sodium conductance reaches a peak about the same time as the peak of the action potential, then it decreases rather rapidly. The K^+ conductance, g_K, increases more slowly, reaches a peak at about the middle of the repolariza-

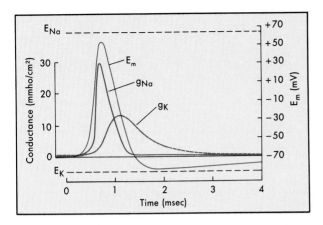

FIGURE 3-5   The action potential (E_m) of a squid giant axon is shown on the same time scale with the associated changes in the conductance of the axon membrane to sodium and potassium ions. (Redrawn from Hodgkin AL and Huxley AF: J Physiol 117:500, 1952.)

tion phase, and then returns more slowly to resting levels.

As described in Chapter 2, the chord conductance equation shows that the membrane potential is a result of the opposing tendencies of the K^+ gradient to bring E_m toward the equilibrium potential for K^+ and the Na^+ gradient to bring E_m toward the equilibrium potential for Na^+. Increasing the conductance of either ion will increase its ability to pull E_m toward its equilibrium potential. The rapid increase in g_{Na} during the early part of the action potential causes the membrane potential to move toward the equilibrium potential for Na^+(+60 mV). The peak of the action potential reaches only about +50 mV because the conductance of K^+ also increases, although more slowly, providing an opposing tendency, and also because g_{Na} quickly decreases toward resting levels. The rapid return of the membrane potential toward the resting potential is caused by the rapid decrease of g_{Na} and the continued increase in g_K. These conductance changes decrease the size of the Na^+ term in the chord conductance equation and increase the size of the K^+ term. During the hyperpolarizing afterpotential, when the membrane potential is actually more negative than the resting potential (more polarized), g_{Na} has returned to baseline levels, but g_K remains elevated above resting levels. Thus E_m is pulled closer to the K^+ equilibrium potential (-100 mV) as long as g_K remains elevated.

ION CHANNELS AND GATES

Hodgkin and Huxley proposed that the ion currents pass through separate Na^+ and K^+ channels, each with distinct characteristics, in the plasma membrane. Recent research supports this interpretation and has determined some of the properties of the channels (Figure 3-6).

To enter the narrowest part of the channel known as the *selectivity filter*, it is believed that K^+ and Na^+ must shed most of their waters of hydration. Tetrodotoxin (TTX) is a specific blocker of the Na^+ channel, and tetraethylammonium ion (TEA^+) blocks the K^+ channel. TTX binds to the extracellular side of the sodium channel. TEA^+ blocks the K^+ channel from the inside surface of the membrane.

The Na^+ channel appears to have both an activation gate and an inactivation gate that account for the changes in g_{Na} during voltage clamp. Digestion of the interior of the squid axon with the proteolytic enzyme pronase does away with the self-inactivation of g_{Na} by destroying the inactivation gate. The gates of the channels appear to be charged peptide chains.

Behavior of Individual Ion Channels

Recently it has been possible to study the behavior of individual ion channels. One way to do this is to

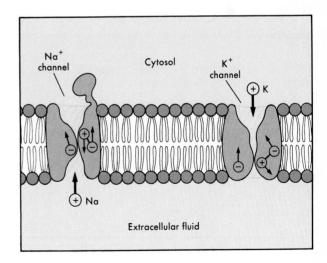

FIGURE 3-6   Schematic representations of the sodium and potassium channels. Movement of the charged residues shown in the channel proteins is believed to be associated with conformational changes that open and close the channels.

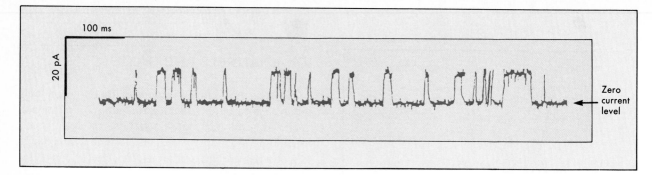

FIGURE 3-7 Ionic current through a single ion channel from rat muscle incorporated into a planar lipid bilayer membrane. The channel opens and closes spontaneously. The fraction of time this channel spends in the open state is a function of calcium ion concentration and membrane potential. (Reproduced from Moczydlowski E and Latorre R: *The Journal of General Physiology*, 1983, vol 82, pp 511-542 by copyright permission of the Rockefeller University Press.)

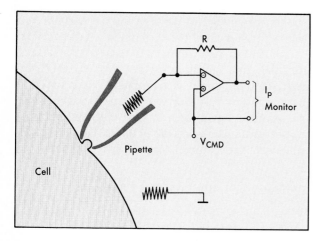

FIGURE 3-8 A patch electrode and circuitry required to record the currents that flow through the small number of ion channels isolated in the electrode. (Redrawn by permission from Sigworth FJ and Neher E: *Nature*, vol 287, pp 447-449. Copyright © 1980 Macmillian Journals Limited.)

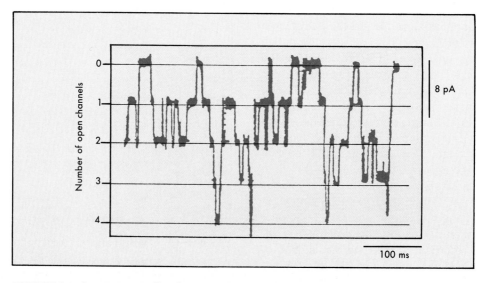

FIGURE 3-9 A current recording from a patch electrode on a muscle cell plasma membrane. The four different current levels show that this particular patch of membrane contains four different ion channels. (Redrawn from Hammill OP et al: Pflugers Arch 391:85, 1981.)

incorporate either purified ion channel proteins or bits of membrane into planar lipid bilayers that separate two aqueous compartments. Then electrodes placed in the aqueous compartments can be used to monitor or impose currents and voltages across the membrane. Under some conditions only one, or a few, ion channels of a particular type may be present in the planar membrane. The ion channels spontaneously oscillate between two conductance states, an open state and a closed state (Figure 3-7).

Another way to study individual ion channels involves the use of so-called patch electrodes. A fire-polished microelectrode is placed against the surface of a cell and suction is applied to the electrode (Figure 3-8). A high-resistance seal is formed around the tip of the electrode. The sealed patch electrode can then be used to monitor the activity of whatever channels

happen to be trapped inside the seal. The patch of membrane can either be studied in situ or be removed from the cell so that the composition of the solution in contact with the intracellular face of the membrane may be manipulated. The patch can even be turned inside out if desired. Sometimes the patch trapped inside the electrode contains more than one functional ion channel of a particular type (Figure 3-9).

Patch electrodes have been used to study Na^+ channels in skeletal muscle cells. In response to a step depolarization of the muscle cell membrane, some of the Na^+ channels show an opening event, some do not open at all, and some open more than once (Figure 3-10). When the currents of a large number of channels are averaged (Figure 3-10, tracing B), the resulting behavior resembles the macroscopic behavior of Na^+ channels discussed earlier. That is, the

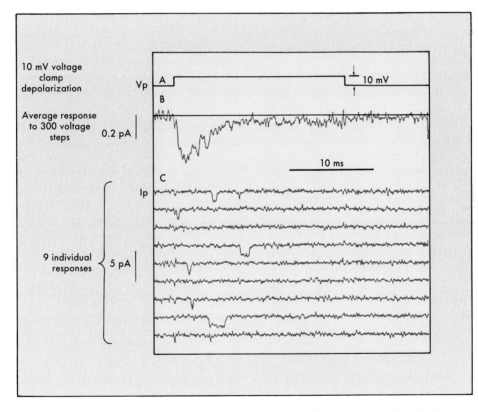

10 mV voltage clamp depolarization

Average response to 300 voltage steps   0.2 pA

9 individual responses   5 pA

FIGURE 3-10   A patch electrode records the currents that flow in a small patch of rat muscle membrane in response to a 10 mV depolarization (trace A). Tetraethylammonium is used to block potassium channels that may be present in the patch. The traces in curves C show responses to nine individual 10 mV depolarizations. The tracing in B is the average of 300 individual responses. Note that this average response resembles the response of large numbers of sodium channels, as seen in conventional voltage clamp experiments. (Redrawn by permission from Sigworth FJ and Neher E: *Nature*, vol 287, pp 447-449. Copyright © 1980 Macmillan Journals Limited.)

"average channel" opens promptly in response to depolarization, then after a short time delay, the channel closes (inactivates), even though the applied depolarization is maintained.

Our knowledge of ion channels is rapidly expanding. The primary amino acid sequence of the Na^+ channel has recently been elucidated by determining the sequence of the DNA that encodes the channel-protein. Although the three-dimensional structure of the Na^+ channel remains to be determined, its intra-membrane domain is known to consist of a number of α-helices that span the membrane and probably surround the ion channel. Groups of charged amino acid residues that may form the activation and inactivation gates have been tentatively identified.

ACTION POTENTIALS IN CARDIAC AND SMOOTH MUSCLE

Cardiac Muscle

An action potential in a cardiac ventricular cell is schematically shown in Figure 3-1. The initial rapid depolarization and overshoot is caused by the rapid entry of Na^+ through channels that are very similar to the Na^+ channels of nerve and skeletal muscle. Because of the rapid kinetics of opening and closing of these channels, they are called *fast Na^+ channels*.

After the initial depolarization and overshoot, the cardiac ventricular action potential has a *plateau phase*. The plateau is caused by another set of channels that are distinct from the fast Na^+ channels. These channels open and close much more slowly than the fast Na^+ channels and are called *slow channels*. The slow channels conduct both Na^+ and Ca^{++} ions, and the Ca^{++} that enters the cell during the plateau phase helps to initiate contraction of the ventricular cell. The repolarization of the ventricular cell is brought about by the closing of the slow channels and by a much delayed opening of K^+ channels. The ionic mechanisms of cardiac action potentials are discussed in more detail in Chapter 16.

Smooth Muscle

Action potentials vary considerably among different types of smooth muscle. Characteristically action potentials in smooth muscle have slower rates of depolarization and repolarization and less overshoot than skeletal muscle action potentials. Smooth muscle cells lack fast Na^+ channels. The depolarizing phase of smooth muscle action potentials is caused primarily by channels that resemble the cardiac slow channels in their kinetics and in conducting both Na^+ and Ca^{++}. The Ca^{++} that enters via the slow channels is often vital for excitation-contraction coupling in smooth muscle. Repolarization is caused by the closing of the slow Na^+/Ca^{++} channels and a simultaneous opening of K^+ channels.

PROPERTIES OF ACTION POTENTIALS

Voltage Inactivation

If a neuron or skeletal muscle cell is partially depolarized, for example, by increasing the concentration of K^+ in the extracellular fluid, its action potential has a slower rate of rise and a smaller overshoot. This is a result of two factors: (1) a smaller electrical force driving Na^+ into the depolarized cell and (2) voltage inactivation of some of the Na^+ channels.

The increase in g_{Na} during the action potential is *self-inactivating*; that is, the inactivation gates close soon after the activation gates open. Once the Na^+ channels are inactivated, the membrane must be repolarized toward the normal resting membrane potential before the channels can be reopened. As the membrane potential is restored toward normal resting levels, more and more of the Na^+ channels again become capable of being activated. Because the action potential mechanism requires a critical density of open Na^+ channels, an action potential may not be generated in response to stimulation when a considerable fraction of Na^+ channels is inactivated because of partial depolarization. This is called *voltage inactivation* of the action potential resulting from voltage inactivation of the Na^+ channels. Voltage inactivation of the Na^+ channels partially accounts for important properties of excitable cells, such as refractoriness and accommodation.

Refractory Periods

During much of the action potential the membrane is completely refractory to further stimulation. This means that no matter how strongly the cell is stimulated, it is unable to fire a second action potential.

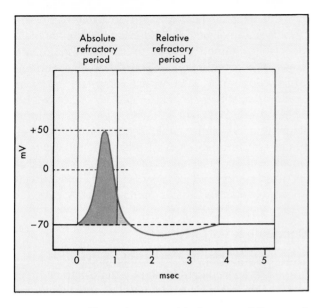

FIGURE 3-11   The action potential of nerve and the associated absolute and relative refractory periods.

This is called the *absolute refractory period* (Figure 3-11). The cell is refractory because the majority of its Na^+ channels are voltage inactivated and cannot be reopened until the membrane is repolarized.

During the latter part of the action potential the cell is able to fire a second action potential, but a stronger than normal stimulus is required. This is the *relative refractory period*. Early in the relative refractory period, before the membrane potential has returned to the resting potential level, some Na^+ channels are voltage inactivated, so a stronger than normal stimulus is required to open the critical number of Na^+ channels needed to trigger an action potential. Throughout the relative refractory period the conductance to K^+ is elevated, which opposes depolarization of the membrane. This also contributes to the refractoriness.

Accommodation to Slow Depolarization

When a nerve or muscle cell is depolarized slowly, the threshold may be passed without an action potential being fired; this is called *accommodation*. Na^+ and K^+ channels are both involved in accommodation. During slow depolarization some of the Na^+ channels that are opened by depolarization have enough time to become voltage inactivated before the threshold potential is attained. If depolarization is slow enough, the critical number of open Na^+ chan-

nels required to trigger the action potential may never be attained. In addition, K^+ channels open in response to the depolarization. The increased g_K tends to repolarize the membrane, making it still more refractory to depolarization.

STRENGTH-DURATION CURVE

A stimulus depolarizes the cell by causing charge to flow across the plasma membrane. The relevant quantity is the *total amount of charge* that flows across the membrane (current × time). A strong stimulus depolarizes the membrane to threshold quickly. A weaker stimulus must be applied longer for the critical amount of charge to flow across the membrane. This often is depicted in the *strength-duration curve*, which is a plot of the stimulus strength versus the minimal time the stimulus must be applied to cause an action potential (Figure 3-12).

A very weak stimulus will not cause an action potential even if applied for a very long time (accommodation). The smallest stimulus strength that can elicit an action potential from a particular preparation is called the *rheobase*. The time needed by a stimulus twice as strong as the rheobase to elicit an action potential is called the *chronaxie* of the preparation. Chronaxie is a useful index of the excitability of a preparation. The larger the chronaxie, the less excitable is the preparation.

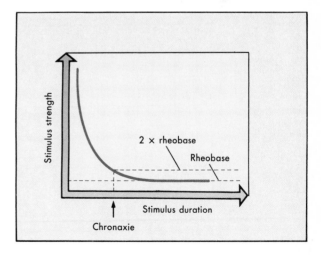

FIGURE 3-12   The strength-duration curve. The ordinate shows the stimulus strength, and the abscissa shows the minimal time a stimulus of that strength must be applied to produce an action potential.

CONDUCTION OF ACTION POTENTIAL

A principal function of neurons is to transfer information by the conduction of action potentials. The axons of the motor neurons of the ventral horn of the spinal cord conduct action potentials from the cell body of the neuron to a number of skeletal muscle fibers. The distance from the motor neuron to one of the muscle fibers it innervates may be longer than 1 m.

Action potentials are conducted along a nerve or muscle fiber by local current flow, just as occurs in electrotonic conduction of subthreshold potential changes. Thus the same factors that govern the velocity of electrotonic conduction also determine the speed of action potential propagation.

The Local Response

Figure 3-13, *A* shows the membrane of an axon or muscle fiber that has been depolarized in a small region. In this region the external aspect of the membrane is negative relative to the adjacent membrane, and the internal face of the depolarized membrane is positively charged relative to neighboring internal areas. The potential differences cause local currents to flow (Figure 3-13, *B*), which depolarize the membrane adjacent to the initial site of depolarization. These newly depolarized areas then cause current flows that depolarize other segments of the membrane still further removed from the initial site of depolarization. This spread of depolarization is called the *local response*. This mechanism of conduction is known as *electrotonic conduction*.

Conduction Velocity

The speed of electrotonic conduction along a nerve or muscle fiber is determined by the electrical properties of the cytoplasm and of the plasma membrane that surrounds the fiber. Fibers that are larger in diameter have a greater conduction velocity. This is principally caused by the decrease in the resistance to conduction in the cytoplasm along the length of the fiber as the radius (and thus the cross-sectional area) of the fiber increases.

Electrotonic Conduction and Decrement

Earlier in this chapter it was noted that the local response dies away to almost nothing over the course of several millimeters (see Figure 3-3). A nerve or muscle fiber has some of the properties of an electrical cable. In a perfect cable the insulation surrounding the core conductor prevents all current loss so that a signal is transmitted along the cable with undiminished strength (Figure 3-14). The plasma membrane of an unmyelinated nerve or muscle fiber serves as the insulation. The membrane has a resistance much higher than the resistance of the cytoplasm, but (partly because of its thinness) the plasma membrane is not a perfect insulator. The higher the ratio of R_m to R_{in}, the better the cell can function as a cable, and the longer the distance that a signal can be transmitted electrotonically without significant decrement. $\sqrt{R_m/R_{in}}$ determines the *length constant* of a cell (see Figure 3-3). The length constant is the distance over which an electrotonically conducted signal falls to 37% (1/e) of its initial strength. A typical length constant for unmyelinated mammalian nerve and muscle fibers is about 1 to 3 mm. Some axons in the human body are about 1 m long, so it is clear that the

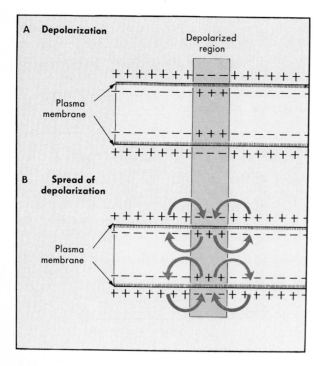

FIGURE 3-13   Mechanism of electrotonic spread of depolarization. **A**, The reversal of membrane polarity that occurs with local depolarization. **B**, The local currents that flow to depolarize adjacent areas of the membrane and allow spreading of the depolarization.

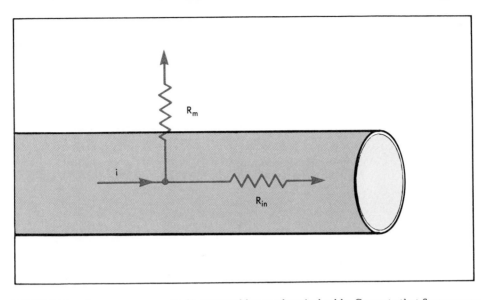

FIGURE 3-14   An axon or a muscle fiber resembles an electrical cable. Currents that flow across the membrane resistance (R_m) are lost from the cable. Currents that flow through the longitudinal resistance (R_{in}) carry the electrical signal along the cable. The larger the ratio R_m/R_{in}, the more efficient is signal transmission along the fiber.

local response cannot conduct a signal over so great a distance.

Action Potential as Self-Reinforcing Signal

Many nerve and muscle fibers are much longer than their length constants. The action potential serves to conduct an electrical impulse with undiminished strength along the full length of those fibers. To do this, the action potential reinforces itself as it is propagated along the fiber. The conduction of the action potential occurs by the mechanism depicted in Figure 3-13. When the areas on either side of the depolarized region reach threshold, these areas also fire action potentials, which locally reverse the polarity of the membrane potential. By local current flow, the areas of the fiber adjacent to these areas are brought to threshold, and then these areas in turn fire action potentials. A cycle of depolarization occurs by local current flow followed by generation of an action potential in a restricted region that then travels along the length of the fiber, with "new" action potentials being generated as they spread. In this way the action potentials are generated as they spread, and the action potential propagates over long distances, keeping the same size and shape.

Since the shape and size of the action potential are usually invariant, only variations in the frequency of the action potentials can be used in the code for information transmission along axons. The maximum frequency is limited by the duration of the absolute refractory period (about 1 msec) to about 1000 impulses/second in large mammalian nerves.

Effect of Myelination on Conduction Velocity

In vertebrates certain nerve fibers are coated with *myelin*; such fibers are said to be *myelinated*. Myelin is formed from multiple wrappings of the plasma membranes of Schwann cells that wind themselves around the nerve fiber (Figure 3-15). The myelin sheath consists of several to more than 100 layers of plasma membrane. Gaps that occur in the sheath every 1 to 2 mm are known as *nodes of Ranvier*. Nodes of Ranvier are about 1 μm wide and are the lateral spaces between adjacent Schwann cells along the axon. Myelin alters the electrical properties of the nerve fiber and results in a great increase in the conduction velocity of the fiber.

A squid giant axon with a 500 μm diameter has a conduction velocity of 25 m·sec^{-1} and is unmyelinated. If conduction velocity were directly proportional to fiber radius, a human nerve fiber with a 10 μm diameter would conduct at 0.5 m·sec^{-1}. With this con-

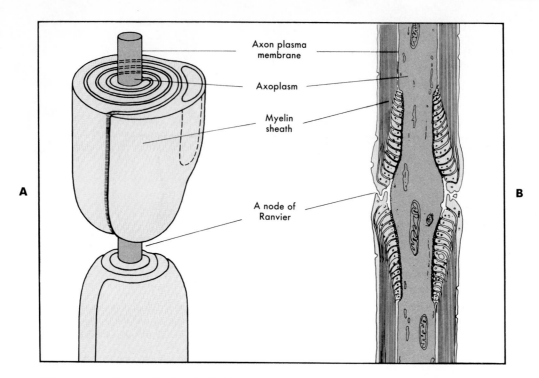

FIGURE 3-15   The myelin sheath. **A**, Schematic drawing of Schwann cells wrapping around an axon to form a myelin sheath. **B**, Drawing of a cross section through a myelinated axon near a node of Ranvier.

duction velocity a reflex withdrawal of the foot from a hot coal would take about 4 seconds. Even though our nerve fibers are much smaller in diameter than squid giant axons, our reflexes are much faster than this. The myelin sheath that surrounds certain vertebrate nerve fibers results in a much greater conduction velocity than that of unmyelinated fibers of similar diameters. A 10 μm myelinated fiber has a conduction velocity about 50 m·sec^{-1}, which is twice that of the 500 μm squid giant axon. The high conduction velocity permits reflexes that are fast enough to allow us to avoid dangerous stimuli. Figure 3-16 shows the large increase in conduction velocity caused by myelination. A myelinated axon has a greater conduction velocity than an unmyelinated fiber that is 100 times larger in diameter. As discussed next, the myelin sheath increases the velocity of action potential conduction by increasing the length constant of the axon, by decreasing the capacitance of the axon, and by restricting the generation of action potentials to the nodes of Ranvier.

Myelination greatly alters the electrical properties of the axon. The many wrappings of membrane around the axon increase the effective membrane resistance, so that R_m/R_{in} and thus the length constant is much greater. Less of a conducted signal is lost through the electrical insulation of the myelin sheath, so that the amplitude of a conducted signal declines less with distance along the axon. The myelin-wrapped membrane has a much smaller electrical capacitance than the naked axonal membrane. For this reason the conduction velocity is greatly increased by myelination. Because of the increase in length constant and in conduction velocity, an action potential is conducted with little decrement and at great speed from one node of Ranvier to the next.

The action potential is regenerated only at the nodes of Ranvier (1 to 2 mm apart), rather than being regenerated at each place along the fiber, as is the case in an unmyelinated fiber. The regeneration of action potentials in myelinated fibers is restricted to the nodes of Ranvier because the internodal plasma membrane cannot produce action potentials. The resistance to the flow of ions across the many layers of Schwann cell membrane that make up the myelin sheath is so high that the ionic currents are effective-

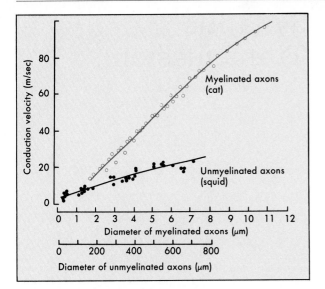

FIGURE 3-16 Conduction velocities of myelinated and unmyelinated axons as functions of axon diameter. Myelinated axons are from cat saphenous nerve at 38° C. Unmyelinated axons are from squid and are at 20° to 22° C. Note that the myelinated axons have faster conduction velocities than unmyelinated axons 100 times greater in diameter. (Data for myelinated axons from Gasser HS and Grundfest H: Am J Physiol 127:393, 1939. Data for unmyelinated axons From Pumphrey RJ and Young JZ: J Exp Biol 15:453, 1938.)

ly localized to the short stretches of naked plasma membrane that occur at the nodes of Ranvier. The action potential is rapidly conducted from node to node and "pauses" to be regenerated at each node. The action potential appears to "jump" from one node of Ranvier to the next, a process called *saltatory conduction*.

Myelinated axons are also more efficient metabolically than unmyelinated axons. The sodium-potassium pump extrudes the sodium that enters and reaccumulates the potassium that leaves the cell during action potentials. In myelinated axons, ionic currents are restricted to the small fraction of the membrane surface at the nodes of Ranvier. For this reason fewer Na^+ and K^+ ions traverse a unit area of fiber membrane, and less ion pumping is required to maintain Na^+ and K^+ gradients.

BIBLIOGRAPHY
Journal Articles

Keynes RD: Ion channels in the nerve-cell membrane, Sci Am 240:126, 1979.

Stevens CF: The neuron, Sci Am 241(3):54, 1979.

Stevens, CF: Studying just one molecule: single channel recording, Trends Pharm Sci 5:131, 1984.

Books and Monographs

Aidley DJ: The physiology of excitable cells, ed 2, Cambridge, 1978, Cambridge University Press.

Davson H: A textbook of general physiology, ed 4, Baltimore, 1970, Williams & Wilkins.

Hodgkin AL: The conduction of the nervous impulse, Springfield, Ill, 1964, Charles C Thomas, Publisher.

Junge D: Nerve and muscle excitation, ed 2, Sunderland, Mass, 1981, Sinauer Associates, Inc.

Kandel ER and Schwartz JH: Principles of neural science, ed 2, New York, 1985, Elsevier Science Publishing Co, Inc.

Katz B: Nerve, muscle, and synapse, New York, 1966, McGraw-Hill Book Co.

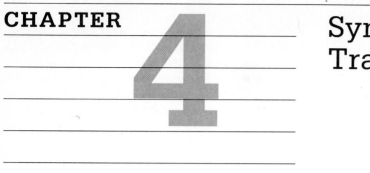

CHAPTER 4

Synaptic Transmission

A *synapse* is a site at which an impulse is transmitted from one cell to another. There are two types: electrical and chemical synapses. At an *electrical synapse* two excitable cells communicate by the direct passage of electrical current between them. This is called *ephaptic* or *electrotonic* transmission. *Gap junctions* link electrotonically coupled cells and provide low-resistance pathways for current flow directly between the cells.

Information is also transferred between excitable cells by means of *chemical synapses*. Chemical synapses may be better suited for the complex modulation of synaptic activity and the integration that occurs at synapses in vertebrate central nervous systems. At a chemical synapse an action potential causes a transmitter substance to be released from the presynaptic neuron. The transmitter diffuses across the extracellular *synaptic cleft* and binds to receptors on the membrane of the postsynaptic cell to change the electrical properties of the postsynaptic membrane. Chemical synapses have *synaptic delay*—the time required for these events to occur. The neuromuscular (or myoneural) junction is a particularly well-studied vertebrate chemical synapse.

Although the nature of the presynaptic and postsynaptic cells, the structure of the synapse, and the transmitter substance vary, chemical synapses have certain characteristics in common.

NEUROMUSCULAR JUNCTIONS

The synapses between the axons of motoneurons and skeletal muscle fibers are called *neuromuscular junctions*, *myoneural junctions*, or *motor endplates*. The neuromuscular junction, the first vertebrate synapse to be well characterized, serves as a model chemical synapse that provides a basis for understanding more complex synaptic interactions among neurons in the central nervous system.

Structure

Near the neuromuscular junction the motor nerve loses its myelin sheath and divides into fine terminal branches (Figure 4-1). The terminal branches of the axon lie in synaptic troughs on the surfaces of the muscle cells. The plasma membrane of the muscle cell lining the trough is thrown into numerous junctional folds. The axon terminals contain many 40 nm smooth-surfaced synaptic vesicles that contain acetylcholine, the neurotransmitter employed at this synapse. The axon terminal and the muscle cell are separated by the *junctional cleft,* which contains a carbohydrate-rich amorphous material.

Acetylcholine receptor molecules are concentrated near the mouths of the junctional folds, and acetylcholinesterase is evenly distributed on the external surface of the postjunctional membrane. The synaptic vesicles in the nerve terminals and specialized release sites (called *active zones*) on the prejunctional membrane are concentrated opposite the mouths of the junctional folds.

Overview of Neuromuscular Transmission

The action potential is conducted down the motor axon to the presynaptic axon terminals. Depolariza-

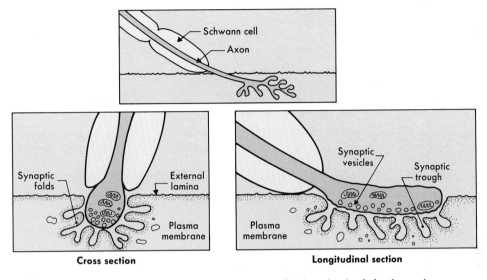

FIGURE 4-1   The structure of the neuromuscular junction in skeletal muscle.

tion of the plasma membrane of the axon terminal transiently opens calcium channels. Ca^{++} from the interstitial fluid flows down its electrochemical potential gradient into the axon terminal. The influx of Ca^{++} causes synaptic vesicles to fuse with the plasma membrane and to empty their acetylcholine into the synaptic cleft by exocytosis. Acetylcholine diffuses across the synaptic cleft and combines with a specific *acetylcholine receptor protein* on the external surface of the muscle plasma membrane of the motor endplate. The combination of acetylcholine with the receptor protein transiently increases the conductance of the postjunctional membrane to Na^+ and K^+. Ionic currents (Na^+ and K^+) result in a transient depolarization of the endplate region. The transient depolarization is called the *endplate potential,* or *EPP* (Figure 4-2). The EPP is transient because the action of acetylcholine is ended by the hydrolysis of acetylcholine to form choline and acetate. The hydrolysis of acetylcholine is catalyzed by the enzyme *acetylcholinesterase,* which is present in high concentration on the postjunctional membrane.

The postjunctional plasma membrane of the neuromuscular junction is not electrically excitable and does not fire action potentials. After it is depolarized, adjacent regions of the muscle cell membrane are depolarized by electrotonic conduction (Figure 4-2,

B). When those regions reach threshold, action potentials are generated. Action potentials are propagated along the muscle fiber at high velocity and initiate the chain of events that leads to muscle contraction (Chapter 11). The steps involved in neuromuscular transmission are listed in the box, and some of these steps are considered next in greater detail.

Synthesis of Acetylcholine

Motoneurons and their axons synthesize acetylcholine. Most other cells are not able to make acetylcholine. The enzyme *choline-O-acetyltransferase* in the motoneuron catalyzes the condensation of acetyl coenzyme A (acetyl CoA) and choline. Acetyl CoA is produced by the neuron, as it is by most cells. However, choline cannot be synthesized by the motoneuron and is obtained by active uptake from the extracellular fluid. The plasma membrane of the motoneuron has a transport system that can accumulate choline against a large electrochemical potential gradient.

Quantal Release of Transmitter

The amount of acetylcholine released by the prejunctional nerve ending does not vary continuously;

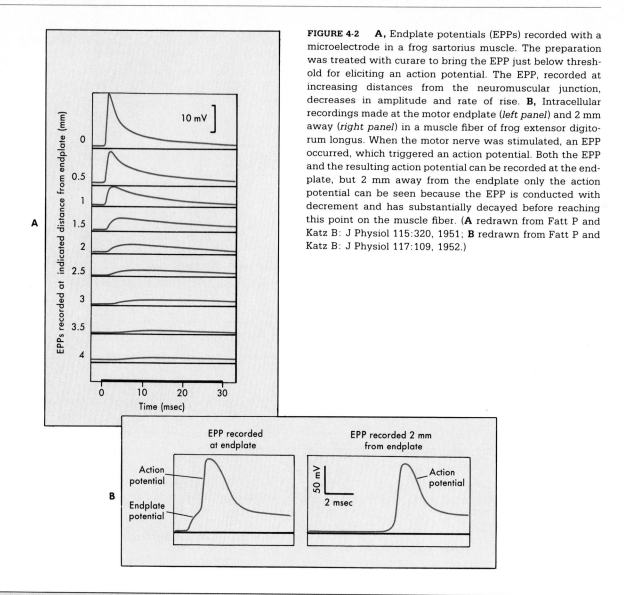

FIGURE 4-2   **A,** Endplate potentials (EPPs) recorded with a microelectrode in a frog sartorius muscle. The preparation was treated with curare to bring the EPP just below threshold for eliciting an action potential. The EPP, recorded at increasing distances from the neuromuscular junction, decreases in amplitude and rate of rise. **B,** Intracellular recordings made at the motor endplate (*left panel*) and 2 mm away (*right panel*) in a muscle fiber of frog extensor digitorum longus. When the motor nerve was stimulated, an EPP occurred, which triggered an action potential. Both the EPP and the resulting action potential can be recorded at the endplate, but 2 mm away from the endplate only the action potential can be seen because the EPP is conducted with decrement and has substantially decayed before reaching this point on the muscle fiber. (**A** redrawn from Fatt P and Katz B: J Physiol 115:320, 1951; **B** redrawn from Fatt P and Katz B: J Physiol 117:109, 1952.)

Summary of Events Occurring During Neuromuscular Transmission

Action potential in presynaptic motor axon terminal
↓
Opening of Ca^{++} channels and entry of Ca^{++} into axon terminal
↓
Release of acetylcholine from synaptic vesicles into synaptic cleft
↓
Diffusion of acetylcholine to postjunctional membrane
↓
Combination of acetylcholine with specific receptor protein of postjunctional membrane
↓
Increase in conductance of postjunctional membrane to Na^+ and K^+ causes EPP
↓
Depolarization of muscle membrane adjacent to endplate initiates action potential

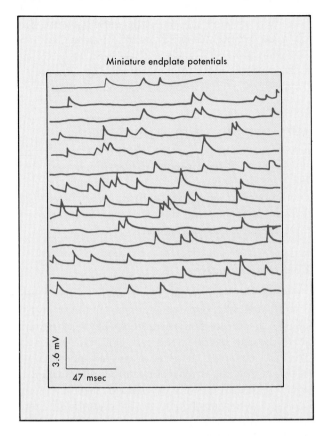

Miniature endplate potentials

3.6 mV

47 msec

FIGURE 4-3 Spontaneous miniature endplate potentials (MEPPs) recorded at a neuromuscular junction in a fiber of frog extensor digitorum longus. (Redrawn from Fatt P and Katz B: Nature 166:597, 1950.)

rather, the amount varies in steps, with each step corresponding to the release of one synaptic vesicle. The amount of acetylcholine contained in one vesicle corresponds to a *quantum* of acetylcholine.

Even if the motoneuron is not stimulated, small depolarizations of the postjunctional muscle cell occur spontaneously. These small spontaneous depolarizations are known as *miniature endplate potentials*, or *MEPPs* (Figure 4-3). They occur at random times with a frequency that averages about 1 per second. Each MEPP depolarizes the postjunctional membrane by only about 0.4 mV on average. The MEPP has the same time course as an EPP that is evoked by an action potential in the nerve terminal. The MEPP is similar to the EPP in its response to most drugs. The EPP and MEPP are both prolonged by drugs that inhibit acetylcholinesterase, and both are similarly depressed by compounds that compete with acetylcholine for binding to the receptor protein. The frequency of MEPPs varies in time, but their amplitudes

are within a relatively narrow range (Figure 4-3). A MEPP is caused by the spontaneous release of one quantum of transmitter into the junctional cleft.

Action of Cholinesterase and Reuptake of Choline

Acetylcholinesterase is concentrated on the external surface of the postjunctional membrane and in the basal lamina. The drugs *eserine* and *edrophonium* are inhibitors of the enzyme and are called *anticholinesterases*. In the presence of an anticholinesterase, the EPP is larger and dramatically prolonged.

The motoneuron cannot synthesize choline, so the reuptake from the synaptic cleft provides choline needed for the resynthesis of acetylcholine. *Hemicholiniums* are drugs that block the choline transport system and inhibit choline uptake. Prolonged treatment with hemicholiniums depletes the store of transmitter and ultimately decreases the acetylcholine content of the quanta.

Ionic Mechanism of EPP

The cation channels that acetylcholine opens in the postjunctional membrane differ from the cation channels of nerve and muscle in that they are independent of the membrane potential. The postjunctional channels are gated by the action of acetylcholine rather than by the transmembrane potential.

The membrane potential (E_m) may be determined primarily by the membrane conductances to K^+ and Na^+, as shown by the chord conductance equation:

$$E_m = \frac{g_K}{g_K + g_{Na}}E_K + \frac{g_{Na}}{g_K + g_{Na}}E_{Na} \qquad (1)$$

Because acetylcholine increases the permeability of the postsynaptic membrane to both Na^+ and K^+, the membrane potential of the endplate should be between E_K and E_{Na} in the presence of acetylcholine. For a muscle cell with a normal resting membrane potential (about -90 mV), this results in a depolarization of the membrane.

Acetylcholine Receptor Protein

The acetylcholine receptor protein has been studied intensively. Development of methods for isolating and purifying hydrophobic membrane proteins and the availability of snake venom neurotoxins that bind

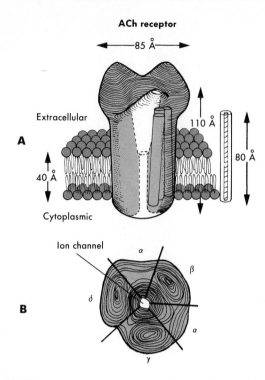

ACh receptor

FIGURE 4-4   A model of the structure of the acetylcholine receptor protein. **A,** Viewed from the side and, **B,** viewed looking down on the acetylcholine receptor from the cytoplasmic surface. The closed curves are electron density profiles. The five subunits surround a central ion channel. (Redrawn from Kistler J, et al: Biophys J 37:371, 1982).

very tightly to the acetylcholine receptor have been essential in these studies.

α-Bungarotoxin, from the venom of the Formosan krait (a relative of the cobra), binds to the acetylcholine receptor almost irreversibly. Binding of radioactively labeled α-bungarotoxin by neuromuscular junctions suggests that there are 10^7 to 10^8 binding sites per motor endplate. In neuromuscular junctions the acetylcholine binding sites concentrated near the mouths of the postjunctional folds have a density of about 20,000/μm^2. This suggests that the receptor molecules are tightly packed.

The acetylcholine receptor protein is an integral membrane protein and is deeply embedded in the hydrophobic lipid matrix of the postjunctional membrane. Cholinesterase, on the other hand, is loosely associated with the surface of the postjunctional membrane by hydrophilic interactions.

The acetylcholine receptor protein consists of five subunits (Figure 4-4), two of which are identical, so

that there are four different polypeptide chains. The duplicated subunit is the α subunit (molecular weight [MW] 40,000). The acetylcholine binding sites are located on the α subunits. The other subunits are β (MW 50,000), γ (MW 60,000), and δ (MW 65,000).

SYNAPSES BETWEEN NEURONS

Chemical transmission between neurons has many of the same properties that characterize the neuromuscular junction. Electrical synapses are also present in the central nervous systems of animals, from invertebrates to mammals.

Electrical Synapses

At an electrical synapse a change in the membrane potential of one cell is transmitted to the other cell by the direct flow of current. Because current flows directly between the two cells that make an electrical synapse, there is essentially no synaptic delay. In general, electrical synapses allow conduction in both directions. In this respect they differ from chemical synapses, which are obligatorily unidirectional. Certain electrical synapses conduct more readily in one direction than in the other; this property is called *rectification*.

Cells that form electrical synapses typically are joined by *gap junctions*. Gap junctions are plaquelike structures in which the plasma membranes of the coupled cells are very close (less than 3 nm). Freeze-fracture electron micrographs of gap junctions display regular arrays of intramembrane protein particles. The intramembrane particles consist of six subunits surrounding a central channel that is accessible to water. The hexagonal array is called a *connexon*. Each of the six subunits is a single protein (one polypeptide chain) called *connexin* (MW 25,000). At the gap junction the connexons of the coupled cells are aligned to form channels (Figure 4-5, *A*). The channels allow the passage of water-soluble molecules up to MWs of 1200 to 1500 from one cell to the other. Such channels are the pathways for electrical current flow between the cells.

Cells that are electrically coupled may become uncoupled by closing of the connexon channels. The channels may close in response to increased intracellular Ca^{++} or H^+ in one of the cells or in response to depolarization of one or both of the cells. A model for

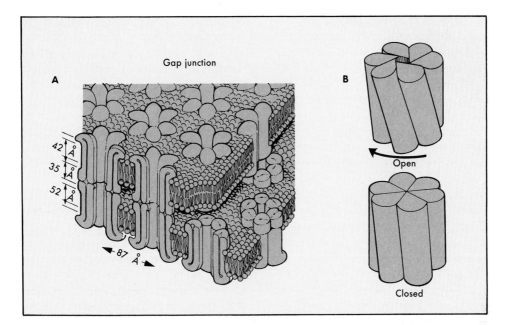

FIGURE 4-5   A, A model for the structure of the gap junction channels. Each plasma membrane contains connexons, each of which consists of an hexagonal array of six connexin polypeptides. The connexons of the two membranes are aligned at the gap junction to form channels between the cytosolic compartments of the two cells. **B,** A model of the opening and closing of the gap junction channel. The individual subunits of the connexon are proposed to twist relative to one another to open and close the central channel. (**A,** Redrawn from Makowski L et al: J Cell Biol 74:629, 1977, by copyright permission of The Rockefeller University Press. **B,** Redrawn from Unwin PNT and Zampighi G: Nature 283:45, 1980, by permission from Macmillan Journals Limited.)

the mechanism of closing the channels is shown in Figure 4-5, *B*.

Electrical synapses are widespread in the peripheral and central nervous systems of invertebrates and vertebrates. Electrical synapses are particularly useful in reflex pathways in which rapid transmission between cells (little synaptic delay) is necessary or when the synchronous response of a number of neurons is required. Among the many nonneuronal cells that are coupled by gap junctions are hepatocytes, myocardial cells, intestinal smooth muscle cells, and the epithelial cells of the lens.

Chemical Synapses

When one neuron makes a chemical synapse with another, the presynaptic nerve terminal characteristically broadens to form a *terminal bouton*. At the synapse itself the presynaptic and postsynaptic membranes are close to each other and lie parallel to one another. Substantial structures stabilize the synapse,

so that when nervous tissue is disrupted, the relationship of the presynaptic and postsynaptic membranes at the synapse is often preserved.

Because of the structure and organization of chemical synapses, conduction is necessarily one way. One-way conduction of chemical synapses contributes to the organization of central nervous systems of vertebrates. The synaptic delay at chemical synapses is about 0.5 msec. Synaptic delay is mainly caused by the time required for the release of transmitter. The time required for calcium channels to open in response to depolarization of the presynaptic terminal is a major component of synaptic delay. In polysynaptic pathways synaptic delay accounts for a significant fraction of the total conduction time.

At chemical synapses the transmitter released by the presynaptic neurons alters the conductance of the postsynaptic plasma membrane to one or more ions. A change in the conductance of the postsynaptic membrane to an ion that is not in equilibrium across the membrane alters the current carried by that ion, which causes a change in the membrane potential of the postsynaptic cell. In most cases transmitters produce their effect by *increasing* the conductance of the postsynaptic membrane to one or more ions. However, some transmitters in invertebrates (and perhaps in vertebrates as well) act by *decreasing* the postsynaptic conductance to specific ions.

The part of the membrane of the postsynaptic neuron that forms the synapse is specialized for chemical sensitivity rather than electrical sensitivity. Action potentials are not produced at the synapse. The change in membrane potential, whether depolarization or hyperpolarization, that occurs at the synapse is conducted electrotonically over the membrane of the postsynaptic neuron to the *axon hillock–initial segment* region. The axon hillock–initial segment has a lower threshold than the rest of the plasma membrane of the postsynaptic cell. An action potential will be generated at that site if the sum of all the inputs to the cell exceeds threshold. Once the action potential has been generated, it is conducted back over the surface of the soma of the postsynaptic cell and is propagated along its axon.

Input-Output Relations

The neuromuscular junction is representative of a very simple type of synapse in which one action potential in the presynaptic cell (the input) elicits a single action potential in the postsynaptic cell (the output). In other types of synapses the output may differ from the input. Synapses can be classified as one-to-one, one-to-many, or many-to-one, based on the relationship between input and output.

In a *one-to-one synapse*, as in the neuromuscular junction, the input and output are the same. A single action potential in the presynaptic cell evokes a single action potential in the postsynaptic cell. Because the output is the same as the input, no integration can occur at this type of synapse.

In a *one-to-many synapse* a single action potential in the presynaptic cell elicits many action potentials in the postsynaptic cell. One-to-many synapses are not common; one example is the synapse of motoneurons on Renshaw cells in the spinal cord. One action potential in the motoneuron induces the Renshaw cell to fire a burst of action potentials.

In a *many-to-one synaptic arrangement* one action potential in the presynaptic cell is not enough to make the postsynaptic cell fire an action potential.

The nearly simultaneous arrival of presynaptic action potentials in several input neurons that synapse on the postsynaptic cell is necessary to depolarize the postsynaptic cell to threshold. The spinal motoneuron has this type of synaptic organization. One hundred or more presynaptic axons synapse on each spinal motoneuron (Figure 4-6). Some of these are excitatory inputs that depolarize the postsynaptic cell and bring it closer to its threshold. Other inputs are inhibitory and hyperpolarize the motoneuron; they take it farther away from threshold.

The changes in postsynaptic potential caused by an action potential in a single presynaptic input are about 1 to 2 mV. Thus no one excitatory input can bring the motoneuron to threshold. A transient depolarization of the postsynaptic neuron evoked by an action potential in a presynaptic axon is called an *excitatory postsynaptic potential (EPSP)* (Figure 4-7). The transient hyperpolarization elicited by an action potential in an inhibitory input is called an *inhibitory postsynaptic potential (IPSP)* (Figure 4-7). At any instant the postsynaptic cell integrates the various inputs. If the momentary sum of the inputs depolarizes the postsynaptic cell to its threshold, it will fire an action potential. This is integration at the level of a single postsynaptic neuron.

Summation of Synaptic Inputs

The summation (or integration) of inputs can occur by either spatial summation or temporal summation (Figure 4-8). *Spatial summation* occurs when two separate inputs arrive simultaneously. The two postsynaptic potentials are added so that two simultaneous excitatory inputs will depolarize the postsynaptic cell about twice as much as either input alone. However, if one EPSP and one IPSP occur simultaneously, they tend to cancel one another. Even inputs at synapses at opposite ends of the postsynaptic cell body act in this way. The postsynaptic potentials (EPSPs and IPSPs) are conducted rapidly over the entire cell membrane of the postsynaptic cell body with almost no decrement. This is because cellular dimensions (less than 100 μm) are much smaller than the length constant (about 1 to 2 mm) for electrotonic conduction. Synaptic potentials that originate in fine dendritic branches decrease in magnitude as they are conducted to the cell body.

Temporal summation occurs when two or more action potentials in a single presynaptic neuron are

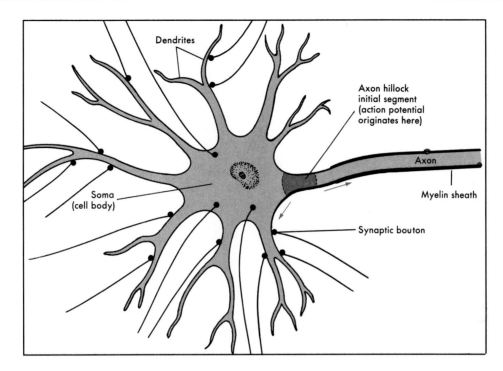

FIGURE 4-6   A spinal motor neuron with multiple synapses on both soma and dendrites. The axon hillock-initial segment has the lowest threshold, and as a result, action potentials tend to originate here.

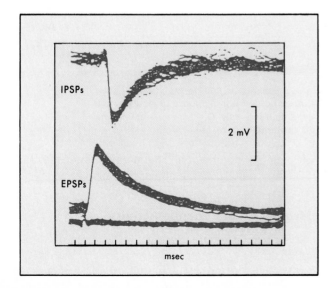

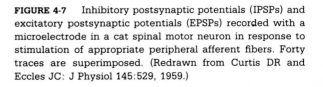

FIGURE 4-7   Inhibitory postsynaptic potentials (IPSPs) and excitatory postsynaptic potentials (EPSPs) recorded with a microelectrode in a cat spinal motor neuron in response to stimulation of appropriate peripheral afferent fibers. Forty traces are superimposed. (Redrawn from Curtis DR and Eccles JC: J Physiol 145:529, 1959.)

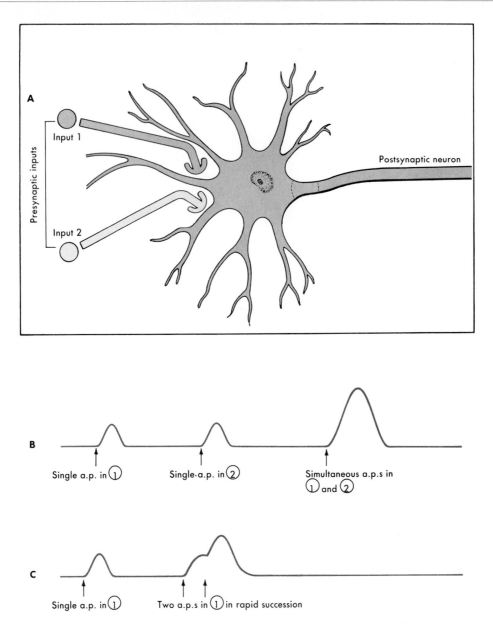

FIGURE 4-8 **A,** Spatial and temporal summation at a postsynaptic neuron with two synaptic inputs (*1* and *2*). **B,** Spatial summation. The postsynaptic potential in response to single action potentials in inputs *1* and *2* occurring separately and simultaneously. **C,** Temporal summation. The postsynaptic response to two impulses in rapid succession in the same input.

fired in rapid succession, so that the resulting postsynaptic potentials overlap in time. A train of impulses in a single presynaptic neuron can cause the potential of the postsynaptic cell to change in a stepwise manner, each step caused by one of the presynaptic impulses.

Integration at the spinal motoneuron takes place because many positive and negative inputs impinge on a single motoneuron. This permits fine control of the firing pattern of the spinal motoneuron.

Modulation of Synaptic Activity

The responses of a postsynaptic neuron to single stimulations of a particular presynaptic neuron are relatively constant in magnitude and time course.

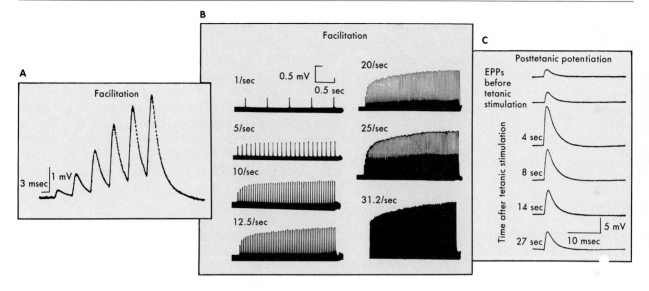

FIGURE 4-9 **A,** Facilitation at a neuromuscular junction. EPPs at a neuromuscular junction in toad sartorius muscle were elicited by successive action potentials in the motor axon. Neuromuscular transmission is depressed by 5 mM Mg^{++} and 2.1 μM curare, so that action potentials do not occur. **B,** EPPs at a frog neuromuscular junction elicited by repetitively stimulating the motor axon at different frequencies. Note that facilitation fails to occur at the lowest frequency of stimulation (1/sec) and that the degree of facilitation increases with increasing frequency of stimulation in the range of frequency employed. Neuromuscular transmission was inhibited by bathing the preparation in 12 to 20 mM Mg^{++}. **C,** Posttetanic potentiation at a frog neuromuscular junction. The top two traces indicate control EPPs in response to single action potentials in the motor nerve. The subsequent traces indicate the responses to single action potentials following tetanic stimulation (50 impulses/sec for 20 seconds) of the motor nerve. The time interval between the end of tetanic stimulation and the single action potential is shown on each trace. The muscle was treated with tetrodotoxin to prevent generation of action potentials. (**A** redrawn from Belnave RJ and Gage PW: J Physiol 266:435, 1977; **B** redrawn from Magelby KL: J Physiol 234:327, 1973; **C** redrawn from Weinrich D: J Physiol 212:431, 1971.)

However, when a presynaptic cell is stimulated repeatedly at relatively high frequency, the postsynaptic response may depend on the frequency and duration of the presynaptic stimulation. When a presynaptic axon is stimulated repeatedly, the postsynaptic response may grow with each stimulation. This phenomenon is called *facilitation* (Figure 4-9, *A*). As shown in Figure 4-9, *B*, the extent of facilitation depends on the frequency of presynaptic impulses. Facilitation dies away rapidly, within tens to hundreds of milliseconds after stimulation stops.

When a presynaptic neuron is stimulated tetanically (many stimuli at high frequency) for several seconds, a longer-lived enhancement of postsynaptic response occurs called *posttetanic potentiation* (Figure 4-9, *C*). Posttetanic potentiation persists much longer than facilitation; it lasts tens of seconds to several minutes after cessation of tetanic stimulation.

An enhancement of synaptic efficacy may occur that is intermediate in time course between facilitation and posttetanic potentiation. This enhancement is called *augmentation*, and it persists for about 10 seconds after repetitive stimulation is ended.

Facilitation, augmentation, and posttetanic potentiation are the result of the effects of repeated stimulation on the presynaptic neuron. These phenomena do not involve a change in the sensitivity of the postsynaptic cell to transmitter. With repeated stimulation, an increased number of quanta of transmitter are released. Increased levels of intracellular calcium enhance transmitter release during repetitive stimulation, but other intracellular events are involved as well

When a synapse is repetitively stimulated for a long time, a point is reached at which each successive presynaptic stimulation elicits a smaller postsynaptic response. This phenomenon is called *synaptic fatigue* (neuromuscular depression at the motor endplate). The postsynaptic cell at a fatigued synapse responds normally to transmitter applied from a micropipette; thus the defect is presynaptic. In some cases a decrease in quantal content (the amount of transmitter per synaptic vesicle) has been implicated in synaptic fatigue. A fatigued synapse typically recovers in a few seconds.

Ionic Mechanisms of Postsynaptic Potentials in Spinal Motoneurons

Much of our current knowledge of synaptic mechanisms in the mammalian central nervous system is derived from studies of cat spinal motoneurons.

EPSP The EPSP (see Figure 4-7) of the cat spinal motoneuron is caused by a transient increase of the conductance of the postsynaptic membrane to both Na^+ and K^+. This was demonstrated by injecting Na^+ or K^+ into the cell to raise the intracellular concentration of that particular ion. Injection of either Na^+ or K^+ diminishes the EPSP because, when the conductance increase occurs, there is a smaller tendency for Na^+ to flow in and depolarize the cell after Na^+ injection, and a greater tendency for K^+ to flow out and oppose depolarization after K^+ injection.

Injection of Cl^- into the cell does not alter the EPSP; thus, Cl^- apparently is not involved in the EPSP.

IPSP The IPSP (see Figure 4-7) of cat spinal motoneurons is caused by an increased Cl^- conductance of the postjunctional membrane. At rest there is a net tendency for Cl^- to enter the cell. The increase in Cl^- conductance, as the result of transmitter release at the inhibitory synapse, allows Cl^- to enter the postsynaptic cell and hyperpolarize it. Injecting Cl^- into the cell or hyperpolarizing it decreases the net tendency for Cl^- to enter the cell and decreases the size of the IPSP. Injection of Na^+ or K^+ produces no change in the IPSP, suggesting that neither Na^+ nor K^+ is involved in the IPSP of spinal motoneurons. In certain other cell types the IPSP appears to be caused by an increased K^+ conductance.

Presynaptic Inhibition Inhibitory interactions are vital in stabilizing the central nervous system. Another type of inhibition is called presynaptic inhibition. If an inhibitory input to a spinal motoneuron is stimulated tetanically and then an excitatory input is stimulated once, the EPSP elicited by stimulating the excitatory input may be reduced in magnitude after the inhibitory volley. This is believed to occur by a mechanism in which axon collaterals of the inhibitory

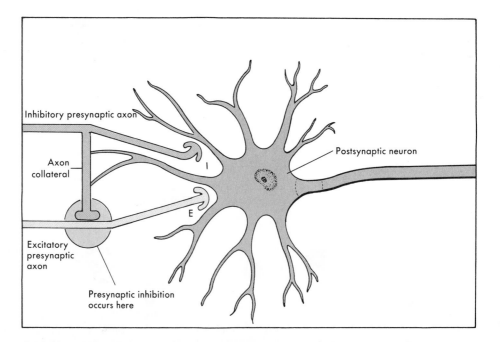

FIGURE 4-10   Presynaptic inhibition. Axon collaterals of the inhibitory axon (*I*) synapse on the excitatory axon terminal (*E*). An action potential in the inhibitory axon depolarizes the excitatory axon terminal. The depolarized excitatory axon terminal will release less transmitter in response to an action potential in the excitatory neuron.

axons synapse on the excitatory nerve terminals (Figure 4-10). An action potential in the inhibitory nerve depolarizes (for a rather long time) the excitatory nerve terminal. This brings the excitatory nerve terminal closer to threshold. The partly depolarized excitatory terminal will release less transmitter in response to an action potential. The smaller release of transmitter diminishes the EPSP. The phenomenon of decreased transmitter release from a partially depolarized nerve terminal is well known at the neuromuscular junction.

TRANSMITTERS IN THE NERVOUS SYSTEM

Identification of Transmitter Substances

Several compounds have been proposed to function as neurotransmitters. Such compounds are called candidate or putative neurotransmitters. *Candidate neurotransmitters* are usually concentrated in specific neurons or in specific neuronal pathways. Microapplication of *putative transmitters* to particular areas of the central nervous system (CNS) may evoke specific responses. Correlation of information about the localization of the putative transmitter with knowledge of the location of neurons that respond to this substance, as well as the ways in which they respond, allows intelligent speculation about the functions of a putative neurotransmitter.

It is often difficult to prove that a substance is the transmitter at a particular synapse. A putative transmitter (X) must satisfy the following criteria before it is accepted as a proven transmitter at a particular synapse:

1. The presynaptic neurons must contain X and must be able to synthesize it.
2. X must be released by the presynaptic neurons on appropriate stimulation.
3. Microapplication of X to the postsynaptic membrane must mimic the effects of stimulation of the presynaptic neuron.
4. The effects of presynaptic stimulation and of microapplication of X should be altered in the same way by pharmacologic agents.

Our knowledge of neurotransmitters has increased greatly in recent years. Some transmitters have rapid and transient effects on the postsynaptic cell. Other transmitters have effects that are much slower in onset and may last for minutes or even hours. Most of the candidate neurotransmitters that have been discovered so far fall into three major chemical classes: amines, amino acids, and oligopeptides.

CNS Transmitters

Acetylcholine As discussed previously, acetylcholine is the transmitter used by all motor axons that arise from the spinal cord. Acetylcholine plays a central role in the autonomic nervous system; it is the transmitter for all autonomic preganglionic neurons and also for postganglionic parasympathetic fibers. The Betz cells of the motor cortex use acetylcholine as their transmitter. The basal ganglia, which are involved in the control of movement, contain high levels of acetylcholine, and acetylcholine is believed to be a transmitter there. In addition, acetylcholine may be the transmitter in a large number of central pathways.

Biogenic Amine Transmitters Among the amines that may serve as neurotransmitters are norepinephrine, epinephrine, dopamine, serotonin, and histamine.

Dopamine, norepinephrine, and *epinephrine* are *catecholamines,* and they share a common biosynthetic pathway that starts with the amino acid tyrosine. Tyrosine is converted to L-dopa by tyrosine hydroxylase. L-Dopa is converted to dopamine by a specific decarboxylase. In dopaminergic neurons the pathway stops here. Noradrenergic neurons have another enzyme, dopamine β-hydroxylase, that converts dopamine to norepinephrine. Other cells add a methyl group to norepinephrine to produce epinephrine. Norepinephrine is the primary transmitter for postganglionic sympathetic neurons.

Neurons that contain high levels of dopamine are prominent in the midbrain regions known as the substantia nigra and the ventral tegmentum. Some of the axons of these neurons travel to the forebrain, where they may play a role in emotional responses. Other dopaminergic axons terminate in the corpus striatum, where they may control complex movements. The degeneration of dopaminergic synapses in the corpus striatum occurs in Parkinson's disease and may be a major cause of the muscular tremors and rigidity that characterize this disease.

Serotonin (5-hydroxytryptamine)-containing neurons are present in high concentration in certain nuclei located in the brainstem. Serotonergic neurons may be involved in temperature regulation, sensory perception, onset of sleep, and control of mood.

Histamine is present in certain neurons in the hypothalamus. The functions of these presumably histaminergic neurons are not yet known.

Amino Acid Transmitters *Glycine,* the simplest amino acid, is an inhibitory neurotransmitter released by certain spinal interneurons.

Glutamate and *aspartate,* dicarboxylic amino acids, have strong excitatory effects on many neurons in the brain. Glutamate and asparate are the most prevalent excitatory transmitters in the brain.

γ-Aminobutyric acid (GABA) is not incorporated into proteins, nor is it present in all cells (as are the other naturally occurring amino acids). GABA is produced from glutamate by a specific decarboxylase present only in the CNS. Among the cells that contain GABA are some cells in the basal ganglia, the cerebellar Purkinje cells, and certain spinal interneurons. In all known cases GABA functions as an *inhibitory transmitter.* It is the most common transmitter in the brain. GABA may be the neurotransmitter at as many as one third of the synapses in the brain.

Neuroactive Peptides Certain cells release peptides that act at very low concentrations to excite or inhibit neurons. To date, about 25 of these so-called neuropeptides, ranging from 2 to about 40 amino acids long, have been identified; most are listed in the box. It is likely that more neuropeptides will soon be added to this list.

Neuropeptides typically affect their target neurons at lower concentrations than the classic neurotransmitters discussed previously, and the neuropeptides usually act longer. Several neuropeptides are more familiar as hormones. Neuropeptides may act as hormones, as neurotransmitters, or as neuromodulators. A *hormone* is a substance that is released into the blood and that reaches its target cell via the circulation. A *neurotransmitter* or *neuromodulator* is released near the surface of its target cell and diffuses to the target cell. Neurotransmitters, as discussed earlier, act to change the conductance of the target cell to one or more ions, and in that way they change the membrane potential of the target cell. A neuromodulator *modulates* synaptic transmission. The neuromodulator may act presynaptically to change the amount of transmitter released in response to an action potential, or it may act on the postsynaptic cell to modify its response to transmitter. A number of neuropeptides act as true transmitters at particular

Neuroactive Peptides

GUT-BRAIN PEPTIDES

Vasoactive intestinal polypeptide (VIP)
Cholecystokinin octapeptide (CCK-8)
Substance P
Neurotensin
Methionine enkephalin
Leucine enkephalin
Motilin
Insulin
Glucagon

HYPOTHALAMIC-RELEASING HORMONES

Thyrotropin-releasing hormone (TRH)
Luteinizing hormone–releasing hormone (LHRH)
Somatostatin (growth hormone releasing–inhibiting factor, or SRIF)

PITUITARY PEPTIDES

Adrenocorticotropin (ACTH)
β-Endorphin
α-Melanocyte-stimulating hormone (α-MSH)

OTHERS

Dynorphin
Angiotensin II
Bradykinin
Vasopressin
Oxytocin
Carnosine
Bombesin

Modified from Snyder SH: Science 209:976, 1980. Copyright 1980 by American Association for the Advancement of Science.

synapses and as neuromodulators at other synapses.

In several instances, some neuropeptides coexist in the same nerve terminals with classic transmitters (Table 4-1). In some of these cases the neuropeptide is released along with the transmitter in response to nerve stimulation. Whether the transmitter and the neuropeptide coexist in the same synaptic vesicles, or are packaged in separate vesicles, is not yet clear.

Synthesis of Neuropeptides. Most classic neurotransmitters are synthesized in nerve terminals by pathways that involve soluble enzymes and simple precursors. Neuropeptides are synthesized in the cell

Table 4-1 Examples of the Coexistence of a Classic Transmitter and Neuropeptide Within the same Nerve terminal*

Transmitter	Neuropeptide
Acetylcholine	Vasoactive intestinal peptide (VIP)
Norepinephrine	Somatostatin
	Enkephalin
	Neurotensin
Dopamine	Cholecystokinin (CCK)
	Enkephalin
Epinephrine	Enkephalin
Serotonin	Substance P
	Thyrotropin-releasing hormone (TRH)

Reprinted by permission of the publisher from *Chemical messengers: small molecules and peptides* by Schwartz JH. In Kandel ER and Schwartz JH, editors: Principles of neural science. Copyright © 1981 by Elsevier Science Publishing Co, Inc.
*Evidence for the coexistence of a classic transmitter substance with a neuroactive peptide has been reported for these combinations. With the information thus far available, it is not yet possible to determine the specificity of the pairs and their physiological significance.

body. They are encoded by the cell's DNA and transcribed into messenger RNA, which is translated on polyribosomes bound to the endoplasmic reticulum. *Secretory vesicles,* containing the neuropeptide, are released from the mature face of the Golgi complex. Secretory vesicles are moved by fast axonal transport to the axon terminals, where they are known as *synaptic vesicles.*

Some neuropeptides are synthesized as *preprohormones.* Cleavage of the signal sequence converts the preprohormone to a *prohormone.* Proteolytic cleavage of the prohormone may then release one or more active peptides. In some cases one prohormone may contain several active peptide sequences. For example, the prohormone of the opioid peptide, β-endorphin, is a 31,000-dalton polypeptide that contains several active sequences. One cleavage of the prohormone releases adrenocorticotropic hormone (ACTH) and β-lipotropin. Cleavage of ACTH releases a hormone, melanocyte-stimulating hormone (α-MSH), and cleavage of β-lipotropin releases β-MSH and a number of active β-endorphins.

Opioid Peptides *Opiates* are drugs that are derived from the juice of the opium poppy. Opiates are useful therapeutically as powerful analgesics. They exert their analgesic affect by binding to specific opiate receptors. The binding of opiates to their receptors is stereospecifically inhibited by a morphine derivative called *naloxone.* Compounds that do not derive from the opium poppy, but that exert direct effects by binding to opiate receptors, are called *opioids.* Operationally, opioids are defined as direct-acting compounds whose effects are stereospecifically antagonized by naloxone.

The three major classes of endogenous opioid peptides in mammals are enkephalins, endorphins, and dynorphin. *Enkephalins* are the simplest opioids; they are pentapeptides. *Dynorphin* and the *endorphins* are somewhat longer peptides that share one or the other of the enkephalin sequences at their *N*-terminal ends.

Opioid peptides are widely distributed in neurons of the CNS and intrinsic neurons of the gastrointestinal tract. Opioid peptides are found in vesicles that resemble synaptic vesicles. The endorphins are discretely localized in particular structures of the CNS, whereas the enkephalins and dynorphins are more widely distributed.

Nonopioid Neuropeptides Most of the known neuropeptides are not opioids.

Substance P, a peptide of 11 amino acids, is present in specific neurons in the brain, in primary sensory neurons, and in plexus neurons in the wall of the gastrointestinal tract. Substance P is a member of a group of neuropeptides called tachykinins (most of the other known tachykinins are present in amphibians). Substance P was the first so-called gut-brain peptide to be discovered. The wall of the gastrointestinal tract is richly innervated with neurons that form networks or plexuses. The intrinsic plexuses of the gastrointestinal tract exert primary control over its motor and secretory activities (see Part VI). These enteric neurons contain many of the neuropeptides, including substance P, that are found in the brain and spinal column.

Substance P is the suspected transmitter at synapses made by primary sensory neurons (their cell bodies are in the dorsal root ganglia) with spinal interneurons in the dorsal horn of the spinal column. Enkephalins act to decrease the release of substance P at these synapses, thereby inhibiting the pathway for pain sensation at the first synapse in the pathway. Opioids also have inhibitory effects in structures in the brain involved in the perception of pain.

Vasoactive intestinal polypeptide (VIP) is a member of a family of neuropeptides related to the hormone *secretin.* VIP was first discovered as a gastrointestinal hormone, but it is now known to be a neuropeptide as well. *Secretin* (27 amino acids) and *glucagon* (29 amino acids) have 14 amino acids in common at similar positions. VIP (28 amino acids) and *gastric inhibitory peptide (GIP,* 43 amino acids) have extensive sequence homology with secretin and glucagon. These four neuropeptides probably have a common ancestor peptide and arose in evolution by gene duplication.

VIP is widely distributed in the CNS and in the intrinsic neurons of the gastrointestinal tract. In neurons in the brain VIP has been localized in synaptic vesicles. VIP appears to function as an inhibitory transmitter to vascular and nonvascular smooth muscle and as an excitatory transmitter to glandular epithelial cells.

Secretin, glucagon, and GIP are molecules whose function as hormones has been well characterized. These peptides have also been found in particular neurons in the CNS, but their CNS functions remain undetermined.

Cholecystokinin (CCK) is a member of a group of neuropeptides that includes gastrin and cerulein, which have similar *C*-terminal sequences. CCK is a well-known gastrointestinal hormone that elicits contraction of the gallbladder (see Chapter 30). CCK(39) is cleaved near its *N* terminus to produce CCK(33), the physiologic form in the gastrointestinal tract. The *N*-terminal octapeptide, CCK(8), is present in particular neurons of the CNS.

Neurotensin is one of the most recently discovered gut-brain neuropeptides. It is present in enteric neurons and in the brain. When injected into cerebrospinal fluid at low concentrations, neurotensin lowers body temperature. Thus neurotensin may function in temperature regulation.

Other Neuromodulators

Some important neuromodulators are not peptides. Purines and purine nucleotides and nucleosides function as neuromodulators in the central, autonomic, and peripheral nervous systems. Substances that serve as classic neurotransmitters may also act as neuromodulators. In some cases the transmitter binds to receptors on the presynaptic neuron that released it, thereby regulating its own release.

BIBLIOGRAPHY
Journal Articles

Bloom FE: Neuropeptides, Sci Am 245:148, 1981.

Gregory RA, editor: Regulatory peptides of gut and brain, Br Med Bull 38:219, 1982.

Hughes J, editor: Opioid peptides, Br Med Bull 39:1, 1983.

Iversen LL: The chemistry of the brain, Sci Am 241:134, 1979.

Llinas RK: Calcium in synaptic transmission, Sci Am 247:56, 1982.

Books and Monographs

Aidley DJ: The physiology of excitable cells, ed 2, Cambridge, 1978, Cambridge University Press.

Cooper JR et al: The biochemical basis of neuropharmacology, ed 4, New York, 1982, Oxford University Press, Inc.

Eccles JC: The physiology of synapses, Berlin, 1964, Springer Verlag.

Kandel ER, and Schwartz JH: Principles of neural science, ed 2, New York, 1985, McGraw-Hill Book Co.

Katz B: Nerve, muscle, and synapse, New York, 1966, McGraw-Hill Book Co.

Kuffler SW et al: From neuron to brain, ed 2, Sunderland, Mass, 1984, Sinauer Associates, Inc.

NERVOUS SYSTEM

WILLIAM D. WILLIS, JR.

PART II

CHAPTER 5

Cellular Organization of the Nervous System

The nervous system is a communication network that allows an organism to interact in appropriate ways with the environment. This system has sensory components that detect environmental events, integrative components that process sensory data and information stored in memory, and motor components that generate movements and other activity. The nervous system can be subdivided into peripheral and central parts, each with a number of further subdivisions.

The functional unit of the nervous system is the *neuron,* whose dendrites and axon make synaptic connections with other neurons to form a communications network. Neuroglia assist neurons, for example, by providing myelin sheaths that speed the conduction of nervous impulses along axons. Neural activity is coded, and information is passed from one neuron to the next by synaptic transmission. Axons not only transmit information but also move chemical substances by axonal transport. Damage to axons may lead to an axonal reaction in the cell bodies, wallerian degeneration of the distal part of the axon, and sometimes transneuronal degeneration. Axonal transport and the reactions to injury can be used experimentally to trace neural pathways. Regeneration of neurons is more effective in peripheral than in central neurons.

Learning and memory permit behavior to change appropriately in response to environmental challenges based on past experience. Other systems, such as the endocrine and immune systems, share these functions, but the nervous system is specialized for them.

Excitability is a cellular property of neurons that enables them to perform their functions. Excitability is manifested by such electrical events as nerve impulses (or *action potentials*), *receptor potentials,* and *synaptic potentials* (see Chapter 4). Chemical events often accompany these electrical ones.

Sensory detection is accomplished by special nerve cells called *sensory receptors.* Various forms of energy are sensed, including light, sound and other mechanical events, chemicals, temperature gradients, and in some animals electrical fields.

Information processing depends on *intercellular communication* in neural circuits, which is accomplished by nerve cells as they respond to and generate chemical signals. The mechanisms involved require both electrical and chemical changes.

Behavior may be covert, as in *cognition* or *memory,* but it is often readily observable as a motor act, such as a *movement* or an *autonomic response.* In humans a particularly important set of behaviors are those involved in *language.*

GENERAL FUNCTIONS OF THE NERVOUS SYSTEM

Functions of the nervous system include *sensory detection, information processing,* and *behavior.*

ORGANIZATION OF THE NERVOUS SYSTEM

The nervous system consists of a highly complex aggregation of cells, part of which forms a communication network and another part, a supportive matrix. The communication network is formed by neurons. The cells involved in communication are specialized for receiving and making decisions on information and for transmitting signals to other neurons or to effector cells. The human brain contains approximately 10^{12} neurons. The supportive cells of the nervous system include the *neuroglia* (meaning "nerve glue"). These cells help maintain an appropriate local environment for neurons, or they ensheath axons to increase the speed of nerve impulse propagation. The human brain has 10 times as many neuroglia as neurons.

The nervous system of most animals can be subdivided into a *peripheral nervous system* and a *central nervous system.* The emphasis here is on the mammalian nervous system, especially that of humans.

Peripheral Nervous System

The peripheral nervous system provides an interface between the central nervous system and the environment, including both the external world and the body apart from the nervous system. The peripheral nervous system includes a sensory component, formed by *sensory receptor organs* and *primary afferent neurons,* and motor components to command effector organs to perform muscular or glandular activity. Motor components include *somatic motor fibers* and *autonomic ganglia* and *autonomic motor fibers.* Autonomic neurons can be further subdivided into *sympathetic, parasympathetic,* and *enteric neurons.* Somatic motor fibers cause contraction of the skeletal muscle fibers. Autonomic motor fibers excite or inhibit cardiac or smooth muscle and glands. The actions of the sympathetic nervous system include preparing the organism for emergency action, whereas the parasympathetic and enteric nervous systems promote more routine activities, such as digestion.

Central Nervous System

The central nervous system (CNS) includes the *spinal cord* and the *brain* (Figure 5-1). The brain can be further subdivided into five regions, based on embryologic development: the myelencephalon, metencephalon, mesencephalon, diencephalon, and telencephalon. In the adult brain the *myelencephalon* includes the medulla; the *metencephalon,* the pons and cerebellum; the *mesencephalon,* the midbrain;

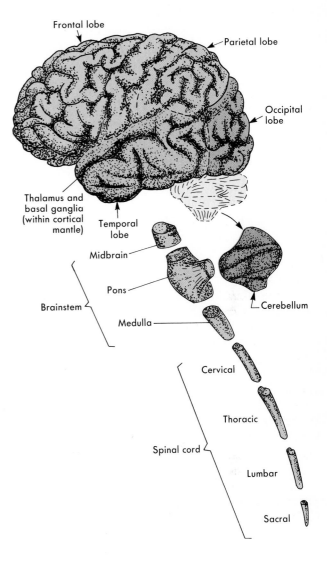

FIGURE 5-1   Exploded view showing the major components of the central nervous system. Also shown are the four major divisions of the cerebral cortex: the occipital, parietal, frontal, and temporal lobes.

and the *telencephalon,* the various lobes of the cerebral cortex. The *diencephalon* includes the thalamus and the telencephalon, the basal ganglia; these are hidden from view in Figure 5-1.

COMPOSITION OF NERVOUS TISSUE

Nervous tissue consists largely of neurons and neuroglia. Neurons are responsible for communica-

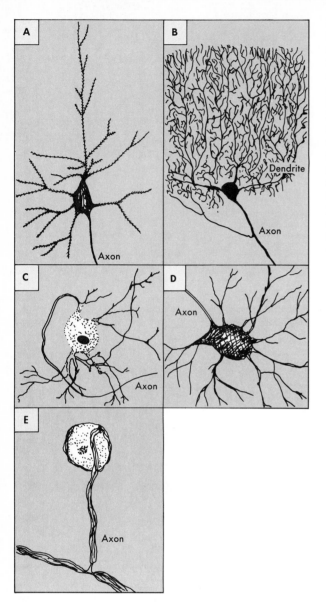

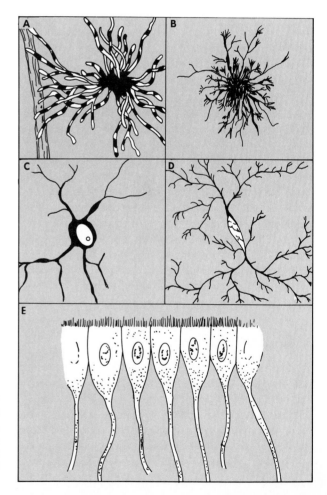

FIGURE 5-2   Various forms of neurons. **A,** Neuron characterized by a cell body that has a roughly pyramidal shape. This type of neuron, called a pyramidal cell, is typical of the cerebral cortex. Note the many spinous processes lining the surface of the dendrites. **B,** Cell type first described by the Czechoslovakian neuroanatomist Purkinje and since known as the Purkinje cell. Purkinje cells are characteristic of the cerebellar cortex. The cell body is pear shaped, with a rich dendritic plexus originating from one end and the axon from the other. The fine branches of the dendrites are covered with spines (not shown). **C,** A sympathetic postganglionic motoneuron. **D,** An α-motoneuron of the spinal cord. Both **C** and **D** are multipolar neurons with radially arranged dendrites. **E,** A sensory dorsal root ganglion cell; no dendrites are present. The axon branches into a central and a peripheral process. Because the axon results from fusion of two processes during embryonic development, these cells are described as pseudounipolar neurons rather than unipolar.

FIGURE 5-3   Different types of neuroglial cells of the central nervous system. **A,** Fibrous atrocyte; **B,** protoplasmic astrocyte. Note the glial foot processes in association with a capillary in **A**. **C,** An oligodendrocyte. Each of the processes is responsible for the production of one or more myelin sheath internodes about central axons. **D,** Microglial cell. **E,** Ependymal cells.

tions. Neurons that communicate with the periphery include *sensory receptor cells* and *somatic* and *autonomic motoneurons.* Other neurons perform integrative tasks through activity in central neural networks. Not surprisingly, given the many different roles that neurons play, neurons display a great variety of shapes and sizes (Figure 5-2).

Support for neurons is provided by *neuroglial cells* (Figure 5-3). These include *Schwann cells* and *satellite cells* in the peripheral nervous system and *astrocytes* and *oligodendroglia* (oligodendrocytes) in the CNS. *Microglia* and *ependymal cells* are also considered neuroglia.

Myelin is formed by Schwann cells and oligodendroglia and allows nerve impulses to be conducted rapidly. Satellite cells encapsulate dorsal root and cranial nerve ganglion cells and regulate their microenvironment. Astrocytes play a similar role in the CNS, although astrocytes contact only a part of the surface of central neurons. However, their processes surround groups of synaptic endings and isolate them from adjacent synapses. Astrocytes have foot processes that contact capillaries and the connective tissue at the surface of the CNS (Figure 5-3). These foot processes may help limit the free diffusion of substances into the CNS.

Microglia are phagocytes that remove the products of cellular damage from the CNS. They are probably derived from the circulation.

Ependymal cells form an epithelium that separates the CNS from the ventricles, a series of cavities within the brain; these cavities contain *cerebrospinal fluid (CSF).* Many substances diffuse readily across the ependyma between the extracellular space of the brain and the CSF. CSF is secreted by specialized ependymal cells of the *choroid plexuses,* which are found in certain ventricles.

Delivery of nutrients and removal of wastes are accomplished by the *vascular system.* Capillaries and other blood vessels are abundant in nervous tissue. Diffusion of many substances between the blood and the CNS is limited by the *blood-brain barrier.* This barrier is formed chiefly by tight junctions between capillary endothelial cells.

The external surface of the CNS is covered by several layers of connective tissue. These layers form the *pia mater, arachnoid,* and *dura mater.*

MICROSCOPIC ANATOMY OF THE NEURON

Most neurons have the following parts: a cell body, or *soma;* one or more dendrites; and an axon.

The cell body (Figure 5-4) contains the *nucleus* and *nucleolus* of the neuron. It possesses a well-developed biosynthetic apparatus for the manufacture of membrane constituents, synthetic enzymes, and other chemical substances needed for the specialized functions of the nerve cell. The neuronal biosynthetic apparatus includes *Nissl bodies,* which are stacks of rough endoplasmic reticulum, and a prominent *Golgi apparatus.* The soma also contains numerous *mitochondria* and cytoskeletal elements, including *neurofilaments* and *microtubules. Lipofuscin* is a pigment formed from incompletely degraded membrane components and accumulates in some neurons. A few groups of neurons in the brainstem contain melanin pigment.

The *dendrites* are extensions of the cell body. In some neurons dendrites may be about 1 mm in length, and they account for more than 90% of the surface area of many neurons. The proximal dendrites (near the cell body) contain Nissl bodies and parts of the Golgi apparatus. However, the main cytoplasmic organelles in dendrites are microtubules and neurofilaments.

The axon arises from the soma (or sometimes from a dendrite) in a specialized region called the axon hillock. The axon hillock and axon differ from the soma and proximal dendrites because they lack rough endoplasmic reticulum and free ribosomes, as well as the Golgi apparatus. The axon contains smooth endoplasmic reticulum and a prominent cytoskeleton. Axons may be short and, as with dendrites, terminate near the soma *(Golgi type 1 neurons),* or they may be long *(Golgi type 2 neurons)* and extend to a meter or more.

Axons may be ensheathed or bare. In the peripheral nervous system, axons are always ensheathed by Schwann cells. Many axons are surrounded by a spiral, multilayered wrapping of Schwann cell membrane called a *myelin sheath.* In the CNS, *myelinated* axons are ensheathed by oligodendroglia (Figure 5-5); other axons are *unmyelinated.* In the peripheral nervous system, unmyelinated axons are embedded in Schwann cells but are not wrapped in myelin (Figure 5-6). In the CNS, unmyelinated axons are bare.

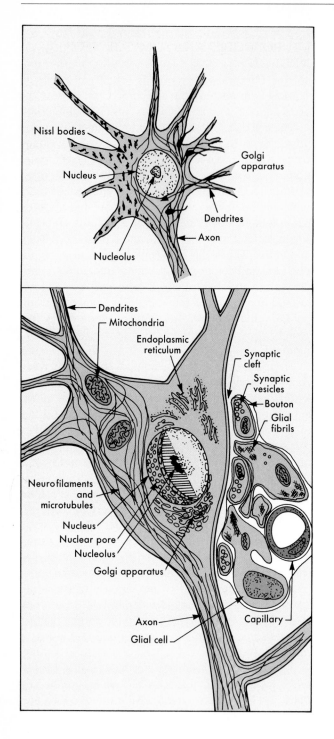

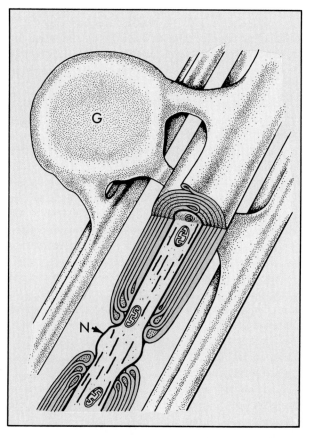

FIGURE 5-4   Organelles of the neuron. The smaller drawing on top shows the organelles typical of a neuron, as seen with the light microscope. The portion of the illustration to the left of the soma represents structures seen with a Nissl stain. These include the nucleus and nucleolus, Nissl bodies in the cytoplasm of the cell body and proximal dendrites, and as a negative image, the Golgi apparatus. The absence of Nissl bodies in the axon hillock and axon is also shown. To the right of the soma are structures seen with a heavy-metal stain; these include neurofibrils. The appropriate heavy-metal stain may demonstrate the Golgi apparatus (not shown). On the surface of the neuron several synaptic endings are indicated, as stained by the heavy metal. The large drawing shows structures visible at the electron microscopic level. The nucleus, nucleolus, chromatin, and nuclear pores are represented. Mitochondria, rough endoplasmic reticulum, Golgi apparatus, neurofilaments, and microtubules are in the cytoplasm. Along the surface membrane are such associated structures as synaptic endings and astrocytic processes.

FIGURE 5-5   The myelin sheath in the CNS. Each oligodendroglial process forms the internode for one axon. *G*, oligodendroglial cell; *N*, node of Ranvier.

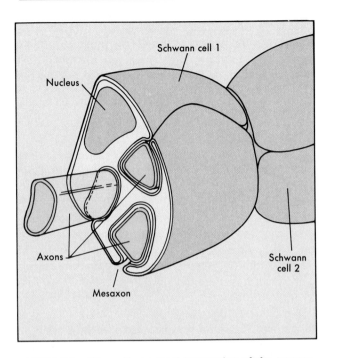

FIGURE 5-6   Three-dimensional impression of the appearance of a Remak's bundle. The cut face of the bundle is seen to the left. One of the three unmyelinated axons is represented as protruding from the bundle. A mesaxon is indicated, as is the nucleus of the Schwann cell. To the right, the junction of adjacent Schwann cells is depicted.

TRANSMISSION OF INFORMATION

A major role of axons is to transmit information from the region of the cell body and dendrites of a neuron to synapses on other neurons or effector cells. The information is generally transmitted as a series of nerve impulses.

The speed of transmission of information depends partly on the *conduction velocity* of the axon. Conduction velocity in turn depends on the diameter of the axon and whether it is unmyelinated or myelinated. Unmyelinated axons are generally less than $1\mu m$ in diameter and conduct at speeds less than 2.5 $m \cdot sec^{-1}$. About 1 second would be required for a signal in an unmyelinated axon supplying a sensory receptor in a person's foot and having a conduction velocity of 1 $m \cdot sec^{-1}$ to reach the spinal cord. Myelinated axons have diameters of 1 to 20 μm and conduct at speeds of 3 to 120 $m \cdot sec^{-1}$. A spinal motoneuron with an axon that conducts at 100 $m \cdot sec^{-1}$ would be able to trigger the contraction of a toe muscle in about 10 msec.

In the CNS certain neurons that lack axons *(amacrine cells)* signal information by electrical current flow rather than by generating action potentials. This current flow produces a *local potential,* which decays over a short distance (millimeters to hundreds of micrometers, depending on the length constant of the neuron involved). Local potentials differ from action potentials in that they are nonpropagating and cannot spread over long distances. In contrast, action potentials can propagate over long distances along axons.

Signaling by local potentials is also characteristic of sensory receptors, which produce *receptor potentials,* and of communications between nerve cells, *synaptic potentials.*

Coding

Information conveyed by axons may be coded in several ways. Sets of neurons may be dedicated to a general function, such as a particular sensory modality. For example, the visual pathway includes the retina, optic nerve and tract, the lateral geniculate nucleus of the thalamus, and the visual part of the cerebral cortex. The normal means of activating the visual system is by light striking the retina, but mechanical or electrical stimulation of the visual system will also produce a visual response, although a distorted one. Thus neurons of the visual system can be regarded as a *labeled line,* which, when activated, causes a visual sensation. Other sensory systems provide further examples of labeled lines.

A second way in which information is coded by the nervous system is through *spatial maps.* The body surface may be mapped by an array of neurons in a sensory or a motor system. This is termed a *somatotopic map* (or, for humans, a "homunculus"; see Chapter 6). With the visual system a *retinotopic map* exists. In the auditory system frequency of sound is represented in a *tonotopic map.*

A third method for coding information is by *patterns of nerve impulses.* Axons transmit a sequence of nerve impulses that result in synaptic transmission of information to a new set of neurons. The information that is communicated is coded in terms of the structure of the nerve impulse trains. Several different types of nerve impulse codes have been proposed. A common code is likely to depend on the *mean discharge frequency.* Other candidate codes depend on the *time of firing,* the *temporal pattern,* and the *duration of bursts.* Still other codes have been proposed.

Synaptic Transmission

Neurons communicate with each other at specialized junctions called *synapses* (see Chapter 4). Typically, synapses are formed between the terminals of the axon of one neuron and the dendrites of another (Figure 5-7); these are called *axodendritic synapses.* However, many other types of synapses occur, including *axosomatic, axoaxonal,* and *dendrodendritic.* The synapse between a motoneuron and a skeletal muscle fiber is called an *endplate* or *neuromuscular junction.*

Axonal Transport

Many axons are too long to allow the movement of substances from the soma to the synaptic endings simply by diffusion. Certain membrane and cytoplasmic components that originate in the biosynthetic apparatus of the soma and proximal dendrites must be distributed along the axon and especially to the presynaptic elements of synapses to replenish secreted or inactivated materials. A special transport mechanism, called axonal transport, accomplishes this distribution.

Several types of axonal transport exist. Certain membrane-bound organelles and mitochondria are transported relatively rapidly by fast axonal transport. Substances (e.g., proteins) that are dissolved in cytoplasm are moved by slow axonal transport. In mammals fast axonal transport proceeds as rapidly as 400 mm/day, whereas slow axonal transport occurs at about 1 mm/day. This means that synaptic vesicles can travel from a motoneuron in the spinal cord to a neuromuscular junction in a person's foot in about 2½ days, whereas the movement of many soluble proteins over the same distance takes nearly 3 years.

Axonal transport requires metabolic energy and involves calcium ions. The cytoskeleton, particularly microtubules, provides a system of guide-wires along which membrane-bound organelles move (Figure 5-8). These organelles may attach to microtubules through a linkage similar to that between the thick and thin filaments of skeletal muscle fibers; calcium triggers the movement of the organelles along the microtubules.

Axonal transport occurs in both directions. Transport from the soma toward the axonal terminals is called *anterograde axonal transport* (Figure 5-9, *A*). This process allows the replenishment of synaptic vesicles and enzymes responsible for neurotransmitter synthesis in synaptic terminals. Transport in the opposite direction is *retrograde axonal transport* (Figure 5-9, *B*). This process returns synaptic vesicle membrane to the soma for lysosomal degradation. Marker substances, such as the enzyme *horseradish peroxidase*, can be transported anterogradely or retrogradely and can be used experimentally to trace neural pathways.

REACTIONS TO INJURY

Injury to nervous tissue elicits responses by neurons and neuroglia. Severe injury causes cell death. Once a neuron is lost, it cannot be replaced because neurons are postmitotic cells.

When an axon is transected, the soma of a neuron

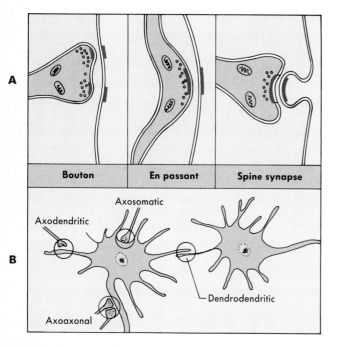

FIGURE 5-7   A, Several types of synaptic configurations. *Left,* The synaptic bouton, which consists of the distended terminal of an axon ending on a postsynaptic structure. *Center,* An en passant ending. The presynaptic element does not terminate; instead synaptic specializations exist along the course of an axon. *Right,* An example of a synapse with a postsynaptic specialization, a dendritic spine. The spine synapse involves either a synaptic bouton or an en passant synapse presynaptically and a protrusion of the postsynaptic element. The spine may contain a special organelle, a cisternal structure called a spine apparatus (not illustrated). **B,** Types of synapses according to the parts of neurons contributing to the presynaptic and postsynaptic elements.

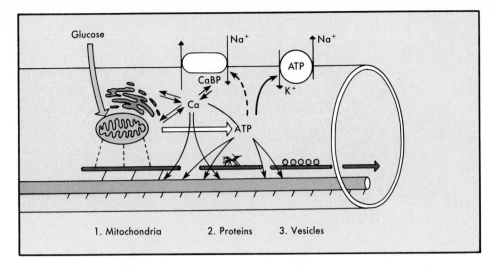

FIGURE 5-8   Axonal transport has been proposed to depend on the movement of transport filaments. Energy is required and is supplied by glucose. Mitochondria control the level of cations in the axoplasm by supplying adenosine triphosphate *(ATP)* to the ion pumps. An important cation for axonal transport is calcium. Transport filaments (black bars at bottom of drawing) move along the cytoskeleton (microtubules, *M,* or neurofilament, *NF*) by means of crossbridges. Transported components attach to the transport filaments.

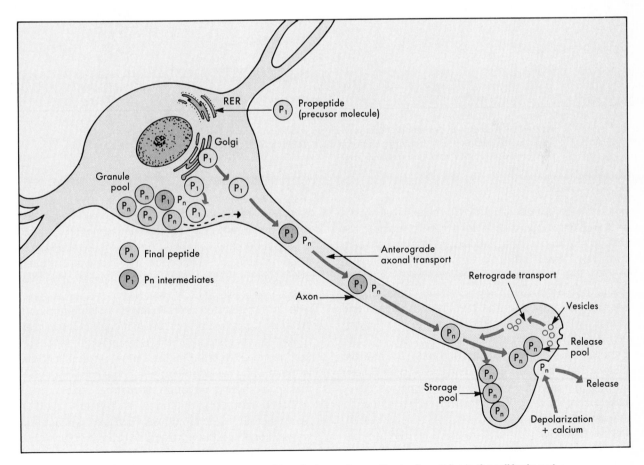

FIGURE 5-9   A, Axonal transport and its relation to the synthesis of peptides in the cell body and their release from terminals. *RER,* Rough endoplasmic reticulum.   *Continued.*

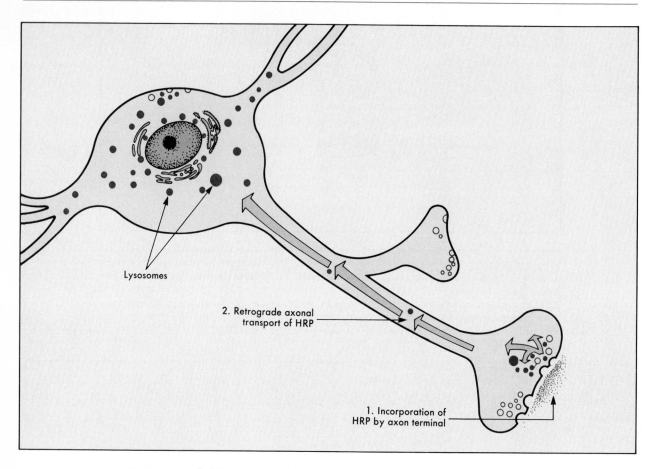

Lysosomes

2. Retrograde axonal
transport of HRP

1. Incorporation of
HRP by axon terminal

FIGURE 5-9 cont'd.   B, Schematic summary of incorporation, retrograde axonal transport, and lysosomal accumulation of horseradish peroxidase *(HRP)* in neurons. Anterograde axonal transport of HRP from the soma is not illustrated.

may show the axonal reaction. Normally, Nissl bodies stain well with basic aniline dyes, which attach to the ribonucleic acid of the ribosomes (Figure 5-10, *A*). During the axonal reaction the cisterns of the rough endoplasmic reticulum become distended with the products of protein synthesis. The ribosomes become disorganized, and thus the Nissl bodies are stained weakly by basic aniline dyes. This alteration in staining is termed *chromatolysis* (Figure 5-10, *C*). The soma may also become swollen and rounded, and the nucleus may assume an eccentric position. These morphologic changes reflect the cytologic processes that accompany protein synthesis. The damaged neuron is repairing itself.

The axon distal to the transection dies (Figure 5-10, *B*). Within a few days the axon and all the synaptic endings formed by the axon disintegrate. If the

axon was myelinated, the myelin sheath fragments and is eventually phagocytized and removed. However, the neuroglial cells that formed the myelin sheath remain viable. This sequence of events was originally described by Waller and is called *wallerian degeneration*.

If the axons that provide the sole or predominant synaptic input to a neuron or an effector cell are interrupted, the postsynaptic cell may undergo degeneration and even death. The best known example of this is the atrophy of skeletal muscle fibers following interruption of their innervation by motoneurons.

These pathologic changes have been useful in neuroanatomic investigations to trace neural pathways. For example, retrograde chromatolysis has been used to reveal groups of neurons whose axons have been deliberately interrupted. The projection

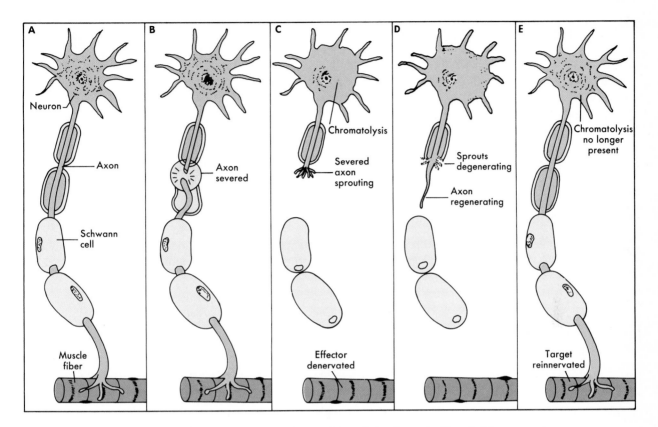

FIGURE 5-10 **A,** Normal motoneuron innervating a skeletal muscle fiber. **B,** Motor axon has been severed, and the motoneuron is undergoing chromatolysis. **C,** This is associated in time with sprouting and, in **D,** with regeneration of the axon. The excess sprouts degenerate. **E,** When the target cell is reinnervated, chromatolysis is no longer present.

target of axons can be determined by following the course of interrupted axons undergoing wallerian degeneration. Synaptic targets can also be mapped if neurons undergo transneuronal degeneration after an axonal bundle is transected.

Regeneration

After an axon is lost through injury, many neurons can regenerate a new axon. The proximal stump of the damaged axon develops *sprouts* (Figure 5-10, *C*). In the peripheral nervous system these sprouts elongate and grow along the path of the original nerve if this route is available. The Schwann cells in the distal stump of the nerve not only survive the wallerian degeneration, but they also proliferate and form rows along the course previously taken by the axons.

Growth cones of the sprouting axons find their way along the rows of Schwann cells and may eventually reinnervate the original peripheral target structures (Figure 5-10, *D* and *E*). The Schwann cells then remyelinate the axons. The rate of regeneration is limited by the rate of slow axonal transport to about 1 mm·day^{-1}.

In the CNS transection of axons also results in sprouting. Proper guidance for the sprouts is lacking, however, because the oligodendroglia do not form a path along which the sprouts can grow. This may occur because a single oligodendroglial cell myelinates many central axons, whereas a given Schwann cell provides myelin for only a single axon in the periphery. Alternately, different chemical signals may affect peripheral and central attempts at regeneration differently. Another obstacle is the formation of glial scars by astrocytes.

ENVIRONMENT OF THE NEURON

The local environment of most neurons is controlled so that neurons are normally protected from extreme variations in the composition of the extracellular fluid that bathes them. This control is provided by regulating the circulation of the CNS (see Chapter 24), the presence of a blood-brain barrier, the buffering function of astrocytes, and the exchange of substances with the CSF.

Fluid Compartments of the Cranium

The cranial cavity contains the brain, blood, and CSF. The brain weighs about 1350 g, of which approximately 15%, or 200 ml, is extracellular fluid. The intracranial blood volume is about 100 ml, and the cranial CSF volume another 100 ml. Thus the extracellular fluid space in the cranial cavity totals approximately 400 ml.

Blood-Brain Barrier

The movement of large molecules and highly charged ions from the blood into the brain and spinal cord is severely restricted (Figure 5-11). The restriction is at least partly caused by the presence of tight junctions between the capillary endothelial cells of the CNS. Astrocytes may also help limit the movements of certain substances. For example, astrocytes can take up potassium ions and thus regulate the K^+

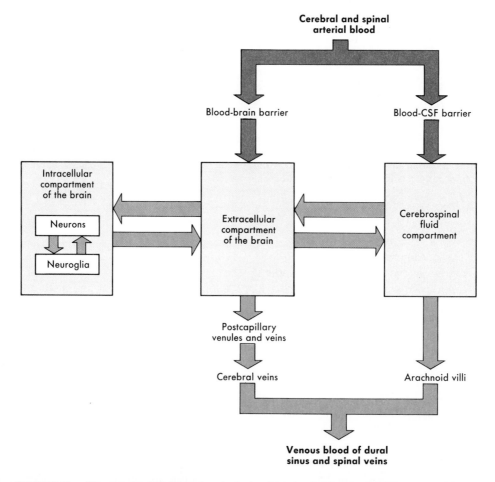

FIGURE 5-11   The structural and functional relationships involved in the blood-brain and blood-CSF barriers. Substances entering the neurons and glial cells (i.e., intracellular compartment) must pass through the cell membrane. Arrows indicate direction of fluid flow under normal conditions.

concentration in the extracellular space. Some substances are removed from the CNS by transport mechanisms.

Cerebrospinal Fluid

The extracellular fluid within the CNS communicates directly with the CSF. Thus the composition of the CSF is an indication of the composition of the extracellular environment of neurons in the brain and spinal cord. The main constituents of CSF in the lumbar cistern are shown in Table 5-1. For comparison, the concentrations of the same constituents in the blood are also given. The CSF has a lower concentration of K^+, glucose, and protein but a greater concentration of Na^+ and Cl^- than does blood. Furthermore, CSF contains practically no blood cells. The increased concentrations of Na^+ and Cl^- allow the CSF to be isotonic to blood, despite the much lower concentration of protein in the former.

The CSF is formed largely by the *choroid plexuses,* which are covered by specialized ependymal cells and are located in certain parts of the ventricular system of the brain. The ventricular system includes the two *lateral ventricles* in the telencephalon, the *third ventricle* of the diencephalon, and the *fourth ventricle* of the met- and myelencephalon. The lateral ventricles connect to the third ventricle by way of the *interventricular foramina,* and the third ventricle connects with the fourth through the *cerebral aqueduct.* Choroid plexuses are found in the lateral ventricles, third ventricle, and fourth ventricle (Figure 5-12).

The CSF escapes from the fourth ventricle through openings in its connective tissue roof (Figure 5-12). These openings are the *unpaired median aperture* and the *paired lateral apertures.* After leaving the ven-

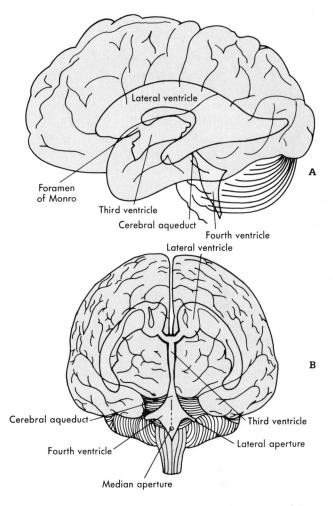

FIGURE 5-12   The ventricular system in situ as seen from the side (**A**) and from the front (**B**).

tricular system, the CSF circulates through the *subarachnoid spaces* surrounding the brain and spinal cord. Part of the CSF is removed by bulk flow through the valvular *arachnoid villi* into the dural venous sinuses.

The volume of the CSF within the cerebral ventricles is approximately 35 ml, and that in the subarachnoid spaces of the brain and spinal cord is about 100 ml. The rate at which CSF is produced is about 0.35 ml·min^{-1}. This allows the CSF to be turned over approximately four times daily.

The pressure in the CSF column is about 120 to 180 mm H_2O when the person is recumbent. The rate at which CSF is formed is relatively independent of the pressure in the ventricles and subarachnoid space

Table 5-1	Constituents of Cerebrospinal Fluid and Blood	
Constituent	**Lumbar CSF**	**Blood**
Na^+ (mEq·L^{-1})	148	136-145
K^+ (mEq·L^{-1})	2.9	3.5-5
Cl^- (mEq·L^{-1})	120-130	100-106
Glucose (mg·dl^{-1})	50-75	70-100
Protein (mg·dl^{-1})	15-45	$6\text{-}8 \times 10^3$
pH	7.3	7.4

and of the systemic blood pressure. However, the absorption rate of CSF is a direct function of CSF pressure.

Obstruction of the circulation of CSF leads to increased CSF pressure and *hydrocephalus*. In hydrocephalus the ventricles become distended, and if the increase continues, brain substance is lost. When the obstruction is within the ventricular system or in the roof of the fourth ventricle, the condition is called *noncommunicating hydrocephalus*. If the obstruction is in the subarachnoid space or arachnoid villi, it is known as *communicating hydrocephalus*.

BIBLIOGRAPHY
Journal Articles

Bray GM et al: Interactions between axons and their sheath cells, Annu Rev Neurosci 4:127, 1981.

Grafstein B and Forman DS: Intracellular transport in neurons, Physiol Rev 60:1167, 1980.

Pardridge WM: Brain metabolism: a perspective from the blood-brain barrier, Physiol Rev 63:1481, 1983.

Schwartz JH: Axonal transport: components, mechanisms, and specificity, Annu Rev Neurosci 2:467, 1979.

Books and Monographs

Bullock TH et al: Introduction to nervous systems, San Francisco, 1977, WH Freeman and Co.

Cajal SR: Degeneration and regeneration of the nervous system, New York, 1959, Hafner Publishing Co.

Cajal SR: The neuron and the glial cell, Springfield, Ill, 1984, Charles C Thomas, Publisher.

Heimer L and Robards MJ, editors: Neuroanatomical tract-tracing methods, New York, 1981, Plenum Press.

Kandel ER and Schwartz JH: Principles of neural science, ed 2, New York, 1985, Elsevier Science Publishing Co, Inc.

Millen JW and Woollam DHM: The anatomy of the cerebrospinal fluid, New York, 1962, Oxford University Press.

Shephard GM: Neurobiology, New York, 1983, Oxford University Press.

Whitfield IC: Neurocommunications: an introduction, New York, 1984, John Wiley & Sons, Inc.

Willis WD and Grossman RG: Medical neurobiology, ed 3, St Louis, 1981, The CV Mosby Co.

General Sensory System

The nervous system can be regarded as a complex of several different subsystems with different functional roles. These subsystems interact, however, and their activity leads to unified behavior.

Some of the neural subsystems are concerned with sensory functions. The general sensory system, or *somatosensory system,* analyzes sensory events relating to mechanical, thermal, or chemical stimulation of the body and face. If both visceral and somatic organs are included, the term *somatovisceral sensory system* is perhaps more appropriate. The special sensory systems include the *visual system,* which analyzes patterns of light detected by the eye; the *auditory system,* which interprets sounds that impinge on the ear; the *vestibular system,* which responds to the position of the head in space; and the *chemical sensory systems,* which are special sensory apparatuses for taste and olfaction.

PRINCIPLES OF SENSORY PHYSIOLOGY

Transduction

Sensory systems are designed to respond to the environment. Useful features of the environment are detected by sensory receptors, which then provide that information to the central nervous system (CNS). The interaction of environmental energy with a sensory receptor is called a stimulus. The effect of the stimulus on the sensory receptor may lead to a response. The process that enables a sensory receptor to respond usefully to a stimulus is called *sensory transduction.*

Environmental events that lead to sensory trans-

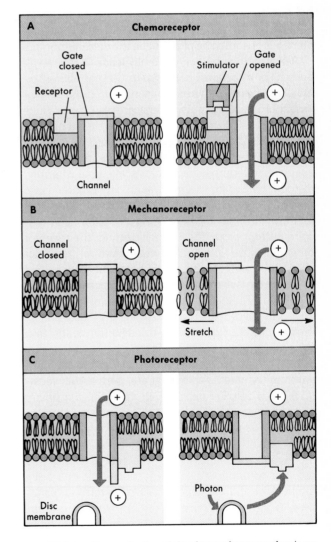

FIGURE 6-1 Conceptual models of transducer mechanisms in three types of receptors. **A,** Chemoreceptor; **B,** mechanoreceptor; **C,** vertebrate photoreceptor. (See text.)

duction can involve mechanical, thermal, chemical, or other forms of energy, depending on the sensory apparatus. Although humans cannot sense electrical and magnetic fields, other animals, such as fish, can respond to such stimuli. Figure 6-1 shows how different types of stimuli can alter the membrane properties of sensory receptor neurons specialized to transduce such stimuli. In Figure 6-1, *A*, a chemoreceptor responds when a molecule of a chemical stimulant reacts with a receptor molecule on the sensory receptor, resulting in the opening of an ion channel and a consequent influx of ionic current. In Figure 6-1, *B*, the ion channel of a mechanoreceptor is opened in response to the application of a mechanical force along the membrane. In Figure 6-1, *C*, the ion channel of a photoreceptor is open in the dark but closes when a photon is absorbed by pigment on the disc membrane.

Sensory transduction generally leads to the development of a *receptor potential* in the peripheral terminal of a primary afferent sensory neuron. A receptor potential is usually a depolarizing event that results from inward current flow, and it brings the membrane potential of the sensory receptor toward or past the threshold needed to trigger a nerve impulse. For example, in Figure 6-2 a mechanical stimulus *(arrow)* distorts the ending of a mechanoreceptor and causes inward current flow at the terminal and longitudinal and outward current flow along the axon. The outward current produces a depolarization, the receptor potential *(solid line in B)*, which may or may not exceed threshold for an action potential *(dashed line in B)*. In this case the action potential is generated at a trigger zone in the first node of Ranvier of the afferent fiber. However, in photoreceptors the cessation of inward current flow during transduction leads to a hyperpolarization of the receptor.

In some sensory receptor organs the primary afferent fiber terminates on a separate, peripherally located sensory cell. For example, in the cochlea, primary afferent fibers end on hair cells. Sensory transduction in such sense organs is made more complex by this arrangement. In the cochlea a receptor potential is produced in the hair cells in response to sound. The receptor potential is a depolarization of the hair cell's membrane, and the depolarization liberates an excitatory neurotransmitter onto the primary afferent terminal. This produces a *generator potential,* which in turn depolarizes the primary afferent fiber. This depolarization brings the membrane potential of the pri-

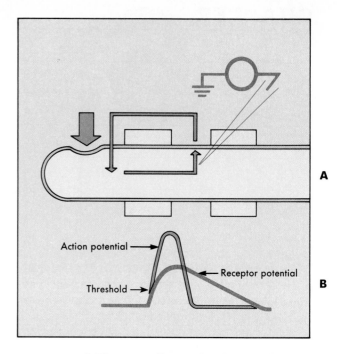

FIGURE 6-2   A, The current flow produced by stimulation of a mechanoreceptor at the site indicated by the arrow and an intracellular recording from a node of Ranvier. **B,** The receptor potential produced by the current *(red line)* and an action potential that may be superimposed on the receptor potential if the latter exceeds threshold *(blue line).*

mary afferent fiber toward or beyond the threshold for firing nerve impulses.

Sensory receptors have the property of *adaptation* to maintained stimuli. A long-lasting stimulus may produce either a prolonged repetitive discharge or a short-lived response (one or a few discharges), depending on whether the sensory receptor is slowly or rapidly adapting. Different adaptation rates result because a prolonged stimulus may produce either a maintained or a transient receptor potential in the sensory receptor. The functional implication of the adaptation rate is that different temporal features of a stimulus can be analyzed by receptors with different adaptation rates. For example, during an indentation of the skin a slowly adapting receptor may respond repetitively at a rate proportional to the amount of indentation (Figure 6-3). On the other hand, rapidly adapting receptors in the skin respond best to transient mechanical stimuli. The information signaled may reflect stimulus velocity or acceleration rather than the amount of skin indentation.

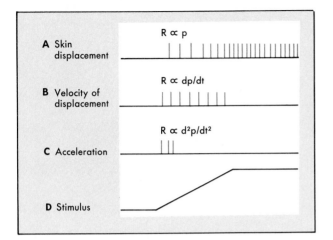

FIGURE 6-3   Responses of slowly and rapidly adapting mechanoreceptors to displacement of the skin. The discharges of the primary afferent fibers supplying the receptors in response to the ramp and hold stimulus (shown at the *bottom*) are termed the response *(R)*. **A,** *R* is proportional to skin position *(p)*. The receptor is slowly adapting and signals skin displacement. **B,** *R* is a function of the velocity of displacement *(dp/dt)*. **C,** *R* is a function of the acceleration (d^2p/dt^2). These receptors are rapidly adapting, but signal different, dynamic features of the stimulus.

Receptive Fields

The relationship between the location of a stimulus and the activation of particular sensory neurons is a major theme in sensory physiology. The receptive field of a sensory neuron is the region that, when stimulated, affects the discharge of the neuron. For example, a sensory receptor might be activated by indentation of only a small area of skin. That area is the *excitatory* receptive field of the sensory receptor. A neuron in the CNS might be excited by stimulation of a receptive field several times as large. The receptive fields of sensory neurons of the CNS are typically larger than those of sensory receptors, since the central neurons receive information from many sensory receptors, each with a slightly different receptive field. The location of the receptive field is determined by the location of the sensory transduction apparatus responsible for signaling information about the stimulus to the sensory neuron.

Generally the receptive fields of sensory receptors are excitatory. However, a central sensory neuron can have either an excitatory or an *inhibitory* receptive field (Figure 6-4). Inhibition results from data processing in sensory neural circuits and is mediated by inhibitory interneurons.

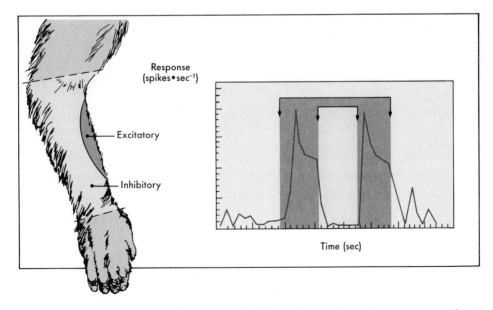

FIGURE 6-4   Excitatory and inhibitory receptive fields of a central somatosensory neuron located in the SI (primary) somatosensory cerebral cortex. The excitatory receptive field is on the forearm and is surrounded by an inhibitory receptive field. The graph shows the response to an excitatory stimulus and the inhibition of that response by a stimulus applied in the inhibitory field.

Sensory Coding

Sensory neurons encode stimuli. In the process of sensory transduction, one or more aspects of the stimulus must be encoded in a way that can be interpreted by the CNS. The encoded information is an abstraction based on the responses of sensory receptors to the stimulus and on information processing within the sensory pathway. Some of the aspects of stimuli that are encoded include modality, spatial location, threshold, intensity, frequency, and duration. Other aspects are presented in reference to particular sensory systems.

A sensory modality is a readily identified class of sensation. For example, maintained mechanical stimuli applied to the skin result in a sensation of touch-pressure, and transient mechanical stimuli may evoke a sensation of flutter-vibration. Other cutaneous modalities include cold, warm, and pain. Vision, audition, position sense, taste, and smell are examples of noncutaneous modalities. Coding for modality is signaled by labeled-line sensory channels in most sensory systems (see Chapter 5). A *labeled-line sensory channel* consists of a set of neurons devoted to a particular sensory modality.

The location of a stimulus is often signaled by the activation of the particular population of sensory neurons whose receptive fields are affected by the stimulus (Figure 6-5, *A*). In some cases an inhibitory receptive field or a contrasting border between an excitatory and an inhibitory receptive field can have localizing value. Resolution of two different adjacent stimuli may depend both on excitation of partially separate populations of neurons and on inhibitory interactions (Figure 6-5, *B*).

A *threshold stimulus* is the weakest that can be detected. For detection, a stimulus must produce receptor potentials large enough to activate one or more primary afferent fibers. Weaker intensities of stimulation can produce subthreshold receptor potentials; however, such stimuli would not excite central sensory neurons. Furthermore, the number of primary afferent fibers that need to be excited for sensory detection depends on the requirements for *spatial* and *temporal summation* in the sensory pathway (see Chapter 4). Thus a stimulus at threshold for detection may be much greater than threshold for activation of the most responsive primary afferent fibers. Conversely, a stimulus that excites some primary afferent fibers may not lead to perception of that stimulus.

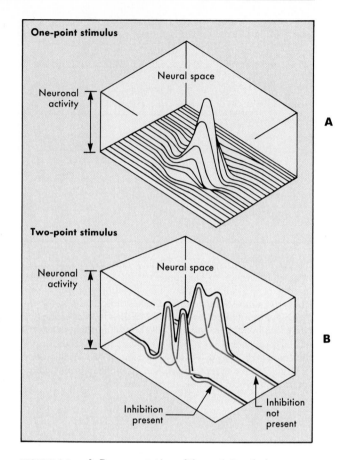

FIGURE 6-5   **A,** Representation of the activity of a large population of neurons distributed three-dimensionally in neural space. The activity is in response to stimulation of a point on the skin. Note that an excitatory peak is surrounded by an inhibitory trough; these are determined by the excitatory and inhibitory fields of sensory neurons in the central pathways. **B,** The activity in response to stimulation of two adjacent points on the skin. Note that the sum of the activity *(solid line)* is separated better into two peaks when inhibition is present than when it is not.

Stimulus intensity may be encoded by the mean frequency of discharge of sensory neurons. The relationship between stimulus intensity and response can be plotted as a stimulus-response function. For many sensory neurons, the stimulus-response function approximates an exponential curve (Figure 6-6). The general equation for such a curve is

$$\text{Response} = \text{Stimulus}^n \times \text{Constant} \qquad \textbf{(1)}$$

The exponent, n, can be less than, equal to, or greater than 1. Stimulus-response functions with fractional exponents are found for many mechanoreceptors (Figure 6-6). Thermoreceptors have linear stimulus-

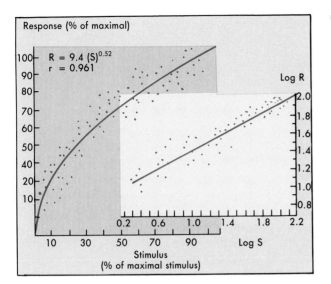

FIGURE 6-6 Stimulus-response function for slowly adapting cutaneous mechanoreceptors. The rate of discharge is plotted against stimulus strength (normalized to maximal). The plots are on linear and on log-log scales. The stimulus-response function is $R = 9.4(S)^{0.52}$.

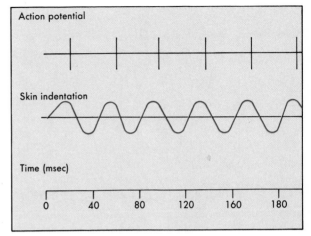

FIGURE 6-7 Coding for the frequency of stimulation. Discharge of a rapidly adapting cutaneous mechanoreceptor in phase with a sinusoidal stimulus. The action potentials are shown at the top and the stimulus in the middle trace.

response curves. Nociceptors may have linear or positively accelerating stimulus-response functions; that is, the exponent for these curves is 1 or more.

Another way in which stimulus intensity is encoded is by the number of sensory receptors activated. A stimulus that is threshold for perception may activate just one or a few primary afferent fibers, whereas a strong stimulus may excite many similar receptors. Central neurons that receive input from a particular class of sensory receptor would be more powerfully activated as more primary afferents are caused to discharge, and a greater activity in central sensory neurons results in the perception of a stronger stimulus.

Stimuli of different intensities may activate different sets of receptors. For example, a weak mechanical stimulus applied to the skin might activate only mechanoreceptors, whereas a strong mechanical stimulus might activate both mechanoreceptors and nociceptors. In this case the sensation evoked by the stronger stimulus would be more intense, and the quality would be different.

Stimulus frequency can be encoded by the intervals between the discharges of sensory neurons. Sometimes the interspike intervals correspond exactly to the intervals between stimuli (Figure 6-7), but in other cases a given neuron may discharge at intervals that are multiples of the interstimulus interval.

Stimulus duration may be encoded in slowly adapting sensory neurons by the duration of enhanced firing. In rapidly adapting neurons the beginning and end of a stimulus may be signaled by transient discharges.

Sensory Pathways

A sensory pathway can be viewed as a set of neurons arranged in series (Figure 6-8). First-, second-, third-, and higher-order neurons serve as sequential elements in a given sensory pathway. Furthermore, several parallel sensory pathways are often involved in transmitting similar sensory information. The *first-order neuron* in a sensory pathway is the primary afferent neuron. The peripheral endings of this neuron form a sensory receptor (or receive input from an accessory sensory cell, such as a hair cell), and thus the neuron responds to a stimulus and transmits encoded information to the CNS. The primary afferent neuron often has its soma in a dorsal root ganglion or a cranial nerve ganglion.

The *second-order neuron* is likely to be located in the spinal cord or brainstem. It receives information from first-order neurons and transmits information to the thalamus. The information may be transformed at the level of the second-order neuron by local neural processing circuits. The ascending axon of the sec-

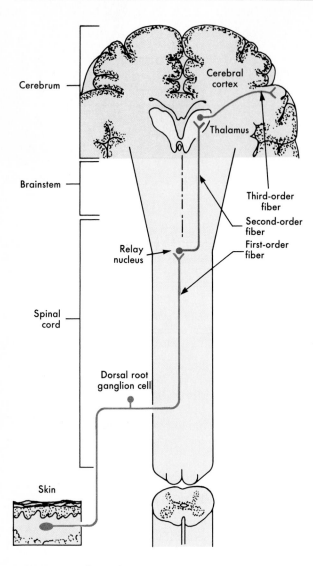

FIGURE 6-8   General arrangement of sensory pathways. First-, second-, and third-order neurons are shown. Note that the axon of the second-order neuron crosses the midline, so that sensory information from one side of the body is transmitted to the opposite side of the brain.

ond-order neuron typically crosses the midline, and thus sensory information that originates on one side of the body reaches the contralateral thalamus.

The *third-order neuron* is in one of the sensory nuclei of the thalamus. Again, local circuits may transform information from second-order neurons before the signals are transmitted to the cerebral cortex.

Fourth-order neurons in the appropriate sensory receiving areas of the cerebral cortex and *higher-*

order neurons in the same and other cerebral cortical areas process the information further. At some undetermined site the sensory information results in perception, which is a conscious awareness of the stimulus.

SOMATOVISCERAL SENSORY SYSTEM

The somatovisceral sensory system includes sensory units that have sensory receptor organs in the skin, muscle, joints, and viscera. Information arising from these sensory receptors reaches the CNS by way of first-order neurons, which are the primary afferent neurons. The cell bodies of the primary afferent neurons are generally located in dorsal root ganglia or cranial nerve ganglia. Each ganglion cell gives off a neurite that bifurcates into a peripheral process and a central process. The *peripheral process* has the structure of an axon and terminates peripherally as a sensory receptor. The *central process* is also an axon and enters the spinal cord through a dorsal root or the brainstem through a cranial nerve. The central process typically gives rise to numerous collateral branches that end synaptically on several second-order neurons.

The processing of somatovisceral sensory information involves a number of CNS structures, including the spinal cord, brainstem, thalamus, and cerebral cortex. The ascending pathways arise from second-order neurons that are located in the spinal cord and brainstem and that project to the contralateral thalamus. The most important ascending somatosensory pathways that carry somatovisceral information from the body are the dorsal column–medial lemniscus path and the spinothalamic tract. The main somatosensory projection that represents the face is the trigeminothalamic tract. Ancillary somatovisceral pathways include the spinocervicothalamic path, the postsynaptic dorsal column path, the dorsal spinocerebellar tract, the spinoreticular tract, and the spinomesencephalic tract.

The somatovisceral sensory system can be regarded as a **general sensory system.** The sensory modalities mediated by the somatovisceral sensory system include touch-pressure, flutter-vibration, position sense, joint movement, thermal sense, pain, and visceral distension.

Receptors

The somatovisceral sensory system includes various types of sensory receptor organs in the skin, muscle, joints, and viscera.

Cutaneous receptors can be subdivided according to the type of stimulus to which they respond. The major types of cutaneous receptors include mechanoreceptors, thermoreceptors, and nociceptors. *Mechanoreceptors* respond to such mechanical stimuli as stroking or indenting the skin and can be rapidly adapting or slowly adapting. Rapidly adapting cutaneous mechanoreceptors include *hair follicle receptors* in the hairy skin, *Meissner's corpuscles* in the

nonhairy (glabrous) skin, and *pacinian corpuscles* in subcutaneous tissue (Figure 6-9, *A*). Hair follicle receptors and Meissner's corpuscles respond best to stimuli repeated at rates of about 30 to 40 Hz, whereas pacinian corpuscles prefer stimuli repeated at approximately 250 Hz. Slowly adapting cutaneous mechanoreceptors include *Merkel's cell endings* and *Ruffini's corpuscles* (Figure 6-9, *B*). Merkel's cell receptors have punctate receptive fields, whereas Ruffini's corpuscles can be activated by stretching the skin some distance from the receptor terminals. The axons of all these receptor types are myelinated.

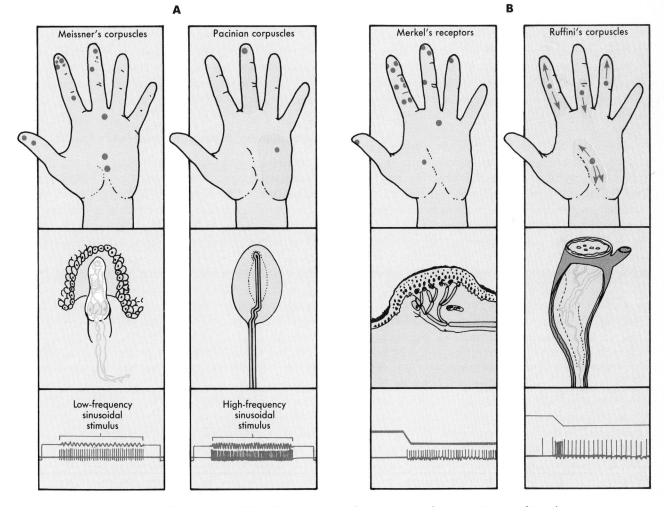

FIGURE 6-9   The receptive fields of several types of cutaneous mechanoreceptors are shown in the top row of drawings. **A,** Rapidly adapting mechanoreceptors: Meissner's corpuscles and pacinian corpuscles. **B,** Slowly adapting mechanoreceptors: Merkel's receptors and Ruffini's corpuscles. The second row of drawings shows the morphology of the receptors; the third row, the responses to sinusoidal stimuli **(A)** or to step indentations of the skin **(B).**

The two types of *thermoreceptors* in the skin are cold receptors and warm receptors. Both classes are slowly adapting, although they also discharge phasically when skin temperature is changed rapidly. These are among the few receptor types that discharge spontaneously under normal circumstances. Cold receptors are supplied by small myelinated axons and warm receptors by unmyelinated axons.

Nociceptors respond to stimuli that threaten or actually produce damage. There are two major classes of cutaneous nociceptors: the *A-δ mechanical nociceptors* and the *C polymodal nociceptors*. A-δ mechanical nociceptors are supplied by finely myelinated (or A-δ) axons, whereas C polymodal nociceptors are supplied by unmyelinated (or C) fibers. The A-δ mechanical nociceptors respond to strong mechanical stimuli, such as pricking the skin with a needle or crushing the skin with forceps. They typically do not respond to noxious thermal or chemical stimuli unless they are previously sensitized. C polymodal nociceptors, on the other hand, respond to several types of noxious stimuli, including mechanical, thermal, and chemical.

Skeletal muscle also contains several types of sensory receptors. These are chiefly mechanoreceptors and nociceptors, although some muscle receptors may possess thermo- or chemosensitivity. The best studied muscle receptors are the *stretch receptors,* which include *muscle spindles* and *Golgi tendon organs.* Although these play an important role in proprioception, they may be more important in motor control. Therefore their structure and function are discussed in Chapter 8.

Other sensory receptors in muscle include nociceptors. These respond to pressure applied to the muscle and to release of metabolites, especially during ischemia. Muscle nociceptors are supplied by medium-sized and small myelinated (group II and III) axons or by unmyelinated (group IV) afferent fibers.

Joints are associated with several types of sensory receptors, including rapidly and slowly adapting mechanoreceptors and nociceptors. The rapidly adapting mechanoreceptors are pacinian corpuscles, which respond to mechanical transients, including vibration. The slowly adapting joint receptors are Ruffini's endings, which respond best to movements of a joint to extremes of flexion or extension; these endings signal pressure or torque applied to the joint. Joint mechanoreceptors are innervated by medium-sized (group II) afferent fibers. Joint nociceptors are activated by probing a joint capsule or by hyperextension or hyperflexion, although many articular nociceptors fail to respond to joint movements under normal conditions. If sensitized by inflammation, however, they can respond to stimuli that are normally innocuous, including movements or weak pressure stimuli. Joint nociceptors are innervated by finely myelinated (group III) or unmyelinated (group IV) primary afferent fibers.

Viscera are supplied with sensory receptors. These are usually involved in reflexes and have little to do with sensory experience. However, some visceral mechanoreceptors are responsible for the sensation of distension, and visceral nociceptors produce visceral pain. Pacinian corpuscles are present in the mesentery and in the capsules of visceral organs such as the pancreas; these presumably signal mechanical transients. Whether some forms of visceral pain result from overactivity of the mechanoreceptor afferents is still controversial. Some viscera, however, clearly have specific nociceptors.

Dermatomes, Myotomes, and Sclerotomes

Primary afferent fibers in the adult are distributed systematically, as determined during embryologic development. The mammalian embryo becomes segmented, and each body segment is called a *somite*. A somite is innervated by an adjacent segment of the spinal cord or, in the case of a somite of the head, by a cranial nerve. The portion of a somite destined to form skin is called a *dermatome*. Similarly, the part of a somite that will form muscle is a *myotome,* and the part that will form bone, a *sclerotome*. Viscera are also supplied by particular segments of the spinal cord or particular cranial nerves.

Many dermatomes become distorted during development, chiefly because of the way the upper and lower extremities are formed and because humans maintain an upright posture. However, the sequence of dermatomes can be understood if pictured on the body in a quadrupedal position (Figure 6-10).

Although a dermatome receives its densest innervation from the corresponding spinal cord segment, the dermatome is also supplied by several adjacent spinal segments. Thus transection of a dorsal root produces little sensory loss in the corresponding dermatome. Anesthesia of any given dermatome requires interruption of several successive dorsal roots.

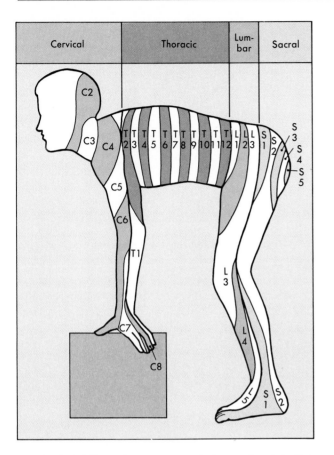

FIGURE 6-10   Dermatomes represented on a drawing of a person assuming a quadrupedal position.

Spinal Roots

Axons of the peripheral nervous system enter or leave the CNS through the spinal roots (or through cranial nerves). The dorsal root on one side of a given spinal segment is composed entirely of the central processes of dorsal root ganglion cells. The ventral root consists chiefly of motor axons, including α–motor axons, γ–motor axons, and at certain segmental levels autonomic preganglionic axons. Ventral roots also contain many primary afferent fibers, whose role is still unclear.

Just before they penetrate the spinal cord, the large myelinated fibers move to a medial position in the dorsal root, whereas the fine myelinated and unmyelinated fibers move to a lateral position. The large, medially placed afferent fibers enter the dorsal funiculus and bifurcate, sending one branch rostrally and another branch caudally. These branches travel through several segments (some even ascend to the medulla) and give off collaterals that pass ventrally into the gray matter of the spinal cord. The fine myelinated and unmyelinated primary afferent fibers enter the dorsolateral (Lissauer's) fasciculus, where they bifurcate and send branches rostrally and caudally for a short distance in this fiber bundle. Collaterals from these branches enter and terminate in the spinal cord gray matter. The endings of the large and small primary afferent fiber projections to the spinal cord are located in distinctly different regions of the gray matter. The spinal cord gray matter can be subdivided into 10 laminae (Figure 6-11). The dorsal horn is composed of laminae I to VI. The intermediate region is equivalent to parts of laminae VI and VII. The ventral horn in the cervical and lumbar enlargements is subdivided into a medial zone, lamina VIII, which is part of the commissural nucleus, and a lateral zone, lamina IX, which is the motor nucleus supplying the limb muscles. A component of lamina IX is also found medially (the motor nucleus for the axial musculature). Lamina VII extends ventrally between lamina VIII and the lateral component of lamina IX. Lamina X surrounds the central canal.

The large myelinated primary afferent fibers entering through the medial part of the dorsal root (Figure 6-11) have terminals in the deeper laminae of the dorsal horn (III to VI), the intermediate region, and the ventral horn. The small myelinated and unmyelinated fibers that enter through the lateral part of the dorsal root have terminals in lamina I, lamina II, and part of lamina V (Figure 6-11).

Trigeminal Nerve

The arrangement for primary afferent fibers that supply the face is comparable to that for fibers supplying the body. Peripheral processes of neurons in the trigeminal ganglion pass through the ophthalmic, maxillary, and mandibular divisions of the trigeminal nerve to innervate dermatome-like regions of the face. The trigeminal nerve also innervates the oral and nasal cavities and the dura mater.

The large myelinated fibers supplying mechanoreceptors of the skin and structures of the oral and nasal cavities synapse in the main sensory nucleus of the trigeminal nerve. Small myelinated and unmyelinated primary afferent fibers of the trigeminal nerve terminate in the nerve's spinal nucleus. Primary afferent fibers from stretch receptors have their cell bodies in

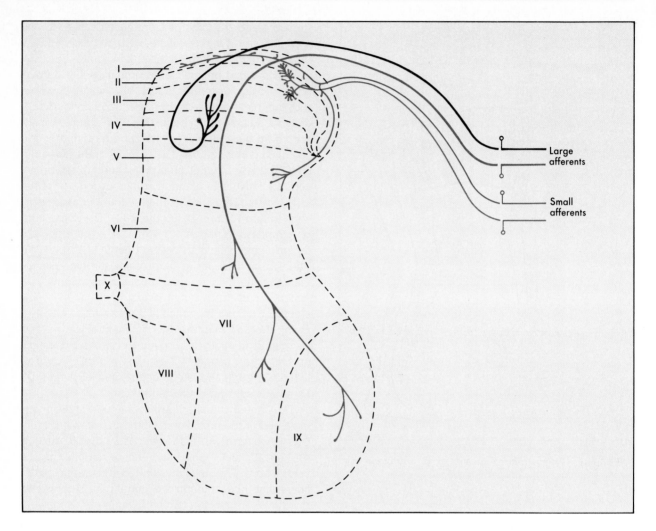

FIGURE 6-11   Distribution of large and small primary afferent fibers in the spinal cord. Terminals of two large primary afferent fibers are shown. One, from a hair follicle receptor, synapses in laminae III to V. The other, from a muscle spindle, synapses in laminae VI, VII, and IX. Terminals from two fine afferent fibers are also shown. The small myelinated fiber (A-δ), from a cutaneous mechanical nociceptor, ends in laminae I and V, whereas the unmyelinated (C) afferent fiber synapses in laminae I and II.

the mesencephalic nucleus of the trigeminal nerve. This is an exceptional arrangement, since all other primary afferent cell bodies in the somatovisceral sensory system are in peripheral ganglia. The central processes synapse in the motor nucleus of the trigeminal nerve.

Dorsal Column–Medial Lemniscus Pathway

The ascending branches of many large myelinated nerve fibers travel rostrally in the dorsal column all the way to the medulla. Axons that innervate sensory receptors of the lower extremity and the lower trunk ascend in the gracile fasciculus, whereas fibers from receptors of the upper extremity and upper trunk ascend in the cuneate fasciculus. These axons are the first-order neurons of the dorsal column path.

The neurons of the dorsal column nuclei respond in much the same way as the primary afferent fibers that synapse on them. Some behave as rapidly adapting receptors, responding to hair movement or to mechanical transients applied to the glabrous skin.

Others discharge at high frequencies when vibratory stimuli are applied to their receptive fields and thus resemble pacinian corpuscles. Still other neurons in the dorsal column nuclei have slowly adapting responses to cutaneous stimuli. In the cuneate nucleus many neurons are activated by stretching muscles. The main differences between the responses of dorsal column neurons and the primary afferent fibers are that dorsal column neurons:

1. Have larger receptive fields, since more than one primary afferent fiber synapses on a given dorsal column neuron

2. Sometimes respond to more than one class of sensory receptor because of convergence of several different types of primary afferent fibers on the second-order neuron

3. Often have inhibitory receptive fields mediated through interneuronal circuits in the dorsal column nuclei

Other Somatosensory Pathways of the Dorsal Spinal Cord Three other pathways that carry somatosensory information ascend in the dorsal part of the spinal cord on the same side as the afferent input: (1) the spinocervical tract, (2) the postsynaptic dorsal column pathway, and (3) the dorsal spinocerebellar tract.

The cells of origin of the *spinocervical tract* receive input largely from cutaneous mechanoreceptors, although some of these cells are activated by nociceptors as well. The cells of origin of the *postsynaptic dorsal column path* receive information similar to that reaching spinocervical tract neurons. The information conveyed by these cells is eventually conveyed to the ventral posterolateral nucleus of the thalamus.

The *dorsal spinocerebellar tract* responds to input from muscle and joint receptors of the lower extremity. The main destination of the tract is the cerebellum, but it also provides proprioceptive information from the leg to the ventral posterolateral nucleus of the thalamus after a relay in the medulla. Proprioceptive information from the arm is signaled by the dorsal column path.

Sensory Functions of the Dorsal Spinal Cord Pathways The sensory qualities mediated by dorsal spinal cord pathways include flutter-vibration, touch-pressure, joint movement and position sense, and visceral distension. Each of these qualities of sensation depends on activity in a set of sensory neurons that collectively form a labeled-line sensory channel. A sensory channel may involve several parallel ascending pathways, and it includes particular primary afferent neurons and sensory processing mechanisms at spinal cord, brainstem, thalamic, and cerebral cortical levels.

Flutter-vibration is a complex sensation. Flutter refers to recognition of events that have low-frequency components. The sensory receptors detecting flutter include hair follicles and Meissner's corpuscles. Ascending sensory tracts that convey the information needed for flutter sensation include the dorsal column–medial lemniscus pathway, the spinocervical tract, and the postsynaptic dorsal column path. In addition, as mentioned later, the spinothalamic tract in the ventral part of the cord is partly responsible for flutter sensation.

High-frequency vibration is primarily detected by pacinian corpuscles. Pacinian corpuscle afferents ascend in the dorsal columns.

Touch-pressure sensation involves the recognition of skin indentation by Merkel's cell and Ruffini's receptors. The ascending pathways that convey information from these receptors include the dorsal column–medial lemniscus path and the postsynaptic dorsal column tract.

The senses of *joint movement* and *joint position* are complex and depend on sensory information that arises from muscle, joint, and cutaneous receptors. For some joints, such as the knee, the most important information is derived from muscle spindles in the muscles that move the joint. In other joints, however, such as those of the digits, Ruffini's endings and joint receptors also contribute. All the information required from the upper extremity ascends in the dorsal column–medial lemniscus path, but part of the information conveyed from the lower extremity depends on the dorsal spinocerebellar tract.

The sense of *distension of certain viscera,* such as the urinary bladder, depends on information transmitted by the dorsal column–medial lemniscus pathway.

Higher Processing of Tactile and Proprioceptive Information

As mentioned earlier, the medial lemniscus synapses in the ventral posterolateral nucleus of the thalamus. Neurons of the ventral posterolateral nucleus are the third-order neurons not only of the dorsal column–medial lemniscus pathway, but also of the spinocervical tract, the postsynaptic dorsal column path, and the dorsal spinocerebellar tract (as well as the spinothalamic tract, as discussed later).

The responses of many neurons in the ventral posterolateral nucleus resemble those of the first- and second-order neurons of the dorsal column-medial lemniscus pathway. The responses may be dominated by a particular type of receptor, and the receptive field may be small, although larger than that of a primary afferent fiber. Thalamic neurons often have inhibitory receptive fields. A notable difference between neurons in the ventral posterolateral nucleus and neurons at lower levels of the dorsal column–medial lemniscus path is that the excitability of the thalamic neurons depends on the stage of the sleep-wake cycle and on the presence or absence of anesthesia.

The primary somatosensory receiving area (SI) of the cortex has several subdivisions. Brodmann described different cortical regions based on the arrangements of cortical neurons as seen in Nissl-stained preparations. The postcentral gyrus includes Brodmann's areas 3, 1, and 2. Area 3 has since been subdivided into areas 3a and 3b. Cutaneous input dominates in areas 3b and 1, whereas muscle and joint input dominates in areas 3a and 2. Thus quite different cortical zones are involved in the processing

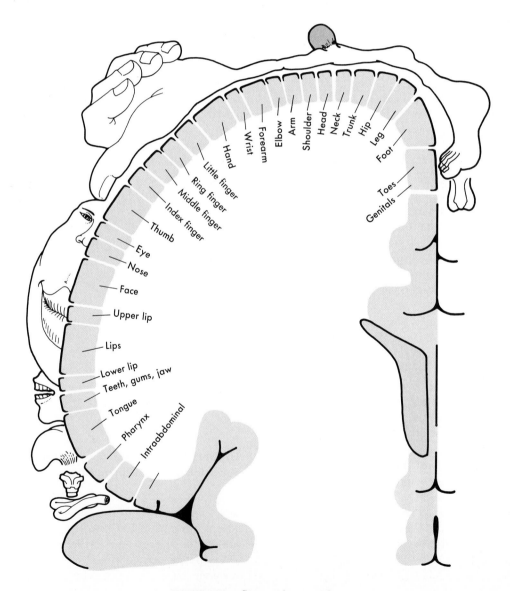

FIGURE 6-12   Sensory homunculus.

of tactile and proprioceptive information. The inputs to these cortical areas are from distinct parts of the ventral posterolateral nucleus. The cutaneous input is from the core of the ventral posterolateral nucleus, whereas the muscle and joint input is from a "shell" region.

Within any particular area of the SI cortex, all the neurons along a line perpendicular to the cortical surface have similar response properties and receptive fields. The SI cortex is thus said to have a *columnar organization*. A comparable columnar organization has also been demonstrated for other primary sensory receiving areas, including the primary visual and auditory cortices. Nearby cortical columns in the SI cortex may process information for different sensory modalities. For example, the cutaneous information reaching one cortical column in area 3b may come from rapidly adapting mechanoreceptors, whereas that reaching a neighboring column might be from slowly adapting mechanoreceptors.

The location of cortical columns in the SI region is

related systematically to the location of the receptive fields on the body surface. This relationship is called a *somatotopic organization*. The body surface is mapped in the SI cortex. The lower extremity is represented in the human in the medial aspect and the apex of the postcentral gyrus, whereas the upper extremity is mapped on the dorsolateral aspect of the postcentral gyrus and the face is mapped just dorsal to the lateral fissure. This somatotopic map for humans is called a "homunculus" (Figure 6-12). The somatotopic organization of the SI cortex is responsible for the coding of stimulus location. The somatotopic organization at the cortical level reflects the same organization at lower levels of the somatosensory system, including the dorsal column nuclei and the ventral posterolateral nucleus of the thalamus.

Besides being responsible for the initial processing of somatosensory information, the SI cortex also begins higher-order processing, such as *feature extraction*. For example, certain neurons in area 1 respond preferentially to a stimulus moving in one

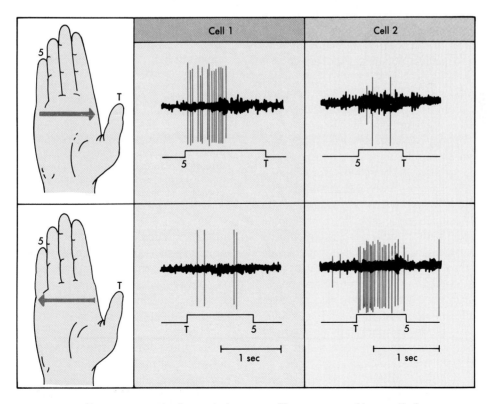

FIGURE 6-13   Feature extraction by cortical neurons. The responses of two cortical neurons are shown to stimuli moved across the palm. Cell 1 was excited strongly by movement of the stimulus from the fifth digit toward the thumb, but only weakly by movement of the stimulus in the opposite direction. Cell 2 showed the converse.

direction across the receptive field but not in the opposite direction (Figure 6-13). Such neurons might contribute to the perceptual ability to recognize the direction of an applied stimulus.

Spinothalamic Tract

The spinothalamic tract originates from spinal cord neurons that project mainly to the contralateral thalamus. The decussation of the axon of a spinothalamic tract cell is within the same segment as the cell body, and the axon ascends in the lateral funiculus. The spinothalamic tract terminates in several nuclei of the thalamus, including the ventral posterolateral nucleus and the central lateral nucleus (one of the intralaminar nuclei).

The cells of origin of the spinothalamic tract are found chiefly in spinal cord laminae I and V. Most spinothalamic tract cells receive an excitatory input from nociceptors of the skin, but many can also be excited by noxious stimulation of muscle or viscera. Effective stimuli include noxious mechanical, thermal, and chemical stimuli. Some spinothalamic neurons are excited by activity in cold or warm thermoreceptors or sensitive mechanoreceptors. Thus different spinothalamic tract cells respond in a manner appropriate for signaling noxious, thermal, or mechanical events.

Some spinothalamic tract cells receive a convergent excitatory input from several different classes of sensory receptors. For example, a given spinothalamic tract neuron may be activated weakly by tactile stimuli but more powerfully by noxious stimuli (Figure 6-14). Such neurons are called *wide-dynamic-range cells* because they are activated by stimuli having a great range of intensities. Wide-dynamic-range neurons mainly signal noxious events, the weak response to tactile stimuli perhaps being ignored by higher centers. However, in pathologic conditions these neurons may be activated sufficiently by normally innocuous stimuli to evoke a sensation of pain. This would explain some pain states in which activation of mechanoreceptors causes pain *(allodynia)*. Other spinothalamic tract cells are activated only by noxious stimuli. Such neurons are often called *nociceptive-specific* or *high-threshold cells* (Figure 6-14).

Spinothalamic tract cells often have inhibitory receptive fields. Inhibition may result from weak mechanical stimuli, but usually the most effective

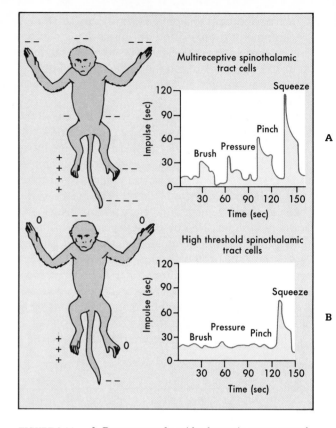

FIGURE 6-14   A, Responses of a wide-dynamic-range or multireceptive spinothalamic tract cell. **B,** Responses of a high-threshold spinothalamic tract cell. The figures indicate the excitatory *(plus signs)* and inhibitory *(minus signs)* receptive fields. The graphs show the responses to graded intensities of mechanical stimulation. *Brush* is with a camel's hair brush repeatedly stroked across the receptive field. *Pressure* is applied by attachment of an arterial clip to the skin. This is a marginally painful stimulus to a human. *Pinch* is by attachment of a stiff arterial clip to the skin and is distinctly painful. *Squeeze* is by compressing a fold of skin with forceps and is damaging to the skin.

inhibitory stimuli are noxious ones. The nociceptive inhibitory receptive fields may be very large and include most of the body and face (Figure 6-14). Such receptive fields may account for the ability of a variety of physical manipulations to suppress pain, including transcutaneous electrical nerve stimulation and acupuncture.

The **gate control theory of pain** explains how innocuous stimuli may inhibit the responses of dorsal horn neurons that transmit information about painful stimuli to the brain. In this theory pain transmission is prevented by innocuous inputs mediated by large myelinated afferent fibers, whereas pain transmission is enhanced by inputs carried over fine afferent fibers. The inhibitory interneurons of lamina II serve as a gating mechanism. The circuit diagram originally proposed has been criticized, but the basic notion of a gating mechanism is still viable.

Noxious stimuli applied to large areas of the body can also inhibit the discharges of nociceptive dorsal horn neurons by inhibitory pathways that descend from the brainstem (see later description of the endogenous analgesia system).

Many of the spinothalamic tract cells that project to medial thalamic nuclei, such as the central lateral nucleus, have very large receptive fields that often include much of the surface of the body and face. The large receptive fields of these spinothalamic neurons suggest that they function to trigger motivational-affective responses to painful stimuli, rather than sensory discrimination.

Other Somatosensory Pathways of the Ventral Spinal Cord Two other pathways that transmit somatosensory information ascend in the ventral part of the spinal cord: the spinoreticular tract and the spinomesencephalic tract.

The cells of origin of the *spinoreticular tract* are often difficult to activate, but when receptive fields are found, these are generally large, sometimes bilateral, and the effective stimuli include noxious ones.

The reticular formation is involved in attentional mechanisms and arousal. Ascending projections from the reticular formation to the intralaminar complex of the thalamus and from there to wide areas of the cerebral cortex are presumably part of the neural mechanism for these actions. In addition, reticulospinal projections presumably contribute to the descending inhibitory system.

Many cells of the *spinomesencephalic tract* respond to noxious stimuli, and the receptive fields are generally small. The terminations of the tract are in several midbrain nuclei, including the *periaqueductal gray,* which is an important component of the endogenous analgesia system (see later discussion). Information from the midbrain is also relayed to the *amygdala,* a part of the limbic system. This may provide one pathway by which noxious stimuli can trigger emotional responses. Motivational-affective responses may also result from activation of the periaqueductal gray and midbrain reticular formation. The latter is an important part of the arousal system, and stimulation in the periaqueductal gray causes vocalization and aversive behavior.

Sensory Functions of the Ventral Spinal Cord Pathways The sensory modalities mediated by ventral spinal cord pathways include a contribution to flutter, but the most important functions are pain and thermal sensations. Although the most essential pathway for flutter-vibration is the dorsal column-medial lemniscus path, the spinothalamic tract can provide sufficient information for flutter sensation. However, the spinothalamic tract alone is insufficient for the recognition of the direction of a tactile stimulus, and discrimination is less than when the dorsal column—medial lemniscus path is intact. Thermal sense depends on input from cold and warm receptors to spinothalamic tract neurons in lamina I.

Pain resulting from stimulation of nociceptors is mediated partly by spinothalamic tract cells and partly by the spinoreticular and spinomesencephalic tracts. Pain is a complex phenomenon and includes both sensory-discriminative and motivational-affective components. That is, pain is a sensory experience that is accompanied by emotional responses and by somatic and autonomic motor adjustments. Presumably the sensory-discriminative component of pain depends on the spinothalamic tract projection to the ventral posterolateral nucleus and the further transmission of nociceptive information to the primary (SI) and secondary (SII) regions of the cerebral cortex. Sensory processing at these and higher levels of the cortex results in the perception of the quality of pain (e.g., pricking, burning, and aching), the location of the painful stimulus, the intensity of the pain, and its duration.

The motivational-affective responses to painful stimuli include attention and arousal, somatic and autonomic reflexes, endocrine responses, and emotional changes. These collectively account for the unpleasant nature of painful stimuli. The motivational-affective responses apparently depend on several ascending pathways, including the component of the spinothalamic tract that projects to the medial thalamus, the spinoreticular tract, and the spinomesencephalic tract. As indicated previously, these pathways have access to attentional, orientational, and arousal systems, as well as to the limbic system.

Pain that originates from the skin is generally well localized, presumably because spinothalamic tract cells have relatively discrete cutaneous receptive fields. Also, the ascending system through which they signal is somatotopically organized. However, pain that originates from deep structures, including muscle and viscera, is poorly localized and is often mistakenly attributed (referred) to superficial structures. An example of this is angina pectoris, which is the pain resulting from ischemia of the heart. Frequently, ischemic heart pain is referred to the inner aspect of the left arm. This area of pain referral is in the T1 dermatome, which corresponds to the spinal cord level of the sensory innervation of the heart.

One explanation of *referred pain* is that many spinothalamic neurons receive excitatory input not only from the skin but also from muscle and viscera. The spinal cord segments that innervate the dermatomes containing the cutaneous receptive field of the cell correspond well to the segments innervating the muscle or viscus. The activity in a population of spinothalamic tract cells may be interpreted as originating from the skin, based on learning this association during childhood. Subsequently, activation of these neurons by pathologic input from visceral nociceptors might be misinterpreted as resulting from stimulation of superficial parts of the body.

Pain sometimes occurs in the absence of nociceptor stimulation. This is most likely to happen after damage to peripheral nerves or to the parts of the CNS involved in transmitting nociceptive information. The mechanism of such pain caused by neural damage is poorly understood.

The Trigeminal System

Sensory processing for the face, oral cavity, and dura mater is organized in a fashion similar to that for the body. Neurons in the main sensory nucleus are apparently responsible for flutter-vibration and touch-pressure sensations from the face, and neurons in the subnucleus caudalis of the spinal nucleus mediate pain and temperature sensations that originate from the face, oral cavity, and dura mater. Pain in the trigeminal distribution is of particular importance, since this includes both tooth pain and headaches.

Centrifugal Control of Somatovisceral Sensation

Sensory experience is not just the passive detection of environmental events; instead, it more often depends on exploration of the environment. Tactile cues are sought by moving the hand over a surface. Visual cues result from scanning visual targets with

the eyes. Thus sensory information is often received as a result of activity in the motor system. Furthermore, sensory transmission in pathways to the sensory centers of the brain is regulated by descending control systems. This allows the brain to control its input by filtering the incoming sensory messages. Important information can be attended to and unimportant information ignored.

The tactile and proprioceptive somatosensory pathways are regulated by descending pathways originating in the motor and SI regions of the cerebral cortex. For example, corticobulbar projections to the dorsal column nuclei help control sensory input over the dorsal column–medial lemniscus path.

Of particular interest is the descending control system that regulates the transmission of nociceptive information. The system presumably serves to reduce excessive pain under certain circumstances. It is well known that soldiers on the battlefield, athletes in competition, accident victims, and others facing stressful circumstances often feel little or no pain at the time a wound occurs or a bone is broken. At a time after the stressful events, however, pain may be severe. Although the descending regulatory system that controls pain is part of the more general centrifugal system that modulates all forms of sensation, the pain control system is so important medically that it is distinguished as a special *endogenous analgesia system*.

Several descending pathways contribute to the endogenous analgesia system. The sensorimotor cortex modulates the activity of trigeminothalamic and spinothalamic tract neurons by way of corticobulbar and corticospinal projections. The activity of these descending pathways may either inhibit or excite spinothalamic tract neurons. The raphe nuclei and periaqueductal gray give rise to direct and indirect projections to the medullary and spinal dorsal horns, and these pathways inhibit nociceptive neurons, including trigeminothalamic and spinothalamic tract cells (Figure 6-15, *A*).

The endogenous analgesia systems can be subdivided into those that use one of the endogenous opioids and those that do not. The endogenous opioid substances are neuropeptides that activate one of the several forms of opiate receptors. Some of the endogenous opioids include enkephalin, dynorphin, and β-endorphin. Opioid analgesia can generally be antagonized with the narcotic antagonist, naloxone. Therefore naloxone is used as a test of whether or not anal-

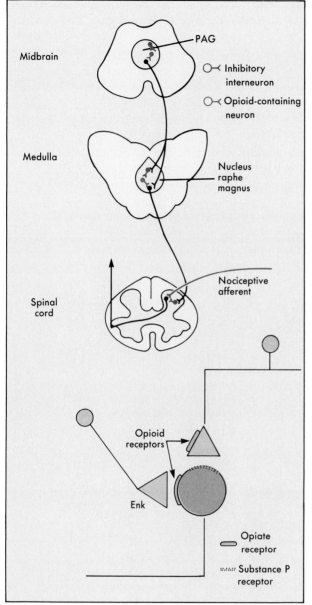

FIGURE 6-15   A, Some of the neurons thought to play a role in the endogenous analgesia system. Neurons in the periaqueductal gray *(PAG)* activate a raphe-spinal pathway, which in turn inhibits spinothalamic tract cells. Interneurons containing opioid substances are involved in the system at each level. **B,** Possible pre- and postsynaptic sites of action of enkephalin *(Enk)*. The presynaptic action might prevent the release of substance P *(Sub P)* from nociceptors.

gesia is mediated through an opioid mechanism.

The opioid-mediated endogenous analgesia system can be activated by the exogenous administration of morphine and other opiate drugs. Thus one of the oldest known medical treatments for pain depends on the triggering of a sensory control system. Opiates typically inhibit neural activity in nociceptive pathways. Two sites of action have been proposed for opiate inhibition: presynaptic and postsynaptic (Figure 6-15, *B*). The presynaptic action of opiates on nociceptive afferent terminals is thought to prevent the release of excitatory transmitters, such as substance P. The postsynaptic action produces an inhibitory postsynaptic potential. How can an inhibitory neurotransmitter cause the activation of descending pathways? One hypothesis suggests that the descending analgesia system is under tonic inhibitory control by inhibitory interneurons in both the midbrain and the medulla. The action of opiates would inhibit the inhibitory interneurons, resulting in a disinhibition of the descending analgesia pathways.

Some endogenous analgesia pathways operate by neurotransmitters other than opioids and thus are unaffected by naloxone. One way of engaging a non-opioid analgesia pathway is through stress. The analgesia so produced is called *stress-induced analgesia*.

Many neurons in the raphe nuclei use serotonin as a neurotransmitter. Serotonin is able to inhibit nociceptive neurons and presumably plays an important role in the endogenous analgesia system. Other brainstem neurons release catecholamines, such as norepinephrine and epinephrine. Catecholamines also inhibit nociceptive neurons, and therefore catecholaminergic neurons may contribute to the endogenous analgesia system. Undoubtedly many other substances will be shown to play a role in the analgesia system. Furthermore, evidence now shows that endogenous opiate antagonists exist that could prevent opiate analgesia.

BIBLIOGRAPHY
Journal Articles

Akil H et al: Endogenous opioids: biology and function, Annu Rev Neurosci 7:223, 1984.

Amit Z and Galina ZH: Stress-induced analgesia: adaptive pain suppression, Physiol Rev 66:1091, 1986.

Basbaum AI and Fields HL: Endogenous pain control systems: brainstem spinal pathways and endorphin circuitry, Annu Rev Neurosci 7:309, 1984.

Besson JM and Chaouch A: Peripheral and spinal mechanisms of nociception, Physiol Rev 67:67, 1987.

Kaas JH: What, if anything, is SI? Organization of first somatosensory area of cortex, Physiol Rev 63:206, 1983.

McCloskey DI: Kinesthetic sensibility, Physiol Rev 58:763, 1978.

Snyder SH and Childers SR: Opiate receptors and opioid peptides, Annu Rev Neurosci 2:35, 1979.

Books and Monographs

Boivie JJG and Perl ER: Neural substrates of somatic sensation. In Hunt CC, editor: Neurophysiology, MTP Physiology series one, vol 3, 1975, International Review of Science, Baltimore, University Park Press.

DeGroot J and Chusid JG: Correlative neuroanatomy, East Norwalk, Conn, 1988, Appleton & Lange.

Kandel ER and Schwartz JH: Principles of neural science, ed 2, New York, 1985, Elsevier Science Publishing Co, Inc.

Kenshalo DR: The skin senses, Springfield, Ill, 1968, Charles C Thomas, Publisher.

Shepherd GM: Neurobiology, New York, 1983, Oxford University Press.

Uttal WR: The psychobiology of sensory coding, New York, 1973, Harper & Row, Publishers, Inc.

Willis WD: Control of nociceptive transmission in the spinal cord, Berlin, 1982, Springer-Verlag.

Willis WD: The pain system, Basel, 1985, Karger.

Willis WD and Coggeshall RE: Sensory mechanisms of the spinal cord, New York, 1978, Plenum Press.

Willis WD and Grossman RG: Medical neurobiology, ed 3, St Louis, 1981, The CV Mosby Co.

Special Senses

An important evolutionary trend was *encephalization,* in which special sensory organs in the heads of animals developed, along with appropriate neural systems in the brain. These special sensory systems, which included the visual, auditory, olfactory, and gustatory systems, allowed the animal to detect and analyze light, sound, and chemical signals in the environment. In addition, the vestibular system evolved to signal the position of the head.

VISUAL SYSTEM

The visual system detects and interprets photic stimuli. In vertebrates effective photic stimuli are electromagnetic waves between 400 and 700 nm long, or visible light. Light enters the eye and impinges on *photoreceptors* of a specialized sensory epithelium, the retina. The photoreceptors are the rods and cones. Rods have low thresholds for detecting light and thus operate best under conditions of reduced lighting *(scotopic vision).* However, rods neither provide well-defined visual images nor contribute to color vision. Cones, by contrast, are not as sensitive as rods to light but operate best under daylight conditions *(photopic vision).* Cones are responsible for high visual acuity and color vision. Information processing within the retina is done by the retinal interneurons, and the output signals are carried to the brain by axons of the retinal ganglion cells through the optic nerves and optic tracts. The main visual pathway is through the lateral geniculate nucleus of the thalamus to the visual receiving areas of the cerebral cortex (Figure 7-1). Also, extrageniculate visual

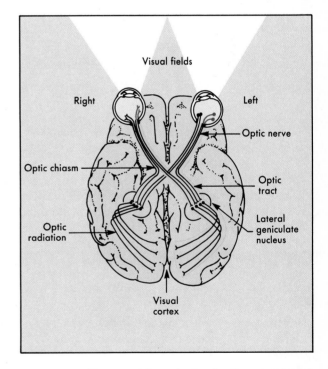

FIGURE 7-1 Diagram of the main visual pathway as viewed from the base of the brain.

pathways involve the superior colliculus, pretectum, and hypothalamus. These participate in orientation of the eyes, control of the pupil, and control of circadian rhythms, respectively.

Structure of the Eye

The wall of the eye is formed of three concentric layers (Figure 7-2). The outer layer is the fibrous coat,

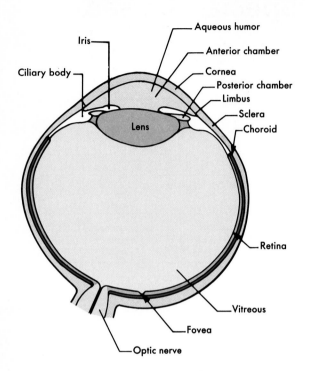

FIGURE 7-2   Drawing of the right eye as viewed from above.

which includes the transparent cornea, with its epithelium, the conjunctiva, and the opaque sclera. The middle layer is the vascular coat, which includes the iris and the choroid. The iris contains both radially and circularly oriented smooth muscle fibers, which comprise the pupillary dilator and sphincter; the iris forms a diaphragm to control pupil size. The dilator is activated by the sympathetic nervous system and the sphincter by the parasympathetic nervous system (oculomotor nerve). The choroid is rich in blood vessels that supply the outer layers of the retina. The inner retinal layers are nourished by tributaries of the central artery and veins of the retina; these vessels course with the optic nerve. The inner layer of the eye is the neural coat, the retina. The functional part of the retina covers the entire posterior eye except for the blind spot, which is the optic nerve head. Visual acuity is highest in the central part of the retina, the *macula lutea*. The fovea is a pitlike depression in the middle of the macula where visual targets are focused; it is the fixation point (Figure 7-2).

Besides the retina, the eye contains a lens to focus light on the retina, pigment to reduce light scatter, and fluids called aqueous and vitreous humor that help maintain the shape of the eye. Externally

attached extraocular muscles aim the eye toward an appropriate visual target.

The lens is held in place behind the iris by the suspensory ligaments (or zonule fibers), which attach to the wall of the eye at the ciliary body (Figure 7-2). When the ciliary muscles are relaxed, the tension exerted by the suspensory ligaments tends to flatten the lens. When the ciliary muscles contract, tension on the suspensory ligaments is reduced, allowing the lens to assume a more spheric shape because of its elastic properties. The ciliary muscles are activated by the parasympathetic nervous system by way of the oculomotor (third cranial) nerve.

Light scattering within the eye is minimized by pigment. The choroid contains an abundance of pigment. In addition, the outermost layer of the retina is a pigment-containing epithelium. The pigment cells of the retinal pigment layer can adjust the length of pigment-containing processes that extend between retinal photoreceptors. This allows fine control of the lengths of the photoreceptors that are potentially exposed to scattered light. The pigment cells are also involved in the turnover of photoreceptor outer segments.

The space around the iris is filled with aqueous humor. This is a clear fluid resembling cerebrospinal fluid. It is actively secreted by the ciliary processes, which form an epithelium that is posterior to the iris and protrudes into a space called the posterior chamber (Figure 7-2). The aqueous humor circulates through the posterior chamber, out the pupil, and into the anterior chamber. It is then reabsorbed through Fontana's spaces into Schlemm's canal and returned to the venous circulation. Imbalance in the secretion and reabsorption of aqueous humor can increase the pressure in the eye, a condition that threatens the viability of the retina. This malady is known as *glaucoma*.

The space behind the lens contains a gelatinous material, the vitreous humor. The vitreous humor turns over very slowly.

The extraocular muscles insert on the sclera from their origins on the bony orbit. Details concerning the organization and operation of the eye movement control system are given later (see Chapter 8).

Physiologic Optics

The eye is often compared with a camera. Both are devices that capture images by using a lens system to focus light on a photosensitive surface. The quality of

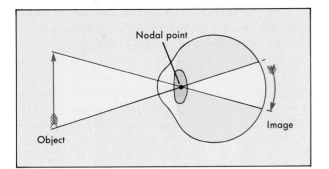

FIGURE 7-3   Image formation in the eye. The image is reversed as rays of light pass through the nodal point of the lens.

the image is enhanced by use of a diaphragm to reduce the effect of spheric aberrations of the lens and to increase depth of field. The diaphragm also controls the amount of entering light.

The eye, as with the camera, produces an image of an object. It is important to realize that the image is inverted (Figure 7-3). The inversion is caused by the light rays from the object crossing at a nodal point within the lens. The image is inverted both from side to side and from above downward.

The ability of a lens to bend light is called its *refractive power*. The unit of refractive power is the diopter. For an image to be in focus on the retina, light coming from any point on the object and passing through the cornea and lens of the eye must be refracted just enough so that it falls on a corresponding point on the retina. The cornea is the main refractive surface of the eye, having a refractive power of 43 diopters. However, the lens is crucial for focusing images on the retina, because its refractive power can be varied (from 13 to 26 diopters). Changes in the refractive power of the lens are produced by changes in the shape of the lens through relaxation or contraction of the ciliary muscles. *Accommodation* is the process by which contraction of the ciliary muscle causes the lens to become more rounded. The result of accommodation is that images of nearby objects are brought into focus on the retina.

During aging the lens tends to lose its elasticity. This loss reduces the ability of the eye to accommodate. This visual disturbance is called *presbyopia*. Other common defects in focusing ability are myopia and hypermetropia. In *myopia* (nearsightedness) images are focused in front of the retina because the eye is disproportionately long for the refractive system. In *hypermetropia* (farsightedness) images are

focused behind the retina because the eye is short relative to the refractive system. *Astigmatism* is the result of asymmetric focusing, usually because the cornea lacks radial symmetry.

Retina

The outermost of the 10 retinal layers is the pigment epithelium. The pigment cells capture stray light. They also phagocytize photoreceptor membrane shed from the outer segments of the rods and cones. Substances that move between the photoreceptors and the blood vessels within the choroid must pass through the pigment cell layer. Interactions between pigment cells and photoreceptor cells are very important in visual function.

Individual photoreceptor cells can be subdivided into three regions: the outer segment, the inner segment, and the synaptic terminal (Figure 7-4). The outer segment contains a stack of membranous discs that are rich in photopigment. The inner segment

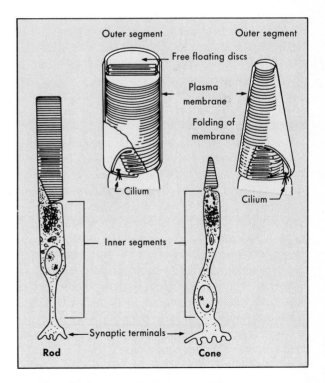

FIGURE 7-4   Drawings showing the structure of rods and cones. The inner and outer segments and the synaptic terminals are shown, as are details of the membranous discs in the outer segments.

connects with the outer segment by way of a modi-fied cilium containing nine pairs of microtubules, but lacking the two pairs of central microtubules seen in most cilia. The inner segment of the photoreceptor cell contains the nucleus, mitochondria, and other organelles. The synaptic ending contacts one or more bipolar cells.

A rod is so sensitive to light that it can respond to a single photon. The greater sensitivity of rods than cones is partly caused by rods' long outer segments. Consequently rods contain more photopigment, which is arranged in a monomolecular layer on each outer segment disc. The pigment is rhodopsin, which is composed of a chromophore, retinal, and a protein, opsin. Retinal is the aldehyde form of vitamin A. In the dark, retinal is bound to opsin in the 11-*cis*-retinal form. Absorption of light causes a change to the all-*trans*-retinal form, which no longer binds to opsin. Before the photopigment can be regenerated, the all-*trans*-retinal must be transported to the pigment cell layer, reduced, isomerized, and esterified.

Cones also contain 11-*cis*-retinal attached to an opsin. However, three different cone opsins are found in three different types of cones, each sensitive to a different part of the visible light spectrum. One cone type responds best to blue light (420 nm), another to green (531 nm), and the third to red (558 nm). The presence of three types of cones gives the retina a mechanism for trichromatic color vision. Light causes a series of changes in the photopigment of cones; these changes resemble the sequence in rods, but the reactions and recovery are quicker.

Color vision requires at least two photopigments. A single pigment absorbs light over much of the spectrum but absorbs best at a particular wavelength. The amount of light absorbed depends on its wavelength and intensity. Light of one wavelength and a given intensity could produce the same effect on a particular photoreceptor as another light of a different wavelength and intensity. Therefore the signal is ambiguous because intensity can substitute for wavelength. However, with at least two different photoreceptors having two different pigments, it is possible to distinguish between different wavelengths, if the intensity of the light falling on both photoreceptors is the same. Three different photoreceptor types reduce the ambiguity even more.

Color blindness is often based on a genetic defect resulting in the loss of one or more of the cone mechanisms or in a change in the absorption spectra of one

or more photopigments. People are normally *trichromats* because they have three cone mechanisms. *Protanopia* is the loss of the long wavelength system; *deuteranopia,* the loss of the medium wavelength system; and *tritanopia,* the loss of the short wavelength system. Any of these causes a person to be a *dichromat. Monochromats* generally lack all three cone mechanisms, but in rare instances they lack two.

The cones are most concentrated in the fovea, where all the photoreceptors are cones (Figure 7-5). This is the region of the retina that provides the greatest visual acuity. In the fovea the retina is thinned to just the outer four layers; thus the image here is of the highest quality. Rods are most concentrated in the parafoveal region.

No photoreceptors exist in the optic disc (Figure

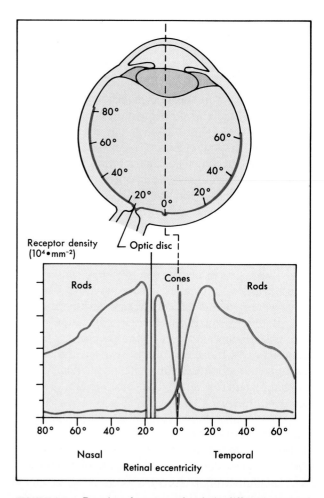

FIGURE 7-5   Density of cones and rods in different parts of the retina.

7-5), which is the site where the ganglion cell axons collect to leave the eye as the optic nerve. The optic disc is therefore a blind spot. The optic disc is in the medial retina. Therefore the part of the visual field that would be imaged on the blind spot would be on the temporal side of the field of vision of that eye. The blind spot is not noticed in binocular vision, since the region of the visual field that fails to be seen by the blind spot in one eye is seen by the opposite eye because the light falls on the temporal side of that retina.

Information Processing in the Retina The most direct route for information flow through the retina is from photoreceptors to bipolar cells and then to ganglion cells. The ganglion cells provide the output of the retina to the thalamus.

The neural pathways in the retina can be subdivided into rod pathways and cone pathways. There is more convergence from photoreceptors onto bipolar cells in the rod than in the cone pathways. This convergence enhances the sensitivity of the rod pathways. Cone pathways display much less convergence, in keeping with their role in visual acuity.

Because the photoreceptor cells and many of the retinal interneurons have short processes, action potentials are not required for transmitting information to the next cell in the circuit. Instead, local potentials cause changes in neurotransmitter release, which in turn provides for information transfer. In darkness photoreceptor cells have open sodium channels, which result in a dark current and consequently a tonic release of neurotransmitter onto bipolar cells and horizontal cells (Figure 7-6). When light is absorbed, the sodium channels are closed, which leads to a hyperpolarization of the photoreceptor cells and a decrease in the release of transmitter, probably glutamate.

This process involves an amplification mechanism that depends on a second messenger system. It appears that cyclic guanosine monophosphate (cGMP) maintains the sodium channels in an open configuration (Figure 7-6). Light activates a G-protein, called transducin, in the photoreceptor membrane, by the conversion of guanosine triphosphate (GTP) to guanosine diphosphate (GDP). Transducin in turn activates a phosphodiesterase, which hydrolyzes cGMP. This causes the sodium channels to close and the membrane to hyperpolarize.

The receptive field of a photoreceptor is a small circular area that is coextensive with the area of the

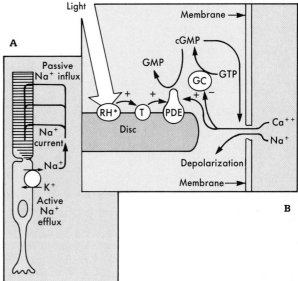

FIGURE 7-6 **A,** The dark current in a photoreceptor caused by passive influx of sodium ions. The sodium ions are returned to the extracellular space by pumping. Light closes the sodium channels and thus reduces the dark current. **B,** The second messenger system underlying phototransduction. When light reacts with rhodopsin (Rh*), the G-protein transducin (T) is activated. This in turn activates phosphodiesterase (PDE), which breaks down cyclic guanosine monophosphate (cGMP) into GMP. The dark current depends on cGMP, and thus a fall in cGMP concentration reduces the dark current. A decrease in dark current causes a hyperpolarization of the photoreceptor. GTP, Guanosine triphosphate; GC, guanylate cyclase.

retina occupied by the photoreceptor. Bipolar cells are of two types: on-center and off-center (Figure 7-7). An on-center bipolar cell is depolarized when light is shined in the center of its receptive field and hyperpolarized when light is shined in an annulus around the center of the receptive field. An off-center bipolar cell behaves in the converse manner. It is unclear exactly how these responses are generated. However, the interactions involve photoreceptors and horizontal cells, and inhibitory mechanisms are probably important.

Ganglion cells, as with bipolar cells, may have center-surround antagonistic receptive fields (Figure 7-7) or, as with amacrine cells, may have large receptive fields. The type of receptive field presumably reflects the dominant input. Ganglion cells can be classified as X-cells, Y-cells, and W-cells. Both X- and Y-cells have center-surround receptive fields. X-cells have smaller receptive fields than Y-cells, respond more

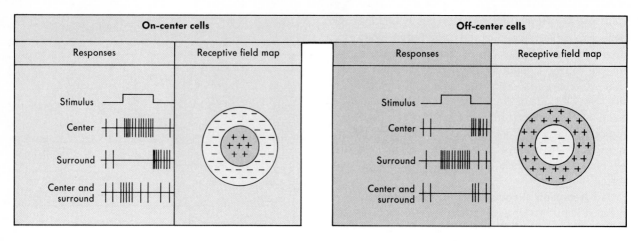

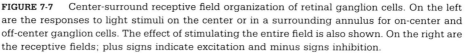

FIGURE 7-7   Center-surround receptive field organization of retinal ganglion cells. On the left are the responses to light stimuli on the center or in a surrounding annulus for on-center and off-center ganglion cells. The effect of stimulating the entire field is also shown. On the right are the receptive fields; plus signs indicate excitation and minus signs inhibition.

tonically to stimuli, have slower axons, and sum multiple responses in a linear fashion. Y-cells respond to complex stimuli in a difficult-to-predict, nonlinear fashion. X-cells respond differently to different wavelengths of light, whereas Y-cells are insensitive to differences in wavelength. W-cells are more heterogeneous. They may have center-surround receptive fields, but most have large, diffuse receptive fields, and their axons are very slowly conducting. The central pathways of X- , Y- and W-cells differ, and these neurons make different contributions to vision.

Central Visual Projections of the Retina The optic nerves from the two eyes converge at the optic chiasm (see Figure 7-1). Part of the optic nerve fibers decussates in the chiasm and part continues posteriorly on the same side as the eye of origin. Crossing fibers originate from the nasal hemiretinas of the two eyes. The uncrossed fibers originate from the temporal hemiretinas. Because of this arrangement, each optic tract contains both uncrossed and crossed fibers.

Lateral Geniculate Body

Most of the lateral geniculate nucleus (LGN) neurons project to the visual cortex; however, some are interneurons. A given LGN neuron receives a dominant input from one or a few retinal ganglion cells, and the responses resemble those of the ganglion cells. Thus the LGN neurons can be classified as X- or Y-cells, and they have on- or off-center receptive fields. However, LGN neurons are subject to inputs from regions other than the retina, including the visual cortex, several brainstem nuclei, and the reticular nucleus of the thalamus. Inhibitory actions originating from the brainstem or reticular nucleus can prevent visual signals from reaching the cortex or can reduce these signals. Cortical input can result in excitation or inhibition of LGN projection neurons. In effect, the LGN serves as a filter for visual information before it accesses the visual cortex.

Visual Field Deficits Caused by Lesions of the Visual Pathway

Interruption of an optic nerve causes blindness in that eye or a *scotoma* (partial blindness in one visual field). A lesion affecting the optic chiasm causes loss of vision in the temporal fields of both eyes, a condition called *bitemporal hemianopsia*. Destruction of an optic tract causes loss of vision in the contralateral half of the visual field of each eye or *contralateral homonymous hemianopsia*. A similar visual field deficit results from destruction of a lateral geniculate body or the entire optic radiation or visual cortex gyri. Macular vision may be spared in cortical lesions, perhaps because of the very large size of the macular representation.

Striate Cortex

The optic radiation ends chiefly in layer IV of Brodmann's area 17. The dense axon terminals form a white stripe (the line of Gennari) that can be seen grossly and that gives rise to the name *striate* for this region of cortex. Projections from magnocellular and parvocellular layers of the LGN are separate, and axons carrying information from the two eyes end in alternating patches of cortex called *ocular dominance columns* (Figure 7-8, *B*). Recordings from neurons in area 17 reveal that usually a given cell receives input from both eyes, although one eye is dominant.

The retina is mapped onto the striate cortex. The macular region is represented at and for a distance anterior to the occipital pole. The remainder of the retina is represented still more anteriorly along the medial aspect of the occipital lobe. More neural representation is devoted to the macula than for the rest of the retina because of the requirements for visual acuity.

Most neurons in the striate cortex respond best to elongated stimuli. A rectangular visual target or an edge evokes a much more vigorous response than a small spot. The orientation of the stimulus is an important factor. Neurons in a region of striate cortex perpendicular to the cortical surface will all respond best to elongated stimuli having the same orientation (Figure 7-8, *B*). These neurons form orientation columns.

Higher Processing of Visual Information

The striate cortex receives visual information from the LGN and begins the analysis of that information. The striate cortex has connections with many other areas of cerebral cortex known as the *extrastriate visual cortex*, which participate in the further processing of visual information. These cortical areas are interconnected with the pulvinar, another nucleus of the thalamus.

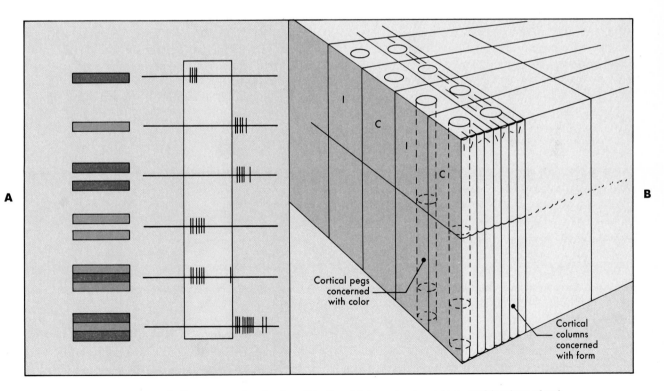

FIGURE 7-8 **A,** Responses of a simple cell in the striate cortex to various combinations of red and green bars. The cell responded best to a red bar flanked by two green bars. **B,** Diagram of the columns in the visual cortex. Ocular dominance columns are indicated by *I* (ipsilateral eye) and *C* (contralateral eye) and orientation columns by the short bars at various angles. The cortical pegs contain neurons that have double-opponent color fields.

The X-cell pathway includes the striate cortex and the inferotemporal area. Some neurons in the latter region respond only to such highly specific visual stimuli as recognition of a face. Thus the inferotemporal region may be responsible for the analysis of fine spatial detail. Other visual areas process color and motion.

Stereopsis is binocular depth perception. It depends on slight differences in the images in the two eyes, so that a given cortical neuron will have its receptive field at points on the two retinas that are slightly out of correspondence.

Color vision depends on discrimination of wavelengths of light. Retinal ganglion cells and LGN neurons may respond selectively to one wavelength and be inhibited by another. These cells are called *spectral opponent neurons.* An example would be a neuron that is excited by red light shined in the center of its receptive field and by green light in the surrounding part of the field (Figure 7-8, *A*). Spectral opponent cells belong to the X-cell category (Y-cells are insensitive to the spectral aspects of light). Neurons in the cortex are able not only to discriminate wavelength but also to take into account brightness, and thus they permit the perception of true color. These neurons are concentrated in *cortical pegs,* which are sets of neurons within the ocular dominance columns (Figure 7-8, *B*).

Superior Colliculus

The superior colliculus is a layered midbrain structure that serves as a visual center and as a coordination center for reflexes related to orientation. The dorsal three layers are involved in visual processing, whereas the deeper four layers also process other sensory inputs, including somatosensory stimuli.

Retinal ganglion cells project to the upper layers of the superior colliculus. The ganglion cells include both Y-cells and W-cells. Most of the projection arises from the nasal retina and is crossed. The superior colliculus also receives a projection from the cerebral cortex. The cortical neurons involved in this projection are activated by Y-cells. Thus the visual input to the superior colliculus is concerned with the functions of Y- and W-cells but not of X-cells. The output of the upper layers of the superior colliculus influences much of the visual processing in the cortex. Experiments in animals suggest that the superior colliculus is important in determining the location of

objects in visual space, whereas the cortex determines what the objects are.

The deep layers of the superior colliculus are considered with the motor system (see Chapter 8).

AUDITORY SYSTEM

The auditory system is designed to analyze sound. Audition is important not only for recognition of environmental cues, but also for communication, especially language in humans.

Sound

Sound is produced by alternating waves of pressure in the air. Sound waves are composed of the sum of a set of sinusoidal waves of the appropriate amplitudes, frequencies, and phase. Thus sound can be regarded as a mixture of pure tones. The human acoustic system acts as a filter that is sensitive to pure tones within a range of frequencies from approximately 20 to 15,000 Hz. Threshold varies with frequency. Sound intensity is measured in decibels (dB), which are expressed in terms of a reference level of sound pressure (P_r), often 0.002 dyne/cm^2, the threshold for hearing. The formula for sound intensity is

$$\text{Sound pressure (decibels)} = 20 \log(P/P_r) \qquad \textbf{(1)}$$

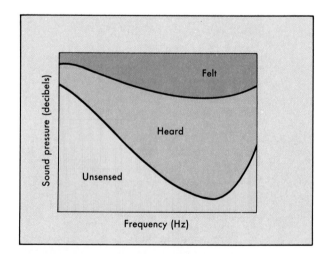

FIGURE 7-9   Sound levels for human hearing as a function of frequency. Below the range for hearing, sound is not sensed; above the hearing range, it is detected by both the auditory and the somatosensory systems.

The ear is most sensitive to tones from 1000 to 3000 Hz. At these frequencies, threshold is by definition 0 dB. Threshold is higher at frequencies less than 1000 Hz and greater than 3000 Hz (Figure 7-9). For example, threshold at 100 Hz is approximately 40 dB. Speech has an intensity of about 65 dB. Damage to the acoustic apparatus can be produced by sounds that exceed 100 dB, and discomfort results from sound pressures exceeding 120 dB.

Structure of the Ear

The ear can be subdivided into the external ear, the middle ear, and the inner ear. The external ear includes the pinna and the external auditory meatus, which leads by way of the auditory canal to the outer surface of the tympanic membrane (Figure 7-10). The auditory canal contains glands that secrete cerumen, a wax that guards the ear from invasion by insects.

The middle ear is a cavity that extends deep to the tympanic membrane. It contains a chain of ossicles, the malleus, incus, and stapes (Figure 7-10, *A*), which connect the tympanic membrane with another membrane that covers the oval window, an opening into the inner ear (Figure 7-10, *B*). A second opening

between the middle and inner ears, also covered by a membrane, is the round window. The middle ear contains two muscles, the tensor tympani and the stapedius; the former attaches to the malleus and the latter to the stapes. Contraction of the middle ear muscles dampens movements of the ossicular chain. The eustachian tube provides an opening from the middle ear to the nasopharynx. This permits pressure differences between the environment and the middle ear to be equalized.

The inner ear is a cavity within the temporal bone and contains the cochlea and the vestibular apparatus (Figure 7-10). The cochlea is the organ of hearing formed by elements of both the bony labyrinth and the membranous labyrinth. The space in the bony labyrinth just inside the oval window is the vestibule. The cochlea is a coiled structure formed by subdivision of the bony labyrinth into two compartments. The partition between the compartments is formed by a component of the membranous labyrinth; this component is called the *cochlear duct,* or *scala media.* The portion of bony labyrinth in continuity with the vestibule is the *scala vestibuli.* This extends along the two and a half turns of the human cochlea to the end of the cochlear duct. At this point the scala vestibuli

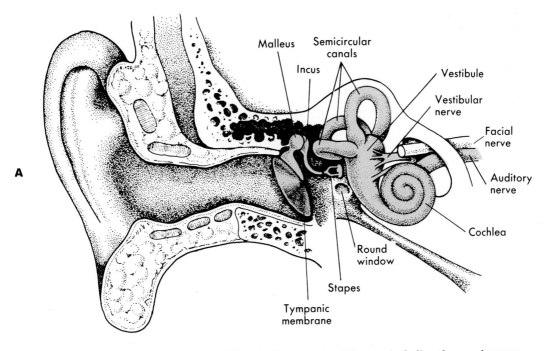

FIGURE 7-10   Structure of the cochlea. **A,** Components of the ear, including the membranous labyrinth.

Continued.

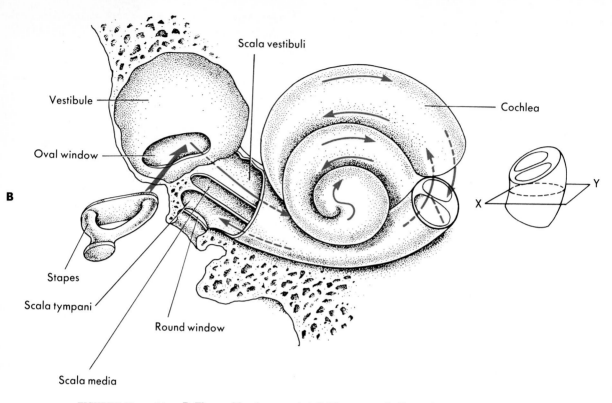

B

Scala vestibuli

Vestibule

Oval window

Cochlea

Y

X

Stapes

Scala tympani

Round window

Scala media

FIGURE 7-10, cont'd.   B, The cochlea in more detail. The arrows indicate the path of fluid movements that would result from movement of the stapes into the oval window.

connects with the scala tympani by way of a space called the *helicotrema*. The scala tympani spirals back to the bony interface with the middle ear and ends at the round window. The base of the cochlea is near the oval and round windows and the apex at the helicotrema. The bony core of the cochlea is the *modiolus*.

The cochlear duct is a tube and part of the membranous labyrinth (Figure 7-11, *A*). The basilar membrane forms the base of the cochlear duct and can be regarded as the main partition between the scalae vestibuli and tympani. The basilar membrane is narrowest near the base of the cochlea and widest near the helicotrema. The basilar membrane is attached internally to a ledge, the *spiral lamina,* that arises from the modiolus. Externally the basilar membrane is anchored to the wall of the cochlea by the spiral ligament. Contained within the spiral ligament is a vascular structure, the *stria vascularis*. The roof over the cochlear duct is formed by *Reissner's membrane*. The cochlear duct contains *endolymph,* a fluid having a high concentration of potassium ions; the endolymph is secreted by the stria vascularis. The bony labyrinth

contains *perilymph,* which resembles cerebrospinal fluid.

The *organ of Corti* is the sense organ for hearing (Figure 7-11, *B*). It lies within the cochlear duct along the basilar membrane. The organ of Corti consists of hair cells, the *tectorial membrane,* a stiff framework, and several types of supportive cells. The *stereocilia* of the hair cells contact the tectorial membrane. The hair cells are innervated by primary afferent fibers and also by efferent fibers of the cochlear nerve. The cell bodies of the primary afferent fibers are in the *spiral ganglion,* which is contained in the modiolus. The spiral ganglion cells are bipolar neurons whose peripheral processes reach the hair cells through the spiral lamina. The central processes join the cochlear nerve, which projects into the brainstem.

FIGURE 7-11   A, The organ of Corti within the cochlear duct (scala media). **B,** Enlargement of the area outlined by the dashed rectangle.

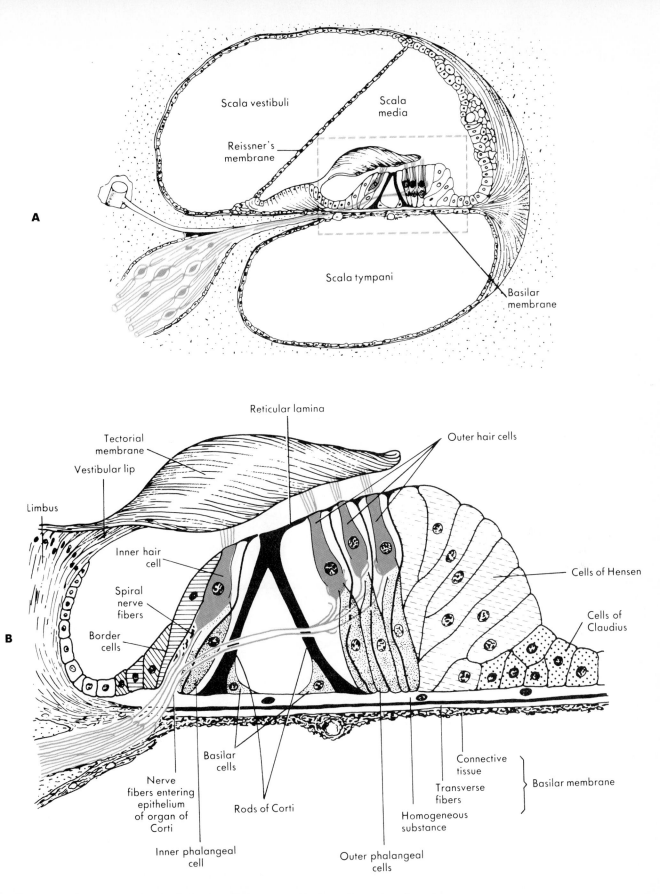

A

Scala vestibuli

Scala media

Reissner's membrane

Scala tympani

Basilar membrane

B

Reticular lamina

Tectorial membrane

Vestibular lip

Limbus

Inner hair cell

Spiral nerve fibers

Border cells

Outer hair cells

Cells of Hensen

Cells of Claudius

Nerve fibers entering epithelium of organ of Corti

Inner phalangeal cell

Basilar cells

Rods of Corti

Outer phalangeal cells

Homogeneous substance

Transverse fibers

Connective tissue

Basilar membrane

FIGURE 7-11   For legend see opposite page.

Sound Transduction

The external ear acts as a filter that is tuned to frequencies between 800 and 6000 Hz. The pinna serves little function in humans, although it is important in many animals. Pressure waves reaching the tympanic membrane cause it and the ossicular chain to vibrate at the frequency of the sound. The ossicular chain in turn causes an oscillation of the oval window and of the fluids within the cochlea. The round window completes the hydraulic pathway.

The middle ear mechanism serves as an impedance matching device to couple airborne sound waves with those conducted through the cochlear fluids (Figure 7-12, A). If sound waves were to be conducted directly from air to the oval window, most of the energy would be reflected and lost. With the

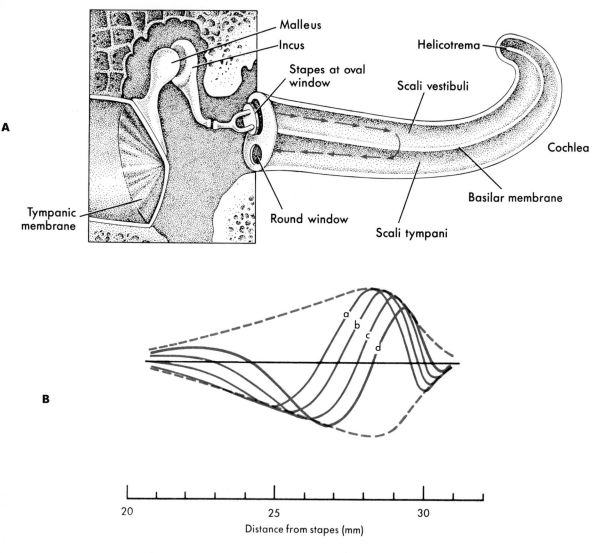

FIGURE 7-12   **A,** The impedance matching arrangement of the ear. The tympanic membrane and ossicular chain oscillate in response to sound waves in the air. The movements of the stapes in the oval window produce comparable oscillations of the fluid columns within the cochlea. The distance along the basilar membrane at which oscillations are maximal depends on the frequency of the sound. The largest displacements of the basilar membrane are near the base of the cochlea for high frequencies and near the apex for low frequencies. **B,** The traveling wave produced by a 200 Hz sound at four different times. The dashed line is the envelope of the peaks of the successive positions of the wave, showing a maximal deflection of the basilar membrane about 29 mm from the stapes.

mechanical advantage provided by the ratio of the area of the tympanic membrane to that of the oval window, plus that provided by the lever action of the ossicular chain, only 10 to 15 dB are lost in the impedance matching process of the ear.

Within the cochlear duct, the maximal amplitude of the oscillations extends for various distances along the basilar membrane; the distance depends on the frequency of the sound (Figure 7-12, B). Although much of the basilar membrane oscillates in a traveling wave in response to a particular frequency of sound, high frequencies result in movements that are largest in the basal part of the cochlea, whereas low frequencies induce movements that are largest near the apex of the cochlea.

As the basilar membrane oscillates, the stereocilia of the hair cells in the organ of Corti are subjected to shear forces at their junctions with the tectorial membrane (Figure 7-13). When the stereocilia are bent in a direction toward the longest cilia, a hair cell will become depolarized because of an increased conductance of the apical membrane to cations. This depolarization is a receptor potential and causes the release of an excitatory transmitter that produces a generator potential in the primary afferent nerve fibers synapsing on the hair cell. As the oscillations of the basilar membrane move in the opposite direction, the membrane potential of the hair cell is hyperpolarized and less transmitter is released. The generator potential in the primary afferent terminals is thus an oscillatory one, and if its amplitude is sufficient during the depolarizing phases, it will trigger action potentials in the primary afferent nerve fiber.

The difference in potential between the endo-

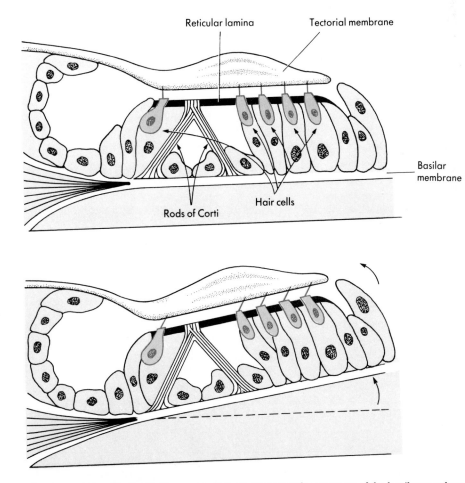

FIGURE 7-13   Transduction in the organ of Corti. An upward movement of the basilar membrane causes the development of shear forces between the stereocilia of the hair cells and the tectorial membrane, resulting in displacement of the cilia.

lymph and the intracellular fluid of the hair cells is unusually high. This potential difference is an important factor in the sensitivity of the auditory system. If the perilymph is considered the reference potential, the endolymph has a positive steady potential of about 85 mV. This is called the *endocochlear potential* and is the result of electrogenic pumping by the stria vascularis. The resting potential of the hair cells is approximately −85 mV with reference to the perilymph. Because of the positive potential in the endolymph, however, the transmembrane potential across the apical membrane of the hair cells can be as great as 170 mV. This increases the ionic driving forces across the transducer membrane.

An oscillatory potential, called the *cochlear microphonic potential,* can be recorded from the bony labyrinth of the cochlea. This potential results from the current flow associated with the activity of the hair cells in response to sound. The cochlear microphonic potential has the frequency of the sound stimulus, and its amplitude is graded with the sound intensity.

Cochlear nerve fibers that innervate hair cells at different points along the length of the organ of Corti are tuned to different frequencies of sound (Figure 7-14). The tuning properties of the primary afferent fibers can be demonstrated by constructing tuning curves that relate the threshold for activation of the fiber to the frequencies of sound stimuli. The frequency that activates the fiber at the lowest intensity is called the *characteristic frequency* of the fiber. Cochlear nerve fibers that innervate the organ of Corti near

the base of the cochlea have high characteristic frequencies, whereas those innervating the apex have low characteristic frequencies. The organ of Corti is thus organized *tonotopically.*

For the lower part of the frequency range detected by the cochlea (less than 4000 Hz), the discharges of a given cochlear nerve fiber show *phase locking.* That is, they occur consistently at a particular phase of the sound oscillation. The discharges of a population of afferent nerve fibers could actually signal the stimulus frequency. This is *volley coding* of acoustic signals. However, cochlear afferent fibers with higher characteristic frequencies do not show phase locking. Coding in these depends on *place coding;* the afferent fibers that innervate regions near the base of the cochlea signal frequencies according to the site innervated. *Intensity coding* depends on the number of discharges evoked by sounds of different intensities and presumably also on the number of neurons that discharge.

Central Auditory Pathway

Branches of individual primary afferent fibers of the cochlear nerve synapse in the cochlear nuclei. The cochlear nuclei project to several brainstem nuclei, including the superior olivary complex and the inferior colliculus. Many of the axons from the cochlear nuclei cross the midline in the trapezoid body to innervate the contralateral superior olivary complex or to ascend in the lateral lemniscus. The superior olivary complexes connect with each other and also

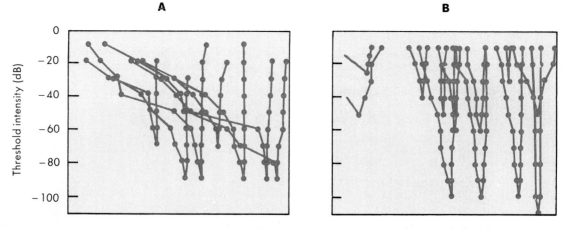

FIGURE 7-14   Tuning curves for neurons in the auditory pathway. **A,** Tuning curves for excitation of seven different neurons in the cochlear nerve. **B,** Tuning curves for 12 different neurons in the inferior colliculus.

project rostrally. Fibers from the inferior colliculus project through the brachium of the inferior colliculus to end in the medial geniculate nucleus. The latter projects through the auditory radiation to the primary auditory cortex, A1, which in the human is in the transverse temporal gyri (Brodmann's areas 41 and 42). As in the visual system, several additional cortical areas also contribute to auditory processing.

Central Auditory Processing

The superior olivary complex is concerned with **sound localization.** Neurons in the medial superior olivary nuclei compare the arrival times of sound in the two ears, whereas neurons in the lateral superior olivary nuclei compare differences in the intensity of sounds reaching the two ears. A sound originating from a source located to the left will reach the left ear first, and the head will provide an acoustic shield that lowers the intensity of the sound that reaches the right ear. By making use of these binaural cues, signals from the superior olivary nuclei allow the central auditory pathways to judge the location of the sound source.

Binaural processing occurs also in the cortex, as shown by the presence of *summation* and *suppression columns* in the auditory cortex. The responses of the neurons in these columns depend on whether sounds are introduced into the left or right ear or both. In summation columns neurons respond better when sound reaches both ears rather than only one. Neurons of suppression columns respond better to sound in one ear than simultaneously in both.

Frequency analysis within the central auditory pathways is reflected in the tonotopic maps characteristic of many auditory structures. The tonotonic map of the cochlea is also reflected in tonotopic maps in the cochlear nuclei, inferior colliculus, medial geniculate nucleus, and several regions of the auditory cortex.

The bilateral organization of the central auditory pathways is the reason that neurologic lesions of the brainstem at levels rostral to the cochlear nuclei do not produce unilateral deafness (although large unilateral lesions of the auditory cortex do interfere with the localization of sounds in space). Unilateral deafness, for example, implies a defect in the sound conduction system (e.g., in the tympanic membrane or the ossicle chain) or in the initial stages of the auditory pathway (organ of Corti, cochlear nerve, or coch-

lear nuclei). These conditions are called *conduction deafness* and *nerve deafness,* respectively. The degree of deafness depends on the severity and the location of the lesion. For example, loud high-frequency sounds can damage part of the organ of Corti and lead to deafness only for high-frequency sound.

The degree of deafness and the frequencies affected can be determined by *audiometry,* which tests the patient in each ear with pure tones of different frequencies and intensities. By comparing auditory thresholds for different sample frequencies with those expected in normal subjects, deficits can be described in terms of decibel losses for a certain range of frequencies or for the entire frequency spectrum.

VESTIBULAR SYSTEM

The sensory role of the vestibular system is a form of proprioception. The vestibular apparatus detects head movements and the position of the head in space. To accomplish this, it uses two sets of sensory epithelia to transduce *angular* and *linear accelerations* of the head. The vestibular apparatus is part of the membranous labyrinth of the inner ear.

Structure of the Vestibular Apparatus

The vestibular apparatus is contained within the bony labyrinth, but unlike the cochlea, its function depends mainly on the membranous labyrinth. The vestibular apparatus is connected with the cochlear duct, contains endolymph, and is surrounded by perilymph. The vestibular apparatus includes three pairs of semicircular canals on each side: the **anterior, posterior,** and **horizontal canals** (Figure 7-15). The anterior and posterior canals are oriented in vertical planes that are perpendicular to each other, as well as perpendicular to the plane of the horizontal canals. Thus the canals are well positioned to sense events in the three dimensions of space. The superior canal on one side is parallel to the posterior canal on the other side; the horizontal canals are in the same plane.

Each of the semicircular canals has a dilation, called an *ampulla.* Within the ampulla is a sensory epithelium, known as an *ampullary crest* (Figure 7-16). The apical surface of each of the hair cells of the sensory epithelium has both stereocilia and a single *kinocilium* (unlike cochlear hair cells, which lack

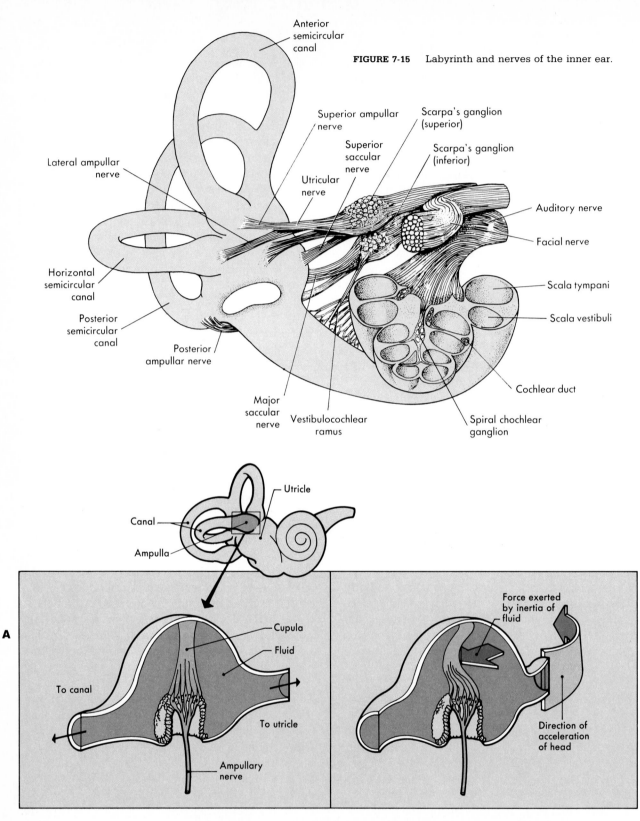

FIGURE 7-15 Labyrinth and nerves of the inner ear.

Anterior semicircular canal

Superior ampullar nerve

Scarpa's ganglion (superior)

Superior saccular nerve

Scarpa's ganglion (inferior)

Lateral ampullar nerve

Utricular nerve

Auditory nerve

Facial nerve

Horizontal semicircular canal

Scala tympani

Scala vestibuli

Posterior semicircular canal

Posterior ampullar nerve

Cochlear duct

Major saccular nerve

Vestibulocochlear ramus

Spiral chochlear ganglion

Utricle

Canal

Ampulla

A

B

Cupula

Fluid

To canal

Force exerted by inertia of fluid

To utricle

Direction of acceleration of head

Ampullary nerve

FIGURE 7-16 **A,** The relationship of the cupula to the ampulla when the head is stationary. **B,** The displacement of the cupula when the head is rotated.

kinocilia). The arrangement of the kinocilium with respect to the stereocilia gives a functional polarity to the vestibular hair cell. The cilia are all oriented in the same way relative to the axis of the semicircular duct. The cilia contact a gelatinous mass, the *cupula,* which extends across the ampulla and occludes it completely. Pressure shifts in the endolymph produced by angular accelerations of the head distort the cupula (Figure 7-16) and bend the cilia of the crista.

The semicircular canals connect with the *utricle,* one of the otolith organs. The sensory epithelium of the utricle is the *macula utriculi,* which is oriented horizontally along the floor of the utricle. The *otolithic membrane* is a gelatinous mass that contains numerous otoliths formed from crystals of calcium carbonate. The hair cells of the macula are oriented in relation to a groove, called the *striola,* along the length of the macula. The kinocilia in the utricle are on the striola side of the hair cells. The saccule is a separate part of the membranous labyrinth, and the macula sacculi is oriented vertically. Linear accelerations of the head cause the otolithic membranes to shift with respect to the hair cells. This shift results in bending of the cilia and sensory transduction. Angular accel-

erations do not affect the otolithic membrane substantially because the otolithic membranes do not protrude into the endolymph.

Vestibular Transduction

When the stereocilia on the vestibular hair cells are bent toward the kinocilium, the hair cell is depolarized because of an increased conductance of the hair cell membrane to cations. Bending of the cilia in the opposite direction leads to hyperpolarization. When vestibular hair cells are depolarized, they release more neurotransmitter (probably an excitatory amino acid such as glutamate), and when they are hyperpolarized, they release less. The neurotransmitter excites primary afferent fibers that end on the hair cells. In the absence of overt stimuli, vestibular primary afferent fibers are spontaneously active. Thus vestibular stimuli modulate afferent activity (Figure 7-17). The activity either increases or decreases, depending on the direction in which the cilia are bent.

In the ampullary crest of the horizontal semicircular duct, the kinocilia are arranged so that they are on the utricular side of the ampulla (Figure 7-18). If the

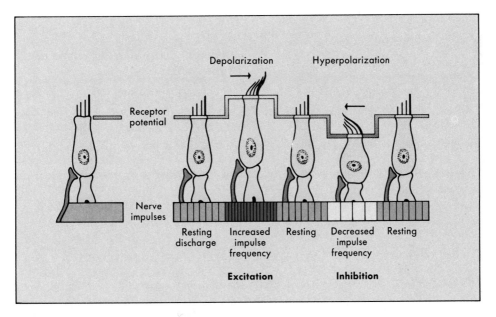

FIGURE 7-17   Directional selectivity of hair cells. Bending of the stereocilia toward the kinocilium depolarizes the hair cell and increases the firing rate in its afferent fiber. Bending of the stereocilia away from the kinocilium hyperpolarizes the hair cell and decreases the firing rate in its afferent fiber.

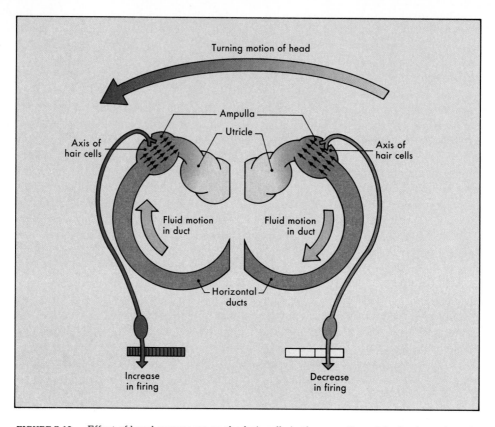

FIGURE 7-18   Effect of head movement on the hair cells in the ampullae of the horizontal semicircular ducts. The functional polarity of the hair cells is indicated by the arrows.

head is rotated to the left, inertial forces will cause the endolymph to shift relatively to the right in both the horizontal canals. In the left ear this means that the stereocilia of the hair cells of the left horizontal canal will bend toward the kinocilium (toward the utricle), and the primary afferent nerve fibers that supply the left ampullary crest will show an increased discharge. Conversely, the stereocilia in the crest of the right horizontal duct will bend away from their kinocilia (away from the utricle), and thus the discharge of the primary afferent fibers of this crest will be reduced.

The orientation of the kinocilia in the macula utriculi is toward the striola. The orientation in the macula sacculi is away from the striola. That is, hair cells on the two sides of the striola are functionally polarized in opposite directions. The changes produced in the discharges of vestibular afferents from a macula by linear acceleration of the head differ for different hair cells. The pattern of input to the central nervous system is analyzed and interpreted by the central vestibular pathways in terms of head position.

Central Vestibular Pathways and Vestibular Sensation

Primary afferent fibers from the vestibular apparatus reach the brainstem by way of the vestibular nerve. The cell bodies of the primary afferent fibers are in the vestibular ganglion. Most of the afferents terminate in the superior, lateral, medial, and inferior vestibular nuclei, but some end in the cerebellum. The main projections from the semicircular ducts are to the superior and medial vestibular nuclei, from the utricle to the lateral and inferior vestibular nuclei, and from the saccule to the inferior vestibular nucleus.

The projections of the vestibular nuclei include interconnections with the cerebellum and reticular formation, the motor nuclei supplying the eye muscles, and the spinal cord. These are very important for the vestibular control of eye and head movements and posture. The reticular formation is also involved in autonomic adjustments in response to vestibular signals.

CHEMICAL SENSES

The chemical senses include taste *(gustation)* and smell *(olfaction)*. They permit detection of chemical substances in food, water, and the atmosphere. Humans are less adept at chemical detection than many animals, but the chemical senses contribute substantially to the affective aspects of life, and their malfunction may be significant in disease.

Taste

The human gustatory system recognizes many different taste stimuli. However, these can generally be classified as one of four primary taste qualities: sweet, salty, sour, and bitter.

The sensory receptors for taste are the *taste buds.*

Most taste buds are on the tongue, but some are on the palate, pharynx, larynx, and upper esophagus. Taste buds occur in groups on papillae (Figure 7-19). *Fungiform papillae* are mushroomlike structures, several hundred of which are present on the anterior two thirds of the tongue. The taste buds of the fungiform papillae respond mainly to sweet and salty substances but also to sour. The taste buds on fungiform papillae are innervated by the chorda tympani branch of the facial (seventh cranial) nerve. *Foliate papillae* are folded structures on the posterior edge of the tongue, and their taste buds respond best to sour stimuli. *Circumvallate papillae* are large, round structures encircled by a depression; they are on the posterior tongue and respond to bitter substances. The foliate and circumvallate papillae are innervated by the glossopharyngeal (ninth cranial) nerve. Taste

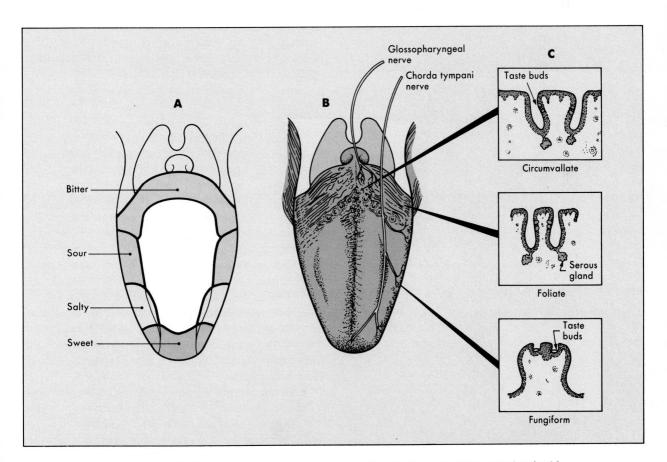

FIGURE 7-19   Peripheral sensory apparatus for gustation. **A,** Taste qualities associated with different regions of the tongue. **B,** The innervation of taste buds in the anterior two thirds and posterior one third of the tongue by the facial and glossopharyngeal nerves. **C,** Arrangement of the taste buds on the three types of papillae.

buds in the region of the epiglottis and upper esophagus are supplied by the superior laryngeal branch of the vagus (tenth cranial) nerve.

A taste bud consists of a group of some 50 gustatory receptor cells in association with supporting cells and basal cells (Figure 7-20). The gustatory cells are continuously turned over and are replaced by differentiation of supporting cells from basal cells. The apical membranes of the gustatory cells have microvilli that protrude into a taste pore, where they come into contact with saliva.

Receptor molecules on the microvilli recognize chemical substances in the saliva. The gustatory cells are in synaptic contact with primary afferent nerve terminals. Gustatory signals apparently evoke a receptor potential in the gustatory cell, which leads to transmitter release, a generator potential, and a coded pattern of nerve impulses in the primary afferent fiber. Individual gustatory cells do not appear to be completely selective for a particular primary taste. Rather, they respond best to one type of taste stimulus and less well to others. The recognition of a particular taste quality depends on the activity of a population of gustatory cells. This is a *modified labeled-line system.*

The primary afferent fibers from the taste buds are in the facial, glossopharyngeal, and vagus nerves. The afferent fibers enter the brainstem and travel caudally in the solitary tract, ending in the gustatory part of the nucleus of the solitary tract. Ascending gustatory fibers reach the parvocellular part of the ventral posteromedial nucleus of the thalamus. This nucleus projects to the postcentral gyrus, ending adjacent to the area representing the tongue. An unusual feature of the gustatory projection is that it is ipsilateral rather than crossed. This pathway appears to mediate taste perception, since this is impaired following lesions of the cortical gustatory areas.

Smell

The human olfactory system can recognize many odors. These are difficult to classify, but there are at least seven primary odors: camphoraceous, musk, floral, peppermint, ethereal, pungent, and putrid.

The sensory receptors for olfaction are located in the olfactory mucosa, a specialized area of about 2.5 cm^2 in each nasal mucosa. The olfactory receptor cells are themselves primary afferent neurons. They have an apical process with cilia that extend into a layer of mucus, in which are dissolved chemical substances that elicit olfactory responses. The base of the olfactory receptor cells gives rise to an axon that projects centrally to end in the olfactory bulb. Associated with the olfactory receptor cells are supporting and basal cells that replace olfactory receptors cells as they turn over.

Olfactory transduction depends on the binding of odorants (dissolved in the mucous layer) to receptor molecules on the cilia of olfactory receptor cells. The resulting receptor potential increases the firing rate of the primary afferent fiber. The firing rate is a function of the concentration of the odorant.

The coding mechanism for odors is a modified

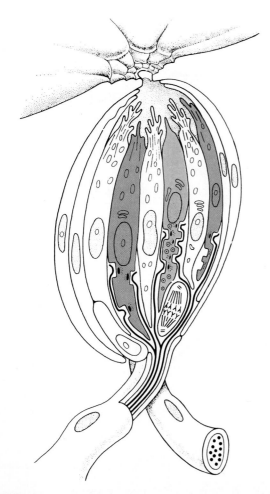

FIGURE 7-20   Taste bud. The receptor cells are darker and the supportive cells lighter.

labeled-line system similar to that for taste. Olfactory receptors respond best to a particular type of odorant and less well to others. Olfactory receptors are grouped according to sensitivity to the class of odorant and are located in different regions of the olfactory mucosa. The central nervous system is presented with a spatially coded input that partly represents odor qualities.

The central olfactory pathway is complex. Olfaction is the most primitive sensation, and thus the organization of the olfactory pathway apparently reflects an early solution to the problem of sensory representation in the brain. Other sensory systems have evolved differently. One unusual feature of the olfactory system is that the primary afferent neurons synapse directly on neurons of the telencephalon, whereas in all other sensory systems, sensory processing occurs at several lower stages before information reaches the telencephalon.

The primary afferent axons from olfactory receptor cells are unmyelinated axons that collect into filaments of the olfactory nerve (Figure 7-21). The olfactory nerve bundles pass through the base of the skull and synapse in the olfactory bulb.

The main efferent projections of the olfactory bulb form the olfactory tract. Terminations are made in a number of structures at the base of the brain. The two olfactory bulbs also interconnect through the anterior

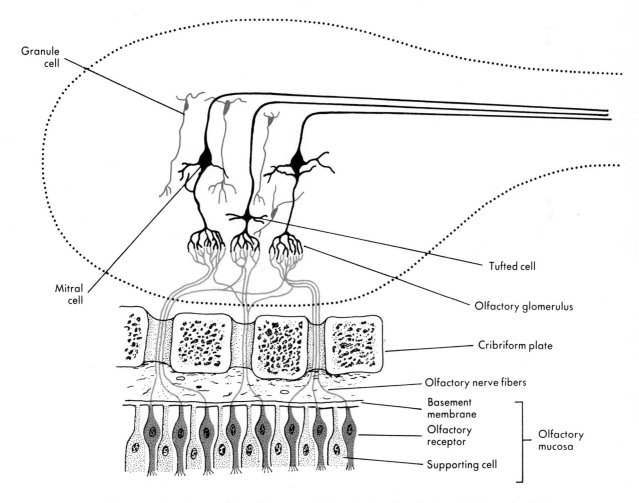

FIGURE 7-21   Initial part of the olfactory pathway, showing the olfactory receptor cells, their projections to the olfactory bulb, and their synapses in the glomeruli with tufted and mitral cells. Also shown are some of the granule cells, which serve as inhibitory interneurons.

commissure. The orbitofrontal region of the neocortex also receives olfactory information by way of the thalamus. Presumably the limbic system projections of the olfactory system are involved in the affective responses to odors, whereas the neocortex is concerned with the discrimination of odors.

Important disturbances in olfaction include *anosmia,* the loss of olfaction on one or both sides (e.g., caused by a tumor compressing the olfactory tract), and *uncinate fits,* seizures originating in the temporal lobe that cause olfactory hallucinations.

BIBLIOGRAPHY
Journal Articles

Allman J et al: Stimulus specific responses from beyond the classical receptive field, Annu Rev Neurosci 8:407, 1985.

Brugge JF and Geisler CD: Auditory mechanisms of the lower brainstem, Annu Rev Neurosci 1:363, 1978.

Getchell TV: Functional properties of vertebrate olfactory receptor neurons, Physiol Rev 66:772, 1986.

Gilbert CD: Microcircuitry of the visual cortex, Annu Rev Neurosci 6:217, 1983.

Hudspeth AJ: Mechanoelectrical transduction by hair cells in the acousticolateralis sensory system, Annu Rev Neurosci 6:187, 1983.

Imig TJ and Morel A: Organization of the thalamocortical auditory system in the cat, Annu Rev Neurosci 6:95, 1983.

Lancet D: Vertebrate olfactory reception, Annu Rev Neurosci 9:329, 1986.

Maunsell JHR and Newsome WT: Visual processing in monkey extrastriate cortex, Annu Rev Neurosci 10:363, 1987.

Moulton DG: Spatial patterning of response to odors in the peripheral olfactory system, Physiol Rev 56:578, 1976.

Patuzzi R and Robertson D: Tuning in the mammalian cochlea, Physiol Rev 68:1009, 1988.

Schwartz EA: Phototransduction in vertebrate rods, Annu Rev Neurosci 8:339, 1985.

Sherman SM and Spear PD: Organization of visual pathways in normal and visually deprived cats, Physiol Rev 62:738, 1982.

Sterling P: Microcircuitry of the cat retina, Annu Rev Neurosci 6:149, 1983.

Stryer L: Cyclic GMP cascade of vision, Annu Rev Neurosci 9:87, 1986.

Travers JB et al: Gustatory neural processing in the hindbrain, Annu Rev Neurosci 10:595, 1987.

Books and Monographs

Dowling JE: The retina, an approachable part of the brain, Cambridge, Mass, 1987, Harvard University Press.

Kandel ER and Schwartz JH: Principles of neural science, ed 2, New York, 1985, Elsevier Science Publishing Co, Inc.

Kuffler SW et al: From neuron to brain, Sunderland, Mass, 1984, Sinauer Associates, Inc.

Shepherd GM: Neurobiology, New York, 1983, Oxford University Press.

Wilson VJ and Jones GM: Mammalian vestibular physiology, New York, 1979, Plenum Press.

CHAPTER

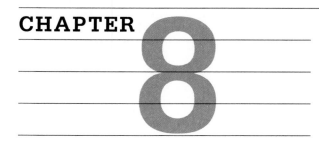

Motor System

The term *motor system* refers to the neural pathways that control the sequence and pattern of contraction of skeletal muscles. Skeletal muscle contractions result in posture, reflexes, rhythmic activity (e.g., locomotion), and voluntary movements. A given motor act may involve several of these. Motor acts make up a substantial part of the readily observable behavior of an organism. Motor behaviors that are especially important in humans include speech and movements of the digits and eyes.

Skeletal muscle fibers are innervated by α-motoneurons located in the spinal cord ventral horn and in cranial nerve motor nuclei. The basic element of motor control is the motor unit, which comprises an α-motoneuron and the skeletal muscle fibers it innervates. Reflex pathways involving muscle stretch receptors and a variety of other sensory receptors are organized within the spinal cord. Reflex circuits form a control system that organizes posture and movement.

The higher motor centers of the brain superimpose commands on activity intrinsic to the spinal cord. These centers include the brainstem, the cerebral cortex, cerebellum, and basal ganglia.

SPINAL CORD MOTOR ORGANIZATION

The Motor Unit

The basic element in motor control is the motor unit (see Chapter 12). This consists of an α-motoneuron, its motor axon, and all the skeletal muscle fibers that it innervates. A muscle unit is the set of skeletal muscle fibers in a motor unit. The discharge of an α-motoneuron will normally result in the contraction of each of the muscle fibers that it supplies, since the endplate potential in skeletal muscle fibers is normally suprathreshold (see Chapter 12). In mammals and other vertebrates no inhibitory synapses exist on skeletal muscle fibers, although they do exist in many invertebrates. This means that all decisions about whether or not a skeletal muscle fiber will contract are normally made by the α-motoneuron. Furthermore, each time an α-motoneuron discharges, the entire muscle unit will contract. This means that the smallest gradation of force that can be generated by a muscle depends on the force of contraction of the weakest muscle units in that muscle.

A given skeletal muscle will contain a number of muscle units. The ratio between the number of α-motoneurons and the total number of skeletal muscle fibers in a muscle is the *innervation ratio*. This gives the number of muscle fibers in the average muscle unit. The number is large for muscles that are used for coarse movements (e.g., 2000 fibers for the gastrocnemius muscle) and small for muscles that produce finely graded movements (e.g., three to six for the eye muscles). The muscle fibers in a muscle unit are distributed widely in a muscle and are separated by fibers belonging to other motor units.

All the skeletal muscle fibers in a muscle unit are of the same histochemical type. That is, they are either all type I, type IIB, or IIA. The contractile properties of these muscle fiber types are summarized in Table 8-1. The motor units that twitch slowly and resist fatigue are classified as S (slow) and have type I fibers. S motor units depend on oxidative metabolism for their

Table 8-1   Muscle Fiber Contractile Properties

Type	Speed	Strength	Fatiguability	Motor Unit
I	Slow	Weak	Fatigue resistant	S
IIB	Fast	Strong	Fatigable	FF
IIA	Fast	Intermediate	Fatigue resistant	FR

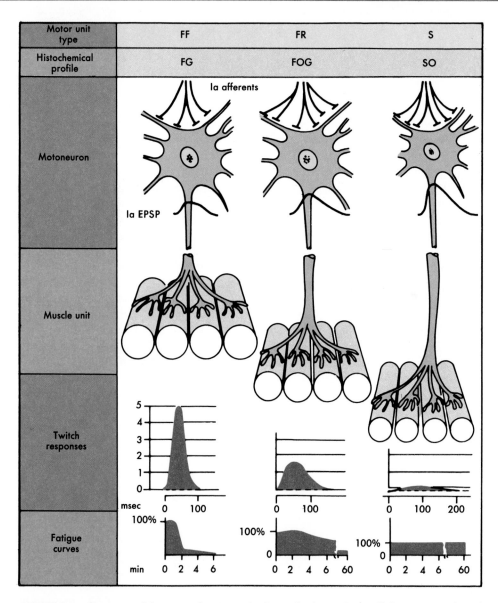

FIGURE 8-1   Summary of features of motor units in a mixed muscle (medial gastrocnemius of cat). Relative sizes are shown for motoneurons, muscle fibers, monosynaptic excitatory postsynaptic potentials evoked by volleys in group Ia afferent fibers, and twitch responses. The densities of shading of the muscle fibers reflect the intensity of histochemically staining for different enzymes. *EPSP*, Excitatory postsynaptic potential. *FG*, Fast glycolytic; *FOG*, Fast oxidative-glycolytic; *SO*, slow oxidative.

energy supply and have weak contractions (Figure 8-1). The motor units with fast twitches are FF (fast, fatiguable) and FR (fast, fatigue resistant). FF motor units have type IIB fibers, use glycolytic metabolism, and have strong contractions but fatigue easily. FR motor units have type IIA fibers and rely on oxidative metabolism; their contractions are of intermediate force, and these motor units resist fatigue (Figure 8-1).

α-Motoneurons

The only way in which the central nervous system can cause skeletal muscle fibers to contract is by evoking discharges in α-motoneurons. Therefore all motor acts depend on neural circuits that eventually impinge on α-motoneurons. This led Sherrington to call α-motoneurons the *final common pathway*.

Motor Nucleus α-Motoneurons are large neurons found in lamina IX of the spinal cord ventral horn and in cranial nerve motor nuclei that supply skeletal muscles. Each muscle or group of synergistic muscles (those having a similar action) has its own motor nucleus. α-Motoneurons that supply a given muscle are generally arranged as a longitudinal column of cells, often extending two to three segments in the spinal cord and several millimeters in the brainstem. The set of α-motoneurons that innervates a muscle is called the *motoneuron pool* of the muscle.

The motor nuclei of different muscles or muscle groups are located in different parts of the ventral horn. That is, motor nuclei have a *somatotopic* organization. Motor nuclei that supply the axial muscles of the body are in the medial part of the ventral horn in the cervical and lumbosacral enlargements (Figure 8-2) and in the most ventral part of the ventral horn in the upper cervical, thoracic, and upper lumbar segments of the spinal cord. The innervation ratio for the motor units in these muscles is large, since the role of the axial muscles includes such gross activities as maintenance of posture, support for limb movements, and respiration.

Motor nuclei that innervate the limb muscles are in the lateral part of the ventral horn in the cervical and lumbosacral enlargements (Figure 8-2). The most distal muscles are supplied by motor nuclei located in the dorsolateral part of the ventral horn, whereas more proximal muscles are innervated by motor nuclei in the ventrolateral ventral horn. The innervation ratios of these muscles are small for the distal muscles and large for the proximal ones.

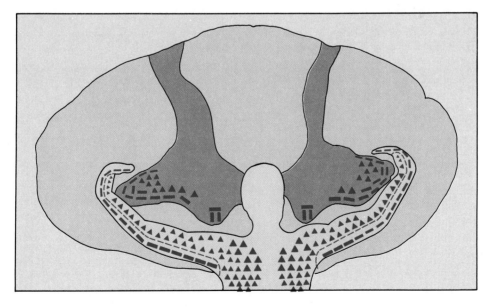

FIGURE 8-2   Somatotopic organization of spinal cord motoneurons. Motoneurons to axial muscles are in the medial ventral horn. In the lateral part of the motor nucleus, motoneurons to more proximal muscles are indicated by the larger symbols. Extensor muscles are supplied by motoneurons indicated by solid rectangles and flexor muscles by motoneurons indicated by triangles.

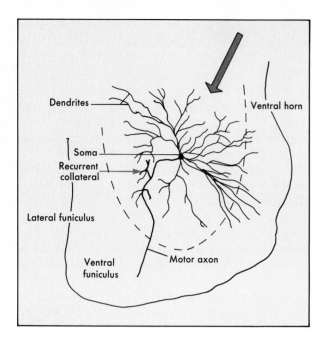

FIGURE 8-3 α-Motoneuron injected intracellularly with horseradish peroxidase. A small arrow indicates a recurrent collateral of the motor axon. The large arrow shows the direction followed by the microelectrode.

Motoneurons The individual α-motoneuron (Figure 8-3) is a cell with a large soma (up to 70 μm in diameter). Each of the 5 to 22 dendrites may be as long as 1 mm. The large myelinated axon has a diameter of 12 to 20 μm, and its conduction velocity is 72 to 120 m · sec^{-1}. The axon of an α-motoneuron is often called an α−motor axon. It arises from an axon hillock on the soma or a proximal dendrite and has a short, unmyelinated initial segment before the myelin sheath begins. The axons of α-motoneurons collect in bundles that leave the ventral horn, pass through the ventral white matter of the spinal cord, and enter a filament of the ventral root. Just before leaving the ventral horn, some α−motor axons give off recurrent collaterals. Recurrent collaterals typically project dorsally and synapse on interneurons, called *Renshaw cells,* in the ventral part of lamina VII.

Synaptic Integration The dendrites and soma of the α-motoneuron are covered with synapses from primary afferent fibers, interneurons, and pathways descending from the brain. Most of the synapses are from interneurons. Approximately half of the surface membrane lies beneath synaptic endings. Some of the synapses are excitatory, whereas others are inhibitory.

The lowest threshold part of the α-motoneuron membrane is thought to be the initial segment, which therefore serves as a trigger zone for the generation of action potentials. Excitatory synaptic currents depolarize all parts of the membrane of the motoneuron, but in terms of the initiation of an action potential, the depolarization of the initial segment is crucial. The *excitatory postsynaptic potentials (EPSPs)* produced by activation of more than one excitatory pathway to a motoneuron may sum (Figure 8-4, *A*), and the summed EPSPs may exceed threshold for discharge *(spatial summation).* Alternatively, repetitive activa-

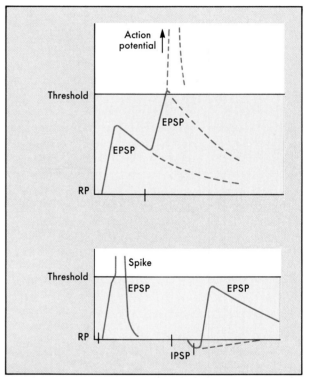

FIGURE 8-4 Synaptic integration. **A,** Summation of two excitatory postsynaptic potentials recorded from an α-motoneuron. The second EPSP exceeds threshold, and so an action potential is triggered. The two EPSPs can be the result of stimulation of separate pathways or activation of a single pathway repetitively. Thus summation can be spatial or temporal. **B,** An EPSP and the action potential it triggers at the left, as well as the results of an interaction between the EPSP and inhibitory postsynaptic potential *(IPSP).* Note that the IPSP prevents the EPSP from triggering the spike. *RP,* Resting potential.

tion of an excitatory pathway can produce *temporal summation* (see Chapter 4). *Inhibitory synaptic currents (IPSPs)* interfere with the excitatory ones and tend to prevent the discharge of an action potential (Figure 8-4, *B*). The interactions of excitatory and inhibitory synaptic currents in determining whether a neuron discharges is termed *synaptic integration*.

The location of a synapse on the membrane of a neuron such as an α-motoneuron may determine the effectiveness of that particular synapse in synaptic integration. Analysis of the passive electrical properties of the α-motoneuron indicates that synaptic currents can reach the initial segment from even the most distant part of the dendritic tree. However, synaptic potentials produced by distal synapses are smaller and slower than those produced by proximal synapses. For example, if a synapse on a dendrite is one length constant (see Chapter 3) away from the initial segment, the size of the membrane potential change in the initial segment will be only about one third (1/e) of that generated in the dendrite. Furthermore, the postsynaptic potential will be slowed considerably.

In many neurons, inhibitory synapses, which prevent action potential generation, tend to be located near the initial segment. Another arrangement is that excitatory and inhibitory synapses from neural pathways with antagonistic functions are located near each other on a given dendrite but away from the initial segment. This allows other pathways to influence the motoneuron independently. In still another arrangement, excitatory synaptic endings of one pathway receive axoaxonal synapses from another. Thus presynaptic inhibition can reduce the effectiveness of one pathway to the motoneuron without altering the excitability of the motoneuron, which can then participate in other pathways.

Action Potential Generation When a motoneuron discharges in response to synaptic excitation, the potentials follow a characteristic sequence. A recording from the soma reveals first an EPSP (Figure 8-5, *A*). Arising from this is a spike potential with two phases: an initial small spike on which is superimposed a slightly delayed but larger spike. The first small spike is believed to represent the action potential generated by the initial segment; it is small in a recording from the soma because of electrotonic decrement. The larger spike is thought to represent invasion of the soma by the action potential. Unseen in

the recording is the axonal action potential that is generated simultaneously and conducted distally from the initial segment toward the muscle supplied by the motoneuron. The activation of a motoneuron in this fashion by an EPSP is called *orthodromic activation* because the sequence is in the normal, or orthograde, direction.

Under experimental conditions an action potential may be initiated in the motor axon and conducted retrogradely to the motoneuron. This is called *antidromic activation* (Figure 8-5, *B*). A recording from the soma of the motoneuron reveals that the antidromic action potential arises directly from the resting membrane potential and has the same initial small spike and slightly delayed large spike as the orthodromic action potential. The missing part is the EPSP. Following the spike is a large, long-lasting *afterhyperpolarization*. This is also present in orthodromic action potentials but is often obscured by the EPSP.

The afterhyperpolarization is a very important feature of the motoneuronal action potential because it helps determine the characteristic firing rate of the neuron. Large α-motoneurons have shorter-lasting afterhyperpolarizations (about 50 msec long) than do small α-motoneurons (about 100 msec long). Therefore large α-motoneurons tend to discharge at rates of about 20 Hz, whereas small motoneurons discharge at approximately 10 Hz.

Muscle fibers contract with a twitch when a motoneuron discharges once. However, repetitive discharges of a motoneuron result in a tetanic contraction (see Chapter 12) of the muscle (Figure 12-8). The contractile force of a tetanic contraction increases with the rate of discharge of the motoneuron up to a limit imposed by the properties of the muscle. When the tetanic contraction is submaximal, muscle force increases with each motoneuronal discharge. This condition is called an *unfused tetanus*. When the tetanic contraction is maximal, the tetanus becomes fused. The motoneuronal firing rate that produces a *fused tetanus* in the muscle unit causes the greatest contractile force possible for that motor unit. A higher firing rate produces no greater action.

The characteristic firing rates of α-motoneurons match the mechanical properties of the skeletal muscle fibers. For example, large motoneurons fire at fast rates and innervate fast-twitch muscle fibers; that is, the motor units of large motoneurons are either the FF or the FR type (Table 8-1). Small motoneurons fire at

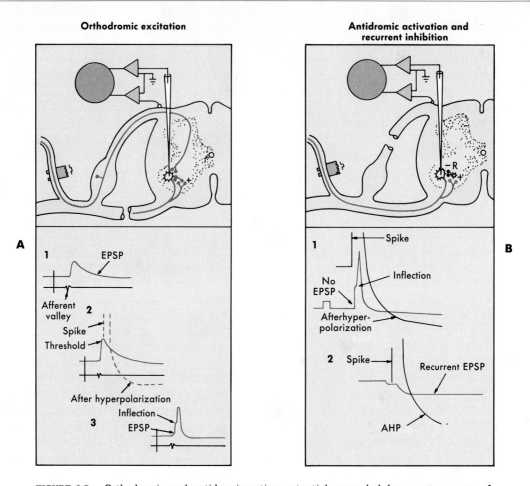

FIGURE 8-5   Orthodromic and antidromic action potentials recorded from motoneurons. **A,** Monosynaptic EPSP *(1)*, a larger EPSP that reaches threshold on some trials *(2)*, and a low-gain recording of an orthodromic action potential *(3)*. In *3* the arrows indicate the EPSP *(lower)* and the inflection between the initial segment and the soma–dendritic spikes *(upper)*. The upper drawing shows the recording arrangement and the interruption of the ventral root to prevent antidromic activation. **B,** Recordings in *1* shows an antidromic action potential in a motoneuron at high gain (spike truncated) and at low gain. Note the inflection on the rising phase of the spike. Also note that the spike is succeeded by a large afterhyperpolarization. This is best seen in the high-gain record. In *2* most of the records are with the stimulus subthreshold for the motor axon, and thus the potentials recorded are IPSPs caused by the activity of Renshaw cells excited by other motor axons. The lower drawing shows the experimental arrangement. Note that the dorsal root is cut to prevent orthodromic excitation, *R,* Renshaw cell.

slow rates and innervate slow-twitch muscle fibers; the motor units of small motoneurons are the S type.

The contraction of a muscle is regulated by the nervous system in two ways. The first is by the firing rate of α-motoneurons. As already indicated, the effects of changing the firing rate are limited by the firing rate at which a tetanus in a given motor unit becomes fused. The second means of regulating mus-

cle tension is by changing the number of active α-motoneurons. The activation of additional motoneurons is called *recruitment.*

The recruitment of α-motoneurons is orderly. Small motoneurons are usually recruited more easily than large motoneurons. This may be related to differences in the membrane properties of small and large motoneurons or may reflect the synaptic organization that controls their discharges. This difference in the effective excitability of small and large α-motoneurons is called the *size principle*. Not only are the small motoneurons recruited before the large ones during excitation, but the activity of the small motoneurons persists longer than that of the large motoneurons during inhibition. Because the motor axons of large motoneurons have greater diameters than those of small motoneurons, the action potentials recorded extracellularly from the ventral root are greater for large than for small motoneurons (Figure 8-6). This allows an evaluation of recruiting sequence by recordings from the ventral root.

Because of the progressive and orderly recruitment of small and then large α-motoneurons, a weak activation of a motoneuronal pool will discharge only the small α-motoneurons. This activity will produce a

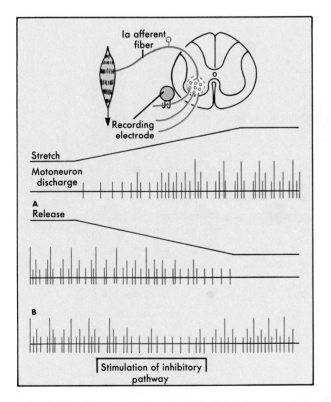

FIGURE 8-6   The size principle and motoneuron recruitment. The drawing shows the experimental arrangement. **A,** Stretching the muscle is shown to activate several motoneurons. The motor axon with the smallest action potential in the ventral root filament is activated first, then progressively larger units begin to discharge. The converse occurs when the muscle is released from the stretch: the large units stop firing first. **B,** An inhibitory input causes cessation of discharge of the larger units but not of the small unit.

weak, slow contraction of slow-twitch muscle fibers. This type of muscle activity is suited to the maintenance of posture and to slow movements such as walking. The recruitment of large α-motoneurons will activate powerful, fast-twitch muscle fibers. The contractions of these add to the initial force evoked by the slow-twitch fibers, and the resulting movements are appropriate to vigorous activity such as running and jumping.

Muscle Stretch Receptors

Skeletal muscles and their tendons contain specialized sensory receptors, called *stretch receptors,* that discharge when the muscles are stretched. These receptors include the muscle spindle and the Golgi tendon organ. These receptors are involved in sensory experience and contribute to proprioception (see Chapter 6). However, they are discussed here because of their importance in motor control.

The most complex muscle receptor is the muscle spindle. Muscle spindles are composed of elongated bundles of narrow muscle fibers, called *intrafusal muscle fibers,* enclosed within a connective tissue capsule. The spindles are richly innervated with both sensory and motor endings. Most of the muscle spindle lies freely within the space between the regular, or extrafusal, muscle fibers, but its distal ends merge with connective tissue in the muscle. This parallel arrangement is important for the operation of the muscle spindle. When the whole muscle is stretched, the muscle spindle is also stretched (Figure 8-7, *B*). However, when the extrafusal fibers of the muscle contract, the muscle spindle will be unloaded, unless the intrafusal muscle fibers also contract (Figure 8-7, *A*).

The intrafusal muscle fibers are of two main types, called *nuclear bag* and *nuclear chain* fibers because of the arrangement of their nuclei (Figure 8-8). Nuclear bag fibers are larger than nuclear chain fibers and have a cluster of nuclei near their midpoint (resembling a "bag" of oranges). Nuclear chain fibers have a single row of nuclei near their midpoint.

The sensory endings in a muscle spindle are of two types (Figure 8-8): a *primary ending* and one or more *secondary endings*. The primary ending has annulospiral-shaped terminals on both nuclear bag and nuclear chain intrafusal muscle fibers and is innervated by a large, myelinated nerve fiber, called a *group Ia fiber*. Secondary endings have spraylike ter-

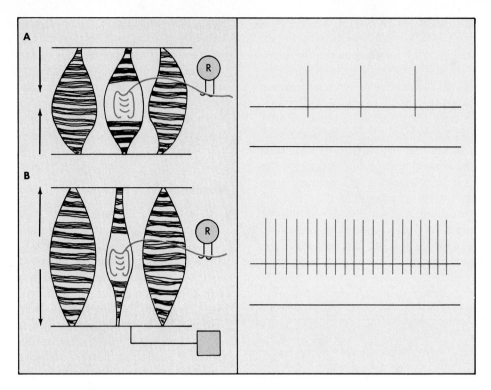

FIGURE 8-7   **A,** Extrafusal muscle fibers in a muscle contract, unloading the muscle spindle and reducing the discharge of an afferent fiber from the spindle. **B,** The muscle is stretched, activating the afferent fiber, *R,* Recording electrode.

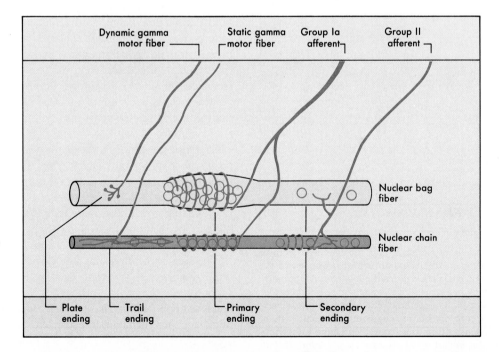
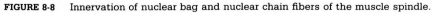

FIGURE 8-8   Innervation of nuclear bag and nuclear chain fibers of the muscle spindle.

minals that are primarily on nuclear chain fibers. The secondary endings are supplied by medium-sized *(group II)* primary afferent nerve fibers.

γ-Motoneurons provide the motor innervation of muscle spindles (Figure 8-8). The endings may be small endplates or elongated trail endings. There are two types of γ-motoneurons: *dynamic γ-motoneurons* innervate chiefly the nuclear bag intrafusal muscle fibers, and *static γ-motoneurons* supply mainly the nuclear chain fibers.

Primary endings respond to maintained muscle stretch with a slowly adapting discharge that has both a phasic and a static component (Figure 8-9). The *phasic response* signals the rate of stretch of the muscle, and the *static response* signals the length of the muscle. Secondary endings have only a static response and thus signal muscle length. The phasic response probably results from nuclear bag fibers showing an elastic rebound after an initial rapid elongation during muscle stretch. Thus the annulospiral

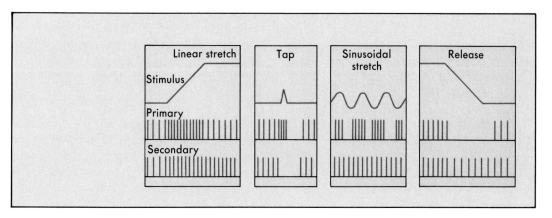

FIGURE 8-9 Responses of primary and secondary endings of a muscle spindle to linear stretch, tap, sinusoidal stretch, and release of the muscle.

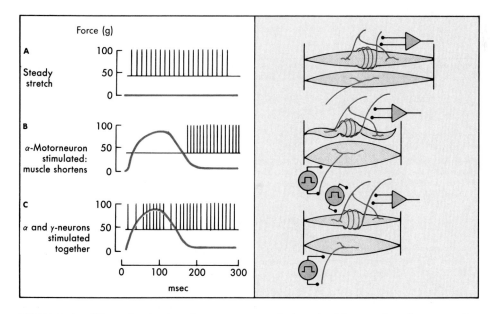

FIGURE 8-10 Effect of activation of γ-motoneurons. **A,** Stretch of the muscle activates an afferent fiber supplying a muscle spindle. **B,** The discharge stops during a muscle contraction produced by activity in α-motoneurons. **C,** The effect of unloading is avoided because α- and γ-motoneurons are co-activated.

coils of the primary ending on the nuclear bag fibers are separated widely at the onset of muscle stretch, and they then creep back toward a closed position as the muscle reaches its new length. The static response may be attributed to the nuclear chain fibers not rebounding; instead they elongate in proportion to the stretch, and the secondary endings are distorted accordingly.

γ-Motoneurons regulate the sensitivity of muscle spindles to muscle stretch (Figure 8-10). They can also prevent the unloading effect of muscle contraction by contracting the intrafusal muscle fibers during or just before contraction of the extrafusal muscle fibers. Dynamic γ-motoneurons enhance the phasic responses of the primary endings, and static γ-motoneurons increase the static responses of both primary and secondary endings. The central nervous system can thus regulate phasic and static responses of muscle spindles independently.

Another type of muscle stretch receptor is the *Golgi tendon organ* (Figure 8-11). These receptors are found in tendons, in the connective tissue inscriptions within skeletal muscles, and around joint capsules. The terminals of Golgi tendon organs interdigitate with bundles of collagen fibers, an arrangement that allows the application of mechanical force to the terminals when the muscle is either contracted or stretched. Golgi tendon organs are therefore arranged

in series with the skeletal muscle. They signal force in the muscle and its tendon. Golgi tendon organs are supplied by large, myelinated primary afferent nerve fibers, called *group Ib fibers*.

Spinal Cord Interneurons

As mentioned earlier, most of the synapses on α-motoneurons originate from spinal cord interneurons. Interneurons, by definition, are neurons interposed between primary afferent neurons and motoneurons. Interneurons whose processes are confined to the spinal cord are often called *propriospinal neurons.*

Most spinal cord interneurons are located in the dorsal horn. Many of these are involved in sensory processing and contribute directly or indirectly to the transmission of sensory information to the brain. However, neurons in the dorsal horn also project to the intermediate nucleus and ventral horn and affect the discharges of motoneurons. Furthermore, axons in pathways descending from the brain only rarely terminate directly on motoneurons. Axons in descending pathways usually end on interneurons and alter motor output by changing the level of activity in spinal cord circuits.

Various types of interneurons involved in motor control have been well characterized. The Renshaw cell has already been mentioned. Renshaw cells are

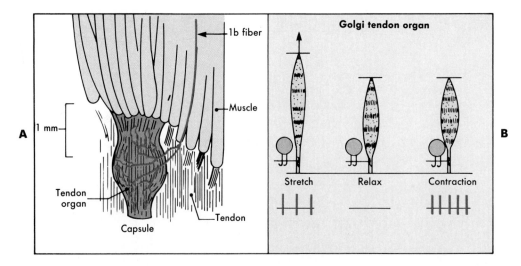

FIGURE 8-11   A, The structure of a Golgi tendon organ and its relationship to the tendon of a muscle. **B,** the recordings at the bottom show that a Golgi tendon organ can be activated by either muscle stretch or contraction. The activity of Golgi tendon organs signals muscle tension.

inhibitory interneurons located in the part of lamina VII that protrudes ventrally between the lateral part of lamina IX and lamina VIII. Recurrent collaterals from α–motor axons synapse on Renshaw cells. When the motor axons discharge, they release acetylcholine at the synapses on Renshaw cells and excite these cells. The Renshaw cells in turn synapse on and inhibit α-motoneurons; thus, when motoneurons discharge, this causes an inhibitory feedback by way of Renshaw cells. This is called *recurrent inhibition* (see Figure 8-5, *B*).

Another well-studied interneuron is the group Ia inhibitory interneuron. These interneurons are located in the dorsal part of lamina VII. They are excited monosynaptically by group Ia primary afferent fibers from the primary endings of muscle spindles. The term *monosynaptic* implies a neural pathway in which only one synapse intervenes between one element of the pathway and the next. Group Ia inhibitory interneurons in turn synapse on the α-motoneurons supplying the muscle or muscle group that serves as the antagonist to the muscle giving rise to the group Ia afferent fibers. Thus a disynaptic pathway is formed involving group Ia fibers from one muscle group, inhibitory interneurons, and α-motoneurons to the antagonist muscle group (Figure 8-12).

Spinal Cord Reflexes

A reflex is a relatively simple, stereotyped motor response to a defined sensory input. Some of the reflexes mediated by spinal cord circuits are described here. However, many other reflexes are also organized at the level of either the spinal cord or the brain.

Stretch Reflex A particularly important spinal reflex is the stretch reflex (sometimes called the *myotatic reflex*). Stretching a muscle causes a reflex contraction of that muscle and a reflex relaxation of the antagonistic muscles. The stretch reflex has two components, the phasic and the tonic stretch reflexes. The *phasic* stretch reflex is elicited by stretching the muscle quickly. This is often done clinically by tapping the tendon of the muscle with a reflex hammer. The *tonic* stretch reflex results from a slower stretch of a muscle, such as occurs during passive movement of a joint. The tonic stretch reflex is important in the maintenance of posture.

Hinge joints, such as the knee and the ankle, are extended or flexed by extensor and flexor muscles, which are antagonists because they produce opposite movements of the joint. A set of extensor and flexor muscles about a joint is called a *myotatic unit*. A phasic stretch reflex produced by stretching an extensor muscle results in contraction of the extensor muscle and relaxation of the flexor muscle. Converse-

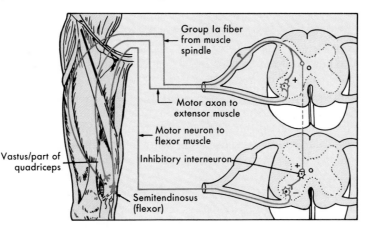

FIGURE 8-12   Reflex pathway for the stretch reflex. The illustration is for the quadriceps stretch reflex, but similar connections would be made for other muscles, including flexor muscles. A muscle spindle is shown to be supplied by a group Ia fiber that enters the spinal cord through a dorsal root and makes monosynaptic excitatory connections with an α-motoneuron to the quadriceps muscle in the same segment and a group Ia inhibitory interneuron in another segment. The inhibitory interneuron synapses with an α-motoneuron to the antagonistic flexor muscle, semitendinosus. *E,* Extensor motoneuron; *F,* flexor motoneuron.

ly, a phasic stretch reflex of the flexor muscle involves a concomitant relaxation of the extensor muscle. The reciprocal organization of the stretch reflex pathways is called *reciprocal innervation.*

The neural basis of a spinal reflex is the *reflex arc,* a circuit that includes a set of primary afferent fibers, interneurons, and α-motoneurons. The reflex arc for the phasic stretch reflex of a particular muscle includes (1) group Ia afferent fibers from primary endings of muscle spindles located within that muscle, (2) a monosynaptic excitatory connection of these afferents with α-motoneurons innervating the muscle, and (3) a disynaptic inhibitory pathway involving group Ia inhibitory interneurons that synapse with α-motoneurons innervating the antagonistic muscles (Figure 8-12).

Group Ia afferent fibers from muscle spindles in synergistic muscles (i.e., muscles that have a similar function) also participate in the excitation of α-motoneurons to a particular muscle, but generally the excitation is not sufficiently powerful to cause the motoneurons to discharge.

The reflex arc for the tonic stretch reflex involves the same connections just described for the phasic stretch reflex. However, group II afferent fibers from secondary endings of muscle spindles also contribute and make monosynaptic excitatory connections with the α-motoneurons that supply the muscle containing the muscle spindles.

As previously mentioned, the sensitivity of the primary and secondary endings of muscle spindles is controlled by dynamic and static γ-motoneurons. Activation of γ-motoneurons can result in a sufficiently strong excitatory input in group Ia afferent fibers to cause discharges in α-motoneurons. The pathway linking γ- to α-motoneurons by way of group Ia primary afferent fibers is called the γ-*loop.* However, the group Ia fibers in humans discharge after contractions of skeletal muscle. This pattern of discharge indicates that γ-motoneurons do activate muscle spindles during voluntary movements, but at about the same time as the activation of α-motoneurons. Presumably the shortening of the muscle spindles functions to prevent unloading. Thus voluntary and other movements depend on co-activation of α- and γ-motoneurons.

Inverse Myotatic Reflex An important reflex whose afferent limb is group Ib afferent fibers from Golgi tendon organs is sometimes called the inverse myotatic reflex. Group Ib afferent fibers from extensor

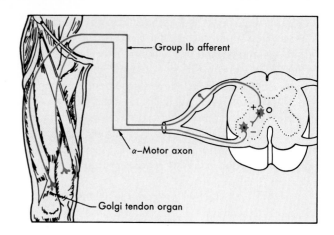

FIGURE 8-13   Pathway for the inverse myotatic reflex produced by Golgi tendon organs. A Golgi tendon organ is shown in the patellar tendon. Its group Ib afferent fiber enters the spinal cord through a dorsal root to terminate on an inhibitory interneuron that synapses on a motoneuron to the quadriceps muscle. *E,* Extensor motoneuron.

muscles synapse monosynaptically on inhibitory interneurons (Figure 8-13). The inhibitory interneurons in turn synapse on the α-motoneurons that supply the same and other extensor muscles in the limb. Flexor muscles are relatively unaffected by this pathway. This pathway is activated by increases in muscle force, which can be produced either by stretch or by contraction of the muscle, rather than by stretch alone. Therefore the word ''myotatic'' (which refers to muscle stretch) is probably inappropriate.

The stretch reflex and the group Ib (inverse myotatic) reflex are examples of negative feedback loops. In a negative feedback loop the output of the system is compared with the desired output (Figure 8-14). Any difference (error) is fed back to the input so that a corrective action can be made. The variable regulated by a negative feedback system is the controlled variable. In the stretch reflex the controlled variable is muscle length; in the group Ib pathway it is muscle force.

The concurrent regulation of muscle length and muscle force makes it possible to control muscle stiffness. Muscle has mechanical properties that resemble those of a spring. When a spring is slack, it exerts no force. When the spring is lengthened beyond a threshold point, known as the *set point* (or *resting*

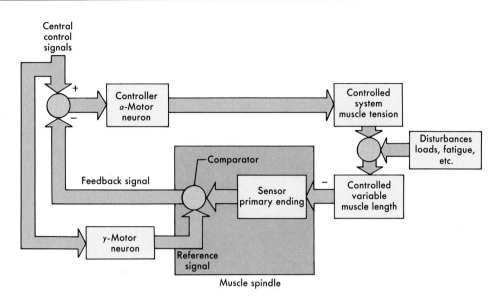

FIGURE 8-14   Diagram showing the operation of the stretch reflex as a negative feedback system that regulates muscle length. Length is determined by the level of muscle tension set by activity in α-motoneurons. Muscle length is detected by the muscle spindle, which has a sensitivity set by γ-motoneurons. If the muscle length is increased, such as by a changed load, feedback conveyed by the primary endings will increase the discharges of α-motoneurons, which reduces muscle length through further contraction of the muscle.

length), each increment of stretch is associated with the development of an increment of force. The relationship between length and force in an ideal spring is linear, and the slope of the curve is the stiffness of the spring. The stiffness of different springs varies, being greater if more force is produced for a given increment of stretch.

Muscles also have characteristic length-force curves (see Chapter 11). These depend on whether they are determined with the muscle relaxed or activated by nerve stimulation, and such curves can be highly nonlinear. If a muscle is passively stretched, the muscle will develop force as it is stretched beyond the set point (resting length). When the nerve to the muscle is stimulated, the length-force curve shifts, now having a lower set point and a greater slope. The increased slope indicates that contraction has increased the stiffness of the muscle.

In joints whose position is controlled by sets of agonist and antagonist muscles, a particular joint angle *(equilibrium point)* can actually be attained in several different ways. Because the myotatic unit is reciprocally innervated, neural circuitry is available to cause one muscle to be activated and the antagonist to be relaxed. The combination of these two events determines the equilibrium point of the joint. Another possibility is the co-contraction of the agonist and antagonist muscles. Although this mechanism requires more energy than the one using reciprocal innervation, co-contraction can provide stability in case of unanticipated changes in load because the stiffness of the joint is increased. One typically performs a new task using co-contraction until the task is learned, when co-contraction is replaced by the strategy of relaxing the antagonist.

Flexion Reflex Other important reflexes, such as the flexion reflex, also operate at the spinal cord level. In the flexion reflex the physiologic flexor muscles of one or more joints in a limb contract and the physiologic extensor muscles relax. The physiologic flexor muscles are those that tend to withdraw the limb from a noxious stimulus. The flexion reflex has several different uses. The flexor withdrawal reflex consists of a defensive removal of a limb from a threatening or damaging stimulus. This reflex takes precedence over other reflexes. It may be accompanied by a *crossed extensor reflex,* which involves the contraction of the extensor muscles and the relaxation of the flexor muscles of the contralateral limb. The crossed extension serves as a postural adjustment to compensate for the loss of the antigravitational support by the limb that flexes. In quadrupeds the converse pattern

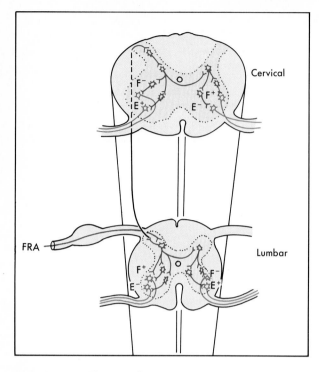

FIGURE 8-15   Flexion reflex pathway. The receptive field of flexion reflex afferents on the lower extremity is not shown. Activation of these leads to input over dorsal roots of the lumbar enlargement. By way of polysynaptic connections, flexor motoneurons of the lower extremity on the side stimulated are activated and extensor motoneurons inhibited. This results in flexion of the lower extremity. Crossed connections cause activation of extensor motoneurons and inhibition of flexor motoneurons to the opposite lower extremity, the crossed extensor reflex. A reverse pattern of reflex activity may be produced in the upper extremities. *E*, Extensor motoneuron; *F*, flexor motoneuron; *FRA*, Flexion reflex afferents.

can occur in the other pair of limbs. The flexion reflex is also involved in locomotion and in the scratch reflex.

The flexion reflex is initiated by the flexion reflex afferent fibers, which supply high-threshold muscle and joint receptors, many cutaneous receptors, and also nociceptors. It seems likely that the lower-threshold receptors help modulate locomotion and that the nociceptors are crucial for evoking the flexor withdrawal reflex. The flexion reflex pathway from primary afferent fibers to the motoneurons is polysynaptic and involves both excitatory and inhibitory interneurons (Figure 8-15). There are both uncrossed and crossed components of the pathway. The motoneurons involved are those appropriate to the reflex movements already described.

REGULATION OF DESCENDING MOTOR PATHWAYS

The spinal cord mediates simple reflex actions and contains the pattern generator for locomotion. More complex motor behaviors are initiated by pathways that originate in the brain. In some cases descending pathways have direct access to α-motoneurons and cause muscle contractions if the motoneurons discharge. More typically, however, motor pathways from the brain alter joint position indirectly by imposing commands on the reflex circuitry of the spinal cord (or brainstem). These commands change the set points of the muscles around the joint by changing muscle stiffness.

The motor system is organized hierarchically. The spinal cord motor apparatus is under the control of both the brainstem and higher brain centers. Several brainstem centers influence spinal reflexes. Postural reflexes are initiated by the vestibular system and by stretch receptors in the neck. The locomotor pattern generator is controlled by the midbrain locomotor center. A special set of motor centers is responsible for the control of eye position. Voluntary movements are initiated by way of the corticospinal and corticobulbar tracts. The patterns of discharge in these pathways are determined by motor programs developed in various cortical centers. The activity of the voluntary motor pathways and the descending pathways from the brainstem is regulated by the cerebellum and basal ganglia.

ORGANIZATION OF DESCENDING MOTOR PATHWAYS

The descending motor pathways have traditionally been subdivided into pyramidal and extrapyramidal components. The pyramidal system includes the corticospinal and corticobulbar tracts; pyramidal refers to the presence of these tracts in the medullary pyramid. The pyramidal system is the main pathway for mediating voluntary movements of the distal parts of the extremities, as well as for mimetic movements of the face muscles and movements of the tongue. The extrapyramidal system originally referred to motor pathways other than the corticospinal and corticobulbar tracts. At present, however, extrapyramidal is most usefully applied to motor disorders associated with lesions involving the basal ganglia, without ref-

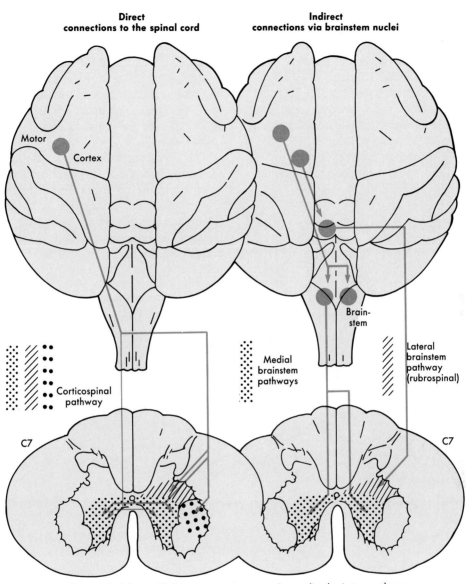

Direct connections to the spinal cord

Indirect connections via brainstem nuclei

Motor

Cortex

Brain-stem

Corticospinal pathway

Medial brainstem pathways

Lateral brainstem pathway (rubrospinal)

C7

C7

Corticospinal (pyramidal) tract

Descending brainstem pathways

FIGURE 8-16   Lateral and medial motor control systems. Descending motor pathways in the lateral funiculus include the lateral corticospinal tract *(left)* and rubrospinal tract *(right)*. The lateral corticospinal tract projects directly to motoneurons innervating distal muscles, as well as to interneurons controlling these motoneurons. The rubrospinal tract projects onto lateral interneurons. Medial pathways include the ventral corticospinal tract *(left)* and several pathways from the medial brainstem *(right)*. These pathways end in the medial ventral horn and control motoneurons to axial and proximal muscles.

erence to the particular motor pathways affected.

A helpful classification of descending motor pathways is based on their site of termination in the spinal cord. One set of pathways ends on α-motoneurons in the lateral part of lamina IX or on the interneurons that project to these (Figure 8-16). This *lateral system*

of descending pathways controls muscles of the distal and part of the proximal limbs. These muscles subserve fine movements used in manipulation and other precise actions, especially of the digits. A parallel control system ends in the brainstem and regulates the part of the facial motor nucleus that supplies the muscles of the lower part of the face, as well as the hypoglossal nucleus that innervates the tongue. The other set of pathways ends on motoneurons in the medial part of lamina IX or on interneurons that project to these (Figure 8-16). This *medial system* of descending pathways controls axial and girdle muscles, as well as most cranial nerve motor nuclei. The muscles of the body regulated by the medial system contribute importantly to posture, balance, and locomotion. Those in the head are involved in such activities as closure of the eyelids, chewing, swallowing, and phonation.

Lateral System

The lateral system includes two pathways from the brain to the spinal cord: the lateral corticospinal tract and the rubrospinal tract. In addition, the part of the corticobulbar tract that controls the lower face and the tongue can be considered part of the lateral system.

The motor cortex has a somatotopic organization (Figure 8-17) resembling that of the somatosensory cortex (see Chapter 6). The component of the lateral

corticospinal tract that controls the upper extremity originates from the dorsolateral aspect of the precentral gyrus (arm representation), and most of the fibers terminate in the cervical enlargement. The component controlling the lower extremity arises from the vertex and medial part of the postcentral gyrus (leg area) and passes caudally in the spinal cord in the lateral part of the tract to end in the lumbosacral enlargement.

The lateral corticospinal tract synapses in the dorsal horn, intermediate region, and ventral horn. The terminations in the dorsal horn come from corticospinal neurons in the postcentral gyrus, thus permitting the somatosensory cortex to regulate sensory transmission in the spinal cord. Most of the terminations of axons from the motor areas of the cerebral cortex are on interneurons in the base of the dorsal horn and in the intermediate region. However, some monosynaptic excitatory connections with α-motoneurons exist in lamina IX. These are especially prominent on α-motoneurons to the muscles of the hand. Such connections allow a direct influence by the motor cortex on fine hand movements.

The rubrospinal tract originates from the red nucle-

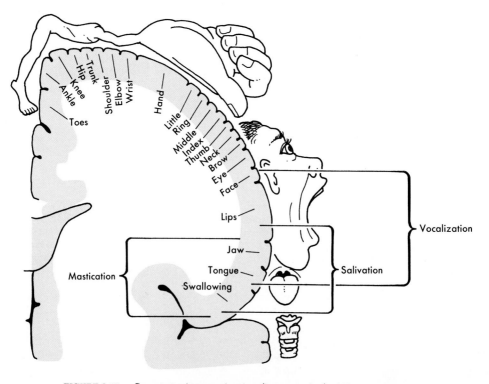

FIGURE 8-17   Somatotopic organization (homunculus) of the motor cortex.

us. The red nucleus has a somatotopic organization, with different components projecting to the cervical and lumbosacral enlargements. The tract decussates as it leaves the ventral aspect of the red nucleus and descends through the brainstem and the dorsal part of the lateral funiculus of the spinal cord. Terminations are in the base of the dorsal horn and intermediate region. The tract ends on interneurons and thus regulates movements indirectly by actions on reflex pathways. Neurons of the red nucleus receive a prominent corticorubral projection, and therefore the rubrospinal pathway is strongly influenced by motor commands originating in the cerebral cortex. The rubrospinal tract appears to be much less significant in humans than in experimental animals.

Medial System

The medial system includes the ventral corticospinal tract, much of the corticobulbar tract, and several pathways descending from the brainstem, including the tectospinal tract, the lateral and medial vestibulospinal tracts, and the pontine and medullary reticulospinal tracts. The descending projections of monoaminergic nuclei in the brainstem form an additional modulatory system.

The ventral corticospinal tract originates from separate neurons of the motor areas of the cortex than from those projecting in the lateral corticospinal tract. The ventral corticospinal tract can influence the activity of motoneurons to both sides of the body. This is an important arrangement, since the axial muscles of both sides often function together.

Much of the corticobulbar tract can be considered to belong to the medial system. The corticobulbar projections of the medial system provide a bilateral innervation to the trigeminal motor nucleus, the component of the facial nucleus that supplies the frontal and orbicularis oculi muscles, the nucleus ambiguus, and the spinal accessory nucleus. The cortical projection from the frontal eye fields to the gaze centers (see later discussion) can also be regarded as a component of the corticobulbar projection because it exerts bilateral motor control, in this case of conjugate eye movements.

The tectospinal tract originates from neurons of the deep layers of the superior colliculus. The tectospinal tract helps control head movements. A tectobulbar tract also helps control eye position.

The lateral and medial vestibulospinal tracts arise,

respectively, from two of the vestibular nuclei, the lateral and medial vestibular nuclei. The lateral vestibulospinal tract enhances the activity of extensor (antigravity) muscles and thus functions in postural adjustments to vestibular signals. The medial vestibulospinal tracts are influenced primarily by the semicircular ducts (see later discussion); these tracts cause postural adjustments of the neck and upper limbs in response to angular accelerations of the head.

The pontine and medullary reticulospinal tracts arise from neurons of the pontine and medullary reticular formation. The main action of the pontine reticulospinal tract is the excitation of extensor (antigravity) motoneurons. The medullary reticulospinal tract inhibits several reflexes and also sensory transmission.

BRAINSTEM CONTROL OF POSTURE AND MOVEMENT

A hierarchic organization of the motor system can be demonstrated by the effects of lesions at different levels of the neuraxis. A lesion can result in a particular effect either (1) by abolishing functions subserved by a structure whose influence is removed by the lesion or (2) by allowing an action to appear through removal of an inhibitory influence. Such latter responses are called *release phenomena*. Lesions that are particularly instructive include spinal cord transection and decerebration.

Spinal Cord Transection

Transection of the cervical spinal cord below the phrenic nucleus has several important effects on the motor system, in addition to a complete loss of sensation from the body. Assuming that respiration is spared, the most important motor loss is of voluntary movement. Immediately after transection, especially in humans, a period of spinal shock may follow in which reflex activities are absent. This is presumably caused by the loss of the excitatory actions of descending pathways from the brain. After a time, up to months in humans, hyperactive stretch and flexion reflexes develop. These changes may reflect loss of descending inhibition and also rearrangements in spinal cord circuits, possibly including sprouting of primary afferent fibers and the formation of new syn-

A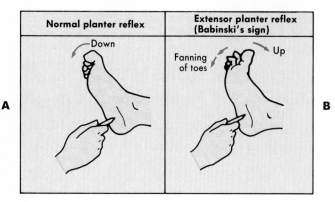

B

FIGURE 8-18   Babinski's sign. **A,** The normal response to stroking the plantar surface of the foot. **B,** Babinski's sign (extensor plantar reflex) in a person with an interruption of the corticospinal tract.

aptic connections within the spinal cord. Other release phenomena include the appearance of pathologic reflexes, such as *Babinski's sign* (Figure 8-18), resulting from interruption of the lateral corticospinal tracts. Although locomotion is not regained in humans, a capability for locomotion reappears in some animal species because the locomotor pattern generator is contained within the neural circuitry of the spinal cord. In animals with chronic spinal transections locomotion is triggered by afferent signals rather than by the midbrain locomotor center.

Hyperactive stretch reflexes may result in *clonus,* an alternating sequence of contractions of extensor and then flexor muscles. *Mass reflexes* are associated with hyperactive flexion reflexes and are characterized by flexion of one or both limbs and evacuation of the bladder and bowel.

Decerebrate Rigidity

Transection of the brainstem at a midbrain level results in a condition known as decerebrate rigidity. This develops immediately as the brainstem is transected, and in animals the ''rigidity'' is expressed as an exaggerated extensor (antigravity) posture caused by hyperactive stretch reflexes. The term *decerebrate rigidity* is unfortunate, since the condition more closely resembles spasticity than the rigidity that results from basal ganglion disease (see later discussion). Activation of the γ-loop is important; the ''rigidity'' is lost after transection of the dorsal roots (which would interrupt input from muscle spindle afferents). The vestibular system is also involved in decerebrate rigidity; destruction of the lateral vestibular nuclei reduces or eliminates the extensor posture.

Postural Reflexes

Various reflexes assist in postural adjustments that occur as the head is moved or the neck is bent. The receptors triggering these reflexes include the vestibular apparatus and stretch receptors in the neck. The visual system also contributes to postural adjustments, but the reflexes described here are elicited in the absence of visual cues.

Angular accelerations of the head activate the sensory epithelia of the semicircular ducts and elicit the acceleratory reflexes. These reflexes cause movements of the eyes, neck, and limbs that tend to oppose changes in position. For example, if the head is turned to the left (Figure 8-19), the eyes will be reflexly rotated a similar degree to the right. This reflex action will help maintain stability of the visual field. The movements of the two eyes are *conjugate,* meaning that the eyes move together in the same direction and through the same angle. When the head rotation exceeds the range of eye movement, the eyes are quickly deflected to the left and another visual target is found. If the head continues to rotate to the left, there will be an alternation of slow eye movements to the right, followed by rapid eye movements to the left. These alternating slow and fast eye movements are called *nystagmus.*

The same stimulus, rotation of the head to the left, will also tend to cause an increased contraction of the extensor (antigravity) muscles on the left, resulting in resistance to any tendency to fall to the left as the head rotation continues.

The neural mechanism underlying the acceleratory reflexes depends on stimulation of the sensory epithelia in the semicircular canals. In the case of rotation of the head in a plane parallel to the ground, the semicircular canals primarily involved are the horizontal ones (Figure 8-19). The inertia of the endolymph within the horizontal semicircular canals causes the endolymph to lag behind as the head rotates, producing a relative shift of the endolymph. This will deflect the cupulas of the ampullae of the horizontal canals and cause the stereocilia of the hair cells in the ampullary crest to bend. The hair cells of one duct will depolarize, causing an increased discharge of the vestibular afferent fibers. The opposite will occur in the other horizontal canal. The mismatch in input from the left

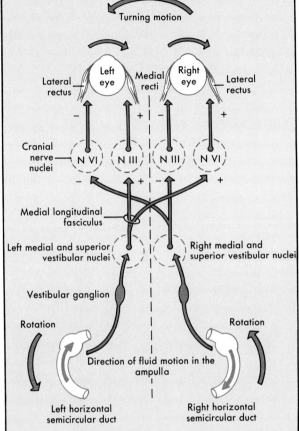

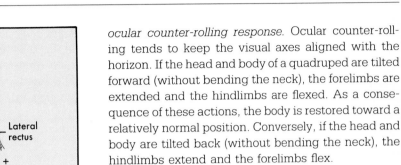

FIGURE 8-19 Neural circuit for the vestibuloocular reflex, an acceleratory reflex. The horizontal semicircular ducts, brainstem pathways, and eyes are viewed from above. Rotation of the head to the left is indicated by the arrow at the top and those next to the drawings of the semicircular ducts. Fluid movement within the ducts in indicated by the arrows in the opposite direction. The neural activity produced by the vestibular system causes the eyes to move conjugately to the right. As the eyes reach the limit of their movement, they are quickly returned to the left.

and right canals to the brainstem will result in reflex discharges that tend to counteract the positional changes resulting from head rotation.

Several reflexes can be elicited by linear accelerations of the head and activation of the otolith organs. If an animal is dropped, the stimulation of the utricles leads to extension of the forelimbs, the *vestibular placing reaction*. The response prepares the animal for landing. If the head is tilted, the otolith organs cause the eyes to rotate in the opposite direction, the *ocular counter-rolling response*. Ocular counter-rolling tends to keep the visual axes aligned with the horizon. If the head and body of a quadruped are tilted forward (without bending the neck), the forelimbs are extended and the hindlimbs are flexed. As a consequence of these actions, the body is restored toward a relatively normal position. Conversely, if the head and body are tilted back (without bending the neck), the hindlimbs extend and the forelimbs flex.

The *tonic neck reflexes* are another class of positional reflexes. The neck muscles contain the largest concentration of muscle spindles of any muscles of the body; these are presumably responsible for the tonic neck reflexes. In the absence of vestibular reflexes (head position normal), if the neck of a quadruped is extended relative to the body, the forelimbs extend and the hindlimbs flex. The converse happens if the neck is flexed relative to the body. These changes are opposite to those expected from the vestibular reflexes. If the neck is turned to the left, the extensor muscles in the limbs on the left will contract more, and the flexor muscles in the limbs on the right will relax.

The *righting reflexes* tend to restore the position of the head and body in space to normal. The vestibular apparatus and the neck stretch receptors are involved in the righting reflexes, as are mechanoreceptors in the body wall.

Locomotion

As mentioned earlier, the pattern generator for locomotion is contained within the neural circuitry of the spinal cord. Actually, separate pattern generators exist for each limb. The activity of these is coupled so that the movements of the limbs during locomotion are coordinated.

The pattern generators for locomotion and for other rhythmic activities (e.g., respiration) are regarded as *biologic oscillators*. Many biologic oscillators operate on the basis of reciprocal inhibition of circuits, called *half-centers*, that control antagonistic muscles. Excitation of an extensor muscle by one half-center is accompanied by reciprocal inhibition of the half-center for the antagonistic flexor muscle. When the excitation of the extensor muscle decreases, there is less inhibition of the antagonist, which can then be activated by the second half-center. This results in reciprocal inhibition of the first half-center. The details of this mechanism and the factors causing switching between the two half-centers varies with the particular oscillator being considered.

The locomotor pattern generator is normally activated by commands that descend from the brainstem. Neurons in a circumscribed region, called the *midbrain locomotor center* (Figure 8-20), play a crucial role in the initiation of locomotion. The midbrain locomotor center activates neurons located in the pontomedullary reticular formation and involved in transmitting the descending commands. The locomotor pattern generator converts tonic activity in the descending pathways into rhythmic discharges of motoneurons to the muscles involved in locomotor activity.

The midbrain locomotor center can be brought into action by voluntary commands that originate in the motor regions of the cerebral cortex. It can also be engaged by afferent signals. The activity of the locomotor pattern generator in the spinal cord is also influenced by afferent signals. These signals modify the ongoing motor program so that motor performance is altered in accord with environmental demands.

Control of Eye Position

Movements of the eyes are generally conjugate (in the same direction), although sometimes they are convergent or divergent when they are targeted on nearby objects (as in reading) or on distant objects.

A rapid conjugate movement of the eyes is called a *saccade*. Usually a saccade causes a visual target to

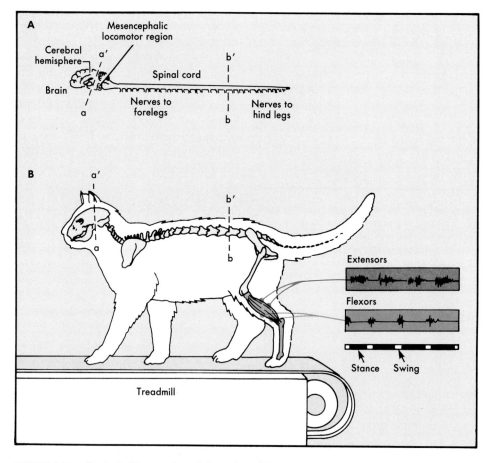

FIGURE 8-20   Control of locomotion. **A,** Location of the mesencephalic locomotor center, which can trigger locomotion even in the decerebrate state (transection at a'-a). **B,** Locomotion can also be elicited by afferent input produced by a treadmill after the spinal cord is transected at b'-b. Electromyographic recordings from extensor and flexor muscles of the hindlimb show bursts of activity during the stance and swing stages of the step cycle.

be imaged on the fovea. However, saccades can be made in the dark.

Once the eyes have located a visual target, fixation is maintained by smooth pursuit movements. Actually, during fixation the eyes drift somewhat and are returned to the target by microsaccades. Without these small movements, the retina would adapt and lose vision of the target. Smooth pursuit movements do not take place in the dark, since they require a visual target.

The *vestibuloocular reflex* is a response to angular acceleration of the head (see Figure 8-19). The eyes move in the direction opposite to the head movement and are usually exactly compensatory so that the visual target is kept on the fovea. It is possible to change the gain of the vestibuloocular reflex by suitable training.

Misalignment of the two visual axes can cause double vision, or *diplopia*. Such misalignment, or *strabismus,* can be caused by weakness of the muscles of one eye, such that its visual axis will differ from that of the other eye. Over time the misaligned eye may lose visual acuity, a condition called *amblyopia*.

Several central nervous system structures influ-

ence eye position. These structures include the gaze centers, the vestibular system, the superior colliculus, and the frontal and occipital lobes of the cerebrum. Horizontal eye movements are organized by the *pontine horizontal gaze center* located near the abducens nucleus. There is also a midbrain vertical gaze center in the pretectum.

Neurons with different response properties have been found in the horizontal gaze center (Figure 8-21). Burst cells discharge rapidly just before saccades, which burst cells are thought to initiate. Tonic cells discharge during slow pursuit movements and fixation. Burst-tonic cells show a burst discharge during saccades and tonic activity during fixation. Pause cells stop firing during saccades and seem to inhibit burst cells. Saccades occur when pause cells stop firing, resulting in a release of activity in burst cells and a discharge of eye muscle motoneurons. Feedback when the eye is on target inhibits the burst cells and reactivates the pause cells.

Only the vestibular nuclei and the gaze centers send direct projections to the motor nuclei supplying the eye muscles. The superior and medial vestibular nuclei project axons rostrally in the medial longitudi-

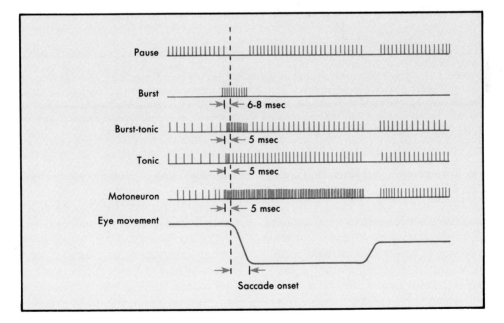

FIGURE 8-21   Types of neurons found in the pontine horizontal gaze center. Pause cells are thought normally to inhibit burst cells. A saccade begins with cessation of activity in pause cells and an explosion of activity in burst cells. Slightly later, burst-tonic and tonic cells discharge. Motoneurons to the muscles involved in the saccade are excited in the burst-tonic fashion, which causes the eye muscle to contract quickly and then to maintain its contraction. This results in the saccadic eye movement.

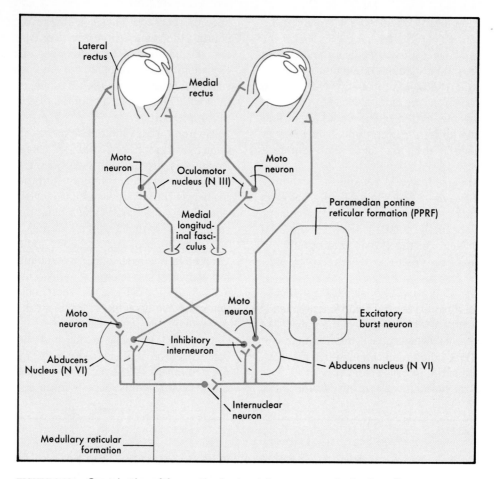

FIGURE 8-22   Organization of the pontine horizontal gaze center. Activation of burst neurons in the right gaze center results in direct excitation of abducens motoneurons on the right and oculomotor motoneurons to the medial rectus muscle on the left through an interneuronal path by way of the medial longitudinal fasciculus. Concurrently, the contralateral saccadic mechanism is inhibited by way of the reticular formation.

nal fasciculus to the abducens, trochlear, and oculomotor nuclei (see Figure 8-19). The circuitry is reciprocally organized, so that a signal causing an eye movement will excite a set of agonistic motoneurons and inhibit the antagonistic ones. Similarly, neurons of the pontine horizontal gaze center send (1) projections directly to abducens motoneurons on the same side to excite them, causing abduction of the ipsilateral eye, and (2) ascending projections through the medial longitudinal fasciculi to excite motoneurons of the contralateral medial rectus muscle (Figure 8-22). Appropriate connections are made to inhibit the motoneurons of the antagonistic muscles.

Neurons located in the deep layers of the superior colliculus and activated by visual, auditory, and somatosensory stimuli project to the horizontal gaze center. They produce saccadic eye movements that are part of an orientation response to a novel or threatening stimulus.

The frontal eye fields in the premotor region of the frontal lobe trigger voluntary saccadic eye movements by way of a projection to the contralateral pontine horizontal gaze center. The occipital eye fields are involved in smooth pursuit movements, optokinetic nystagmus, and visual fixation. Adjustments for near vision include vergence, pupillary constriction, and rounding of the lens. They are also organized in this region. The occipital eye fields are connected with the superior colliculus and the pretectal region and influence the vertical and horizontal gaze centers.

CORTICAL CONTROL OF VOLUNTARY MOVEMENT

The corticospinal and corticobulbar tracts are the most important pathways used in the initiation of voluntary movements. The lateral corticospinal tract and the comparable part of the corticobulbar tract control the fine movements produced by muscles of the contralateral distal extremities, face, and tongue. The ventral corticospinal tract, part of the corticobulbar tract, as well as more indirect pathways such as the corticorubrospinal and corticoreticulospinal tracts, provide for postural support of voluntary movements.

Corticospinal and corticobulbar neurons do not operate in isolation. Their discharges represent decisions based on inputs from many sources. The motor cortex receives projections from the ventral lateral nucleus of the thalamus, the postcentral gyrus, the posterior parietal cortex, the supplementary motor cortex, and the premotor cortex. The ventral lateral thalamic nucleus is part of the circuitry by which the cerebellum and basal ganglia regulate movements (see later discussion). The postcentral gyrus processes and then transmits somatosensory information to the motor cortex. This provides feedback about movements and about contacts between the skin and objects being explored. The posterior parietal cortex, supplementary motor cortex, and premotor cortex help program movements.

Motor Programs

Voluntary movements require contractions and relaxations in the proper sequence, not only of the muscles directly involved in the movements, but also of the appropriate postural muscles. Therefore a mechanism is needed for programming these complex events. The cortical areas thought to be responsible for cortical motor programs include the posterior parietal lobe, the supplementary motor cortex, and

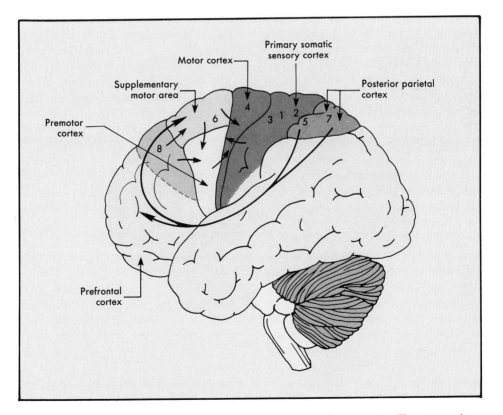

FIGURE 8-23   Cortical regions involved in the programming of movements. The arrows show some of the interconnections of these regions. The numbers refer to Brodmann's areas.

the premotor cortex (Figure 8-23).

The posterior parietal lobe (Brodmann's area 5 and 7) receives somatosensory information from the postcentral gyrus and visual information from the occipital cortex. The posterior parietal cortex connects with the supplementary motor cortex and the premotor cortex (Figure 8-23) and is important for processing sensory information that leads to goal-directed movements. A lesion of the posterior parietal cortex will cause deficits in visually guided movements. Humans may develop a *neglect syndrome* (especially if the lesion is in the nondominant hemisphere for speech), in which a patient will be unable to recognize objects placed in the contralateral hand and unable to draw three-dimensional objects accurately. In the posterior parietal cortex of monkeys, neurons have been recorded that discharge only when the animal reaches for something in the immediate surroundings (arm projection neurons) or when the hand is used for exploration (hand manipulation neurons). Other neurons discharge only when the eyes are moved to a target and the animal simultaneously reaches for the object (hand-eye coordination neurons).

Electrical stimulation of the supplementary motor cortex produces complex, often bilateral, movements. Recordings from the supplementary cortex show that neurons in this region discharge in association with complex activity, such as movements of either arm. No relationship exists between the discharges and the details of the movement, suggesting that the activity of the supplementary motor cortex is only indirectly involved in producing a particular movement. Lesions of the supplementary motor cortex cause deficits in orientation during movements and impairment of bilateral coordination.

The premotor cortex receives input from the part of the ventral lateral nucleus of the thalamus controlled by the cerebellum. It also receives input from the posterior parietal lobe and the supplementary motor cortex. The premotor cortex projects to the motor cortex and also to the spinal cord and brainstem. It is thought to be especially involved in the control of axial and proximal muscles, providing the appropriate postural background for movements of the distal muscles. A lesion of the premotor cortex in humans or monkeys releases the grasp response, in which touching the palm or extending the fingers elicits a grasping movement of the hand.

Motor Cortex

The motor cortex is recognizable in the microscope by the presence of the giant pyramidal Betz cells. However, many more projections of this area arise from small and medium-sized pyramidal cells than from the Betz cells. The corticospinal and corticobulbar tracts originate from pyramidal cells in layer 5 of the motor cortex, as well as from other cortical regions, including the premotor and supplementary motor cortex and the postcentral gyrus.

The somatotopic organization of the motor cortex has already been described (see Figure 8-17).

The motor cortex controls both distal and proximal muscles. However, the corticospinal and corticobulbar projections of the lateral system are especially important for the activation of distal muscles of the contralateral upper and lower extremities and of the contralateral lower face and tongue. A lesion that interrupts the corticospinal and corticobulbar tracts eliminates movements of distal muscles, but other pathways can still be used to activate proximal and axial muscles.

The lateral corticospinal tract makes monosynaptic excitatory connections with α-motoneurons, especially those that are located in the dorsolateral part of the ventral horn and that innervate distal muscles. The same pathway also excites γ-motoneurons, probably through an interneuronal pathway. Thus when the lateral corticospinal tract commands a voluntary movement, it co-activates α- and γ-motoneurons. In addition, interneurons in various reflex pathways are activated, so that the corticospinal tract influences spinal cord activity by regulating reflex transmission.

Recordings from neurons in the motor cortex during learned movements (Figure 8-24) reveal that some pyramidal tract neurons discharge just before a particular phase of the movement (e.g., flexion or extension of a joint). The activity encodes the force exerted by the muscles involved in the movement rather than encoding the position of the joint. Some neurons code for the rate at which force is developed, whereas others code for the steady-state force.

Neurons in the motor cortex receive input from somatosensory receptors in the skin, muscles, or joints. The receptive field of a given cortical neuron is related to the muscles activated from the same area of cortex. The somatosensory information presumably reaches the motor cortex by way of the somatosenso-

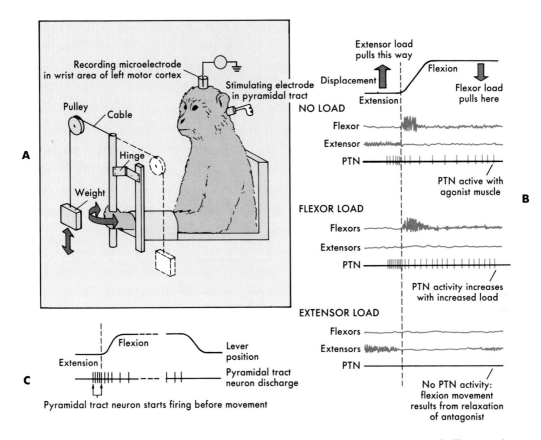

FIGURE 8-24   Recordings from corticospinal tract neurons during movements. **A,** The experimental arrangement. **B,** The recordings of joint position, electromyographic activity in flexor and extensor muscles, and the activity of a corticospinal neuron. *PTN,* Pyramidal tract neuron. The PTN starts to discharge before the movement (**B** and **C**). When a load is placed on the flexors, the discharge of the PTN cell increases, whereas the activity stops when a load is placed on the extensors. Since the movements are the same, the activity of the cell codes for force rather than for position.

ry cortex, although a more direct pathway through the thalamus is also possible. The interaction between sensory feedback and motor cortical output is important for exploratory behavior. The *tactile placing reaction* is mediated by the motor cortex. When the dorsal surface of the paw of an animal held off the ground makes contact with a surface, the limb is flexed and then extended to establish support for the body.

Interruption of the Pyramidal Tract

When the corticospinal and corticobulbar tracts are completely interrupted, the distal muscles of the contralateral upper and lower extremities and mus-

cles of the contralateral lower face and tongue are paralyzed *(hemiplegia).* Unless the lesion is restricted to these tracts, the deficit is a spastic paralysis. Spasticity usually accompanies hemiplegia produced by a lesion of the internal capsule or at other levels of nervous system, since the corticoreticulospinal pathway is interrupted along with the pyramidal tract. Spasticity is also present in spinal cord injuries when transection at an upper cervical level causes paralysis of all four extremities *(quadriplegia)* and transection below the cervical enlargement causes paralysis of both lower extremities *(paraplegia).*

Spastic paralysis is associated with an increase in muscle tone and increased phasic stretch reflexes. The latter may lead to clonus, such as in the ankle, in response to a brisk passive movement. Interruption of the corticospinal tract at any level causes an important release response, Babinski's sign (see Figure 8-18).

CEREBELLAR REGULATION OF POSTURE AND MOVEMENT

The cerebellum assists in the performance of coordinated movements by receiving sensory information about the status of movements and then adjusting the activity of the various descending motor pathways to optimize performance. These functions improve with practice, and thus the cerebellum is involved in the learning of motor skills. Destruction of the cerebellum produces no sensory deficits; therefore it has no essential role in sensation.

Organization of the Cerebellum

Afferent fibers from other parts of the central nervous system approach the cerebellar cortex through the cerebellar white matter. Two types of afferent fibers are found: mossy fibers and climbing fibers (Figure 8-25). *Mossy fibers* originate from a variety of sources, but all *climbing fibers* are derived from the contralateral inferior olivary nucleus. In the cerebellar cortex the mossy fibers synapse in the granular layer on the dendrites of granule cells. There is considerable divergence, since a given mossy fiber branches repeatedly and synapses on many different granule cells. An individual climbing fiber synapses at many sites on the soma and dendritic tree of one or a few Purkinje cells. Thus the climbing fiber pathways show little divergence.

The granule cell axons form bundles of parallel fibers that synapse on the dendrites of Purkinje cells and of several classes of interneurons: Golgi cells, basket cells, and stellate cells. *Granule cells* are the only excitatory interneurons in the cerebellar cortex. The mossy fiber–granule cell pathway and the climbing fiber pathway are able to excite Purkinje cells, and so these may be regarded as the excitatory circuits of the cerebellar cortex. Mossy fiber–granule cell excitation typically elicits single action potentials in a Purkinje cell (simple spike response), whereas a climbing fiber evokes a high-frequency burst of action potentials in a Purkinje cell (complex spike).

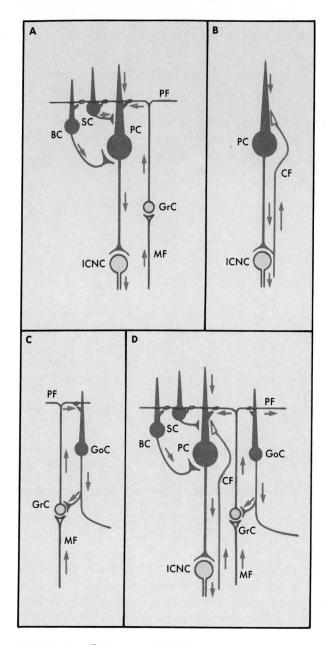

FIGURE 8-25   Excitatory and inhibitory circuits in the cerebellar cortex. Excitatory neurons are shown in outline; inhibitory neurons are dark. **A,** Connections made by mossy fibers through granule cells to Purkinje cells, stellate cells, and basket cells. Purkinje cells inhibit neurons of the deep cerebellar nuclei. **B,** The climbing fiber input to a Purkinje cell. **C,** The excitation of a Golgi cell by the mossy fiber through the granule cell path, with inhibition of granule cells by the Golgi cell. **D,** Combination of these circuits. *PC*, Purkinje cell; *SC*, stellate cell; *BC*, basket cell; *PF*, parallel fiber; *GrC*, granule cell; *MF*, mossy fiber; *ICNC*, deep cerebellar nuclear cell; *CF*, climbing fiber; *GoC*, Golgi cell.

The Golgi cells, basket cells, and stellate cells are all inhibitory interneurons of the cerebellar cortex. The Golgi cells inhibit granule cells; the basket cells inhibit Purkinje cell somata; the stellate cells inhibit Purkinje cell dendrites. All these inhibitory interneurons are activated by the mossy fiber–granule cell pathway.

Most of the discharges that can be recorded from Purkinje cells are simple spikes. The firing rate of Purkinje cells is modulated in response to sensory input or in relation to movements. Complex spikes occur at a low rate, approximately 1 Hz; they are too infrequent to play a role in the regulation of ongoing movements. Evidence suggests, however, that complex spikes affect the responsiveness of Purkinje cells to mossy fiber inputs. Therefore the effects of climbing fibers on Purkinje cell responsiveness may be important in motor learning.

A surprising observation is that although Purkinje cells are the only output neurons of the cerebellar cortex, their synaptic actions are inhibitory. Most Purkinje cells project to the deep cerebellar nuclei, but some synapse in the lateral vestibular nucleus. The cells of the deep cerebellar nuclei are tonically active, receiving excitatory input from collaterals of the mossy fibers and climbing fibers. Thus the inhibitory action of Purkinje cells modulates the discharges of the neurons of the deep cerebellar nuclei.

Functional Systems of the Cerebellum

The cerebellum can be considered on phylogenetic and functional grounds to be composed of three major components: the archicerebellum, the paleocerebellum, and the neocerebellum.

The *archicerebellum* is the earliest part of the cerebellum to evolve and is related in function primarily to the vestibular system. The archicerebellum is thus often referred to as the *vestibulocerebellum*. It corresponds in the human to the flocculonodular lobe and parts of the vermis in addition to the nodule. The archicerebellum helps control axial muscles and thus balance and also coordinates head and eye movements. A lesion of the archicerebellum can result in a "drunken" stagger called an *ataxic gait* and also in nystagmus.

The *paleocerebellum* receives somatotopically organized information from the spinal cord, and so it is often called the *spinocerebellum*. The paleocerebellum regulates both movement and muscle tone. Lesions of the paleocerebellum produce deficits in coordination similar to those seen after damage to the neocerebellum.

The *neocerebellum* is the dominant component of the human cerebellum. It occupies the hemispheres of the cerebellum. The input is from wide areas of the cerebral cortex, and so this region is sometimes called the *cerebrocerebellum*. The neocerebellum modulates the output of the motor cortex. Since one side of the neocerebellum controls activity in the contralateral cortex, and since the contralateral cortex influences movements of the ipsilateral limbs, the neocerebellum regulates motor activity of the same side of the body.

The neocerebellum probably interacts with neurons of the premotor cortex in programming movements. Lesions of the neocerebellum affect chiefly the distal limbs. The deficits include delayed initiation of movements, ataxia of the limbs (incoordination), and reduced muscle tone. The limb ataxia results in *asynergy* (lack of synergy in movements), *dysmetria* (inaccurate movements), *intention tremor* (oscillations at the end of a movement), and *dysdiadochokinesia* (irregular performance of pronation and supination movements of the forearm). Reduced muscle tone leads to pendular phasic stretch reflexes in the lower extremity. Bilateral lesions of the neocerebellum may result in *dysarthria* (slow, slurred speech; synonymous with scanning speech).

REGULATION OF POSTURE AND MOVEMENT BY THE BASAL GANGLIA

As with the neocerebellum, the basal ganglia help regulate the activity of the motor cortex. Unlike the archi- and paleocerebellum, the basal ganglia exert only a minor influence on descending motor pathways other than the corticospinal and corticobulbar tracts. Judging from the effects of lesions, the role of the basal ganglia is often opposed to that of the neocerebellum.

Organization of the Basal Ganglia

The basal ganglia are the deep nuclei of the telencephalon. They include the caudate nucleus and putamen (neostriatum) and the globus pallidus (pale-ostriatum). The caudate nucleus and putamen are often collectively called the *striatum* because of the "striations" formed by fibers of the anterior limb of the internal capsule that pass between these nuclei in the human. The role of the basal ganglia in motor control has been inferred more on the basis of the effects of disorders of the basal ganglia than from experimental evidence.

Disturbances Caused by Basal Ganglia Diseases

Different basal ganglia diseases can produce various motor disturbances. These can be categorized as disorders of movement and disorders of posture. Movement disturbances include abnormal movements, such as *tremor* (rhythmic, "pilling-rolling" oscillations at rest), *chorea* (rapid flicking movements), *athetosis* (slow, writhing movements of limbs), *ballism* (violent, flailing movements), and *dystonia* (slow, twisting movements of the torso), and movements that are delayed in initiation and slow to reach completion, or *bradykinesia*. The disorder of posture produced in basal ganglion disease is *rigidity*. The rigidity of basal ganglion disease may be of the cogwheel type; as a joint is moved, resistance occurs throughout the range of the movement, although there may be repeated alterations in the amount of resistance. These alterations produce a ratchetlike effect as the joint is passively moved. Alternately, a leadpipe rigidity may occur, in which resistance is constantly present through the range of motion of the joint. The rigidity of basal ganglion disease should be distinguished from "decerebrate rigidity," which is more similar to spasticity.

Parkinson's disease is caused by a lesion of the substantia nigra and is characterized by tremor, rigidity, and bradykinesia. The loss of dopaminergic projections to the striatum is thought to be crucial. Hemiparkinsonism results when one substantia nigra is affected; the manifestations are contralateral, since these are caused by inappropriate regulation of the corticospinal tract.

Destruction of part of the subthalamic nucleus on one side results in a hemiballism, characterized by ballistic movements on the contralateral side.

Huntington's chorea is a genetic disorder in which there is loss of striatopallidal and striatonigral neurons containing γ-aminobutyric acid; cholinergic striatal interneurons also are lost. Loss of inhibitory input to the globus pallidus is thought to underlie the choreiform movements characteristic of the disease. The cerebral cortex also degenerates, leading to severe mental deterioration.

In cerebral palsy athetosis often occurs because of damage to the striatum and globus pallidus.

BIBLIOGRAPHY
Journal Articles

Alexander GE et al: Parallel organization of functionally segregated circuits linking basal ganglia and cortex, Annu Rev Neurosci 9:357, 1986.

Burke RE: Motor unit properties and selective involvement in movement, Exerc Sport Sci Rev 3:31, 1975.

Cullheim S and Kellerth JO: Combined light and electron microscopic tracing of neurons, including axons and synaptic terminals, after intracellular injection of horseradish peroxidase, Neurosci Lett 2:307, 1976.

Ito M: Cerebellar control of the vestibulo-ocular reflex around the flocculus hypothesis, Annu Rev Neurosci 5:275, 1982.

Lisberger SG et al: Visual motion processing and sensorimotor integration for smooth pursuit eye movements, Annu Rev Neurosci 10:97, 1987.

Penney JB and Young AB: Speculations on the functional anatomy of basal ganglia disorders, Annu Rev Neurosci 6:73, 1983.

Shik ML and Orlovsky GN: Neurophysiology of locomotor automatism, Physiol Rev 56:465, 1976.

Sparks DL: Translation of sensory signals into commands for control of saccadic eye movements: role of primate superior colliculus, Physiol Rev 66:118, 1986.

Wilson VJ and Peterson BW: Peripheral and central substrates of vestibulospinal reflexes, Physiol Rev 58:80, 1978.

Wise SP: Premotor cortex: past, present, and preparatory, Annu Rev Neurosci 8:1, 1985.

Wurtz RH and Albano JE: Visual-motor function of the primate superior colliculus, Annu Rev Neurosci 3:189, 1980.

Books and Monographs

Brooks VB: The neural basis of motor control, New York, 1986, Oxford University Press.

Burke RE: Motor units: anatomy, physiology, and functional organization. In Handbook of physiology, section 1: The nervous system, vol II: Motor control, part 1, Bethesda, Md, 1981, American Physiological Society.

Eccles JC et al: The cerebellum as a neuronal machine, New York, 1967, Springer Publishing Co.

Kandel ER and Schwartz JH: Principles of neural science, ed 2, New York, 1985, Elsevier Science Publishing Co, Inc.

Matthews PBC: Mammalian muscle receptors and their central actions, Baltimore, 1971, Williams & Wilkins.

Phillips CG and Porter R: Corticospinal neurons: their role in movement, New York, 1977, Academic Press.

Roberts TDM: Neurophysiology of postural mechanisms, ed 2, London, 1978, Butterworth Publishers.

Willis WD and Grossman RG: Medical neurobiology, St Louis, ed 3, 1981, The CV Mosby Co.

Wilson VJ and Jones GM: Mammalian vestibular physiology, New York, 1979, Plenum Press.

Autonomic Nervous System and Its Control

The autonomic nervous system is a motor system concerned with the regulation of smooth muscle, cardiac muscle, and glands. The autonomic nervous system is not directly accessible to voluntary control. Instead, it operates in an "automatic" fashion on the basis of autonomic reflexes and central control. A major function of the autonomic nervous system is homeostasis, which is the maintenance of the internal environment in an optimal state. For instance, the autonomic nervous system, in cooperation with the somatic motor system, helps keep the body temperature of homeothermic animals relatively constant. Another important role of the autonomic nervous system is to make the appropriate adjustments in smooth muscle tone, cardiac muscle activity, and glandular secretion for different behaviors. For example, the autonomic activity that can be observed during digestion is very different from that during a sprint.

The autonomic nervous system proper consists of the sympathetic nervous system, the parasympathetic nervous system, and the enteric nervous system. The final common pathway in the sympathetic and parasympathetic nervous systems consists of a sequence of two types of motoneurons: **preganglionic neurons,** whose cell bodies are located in the spinal cord or brainstem, and **postganglionic neurons,** whose cell bodies are in peripheral autonomic ganglia. Accompanying the autonomic motor axons in visceral nerves are visceral afferent fibers, which enter the central nervous system in dorsal roots or in cranial nerves. Visceral afferent fibers form the afferent link in many autonomic reflexes. Somatic afferent fibers can also trigger somatovisceral reflexes that employ an autonomic output. The enteric nervous

system is a peripheral reflex network located in the wall of the gastrointestinal tract. Afferent neurons, postganglionic motoneurons, and interneurons are present in the enteric ganglia of the Meissner's submucosal and Auerbach's myenteric plexuses (see Chapter 29).

The peripheral component of the autonomic nervous system is under the control of several central nervous system structures; important ones include the brainstem reticular formation, the hypothalamus, and the limbic system. However, other motor system structures of the brain, including the cerebellum and the basal ganglia, also strongly influence autonomic activity.

ORGANIZATION OF THE AUTONOMIC NERVOUS SYSTEM

Sympathetic System

The sympathetic nervous system is a widely distributed motor system. It reaches not only the viscera contained in the body cavities, but also the skin and muscles of the body wall.

The cell bodies of the preganglionic neurons are located in the thoracic and upper lumbar spinal cord (T1 to about L2) in the intermediolateral and intermediomedial cell columns. The motor axons of the sympathetic preganglionic neurons leave the spinal cord in the T1 to L2 ventral roots. The motor axons are small myelinated B fibers or in some cases unmyelinated C fibers. They pass from the spinal nerves into the white communicating rami.

When the sympathetic preganglionic axons in a given white ramus reach the sympathetic paravertebral ganglion of the same segment, they may (1) synapse in that ganglion, (2) turn rostrally or caudally in the sympathetic chain and synapse in a paravertebral ganglion at another segmental level, or (3) continue through a splanchnic nerve to synapse in a prevertebral ganglion (Figure 9-1). In this way preganglionic axons that originate from motoneurons limited to spinal cord segments T1 to L2 are able to synapse on postganglionic neurons located in the entire chain of paravertebral sympathetic ganglia (including the superior, middle, and inferior cervical sympathetic ganglia and the ganglia below L2 that do not receive white communicating rami), as well as in the prevertebral ganglia of the abdominal cavity (Figure 9-2). Preganglionic axons also directly innervate the chromaffin cells of the adrenal medulla, which are developmentally comparable to sympathetic ganglion cells.

Sympathetic postganglionic axons are unmyelinated fibers that originate from ganglion cells of the sympathetic ganglia. Axons from ganglion cells in the paravertebral ganglia may distribute either to the body wall or to the viscera in the body cavities (Figure

9-2). If they are destined for the body wall, they pass from a ganglion into a spinal nerve by way of a gray communicating ramus. Gray rami are found on all ganglia of the sympathetic chain and connect with the appropriate spinal nerves. Sympathetic postganglionic axons destined for viscera in the body cavities enter splanchnic nerves and distribute to their targets. These axons may traverse prevertebral ganglia without synapsing en route to their targets. Postganglionic axons originating in prevertebral ganglia distribute to their targets through the sympathetic plexuses near the target organs.

Parasympathetic System

The parasympathetic nervous system is less widely distributed than the sympathetic nervous system. A parasympathetic supply exists for various structures in the head and neck, but much of the distribution is to the viscera contained in the body cavities. No parasympathetic outflow reaches the skin or muscles of the body wall or extremities.

As with the sympathetic nervous system, the parasympathetic outflow involves a sequence of pre- and postganglionic neurons (Figure 9-3). The cell bodies of the parasympathetic preganglionic neurons are located either in the brainstem or in the sacral spinal cord (S2 to S4). Cranial nerve nuclei that contain preganglionic parasympathetic neurons include the Edinger-Westphal nucleus (cranial nerve III), superior salivatory nucleus (cranial nerve VII), inferior salivatory nucleus (cranial nerve IX), and nucleus ambiguus, and dorsal motor nucleus of the vagus (cranial nerve X). Sacral parasympathetic preganglionic neurons are located in the sacral preganglionic nucleus. No lateral horn exists in the sacral spinal cord.

The cranial parasympathetic preganglionic axons leave the brainstem in the appropriate cranial nerves and synapse on ganglion cells in the cranial parasympathetic ganglia (Figure 9-3; III: ciliary ganglion; VII: sphenopalatine and submaxillary ganglia; IX: otic ganglion) or on ganglia in or near the walls of target viscera in the thoracic and abdominal cavities (X). For the gastrointestinal tract the vagal preganglionic axons synapse on neurons belonging to the enteric nervous system (see following discussion). The sacral parasympathetic preganglionic axons distribute to the abdominal cavity and pelvis and synapse on ganglion cells located in the walls of viscera in these

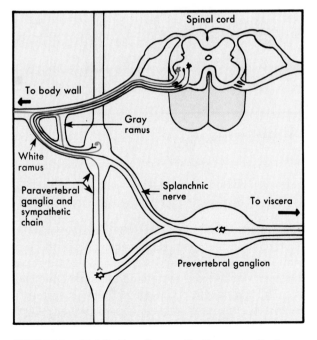

FIGURE 9-1   Distribution of sympathetic preganglionic projections to para- and prevertebral ganglia.

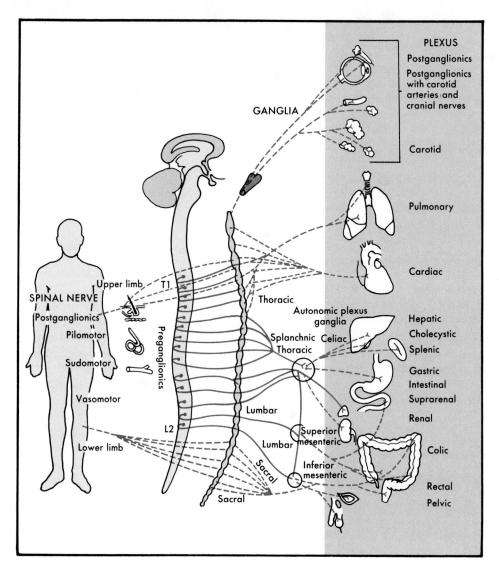

FIGURE 9-2   Sympathetic nervous system.

regions. The splenic flexure is the boundary between the gastrointestinal organs supplied by the vagus nerve and those by sacral parasympathetics. Parasympathetic postganglionic neurons directly innervate their nearby target organs.

The *enteric nervous system* is a miniature nervous system within the wall of the gastrointestinal tract (see Chapter 29). Reflex networks in this system organize gut movements that can occur even when the gut is removed from the body. Afferent neurons, interneurons, and motoneurons are included in the system. Parasympathetic and sympathetic connections to the enteric nervous system permit autonomic control. The component of the enteric nervous system in Auerbach's myenteric plexus controls the activity of the muscular layers, and that in Meissner's submucosal plexus controls the muscularis mucosae and intestinal glands.

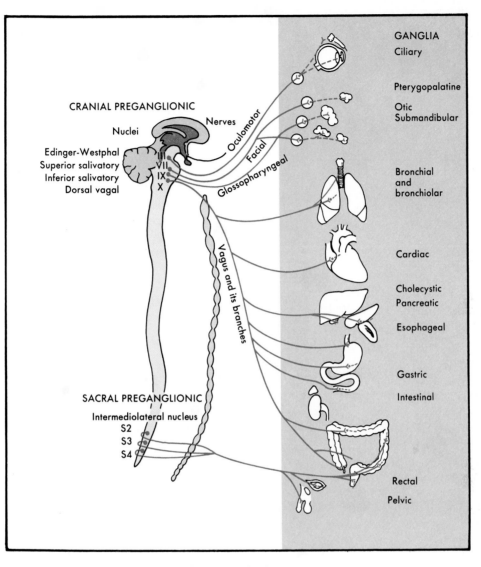

FIGURE 9-3   Parasympathetic nervous system.

AUTONOMIC FUNCTIONS

The sympathetic and parasympathetic nervous systems act continuously to adjust activity in smooth muscles, cardiac muscle, and glands, often by exerting a reciprocal control. The enteric nervous system regulates digestion. When an emergency situation leads to dramatic aggressive or defensive behavior (fight or flight), sympathetic action is prominent. For example, the pupils dilate, the skin blanches, piloerector muscles contract, the saliva becomes thick, the heart rate accelerates, the blood pressure is elevated, and blood flow is directed away from the abdominal viscera and to the skeletal musculature. The sympathetic nervous system also actively regulates visceral function under normal circumstances. The parasympathetic nervous system often acts contrary to the sympathetic nervous system when a giv-

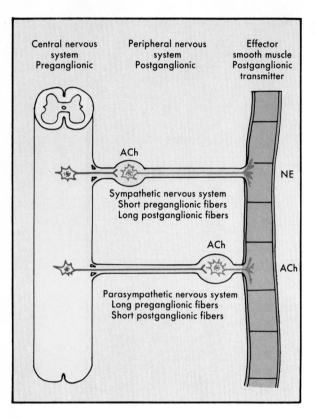

FIGURE 9-4 Transmitters of autonomic ganglia and postganglionic synapses. *Ach,* Acetylcholine.

en organ is innervated by both systems. However, it is more appropriate to view concurrent control of organs by activity in the sympathetic and parasympathetic nervous systems as a means of coordinating visceral activity.

The preganglionic neurons of the autonomic nervous system, as with α-motoneurons of the somatic motor system, use *acetylcholine* as their neurotransmitter (Figure 9-4). The acetylcholine receptors on postganglionic neurons, as with those of skeletal muscle fibers, are of the *nicotinic* type. Nicotinic receptors are activated by low doses of nicotine and blocked by curare. Parasympathetic and some sympathetic postganglionic neurons also use acetylcholine. However, the receptors on target organs are of the *muscarinic* type. Muscarinic receptors are activated by muscarine and blocked by atropine. The cholinergic sympathetic fibers include those to sweat glands and the vasodilator fibers in skin and skeletal muscle.

Sympathetic postganglionic neurons generally use

norepinephrine as their neurotransmitter (Figure 9-4). Receptors for norepinephrine include α- and β-adrenergic receptors. α-Receptors are more powerfully activated by norepinephrine than by isoproterenol; the converse is true of β-receptors. Agents such as phenoxybenzamine block α-receptors, whereas drugs such as propranolol block β-receptors. α-Receptors can be further divided into α_1 and α_2 subtypes, and β-receptors into β_1 and β_2.

The adrenal medulla is supplied by sympathetic preganglionic axons, which release acetylcholine as their neurotransmitter. The chromaffin cells of the adrenal medulla are developmentally similar to sympathetic postganglionic neurons, and they secrete epinephrine and norepinephrine into the circulation, where these agents act as hormones. In humans the ratio of epinephrine to norepinephrine is 4:1.

Neurons of the enteric nervous system release not only acetylcholine and norepinephrine, but also serotonin, adenosine triphosphate (ATP), and a variety of peptides as neurotransmitters and neuromodulators.

CONTROL OF AUTONOMIC FUNCTION

The operation of the autonomic nervous system is regulated hierarchically in much the same way as the somatic motor system. The most direct neural control of many organs is by means of autonomic reflexes. However, these are often regulated by descending pathways from the brainstem. In addition, autonomic function is controlled by higher autonomic centers, including the hypothalamus and other parts of the limbic system.

Autonomic reflexes are mediated by neural circuits in the spinal cord and brainstem. The afferent limbs of these reflex pathways include both visceral and somatic afferent fibers. The pathways involve interneurons that receive a convergent input from visceral and somatic sensory receptors. The efferent limbs are formed by sympathetic and parasympathetic pre- and postganglionic neurons. The actions of the two autonomic systems are generally reciprocal.

Brainstem pathways that regulate the activity of autonomic preganglionic neurons originate from several sites, including the reticular formation, raphe nuclei, and locus ceruleus complex. These brainstem structures receive information about the visceral activities they regulate by way of ascending tracts. Some autonomic functions depend strongly on these brainstem pathways. For example, micturition and defecation depend on the integrity of pathways that interconnect the sacral spinal cord and the pons. Transection of the spinal cord causes a transient loss of bladder and bowel function. Partial return of function occurs as spinal reflexes become reorganized.

Higher centers that regulate autonomic function include the hypothalamus and other components of the limbic system. Limbic structures are interconnected with nonlimbic parts of the nervous system, including the neocortex, cerebellum, and basal ganglia. The hypothalamus projects to the brainstem (e.g., to the reticular formation) and to the spinal cord. The limbic system controls motivation both through neural pathways and indirectly through the endocrine system.

FUNCTIONS OF THE HYPOTHALAMUS

The hypothalamus has several broadly defined functions, which include the regulation of homeostasis, motivation, and emotional behavior. The means whereby these functions are mediated is through hypothalamic control of autonomic and endocrine activity, as well as by interactions between the hypothalamus and other parts of the limbic system.

If the hypothalamus is stimulated electrically, particular regions are shown to be related to particular autonomic responses. For example, stimulation in the lateral and posterior hypothalamus produces responses mediated by the sympathetic nervous system. Stimulation in the anterior hypothalamus activates parasympathetic output. The responses include changes in heart rate and blood pressure. Similarly, lesions of the hypothalamus can disrupt particular functions. For instance, lesions in the preoptic region and anterior hypothalamus interfere with the ability of the body to lose heat by sweating and cutaneous vasodilation. On the other hand, lesions of the posterior hypothalamus result in hypothermia. Other global functions controlled by the hypothalamus include food and water intake, emotional behavior, and regulation of the immune system.

The neurons in some hypothalamic nuclei release peptides, either as hormones or as neuromodulator substances. Such neurons are classified as neuroendocrine cells. Neuroendocrine structures include the paraventricular and supraoptic nuclei, which give rise to the hypothalamohypophyseal tract from the hypothalamus to the posterior pituitary gland. This tract releases the peptide hormones oxytocin and vasopressin into the circulation (see Chapter 39). The paraventricular nucleus also sends peptide-containing axons to various sites within the central nervous system, including the solitary nucleus, dorsal motor nucleus of the vagus, and intermediolateral cell column of the spinal cord. Oxytocin and vasopressin apparently are used both as hormones and as neuromodulators of autonomic function.

Neuroendocrine cells in a number of hypothalamic nuclei secrete hormones into the portal system that supplies the anterior pituitary gland (see Chapter 39). These hormones cause the release or inhibit the release of pituitary hormones into the circulation, and they are very important in endocrine regulation. As in the case of oxytocin and vasopressin, the same hypothalamic hormones can be used as neuromodulatory substances at synaptic terminals within the central nervous system.

THE LIMBIC SYSTEM

The limbic system includes the limbic lobe of the telencephalon, as well as the hypothalamus and several midbrain nuclei. The limbic components of the telencephalon include the cingulate, parahippocampal, and subcallosal gyri, as well as the hippocampal formation (hippocampus, dentate gyrus, and subiculum).

Functions of the Limbic System

Bilateral removal of several temporal lobe structures, including the amygdaloid nuclei, results in a complex set of changes in behavior called the *Klüver-Bucy syndrome*. Animals previously wild become tame; they develop a pronounced tendency to put objects into their mouths; and they become sexually hyperactive. These changes result chiefly from damage to the amygdaloid nuclei.

Thus the functions of the limbic system include regulation of aggressive behavior and sexuality. More generally the limbic system appears to be concerned with motivational states, which in turn are vital for survival of both the individual and the species.

The hippocampus appears to be important for recent memory (see also Chapter 10). Memories are stored in the following sequence: short-term memory, recent memory, and long-term memory. Short-term memory is easily disrupted and is presumed to depend on ongoing neural events. Long-term memory seems to result from a permanent functional or structural change in the nervous system. Lesions of the hippocampus may not interfere with either short- or long-term memory but may prevent the process by which short-term memories are permanently stored. The process of recollection of memories may also be disrupted, resulting in amnesia.

BIBLIOGRAPHY
Journal Articles

Gershon MD: The enteric nervous system, Annu Rev Neurosci 4:227, 1981.

Hayward JN: Functional and morphological aspects of hypothalamic neurons, Physiol Rev 57:574, 1977.

Jaenig W and McLachlan E: Organization of lumbar spinal outflow to distal colon and pelvic organs, Physiol Rev 67:1332, 1987.

Simon E et al: Central and peripheral thermal control of effectors in homeothermic temperature regulation, Physiol Rev 66:235, 1986.

Smith OA and DeVito JL: Central neural integration for the control of autonomic responses associated with emotion, Annu Rev Neurosci 7:43, 1984.

Books and Monographs

Bannister R: Autonomic failure, Oxford, 1983, Oxford University Press.

Carpenter MB and Sutin J: Human neuroanatomy, ed 8, Baltimore, 1983, Williams & Wilkins.

Kandel ER and Schwartz JH: Principles of neural science, ed 2, New York, 1985, Elsevier Science Publishing Co, Inc.

Newman PP: Visceral afferent functions of the nervous system, Baltimore, 1974, Williams & Wilkins.

Willis WD and Grossman RG: Medical neurobiology, ed 3, St Louis, 1981, The CV Mosby Co.

CHAPTER 10

Higher Functions of the Nervous System

The central nervous system is responsible for the higher functions that characterize humans. These functions include consciousness, thought, perception, learning, memory, and language. This chapter first considers the electroencephalogram and evoked potentials. States of consciousness, including the sleep-wake cycle, are generally studied with the help of these neurophysiologic techniques.

Learning and memory depend on alterations in neural function and even in structure. The human brain is actually two brains. One hemisphere is dominant for certain functions, including handedness and speech, and the opposite hemisphere is dominant for others, such as spatial relations and music.

THE ELECTROENCEPHALOGRAM

The higher functions of the human brain are expressed on the background of continuous thalamocortical interactions. Neurons of all the nuclei of the dorsal thalamus project to the cerebral cortex, and the cerebral cortex projects back to the dorsal thalamus.

Recordings from the surface of the cerebral cortex (electrocorticogram) or from the scalp (electroencephalogram, or EEG) reveal the incessant oscillations of extracellular potentials caused by membrane potential oscillations in large numbers of cortical neurons in response to the rhythmic alterations of activity in thalamocortical circuits. On a single neuron level, activity corresponding to the EEG consists of alternating excitatory and inhibitory postsynaptic potentials. The

excitatory potentials often result in discharges of cortical neurons.

The normal EEG can be described in terms of its frequency composition. Several characteristic frequency ranges can be recognized (Figure 10-1). These are called *alpha waves* (8 to 13 Hz), *beta waves* (more than 13 Hz), *theta waves* (4 to 7 Hz), and *delta waves* (less than 4 Hz). Other transient waves are also seen. The dominant frequency depends on several factors, including age, state of consciousness, recording site, the action of drugs, and the presence of disease. During the early years of life the EEG is dominated by low frequencies. In mature individuals at rest with eyes closed the EEG recorded from the posterior region of the brain shows an alpha rhythm, whereas that recorded from the anterior part of the brain has a beta rhythm. If the individual is aroused, lower-voltage, higher-frequency beta rhythms take the place of the synchronized, lower-frequency alpha waves (Figure 10-1). Slower waves of theta and delta frequencies are associated with deeper levels of sleep. One definition of brain death is a persistent isoelectric EEG in the absence of depressant drugs.

EVOKED POTENTIALS

The EEG represents spontaneous activity that is not linked to a particular event. Similar activity can be evoked in response to a stimulus that activates a neural pathway to the thalamus or circuits within the cerebral cortex. Such stimulus-linked activity recorded from the cortex is called a cortical evoked poten-

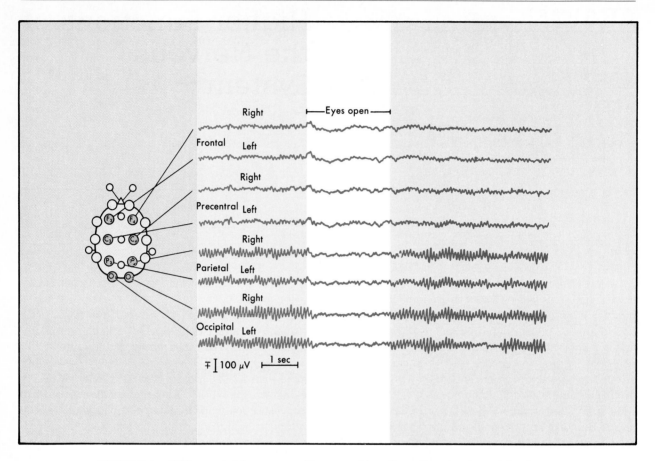

FIGURE 10-1   EEG recorded from a normal human subject. Recordings are from eight sites on the scalp (the indifferent electrodes on the earlobes are coupled). In the resting condition an alpha rhythm is prominent over the parietal and occipital lobes. When the eyes are opened, the alpha rhythm is blocked and replaced by a beta rhythm.

tial. Evoked potentials can easily be produced in human subjects by stimulating a peripheral nerve with electric shocks, the retina with flashes of light, or the ear with an acoustic stimulus such as a click. The waveform that is recorded is largest over the appropriate region of the brain.

STATES OF CONSCIOUSNESS

Mental processes occur in the brain of conscious subjects. Consciousness is not understood, but it is required for perception, thought, and the use of language. In disease, consciousness may be impaired to various degrees (lethargy, stupor), or even lost (coma). The level of consciousness can be tested by observing such behavioral events as speech, motion

of the limbs, and eye movements. In coma, vocalization is absent, the response to painful stimulation is minimal, and the eyes are closed and do not move spontaneously.

The conscious state appears to depend on an interaction between the brainstem reticular formation and thalamocortical circuits. When consciousness is depressed, the EEG becomes more synchronous and slowed in frequency. During behavioral arousal, as in response to a painful stimulus, the EEG changes to a low-voltage, high-frequency pattern (EEG arousal).

Sleep

Sleep is an alteration of, rather than a loss of, consciousness. This is shown by the ease with which sleep is interrupted by significant environmental events, such as a baby's cry. Sleep has a circadian rhythm, as well as a more rapid oscillation. The sleep rhythm, along with many other biologic rhythms, is normally entrained by the light-dark cycle. When an individual rapidly changes location to a different time zone, it takes days for the circadian rhythms to be reentrained. The sleep disturbance and other disorders that result are collectively known as "jet lag."

The various stages of sleep are characterized by different types of motor and autonomic, EEG, and psychologic activity. The major distinction is between rapid eye movement (REM) sleep and non-REM sleep. When an individual falls asleep, the initial stage is non-REM sleep. The EEG becomes more synchronized and slows (Figure 10-2). There are four levels of non-REM sleep. Over time the depth of non-REM sleep increases; that is, the EEG becomes progressively slower, and the person becomes more difficult to arouse. In addition, muscle tone and reflex activity are depressed, blood pressure drops, the heart rate slows, and the pupils constrict. The amount of time spent in the deepest stage of non-REM sleep decreases with age, and this stage may entirely disappear after age 60.

After about 90 minutes sleep lightens and changes to a period of REM sleep that lasts approximately 20 minutes. During REM sleep the EEG becomes desynchronized and has a low voltage; the pattern is similar to that seen during arousal (Figure 10-2). Tone in many muscles disappears, and reflexes are inhibited. Interrupting this tonic inhibition are phasic motor events, including rapid movements of the eyes and brief contractions of other muscles. *Transient waves* (pontine-geniculate-occipital waves) can be recorded from the brainstem and occipital cortex at the times of the REMs. Autonomic events include irregular respiration, reduced blood pressure interrupted by episodes of hypertension, and penile erection in males. Dreams tend to occur in REM sleep, although nightmares usually occur during deep non-REM sleep. It is difficult to awaken individuals from REM sleep, but spontaneous awakening often occurs. REM sleep recurs about six times during a night. The proportion of time spent in REM sleep is greatest in the fetus and newborn, but it declines sharply during early infancy and then further with aging.

Sleep appears to be triggered by an active mechanism that involves the reticular formation and monoamine neurons in the brainstem. Some of the neurotransmitters associated with sleep include serotonin, norepinephrine, and acetylcholine. Several sleep-inducing peptides have been discovered as well. Sleep disorders are common and can be difficult to

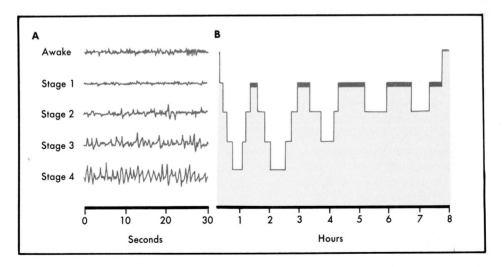

FIGURE 10-2   Stages of sleep and changes during the course of a night. **A,** EEG recordings during the waking state and during progressively deeper levels of non-REM sleep. The EEG in REM sleep would resemble that shown for the awake individual. **B,** Different sleep stages experienced during a typical night for a young adult. The black bars represent periods of REM sleep.

diagnose and manage. *Insomnia* is a chronic condition in which insufficient sleep occurs. This may occur because of a disruption in the circadian rhythms, as in jet lag, or simply because of the aging process. The most common cause is emotional upset. Drugs have profound effects on the sleep-wake cycle, and thus sleep disorders typically result from the misuse of drugs. *Enuresis* (bed-wetting) and *somnambulism* (sleepwalking) occur in some individuals during deep non-REM sleep; their incidence tends to decrease with age. *Narcolepsy* is an uncommon condition characterized by sleep attacks. In narcolepsy, sleep is initiated in the REM stage and is accompanied by loss of muscle tone *(cataplexy)*. Other symptoms may be sleep paralysis (motor inhibition) and hypnogogic hallucinations (similar to the dreams in REM sleep).

Attention

Attention is the process by which perception is directed at particular events. It involves orientation to stimuli that are potentially significant, such as novel stimuli or stimuli likely to lead to a reward or punishment. Attention to a stimulus is associated with a cortical evoked potential and EEG arousal. Repeated application of a stimulus may result in the loss of these changes *(habituation)*. Application of a threatening stimulus enhances attention; this process is termed *sensitization*.

Epilepsy

Epilepsy refers to disease states characterized by behavioral and EEG seizures. The seizures may be partial or generalized. In *partial seizures* only part of the brain shows abnormal activity, and consciousness is retained. In *generalized seizures* large regions of the brain are involved, and consciousness is lost.

Partial seizures may originate in a damaged area of the motor cortex. Such seizures are characterized by contractions of muscles in the somatotopically appropriate region on the contralateral side and a focal EEG *spike train* (an EEG spike is a synchronous wave resulting from simultaneous activity in many neurons). The seizures often spread to adjacent areas of cortex. This spread results in a "march" of convulsive activity to other contralateral parts of the body. For example, the seizure may start with contractions of the fingers, but the movements may then spread to the arm, shoulder, and face and lower extremity. Partial seizures may also originate from the somatosensory cortex and produce focal sensory experiences contralaterally. *Psychomotor seizures* are partial seizures that originate in the limbic lobe. These are characterized by semipurposeful movements, changes in consciousness, hallucinations, and illusions. A common hallucination is an unpleasant odor.

Generalized seizures include grand mal and petit mal seizures. *Grand mal* attacks may be preceded by an aura. Consciousness is soon lost, followed by tonic and then clonic movements bilaterally. *Petit mal* attacks are brief losses of consciousness, accompanied by a characteristic EEG pattern.

LEARNING AND MEMORY

Learning is a process by which behavior is modified on the basis of experience. Memory is the storage of information that has been learned. There are several stages of memory, including short-term memory, recent memory, and long-term memory. Short-term memory appears to depend on ongoing neural activity because it is easily disrupted (e.g., by head trauma). Recent memory refers to the process by which information in short-term memory is transformed into long-term memory. This process seems to depend on activity transmitted by the hippocampal formation, since damage to the hippocampus and related structures prevents the consolidation of short-term into long-term memory. Long-term memory apparently depends on permanent changes in widely distributed sets of neurons. The changes may include morphologic as well as functional changes. In addition to memory stores, there must be mechanisms for accessing these stores, retrieving the information, recalling it to consciousness, comparing it to other information, and using the information for decisions.

Little is known about the mechanisms of learning and memory. Experiments on the simple nervous system of invertebrates have shed some light on the neural basis of simple forms of learning. Habituation and sensitization are examples of *nonassociative learning* because these do not require learning an association between two events. In habituation a response to a particular stimulus diminishes with repetition of the stimulus. Habituation is thus the process of learning that a stimulus is unimportant. Conversely, sensitization is the process by which the subject learns that a stimulus is important. For example, with repetition of a painful stimulus, an individual quickly learns to respond.

In *associative learning* the relationship between two different stimuli is learned. In *classical conditioning* a conditioned stimulus is paired with an unconditioned stimulus. The latter initially produces an unconditioned response (e.g., food produces salivation in a hungry dog). After conditioning the conditioned stimulus may now produce the same response (e.g., ringing a bell at the time food is presented ultimately causes salivation even if the food is omitted). In *operant conditioning* reinforcement of a response changes the probability of the response. Operant behaviors are not reflexes but rather spontaneous actions. An example of operant conditioning would be an animal's avoidance of a wire grid that induces an electric shock when the animal happens to step on the grid. In this case the conditioning stimulus provides negative reinforcement.

Habituation, sensitization, and classical conditioning have all been demonstrated in invertebrate models, and the neural mechanisms that underlie both nonassociative and associative learning are being studied. A major theme of such work is that synaptic efficacy changes during these simple forms of learning. These changes depend on the activation of second messenger systems. Long-term changes are accompanied by structural as well as functional changes. A parallel experimental approach in mammals involves the enhancement of synaptic transmission for hours after activation of particular pathways in the hippocampus. The mechanisms that underlie this "long-term potentiation" are also under active study.

CEREBRAL DOMINANCE

The two halves of the human brain are not equivalent. In a real sense the human has two brains that communicate with each other by way of the cerebral commissures. The left hemisphere is *dominant* in most individuals with respect to control of the preferred hand (right in most people) and in language. However, the right hemisphere can be considered dominant for other functions (music, spatial relationships). A structural correlate of cerebral dominance is the greater size of the left than of the right planum

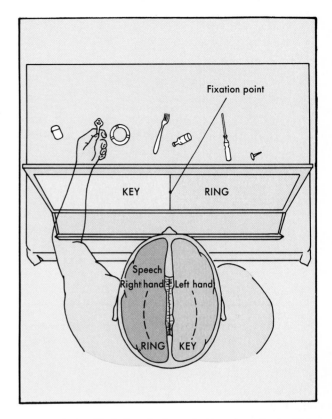

FIGURE 10-3   Technique for investigation of a patient with a disconnection syndrome caused by transection of the corpus callosum. The subject is asked to fix his or her vision on a fixation point at the center of a screen. Pictures are projected on the screen. If the picture is in the left visual field, the image is processed in the right hemisphere (key in this instance). The subject can also reach under the screen to reach objects that can be identified by tactile cues. (See also text.)

temporale (temporal plane, the superior surface of the temporal lobe; see Figure 10-4).

Cerebral dominance has been studied best in patients whose left and right hemispheres have been disconnected by surgical division of the corpus callosum. Visual images can be shown separately to the left and right visual fields of such individuals, and they can be asked to identify objects placed in the right or left hand (Figure 10-3). If a picture of an object, such as a ring, is presented to the left hemisphere, the subject can identify the object verbally as a ring. If a picture of a key is presented to the right hemisphere, the subject cannot identify it verbally. This is because information that reaches the right hemisphere about the key does not access to the lan-

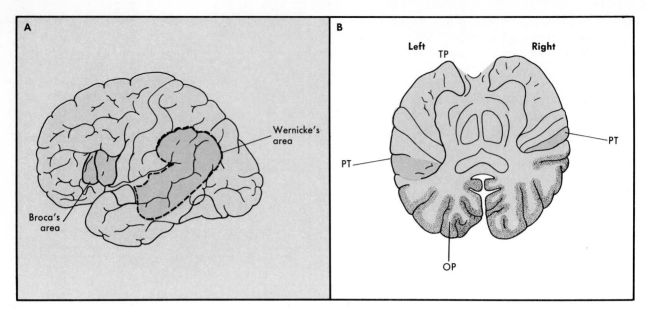

FIGURE 10-4   Areas of the cerebral cortex important for language. **A,** Broca's and Wernicke's areas. **B,** The relative size of the planum temporale *(PT)* on the two sides of the brain. *TP,* Temporal pole; *OP,* occipital pole.

guage centers of the left hemisphere. However, the subject can identify the picture in another way, by picking up a key with the left hand after feeling a group of objects.

Language

Language depends on activity in the left hemisphere in most people. This can be demonstrated by injecting local anesthetic into the carotid circulation on the left while an individual is speaking. The anesthetic stops speech *(aphasia)*. Analysis of aphasia that occurs after damage to the left hemisphere of adults has revealed that several major zones are important for language. One of these is called *Broca's area,* which is located in the inferior frontal gyrus just anterior to the face representation in the motor cortex (Figure 10-4). Damage to Broca's area diminishes the ability of the individual to speak and write. The person understands spoken or written words, and there is not necessarily an impediment to sound production. This type of aphasia is called *Broca's aphasia.*

The other important region for the control of language is *Wernicke's area,* which is in the supramarginal and angular gyri of the temporal lobe and the posterior part of the superior temporal gyrus. Damage here diminishes the comprehension of spoken or written language. However, the person has fluent speech, but much of it is meaningless. This type of aphasia is called *Wernicke's aphasia.*

BIBLIOGRAPHY
Journal Articles

Byrne JH: Cellular analysis of associative learning, Physiol Rev 67:329, 1987.

Damasio AR and Geschwind N: The neural basis of language, Annu Rev Neurosci 7:127, 1984.

Steriade M and Llinas RR: The functional states of the thalamus and the associated neuronal interplay, Physiol Rev 68:649, 1988.

Teyler TJ and DiScenna P: Long-term potentiation, Annu Rev Neurosci 10:131, 1987.

Thompson RF et al: Cellular processes of learning and memory in the mammalian CNS, Annu Rev Neurosci 6:447, 1983.

Books and Monographs

Anderson P and Andersson SA: Physiological basis of the alpha rhythm, New York, 1968, Appleton-Century-Crofts.

Bergamini L and Bergamasco B: Cortical evoked potentials in man, Springfield, Ill, 1967, Charles C Thomas, Publisher.

Creutzfeldt OD et al: Electrophysiology of cortical nerve cells. In Purpura DP and Yahr MD, editors: The thalamus, New York, 1966, Columbia University Press.

Gazzaniga MS and LeDoux JE: The integrated mind, New York, 1978, Plenum Press.

Jones EG: The thalamus, New York, 1985, Plenum Press.

Kandel ER and Schwartz JH: Principles of neural science, ed 2, New York, 1985, Elsevier Science Publishing Co, Inc.

Magoun HW: The waking brain, ed 2, Springfield, Ill, 1963, Charles C Thomas, Pubisher.

Penfield W and Jasper H: Epilepsy and the functional anatomy of the human brain, Boston, 1954, Little, Brown & Co.

Willis WD and Grossman RG: Medical neurobiology, ed 3, St Louis, 1981, The CV Mosby Co.

MUSCLE

RICHARD A. MURPHY

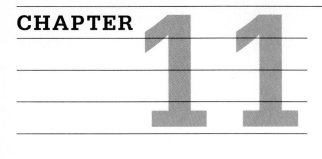

PART III

Molecular Basis of Contraction

Movement, produced by the neuromuscular system, is perhaps the most striking difference between plants and animals. The basis for movement is a biologic energy transformation called *chemomechanical transduction*. In this process most of the body's metabolic production of adenosine triphosphate (ATP) is converted into force or movement by muscle cells. Evolution has led to specialization of muscle cells to minimize the ATP consumption required for specific functions. Nevertheless, the basic molecular process underlying contraction is the same in all muscle cells. The objective of this chapter is to show how muscle cells contract by the *sliding filament–crossbridge mechanism.*

THE CONTRACTILE UNIT

The basic structure involved in contraction consists of organized arrays of insoluble structural proteins. One set of proteins forms a *cytoskeleton* that serves as an anchor and force-transmitting structure for the contractile proteins organized in the *myofilaments.* The simplest form of a contractile unit, termed a *sarcomere,* is found in *striated muscle* cells (Figure 11-1). Enormous numbers of sarcomeres are linked together by the cytoskeleton. *Z-disks* mechanically link sarcomeres end to end. *Intermediate filaments* (polymers of the proteins *desmin* or *vimentin*) connect the Z-disks of adjacent myofibrils within a striated muscle cell. The transverse alignment of sarcomeres and their constituent myofilaments give these cells their striated appearance.

Thin Filaments

Thin filaments are ubiquitous cell structures that always contain *actin* and *tropomyosin* (Figure 11-2). The thin filaments in vertebrate striated muscle are anchored in the Z-disks. The thin filaments of striated muscle also contain *troponin* bound to each tropomyosin molecule. Troponin, a regulatory protein, con-

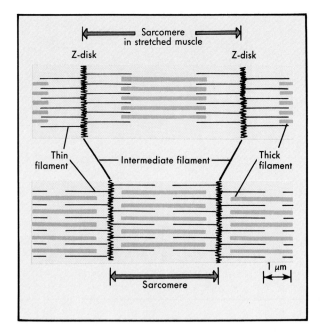
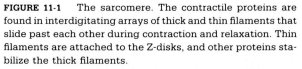

FIGURE 11-1 The sarcomere. The contractile proteins are found in interdigitating arrays of thick and thin filaments that slide past each other during contraction and relaxation. Thin filaments are attached to the Z-disks, and other proteins stabilize the thick filaments.

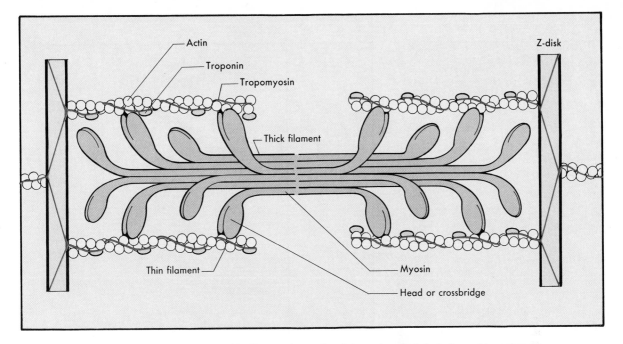

FIGURE 11-2   The core of the thin filament is a twisted, two-stranded chain formed by polymerization of the globular protein actin. Each molecule of the long, rigid tropomyosin binds with six or seven actin monomers plus one troponin. Thick filaments are composed of approximately 300 to 400 myosin molecules. The tails of the very large myosin molecules aggregate to form the filament. The heads protrude as crossbridges that can interact with the thin filament. Note the bipolar structure of the thick filament: the central bare zone lacking crossbridges divides the filament into two halves, where the crossbridges have opposite orientations.

tains Ca^{++}-binding sites that are involved in control of contraction and relaxation.

Thick Filaments

Myosin is a large, complex molecule consisting of tail, neck, and head regions (Figure 11-2). The tails aggregate to form thick filaments, with the neck and head projecting laterally to form a *crossbridge*. Each head contains an actin-binding site and an enzymatic site that can hydrolyze ATP to adenosine diphosphate (ADP) and inorganic phosphate (P_i). Both sites are involved in chemomechanical transduction. The interactions between the crossbridges and the thin filament draw the thin filaments toward the sarcomere's center, thereby causing shortening of the sarcomere as the Z-disks come closer together (Figure 11-1).

THE CROSSBRIDGE CYCLE

A cyclic hydrolysis of ATP occurs when purified myosin and thin filaments are mixed in a solution that approximates the ionic content of the cytoplasm, often called the *myoplasm* when referring to muscle cells (Figure 11-3, *A*). This cycle can be represented by four steps. ATP binds to myosin and is hydrolyzed to form the myosin-ADP-P_i complex. This complex, characterized by a high level of free energy, has a great affinity for actin and rapidly binds to the thin filament *(step 1)*. ADP and P_i are released after myosin attaches to the thin filament, and the myosin head undergoes a conformational change *(step 2)*. The resulting actin-myosin complex has a low level of free energy. In *step 3* the actin-myosin complex binds ATP. The resulting actin-myosin-ATP complex has a low binding affinity, so that the crossbridge dissociates from the thin filament. Internal hydrolysis of the bound ATP regenerates the high-energy myosin-ADP-P_i complex to complete the cycle *(step 4)*.

In this biochemical cycle the large release of free energy occurring in the overall reaction, ATP to ADP plus P_i, is lost as heat. The conversion of part of this energy into mechanical work depends on the sarcomere structure of the muscle cells. The orientation of the myosin heads incorporated into a thick filament is constrained. The preferred orientation of the high-

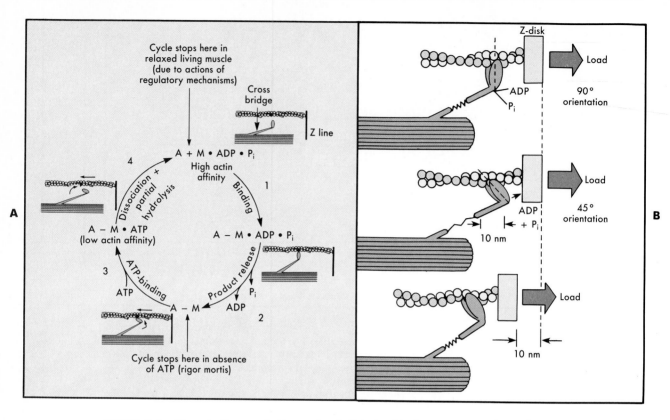

FIGURE 11-3   Steps in the crossbridge cycle. **A,** Relationship between steps *(1 to 4)* in the hydrolysis of adenosine triphosphate *(ATP)* and crossbridge conformations. **B,** Preferred or minimal free energy conformations of attached crossbridges. The transition from the 90- to the 45-degree conformation that occurs on release of adenosine diphosphate *(ADP)* plus inorganic phosphate *(P_i)* generates a force in the crossbridge represented by the stretched spring. This force can be translated into shortening by movement of the thin filament past the thick filament. *A,* Actin; *M,* myosin.

energy complexes (myosin-ADP-P_i and actin-myosin-ADP-P_i) is perpendicular (90 degrees) to the thick filament (Figure 11-3, *B*). However, the preferred (i.e., lowest level of free energy) conformation of the actin-myosin complex occurs when the crossbridge is 45 degrees to the filaments. Thus part of the energy from ATP is translated into conformational changes in the crossbridges. This "bending" of the crossbridges generates forces that draw the thin filaments past the thick filaments toward the center of the sarcomere. The force is transmitted by the cytoskeleton to the ends of the cell, where it exerts a force on the skeleton. A single crossbridge cycle moves the filaments about 10 nm (10^{-8} cm) and develops a minute force whose effect would be lost on crossbridge detachment. However, millions of crossbridges cycling asynchronously can generate great forces and shorten the sarcomere considerably (see Figure 11-1).

Determinants of Crossbridge Cycling

The crossbridge cycle illustrated in Figure 11-3 will continue until all the ATP is consumed and the cycle is arrested (after step 2). This arrest only occurs after the death of the animal, when the ATP supplies are not replenished. This state is characterized by *rigor mortis,* or muscular rigidity.

Resting or relaxed muscle contains detached crossbridges in the myosin-ADP-P_i state and is freely extensible. These crossbridges are prevented from attaching to the thin filaments by Ca^{++}-dependent regulatory systems, which differ among muscle types (see Chapters 12 and 13). Such regulatory systems control the number of crossbridges interacting with the thin filaments.

Crossbridge cycling rates determine how fast a muscle shortens. The maximal shortening velocities occur when no load opposes filament sliding. A load on a muscle cell is transmitted by the cytoskeleton to the sarcomere, where it opposes bending of the cross-bridges. Increasing the load slows crossbridge cycling. Velocities fall to zero when the load prevents the transition from the 90- to the 45-degree conformation. Different types of muscle cells vary in their maximal unloaded shortening velocities. Such functional differences are determined by the isoenzymatic variant of myosin expressed in a particular muscle cell.

THE BIOLOGIC RESPONSE: CHARACTERIZING CONTRACTION

Contracting muscle cells can perform several actions. They may develop a force without shortening, may shorten at various velocities, or may even lengthen while opposing a larger force than they can generate. The response depends on the loading. A simple mechanical analysis of contraction describes the output of the muscle cells and helps show how muscle functions.

Only three variables are needed to describe the output of a muscle completely: force, length, and time (Table 11-1). The analysis is simplified by experimentally holding one of the three variables constant and determining the relationship between the other two. This constraint yields two types of contractions: *isometric* (*iso*, constant or equal; *metric*, length) and *isotonic* (at constant force or load).

Table 11-1 Basic Mechanical Variables in Muscle Contraction

Parameter (Symbol)	Units	Definition
Force (F)	Newton (n)	
Length (L)	Meter (m)	
Time (T)	Second (sec)	
DERIVED VARIABLES		
Velocity (V)	$m \cdot sec^{-1}$	Change in length/ change in time
Work (W)	$n \cdot m$	Force times distance
Power (P)	$n \cdot m \cdot sec^{-1}$	Work/time
Stress (S)	$n \cdot m^{-2}$	Force/cross-sectional area

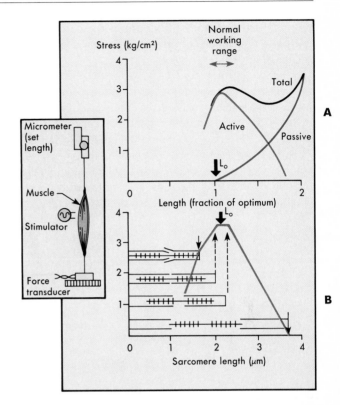

FIGURE 11-4   A, Relaxed muscles exhibit an elastic counterforce as the cell is lengthened; the curve is termed the *passive force-length relationship*. Contracting muscles develop greater forces; this curve is depicted as the *total force-length relationship*. The difference between the total force and the passive force curves represents the force-generating properties of the crossbridges—the curve called the *active force-length relationship*. Insert shows experimental setup. **B,** A sophisticated analysis of single cells or sarcomeres reveals that the generated force depends on the overlap of thick and thin filaments. At an optimal length (L_o), the cell can develop the most force. As illustrated, L_o provides the maximal crossbridge interactions with the thin filament.

Isometric Contractions: the Dependence of Force on Length

A muscle cell develops a characteristic force when maximally stimulated at a fixed length. The force-length relationship depicts this steady-state behavior (Figure 11-4).

Stimulated muscle cells develop no force if they are first stretched to sarcomere lengths greater than 3.7 μm. At shorter lengths **force is proportional to the number of crossbridges that interact with the thin filament in each half sarcomere.** Force generation is also lower at muscle lengths less than the optimal length (L_o); disturbances of the sarcomeric structure and failure of the activation processes are responsible.

The forces generated depend on the size of the muscle cell and the number of filaments. When normalized for size as force per cell cross-sectional area (to give a stress), vertebrate muscle cells generate about 3×10^5 n·m^{-2} (or about 3 kg·cm^{-2}) at their L_o. The passive force-length behavior (Figure 11-4) results from the elasticity of structural proteins. These include cytoskeletal proteins and extracellular connective tissue proteins organized in fibrils of *collagen* and *elastin*.

Isotonic Contractions: the Dependence of Velocity on Load

A muscle lever measures shortening of a muscle at a constant load (Figure 11-5, *A*). The relaxed muscle is adjusted to the length that is optimal for force development (Figure 11-4), and different loads are attached to the lever before the muscle is stimulated. If the load is greater than the muscle can lift, the maximal force (F_o) is developed in an isometric contraction, as just discussed. If the load is somewhat smaller, force development occurs with no shortening until the force developed by the muscle is equal to the load. The muscle then begins to shorten isotonically (Fig-

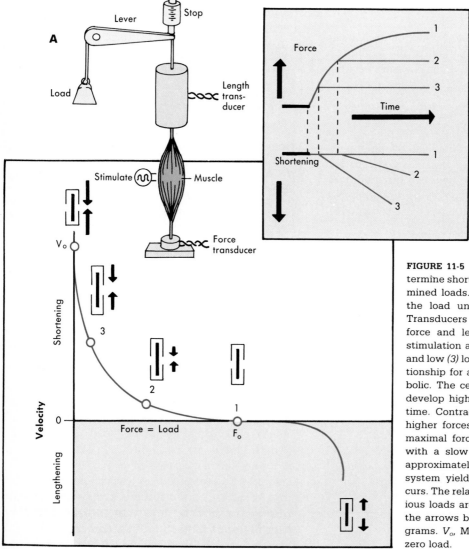

FIGURE 11-5   A, Schematic device to determine shortening in a muscle at predetermined loads. Note the stop that supports the load until the muscle contracts. **B,** Transducers in the apparatus detecting force and length show the response to stimulation at very high *(1),* moderate *(2),* and low *(3)* loads. **C,** The velocity-force relationship for a shortening muscle is hyperbolic. The cell can shorten rapidly or can develop high forces, but not at the same time. Contracting muscles can withstand higher forces than they develop (loads > maximal force, F_o). These are associated with a slow lengthening up to loads of approximately 1.6 F_o, where the contractile system yields and rapid lengthening occurs. The relative rates of movement at various loads are indicated by the lengths of the arrows beside the five sarcomere diagrams. V_o, Maximal shortening velocity at zero load.

ure 11-5, *B*). The slope of the length versus time record gives the shortening velocity. A third contraction with an even lighter load gives a higher shortening velocity. The complete dependence of velocity on load obtained from many contractions is shown in Figure 11-5, *C*.

The velocity-force relationship is the mechanical manifestation of the sum of all the crossbridge interactions in the cell. Shortening velocity is proportional to the average crossbridge cycling rate. Maximal velocities (V_o) occur with zero load. They are determined by the kinetics of the specific isoenzymatic form of myosin synthesized in the cell. A load slows the average cycling rate as it opposes the transition from the 90- to 45-degree conformation (step *2* in Figure 11-3, *A*).

Resisting Imposed Loads

Crossbridge cycling can only lead to force development and shortening. Nevertheless, gravitation or other counterforces can impose large loads that may lengthen contracting muscle cells. A muscle cell can briefly resist loads 60% greater than loads the cell can develop (Figure 11-5, *C*). The crossbridge attachment is strong, and high forces are required to impose a conformational change from the 90-degree state to a 135-degree extended state. The stretched crossbridges are pulled free of the thin filament and subsequently reattach to resist further stretch. Under these physiologically typical conditions the crossbridge cycle, with release of ADP and P_i and binding of ATP, is never completed (see Figure 11-3). The work is done **on** the muscle rather than by the cell, and no ATP cost is incurred.

Energy Cost of Contraction

Crossbridge cycling accounts for most of the body's ATP consumption. Functional specialization among muscle types minimizes this cost. One measure of energy cost is the *efficiency* of contraction, which is defined as the ratio of the mechanical work performed to the chemical energy released by ATP hydrolysis. The efficiency of contraction varies. Work equals force times distance shortened. Work is zero in an isometric contraction (distance shortened equals 0) and in an unloaded contraction (force equals 0). Under these conditions the efficiency is also zero. The greatest efficiency is approximately 45% conversion

of chemical to mechanical energy by the crossbridges. The maximal efficiency occurs with a moderately loaded muscle when the power output is high (Figure 11-6).

The power output and efficiency are low at point *1* in Figure 11-6. At very low loads many crossbridges attach and cycle without exerting their maximal force on the rapidly sliding thin filaments. At high loads (point *3*) a crossbridge is likely to exert its maximal effect in the cycle, but it has a high probability of reattaching at the same point on the very slowly moving thin filament. The optimal loading of about 0.3 F_o (point *2*) gives the highest probability of capturing the maximal amount of energy from ATP hydrolysis.

The efficiency of chemomechanical transduction in the crossbridge cycle is quite high (comparable to modern gasoline engines). Nevertheless, more than half the potential free energy is dissipated as heat, even under optimal conditions. The resulting increase in temperature of muscle cells is a significant physiologic burden during exercise. However, muscles can be used to generate heat during shivering to maintain normal body temperatures in cold environments. The crossbridge cycle is not the only energy-consuming reaction in muscle cells, although it is the only one that yields mechanical work. These

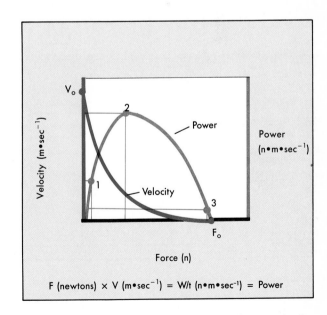

FIGURE 11-6 The power output of a muscle is simply the product of its force *(F)* and velocity *(V)* and can be calculated for each point on the velocity-force relationship.

other reactions lower the overall efficiency of muscular contraction to about 20% to 25%.

Isometric contractions can be characterized energetically in terms of their *economy,* defined as force × time/ATP consumed. The more time a crossbridge spends in the force-generating configuration, the greater is the economy of a contraction. High economies are associated with slow crossbridge cycling rates. Thus a muscle cell can be fast with a high power output and ATP consumption, or it can be slow and economical.

Combining Contractile Units

Large numbers of individual contractile units or sarcomeres are combined in a muscle cell. Furthermore, muscles contain a myriad of muscle cells. The nature of their combination determines the output of the cell or muscle (Figure 11-7).

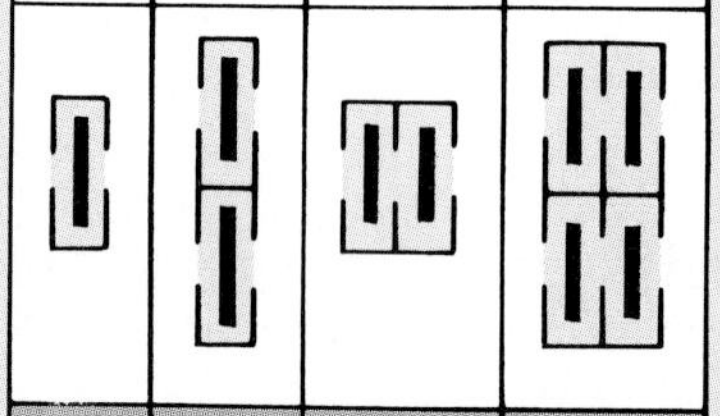

	Alone	Series	Parallel	Both
Absolute values				
Force	X	X	2X	2X
Velocity	X	2X	X	2X
Shortening capacity	X	2X	X	2X
Normalized values				
Force/area	X	X	X	X
Velocity ($L_o \cdot sec^{-1}$)	X	X	X	X
Shortening (% L_o) capacity	X	X	X	X

FIGURE 11-7   Contractile units can be mechanically linked in series and in parallel to give longer or wider cells. These additions that occur with growth directly determine the output of the cell. The effects of combining two contractile units in series and in parallel are shown in the table. Appropriate normalization allows comparisons of cells of different sizes.

BIBLIOGRAPHY
Journal Articles

Bárány M: ATPase activity of myosin correlated with speed of muscle shortening, J Gen Physiol 50(part 2):197, 1967.

Eisenberg E and Hill TL: Muscle contraction and free energy transduction in biological systems, Science 227:999, 1985.

Gordon AM et al: The variation in isometric tension with sarcomere length in vertebrate muscle fibres, J Physiol 184:170, 1966.

Harrington WF and Rodgers ME: Myosin, Annu Rev Biochem 53:35, 1984.

Huxley AF: Muscular contraction, Annu Rev Physiol 50:1, 1988, (prefatory chapter).

Kodama T: Thermodynamic analysis of muscle ATPase mechanisms, Physiol Rev 65:467, 1985.

Swynghedauw B: Developmental and functional adaptation of contractile proteins in cardiac and skeletal muscles, Physiol Rev 66:710, 1986.

Books and Monographs

Carlson FD and Wilke DR: Muscle physiology, Englewood Cliffs, NJ, 1974, Prentice-Hall, Inc.

dos Remedios CG and Barden JA: Actin: structure and function in muscle and non-muscle cells, New York, 1983, Academic Press.

Huxley AF: Reflections of muscle, Princeton, NJ, 1980, Princeton University Press.

Lackie JM: Cell movement and cell behavior, London, 1986, Allen & Unwin.

Peachey LD et al, editors: Handbook of physiology, section 10: Skeletal muscle, Bethesda, Md, 1983, American Physiological Society.

Rüegg JC: Calcium in muscle activation, Berlin, 1986, Springer-Verlag.

Squire JM: The structural basis of muscular contraction, New York, 1981, Plenum Press.

Sugi H and Pollack GH, editors: Cross-bridge mechanism in muscle contraction, Baltimore, 1978, University Park Press.

Muscles Acting on the Skeleton

Vertebrates differ from all other animal groups in possessing an internal skeleton to which the *skeletal muscle* cells are attached. Skeletal muscle cells are sometimes called *striated muscle* because of their appearance. They are also called *voluntary muscle,* which reflects their neural connections to the motor cortex. A muscle cell acting on a skeleton to produce movement has a specific role that dictates many of its properties. These properties, starting with the structure of skeletal muscle, are examined in this chapter. The ways in which crossbridge cycling is controlled to produce contraction and relaxation are then explored. Finally, specializations that allow functional diversity and adaptation are considered.

MUSCULOSKELETAL RELATIONSHIPS

The skeleton serves as a supporting lever system on which most cells act (Figure 12-1). Exceptions include skeletal muscle in the lips and esophagus, where the muscle participates in the voluntary act of swallowing. Characteristically, striated muscle cells in the limbs bridge two joints before they attach to the skeleton via *tendons* or other mechanical connections. Embryonic muscle cells fuse end to end to form the enormous multinucleated, differentiated skeletal muscle cells. Although only the diameter of a fine thread (50 to 100 μm), these multinucleated cells may be many centimeters long.

The relationship between the muscle cells and the skeleton dictates the following important characteristics of skeletal muscle:

1. Each cell acts independently in response to a nerve impulse. The force of contraction can be increased by recruiting more cells.
2. The skeleton bears most gravitational loads. Skeletal muscle cells are normally relaxed.
3. Skeletal muscle cells typically act on the short end of the skeletal lever system (Figure 12-1). Thus they must develop forces that are much greater than the load moved. However, large movements can result from limited cell shortening. The cells are constrained by the skeleton to operate near their optimal length for force development (see Figure 11-4).
4. Most contractions produce movement and do mechanical work (force × distance). The rate of doing work (power = work/time) can be great. Skeletal muscles are characterized by a high efficiency.

Muscles such as the biceps consist of bundles of muscle cells that are linked together and are separated from other muscles by connective tissues. The contraction of some or all of the cells in a muscle yields complex movements. Mammals have more than 400 different muscles, and several are attached to one bone. Discrete movements are the result of coordinated contractions involving many muscles. Muscles may act together as *synergists* to produce the same movement, or they may function as *antagonists* to decelerate a motion. Their summed actions can stabilize a joint or produce a precisely controlled movement. *Flexors* and *extensors* (antagonist muscles that act on the limb joints) are normally both involved in a movement. A surprising variation exists among individuals in the muscles they use to accomplish a specific motion. These patterns reflect differences in neuromuscular learning or training.

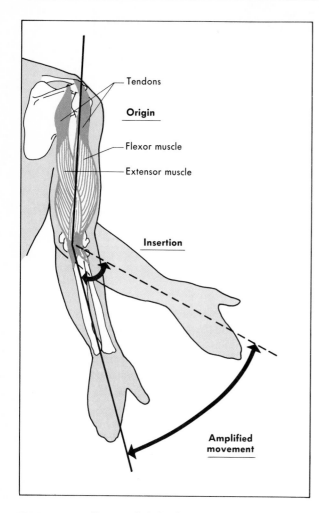

FIGURE 12-1   Groups of skeletal muscle cells form discrete muscles, which span anchor sites on the skeletal lever system. Note how a relatively small shortening of the muscle cells can produce a large movement of the arm, if the force generated is sufficient. Tendons connecting muscle cells to the skeleton contain inextensible collagen fibrils.

STRUCTURE OF SKELETAL MUSCLE

The contractile units, or sarcomeres, of skeletal muscle are linked in series along a *myofibril* (Figure 12-2). The cytoskeleton links the Z-disks of the myofibrils so that the sarcomeres are aligned. The resulting alternating dark and light stripes correspond to regions that contain thick filaments separated by regions containing only thin filaments. These stripes are clearly visible when viewed through the light microscope.

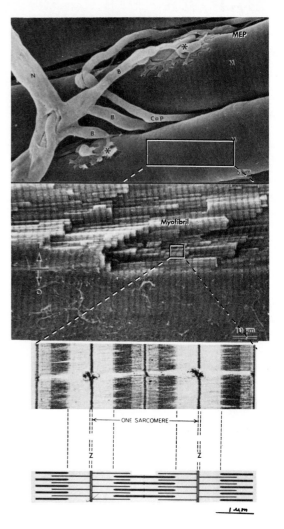

FIGURE 12-2   **A,** Scanning electron micrograph of segments of three mammalian skeletal muscle cells *(M)* showing striations. Each cell receives a branch of a motor nerve in a complex structure termed the motor end plate or neuromuscular junction *(asterisks)*. Note close association of the giant cells with capillaries *(cap)*. **B,** Scanning electron micrograph of skeletal muscle cell that was broken open to show interior packed with large numbers of striated myofibrils. **C,** Higher magnification transmission electron micrograph of thin section through two myofibrils. Individual filaments are difficult to discern at this magnification. **D** shows the structure of the sarcomere arising from thin filaments attached to Z disks interdigitating with a central lattice of thick filaments. (**A** From Desaki J and Uehara Y, 1981. **B,** From Swada H, Ishikawa H, and Yamada E, 1978. **C,** From Huxley, H.E., 1965.)

Looking at the changing scales in Figure 12-2, one can envision the enormous numbers of sarcomeres present in a cell. For example, a 10 cm long cell would have more than 4500 sarcomeres.

The Membranes

Four structurally and functionally distinct membranes in the cell are involved in regulation of contraction.

1. The *neuromuscular junction,* or *motor endplate,* is a specialized region of the plasma membrane (Figure 12-3).

2. The plasma membrane, or *sarcolemma,* along which action potentials are propagated, is the second membrane. Neuromuscular transmission is described in Chapter 4.

3. Tiny openings in the sarcolemma lead into the *transverse-tubular (T-tubular) network* at the level of the Z-disk in mammals. This network defines an extracellular space within the cell (Figure 12-3). The extensive T-tubular network virtually encircles each myofibril. However, the diameter of the tubules is small, and the tiny volume of this extracellular space prevents propagation of action potentials down the T-tubules into the cell. A graded depolarization spreads into the cell via this system when the action potential travels down the sarcolemma. By this means excitation spreads to the level of the myofibrils.

4. The *sarcoplasmic reticulum,* a distinct membrane system, connects intimately with the T-tubules. The sarcoplasmic reticulum surrounds an intracellular compartment that forms a sleeve around each myofibril.

NEUROMUSCULAR FUNCTION

The basic neuromuscular relationships are depicted in Figure 12-4. During early development a motor nerve axon originating from the spinal cord makes contact with a muscle cell. This leads to formation of an endplate. Connections with other growing axons are inhibited so that each muscle cell has one input. Note that individual axons branch in the muscle, and each nerve will control many muscle cells. The resulting functional grouping of a nerve and its associated muscle cells is called a *motor unit.* Motor units may contain only a few muscle cells or up to thousands.

An action potential is elicited in a motor nerve when the sum of the excitatory and inhibitory synaptic inputs to the cell body produces a critical depolarization (see Chapter 3). That action potential leads to the release of sufficient acetylcholine to produce an endplate potential and generate an action potential in

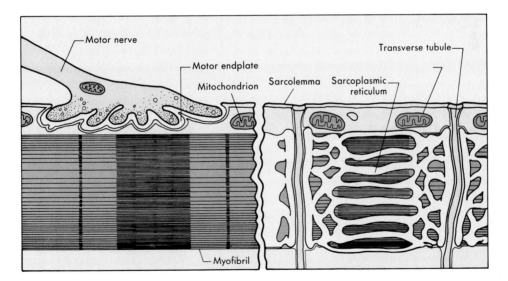

FIGURE 12-3   The membranes of skeletal muscle. The plasma membrane separating the extracellular space from the intracellular space (myoplasm) has three specialized portions: the motor endplate, the sarcolemma, and the transverse tubules (T-tubules) forming a network at the ends of each sarcomere in mammals. The sarcoplasmic reticulum is a distinct membrane system enclosing a separate intracellular compartment surrounding each myofibril and is intimately associated with the T-tubular system.

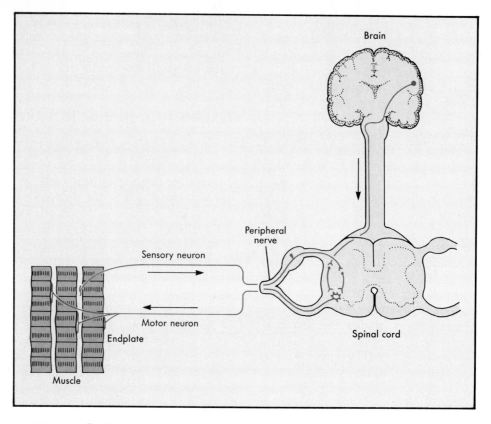

FIGURE 12-4   Basic neuromuscular relationships in skeletal muscle. Motor nerves receive both excitatory and inhibitory input from peripheral sensory receptors involved in reflex arcs and from many levels of the spinal cord, cerebellum, and brain. The cells of individual motor units are scattered among cells in other motor units.

all the muscle cells in the motor unit, followed by synchronous contractions.

REGULATION OF CONTRACTION AND RELAXATION

The process linking the action potential to crossbridge cycling and contraction is called *excitation-contraction coupling*. The events involved are (1) signal transduction at the cell membrane and (2) generation of a second messenger that (3) acts on myofibrillar regulatory mechanisms controlling crossbridge cycling. Excitation-contraction coupling in skeletal muscle is comparatively simple, in the sense that only one event is critical in each of the three steps.

The second messenger that regulates crossbridge cycling in all muscles is Ca^{++}. Skeletal muscle cells are too large and contract too rapidly for Ca^{++} to dif-

fuse through channels in the sarcolemma from the extracellular space to the myofibrils. The cellular compartment enclosed by the sarcoplasmic reticulum contains the Ca^{++} pool involved in activation (Figure 12-5). Signal transduction in skeletal muscle cells is the process by which an action potential triggers Ca^{++} release from the sarcoplasmic reticulum. It appears that the graded depolarization of T-tubules causes conformational changes in the sarcoplasmic reticular membrane by charge coupling. This opens channels in the sarcoplasmic reticulum that allow Ca^{++} ions to diffuse down their electrochemical gradient into the myofibrils. This process is rapid (1 to 2 msec) because the Ca^{++} concentration gradient is huge (about 10^5) and the distances are short (about 1 μm).

The sarcoplasmic reticular membrane contains large amounts of a protein complex that pumps Ca^{++} from the myoplasm back into the sarcoplasmic retic-

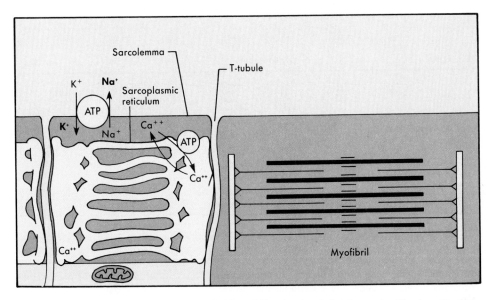

FIGURE 12-5   Signal transduction and Ca^{++} mobilization in skeletal muscle. The transit of an action potential along the sarcolemma causes a brief, graded depolarization of the T-tubular network and a momentary jump in the permeability of the sarcoplasmic reticulum. A myoplasmic Ca^{++} pulse occurs as concentrations rise from less than 10^{-7} M to greater than 10^{-5} M and then rapidly fall to resting values because of active transport back into the sarcoplasmic reticulum.

ulum, accompanied by the hydrolysis of adenosine triphosphate (ATP). Because the membrane potential changes during each action potential are uniform, the release and reuptake of Ca^{++} generate reproducible myoplasmic Ca^{++} pulses (Figure 12-6).

CALCIUM AND CROSSBRIDGE REGULATION

Two basic crossbridge states are possible in skeletal muscle: free and attached. A Ca^{++} switch effectively allows the transition from an **off state,** in which crossbridges cannot attach in a relaxed muscle, to an **on state,** in which attachment and cycling are possible (Figure 12-7).

Troponin, the regulatory protein associated with tropomyosin, has four high-affinity Ca^{++}-binding sites. These are all occupied very rapidly when Ca^{++} is released from the sarcoplasmic reticulum (Figure 12-6). The result is that all the thin filaments are quickly turned on, and the crossbridges cycle until the transport pumps lower the Ca^{++} concentration to the values at which Ca^{++} dissociates from troponin

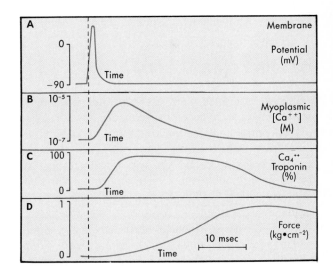

FIGURE 12-6   Excitation-contraction coupling in skeletal muscle. Rapid transit of the action potential **(A)** induces a slower transient increase in the myoplasmic Ca^{++} concentration **(B).** The binding sites on troponin are rapidly saturated **(C),** leading to a conformational change in the thin filament and thus allowing crossbridge attachment, cycling, and force development **(D).** The mechanical twitch begins after a delay of only a few milliseconds and lasts some tens or hundreds of milliseconds, depending on the cell type.

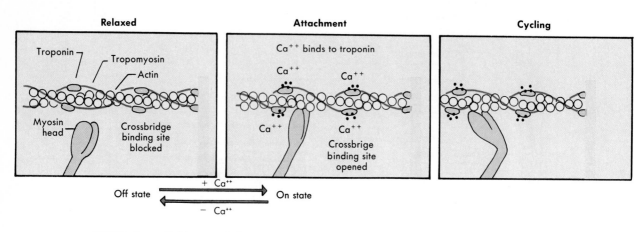

FIGURE 12-7   Ca^{++} can switch on crossbridge cycling by binding to the regulatory protein troponin on the thin filament. The resulting conformational change allows crossbridge attachment and cycling. *A*, Actin; *M*, myosin; *ATP*, adenosine triphosphate; *ADP*, adenosine diphosphate, *P$_i$*, inorganic phosphate.

(Figure 12-6). The thin filaments return to the off conformation, and the cell relaxes. Activation in skeletal muscle is an *all-or-none process:* action potentials give uniform Ca^{++} transients. This rapidly switches the thin filaments on for the same period and leads to a consistent mechanical response, called a *twitch.*

GRADING CONTRACTILE FORCE

Skeletal muscles must generate different forces, sometimes for considerable periods. Two mechanisms control the force generated by a muscle. First, because muscles contain large numbers of motor units, force can be varied over a wide range by recruitment of more and more motor units. The second way to increase force and prolong a contraction is to increase the frequency of firing of the motor nerves (Figure 12-8). Even though all the crossbridges are cycling during a twitch, the maximal force is not attained before the Ca^{++} levels fall and the contractile apparatus is turned off. There is simply not enough time for sufficient crossbridge cycles to generate the full force. Firing the motor nerves at higher frequencies elicits further Ca^{++} transients in time to sum the mechanical responses and produce a *tetanus.* The maximal tetanus force may be three- to fivefold greater than the twitch force.

Several factors contribute to finely graded contrac-

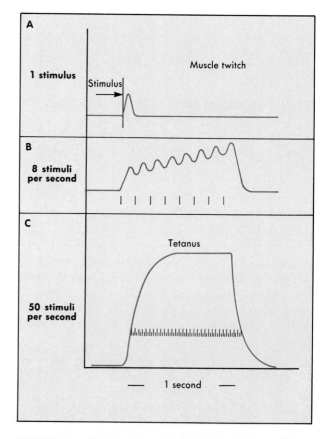

FIGURE 12-8   The force of contraction of a motor unit can be increased by more frequent recruitment so that twitches (**A**) sum in an incomplete (**B**) or complete tetanus (**C**).

tions. The motor units recruited for weak contractions are the smallest, giving small increments in force. **The presence of large numbers of motor units in a muscle and tetanization allow continuous gradation in force generation over a wide range.**

FUNCTION DIVERSITY IN SKELETAL MUSCLE

The transition from rest to contraction in skeletal muscle triggers an extraordinary jump in ATP con-sumption. This must be matched instantaneously by an increase in ATP resynthesis. Muscle, as with all cells, has three pathways to regenerate ATP (Figure 12-9).

Direct phosphorylation of adenosine diphosphate (ADP) by creatine phosphoryltransferase from cre-atine phosphate is not a net synthetic reaction. Cre-atine phosphate serves as a large storage pool of almost instantaneously available high-energy phos-phate. Direct phosphorylation buffers the cellular ATP levels at the onset of contraction while synthetic pathways become active. The cell has only two ways to synthesize ATP, and these have very different characteristics (Figure 12-9).

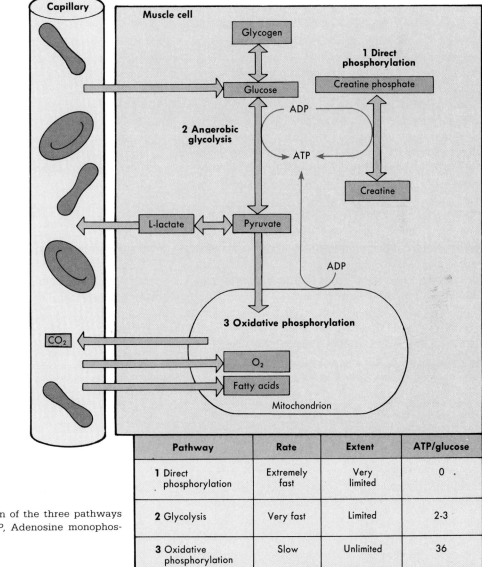

FIGURE 12-9   Comparison of the three pathways for ATP production. *AMP,* Adenosine monophos-phate.

Pathway	Rate	Extent	ATP/glucose
1 Direct phosphorylation	Extremely fast	Very limited	0
2 Glycolysis	Very fast	Limited	2-3
3 Oxidative phosphorylation	Slow	Unlimited	36

Glycolysis can supply ATP at very high rates, even though the yield per mole of glucose is low. However, this pathway fails when the cellular glycogen stores are depleted (seconds to minutes in muscles). In comparison, *oxidative phosphorylation* in the mitochondria generates ATP continuously and very efficiently; it uses oxygen and substrates that diffuse into the cell from the capillaries. The disadvantage is that this pathway is much slower than glycolysis. Oxidative phosphorylation cannot meet the demands of very rapid crossbridge cycling rates.

Skeletal muscle cells are specialized into two main types in humans and other primates. This specialization allows either high contraction velocities or long duration contractions. The two cell classes are differentiated on the basis of whether the gene for a slow or a fast myosin isoenzyme (i.e., having a moderate or a high ATPase activity or cycling rate) is expressed in the cell (Table 12-1). *Slow fibers*, characterized by moderate shortening velocities, consume ATP at moderate rates. Slow fibers have a high blood supply (high capillary density), many mitochondria, and a moderate diameter. These characteristics minimize diffusion distances for oxygen and substrates. Slow fibers are sometimes called *red fibers* because of the distinctive coloration provided by the iron-containing hemoglobin (blood supply) and cytochromes in the mitochondria. If the blood supply is adequate, slow fibers provide great endurance.

The maximal ATP consumption rate of *fast fibers* can only be met by glycolysis. These large, pale cells (few oxygen-binding proteins) have a more extensive sarcoplasmic reticulum so that fast contractions are matched by rapid relaxations. Fast fibers fatigue rapidly with glycogen depletion.

The motor unit rather than the cell is the functional grouping; thus fast and slow motor units exist. These units differ by more than their exclusive composition of either slow or fast fibers (Figure 12-10). In general, slow motor units generate low forces, resulting from both the smaller average fiber diameter and the comparatively few cells. The motor nerve determines the physiologic characteristics of a motor unit. Nerves with small cell bodies and narrow axons can synthesize limited amounts of acetylcholine, and they form small motor units. Small motor axons are also more excitable; relatively few excitatory postsynaptic potentials at the small cell body are needed to depolarize the cell to the critical potential to fire an action potential. The large axons of fast motor units are less excitable.

Most muscles contain large numbers of slow motor units and a few very large, fast motor units. The largest units may contain more than a thousand cells and develop hundreds of grams of force in a large animal. This means that most muscles can carry out various types of activities. For instance, moderate workloads can be sustained with little *fatigue*. The initial motor units recruited are the excitable slow units, which are fatigue resistant. For rapid, forceful contractions with a high power output, the fast motor units are **also** recruited. These maximal efforts cannot be sustained, and the fast units fatigue rapidly (Figure 12-11).

FATIGUE

Surprisingly little is known about the causes of the *general physical fatigue* one experiences after working vigorously. The state of disturbed homeostasis

---

Table 12-1 Fiber Types in Primate Skeletal Muscle

	Slow, Oxidative (Red)	Fast, Glycolytic (White)
Myosin isoenzyme (ATPase rate)	Moderate	Fast
Sarcoplasmic reticular Ca^{++} pumping rate	Moderate	Fast
ATP consumption rate	Moderate	Extremely high
Diameter (diffusion distance)	Moderate	Large
Oxidative capacity: mitochondrial content, capillary density	High	Low
Glycolytic capacity	Moderate	High

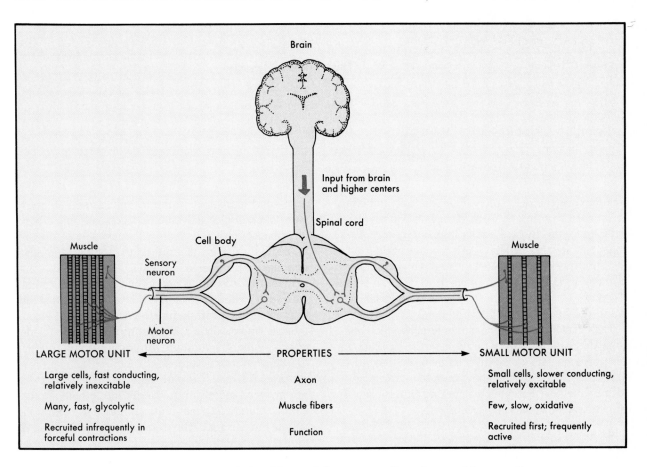

FIGURE 12-10   The characteristics of fast and slow motor units and some of the synaptic connections to the motor axons.

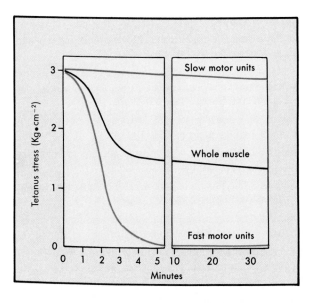

FIGURE 12-11   Fatigue in a muscle that is repeatedly tetanized for part of each second by experimentally stimulating the nerve trunk. Fatigue of the fast motor units occurs with glycogen depletion.

Table 12-2 Effects of Exercise

Type of Training	Example	Major Adaptive Response
Learning/coordination	Typing	Increases the rate and accuracy of motor skills (central nervous system)
Endurance (submaximal, sustained efforts)	Marathon running	Increased oxidative capacity in all involved motor units with limited cellular hypertrophy
Strength (brief, maximal efforts)	Weight lifting	Hypertrophy and enhanced glycolytic capacity of motor units employed

perceived as the discomfort of fatigue occurs before cells cease to contract. Such *cellular fatigue* is illustrated in Figure 12-11. Persons stop using their motor units before the cellular ATP concentration falls (avoiding rigor mortis). Metabolic changes, such as an increased blood lactate concentration and a fall in pH, may contribute to the perception of fatigue but cannot fully explain the phenomenon.

The cellular creatine phosphate and glycogen contents decrease during activity, even in slow motor units. An elevated oxidative metabolism resynthesizes these stores during a few-minute recovery period after exercise. Fatigue persists long after these cellular recovery processes are complete.

GROWTH AND ADAPTATION

Differentiation of skeletal muscle cells depends on their pattern of contractile activity. Motor units that are frequently active express the slow myosin isoenzyme and develop a high oxidative capacity. Motor units that contract infrequently differentiate into the fast fiber phenotype. Thus differentiation depends on innervation.

Skeletal muscle will *atrophy* if not used. Cell diameter and the number of myofibrils decrease. Such changes can be initiated after only 2 days of bed rest. If the motor nerve is destroyed by injury or disease (e.g., poliomyelitis), the denervated muscle cells will first atrophy and most will degenerate within a few months. These processes are reversible if reinnervation occurs. The fiber type will alter if a cell in a formerly fast motor unit is reinnervated by a small motor nerve, and vice versa.

Exercise has great and diverse effects on muscle. These depend on the type of exercise, as summarized in Table 12-2. The responses are adaptive, and they include neuromuscular learning, increased endur-

ance, and increased strength. The learned ability to carry out complex movements, such as riding a bicycle, persists for years even without practice. However, regular exercise is required to maintain the other changes. The increased physical well-being that results from moderate endurance exercise is mainly caused by enhanced respiratory and cardiovascular system capacity rather than the effects on the skeletal muscle fibers. Strength training also has broader effects, which include growth in bones and tendons to bear the greater forces.

Although exercise can have profound effects on the involved motor units, weight lifting will not convert slow units into fast units or induce the formation of new muscle cells. The transformation of slow fibers to fast fibers and the reverse can occur, as shown by cross-innervation experiments. However, no exercise regimen appears to alter activity patterns sufficiently to alter the expression of the myosin isoenzymes.

BIBLIOGRAPHY
Journal Articles

Buchthal F and Schmalbruch H: Motor unit of mammalian muscle, Physiol Rev 60:90, 1980.

Desaki J and Uehara Y: The overall morphology of neuromuscular junctions as revealed by scanning electron microscopy, J Neurocytol 10:107, 1981.

Emerson CP Jr and Bernstein SI: Molecular genetics of myosin, Annu Rev Biochem 56:695, 1987.

Freund H-J: Motor unit and muscle activity in voluntary motor control, Physiol Rev 63:387, 1983.

Huxley HE: The mechanism of muscle contraction, Scientific American 213:18, 1965.

Hwang CL-H: Intramembrane charge movements in skeletal muscle, Physiol Rev 68:1197, 1988.

Martonosi AN: Mechanisms of Ca^{++} release from sarcoplasmic reticulum of skeletal muscle, Physiol Rev 64:1240, 1984.

Rasmussen H and Barrett PQ: Calcium messenger system: an integrated view, Physiol Rev 64:938, 1984.

Sawada H, Ishikawa H, and Yamada, E: High resolution scanning electron microscopy of frog sartorius muscle, Tissue Cell 10:183, 1978.

Books and Monographs

Bagshaw CR: Muscle contraction, New York, 1982, Chapman and Hall.

Fernandez HL and Donoso JA, editors: Nerve-muscle cell trophic communication, Boca Raton, Fla, 1988, CRC Press.

Goldspink DF, editor: Development and specialization of skeletal muscle, Cambridge, 1980, Cambridge University Press.

Kedes LH and Stockdale FE, editors: Cellular and molecular biology of muscle development, New York, 1988, Alan R Liss, Inc.

Keynes RD and Aidley DJ: Nerve and muscle, Cambridge, 1981, Cambridge University Press.

Lüttgau HC and Stephenson GD: Ion movements in skeletal muscle, Chapter 28. In Andreoli TE et al, editors: Physiology of membrane disorders, New York, 1986, Plenum Press.

McMahon TA: Muscles, reflexes, and locomotion, Princeton, NJ, 1984, Princeton University Press.

Netter FH: Musculoskeletal system, part I: Anatomy, physiology and metabolic disorders. In Dingle RV, editor: The Ciba collection of medical illustrations, vol 8, Summit, NJ, 1987, Ciba-Geigy Corp.

Rüegg, JC: Calcium in muscle activation: a comparative approach, Heidelberg, 1986, Springer-Verlag.

Stein RB: Nerve and muscle: membranes, cells, and systems, New York, 1980, Plenum Press.

Woledge RC et al: Energetic aspects of muscle contraction, New York, 1985, Academic Press, Inc.

CHAPTER 13

Muscle in the Walls of Hollow Organs

Most people equate muscle with skeletal muscle. However, muscle is an important constituent of all organs and plays important roles in their function. These organ systems include the heart and vascular system, the airways, the entire gastrointestinal tract, and the urogenital system. Besides playing diverse roles, these muscles also have great medical significance because of their involvement in many diseases, including asthma, hypertension, and atherosclerosis. The muscle in hollow organs is a major target of drug therapy.

Muscle function differs in the absence of a skeleton. This chapter begins with an analysis of this difference, since it underlies the stuctural and functional contrasts between skeletal muscle, the cross-striated muscle in the heart, and the nonstriated or *smooth muscle* present in most of the other hollow organs. This material is covered in four sections devoted to (1) tissue and smooth muscle cell structure and function, (2) signal transduction and crossbridge regulation in smooth muscle, (3) crossbridge function and the mechanics and energetics of smooth muscles, and (4) a brief consideration of the special properties of cardiac muscle.

MUSCLE FUNCTION IN HOLLOW ORGANS

The load on the cells in hollow organs is imposed by the pressure in the organ (compare Figure 13-1 with Figure 12-1). If the muscle cells are relaxed, the volume of the organ will increase with the volume of the contents, until the connective tissue matrix with-

in and around the organ limits further expansion. At this point the pressure increases. The muscle may shorten to empty the organ by briefly increasing the pressure in what is termed a *phasic contraction*. However, the muscle may contract isometrically for long periods to maintain the dimensions of the organ. Such *tonic contractions* typify airways and blood vessels, where smooth muscle determines the resistance to flow and thus the distribution of air within the lungs or blood within the vessels throughout the body. In such cases the muscle serves as an "adjustable skeleton." The *economy* (force × time/adenosine triphosphate [ATP] consumption) is critical in this case. In fact, **the economy of some smooth muscles is more than 300 times that of striated muscle.**

Unlike cells in skeletal muscle, the cells in hollow organs do not bridge two fixed anchor points but are connected to each other. Thus they cannot be recruited to produce force as independent units. Each link in the chain must be equally activated and develop the same force. Furthermore, contraction in one part of an organ will change the pressure throughout, so muscle cell function must be completely coordinated.

CLASSIFICATION AND FUNCTIONAL DIVERSITY

Two classes of skeletal muscle are distinguished primarily on the basis of expressing the fast or slow myosin isoenzyme and secondarily in terms of the dominant metabolic pathway. All smooth muscles

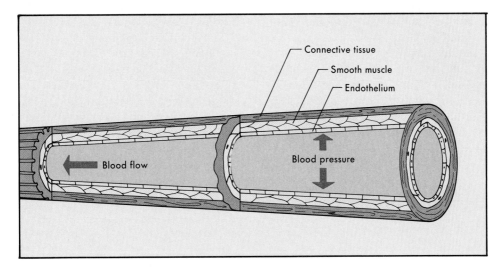

FIGURE 13-1   Functional elements in a simple hollow organ, such as many blood vessels, include (1) a connective tissue network to limit passively organ distension, (2) muscle cells circumferentially arranged to alter vessel circumference and thus resistance to blood flow, and (3) an endothelial lining that interfaces with the organ contents.

appear able to meet ATP requirements for contraction by oxidative phosphorylation, and no functionally important myosin variants are documented. However, smooth muscle from various tissues is characterized by an extraordinary range of distinguishing features. These differences mainly are related to the properties of the cell membranes (innervation, receptors, ion pumps and channels, and junctions) and the ways in which Ca^{++} is mobilized to regulate the crossbridges.

No scheme to classify smooth muscle into characteristic classes has been satisfactory, although it is useful to make the functional distinction between phasic and tonic muscles already noted. A good example of these is the phasic esophageal smooth muscle that functions similar to the striated muscle present in the upper esophagus. This smooth muscle is normally relaxed and contracts phasically during swallowing. Another example is the lower esophageal sphincter at the junction with the stomach. This muscle is tonic and normally contracted; it relaxes only to let food enter the stomach. However, all smooth muscles can exhibit both types of behavior, and many function somewhere between these extremes. This chapter emphasizes the mechanisms that underlie this range of behavior; other chapters describe smooth muscle function in specific organ systems.

STRUCTURE-FUNCTION RELATIONSHIPS IN HOLLOW ORGANS

Organ structure reflects the role of muscle in organ function. The simplest case may be an arteriole that consists of three elements: endothelial cell lining, a single smooth muscle cell encircling the endothelial cells, and connective tissue. Smooth muscle contraction can tonically maintain the diameter to determine flow. It causes blood flow to decrease by shortening or allows flow to increase by relaxation (see Chapter 22).

In the intestinal tract the lining of the tube is a mucosal layer involved in the digestion and absorption of nutrients. The muscle in the walls mixes and propels the contents (see Chapter 29). Two muscle layers are required: (1) a circular layer, as in most blood vessels, to determine circumference, and (2) a longitudinal layer to control length. Coordination depends on an elaborate neural network originating from plexuses between the two muscle layers.

A third category of organ is a sack in which the muscle is normally relaxed. The volume increases with the delivery of the contents. Contraction of the muscle empties the sack, and the movement is regulated by valves or sphincters. Examples are the heart, urinary bladder, and rectum. Complex layers of muscle accomplish large volume changes to empty the organ.

Control and coordination of the muscle depend on three systems: (1) the intrinsic and extrinsic innervation, (2) the blood supply that provides nutrients and circulating hormones, and (3) cell junctions that allow electrical, chemical, and mechanical interactions. The nature and relative importance of these three elements vary enormously.

STRUCTURE OF SMOOTH MUSCLE CELLS

Smooth muscle cells are large (although tiny compared with skeletal muscle cells): typically 2 to 5 μm in diameter and 100 to 400 μm long at the optimal length for force generation (L_o). They have a single central nucleus and taper toward the ends. The cells are comparatively featureless when viewed with a light microscope. Electron microscopy reveals that

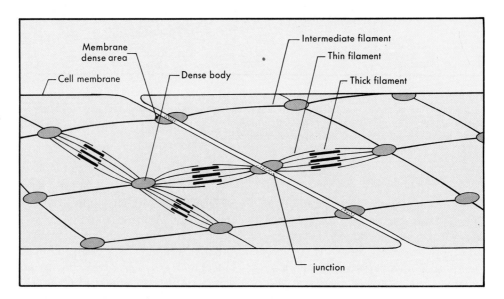

FIGURE 13-2   Possible organization of the cytoskeleton and myofilaments in smooth muscle. Three-dimensional imaging techniques are needed to reconstruct the structure of cells where the filaments are not organized into parallel arrays. Smooth muscle has a high thin-filament density and few thick filaments compared with striated muscle.

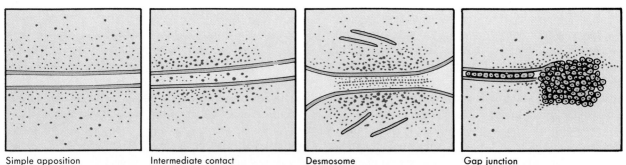

Simple apposition   Intermediate contact   Desmosome   Gap junction

FIGURE 13-3   Junctions between smooth muscle cells serve as mechanical links, and some types allow chemical and/or electrical coupling that provides communication between cells. Junctions also occur between smooth muscle cells and endothelial or other cells involved in integrated organ function.

most of the myoplasm is filled with thin filaments and relatively small numbers of thick filaments. The filaments are aligned in the long axis of the cell when the cell is at L_o, but shortening can lead to angular displacements. The thick filaments occur in small groups of three to five, and the equivalent of a myofibrillar sarcomere may consist of such groups with many interdigitating thin filaments (Figure 13-2). The absence of a precise transverse alignment of thick and thin filaments is responsible for the lack of striations. A cytoskeleton transmits the forces generated. The thin filaments are anchored to the *dense bodies,* or *membrane dense areas,* which are analogous to the Z-disks of striated muscle and contain the same major protein. The dense bodies are connected by intermediate filaments. The detailed organization of the contractile apparatus remains uncertain. However, this structure allows the contractile system to generate force in an arc.

Smooth muscle cells contain a sarcoplasmic reticulum in the form of a network or a fenestrated sheath of tubules close to the sarcolemma. Smooth muscle has no transverse-tubular system, and the sarcoplas-

mic reticulum does not form obvious junctions with the sarcolemma. Diffusion distances are short in the thin cells, and no special association is apparent between the sarcoplasmic reticulum and the myofilaments to give a myofibrillar organization.

Smooth muscle cells are linked to each other by a variety of junctions (Figure 13-3). Some junctions *(desmosomes)* mechanically link the contractile system of adjacent cells. In effect, the contractile apparatus is distributed throughout the tissue (Figure 13-2). Junctions also provide for electrical or chemical signaling if they have low-resistance pathways for ions or small molecules (e.g., *gap junctions*). This type of junction is common in tissues where coordinated phasic contractions occur, since they allow propagation of action potentials from cell to cell.

THE CONTROL SYSTEMS: INPUTS TO THE SARCOLEMMA

Smooth muscle cells are not organized in motor units with a motor nerve. Instead, an integrated and

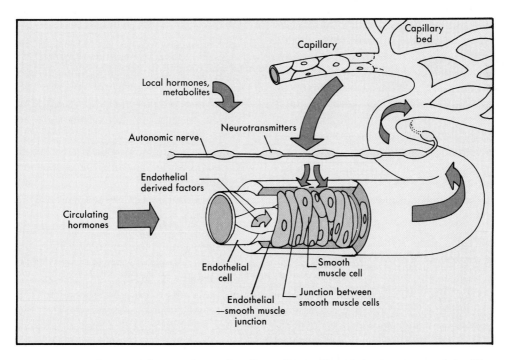

FIGURE 13-4   Inputs to the smooth muscle cell membrane. The relative importance of the different inputs varies widely among tissues (an arteriole is illustrated). Note that circulating hormones can affect vascular smooth muscle via the endothelial cells or can act directly on smooth muscle cells after leaving the circulation in the capillary beds.

coordinated response results from many inputs that can be both excitatory and inhibitory. These inputs include (Figure 13-4): autonomic nerves; circulating hormones or drugs; local hormones, ions, and metabolites; signals from other cell types such as endothelial cells; and signals from coupled smooth muscle cells. Details on the key mechanisms are given in the chapters devoted to specific organ systems.

Nerves are generally the most important input. However, a few smooth muscles receive no innervation, and virtually all continue to function more or less appropriately after the central autonomic connections are severed. Several factors influence neural control. Most smooth muscle tissues have more than one type of innervation, typically parasympathetic and sympathetic. Increasingly, more and more types of nerves and neurotransmitters are discovered, and the neural system of the gastrointestinal tract rivals that of the brainstem in complexity. Dual innervation usually acts reciprocally, with excitatory nerves causing contraction and inhibitory nerves producing relaxation. The neurotransmitter-releasing sites of the nerves may form intimate neuromuscular contacts with each smooth muscle cell, giving it specific control. However, neural contact may be fairly indirect, with long diffusion paths and gradients for the transmitters (Figure 13-4). The action of a particular class of nerve and its neurotransmitter on a specific smooth muscle cell depends on the receptors expressed for that neurotransmitter. Some cells respond to norepinephrine by contracting, whereas others relax, reflecting differences in the receptors and signal transduction mechanisms. These different responses allow coordinated, whole body adjustments. Under stress, for example, epinephrine is released from the adrenal gland. The vascular smooth muscle of the gut will constrict and shunt blood to the cardiac and skeletal muscle, where the vascular beds dilate.

PATTERNS OF Ca^{++} MOBILIZATION

The myoplasmic Ca^{++} concentration $[Ca^{++}]$ in smooth muscle regulates crossbridge interactions as in skeletal muscle. Ca^{++} mobilization from the sarcoplasmic reticulum is a quantal event; release is followed by reuptake, with no control over the steady-state concentration as in skeletal muscle (see Chapter 12). However, the mechanisms that regulate the myoplasmic $[Ca^{++}]$ are much more complex in smooth

FIGURE 13-5   Patterns of Ca^{++} mobilization in phasic **(A)** and tonic **(B)** contractions of smooth muscle. Phasic contractions show stimulus, $[Ca^{++}]$, and force relationships similar to skeletal muscle, although on a much slower time scale. In tonic contractions activation is maintained with continued receptor occupancy by neurotransmitters or other agents **(B)**. An initial Ca^{++} peak representing release from the sarcoplasmic reticulum plus influx from the extracellular pool is not maintained. During the sustained, tonic contraction the $[Ca^{++}]$ remains somewhat elevated and depends on extracellular Ca^{++}. The initial Ca^{++} transient gives fast contractions *(solid lines)* but does not affect the final force attained *(dashed lines)*. Thus Ca^{++} can affect the rate as well as the force of a contraction. This phenomenon, in which crossbridge cycling rates are regulated in a Ca^{++}-dependent manner, does not occur in skeletal muscle. The reasons reflect differences in how Ca^{++} regulates crossbridges in smooth muscle.

muscle. One factor involves the multiplicity of inputs, as already discussed. Another factor is the necessity for smooth muscle to regulate precisely the myoplasmic $[Ca^{++}]$ at submicromolar values during tonic contractions. Regulation of Ca^{++} in smooth muscle involves the sarcolemma as well as the sarcoplasmic reticulum. It also involves another Ca^{++} pool: the extracellular fluid, where the $[Ca^{++}]$ is about 1.6 mM.

Ca^{++} mobilization and regulation is schematized for phasic and tonic contractions in Figure 13-5. Phasic contractions are short and are elicited by brief periods of activation. This could be a propagating action potential or more typically a burst of action potentials; alternately, it could be a brief period of receptor occupancy by an activating neurotransmitter or hormone. The result is a transient increase in myoplasmic $[Ca^{++}]$ and a small contraction or a series of transient increases with a larger contraction analogous to a tetanus (Figure 13-5, *A*). Ca^{++} release from the sarcoplasmic reticulum and influx through the sarcolemma may both contribute to the transient increase. After the stimulus the myoplasmic Ca^{++} is resequestered in the sarcoplasmic reticulum or pumped out of the cell (Figure 13-6).

If the stimulus is prolonged by continuous nerve firing or by delivery of excitatory agents through other means so that receptor occupancy persists, the myoplasmic $[Ca^{++}]$ remains significantly elevated after an initial transient (Figure 13-5, *B*). The resulting contraction is fast and may remain at or near peak levels, despite the fall in the $[Ca^{++}]$. The initial transient

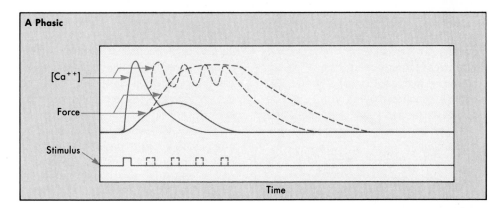

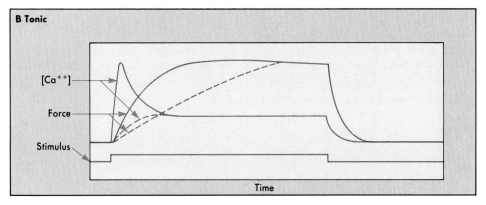

FIGURE 13-5   For legend see opposite page.

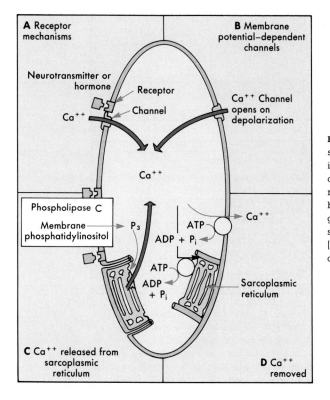

FIGURE 13-6   Mechanisms regulating myoplasmic [Ca^{++}] in smooth muscle. **A,** Receptor-operated channels allow Ca^{++} influx through the sarcolemma. **B,** Membrane potential–dependent Ca^{++} channels also lead to Ca^{++} influx. **C,** Ca^{++} release from the sarcoplasmic reticulum is probably caused by the second messenger, inositol 1,4,5-trisphosphate (IP$_3$) generated at the plasma membrane. **D,** Ca^{++} pumps in the sarcoplasmic reticulum and plasma membrane reduce the [Ca^{++}]. *ATP,* Adenosine trisphosphate; *ADP,* adenosine diphosphate; *P$_i$,* inorganic phosphate.

occurs in the absence of extracellular Ca^{++}, which shows that the transient largerly results from release of Ca^{++} from the sarcoplasmic reticulum. On the other hand, the sustained modest elevation in Ca^{++} is totally dependent on extracellular Ca^{++} (*dashed* line in Figure 13-5, *B*). The same force is generated in a much slower contraction.

SIGNAL TRANSDUCTION MECHANISMS IN THE SARCOLEMMA

How is the myoplasmic $[Ca^{++}]$ regulated by the inputs just discussed? Four mechanisms are involved: (1) membrane potential–dependent Ca^{++} influx from the extracellular space, (2) receptor-activated channels in the sarcolemma, (3) control of sarcoplasmic reticulum Ca^{++} release, and (4) sequestration and extrusion of Ca^{++} by pumps in both membranes (Figure 13-6).

The membrane potential in smooth muscle is the sum of two processes. A major factor is the *Donnan potential*, which reflects the relative permeabilities and concentration gradients for K^+ and Na^+ across the sarcolemma (as in skeletal muscle). However, the Na^+, K^+ pump extrudes three Na^+ ions in exchange for two K^+ ions and thereby results in the transfer of one positive charge out of the cell for each cycle. This process may add -20 mV, to give a normal membrane potential of -50 to -70 mV. Changes in the activity of the Na^+, K^+ pump are responsible for slow oscillations in the membrane potential (Figure 13-7). The ion channels in the sarcolemma that open on depolarization in smooth muscle are primarily permeable to Ca^{++} ions. Some smooth muscles (mainly phasic) generate action potentials at some critical level of depolarization (Figure 13-7, *A* and *B*). Ca^{++} influx is the current-carrying ion in these action potentials. Other smooth muscles (functionally tonic) do not generate action potentials. Nevertheless, these cells have potential-dependent Ca^{++} channels. Graded depolarization caused by reduced electrogenic ion

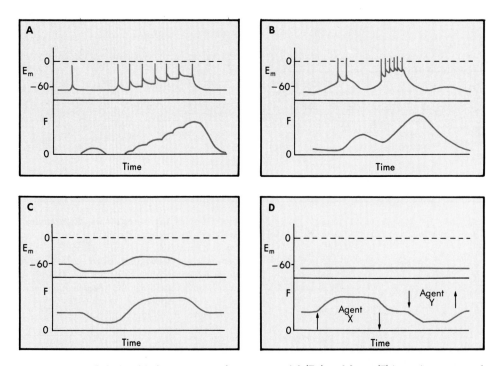

FIGURE 13-7   Relationship between membrane potential *(E_m)* and force *(F)* in various types of smooth muscle. **A,** Action potentials may be generated by pacemaker cells and propagated through the tissue. **B,** Slow waves reflecting oscillations in the Na^+, K^+ pump activity can trigger bursts of action potentials and rhythmic contractile activity. **C,** Tone will vary with E_m in tonic tissues. Action potentials are not normally generated in tonic tissues such as most vascular smooth muscle. **D,** Receptor-mediated mechanisms can alter cell $[Ca^{++}]$ without detectable changes in the E_m.

pumping will increase Ca^{++} influx (Figure 13-7, *C*).

Receptors mediate two paths for Ca^{++} mobilization (see Figure 13-6, *A* and *C*). The binding of an agent to a receptor can increase the Ca^{++} permeability of the sarcolemma by opening *receptor-activated channels*. This allows influx of Ca^{++} from the extracellular pool. Receptor occupancy can also cause Ca^{++} release from the sarcoplasmic reticulum. Such receptor-mediated mechanisms can increase (or decrease) myoplasmic $[Ca^{++}]$ to alter tone without detectable changes in the membrane potential (Figure 13-7, *D*). This is called *pharmacomechanical coupling* to distinguish it from activation mechanisms that involve changes in the membrane potential, or *excitation-contraction coupling*.

A chemical messenger is implicated in Ca^{++} release from the sarcoplasmic reticulum, unlike in skeletal muscle, where charge coupling between the closely associated T-tubular system and the sarcoplasmic reticulum causes release. Phospholipase A is activated by occupancy of specific receptors and hydrolyzes membrane phosphatidylinositol to yield *inositol 1,4,5-trisphosphate*. The latter compound diffuses to the sarcoplasmic reticulum to induce Ca^{++} release.

The last class of mechanisms that affect myoplasmic $[Ca^{++}]$ is the membrane pumps that actively transport Ca^{++} back into the sarcoplasmic reticulum or extrude it into the extracellular space. Drugs, neurotransmitters, or hormones that inhibit contraction or induce relaxation in smooth muscle act by various mechanisms, including reduction of Ca^{++} influx or Ca^{++} release from the sarcoplasmic reticulum, as well as enhancement of Ca^{++} pump activity.

Ca^{++} AND CROSSBRIDGE REGULATION IN SMOOTH MUSCLE

Smooth muscle lacks troponin—the Ca^{++}-binding, thin-filament, regulatory protein of skeletal and cardiac muscle. The available evidence suggests that

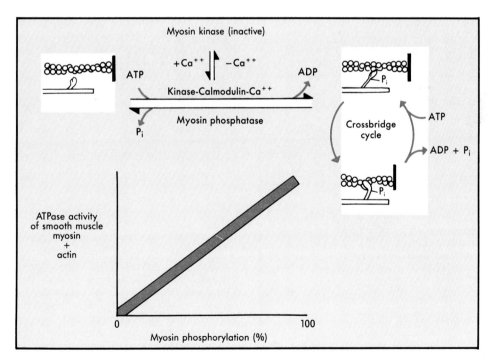

FIGURE 13-8   A covalent regulatory mechanism in which phosphorylation of the crossbridges regulates attachment and cycling in smooth muscle. Phosphorylation is proportional to the $[Ca^{++}]$, which determines the activity of myosin kinase. If the myoplasmic $[Ca^{++}]$ is lowered, Ca^{++} and calmodulin dissociate from myosin kinase, and the crossbridges are dephosphorylated by myosin phosphatase. This scheme explains the dependence of smooth muscle actomyosin ATPase activity in vitro on myosin phosphorylation.

regulation occurs at the crossbridge itself (Figure 13-8). The mechanism is very different from the allosteric regulation of striated muscle; it involves phosphorylation of the crossbridge at a site on its regulatory light chain. This covalent regulatory mechanism uses ATP as the phosphate donor.

In smooth muscle the crossbridges cannot attach to the thin filament and cycle, unless they are phosphorylated by a specific enzyme, *myosin kinase*. The active form of myosin kinase is a complex with Ca^{++}-calmodulin (Figure 13-8). *Calmodulin* is a cytoplasmic protein that has four high-affinity Ca^{++}-binding sites and participates in the activation of several Ca^{++}-dependent enzymes. The phosphorylated crossbridge cycles with no further need for Ca^{++} until it is dephosphorylated by *myosin phosphatase*.

The regulatory scheme illustrated in Figure 13-8 is incomplete, since it postulates that phosphorylation is a simple switch to turn on a crossbridge. The model cannot explain the phenomenon illustrated in Figure 13-5, *B,* in which crossbridge cycling rates depend on Ca^{++}. It was only recently discovered that crossbridge function in smooth muscle has features that are not present in skeletal muscle. These reflect a more complex regulatory process that is considered in the next section.

Contractile System Function

The same two steady-state relationships that describe contractile system function in skeletal muscle are applicable to smooth muscle: the force-length and velocity-load relationships. Force varies with tissue or cell length, as in skeletal muscle, and the maximal stress at L_o is similar or somewhat greater in smooth muscle (Figure 13-9). A significant passive force is present at L_o that reflects the connective tissue component. Although smooth muscle can shorten more than skeletal muscle in the body, other factors such as wall thickening contribute to large changes in organ volume (Figure 13-9).

Although smooth muscle can generate a high stress, its contraction velocities are very low (Figure 13-10). In fact, the specific ATPase activity of phosphorylated smooth muscle myosin is more than 100-fold lower than the fast skeletal muscle myosin isoenzyme. However, when plotted on an appropriate scale, the velocity-stress curves exhibit the characteristic hyperbolic fall in velocities with load. Furthermore, the power output curves and the ability to resist an imposed load of 1.6 times the load that can be generated are the same as illustrated for skeletal muscle in Figure 11-5. The similarity of the force-length and velocity-load relationships provides strong

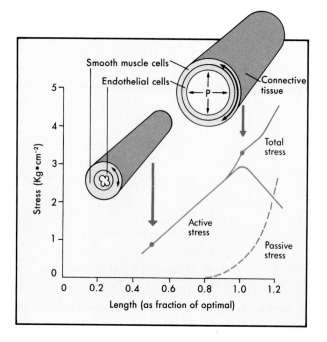

FIGURE 13-9   Force-length and pressure-volume behavior of smooth muscle. The active and passive force-length curves for a strip of smooth muscle are similar to those of skeletal muscle. Diagram representing a small arteriole shows how shortening of the smooth muscle by only 50% can give a 36-fold reduction in lumen area and stop red cell flow. The smooth muscle cannot develop as much force at the shortened length. However, if the blood pressure remains the same, the force on the smooth muscle cells tending to distend the vessel also falls for geometric reasons. This force, indicated by the double-headed arrows, equals P(r/w), where P = pressure, r = radius, and w = wall thickness (Laplace's law). As the vessel radius increases, the load is increasingly resisted by the connective tissues.

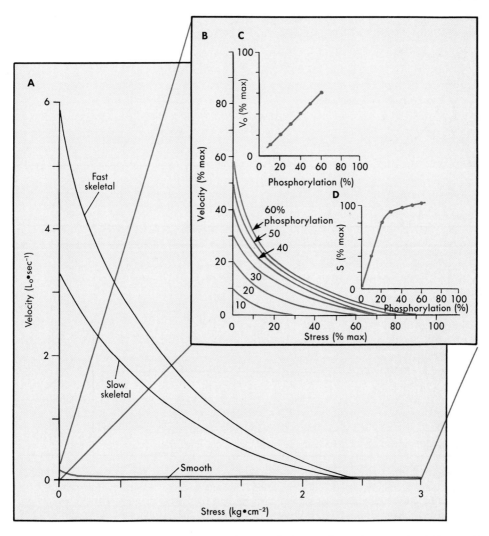

FIGURE 13-10   A, Comparison of velocity-load curves for human fast and slow skeletal fibers and smooth muscle. **B,** Instead of a unique velocity-load curve, smooth muscle has a family of curves determined by the level of crossbridge phosphorylation. **C,** Maximal shortening velocities (intercepts on the ordinate) are directly dependent on phosphorylation. **D,** Active stress (abscissa intercepts) rises rapidly with phosphorylation, and near maximal stress may be generated with only 25% to 30% of the crossbridges in the phosphorylated state.

evidence that the sliding filament–crossbridge mechanism operates in smooth muscle.

Figure 13-10 shows that covalent regulation by crossbridge phosphorylation is associated with new properties. Average crossbridge cycling rates are regulated. Shortening velocities are much higher when the myoplasmic $[Ca^{++}]$ and phosphorylation are elevated, even though force may not increase significantly. Fairly low $[Ca^{++}]$ and phosphorylation support high levels of force. These steady-state relationships

explain the different time courses of a phasic and tonic contraction (Figure 13-11).

Crossbridge Regulation

The properties illustrated in Figures 13-10 and 13-11 confer advantages for muscle in the walls of hollow organs with diverse functions. With appropriate modulation of the myoplasmic $[Ca^{++}]$ (determined by the membranes and their inputs), comparatively rapid phasic contractions are possible. However, tonic con-

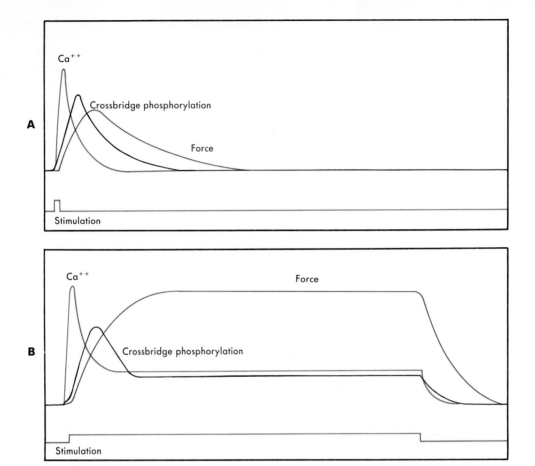

FIGURE 13-11   Time course of events in activation and contraction of smooth muscle. **A,** Phasic contraction following brief period of stimulation. The initial event is a transient increase in Ca^{++}, followed by crossbridge phosphorylation, cycling, and force development. These changes are reversed as myoplasmic $[Ca^{++}]$ is restored to resting levels. **B,** In a tonic contraction produced by sustained stimulation, the $[Ca^{++}]$ and phosphorylation levels remain somewhat elevated after an initial peak (allowing rapid force development). In this case the force (tone) is maintained with reduced crossbridge cycling rates, as manifested by lower shortening velocities, lower ATP consumption, and an overall increase in economy (despite the ATP used for phosphorylation).

tractions are also possible with reduced crossbridge turnover rates. The problem is how to explain this behavior. A very simple modification of the model for crossbridge regulation depicted in Figure 13-8 may explain the characteristics of smooth muscle. This modification is only possible with a covalent regulatory mechanism (Figure 13-12).

Two crossbridge cycles can occur with four crossbridge states. One is a comparatively rapid cycle characteristic of phosphorylated crossbridges (state 2 to 3 to 2) in which one ATP is consumed. A second slow cycle occurs via states 1, 2, 3, and 4 to 1. Two ATP molecules are hydrolyzed in this cycle: one for

chemomechanical transduction and one for phosphorylation. At high Ca^{++} and phosphorylation values, most crossbridges will be phosphorylated and will cycle rapidly. At low Ca^{++} and phosphorylation levels, crossbridges in state 4 predominate because their detachment is slow. ATP used for phosphorylation and dephosphorylation is not captured as mechanical work. Thus covalent regulation reduces the efficiency of contraction in smooth muscle. This disadvantage is offset by the gains in the economy of a tonic contraction when crossbridge turnover is reduced. Total ATP consumption rates are low in smooth muscle, and requirements of the contractile

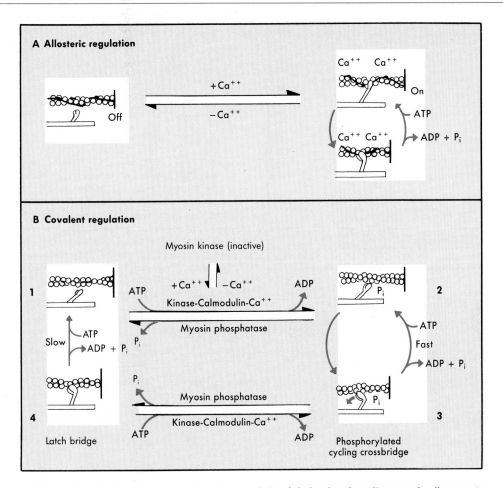

FIGURE 13-12   A, Allosteric regulation characteristic of skeletal and cardiac muscle allows two crossbridge states: free and attached. **B,** Four crossbridge states are present if myosin kinase and phosphatase act on both free and attached crossbridges in smooth muscle. This model assumes that an attached, dephosphorylated crossbridge (state 4) is identical to a phosphorylated crossbridge (state 3), except that its detachment rate is slower. This model predicts the behavior illustrated in Figures 13-5, 13-10, and 13-11).

system are met by oxidative phosphorylation. Smooth muscles do not exhibit fatigue as long as the blood supply is intact.

CARDIAC MUSCLE

The muscle in the heart has unique characteristics (see Chapter 17). The cells are striated and possess a troponin-based regulatory system on their thin filaments. The myosin isoenzymes differ from striated muscle, but the cells have metabolic and contractile properties comparable to those of slow skeletal muscle, with a very high oxidative capacity. However, the heart is a hollow organ. As with smooth muscle cells, cardiac muscle cells are small, with one central nucleus, and are connected by specialized junctions that provide both electrical and mechanical coupling.

This mixture of skeletal and smooth muscle characteristics is well suited for the special function of this organ. The heart is a pump that must contract and relax fairly rapidly; thus it must generate a high power output during shortening to eject the blood. The efficiency of chemomechanical transduction is important, but because force is generated only briefly for each heartbeat, the economy of force maintenance is irrelevant.

Excitation-contraction coupling has some special

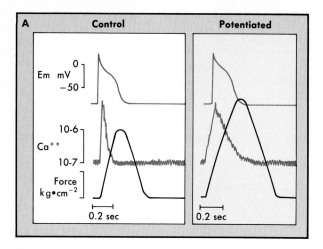

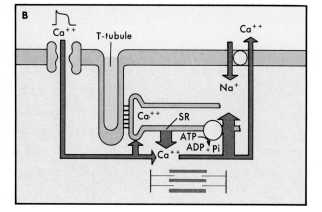

FIGURE 13-13   Regulating the force of the twitch in cardiac muscle. **A,** The Ca^{++} transient triggered by the membrane action potential (E_m) is normally insufficient to saturate all the binding sites on troponin and thus allow all the cross-bridges to cycle. Certain hormones and drugs can enhance Ca^{++} sequestration in the sarcoplasmic reticulum. Subsequent action potentials release more Ca^{++}, and the twitch is potentiated. **B,** Ca^{++} influx during the action potential helps trigger Ca^{++} release from the sarcoplasmic reticulum. The activity of Ca^{++} pumps, exchangers, and channels can be regulated to change the amount of Ca^{++} in the sarcoplasmic reticulum and the size of the Ca^{++} transient. The affinity of troponin for Ca^{++} can also be altered in a coordinated regulatory step.

Table 13-1   Mechanisms for Grading Contractile Force

General Mechanism	Occurrence		
	Skeletal	Cardiac	Smooth*
Recruit more cells (motor units)	+	−	(+)
Sum twitches by increasing stimulation frequency (tetanus)	+	−	+
Alter filament overlap by stretch	(+)	+	+
Vary twitch by changing Ca^{++} transient	−	+	+
Alter Ca^{++} sensitivity of regulatory systems	−	+	?
Tonic depolarization and activation of potential-dependent channels without action potentials	−	−	+
Receptor-activated channels (pharmacomechanical coupling)	−	−	+

*The relative importance of these mechanisms varies greatly with the type of smooth muscle.
Parentheses indicate the possibility of little physiological importance.

features associated with cardiac function (Figure 13-13). The action potential is prolonged because the sarcolemma has channels for both Na^+ and Ca^{++} ions. Influx of Ca^{++} through the sarcolemma delays repolarization. Entry of small amounts of extracellular Ca^{++} triggers release of the main pool of activator Ca^{++} from the sarcoplasmic reticulum (see Chapter 17). The resulting Ca^{++} transient induces the heartbeat (i.e., a twitch). The prolonged action potential followed by a refractory period prevents tetanization of cardiac muscle. Thus the only way that cardiac output can be regulated is to alter either the heart rate or the force of each contraction. Unlike skeletal muscle, the heart has several mechanisms to vary the twitch force (Figure 13-13). One mechanism is to alter the activity of the Ca^{++} pumps and exchangers or the open time of channels in the membrane systems to change the amount of Ca^{++} in the sarcoplasmic reticulum. This is the Ca^{++} available for release following the depolarization. The other mechanism increases the sensitivity of the contractile apparatus to the Ca^{++} released.

Table 13-1 compares the mechanisms for grading the force of contraction in skeletal, cardiac, and smooth muscles.

BIBLIOGRAPHY
Journal Articles

Bagby R: Towards comprehensive three-dimensional model of the contractile system of vertebrate smooth muscle cells, Int Rev Cytol 105:67, 1986.

Gabella G: Structural apparatus for force transmission in smooth muscles, Physiol Rev 64:455, 1984.

Murphy RA: Special topic: contraction in smooth muscle cells, Annu Rev Physiol 51, 1989.

Schwartz SM et al: Replication of smooth muscle cells in vascular disease, Circ Res 58:427, 1986.

Somlyo AP: Excitation-contraction coupling and the ultrastructure of smooth muscle, Circ Res 57:497, 1985.

Somlyo AV et al: Cross-bridge kinetics, cooperativity, and negatively stained cross-bridges in vertebrate smooth muscle, J Gen Physiol 91:165, 1988.

Somlyo AP et al: Inositol trisphosphate, calcium and muscle contraction, Philos Trans R Soc Lond (Biol) 320:399, 1988.

Winegrad S: Regulation of cardiac contractile proteins, Circ Res 55:565, 1984.

Books and Monographs

Abramson DI and Dobrin PB, editors: Vessels and lymphatics in organ systems, Orlando, Fla, 1984, Academic Press, Inc.

Bohr DF et al, editors: Handbook of physiology, section 2: The cardiovascular system, vol II: Vascular smooth muscle, Bethesda, Md, 1980, American Physiological Society.

Grover AK and Daniel EE, editors: Calcium and contractility: smooth muscle, Clifton, NJ 1985, Humana Press.

Grundy D: Gastrointestinal motility: the integration of physiological mechanisms, Lancaster, UK, 1985, MTP Press, Ltd.

Hartshorne DJ: Biochemistry of the contractile process in smooth muscle. In Johnson LR, editor: Physiology of the gastrointestinal tract, ed 2, New York, 1987, Raven Press.

Netter FH: Heart. In Yonkman FF, editor: The Ciba collection of medical illustrations, vol 5, Summit, NJ, 1969, Ciba-Geigy Corp.

Rüegg JC: Calcium in muscle activation: a comparative approach, Berlin, 1986, Springer-Verlag.

Sellers JR and Adelstein RS: Regulation of contractile activity. In Boyer PD and Krebs EG, editors: The enzymes, Orlando, Fla, 1987, Academic Press, Inc.

Siegman MJ et al: Regulation and contraction of smooth muscle, New York, 1987, Alan R Liss, Inc.

Sperelakis N, editor: Physiology and pathophysiology of the heart, ed 2, Boston, 1989, Kluwer Academic Publishers.

CARDIOVASCULAR SYSTEM

ROBERT M. BERNE
MATTHEW N. LEVY

PART IV

CHAPTER 14

Blood and Hemostasis

BLOOD

The main function of the circulating blood is to carry oxygen and nutrients to the tissues and to remove carbon dioxide and waste products. However, blood also transports other substances (e.g., hormones) from their site of formation to their site of action and white blood cells and platelets to where they are needed. In addition, blood aids in the distribution of heat and thus contributes to *homeostasis,* a constancy of the body's internal environment.

Blood is a suspension of red cells, white cells, and platelets in a complex solution *(plasma)* of gases, salts, proteins, carbohydrates, and lipids. The circulating blood volume is about 7% of body weight. Approximately 55% of the blood is plasma whose protein content is 7 g·dl^{-1} (about 4 g·dl^{-1} of albumin and 3 g·dl^{-1} of immunoglobulins).

Blood Components

Erythrocytes

The erythrocytes (red cells) are anuclear, flexible, biconcave disks that transport oxygen to the body tissues (Figure 14-1). The erythrocytes average 7 μm in diameter and 5 million·dl^{-1} in number. They arise from stem cells in the bone marrow, and during maturation they lose their nuclei before entering the circulation, where their average life span is 120 days.

The main protein in erythrocytes is *hemoglobin* (about 15 g·dl^{-1} of blood) which consists of *heme,* an iron containing tetrapyrrole, linked to *globin,* a protein composed of four polypeptide chains (two α and two β in the normal adult). The iron moiety of hemo-

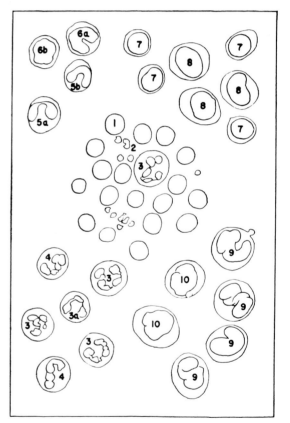

FIGURE 14-1 The morphology of blood cells. *1,* Normal red cells. *2,* Platelets. *3,* Neutrophil, adult. *3a.* Neutrophil, adult (two lobes). *4,* Neutrophil, band form. *5a,* Eosinophil, two lobes. *5b,* Eosinophil, band form. *6a,* Basophil, band form. *6b,* Metamyelocyte, basophilic. *7,* Lymphocyte, small. *8,* Lymphocyte, large. *9,* Monocyte, mature. *10,* Monocyte, young. (From Daland GA: A color atlas of morphologic hematology, Cambridge, Mass, 1951. With permission, Harvard University Press.)

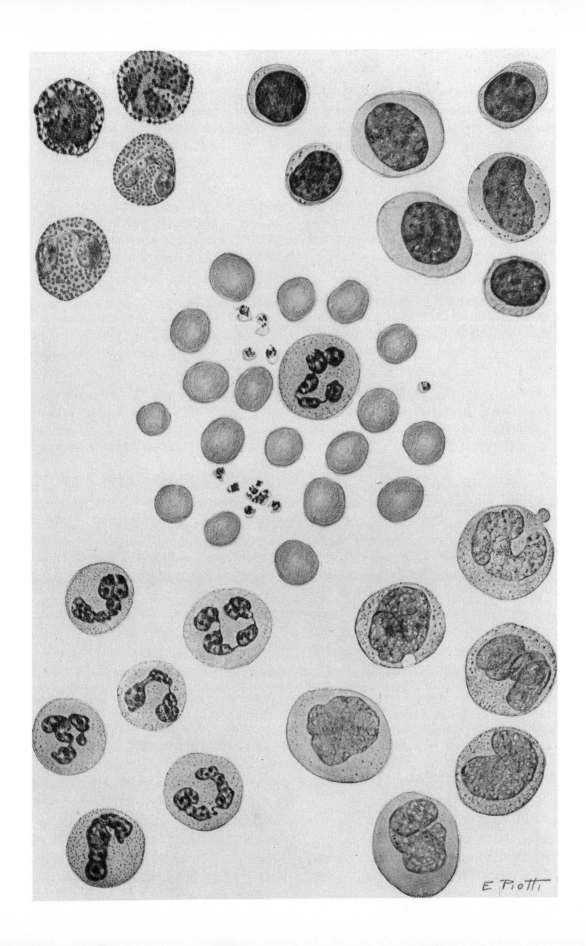

E. Piotti

globin binds loosely and reversibly to oxygen to form *oxyhemoglobin*. The affinity of hemoglobin for oxygen is affected by pH, temperature, and 2,3-diphosphoglycerate concentration. These factors facilitate O_2 uptake in the lungs and its release in the tissues (see Chapter 27). Changes in the polypeptide subunits of globin can also affect the affinity of hemoglobin for O_2 (e.g., fetal hemoglobin has two γ chains instead of two β chains and has greater affinity for O_2), or they can result in disease states, such as sickle cell anemia or thalassemia.

The number of circulating red cells is fairly constant under normal conditions. The production of erythrocytes *(erythropoiesis)* is regulated by the glycoprotein *erythropoietin,* which is secreted mainly by the kidneys. Erythropoietin acts by accelerating the differentiation of stem cells in the bone marrow. Anemia and chronic hypoxia (e.g., living at high altitudes) stimulate erythrocyte production and can produce *polycythemia* (increased number of red cells). When the hypoxic stimulus is removed in subjects with altitude polycythemia, the high red cell concentration in the blood inhibits erythropoiesis.

Leukocytes

There are normally 4000 to 10,000 leukocytes (white blood cells)·μL^{-1} of blood. The leukocytes include granulocytes (65%), lymphocytes (30%), and monocytes (5%). Of the granulocytes, about 95% are neutrophils, 4% eosinophils, and 1% basophils. White blood cells originate from the primitive stem cells in the bone marrow (Figure 14-1). After birth the granulocytes and monocytes continue to originate in the bone marrow, whereas the lymphocytes take origin in lymph nodes, spleen, and thymus.

The granulocytes and the monocytes are motile, nucleated cells that contain *lysosomes,* which in turn contain enzymes capable of digesting foreign material such as microorganisms, damaged cells, and cellular debris. Thus the leukocytes constitute a major defense mechanism against infections. Microorganisms or the products of cell destruction release *chemotactic* substances that attract granulocytes and monocytes. When the migrating leukocytes reach the foreign agents, they engulf them *(phagocytosis)* and then destroy them by action of enzymes that form oxygen-derived free radicals and hydrogen peroxide.

Lymphocytes

The lymphocytes vary in size, have large nuclei, and most lack cytoplasmic granules (Figure 14-1).

The two main types are *B lymphocytes,* which confer humoral immunity, and *T lymphocytes,* which confer cell-mediated immunity. When stimulated by an *antigen,* the B lymphocytes are transformed into *plasma cells,* which synthesize and secrete antibodies (gamma globulin), which are carried by the bloodstream to their site of action. The main T lymphocytes are cytotoxic and are responsible for long-term protection against some viruses, bacteria, and cancer cells. They are also responsible for the rejection of transplanted organs. Other T lymphocytes are *helper T cells,* which activate B cells, and *suppressor T cells,* which inhibit B cell activity. Special B and T lymphocytes, called *memory cells,* "remember" specific antigens. These cells can quickly generate an immune response when subsequently exposed to the same antigen.

Platelets

The platelets are small (3 μm) anuclear cell fragments of *megakaryocytes.* The megakaryocytes reside in the bone marrow, and when mature they break up into platelets and enter the circulation. The platelets are important in hemostasis, as discussed later.

Blood Groups

In humans there are four principal blood groups, designated O, A, B, and AB. The plasma of group O blood contains antibodies to group A, group B, and group AB cells. Group A plasma contains antibodies to group B cells, and group B plasma contains antibodies to group A cells. Group AB plasma has no A, B, O antibodies to cells of group O, A, or B. In blood transfusions crossmatching is necessary to prevent agglutination of donor red cells in the recipient. Thus the recipient must not have antibodies to the red cells of the donor. Because plasma of groups A, B, and AB have no A, B, O antibodies to group O cells, people with group O blood are called *universal donors.* Conversely, persons with AB blood are called *universal recipients* because their plasma has no A, B, O antibodies to cells of the other three groups.

In addition to the ABO blood grouping, there are Rh (rhesus factor)-positive and Rh-negative groups. An Rh-negative person can develop antibodies to Rh-positive cells if exposed to Rh-positive blood. This can occur during pregnancy if the mother is Rh negative and the fetus is Rh positive (inherited from the father). Rh-positive red cells from the fetus can enter the maternal bloodstream at the time of placental separation and induce Rh-positive antibodies in the mother's plasma. The Rh-positive antibodies from the mother can reach the fetus also via the placenta and agglutinate and hemolyze fetal red cells (hemolytic disease of the newborn, called *erythroblastosis fetalis*). Red cell destruction can also occur in Rh-negative individuals who have previously been transfused with Rh-positive blood and have developed Rh antibodies. If these individuals are given a subsequent transfusion of Rh-positive blood, the transfused red cells will be destroyed by the Rh antibodies in their plasma.

HEMOSTASIS

When blood vessels are damaged, three processes act to stem the flow of blood: vasoconstriction, platelet aggregation, and blood coagulation.

Vasoconstriction

Injury to a blood vessel elicits a contractile response of the vascular smooth muscle and thus a narrowing of the vessel. Vasoconstriction in severed arterioles or small arteries can completely obliterate the lumen of the vessel and stop the flow of blood. The contraction of the vascular smooth muscle is probably caused by direct mechanical stimulation as well as by mechanical stimulation of perivascular nerves.

Platelets

Damage to the endothelium of a blood vessel engenders platelet adherence at the site of injury. The adherent platelets release *adenosine diphosphate* and *thromboxane A_2*, which produce adherence of additional platelets. The aggregation of platelets may continue in this manner until some of the small blood vessels become blocked by the mass of aggregated platelets. Extension of the platelet aggregate along the vessel is prevented by the antiaggregation action of *prostacyclin*. This substance is released from the normal endothelial cells in the adjacent, uninjured part of the vessel. Platelets also release *serotonin* (5-hydroxytryptamine), which enhances vasoconstriction, as well as *thromboplastin,* which hastens blood coagulation. When *thrombocytopenia* (reduced number of platelets) occurs, tiny hemorrhages *(petechiae)* may appear in the skin and mucous membranes. Hemorrhage into subcutaneous tissue *(ecchymosis)* after minor trauma is also common in thrombocytopenia.

Blood Coagulation

The clotting of blood is a complex process consisting of sequential activation of various factors that are present in an inactive state in the blood. The cascade of reactions in which one activated factor activates another and so on is depicted in Figure 14-2. Several of the factors are synthesized in the liver, as is vitamin K, which is essential for synthesis of these liver-derived clotting factors.

The key step in blood clotting is the conversion of fibrinogen to fibrin by thrombin. The clot that is formed by this reaction consists of a dense network of fibrin strands in which blood cells and plasma are trapped (Figure 14-3). The two blood coagulation pathways, the *extrinsic pathway* and the *intrinsic pathway,* converge on the activation of factor X, which catalyzes the cleavage of prothrombin to thrombin (Figure 14-2). Blood clotting via the extrinsic pathway is initiated by tissue damage and the release of tissue thromboplastin. Blood clotting via the intrinsic pathway is initiated by exposure of the blood to a negatively charged surface. This can occur within blood vessels when the endothelium is damaged and blood comes in contact with collagen. Alternatively, it can occur outside the body when blood comes in contact with negatively charged surfaces such as glass. If blood is carefully drawn into a syringe coated with silicone, clotting is greatly delayed.

After a clot is formed, the actin and myosin of the platelets trapped in the fibrin mesh interact in a manner similar to that in muscle. The resultant contraction pulls the fibrin strands toward the platelets, thereby extruding the *serum* (plasma without fibrinogen) and shrinking the clot. The process is called *clot retraction.* The function of clot retraction is not clear, but it may serve to approximate the edges of severed blood vessels.

Several cofactors are required for blood coagulation (Figure 14-2), the most important being calcium. If the calcium ions in blood are removed or bound, coagulation will not occur.

Clot Lysis

Blood clots may be liquified *(fibrinolysis)* by a proteolytic enzyme called *plasmin.* Normal blood con-

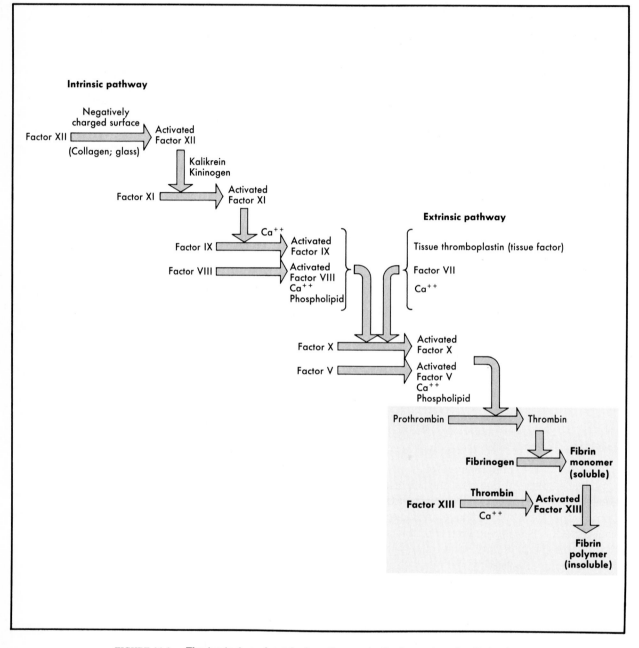

FIGURE 14-2   The intrinsic and extrinsic pathways in the formation of a fibrin clot.

FIGURE 14-3   Human blood clot showing red blood cells immobilized within a network of fibrin threads. The small spheres are platelets. Scanning electron micrograph (×9000). (Reproduced with permission from Shelly WB: Of red cell bondage, JAMA 249:3089. Copyright 1983, American Medical Association.)

tains *plasminogen,* an inactive precursor of plasmin. Activators of the conversion of plasminogen to plasmin are found in tissues, plasma, and urine *(urokinase).* Exogenous plasminogen activators, such as *streptokinase* and *tissue plasminogen activator (tPA),* are used clinically to dissolve intravascular clots, especially in coronary arteries.

Anticoagulants

Blood coagulation can be prevented in vitro by the addition of *citrate* or *oxalate,* which removes the calcium ions from solution. *Heparin,* a sulfated polysaccharide produced by mast cells, is commercially available and is used for rapid anticoagulation. It is used in extracorporeal circuits during open-heart surgery and for prevention of intravascular clot extension. For prolonged anticoagulation, *dicumarol* is used. This drug inhibits the synthesis of vitamin K–dependent factors and is used in treating such conditions as thrombophlebitis (inflammation of a vein associated with an intravascular blood clot).

BIBLIOGRAPHY
Journal Articles

Jackson CM and Nemerson Y: Blood coagulation, Annu Rev Biochem 49:765, 1980.

Shattil SJ and Bennett JS: Platelets and their membranes in hemostasis: physiology and pathophysiology, Ann Intern Med 94:108, 1981.

Books and Monographs

Babior BM and Stossel TP: Hematology: a pathophysiological approach, New York, 1984, Churchill Livingstone Inc.

Eastham RD: Clinical haematology, ed 6, Bristol, 1984, John Wright/PSG Inc.

Erslev AJ and Gabuzda TG: Pathophysiology of blood, ed 3, Philadelphia, 1985, WB Saunders Co.

Ogston D: The physiology of hemostasis, Cambridge, 1983, Harvard University Press.

Ratnoff OD and Forbes CD, editors: Disorders of hemostasis, Orlando, Fla, 1984, Grune & Stratton, Inc.

CHAPTER 15

The Circuitry

The cardiovascular system is made up of a pump, a series of distributing and collecting tubes, and an extensive system of thin vessels that permit rapid exchange of substances between the tissues and the vascular channels. The heart consists of two pumps in series: (1) the right ventricle, to propel blood through the lungs for exchange of oxygen and carbon dioxide; and (2) the left ventricle, to propel blood to all other tissues of the body. Unidirectional flow through the heart is achieved by the appropriate arrangement of effective flap valves. Although the cardiac output is intermittent, continuous flow to the periphery is accomplished by distension of the aorta and its branches during ventricular contraction (systole) and elastic recoil of the walls of the large arteries with forward propulsion of the blood during ventricular relaxation (diastole). Blood moves rapidly through the aorta and its arterial branches. The branches become narrower and their walls become thinner and change histologically toward the periphery. The aorta is predominantly an elastic structure. However, the peripheral arteries are more muscular, and in the arterioles the muscular layer predominates (Figure 15-1).

From the aorta to the small arteries and the arterioles, frictional (viscous) resistance to blood flow is relatively small, and the pressure drop from the root of the aorta to these vessels is also relatively small (Figure 15-2). The arterioles, the "stopcocks" of the vascular tree, are the principal points of resistance to blood flow in the circulatory system. The large resistance offered by the arterioles is reflected by the considerable fall in pressure from arterioles to capillaries. The amount of contraction in the circular muscle of these small vessels regulates tissue blood flow and aids in controlling the arterial blood pressure.

In addition to a sharp reduction in pressure across the arterioles, flow changes from pulsatile to steady. The pulsatile arterial blood flow, caused by the intermittent cardiac ejection, is damped at the capillary level by the combinaton of distensibility of the large arteries and frictional resistance in the arterioles. Many capillaries arise from each arteriole. Therefore the total cross-sectional area of the capillary bed is very large, despite the cross-sectional area of each capillary being less than that of each arteriole. As a result, blood flow slows in the capillaries, analogous to the decrease in flow velocity at the wide regions of a river. Because the capillaries consist of short tubes with walls only one cell thick and because flow velocity is slow, conditions in the capillaries are ideal for the exchange of diffusible substances between blood and tissue.

On its return to the heart from the capillaries, blood passes through venules and then through veins of increasing size. As the heart is approached, the number of veins decreases, the thickness and composition of the vein walls change (Figure 15-1), the total cross-sectional area of the venous channels diminishes, and the velocity of blood flow increases (Figure 15-2). Also, most of the blood in the systemic circulation is located in the venous vessels (Figure 15-2). Conversely, the blood in the pulmonary vascular bed is about equally divided among the arterial, capillary, and venous vessels.

Blood entering the right ventricle from the right atrium is pumped through the pulmonary arterial system at a mean pressure about one-seventh that in the systemic arteries. The blood then passes through the

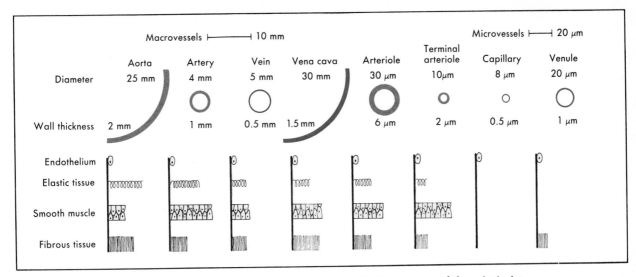

FIGURE 15-1   Internal diameter, wall thickness, and relative amounts of the principal components of the vessel walls of the various blood vessels that compose the circulatory system. Cross sections of the vessels are not drawn to scale because of the huge range from aorta and venae cavae to capillary. (Redrawn from Burton AC: Physiol Rev 34:619, 1954.)

lung capillaries, where carbon dioxide is released and oxygen is taken up. The oxygen-rich blood returns via the pulmonary veins to the left atrium and ventricle to complete the cycle. The systemic and pulmonary circulation systems are diagrammed in Figure 15-3.

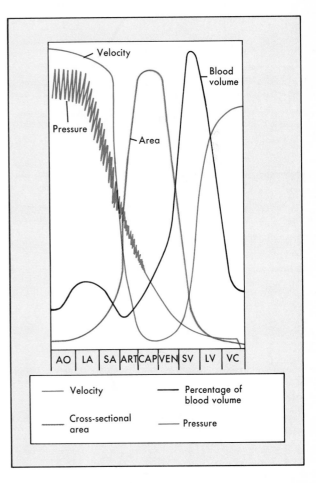

FIGURE 15-2   Pressure, velocity of flow, cross-sectional area, and capacity of the blood vessels of the systemic circulation. The important features are the inverse relationship between velocity and cross-sectional area, the major pressure drop across the arterioles, the maximal cross-sectional area and minimal flow rate in the capillaries, and the large capacity of the venous system. The small but abrupt drop in pressure in the venae cavae indicates the point of entrance of these vessels into the thoracic cavity and reflects the effect of the negative intrathoracic pressure. To permit schematic representation of velocity and cross-sectional area on a single linear scale, only approximations are possible at the lower values. *AO*, Aorta; *LA*, large arteries; *SA*, small arteries; *ART*, arterioles; *CAP*, capillaries; *VEN*, venules; *SV*, small veins; *LV*, large veins; *VC*, venae cavae.

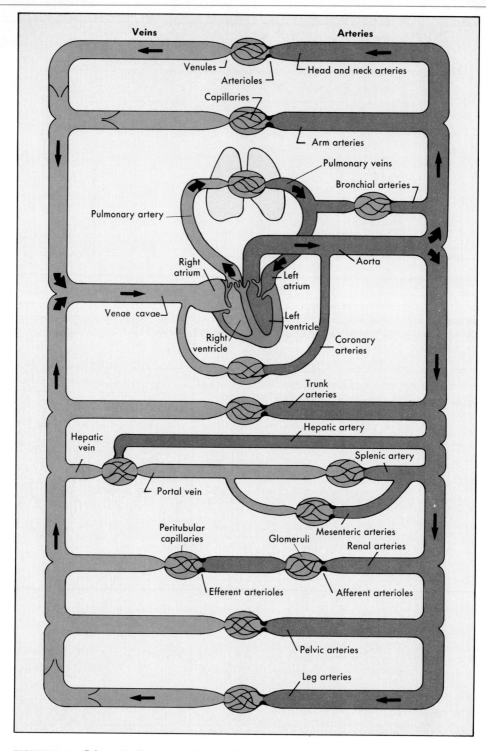

FIGURE 15-3   Schematic diagram of the parallel and series arrangement of the vessels composing the circulatory system. The capillary beds are represented by thin lines connecting the arteries (on the right) with the veins (on the left). The crescent-shaped thickenings proximal to the capillary beds represent the arterioles (resistance vessels). (Redrawn from Green HD: In Glasser O, editor: Medical physics, vol 1, Chicago, 1944, Year Book Medical Publishers, Inc.)

Electrical Activity of the Heart

TRANSMEMBRANE POTENTIALS OF CARDIAC CELLS

The electrical behavior of cardiac cells differs considerably from that of nerve cells or of smooth or skeletal muscle cells (see Chapters 3 and 13). Figure 16-1, A, shows the potential changes recorded from a ventricular muscle cell immersed in an electrolyte solution. When a microelectrode and a reference electrode are placed in the solution near the quiescent cell, no measurable potential difference exists between the two electrodes (a, Fig. 16-1, A).

At b the microelectrode is inserted into the interior of the cell. Immediately a potential difference is recorded across the cell membrane; the potential of the cell interior is about 90 mV lower than that of the surrounding medium. Such electronegativity of the cell interior is also characteristic of skeletal and smooth muscle, of nerve, and indeed of most cells within the body (see Chapter 2).

At c the cell is stimulated and the cell membrane rapidly "depolarizes." Actually the potential difference is reversed (positive overshoot), such that the potential of the interior of the cell exceeds that of the exterior by about 20 mV. The rapid upstroke of the *action potential* is designated *phase 0*. A brief period of partial repolarization (*phase 1*) occurs immediately after the upstroke and is followed by a *plateau (phase 2)* that persists for about 0.2 second. The potential then becomes progressively more negative (*phase 3*) until the resting potential is again attained (at e). The repolarization (*phase 3*) proceeds much more slowly than does the depolarization (phase 0). The interval from the completion of repolarization until the beginning of the next action potential is designated *phase 4*.

The temporal relationships between the electrical activity and the mechanical contraction of cardiac muscle are shown in Figure 16-2. Note that rapid depolarization (phase 0) precedes force development and that repolarization is completed at about the same time as peak force is attained. Thus the duration of contraction parallels the duration of the action potential.

Principal Types of Cardiac Action Potentials

Two main types of action potentials may be recorded in the heart (Figure 16-1). One type, the *fast response* (panel *A*), occurs in atrial and ventricular myocardial fibers and in specialized conducting fibers (*Purkinje fibers*). The other type, the *slow response* (panel *B*), is found in the *sinoatrial (SA) node,* the natural pacemaker region of the heart, and in the *atrioventricular (AV) node,* the specialized tissue that conducts the cardiac impulse from atria to ventricles. Fast responses may be converted to slow responses under certain experimental or pathologic conditions.

As shown in Figure 16-1, the resting membrane potential of the slow response is considerably less negative than that of the fast response. Also, the slope of the upstroke (phase 0) and the amplitude and overshoot of the slow-response action potentials are less than the corresponding values for the fast-response action potentials. The amplitude of the action potential and the rate of rise of the upstroke are

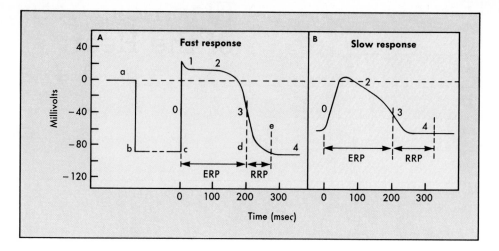

FIGURE 16-1   Changes in transmembrane potential recorded from a fast response **(A)** and slow response **(B)** cardiac fiber in isolated cardiac tissue immersed in an electrolyte solution. **A,** At time *a* the microelectrode was in the solution surrounding the cardiac fiber. At time *b* the microelectrode was introduced into the fiber. At time *c* an action potential was initiated in the impaled fiber. Time *c* to *d* represents the effective refractory period *(ERP),* and time *d* to *e* represents the relative refractory period *(RRP).* **B,** An action potential recorded from a slow response cardiac fiber. Note that compared to the fast response fiber, the resting potential of the slow fiber is less negative, the upstroke (phase *0*) of the action potential is less steep, the amplitude of the action potential is smaller, phase *1* is absent, and the relative refractory period (RRP) extends well into phase *4,* after the fiber has fully repolarized.

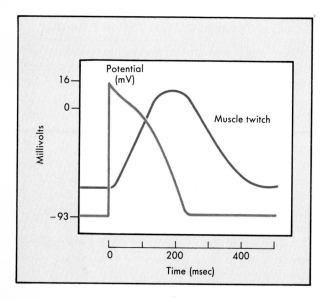

FIGURE 16-2   Time relationships between the mechanical force developed by a thin strip of ventricular muscle and the changes in transmembrane potential. (Redrawn from Kavaler F et al: Bull NY Acad Med 41:592, 1965.)

important determinants of propagation velocity, as described later. Thus, in cardiac tissue characterized by the slow response, conduction velocity is much slower, and impulses are more likely to be blocked than in tissues that display the fast response.

IONIC BASIS OF THE MEMBRANE POTENTIAL

The various phases of the cardiac action potential are associated with changes in the conductivity of the cell membrane, mainly to sodium (Na^+), potassium (K^+), and calcium (Ca^{++}) ions. Each phase of the action potential is associated with a change in conductivity to one or more specific ions.

Resting Potential

The resting potential (phase 4) depends mainly on the conductivity of the cell membrane to K^+. Just as with all other cells in the body, the concentration of

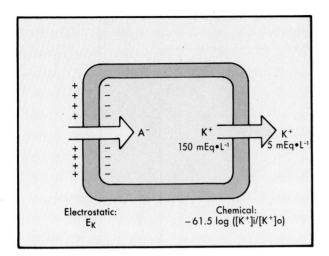

Electrostatic:
E_K

Chemical:
$-61.5 \log ([K^+]i/[K^+]o)$

FIGURE 16-3 The balance of chemical and electrostatic forces acting on a resting cardiac cell membrane, based on 30:1 ratio of the intracellular to extracellular K^+ concentrations and on the presence of a nondiffusible anion (A⁻) inside but not outside the cell.

Table 16-1 Intracellular and Extracellular Ion Concentrations and Equilibrium Potentials in Cardiac Muscle Cells

Ion	Extracellular Concentrations (mM)	Intracellular Concentrations (mM)*	Equilibrium Potential (mV)
Na^+	145	10	70
K^+	4	135	−94
Ca^{++}	2	10^{-4}	132

Modified from Ten Eick RE et al: Prog Cardiovasc Dis 24:157, 1981.
*The intracellular concentrations are estimates of the free concentrations in the cytoplasm.

potassium ions inside a cardiac muscle cell, $[K^+]_i$, greatly exceeds the concentration outside the cell, $[K^+]_o$, (Figure 16-3). The reverse concentration gradient exists for Na^+ and for unbound Ca^{++}. Estimates of the extracellular and intracellular concentrations of Na^+, K^+, and Ca^{++} and of the *Nernst equilibrium potentials* (see Chapter 2) for these ions are compiled in Table 16-1. The K^+ conductance of the resting cell membrane greatly exceeds the Na^+ or Ca^{++} conductance. Thus the chord conductance equation (see Chapter 2) indicates that the resting membrane potential will be determined mainly by the intracellular/extracellular K^+ concentration ratio ($[K^+]_i/[K^+]_o$).

The predictions of the chord conductance equations for the dependence of the resting membrane potential on K^+ have been verified experimentally. Changes in the extracellular concentrations of Na^+ or Ca^{++} scarcely affect the resting potential. However, when $[K^+]_i/[K^+]_o$ is decreased experimentally by raising $[K^+]_o$, the measured value of the resting potential (V_m) approximates that predicted by the Nernst equation for K^+ (Figure 16-4). For extracellular K^+ concentrations above 5 mM, the measured values of V_m correspond closely with the predicted values. The measured levels are slightly less than those predicted by the Nernst equation because of the small but finite value of the Na^+ conductance (g_{Na}). For values of $[K^+]_o$ below 5 mM, the K^+ conductance (g_K) decreases as $[K^+]_o$ is diminished. As g_K decreases,

the effect of g_{Na} on the transmembrane potential becomes relatively more important, as predicted by the chord conductance equation. This change in g_K accounts for the greater deviation, at $[K^+]_o$ levels below 5 mM, of the measured V_m from the value of V_m predicted by the Nernst equation (Figure 16-4).

Fast-Response Action Potential

Genesis of the Upstroke Any process that abruptly decreases the negativity of the resting membrane potential to a critical value (called the *threshold*) will elicit a propagated action potential. The characteristics of fast-response action potentials resemble those shown in Figure 16-1, *A*. The rapid depolarization (phase 0) induced by an effective stimulus is accomplished almost exclusively by a sudden increase in g_{Na}, which leads to an inrush of Na^+. Thus the process responsible for the rapid depolarization of cardiac muscle cells is the same as that responsible for the depolarization of nerve fibers and skeletal muscle cells (see Chapter 3). The amplitude of the cardiac action potential (the magnitude of the potential change during phase 0) varies linearly with the logarithm of $[Na^+]_o$, as shown in Figure 16-5.

In the intact heart certain specialized cells with the property of *automaticity* (described later) can generate action potentials spontaneously. Myocardial cells do not have this property. Therefore a given myocardial cell is brought to the threshold potential when an action potential, originating from some distant *automatic* cell, arrives at that myocardial cell by the process of cell-to-cell conduction.

The approach of an action potential in a nearby cell

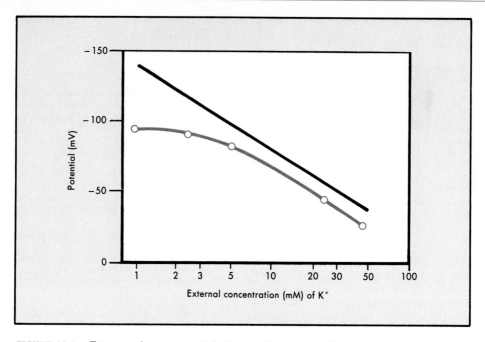

FIGURE 16-4   Transmembrane potential of a cardiac muscle fiber varies inversely with the potassium concentration of the external (extracellular) medium *(red curve)*. The oblique black line represents the change in transmembrane potential predicted by the Nernst equation for E_K. (Redrawn from Page E: Circulation 26:582, 1962.)

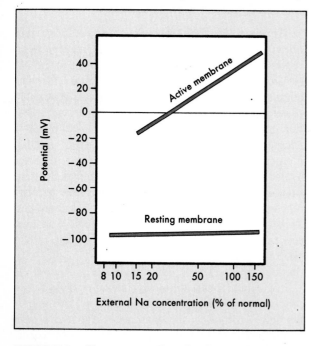

FIGURE 16-5   The concentration of sodium in the external medium is the main determinant of the action potential amplitude *(upper curve)* in cardiac muscle, but it has relatively little influence on the resting potential *(lower curve)*. (Redrawn from Weidmann S: Elektrophysiologie der Herzmuskelfaser, Bern, 1956, Verlag Hans Huber.)

makes V_m less negative for the cell in question. The *activation gates* (see Chapter 3) in some of the specific voltage-sensitive channels (the *fast Na+ channels*) open as V_m becomes less negative; these channels are said to be *activated*. The Na+ conductance (g_{Na}) increases (Figure 16-6) as these channels are activated. Thus Na+ rapidly enters the cell, because (1) g_{Na} has increased, (2) the interior of the cell is negatively charged, and (3) $[Na^+]_o$ greatly exceeds $[Na^+]_i$ (Table 16-1). The entry of the positively charged Na ions decreases the negativity of V_m still more. The entry of additional Na+ thereby opens the activation gates in more fast Na+ channels, in turn causing a further increase in g_{Na}. The process is thus *regenerative* in that an increase in g_{Na} leads to a still greater increase in g_{Na}.

The regenerative increase in g_{Na} accounts for the explosive nature of the inrush of Na+ across the cell membrane and thus for the steepness of the upstroke of the action potential. The inrush of Na+ ceases after only 1 or 2 msec, however, for two principal reasons. First, as V_m approaches the Nernst equilibrium potential for Na+ (Table 16-1), the tendency for Na+ to enter the cell because of the large concentration gradient is partly counterbalanced by the positive

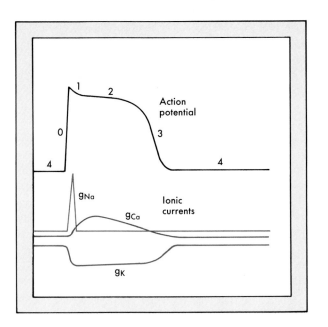

FIGURE 16-6   Changes in the conductances of Na^+ (g_{Na}), Ca^{++} (g_{Ca}), and K^+ (g_K) during the various phases of the action potential of a fast-response cardiac cell. The conductance diagrams indicate directional changes only.

electrostatic charge that exists inside the cell as V_m approaches the peak of the upstroke. This positive electrostatic force tends to repel the influx of Na^+. Second, and even more importantly, very soon after the activation gates in the fast Na^+ channels open, other gates (*inactivation gates;* see Chapter 3) close; that is, the channels are *inactivated.* Thus g_{Na} shortly returns to its preactivation value (Figure 16-6). The activation and inactivation gates are not reset again until the cell membrane is almost fully repolarized. The implications of this resetting are discussed in the section on Cardiac Excitability.

Genesis of the Plateau   In cardiac cells with action potentials that have a prominent plateau, and especially in Purkinje fibers, phase 1 constitutes an early, brief period of limited repolarization that occurs immediately after the upstroke. Phase 1 mainly reflects the initial inactivation of the fast sodium channels (Figure 16-6).

During the plateau (phase 2) of the action potential, Ca^{++} and some Na^+ enter the cell through *slow channels.* Their influx through the slow channels has been called the *slow inward current.* The slow channels differ substantially from the fast Na^+ channels that open during phase 0. The activation, inactiva-

tion, and recovery processes are much slower for the slow channels than for the fast channels. The fast channels may be blocked by tetrodotoxin, whereas the slow channels may be blocked by manganese ions or verapamil, agents known to impede the movement of Ca^{++} into the cell.

The slow channels are activated when V_m reaches their threshold voltage of about -35 mV. Opening of these channels is reflected by an increase in Ca^{++} conductance (g_{Ca}), which begins shortly after the upstroke of the action potential (Figure 16-6). Because of the increase in g_{Ca}, and because the intracellular Ca^{++} concentration is much less than the extracellular concentration (Table 16-1), Ca^{++} enters the cardiac cell throughout the plateau. This Ca^{++} influx is involved in *excitation-contraction coupling,* as described in Chapter 11.

Various factors may influence the slow inward current. This current may be increased by catecholamines, such as epinephrine and norepinephrine. An increase in slow inward current is a crucial step in the mechanism by which catecholamines enhance myocardial contractility. Conversely, drugs such as verapamil, nifedipine, and diltiazem impede the slow inward current; such drugs are known as *calcium channel blocking agents.* By reducing the amount of Ca^{++} that enters the myocardial cells, these drugs diminish the strength of the cardiac contraction (Figure 16-7).

During the plateau of the action potential the concentration gradient for K^+ between the inside and outside of the cell (Table 16-1) is virtually the same as it is during phase 4. Thus the concentration gradient moves K^+ out of the cell. During the plateau the inside of the cell has a slightly positive charge. Thus electrostatic forces also push K^+ out of the cell. The efflux of the positively charged K^+ makes the interior of the cell membrane more negative; that is, it repolarizes the cell membrane, thereby terminating the plateau.

The g_K of the cardiac cell membrane diminishes during phase 2 (Figure 16-6). Consequently the outward current of K^+ is small during the plateau. The outward K^+ current tends to balance the slow inward currents of Ca^{++} and Na^+, thereby helping to sustain the plateau at a V_m close to 0 mV. The effects of altering this balance between the inward currents of Ca^{++} and Na^+ and the outward current of K^+ are demonstrated by administering a calcium channel blocking drug. Figure 16-7 shows that as the concen-

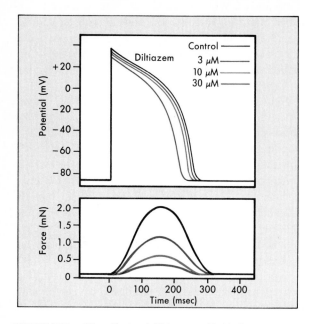

FIGURE 16-7   The effects of diltiazem, a Ca^{++} channel blocking drug, on the action potentials (in millivolts) and isometric contractile forces (in millinewtons) recorded from an isolated papillary muscle of a guinea pig. The tracings were recorded under control conditions and in the presence of diltiazem, in concentrations of 3, 10, and 30 μmol/L. (Redrawn from Hirth C et al: Journal of Molecular and Cellular Cardiology 15:799. Copyright 1983 by Academic Press, Inc [London], Limited.)

tration of diltiazem is increased, the voltage of the plateau becomes less positive and the duration of the plateau diminishes.

Genesis of Repolarization   Final repolarization (phase 3) is achieved by two main processes: (1) an increase in g_K back toward the resting level and (2) a reduction in g_{Ca} (see Figure 16-6). The increase in g_K is induced in part by the elevation of intracellular Ca^{++}, consequent to the inward Ca^{++} current during the plateau. The greater g_K increases the efflux of K^+ from the cell. The outward current of K^+ is no longer balanced by the diminishing slow inward current of Ca^{++} and Na^+, and therefore the charge on the inside of the cell membrane becomes progressively more negative.

The increase in g_K is voltage dependent. As the inside of the cell becomes more negative, g_K becomes greater, and therefore the outward flux of K^+ is accelerated. Thus this rapid phase of repolarization (phase 3) is regenerative, as is the inward current of Na^+ during phase 0. The efflux of K^+ during phase 3 rap-

idly restores the resting level of the membrane potential. The excess Na^+ that had entered the cell mainly during phases 0 and 2 is eliminated by the Na^+/K^+ pump (see Chapter 1), which also pulls back into the cell the K^+ that had exited during phases 2 and 3.

Slow-Response Action Potential

Fast-response action potentials consist of two principal components, a spike (phases 0 and 1) and a plateau (phase 2). In the slow response (see Figure 16-1, *B*) the first component is absent, and the second component accounts for the entire action potential. When the fast Na^+ channels are blocked in myocardial fibers by compounds such as tetrodotoxin, slow responses may be generated in the same fibers under appropriate conditions.

The Purkinje fiber action potentials shown in the first four panels of Figure 16-8 clearly exhibit the two components. In the control tracing (panel *A*) a prominent notch separates the spike from the plateau. In panels *B* to *E*, the concentration of tetrodotoxin was progressively raised to block more and more fast Na^+ channels. The spike is progressively less prominent in panels *B* to *D*, and it disappears entirely in *E*. Thus the tetrodotoxin had a pronounced effect on the spike but only a negligible effect on the plateau. With elimination of the spike (panel *E*), the action potential is typical of the slow response.

Certain cells in the heart, notably those in the SA and AV nodes, are normally slow-response fibers. In such fibers depolarization is achieved by the inward currents of Ca^{++} and Na^+ through the slow channels. These ionic events closely resemble those that occur during the second component of fast-response action potentials. The slow channels in nodal cells can be blocked by Mn^{++} or verapamil, as in fast-response fibers.

CONDUCTION IN CARDIAC FIBERS

An action potential traveling down a myocardial fiber is propagated by local circuit currents, just as in nerve and skeletal muscle fibers (see Figure 3-13). These local currents at the border between the depolarized and polarized sections of a myocardial fiber will tend to depolarize the region of the resting fiber adjacent to the border.

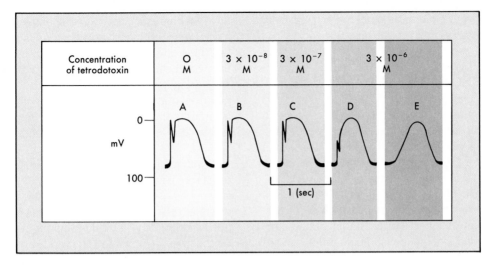

FIGURE 16-8 Effect of tetrodotoxin on the action potential recorded in a calf Purkinje fiber perfused with a solution containing epinephrine and 10.8 mM K^+. The concentration of tetrodotoxin was 0 M in *A*, 3×10^{-8} M in *B*, 3×10^{-7} M in *C*, and 3×10^{-6} M in *D* and *E*; *E* was recorded later than *D*. (Redrawn from Carmeliet E and Vereecke J: Pflugers Arch 313:300, 1969.)

In fast-response fibers the fast Na^+ channels will be activated when the transmembrane potential is suddenly brought to the threshold value of about -70 mV. The inward Na^+ current will then depolarize the cell very rapidly at that site. This portion of the fiber will become part of the depolarized zone, and the border will be displaced accordingly (see Figure 3-13, *B*). This process will be repeated over and over, and the border will move continuously down the fiber as a wave of depolarization.

Local circuits are also responsible for propagation in slow-response fibers. However, the characteristics of the conduction process are different from those of the fast response. The threshold potential is about -40 mV for the slow response, and the conduction velocity is much less than for the fast response. The conduction velocities of the slow responses in the SA and AV nodes are about 0.02 to 0.1 m·sec^{-1}. The fast-response conduction velocities are about 0.3 to 1 m·sec^{-1} for myocardial cells and about 1 to 4 m·sec^{-1} for the specialized conducting fibers in the atria and ventricles. Slow responses are more likely to be blocked by drugs and pathologic processes than are fast responses. Also, slow responses cannot be conducted at as high frequencies as can fast responses.

At any given locus on a fiber, the greater the ampli-tude of the action potential and the greater the rate of change of potential (dV_m/dt) during phase 0, the more rapid will be the conduction down the fiber. The amplitude of the action potential equals the difference in potential between the fully depolarized and the fully polarized regions of the cell interior (see Figure 3-13). The magnitude of the local currents is proportional to this potential difference. Because these local currents shift the potential of the resting zone toward the threshold value, they are the local stimuli that ultimately depolarize the adjacent resting portion of the fiber to its threshold potential. The greater the potential difference between the depolarized and polarized regions, the more efficacious the local stimuli will be, and the more rapidly the wave of depolarization will be propagated down the fiber.

The dV_m/dt during phase 0 (i.e., the slope of the action potential upstroke) is also an important determinant of the conduction velocity. The reason can be appreciated by referring again to Figure 3-13. If the active portion of the fiber depolarizes very gradually (i.e., if the action potential upstroke is not very steep), the local currents over a fixed distance across the border between the depolarized and polarized regions would be relatively small. Thus the resting region adjacent to the active zone would be depolarized very slowly, and consequently each new section of the fiber would require more time to reach threshold. Thus propagation would proceed more slowly.

CARDIAC EXCITABILITY

The *excitability* of a cardiac cell is the ease with which it can be activated. One way to measure the excitability of a cardiac cell is to measure how much electrical current is necessary to induce an action potential. Understanding cardiac excitability is important because of the use of artificial pacemakers and other electrical devices for correcting serious cardiac rhythm disturbances. The excitability characteristics of fast- and slow-response fibers differ considerably.

Fast Response

Once a fast-response action potential has been initiated, the depolarized cell will no longer be excitable until the middle of the period of final repolarization (phase 3). The interval from the beginning of the action potential until the time the fiber can conduct another action potential is called the *effective refractory period*. In the fast response this period extends from the beginning of phase 0 to the time in phase 3 at which V_m has reached about -50 mV (period *c* to *d* in Figure 16-1, *A*). At this value of V_m the electrochemical gates for many of the fast Na^+ channels

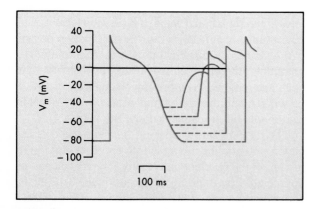

FIGURE 16-9   The changes in action potential amplitude and slope of the upstroke as action potentials are initiated at different stages of the relative refractory period of the preceding excitation. (Redrawn from Rosen MR et al: Am Heart J 88:380, 1974.)

have been reset; that is, the activation gates are closed and the inactivation gates have reopened.

Full excitability is not regained until the cardiac fiber has been fully repolarized (point *e* in Figure 16-1). During period *d* to *e* in Figure 16-1, an action potential may be evoked, but only when the stimulus is stronger than that eliciting a response during phase 4. Period *d* to *e* is called the *relative refractory period*.

When a fast response is evoked during the relative refractory period of an antecedent excitation, its characteristics vary with the membrane potential that exists at the time of stimulation. This dependency on transmembrane potential is illustrated in Figure 16-9. As the fiber is stimulated later and later in the relative refractory period, the amplitude and the rate of rise of the upstroke increase progressively. Presumably the number of fast Na^+ channels that have recovered from inactivation increases as repolarization proceeds during phase 3. Because of the greater amplitude and slope of the evoked action potential, its propagation velocity increases. Once the fiber is fully repolarized, the response is constant no matter when in phase 4 the stimulus is applied. By the end of phase 3 the activation and inactivation gates of all channels are in their final resting positions, and thus excitability remains constant.

Slow Response

The relative refractory period in slow-response fibers frequently extends well beyond phase 3 (see Figure 16-1, *B*). This characteristic has been termed *postrepolarization refractoriness*. Even after the cell has completely repolarized (phase 4), it may require a relatively strong stimulus to evoke a propagated response. Thus the recovery of full excitability is much slower than it is for the fast response.

Until excitability is fully restored, conduction velocity varies with excitability. Impulses arriving at a slow-response fiber early in its relative refractory period are conducted much more slowly than those arriving late in that period. The lengthy refractory periods also account for the greater tendency for conduction to be blocked in slow-response fibers. Even when such fibers are excited at a relatively low repetition rate, the fiber may be able to conduct only a fraction of those impulses.

Effects of Cycle Length

Changes in cycle length alter the action potential duration of cardiac cells and thus change their refractory periods. Consequently the cycle length is often an important factor in initiating or terminating certain arrhythmias. The decreases in action potential duration produced by reductions in cycle length in a Purkinje fiber are shown in Figure 16-10.

This direct correlation between action potential duration and cycle length is ascribable to the effect of the intracellular Ca^{++} concentration on g_K. As already described (see Figure 16-6), Ca^{++} enters the cardiac cell mainly during the action potential plateau. Whenever the cycle length is diminished for a substantial number of heartbeats, the number of plateaus per minute increases, and therefore more Ca^{++} enters the cell each minute. The resultant elevation of intracellular Ca^{++} concentration causes g_K to increase earlier. Thus repolarization begins earlier, and the action potential is abbreviated.

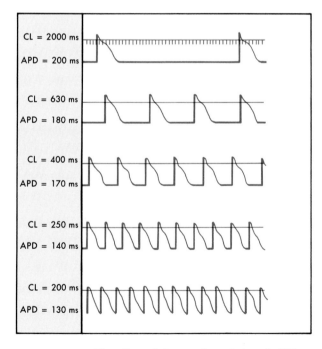

FIGURE 16-10   The effect of changes in cycle length *(CL)* on the action potential duration *(APD)* of canine Purkinje fibers. (Modified from Singer D and Ten Eick RE: Am J Cardiol 2:381, 1971.)

NATURAL EXCITATION OF THE HEART

The properties of *automaticity* (the ability to initiate a heartbeat) and of *rhythmicity* (the frequency and regularity of such pacemaking activity) are intrinsic to cardiac tissue. The heart will continue to beat for a while even when it is completely removed from the body and devoid of any influence from the central nervous system. If the coronary vessels are perfused, the heart will contract rhythmically for many hours. At least some cells in each cardiac chamber can initiate beats; such automatic cells reside mainly in the nodal and specialized conducting tissues. The nervous system affects the frequency at which the heart will beat, and it influences other important cardiac functions as well. However, intact nervous pathways are **not essential** for effective cardiac function.

In the mammalian heart the automatic cells that ordinarily fire at the highest frequency are located in the SA node; this structure is called the *natural pacemaker* of the heart. Other regions of the heart that can initiate beats under special circumstances are called *ectopic foci*, or *ectopic pacemakers*. Ectopic foci may become pacemakers when (1) their own rhythmicity is enhanced, (2) the more rhythmic pacemakers are depressed, or (3) all conduction pathways between the ectopic focus and the more rhythmic foci are blocked.

When the SA node is destroyed, automatic cells in the AV junction usually have the next highest level of rhythmicity, and they become the pacemakers for the entire heart. After some time, which may vary from minutes to days, automatic cells in the atria usually then become dominant. In the dog the most common site for impulse initiation is at the junction of the inferior vena cava and the right atrium.

Purkinje fibers in the specialized conduction system of the ventricles are also automatic. Characteristically, these *idioventricular pacemakers* fire at a very slow rate (about 35 min^{-1}). Ordinarily they do not fire at all, since impulses that originate in the SA node depolarize the Purkinje fibers at a frequency that is much greater than their intrinsic frequency. When the AV junction fails to conduct the cardiac impulse from the atria to the ventricles, the idioventricular pacemakers initiate ventricular contractions at a rate of about 35 min^{-1}. At this low frequency the heart usually cannot pump enough blood to support normal

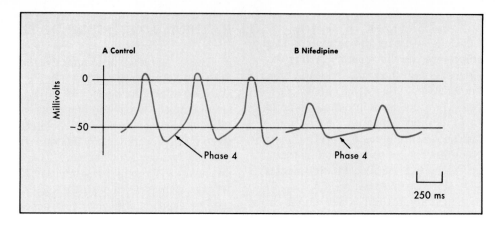

FIGURE 16-11   The effects of nifedipine (5.6×10^{-7} M), a Ca^{++} channel blocking drug, on the transmembrane potentials recorded from a rabbit's SA node cell. (Redrawn From Ning W and Wit AL: Am Heart J 106:345, 1983.)

body function. An artificial pacemaker may be required to correct this disturbance.

Sinoatrial Node

The SA node is the phylogenetic remnant of the sinus venosus of lower vertebrate hearts. In humans it is about 15 mm long, 5 mm wide, and 2 mm thick. It lies in the terminal sulcus on the posterior aspect of the heart at the junction of the superior vena cava and the right atrium.

Typical transmembrane action potentials recorded from an SA node cell are depicted in Figure 16-11, A. Compared with an action potential recorded from a ventricular myocardial cell (see Figure 16-1, A), the minimal potential of the SA node cell is less negative, the upstroke is less steep, the plateau is absent, and repolarization is more gradual. These are all characteristic of the slow response. Tetrodotoxin does not influence the SA nodal action potential, since the upstroke of the action potential is not produced by an influx of Na^+ through the fast channels.

The principal distinguishing feature of an automatic fiber resides in phase 4. In nonautomatic cells (see Figure 16-1) the potential remains constant during phase 4, regardless of whether the cell is a fast- or slow-response fiber. However, an automatic fiber displays a slow depolarization, called the *pacemaker potential,* during phase 4 (Figure 16-11). Depolarization proceeds at a steady rate until the threshold is attained, and then an action potential is triggered.

The firing frequency of an automatic cell is usually varied by changing either the slope of the pacemaker potential or the maximal negativity during phase 4. When the slope of the pacemaker potential decreases (Figure 16-12, A), more time is required for the pacemaker potential to reach threshold, and therefore the firing frequency diminishes. Similarly, if a more negative transmembrane potential is achieved during phase 4 (i.e., the membrane becomes hyperpolarized), more time will again be required for the pacemaker potential to reach threshold (Figure 16-12, B), and therefore the firing frequency also will diminish. A combination of these two mechanisms will also alter the firing frequency.

Ionic Basis of Automaticity Several ionic currents contribute to the slow diastolic depolarization that occurs in cardiac automatic cells. In the automatic cells of the SA node, diastolic depolarization is ascribable to at least three ionic currents: (1) an inward current, i_f, induced by hyperpolarization; (2) the slow inward current, i_{si}; and (3) an outward K^+ current, i_K (Figure 16-13).

The inward current, i_f, is carried mainly by Na^+. This current begins during repolarization (phase 3) of the action potential, as the membrane potential becomes more negative than about -50 mV. The more negative the membrane potential becomes at the end of repolarization, the greater will be the i_f current.

The slow inward current, i_{si}, is carried by Ca^{++} and Na^+. This current begins toward the end of phase

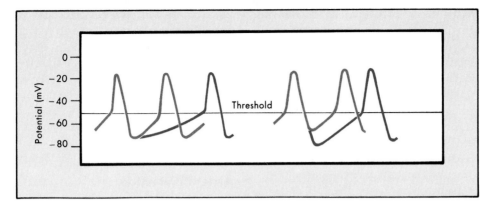

FIGURE 16-12   Effects of changes in the slope of the pacemaker potential or in the maximal diastolic potential on the firing period of an SA nodal cell.

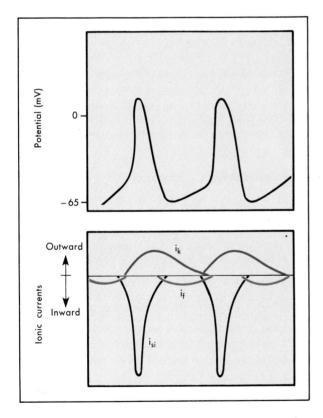

FIGURE 16-13   The transmembrane potential changes *(top panel)* that occur in SA node cells are produced by three principal currents *(bottom panel):* (1) the slow inward current, i_{si}; (2) a hyperpolarization-induced inward current, i_f; and (3) an outward K$^+$ current, i_K. (Redrawn from Brown HF: Physiol Rev 61:644, 1981.)

4, as the transmembrane potential becomes about -55 mV (Figure 16-13). When the slow channels open, Ca^{++} and Na$^+$ enter the cell. The influx of these cations greatly accelerates depolarization, which brings about the upstroke of the action potential. A decrease in the external Ca^{++} concentration or the addition of nifedipine, a calcium channel blocking drug, diminishes the amplitude of the action potential and the slope of the pacemaker potential in SA node cells (Figure 16-11, *B*).

The progressive diastolic depolarization mediated by the two inward currents, i_f and i_{si}, is opposed by a third current, an outward K$^+$ current, i_K. The efflux of K$^+$ tends to repolarize the cell. However, the outward K$^+$ current decays steadily throughout phase 4 (Figure 16-13), and the opposition to the depolarizing effects of the two inward currents gradually decreases.

The ionic basis for automaticity in the AV node pacemaker cells is probably identical to that in the SA node cells. Similar mechanisms probably also account for automaticity in Purkinje fibers, except that the slow inward current is not involved during diastole. Thus the slow diastolic depolarization in Purkinje fibers is mediated principally by the balance between the hyperpolarization-induced inward current, i_f, and the outward K$^+$ current, i_K.

Effects of Autonomic Nerves The transmitters released by the autonomic nerves affect automaticity by altering the ionic currents across the pacemaker cell membranes. Increased sympathetic nervous activity, through the release of norepinephrine, raises the heart rate principally by increasing the slope of the pacemaker potential (Figure 16-12, *A*). Norepinephrine increases all three currents (Figure 16-13) that are involved in SA nodal automaticity. The adrenergically mediated increase in the slope of diastolic depolarization indicates that the increments of i_f and i_{si} exceed the increment of i_K.

Increased parasympathetic activity, through the release of acetylcholine, diminishes the heart rate by increasing the maximal negativity of the pacemaker cell transmembrane potential (Figure 16-12, *B*) and by reducing the slope (Figure 16-12, *A*) of the pacemaker potential. These changes in phase 4 induced by acetylcholine are achieved by an increase in g_K. This change in potassium conductance is mediated by opening specific potassium channels that are controlled by the cholinergic receptors, rather than by opening the regular i_K channels, which are controlled by the voltage across the membrane.

Overdrive Suppression The automaticity of pacemaker cells is temporarily depressed after they are driven at a high frequency. This phenomenon is known as *overdrive suppression*. Because the SA node cells usually fire at a greater frequency than do cells in the other latent pacemaking sites in the heart, the firing of the SA node cells at their greater frequency tends to suppress the automaticity in the other sites. If the SA node cells suddenly stopped firing, the automatic cells in the ectopic sites might remain quiescent for several seconds because of overdrive suppression. Thus the individual might lose consciousness if the SA node suddenly ceased firing, even though numerous automatic cells were present in various ectopic sites in the heart.

The mechanism responsible for overdrive suppression involves the membrane pump (Na^+, K^+-ATPase) that actively extrudes Na^+ from the cell, in partial exchange for K^+ (see Chapter 1). During each depolarization of an automatic Purkinje fiber, for example, a certain quantity of Na^+ enters the cell during phase 0 of the action potential. The more frequently it is depolarized, therefore, the more Na^+ that enters the cell per minute. Under conditions of overdrive the Na^+ pump extrudes this larger quantity of Na^+ more actively from the cell interior. The quantity of Na^+ extruded by the Na^+, K^+-ATPase exceeds the quantity of K^+ that enters the cell. This enhanced activity of the ''electrogenic'' pump hyperpolarizes the cell, since a net loss of cations occurs from the cell interior.

Because of the hyperpolarization, the pacemaker potential requires more time to reach the threshold, as illustrated in Figure 16-12, *B*. Furthermore, when the overdrive suddenly ceases, the Na^+ pump usually does not decelerate instantaneously, but it continues to operate more actively for some time. This excessive extrusion of Na^+ opposes the gradual depolarization of the pacemaker cell during phase 4 and thereby suppresses its automaticity temporarily.

Atrial Conduction

From the SA node the cardiac impulse spreads radially throughout the right atrium (Figure 16-14) along ordinary atrial myocardial fibers, at a conduction velocity of approximately 1 $m \cdot sec^{-1}$. A special pathway, the anterior *interatrial myocardial band* or *Bachmann's bundle*, conducts the impulse most directly from the SA node to the left atrium. Even if this direct pathway is destroyed experimentally, however, conduction proceeds expeditiously from the right to the left atrium along ordinary myocardial fibers. Some of the action potentials that proceed inferiorly through the right atrium ultimately reach the AV node, which is normally the sole source of entry of the cardiac impulse from the atria to the ventricles.

Atrioventricular Conduction

The AV node in adult humans is approximately 22 mm long, 10 mm wide, and 3 mm thick. It is situated posteriorly on the right side of the interatrial septum near the ostium of the coronary sinus.

The AV node in adult humans has been divided into three functional regions: (1) the A-N region, the transitional zone between the atrium and the remainder of the node; (2) the N region, the midportion of the AV node; and (3) the N-H region, the zone in which nodal fibers gradually merge with the *bundle of His,* which is the beginning of the specialized conducting system for the ventricles.

Several features of AV conduction are physiologically and clinically significant. The principal delay in the passage of the impulse from the atrial to the ventricular myocardial cells occurs in the A-N region of the node. The conduction velocity is actually less in the N region than in the A-N region. However, the path length is substantially greater in the A-N than in the N region, which accounts for the greater delay in the A-N than in the N region.

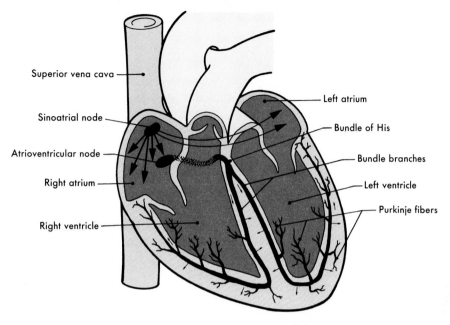

FIGURE 16-14   The conduction system of the heart.

The conduction times through the A-N and N zones account for a considerable fraction of the *P-R interval* (see Figure 16-17), which is the electrocardiographic interval that signifies the delay between atrial and ventricular excitation. Functionally, this delay permits atrial contraction to contribute optimally to ventricular filling.

The action potentials of cells in the N region display the characteristics of the slow response. The resting potential is about −60 mV, the upstroke is not very steep, and the conduction velocity is about 0.05 m·sec^{-1}. Tetrodotoxin, which blocks the fast Na+ channels, has almost no effect on the action potentials in this region. Conversely, diltiazem, a calcium channel blocking drug, decreases the amplitude and duration of the action potentials (Figure 16-15) and depresses AV conduction. The action potentials of cells in the A-N region are intermediate in shape between those of cells in the N region and atria. Similarly, the action potentials of cells in the N-H region are transitional between those of cells in the N region and bundle of His.

Cells in the N region display postrepolarization refractoriness. As the repetition rate of atrial depolarizations is increased, conduction through the AV junction becomes prolonged. Most of that prolongation takes place in the N region. Impulses tend to be

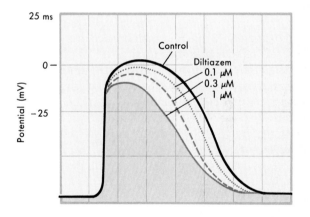

FIGURE 16-15   Transmembrane potentials recorded from a rabbit AV node cell under control conditions and in the presence of the calcium channel blocking drug, diltiazem, in concentrations of 0.1, 0.3, and 1.0 μM·L^{-1}. (Redrawn from Hirth C et al: J Mol Cell Cardiol 15:799, 1983.)

blocked at stimulus repetition rates that are easily conducted in other regions of the heart. If the atria are paced at a high frequency, some fraction of the impulses might be blocked in the AV junction. This tends to protect the ventricles from excessive contraction frequencies. At very high contraction rates, the filling time between contractions might be inadequate, and therefore the heart might not be able to pump a sufficient quantity of blood to the tissues. Thus, when the atria contract at a very rapid rate *(atrial tachycardia),* the ventricles may pump more blood when some of the atrial depolarizations are not conducted through the AV junction.

Autonomic Effects Weak vagal activity may simply prolong the AV conduction time. Stronger vagal activity may cause some or all of the impulses arriving from the atria to be blocked in the node. The delayed conduction or block occurs largely in the N region of the node.

The cardiac sympathetic nerves, on the other hand, are facilitatory. They decrease the AV conduction time and enhance the rhythmicity of the latent pacemakers in the AV junction. The norepinephrine released at the sympathetic nerve terminals increases the amplitude and slope of the upstroke of the AV nodal action potentials, principally in the N region of the node.

Ventricular Conduction

The bundle of His passes subendocardially down the right side of the interventricular septum for approximately 12 mm and then divides into the right and left bundle branches (Figure 16-14). The right bundle branch, a direct continuation of the bundle of His, proceeds down the right side of the interventricular septum. The left bundle branch is considerably thicker than the right. It arises almost perpendicularly from the bundle of His and perforates the interventricular septum.

The bundle branches ultimately subdivide into a complex network of conducting fibers, called *Purkinje fibers,* which ramify over the subendocardial surfaces of both ventricles. Purkinje fibers are the broadest cells in the heart, 70 to 80 μm in diameter, compared with 10 to 15 μm for ventricular myocardial cells. Their large diameter accounts in part for the greater conduction velocity in Purkinje than in myocardial fibers.

The conduction velocity of the action potential over the Purkinje fiber system is the fastest of any tissue within the heart; estimates vary from 1 to 4 m·sec^{-1}. This permits a rapid activation of the entire endocardial surface of the ventricles.

The configuration of the action potentials recorded from Purkinje fibers (see Figure 16-10) is similar to that from ordinary ventricular myocardial fibers (see Figure 16-2). In general, phase 1 is more prominent in

Purkinje fiber action potentials than in those recorded from ventricular fibers, and the duration of the plateau (phase 2) is longer. The prolonged plateau confers a long refractory period on the Purkinje fibers. Thus many premature atrial depolarizations may be conducted through the AV junction, only to be blocked by the Purkinje fibers. This function of protecting the ventricles against the effects of premature atrial depolarizations is especially pronounced at slow heart rates, since the action potential duration and therefore the effective refractory period of the Purkinje fibers vary inversely with the heart rate (see Figure 16-10). Similar directional changes in the refractory period occur in most of the other cells in the heart with changes in rate. However, in the AV node the effective refractory period does not change appreciably over the normal range of heart rates, and it actually increases at very rapid heart rates. Therefore, at high heart rates, it is the AV node rather than the Purkinje fibers that protects the ventricles when impulses arrive at excessive repetition rates.

The first portions of the ventricles to be excited are the interventricular septum and the papillary muscles. The wave of activation spreads into the substance of the septum from both its left and its right endocardial surfaces. Early contraction of the septum makes it more rigid and allows it to serve as an anchor point for the contraction of the remaining ventricular myocardium. Furthermore, early contraction of the papillary muscles prevents eversion of the AV valves into the atria during ventricular systole (see Chapter 17).

The endocardial surfaces of both ventricles are activated rapidly, but the wave of excitation spreads from endocardium to epicardium more slowly (about 0.3 to 0.4 m·sec^{-1}). Because the right ventricular wall is appreciably thinner than the left, the epicardial surface of the right ventricle is activated earlier than that of the left ventricle. Also, apical and central epicardial regions of both ventricles are activated earlier than their respective basal regions. The last portions of the ventricles to be excited are the posterior basal epicardial regions and a small zone in the basal portion of the interventricular septum.

REENTRY

Under appropriate conditions a cardiac impulse may reexcite some region through which it had

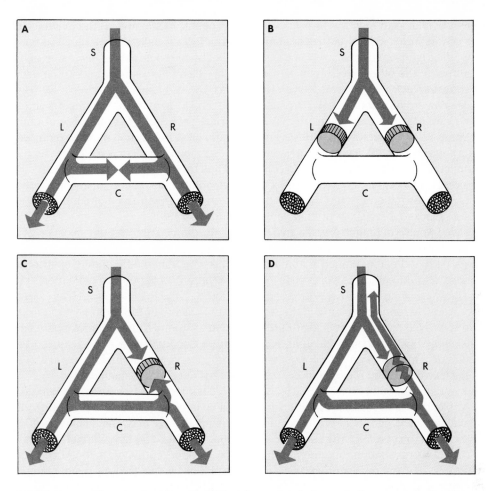

FIGURE 16-16   The role of unidirectional block in reentry. **A,** An excitation wave traveling down a single bundle *(S)* of fibers continues down the left *(L)* and right *(R)* branches. The depolarization wave enters the connecting branch *(C)* from both ends and is extinguished at the zone of collision. Reentry does not occur. **B,** The wave is blocked in the *L* and *R* branches. Reentry does not occur. **C,** The antegrade and retrograde impulses in branch *R* are both blocked. Reentry does not occur. **D,** Unidirectional block exists in branch *R*. The antegrade impulse is blocked. The retrograde impulse arrives later, after the previously refractory segment of tissue in branch R had regained its excitability. The retrograde impulse travels through branch *R* and reexcites bundle *S* to complete the reentry circuit.

passed previously. This phenomenon, known as *reentry,* is responsible for many clinical disturbances of cardiac rhythm.

The conditions necessary for reentry are illustrated in Figure 16-16. In each of the four panels, a single bundle *(S)* of cardiac fibers splits into a left *(L)* and a right *(R)* branch. A connecting bundle *(C)* runs between the two branches. Normally the impulse coming down bundle *S* is conducted along the *L* and *R* branches (panel *A*). As the impulse reaches connecting link *C,* it enters from both sides and becomes

extinguished at the point of collision. The impulse from the left side cannot proceed beyond the point of collision because the tissue beyond is absolutely refractory, since it had just been depolarized from the other direction. Likewise, the impulse cannot pass through bundle *C* from the right side.

It is obvious from Figure 16-16, *B,* that the impulse cannot make a complete circuit if antegrade block exists in the two branches (*L* and *R*) of the fiber bundle. Furthermore, if bidirectional block exists at any point in the loop (for example, branch *R* in panel *C*), the impulse will not be able to reenter.

A necessary condition for reentry is that at some point in the loop, the impulse must be able to pass in one direction but not in the other. This phenomenon is called *unidirectional block.* As shown in Figure 16-16, *D,* the impulse may travel down branch *L* normally. However, the impulse traveling down branch *R* may be blocked in the antegrade direction. The impulse that had been conducted down branch *L* and through the connecting branch *C* may be able to penetrate the depressed region in branch *R* from the retrograde direction, even though the antegrade impulse had been blocked previously at this same site. The antegrade impulse in branch *R* will arrive at the depressed region earlier than the impulse that traverses a longer path and enters branch *R* from the opposite direction. The antegrade impulse may be blocked simply because it arrives at the depressed region during its effective refractory period. If the retrograde impulse is delayed sufficiently because of the longer path, the refractory period may have ended. Then the retrograde impulse can be conducted through the previously depressed region and back to bundle *S,* completing the circuit.

Unidirectional block is a necessary condition for reentry, but not a sufficient one. It is also essential that the effective refractory period of the reentered region be less than the propagation time around the loop. In Figure 16-16, *D,* if the retrograde impulse is conducted through the depressed zone in branch *R,* and if the tissue just beyond is still refractory from the antegrade depolarization, branch *S* will not be reexcited. Therefore the conditions that promote reentry are those that prolong conduction time or shorten the effective refractory period.

ELECTROCARDIOGRAPHY

The electrocardiograph is a valuable instrument because it enables the physician to infer the propagation of the cardiac impulse simply by recording the variations in electrical potential at various loci on the surface of the body. By analyzing the details of these potential fluctuations, the physician gains valuable insight concerning (1) the anatomic orientation of the heart; (2) the relative sizes of its chambers; (3) a variety of disturbances in rhythm and conduction; (4) the extent, location, and progress of ischemic damage to the myocardium; (5) the effects of altered electrolyte

concentrations; and (6) the influence of certain drugs (notably digitalis and its derivatives). The science of electrocardiography is extensive and complex; only the elementary features of the electrocardiogram (ECG) are presented here.

The ECG reflects the temporal changes in the electrical potential between pairs of points on the skin surface. The cardiac impulse progresses through the heart in a very complex three-dimensional pattern. Thus the precise configuration of the ECG varies from person to person, and in any given individual the pattern varies with the anatomic location of the recording electrodes.

In general the pattern consists of P, QRS, and T waves (Figure 16-17). The P-R interval is a measure of the time from the beginning of atrial activation to the beginning of ventricular activation; it normally ranges from 0.12 to 0.20 second. Most of this time involves the passage of the impulse through the AV conduction system. Pathologic prolongations of this interval are associated with disturbances of AV conduction produced by inflammatory, circulatory, pharmacologic, or nervous mechanisms.

The configuration and amplitude of the QRS complex vary considerably among individuals. The duration is usually between 0.06 and 0.10 second. Abnormal prolongation may indicate a block in the normal conduction pathways through the ventricles (e.g., a block of the left or right bundle branch).

During the S-T interval the entire ventricular myocardium is depolarized. Because all the myocardial

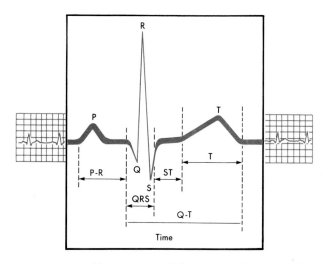

FIGURE 16-17   The important deflections and intervals of a typical scalar electrocardiogram.

cells are at about the same potential, the S-T segment lies on the isoelectric line.

The Q-T interval is sometimes referred to as the period of "electrical systole" of the ventricles; it reflects the duration of the action potentials of the myocardial cells. Its duration is about 0.4 second, but it varies inversely with the heart rate, mainly because the myocardial cell action potential duration varies inversely with the heart rate (see Figure 16-10).

The T wave reflects the repolarization of the ventricular myocardial cells. The T wave is usually deflected in the same direction from the isoelectric line as the major component of the QRS complex. When the T wave and QRS complex do deviate in the same direction from the isoelectric line, it indicates that the propagation of the repolarization process does not follow the same route as the propagation of the depolarization process. Normally depolarization proceeds from endocardium to epicardium, whereas repolarization proceeds in the opposite direction; that is, the epicardial myocytes ordinarily have shorter action potentials than do endocardial myocytes. The reasons for this disparity are not known.

BIBLIOGRAPHY

Journal Articles

Bonke FIM et al: Impulse propagation from the SA-node to the ventricles, Experientia 43:1044, 1987.

Brown HF: Electrophysiology of the sinoatrial node, Physiol Rev 62:505, 1982.

Childers R: The AV node: normal and abnormal physiology, Prog Cardiovasc Dis 19:361, 1977.

Fozzard HA et al: New studies of the excitatory sodium currents in heart muscle, Circ Res 56:475, 1985.

Horackova M: Transmembrane calcium transport and the activation of cardiac contraction, Can J Physiol Pharmacol 62:874, 1984.

Maylie J and Morad M: Ionic currents responsible for the generation of pacemaker current in the rabbit sinoatrial node, J Physiol 355:215, 1984.

Meijler FL and Janse MJ: Morphology and electrophysiology of the mammalian atrioventricular node, Physiol Rev 68:608, 1988.

Noble D: The surprising heart: a review of recent progress in cardiac electrophysiology, J Physiol (Lond) 353:1, 1984.

Rosen MR: Links between basic and clinical cardiac electrophysiology, Circulation 77:251, 1988.

Shamoo AE and Ambudkar IS: Regulation of calcium transport in cardiac cells, Can J Physiol Pharmacol 62:9, 1984.

Ten Eick RE et al: Ventricular dysrhythmia: membrane basis, or of currents, channels, gates, and cables, Prog Cardiovasc Dis 24:157, 1981.

Books and Monographs

Bouman LN and Jongsma HJ: Cardiac rate and rhythm, The Hague, 1982, Martinus Nijhoff Publishers.

Levy MN and Vassalle M: Excitation and neural control of the heart, Bethesda, Md, 1982, American Physiological Society.

Mazgalev T et al: Electrophysiology of the sinoatrial and atrioventricular nodes, New York, 1988, Alan R Liss, Inc.

Mullins LJ: Ion transport in heart, New York, 1981, Raven Press.

Nathan RD: Cardiac muscle: Regulation of excitation and contraction, Orlando, Fla, 1986, Academic Press, Inc.

Noble D and Powell T: Electrophysiology of single cardiac cells, Orlando, Fla, 1987, Academic Press, Inc.

Paes de Carvalho A et al: Normal and abnormal conduction in the heart, Mt Kisco, NY, 1982, Futura Publishing Co, Inc.

Sperelakis N: Physiology and pathophysiology of the heart, ed 2, Boston, 1989, Kluwer Academic Publishers.

Zipes DP and Jalife J: Cardiac electrophysiology and arrhythmias, Orlando, Fla, 1985, Grune & Stratton, Inc.

CHAPTER 17

The Cardiac Pump

It is nearly impossible to contemplate the pumping action of the heart without being struck by its simplicity of design, its wide range of activity and functional capacity, and the staggering amount of work it performs relentlessly over the lifetime of an individual. To understand how the heart accomplishes its important task, it is first necessary to consider the relationships between the structure and the function of its components.

STRUCTURE OF THE HEART IN RELATION TO FUNCTION

Myocardial Cell

Several important morphologic and functional differences exist between myocardial and skeletal muscle cells (see Chapters 12 and 13). However, the contractile elements within the two types of cells are quite similar; each skeletal and cardiac muscle cell is made up of sarcomeres that contain thick filaments composed of myosin and thin filaments composed of actin. As in skeletal muscle, shortening of the cardiac sarcomere occurs by the sliding filament mechanism. Actin filaments slide along adjacent myosin filaments by cycling of the intervening crossbridges, thereby bringing the Z lines closer together.

Skeletal and cardiac muscle show similar length-force relationships. The sarcomere length has been determined with electron microscopy in regions of the ventricles that have been rapidly fixed during systole or diastole. Maximal *developed force* is observed at resting sarcomere lengths of 2.0 to 2.4 μm for car-

diac muscle. At such lengths the thick and thin filaments overlap, and the number of crossbridge attachments is maximal. Developed force of cardiac muscle is less than the maximal value when the sarcomeres are stretched beyond the optimal length, since the overlap of the filaments is less and thus the cycling of the crossbridge is less. At resting sarcomere lengths shorter than the optimal value, the thin filaments that extend from adjacent Z lines overlap each other in the central region of the sarcomere. This arrangement of the thin filaments diminishes contractile force.

In general the length-force relationship for fibers in the papillary muscle also holds true for fibers in the intact heart. This relationship may be expressed graphically, as in Figure 17-1, by substituting ventricular systolic pressure for force, and end-diastolic ventricular volume for myocardial resting fiber (and thus sarcomere) length. The lower curve depicts the ventricular pressure produced by increments in ventricular volume during diastole. The upper curve represents the peak pressure developed by the ventricle during systole at each filling volume. The graph illustrates the *Frank-Starling relationship* of force (or pressure) development by the ventricle as a function of initial fiber length (or initial volume).

Note that the pressure-volume curve in diastole is quite flat at low volumes. Thus large increases in volume can be accommodated with only small increases in pressure. Nevertheless, systolic pressure development is considerable at the lower filling pressures. The ventricle becomes much less distensible with greater filling, however, as evidenced by the sharp rise of the diastolic curve at large intraventricular volumes. Usually ventricular diastolic pressure is less

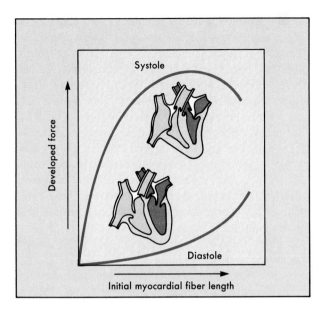

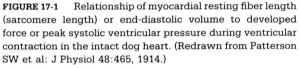

FIGURE 17-1   Relationship of myocardial resting fiber length (sarcomere length) or end-diastolic volume to developed force or peak systolic ventricular pressure during ventricular contraction in the intact dog heart. (Redrawn from Patterson SW et al: J Physiol 48:465, 1914.)

than 7 mm Hg, and the average diastolic sarcomere length is about 2.2 μm. Thus the normal heart operates on the ascending portion of the Frank-Starling curve depicted in Figure 17-1.

A striking difference in the appearance of cardiac and skeletal muscle is that cardiac muscle appears to be a *syncytium* (a single multinucleated cell formed from many fused cells) with branching interconnecting fibers, whereas skeletal muscle cells do not interconnect. However, the myocardium is not a true anatomic syncytium because (1) laterally the myocardial fibers are separated from adjacent fibers by their respective sarcolemmas, and (2) the end of each fiber is separated from its neighbor by dense structures, *intercalated disks,* that are continuous with the sarcolemma (Figure 17-2). Nevertheless, cardiac muscle functions as a syncytium, since a wave of depolarization followed by contraction of the atria or ventricles (an all-or-none response) occurs when a suprathreshold stimulus is applied.

As the wave of excitation approaches the end of a cardiac cell, the spread of excitation to the next cell depends on the electrical conductance of the bound-

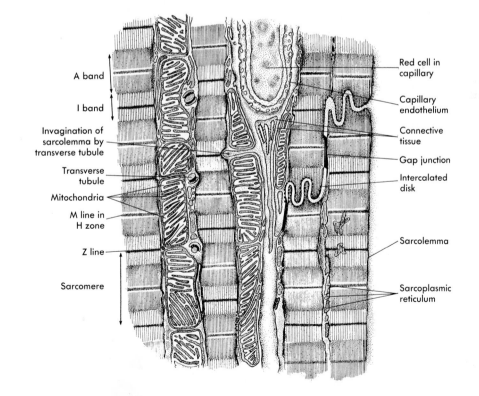

A band
I band
Invagination of sarcolemma by transverse tubule
Transverse tubule
Mitochondria
M line in H zone
Z line
Sarcomere

Red cell in capillary
Capillary endothelium
Connective tissue
Gap junction
Intercalated disk
Sarcolemma
Sarcoplasmic reticulum

FIGURE 17-2   Diagram of an electron micrograph of cardiac muscle. Note the large number of mitochondria and the intercalated disks with nexi (gap junctions), transverse tubules, and longitudinal tubules.

ary between the two cells. Gap junctions (*nexi*) with high conductances are present in the intercalated disks between adjacent cells (Figure 17-2). These gap junctions facilitate the conduction of the cardiac impulse from one cell to the next. Consequently impulse conduction in cardiac tissues progresses more rapidly in a direction parallel to rather than perpendicular to the long axes of the constituent fibers.

Another difference between cardiac and fast skeletal muscle fibers is in the abundance of mitochondria *(sarcosomes)* in the two tissues. Fast skeletal muscle fibers have relatively few mitochondria. Such fibers are called on for relatively short periods of repetitive or sustained contraction. They can metabolize anaerobically and build up a substantial oxygen debt. In contrast, cardiac muscle (1) is richly endowed with mitochondria (Figure 17-2), (2) must contract repetitively for a lifetime, and (3) is incapable of developing a significant oxygen debt. Rapid oxidation of substrates, with the synthesis of adenosine triphosphate (ATP), can keep pace with the myocardial energy requirements because of the large numbers of mitochondria, which contain the respiratory enzymes necessary for oxidative phosphorylation.

To provide adequate oxygen and substrate for its metabolic machinery, the myocardium is also endowed with a rich capillary supply, about one capillary per fiber. Thus diffusion distances are short, and oxygen, carbon dioxide, substrates, and waste material can move rapidly between myocardial cell and capillary. With respect to such exchanges, electron micrographs of myocardium show deep invaginations of the sarcolemma into the fiber at the Z lines (Figure 17-2). These sarcolemmal invaginations constitute the *transverse-tubular*, or *T-tubular, system*. The lumina of these T-tubules are continuous with the bulk interstitial fluid, and they play a key role in excitation-contraction coupling.

A network of sarcoplasmic reticulum consisting of small-diameter sarcotubules surrounds the myofibrils. The sarcoplasmic reticulum releases and takes up calcium, thus playing a key role in myocardial contraction and relaxation.

Excitation-Contraction Coupling The heart requires optimal concentrations of Na^+, K^+, and Ca^{++} to function normally. In the absence of Na^+ the heart is not excitable and will not beat because the action potential of myocardial fibers depends on extracellular Na^+. In contrast, the resting membrane potential is independent of the Na^+ gradient across

the membrane (Figure 16-5). Under normal conditions the extracellular K^+ concentration is about 4 mM. An increase in extracellular K^+, if great enough, produces depolarization, loss of excitability of the myocardial cells, and cardiac arrest in diastole. Ca^{++} is also essential for cardiac contraction. Removal of Ca^{++} from the extracellular fluid decreases contractile force and eventually causes arrest in diastole. Conversely, an increase in extracellular Ca^{++} concentration enhances contractile force, but very high Ca^{++} concentrations induce cardiac arrest in systole *(rigor)*. It is now well documented that the free intracellular Ca^{++} concentration is the factor responsible for the contractile state of the myocardium.

Initially a wave of excitation spreads rapidly along the myocardial sarcolemma from cell to cell via gap junctions. Excitation also spreads into the interior of the cells via the T-tubules. During the plateau (phase 2) of the action potential, Ca^{++} permeability of the sarcolemma increases (see Chapter 16). Ca^{++} flows down its electrochemical gradient and is the main component of the slow inward current. Ca^{++} enters the cell through Ca^{++} channels in the sarcolemma and in the invaginations of the sarcolemma, the T-tubules (Figure 17-3). The primary source of extracellular Ca^{++} is the interstitial fluid (2 mM Ca^{++}). The amount of calcium entering the cell from the extracellular space is not sufficient to induce contraction of the myofibrils, but it serves as a trigger *(trigger Ca^{++})* to release Ca^{++} from the intracellular Ca^{++} stores, the sarcoplasmic reticulum (Figure 17-3). The cytosolic free Ca^{++} concentration increases from a resting level of less than 0.1 μM to levels of 1.0 to 10 μM during excitation, and the Ca^{++} binds to the protein troponin C (see Chapter 11). The Ca^{++}-troponin complex interacts with tropomyosin to unblock active sites between the actin and myosin filaments. This unblocking action allows crossbridge cycling and thus contraction of the myofibrils (systole). Mechanisms that raise the cytosolic Ca^{++} concentration increase the developed force, and mechanisms that lower the Ca^{++} concentration decrease the developed force.

At the end of systole the Ca^{++} influx ceases and the sarcoplasmic reticulum is no longer stimulated to release Ca^{++}. In fact the sarcoplasmic reticulum avidly takes up Ca^{++} by means of an ATP-energized calcium pump that is activated by phosphorylation. The resultant decrease in cytosolic Ca^{++} concentration reverses the binding of Ca^{++} by troponin C. This permits tropomyosin to block again the sites for inter-

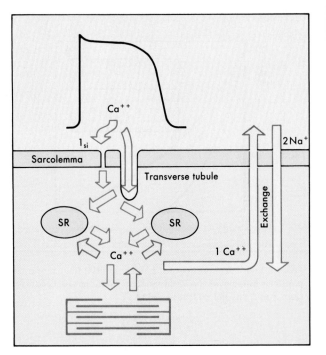

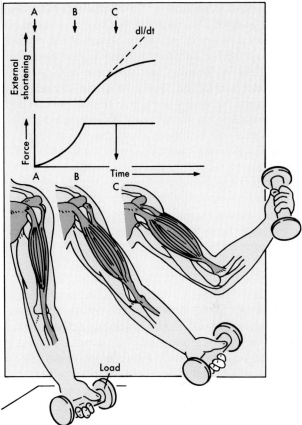

FIGURE 17-3 Schematic diagram of the movements of calcium in excitation-contraction coupling in cardiac muscle. The influx of Ca^{++} from the interstitial fluid during excitation (plateau of action potential) triggers the release of Ca^{++} from the sarcoplasmic reticulum (*SR*). The free cytosolic Ca^{++} activates contraction of the myofilaments (systole). Relaxation (diastole) occurs as a result of uptake of Ca^{++} by the sarcoplasmic reticulum and extrusion of intracellular Ca^{++} by Na^{+}-Ca^{++} exchange.

FIGURE 17-4 Model for a preloaded and afterloaded isotonic contraction of papillary muscle. *A*, Muscle at rest. Preload is represented by partial stretch of the series elastic element. *B*, Partial contraction of the contractile element with stretch of the series elastic element and no external shortening (the isometric phase of the contraction). *C*, Further contraction of the contractile element with external shortening and lifting of the afterload. The tangent (*dl/dt*) to the initial slope of the shortening curve on the right is the velocity of initial shortening. (Redrawn from Sonnenblick EH: The myocardial cell, Philadelphia, 1966, University of Pennsylvania Press.)

action between the actin and myosin filaments, and relaxation (diastole) occurs.

The Ca^{++} that enters the cell to initiate contraction must be removed during diastole. The removal is primarily accomplished by an electroneutral exchange of two Na^{+} for one Ca^{++} (Figure 17-3). Ca^{++} also may be removed from the cell by an electrogenic pump that uses energy to transport three or more Na^{+} for each Ca^{++} across the sarcolemma. A similar electrogenic pump may transport Ca^{++} into the cell. The contributions of Ca^{++} electrogenic pumps to the movement of Ca^{++} across the sarcolemma is controversial; thus this transport mechanism has not been included in Figure 17-3.

Force-Velocity Relationship Velocity and force of contraction depend on the intracellular concentration of free Ca ions. Force and velocity are inversely related, so that with no load, force is negligible and velocity is maximal. **In an isometric contraction,**

where no external shortening occurs, force is maximal and velocity is zero.

The sequence of events in a preloaded and afterloaded isotonic contraction of a papillary muscle is illustrated in Figure 17-4. In the resting state (point *A*) the preload is responsible for the initial stretch. With stimulation the contractile element (interdigitating actin and myosin filaments) begins to shorten and develop force (point *B*). However, the force is not great enough to lift the load, but it does stretch a component of the muscle called the *series elastic element,* whose composition is unclear but may include the muscle fibers. Because the contractile force does not lift the load (no overall shortening), this phase of contraction is termed *isometric*. Stretch of the elastic component of the muscle is represented in the upper half of Figure 17-4 as a progressive rise in force with no external shortening. This stretch of the series elastic element consumes a certain amount of energy. Therefore the total energy used for shortening of the muscle consists of the energy expended during iometric contraction plus that expended in lifting the load. At point *C* the force developed by the contractile component has exceeded the load, and the load has been raised without further stretch of the elastic component. This is represented at the top of Figure 17-4 as external shortening of the muscle without a further increase in force.

The initial slope (dashed tangent) of the shortening curve (Figure 17-4, *top*) depicts the initial rate of shortening (change in length with change in time, dl/dt). If the initial velocity of shortening is plotted against the afterload, the force-velocity curves shown in Figure 17-5 are obtained. The maximal velocity (V_o) may be estimated by extrapolation of the force-velocity curve back to zero load (as indicated by the dotted lines in Figure 17-5); V_o reflects the maximal rate of cycling of the crossbridges.

Contractility represents the performance of the heart at a given preload and afterload. Several indices of contractility have been devised. Contractility of the left ventricle can be assessed by determining the change in developed force at a given preload (end-diastolic pressure) and afterload (aortic pressure). However, it also may be assessed in terms of a change in V_o. Contractility is augmented by certain drugs (e.g., norepinephrine and digitalis) and with an increase in contraction frequency (tachycardia, when applied to the whole heart). The increase in contractility (positive inotropic effect) produced by any of the preceding interventions is reflected by increments in developed force and V_o.

An increase in initial fiber length produces a more forceful contraction, as shown in Figure 17-5. However, this greater developed force is not associated with any change in contractility, as estimated by V_o. It is apparent that with an increase in initial fiber length, greater force may be developed, but the estimated V_o is the same for all three initial lengths. However, some investigators have observed changes in V_o with changes in initial length. Thus, V_o may be in part length dependent. Although the experiments from

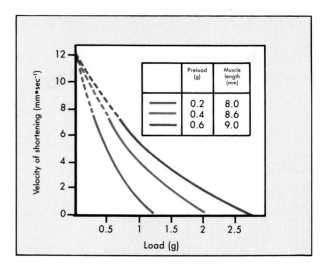

FIGURE 17-5   The effect of increasing initial length of a cat papillary muscle on the force-velocity relationship. (Redrawn from Am J Physiol 202:931, 1962.)

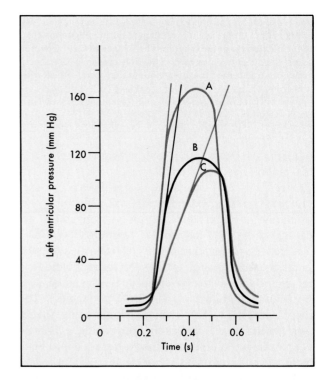

FIGURE 17-6   Left ventricular pressure curves with tangents drawn to the steepest portions of the ascending limbs to indicate maximal dP/dt value. *A,* Control; *B,* hyperdynamic heart, as with norepinephrine administration; *C,* hypodynamic heart, as in cardiac failure.

which the length-force and force-velocity relationships have been derived were carried out on papillary muscles (essentially one dimensional), the findings may be applied qualitatively to the intact heart (three dimensional).

A reasonable index of myocardial contractility can be derived from the contour of ventricular pressure curves (Figure 17-6). A hypodynamic heart is characterized by an elevated end-diastolic pressure, a slowly rising ventricular pressure, and a somewhat reduced ejection phase (curve C). A normal ventricle under adrenergic stimulation shows a reduced end-diastolic pressure, a fast-rising ventricular pressure, and a brief ejection phase (curve B). The slope of the ascending limb of the ventricular pressure curve indicates the maximal rate of pressure development by the ventricle (maximal dP/dt, as illustrated by the tangents to the steepest portion of the ascending limbs of the ventricular pressure curves in Figure 17-6). The slope is maximal during the isovolumic phase of systole. At any given degree of ventricular filling, maximal dP/dt provides an index of the initial contraction velocity and thus of contractility. Similarly, one can obtain an indication of the contractile state of the myocardium from the initial velocity of blood flow in the ascending aorta (the initial slope of the aortic flow curve). The *ejection fraction,* which is the ratio of the volume of blood ejected from the left ventricle per beat *(stroke volume)* to the volume of blood in the left ventricle at the end of diastole, is widely used clinically as an index of contractility.

Cardiac Chambers

The atria are thin-walled, low-pressure chambers that function more as large reservoir conduits of blood for their respective ventricles than as important pumps for ventricular filling. The ventricles are formed by a continuum of muscle fibers that originate from the fibrous skeleton at the base of the heart, primarily around the aortic orifice. These fibers sweep toward the apex at the epicardial surface. They also pass toward the endocardium as they gradually undergo a 180-degree change in direction to lie parallel to the epicardial fibers and form the endocardium and papillary muscles. At the apex of the heart the fibers twist and turn inward to form papillary muscles. At the base of the heart and around the valve orifices they form a thick, powerful muscle that decreases the ventricular circumference to eject blood and narrows

the atrioventricular (AV) valve orifices to aid valve closure. Ventricular ejection is implemented not only by a reduction in circumference, but also by a decrease in the longitudinal axis, accomplished by a descent of the base of the heart. The early contraction of the ventricular apex coupled with approximation of the ventricular walls propels the blood toward the outflow tracts.

Cardiac Valves

The cardiac valves consist of thin flaps of flexible, tough, endothelium-covered fibrous tissue firmly attached at the base to the fibrous valve rings. Movements of the valve leaflets are essentially passive, and the orientation of the cardiac valves is responsible for the unidirectional flow of blood through the heart. There are two types of valves in the heart: the AV and semilunar valves (Figures 17-7 and 17-8).

Atrioventricular Valves The *tricuspid valve* lies between the right atrium and right ventricle and is made up of three cusps, whereas the *mitral valve* lies between the left atrium and left ventricle and has two cusps. The total area of the cusps of each AV valve is approximately twice that of the respective AV orifice, and thus the leaflets overlap considerably in the closed position (Figures 17-7 and 17-8). Attached to the free edges of these valves are fine, strong filaments *(chordae tendineae),* which arise from the powerful papillary muscles of the respective ventricles and prevent eversion of the valves during ventricular systole.

In the normal heart the valve leaflets are relatively close to one another during ventricular filling and provide a funnel for the transfer of blood from atrium to ventricle. This partial approximation of the valve surfaces during diastole is caused primarily by eddy currents behind the leaflets. Also, the chordae tendineae and the papillary muscles are stretched by the filling ventricle and exert tension on the free edges of the valves.

Movements of the mitral valve leaflets throughout the cardiac cycle are shown in an echocardiogram (Figure 17-9). *Echocardiography* consists of sending short pulses of high-frequency sound waves (ultrasound) through the chest tissues and the heart and recording the echoes reflected from the various cardiac structures. The timing and the pattern of the reflected waves provide such information as the diameter of the heart, the ventricular wall thickness,

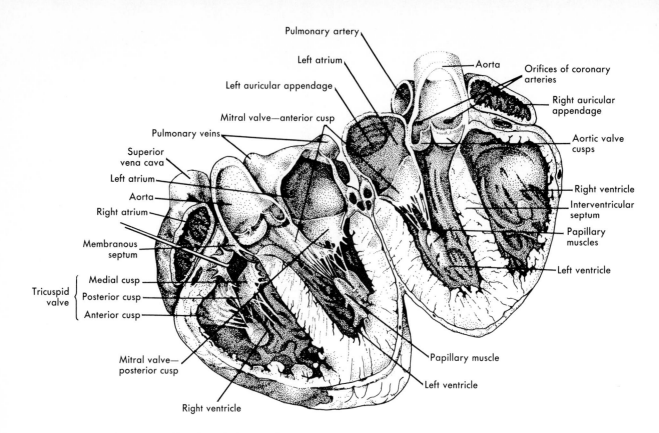

FIGURE 17-7 Drawing of a heart split perpendicular to the interventricular septum to illustrate the anatomic relationships of the leaflets of the AV and aortic valves.

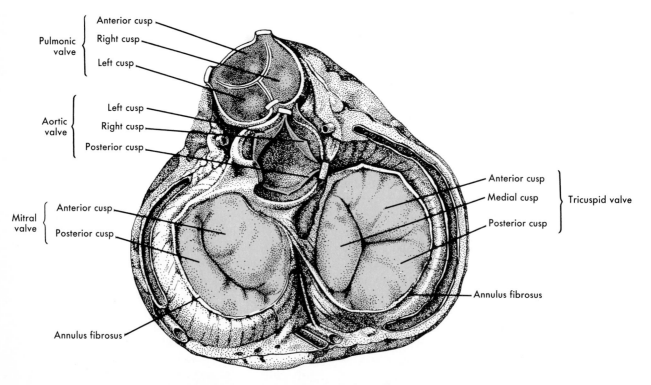

FIGURE 17-8 Four cardiac valves as viewed from the base of the heart. Note how the leaflets overlap in the closed valves.

and the magnitude and direction of the movements of various components of the heart, including the valves.

In Figure 17-9 the echocardiograph transducer is positioned to depict movement of the anterior leaflet of the mitral valve. The posterior leaflet moves in a pattern that is a mirror image of the anterior leaflet, except that in the projection shown in Figure 17-9 the excursions of the leaflet appear much smaller. At point D the mitral valve opens, and during rapid filling (*D* to *E*) the anterior leaflet moves toward the ventricular septum. During the reduced filling phase (*E* to *F*) the valve leaflets float toward each other, but the valve does not close. The ventricular filling contributed by atrial contraction (*F* to *A*) forces the leaflets apart, and a second approximation of the leaflets fol-

lows (*A* to *C*). At point *C* the valve is closed by ventricular contraction. The valve leaflets, which bulge toward the atrium, stay pressed together during ventricular systole (*C* to *D*).

Semilunar Valves The valves between the right ventricle and the pulmonary artery and between the left ventricle and the aorta consist of three cuplike cusps attached to the valve rings (Figures 17-7 and 17-8). At the end of the reduced ejection phase of ventricular systole, blood flow reverses briefly toward the ventricles (shown as a negative flow in the phasic aortic flow curve in Figure 17-10). This flow reversal snaps the cusps together and prevents regurgitation of blood into the ventricles. During ventricular systole the cusps do not lie back against the walls of the pulmonary artery and aorta; rather, they float in the

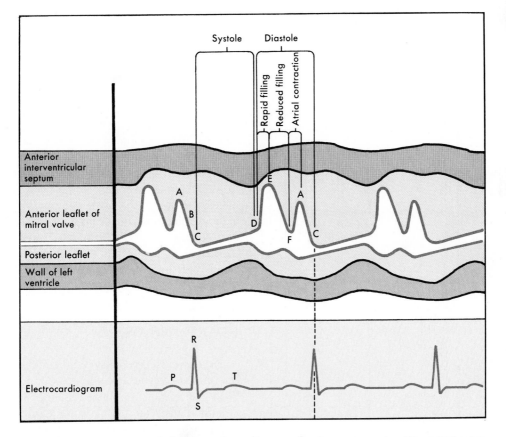

FIGURE 17-9 Drawing made from an echocardiogram showing movements of the mitral valve leaflets (particularly the anterior leaflet) and the changes in the diameter of the left ventricular cavity and the thickness of the left ventricular walls during cardiac cycles in a normal person. *D* to *C*, ventricular diastole; *C* to *D*, ventricular systole; *D* to *E*, rapid filling; *E* to *F*, reduced filling (diastasis); *F* to *A*, atrial contraction. Mitral valve closes at *C* and opens at *D*. Simultaneously recorded electrocardiogram at bottom. (Original echocardiogram courtesy Dr. Sanjiv Kaul.)

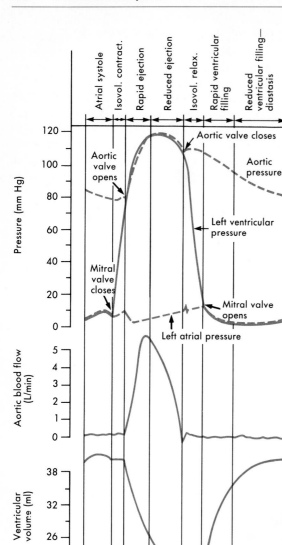

FIGURE 17-10 Left atrial, aortic, and left ventricular pressure pulses correlated in time with aortic flow, ventricular volume, heart sounds, venous pulse, and the electrocardiogram for a complete cardiac cycle in the dog.

bloodstream approximately midway between the vessel walls and their closed position. Behind the semilunar valves are small outpocketings *(sinuses of Valsalva)* of the pulmonary artery and aorta. Eddy currents develop in these sinuses and keep the valve cusps away from the vessel walls. The orifices of the right and left coronary arteries are located behind the right and left cusps, respectively, of the aortic valve. Were it not for the presence of the sinuses of Valsalva and the eddy currents developed therein, the coronary ostia could be blocked by the valve cusps.

The Pericardium

The pericardium is an epithelialized fibrous sac. It closely invests the entire heart and the cardiac portion of the great vessels and is reflected onto the cardiac surface as the epicardium. The sac normally contains a small amount of fluid, which provides lubrication for the continuous movement of the enclosed heart. The pericardium is not very distensible and thus strongly resists a large, rapid increase in cardiac size. Therefore the pericardium helps prevent sudden overdistension of the heart chambers. In contrast to an acute change in intracardiac pressure, progressive and sustained distension of the heart (as occurs in cardiac hypertrophy) or a slow progressive increase in pericardial fluid (as occurs with pericardial effusion) gradually stretches the intact pericardium.

HEART SOUNDS

Four sounds are usually produced by the heart, but only two are ordinarily audible through a stethoscope. With electronic amplification the heart sounds, including the less intense sounds, can be detected and recorded graphically as a phonocardiogram.

The first heart sound is initiated at the onset of ventricular systole (Figure 17-10) and consists of a series of vibrations of mixed, unrelated, low frequencies (a noise). It is the loudest and longest of the heart sounds, has a crescendo-decrescendo quality, and is heard best over the apical region of the heart. The first heart sound is primarily caused by the oscillation of blood in the ventricular chambers and vibration of the chamber walls. The vibrations are engendered in part by the abrupt rise of ventricular pressure with acceleration of blood back toward the atria. However, the sound is produced mainly by sudden tension and recoil of the AV valves and adjacent structures when the blood is decelerated by closure of the AV valves.

The second heart sound, which occurs with closure of the semilunar valves (Figure 17-10), is composed of higher-frequency vibrations (higher pitch), is of shorter duration and lower intensity, and has a more snapping quality than the first heart sound. The second sound is caused by abrupt closure of the semilunar valves, which initiates oscillations of the columns of blood and the tensed vessel walls by the stretch and recoil of the closed valves.

The third heart sound, which is sometimes heard in children with thin chest walls or in patients with left ventricular failure, consists of a few low-intensity, low-frequency vibrations heard best in the region of the apex. It occurs in early diastole and is believed to be the result of vibrations of the ventricular walls caused by abrupt cessation of ventricular distension and deceleration of blood entering the ventricles.

A fourth, or atrial, sound, consisting of a few low-frequency oscillations, is occasionally heard in normal individuals. It is caused by oscillation of blood and cardiac chambers resulting from atrial contraction (Figure 17-10).

Asynchronous valve closures can produce split sounds over the apex of the heart for the AV valves and over the base for the semilunar valves, and deformities of the valves can produce cardiac murmurs.

CARDIAC CYCLE

Ventricular Systole

Isovolumic Contraction The onset of ventricular contraction coincides with the peak of the R wave of the electrocardiogram and the initial vibration of the first heart sound. It is indicated on the ventricular pressure curve as the earliest rise in ventricular pressure after atrial contraction. The interval between the start of ventricular systole and the opening of the semilunar valves (when ventricular pressure rises abruptly) is called isovolumic contraction, since ventricular volume is constant during this brief period (Figure 17-10).

Ejection Opening of the semilunar valves marks the onset of the ejection phase, which may be subdivided into an earlier, shorter phase (rapid ejection) and a later, longer phase (reduced ejection). The *rapid ejection* phase is characterized by (1) the sharp rise in ventricular and aortic pressures that terminates at the peak ventricular and aortic pressures, (2) an abrupt decrease in ventricular volume, and (3) a large aortic blood flow (Figure 17-10). The sharp decrease in the left atrial pressure curve at the onset of ventricular ejection results from the descent of the base of the heart and stretch of the atria. During the *reduced ejection* period, runoff of blood from the aorta to the periphery exceeds ventricular output, and therefore aortic pressure declines. Throughout ventricular systole the blood returning to the atria progressively increases the atrial pressure.

Note that during the first third of ventricular ejection, left ventricular pressure slightly exceeds aortic pressure and flow accelerates (continues to increase), whereas during the last two thirds of ventricular ejection the reverse holds true. This reversal of the ventricular/aortic pressure gradient in the presence of continued flow of blood from the left ventricle to the aorta (caused by the momentum of the forward blood flow) results from the storage of potential energy in the stretched arterial walls, which decelerates the flow of blood into the aorta.

The effect of ventricular systole on left ventricular diameter is shown in an echocardiogram (see Figure 17-9). During ventricular systole (Figure 17-9, *C* to *D*) the septum and the free wall of the left ventricle become thicker and move closer to each other.

At the end of ejection a volume of blood approximately equal to that ejected during systole remains in the ventricular cavities. This *residual volume* is fairly constant in normal hearts. However, it is smaller when heart rate increases or when outflow resistance is reduced, and it is larger when the opposite conditions prevail. An increase in myocardial contractility may decrease residual volume, especially in the depressed heart. In severely hypodynamic and dilated hearts, as in heart failure, the residual volume can become much greater than the stroke volume. In addition to serving as a small adjustable blood reservoir, the residual volume to a limited degree can permit transient disparities between the outputs of the two ventricles.

Ventricular Diastole

Isovolumic Relaxation Closure of the aortic valve produces the incisura on the descending limb of the aortic pressure curve and marks the end of ventricular systole. The period between closure of the semilunar valves and opening of the AV valves is called isovolumic relaxation. It is characterized by a precipitous fall in ventricular pressure without a change in ventricular volume.

Rapid Filling Phase Most of the ventricular filling occurs immediately after the AV valves open. The blood that had returned to the atria during the previous ventricular systole is abruptly released into the relaxing ventricles. This period of ventricular filling is called the *rapid filling phase* (Figure 17-10). The atrial and ventricular pressures decrease despite the increase in ventricular volume because the relaxing ventricles are exerting less and less force on the blood in their cavities.

Diastasis The rapid filling phase is followed by a phase of slow filling called diastasis. During diastasis blood returning from the periphery flows into the right ventricle, and blood from the pulmonary circulation flows into the left ventricle. This small, slow addition to ventricular filling is indicated by gradual increases in atrial, ventricular, and venous pressures and in ventricular volume.

Pressure-Volume Relationship The changes in left ventricular pressure and volume throughout the cardiac cycle are summarized in Figure 17-11. The element of time is not considered in this pressure-volume loop. Diastolic filling starts at *A* and terminates at *C*, when the mitral valve closes. The initial

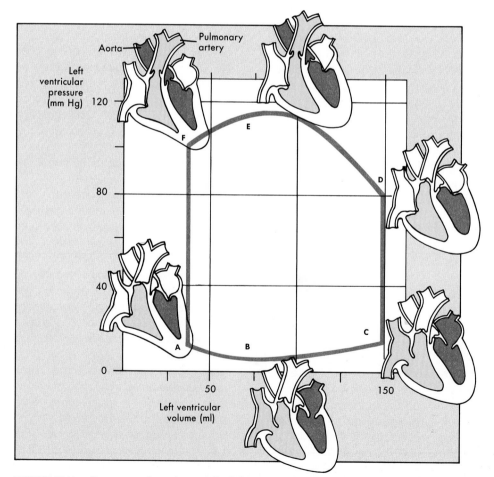

FIGURE 17-11   Pressure-volume loop of the left ventricle for a single cardiac cycle (*A* to *F*).

decrease in left ventricular pressure *(A to B)*, despite the rapid inflow of blood from the atrium, results from progressive ventricular relaxation and increased distensibility. During the remainder of diastole *(B to C)* the increase in ventricular pressure reflects ventricular filling and the passive elastic characteristics of the ventricle. Note that after the initial phase of ventricular diastole, only a small increase in pressure occurs with the increase in ventricular volume *(B to C)*. With isovolumic contraction *(C to D)*, pressure rises steeply and ventricular volume remains constant. At *D* the aortic valve opens. During the first phase of ejection (rapid ejection, *D to E*) the large reduction in volume is associated with a progressive increase in ventricular pressure that is less than the increase occurring during isovolumic contraction. This phase is followed by reduced ejection *(E to F)* and a small decrease in ventricular pressure. The aortic valve closes at *F*; this is followed by isovolumic relaxation *(F to A)*, which is characterized by a sharp drop in pressure and no change in volume. The mitral valve opens at *A* to complete one cardiac cycle.

Atrial Systole The onset of atrial systole occurs soon after the beginning of the P wave of the electrocardiogram (curve of atrial depolarization). The transfer of blood from atrium to ventricle accomplished by the peristalsis-like wave of atrial contraction completes the period of ventricular filling (Figure 17-10). Throughout ventricular diastole, atrial pressure barely exceeds ventricular pressure. This indicates that the resistance of the pathway through the open AV valves during ventricular filling is very low.

Because there are no valves at the junctions of the venae cavae and right atrium or junctions of the pulmonary veins and left atrium, atrial contraction can force blood in both directions. Little blood is pumped back into the venous tributaries during the brief atrial contraction, mainly because of the inertia of the inflowing blood.

Atrial contraction is not essential for ventricular filling. Adequate filling is often observed in patients with atrial fibrillation or complete heart block despite the absence of an effective atrial contribution. However, the contribution of atrial contraction is governed to a great extent by the heart rate and the structure of the AV valves. At slow heart rates, filling practically ceases toward the end of diastasis, and atrial contraction contributes little additional filling. When the heart rate is rapid, diastasis is abbreviated and the atrial contribution can become substantial, especially if the atrium contracts immediately after the rapid filling phase when the AV pressure gradient is maximal. If tachycardia becomes so great that the rapid filling phase is abridged, atrial contraction assumes great importance in propelling blood rapidly into the ventricle during this brief period of the cardiac cycle. Of course, if the period of ventricular relaxation is so brief that filling is seriously impaired, even atrial contraction cannot prevent inadequate ventricular filling. In certain disease states, the AV valves may be severely narrowed *(stenotic)*. Under such conditions atrial contraction may be more essential to ventricular filling than it is in the normal heart.

MEASUREMENT OF CARDIAC OUTPUT

Fick Principle

Adolph Fick contrived the first method for measuring cardiac output in intact animals and humans. The basis for this method, called the Fick principle, is simply an application of the law of conservation of mass. It is derived from the fact that the quantity of O_2 delivered to the pulmonary capillaries via the pulmonary artery plus the quantity of O_2 that enters the pulmonary capillaries from the alveoli must equal the quantity of O_2 that is carried away by the pulmonary veins.

This is depicted schematically in Figure 17-12. The rate, $\dot{q}_1$, of O_2 delivery to the lungs equals the O_2 concentration in the pulmonary arterial blood, $[O_2]_{pa}$, times the pulmonary arterial blood flow, $\dot{V}$, which equals the cardiac output; that is,

$$\dot{q}_1 = \dot{V}[O_2]_{pa} \tag{1}$$

Let $\dot{q}_2$ be the net rate of O_2 uptake by the pulmonary capillaries from the alveoli. At equilibrium, $\dot{q}_2$ equals the O_2 consumption of the body. The rate, $\dot{q}_3$, at which O_2 is carried away by the pulmonary veins equals the O_2 concentration in the pulmonary venous blood, $[O_2]_{pv}$, times the total pulmonary venous flow, which is virtually equal to the pulmonary arterial blood flow, $\dot{V}$; that is,

$$\dot{q}_3 = \dot{V}[O_2]_{pv} \tag{2}$$

From the conservation of mass,

$$\dot{q}_1 + \dot{q}_2 = \dot{q}_3 \tag{3}$$

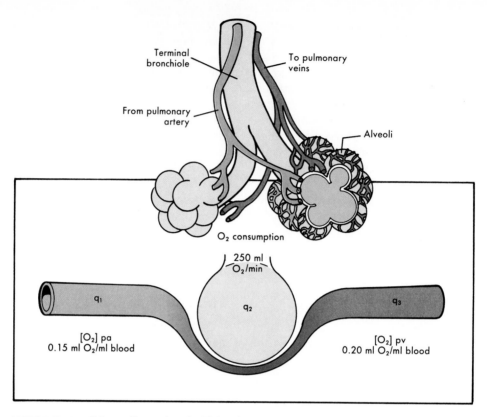

FIGURE 17-12   Schema illustrating the Fick principle for measuring cardiac output. The change in color from pulmonary artery to pulmonary vein represents the change in color of the blood as venous blood becomes fully oxygenated.

Therefore

$$\dot{V}[O_2]_{pa} + \dot{q}_2 = \dot{V}[O_2]_{pv} \qquad \textbf{(4)}$$

Solving for cardiac output,

$$\dot{V} = \dot{q}_2 / ([O_2])_{pv} - [O_2]_{pa}) \qquad \textbf{(5)}$$

Equation 5 is the statement of the Fick principle.

In the clinical determination of cardiac output, O_2 consumption is computed from measurements of the volume and O_2 content of expired air over a given interval. Because the O_2 concentration of peripheral arterial blood is essentially identical to that in the pulmonary veins, $[O_2]_{pv}$, it is determined on a sample of peripheral arterial blood withdrawn by needle puncture. Pulmonary arterial blood actually represents mixed systemic venous blood. Samples for O_2 analysis are obtained from the pulmonary artery or right ventricle through a cardiac catheter.

An example of the calculation of cardiac output in a normal, resting adult is illustrated in Figure 17-12. With an O_2 consumption of 250 ml·min^{-1}, an arterial

(pulmonary venous) O_2 content of 0.20 ml O_2/ml blood, and a mixed venous (pulmonary arterial) O_2 content of 0.15 ml O_2/ml blood, the cardiac output would be 250/(0.20 − 0.15) = 5000 ml·min^{-1}.

The Fick principle is also used for estimating the O_2 consumption of organs in situ, when blood flow and the O_2 contents of the arterial and venous blood can be determined. Algebraic rearrangement reveals that O_2 consumption equals the blood flow times the arteriovenous O_2 concentration difference. For example, if the blood flow through one kidney is 700 ml·min^{-1}, arterial O_2 content is 0.20 ml O_2/ml blood, and renal venous O_2 content is 0.18 ml O_2/ml blood, then the rate of O_2 consumption by that kidney must be 700 × (0.20 − 0.18) = 14 ml·min^{-1}.

Indicator Dilution Technique

The indicator dilution technique has been widely used to estimate cardiac output in humans. A measured quantity of some indicator (a dye or isotope that remains within the circulation) is injected rapidly into a large central vein or into the right side of the heart through a catheter. Arterial blood is continuously drawn through a detector (densitometer or isotope rate counter), and a curve of indicator concentration is recorded as a function of time. The greater the blood flow (cardiac output), the greater is the detection of the injected dye. Currently the most common indicator is a bolus of cold saline, and the cardiac output can be calculated from the change in blood temperature.

BIBLIOGRAPHY

Journal Articles

Allen DG and Orchard CH: Myocardial contractile function during ischemia and hypoxia, Circ Res 60:153, 1987.

Alpert NR et al: Heart muscle mechanics, Annu Rev Physiol 41:521, 1979.

Ballerman BJ and Brenner BM: Role of artrial peptides in body fluid homeostasis, Circ Res 58:619, 1986.

Chapman RA: Control of cardiac contractility at the cellular level, Am J Physiol 245:H535, 1983.

Fabiato A and Fabiato F: Calcium and cardiac excitation-contraction coupling, Annu Rev Physiol 41:473, 1979.

Horackova M: Transmembrane calcium transport and the activation of cardiac contraction, Can J Physiol Pharmacol 62:874, 1984.

Jewell BR: A reexamination of the influence of muscle length on myocardial performance, Circ Res 40:221, 1977.

Katz AM: Influence of altered inotropy and lusitropy on ventricular pressure-volume loops, J Am Coll Cardiol 11:438, 1988.

Langer GA: Sodium-calcium exchange in the heart, Annu Rev Physiol 44:435, 1982.

Sagawa K: The ventricular pressure-volume diagram revisited, Circ Res 43:677, 1978.

Wohlfart B and Nobel MIM: The cardiac excitation-contraction cycle, Pharmacol Ther 16:1, 1982.

Books and Monographs

Fozzard HA: Some experimental studies on Na-Ca exchange in heart muscle. In Nathan RD, editor: Cardiac muscle: the regulation of excitation and contraction, Orlando, Fla, 1986, Academic Press, Inc.

Fozzard HA et al: The heart and cardiovascular system, vols 1 and 2, New York, 1986, Raven Press.

Harvey W: Anatomical studies on the motion of the heart and blood, Springfield, Ill, 1928, Charles C Thomas, Publisher (translated by CD Leake).

Katz AM: Physiology of the heart, New York, 1977, Raven Press.

Katz AM: Role of calcium in contraction of cardiac muscle. In Calcium channel blocking agents in the treatment of cardiovascular disorders, Mt Kisco, NY, 1983, Futura Publishing Co, Inc.

Parmley WW and Talbot L: Heart as a pump. In handbook of physiology, section 2: The cardiovascular system—the heart, vol I, Bethesda, Md, 1979, American Physiological Society.

Sommer JR and Johnson EA: Ultrastructure of cardiac muscle. In Handbook of physiology, section 2: The cardiovascular system—the heart, vol I, Bethesda, Md, 1979, American Physiological Society.

Sperelakis N, editor: Physiology and pathophysiology of the heart, ed 2, Boston, 1989, Kluwer Academic Publishers.

CHAPTER 18

Regulation of the Heartbeat

The quantity of blood pumped by the heart each minute *(cardiac output)* may be varied by changing its frequency of beating *(heart rate)* or the volume it ejects per beat *(stroke volume)*. A discussion of the control of cardiac activity may therefore be subdivided into a consideration of the regulation of pacemaker activity and the regulation of myocardial performance. The control of heart rate is mediated almost exclusively by the autonomic nervous system. This system also regulates myocardial performance, but several mechanical and humoral factors are also important.

CONTROL OF HEART RATE

In normal adults the average heart rate at rest is about 70 min^{-1}, but the rate is signficantly greater in children. During sleep the heart rate diminishes by 10 to 20 min^{-1}, but during exercise or emotional excitement it may accelerate to rates considerably more than 100 min^{-1}. In well-trained athletes at rest the rate is usually about 50 min^{-1}.

The sinoatrial (SA) node is usually under the tonic influence of both divisions of the autonomic nervous system. The sympathetic system increases heart rate, whereas the parasympathetic system decreases it. Changes in heart rate usually involve a reciprocal action of the two divisions of the autonomic nervous system. Thus an increased heart rate is usually achieved by a diminution of parasympathetic activity and a concomitant increase in sympathetic activity; deceleration is usually achieved by the opposite mechanisms.

In healthy resting individuals parasympathetic tone ordinarily predominates. Abolition of parasympathetic influences by giving atropine usually increases the heart rate substantially (Figure 18-1). Conversely, abrogation of sympathetic effects by giving propranolol usually slows the heart only slightly (Figure 18-1). When both divisions of the autonomic nervous system are blocked, the heart rate of adults averages about 100 min^{-1}. The rate that prevails after complete autonomic blockade is called the *intrinsic heart rate*.

Nervous Control

Sympathetic Pathways The cardiac sympathetic fibers originate in the intermediolateral columns of the upper five or six thoracic and lower one or two cervical segments of the spinal cord. These preganglionic fibers emerge from the spinal column through the white communicating branches and enter the paravertebral chains of ganglia. Most of the preganglionic fibers ascend the paravertebral chains and synapse with postganglionic neurons, mainly in the stellate and middle cervical ganglia. Postganglionic sympathetic fibers then join with parasympathetic fibers to form a complex network of mixed efferent nerves to the heart.

Sympathetic fibers from the right and left sides of the body are distributed differentially to the various structures in the heart. In the dog, for example, the fibers on the left side have more pronounced effects on myocardial contractility than on heart rate. In some dogs left cardiac sympathetic nerve stimulation may not affect heart rate, even though it may

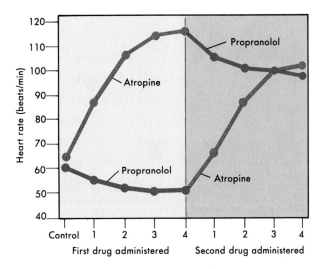

FIGURE 18-1   The effects of four equal doses of atropine (0.04 mg/kg body weight total) and of propranolol (0.2 mg/kg total), given sequentially, on the heart rate of 10 healthy young men (mean age, 21.9 years). In half of the trials atropine was given first *(top curve);* in the other half propranolol was given first *(bottom curve).* (Redrawn from Katona PG et al: J Appl Physiol 52:1652, 1982.)

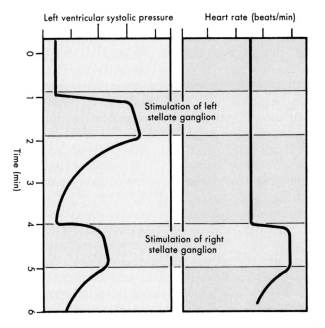

FIGURE 18-2   In an experiment on a canine isovolumic left ventricle preparation, stimulation of the left stellate ganglion did not affect heart rate at all, but it enhanced left ventricular performance substantially. In most experiments however, left stellate stimulation does increase heart rate, but not nearly as greatly as does right stellate stimulation.

enhance ventricular performance profoundly, as shown in Figure 18-2. Conversely, right cardiac sympathetic nerve stimulation increases heart rate more and increases contractile force less than does equivalent stimulation of sympathetic fibers on the left side. This bilateral asymmetry probably also exists in humans.

The effects of sympathetic stimulation decay very gradually after the cessation of stimulation (Figure 18-2). Most of the norepinephrine released from the nerve endings during sympathetic stimulation is taken up again by the nerve terminals, and much of the remainder is carried away by the bloodstream. Relatively little of the released norepinephrine is degraded in the tissues.

The adrenergic receptors in the cardiac tissues are predominantly of the β type; that is, they are responsive to β-adrenergic agonists (e.g., isoproterenol) and are inhibited by specific β-adrenergic blocking agents (e.g., propranolol).

Parasympathetic Pathways The preganglionic parasympathetic fibers to the heart originate in the medulla oblongata, in cells that lie in the dorsal motor nucleus or the nucleus ambiguus. The precise location varies from species to species. Centrifugal vagal fibers pass inferiorly through the neck close to the common carotid arteries and then through the mediastinum to synapse with postganglionic cells located on the epicardial surface or within the walls of the heart itself. Many of the cardiac ganglion cells are located near the SA and atrioventricular (AV) nodes.

The vagal effects on the heart are inhibitory. The right and left vagi are usually distributed differentially to the various cardiac structures. The right vagus nerve affects the SA node predominantly; stimulation decreases the firing rate. The left vagus nerve mainly retards AV conduction and may actually block impulse conduction from atria to ventricles. However, the innervation overlaps considerably; left vagal stimulation inhibits the SA node, and right vagal stimulation impedes AV conduction.

The SA and AV nodes are rich in cholinesterase. Thus the effects of any given vagal impulse are transient because the acetylcholine released at the nerve terminals is hydrolyzed so quickly. Also, parasympathetic effects preponderate over sympathetic effects at the SA node, as shown in Figure 18-3. As the frequency of sympathetic stimulation in an anesthetized dog was increased from 0 to 4 Hz in the absence of

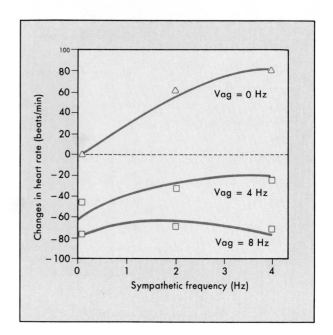

FIGURE 18-3   The changes in heart rate in an anesthetized dog when the vagus and cardiac sympathetic nerves were stimulated simultaneously. The sympathetic nerves were stimulated at 0, 2, and 4 Hz; the vagus nerves at 0, 4 and 8 Hz. The symbols represent the observed changes in heart rate; the curves were derived from the computed regression equation. (Modified from Levy MN and Zieske H: J Appl Physiol 27:465, 1969.)

any concomitant vagal stimulation, the heart rate increased by 80 min^{-1} *(top curve)*. However, during concurrent vagal stimulation at 8 Hz *(bottom curve)*, the same increase in sympathetic stimulation had scarcely any effect on heart rate. The mechanisms responsible for this vagal predominance are discussed later in relation to the neural control of myocardial contractility.

Control by Higher Centers Several higher cerebral centers help regulate cardiac rate, rhythm, and contractility. In the *thalamus,* tachycardia may be induced by stimulating the midline, ventral, and medial groups of nuclei. Variations in heart rate may also be evoked by stimulating the posterior and posterolateral regions of the *hypothalamus.* Hypothalamic centers are also involved in the circulatory responses to alterations in environmental temperature. Experimentally induced temperature changes in the anterior hypothalamus greatly alter heart rate and peripheral resistance. Stimuli applied to the H_2 fields of Forel in the diencephalon elicit a variety of cardiovascular responses, including tachycardia; such changes closely simulate those observed during muscular exercise.

In the *cerebral cortex* the centers that influence cardiac function are located mostly in the anterior half of the brain, principally in the frontal lobe, the orbital cortex, the motor and premotor cortex, the anterior part of the temporal lobe, the insula, and the cingulate gyrus. Cortical and diencephalic centers are undoubtedly responsible for initiating the cardiac reactions that occur during excitement, anxiety, and other emotional states.

Reflex Control

Baroreceptor Reflex Acute changes in blood pressure reflexly alter heart rate. Such changes in heart rate are mediated mainly by the baroreceptors located in the carotid sinuses and aortic arch (see Chapter 22). An example of the changes in heart rate elicited by drug-induced changes in arterial blood pressure in a group of normal human subjects is shown in Figure 18-4. Blood pressure was elevated by infusing phenylephrine, which is a potent vasocon-

strictor, whereas blood pressure was reduced by infusing nitroprusside, a vasodilator. Over the range of induced blood pressures, the cardiac cycle length (reciprocal of heart rate) varied linearly with the arterial blood pressure.

When the arterial blood pressure is in the normal range, moderate alterations in baroreceptor stimulation change heart rate by evoking reciprocal changes in activity in the two divisions of the autonomic nervous system. For example, a moderate increase in arterial blood pressure will lower the heart rate by increasing efferent vagal activity and by decreasing efferent sympathetic activity concurrently. However, when blood pressure is increased by more than about 25 mm Hg, cardiac sympathetic tone is completely suppressed. Thereafter the additional reduction in heart rate produced by any further rise in blood pressure is evoked entirely by increased vagal activity. The converse applies during the development of severe hypotension. Vagal tone virtually disappears after a moderate drop in blood pressure. As the pressure continues to decline, further acceleration of the heart is ascribable solely to a progressive increase in sympathetic activity.

Bainbridge Reflex and Atrial Receptors In 1915 Bainbridge reported that infusions of blood or saline solution increased the heart rate. The heart accelerated regardless of whether the infusions did or did not increase arterial blood pressure. Acceleration was observed whenever central venous pressure rose sufficiently to distend the right side of the heart, and the effect was abolished by bilateral transection of the vagi. Bainbridge postulated that increased cardiac filling elicited tachycardia reflexly, and that the afferent impulses were conducted by the vagi.

Many investigators have confirmed that the heart may accelerate in response to the intravenous administration of fluid. However, the magnitude and direction of the response depend on several factors, especially the prevailing heart rate. When the heart rate is relatively slow, intravenous infusions usually accelerate the heart. When the heart rate is more rapid, however, infusions will ordinarily slow the heart. Acute increases in blood volume not only evoke the Bainbridge reflex, but also activate other reflexes (notably the baroreceptor reflex) that tend to change heart rate in the opposite direction (Figure 18-5). The actual change in heart rate induced by an intravenous infusion is therefore the result of these antagonistic reflex effects.

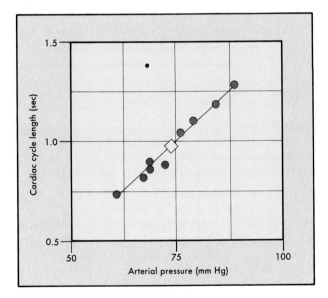

FIGURE 18-4 The relation between cardiac cycle length and diastolic arterial blood pressure in a group of healthy human subjects. The pressure changes were produced by infusions of nitroprusside *(blue circles)* and phenylephrine *(red circles)*. The diamond symbol represents the mean values of cardiac cycle length and arterial blood pressure in these subjects before the drug infusions. (Modified from Eckberg DL et al: J Clin Invest 78:366, 1986, by copyright permission of the American Society for Clinical Investigation.)

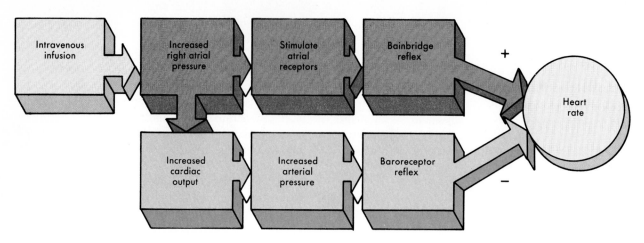

FIGURE 18-5   Intravenous infusions of blood or electrolyte solutions tend to increase heart rate via the Bainbridge reflex and to decrease heart rate via the baroreceptor reflex. The actual change in heart rate induced by such infusions is the result of these two opposing effects.

Sensory receptors that influence heart rate exist in both atria. The receptors are located principally in the venoatrial junctions: in the right atrium at its junctions with the venae cavae and in the left atrium at its junctions with the pulmonary veins. Distension of these receptors sends impulses centripetally in the vagi. The efferent impulses are carried by sympathetic and parasympathetic fibers to the SA node.

Stimulation of the atrial receptors also increases urine flow. A reduction in renal sympathetic nerve activity might be partly responsible for this diuresis. However, the principal mechanism appears to be a neurally mediated reduction in the secretion of vasopressin (antidiuretic hormone) by the posterior pituitary gland (Chapter 39). Furthermore, a peptide with potent diuretic and natriuretic properties is present in atrial tissue. This peptide is called *atrial natriuretic peptide,* and its role in the regulation of water and electrolyte balance is being investigated (see Chapters 33 and 34).

Respiratory Cardiac Arrhythmia Rhythmic variations in cardiac cycle length, occurring at the frequency of respiration, are detectable in most resting individuals and are more pronounced in children. Typically the cycle length decreases during inspiration and increases during expiration (Figure 18-6).

Recordings from the autonomic nerves to the heart and from the phrenic nerves to the diaphragm in animals reveal that the activity increases in the sympathetic nerve fibers during inspiration, whereas the activity increases in the vagal fibers during expiration (Figure 18-7). The acetylcholine released at the vagal

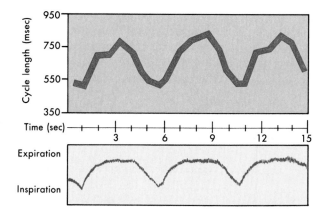

FIGURE 18-6   Respiratory sinus arrhythmia in a resting, unanesthetized dog. Note that the cardiac cycle length increases during expiration and decreases during inspiration. (Modified from Warner MR et al: Am J Physiol 251:H1134, 1986.)

endings is removed so rapidly that the periodic bursts of vagal activity can cause the heart rate to vary rhythmically. Conversely, the norepinephrine released at the sympathetic endings is removed from the tissues much more slowly, thus damping out the effects of the rhythmic variations in release of this transmitter on heart rate. Thus the rhythmic changes in heart rate are ascribable almost entirely to the oscillations in vagal activity. Respiratory sinus arrhythmia is exaggerated when vagal tone is enhanced.

Reflex and central factors both contribute to the genesis of the respiratory cardiac arrhythmia (Figure

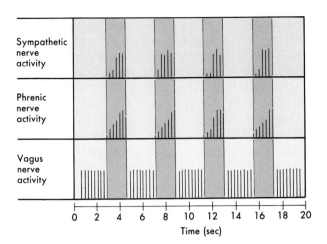

FIGURE 18-7   The respiratory fluctuations in efferent neural activity in the cardiac and respiratory nerves of an anesthetized dog. Note that the sympathetic nerve activity occurs synchronously with the phrenic nerve discharges (which initiate diaphragmatic contraction), whereas the vagus nerve activity occurs between the phrenic nerve discharges. The phrenic discharges mark the inspiratory phase of respiration, whereas the periods between the phrenic discharges denote the expiratory phase. (From Kollai M and Koizumi K: J Auton Nerv Syst. 1:33, 1979.)

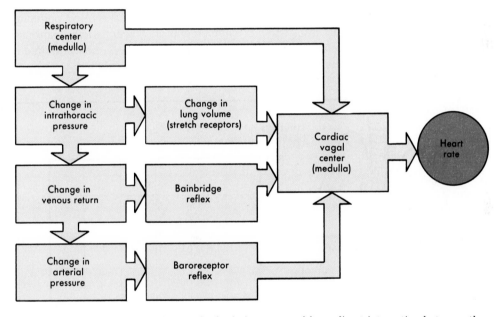

FIGURE 18-8   Respiratory sinus arrhythmia is generated by a direct interaction between the respiratory and cardiac centers in the medulla, as well as by reflexes originating from stretch receptors in the lungs, stretch receptors in the right atrium (Bainbridge reflex), and baroreceptors in the carotid sinuses and aortic arch.

18-8). During inspiration the lung volume increases and the intrathoracic pressure decreases (see Chapter 25). Lung distension stimulates stretch receptors in the lungs, and stimulation of these receptors can reflexly increase heart rate. The reduction in intrathoracic pressure during inspiration increases venous return to the right side of the heart. The resultant distension of the right atrium elicits the Bainbridge

reflex (Figure 18-8). After the time delay required for the increased venous return to reach the left side of the heart, left ventricular output increases and raises systemic arterial blood pressure. This in turn reduces heart rate reflexly through baroreceptor stimulation (Figure 18-8).

The respiratory center in the medulla directly influences the cardiac autonomic centers (Figure 18-8), which are also located in the medulla. In animals that are placed on total heart-lung bypass, the chest is open and the lungs remain collapsed. Rhythmic changes in arterial blood pressure or in central venous pressure are absent in this preparation. Nevertheless, respiratory movements of the rib cage and diaphragm are evident, and cyclic heart rate changes accompany the respiratory movements. These rhythmic changes in heart rate are almost certainly induced by a direct interaction between the respiratory and cardiac centers in the medulla.

Chemoreceptor Reflex The cardiac response to peripheral chemoreceptor stimulation merits special consideration because it illustrates the complexity that may be introduced when one stimulus excites two organ systems simultaneously. In intact animals stimulation of the carotid chemoreceptors consistently increases ventilatory rate and depth (see Chapter 28) but ordinarily has little effect on heart rate. The small, bidirectional changes in heart rate are related to the degree of enhancement of pulmonary ventila-

tion, as shown in Figure 18-9. When respiratory stimulation is relatively mild, heart rate usually diminishes; when the increase in pulmonary ventilation is more pronounced, heart rate usually increases.

The cardiac response to peripheral chemoreceptor stimulation is the result of primary and secondary reflex mechanisms (Figure 18-10). The **primary** reflex effect of carotid chemoreceptor excitation on the SA node is inhibitory. The **secondary** effects of respiratory excitation are facilitatory, and they vary with the extent of the respiratory stimulation.

An example of the primary inhibitory influence is displayed in Figure 18-11. In this experiment on an anesthetized dog the lungs were completely collapsed, and blood oxygenation was accomplished by an artificial oxygenator. When the carotid chemoreceptors were stimulated by suddenly reducing the oxygen saturation of the blood perfusing the carotid chemoreceptors *(lower panel),* an intense bradycardia ensued. Such effects on heart rate are mediated primarily by efferent vagal fibers.

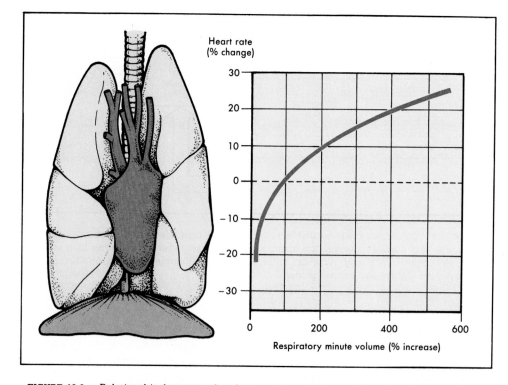

FIGURE 18-9   Relationship between the change in heart rate and the change in respiratory minute volume during carotid chemoreceptor stimulation in spontaneously breathing cats and dogs. When respiratory stimulation was relatively slight, heart rate usually diminished; when respiratory stimulation was more pronounced, heart rate usually increased. (Modified from Daly M deB and Scott MJ: J Physiol 144:148, 1958.)

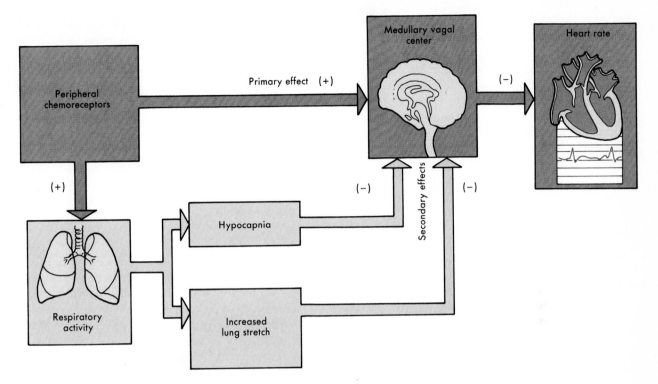

FIGURE 18-10   The primary effect of stimulation of the peripheral chemoreceptors on heart rate is to excite the cardiac vagal center in the medulla and thus to decrease heart rate. Peripheral chemoreceptor stimulation also excites the respiratory center in the medulla. This effect produces hypocapnia and increases lung inflation, both of which secondarily inhibit the medullary vagal center. Thus these secondary influences attenuate the primary reflex effect of peripheral chemoreceptor stimulation on heart rate.

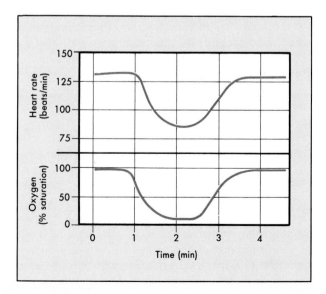

FIGURE 18-11   Changes in heart rate during carotid chemoreceptor stimulation in an anesthetized dog on heart-lung bypass. The lungs are permanently deflated in this preparation, and respiratory gas exchange is accomplished by an artificial oxygenator. The lower tracing represents the oxygen saturation of the blood perfusing the carotid chemoreceptors. Note that when the oxygen saturation was diminished, the heart rate decreased. The blood perfusing the remainder of the animal, including the myocardium, was fully saturated with oxygen throughout the experiment. (Modified from Levy MN et al: Circ Res 18:67, 1966.)

The pulmonary hyperventilation usually evoked by carotid chemoreceptor stimulation (Chapter 28) influences heart rate secondarily, both by initiating more pronounced pulmonary inflation reflexes and by producing hypocapnia (Figure 18-10). Each of these influences accelerates the heart and depresses the primary cardiac response to chemoreceptor stimulation. Thus, when pulmonary hyperventilation is not prevented, carotid chemoreceptor stimulation usually affects heart rate minimally.

Ventricular Receptor Reflexes Sensory receptors located near the endocardial surfaces of the ventricular walls initiate reflex effects similar to those elicited by the arterial baroreceptors. Excitation of these endocardial receptors diminishes the heart rate and peripheral resistance. The receptors discharge in a pattern that parallels the changes in ventricular pressure. Impulses that originate in these receptors are transmitted to the medulla by myelinated vagal fibers.

Other sensory receptors have been identified in the epicardial regions of the ventricles. These receptors discharge in patterns that are not related to the changes in ventricular pressure. These ventricular receptors are excited by a variety of mechanical and chemical stimuli, and their physiologic functions are not clear.

REGULATION OF MYOCARDIAL PERFORMANCE

Intrinsic Regulation

Just as the heart can initiate its own beat in the absence of any nervous or hormonal control, the myocardium can adapt to changing hemodynamic conditions by mechanisms that are intrinsic to cardiac muscle itself. Experiments on denervated hearts reveal that this organ adjusts remarkably well to stress. For example, racing greyhounds with denervated hearts perform almost as well as those with intact innervation. Their maximal running speed is only 5% less after complete cardiac denervation. In these dogs the fourfold increase in cardiac output that occurs when they run is achieved principally by an increase in stroke volume. In normal dogs the increase in cardiac output with exercise is accompanied by a proportionate increase of heart rate; stroke volume does not change much (see Chapter 46). The

cardiac adaptation in the denervated animals is not achieved entirely by intrinsic mechanisms; circulating catecholamines contribute significantly. If the β-adrenergic receptors are blocked by propranolol in greyhounds with denervated hearts, their racing performance is severely impaired.

The intrinsic cardiac adaptation that has received the greatest attention involves changes in the resting length of the myocardial fibers. This adaptation is designated *Starling's law of the heart,* or the *Frank-Starling mechanism.* The mechanical and structural bases for this mechanism have been explained in Chapters 11 and 17. However, certain other intrinsic mechanisms also help to regulate myocardial performance, and these mechanisms do not necessarily involve any changes in resting length. The principal examples involve the alterations in performance that are elicited by varying the interval between contractions.

Frank-Starling Mechanism The Frank-Starling mechanism has been studied primarily under two conditions: (1) on isolated hearts and (2) in more intact preparations.

Isolated heart. In 1895 Frank described the response of the isolated heart of the frog to alterations in the load (i.e., the preload) on the myocardial fibers just before contraction. He observed that as the preload was increased, the heart responded with a more forceful contraction.

In 1914 Starling described the intrinsic response of the canine heart to changes in right atrial and aortic pressure in the isolated heart-lung preparation. In this preparation the right ventricular filling pressure is varied by altering the height of a reservoir connected to the right atrium. The right ventricle then pumps this blood through the pulmonary vessels. The trachea is cannulated, and the lungs are artificially ventilated.

The aorta is ligated distal to the arch, and a cannula is inserted into the brachiocephalic artery. Blood is pumped by the left ventricle into the aortic arch. It then flows through the cannula, through some external tubing, and finally back to the reservoir connected to the right atrium. A resistance device in the external tubing allows the investigator to control the aortic pressure (i.e., the left ventricular afterload). The combined volumes of the two ventricles can be recorded by a special device.

The response of the isolated heart to a sudden increase of right atrial pressure is shown in Figure

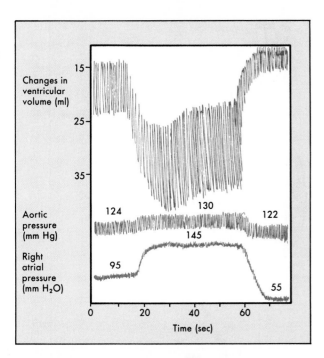

FIGURE 18-12   Changes in ventricular volume in a heart-lung preparation when the venous reservoir was suddenly raised (right atrial pressure increased from 95 to 145 mm H_2O) and subsequently lowered (right atrial pressure decreased from 145 to 55 mm H_2O). Note that an increase in ventricular volume is registered as a downward shift in the volume tracing. (Redrawn from Patterson SW, Piper H, and Starling EH: J Physiol 48:465, 1914.)

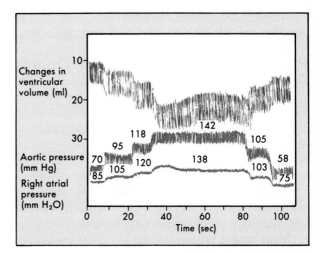

FIGURE 18-13   Changes in ventricular volume, aortic pressure, and right atrial pressure in a heart-lung preparation when peripheral resistance was raised and subsequently lowered in several steps. Note that an increase in ventricular volume is registered as a downward shift in the volume tracing. (Redrawn from Patterson SW, Piper H, and Starling EH: J Physiol 48:465, 1914.)

18-12. Aortic pressure was permitted to increase only slightly. In the top tracing an increase of ventricular volume is registered as a downward deflection. Thus the upper border of the tracing represents the systolic volume, the lower border indicates the diastolic volume, and the width of the tracing reflects the stroke volume.

For several beats after the rise in right atrial pressure, the ventricular volume progressively increased. This indicates that a disparity must have existed between ventricular inflow during diastole and ventricular output during systole; that is, during a given systole the ventricles did not expel as much blood as had entered them during the preceding diastole. This progressive accumulation of blood dilated the ventricles and lengthened the individual myocardial fibers in the walls of the ventricles.

The increased diastolic fiber length somehow facilitates ventricular contraction and enables the ventricles to pump a greater stroke volume. Diastolic fiber length continues to increase until, at equilibrium, the cardiac output exactly matches the augmented filling volume. An optimal fiber length apparently exists, beyond which contraction is actually impaired (Chapter 11). Therefore excessively high filling pressures may depress rather than enhance the pumping capacity of the ventricles by overstretching the myocardial fibers.

Changes in diastolic fiber length also permit the isolated heart to compensate for an increase in afterload. In the experiment depicted in Figure 18-13, the arterial resistance was abruptly raised in three steps, whereas venous return was held constant. Each rise in resistance increased the arterial pressure and ventricular volume. With each abrupt elevation of arterial pressure, which reflects the afterload, the left ventricle was at first unable to pump a normal stroke volume. Because venous return to the right atrium was held constant, the ventricular filling volume exceeded the diminished stroke volume. This disparity between ventricular filling and emptying augmented the ventricular diastolic volume and therefore increased the length of the myocardial fibers. This change in end-diastolic fiber length finally enabled the ventricle to pump a stroke volume equal to the control stroke volume, but against a greater afterload.

When cardiac compensation involves ventricular dilation, the force required by each myocardial fiber to generate a given intraventricular systolic pressure must be appreciably greater than that developed by the fibers in a ventricle of normal size. The relationship between wall tension and cavity pressure resembles that for cylindric tubes (Laplace's law, Chapter 21), in that for a constant internal pressure, wall tension varies directly with the radius. Consequently the myocardial fibers in a dilated heart must develop considerably more tension than do those in a normal-sized heart. Therefore these fibers require considerably more oxygen to perform a given amount of external work than do those in a normal-sized heart.

More intact preparations. The major problem of assessing the role of the Frank-Starling mechanism in intact animals and humans is the difficulty of measuring end-diastolic myocardial fiber length. The Frank-Starling mechanism has been represented graphically by plotting some index of ventricular performance along the ordinate and some index of fiber length along the abscissa. The most commonly used indices of ventricular performance are cardiac output, stroke volume, and stroke work. The indices of fiber length include ventricular end-diastolic volume, ventricular end-diastolic pressure, ventricular circumference, and mean atrial pressure.

The Frank-Starling mechanism is better represented by a family of *ventricular function curves* rather than by a single curve. To construct a given ventricular function curve, blood volume is altered over a wide range of values, and stroke work and end-diastolic pressure are measured at each step. Similar observations are then made during the desired experimental intervention. For example, the ventricular function curve obtained during a norepinephrine infusion lies above and to the left of a control ventricular function curve (Figure 18-14). It is evident that, for a given level of left ventricular end-diastolic pressure, the left ventricle performs more work during a norepinephrine infusion than during control conditions. Thus a shift of the ventricular function curve to the left usually signifies an improvement of ventricular **contractility**; a shift to the right usually indicates an impairment of contractility and a consequent tendency toward **cardiac failure**.

A shift in a ventricular function curve does not uniformly indicate a change in contractility, however. *Contractility* is a measure of cardiac performance at a given level of preload and afterload. The end-diastolic

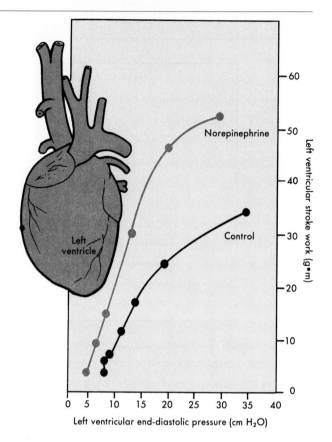

FIGURE 18-14   A constant infusion of norepinephrine in a dog shifts the ventricular function curve to the left. This shift signifies an enhancement of ventricular contractility. (Redrawn from Sarnoff SJ et al: Circ Res 8:1108, 1960.)

pressure is usually a good index of preload, whereas the aortic systolic pressure is a good index of afterload. In assessing myocardial contractility, the cardiac afterload must be held constant as the end-diastolic pressure is varied over a range of values.

The Frank-Starling mechanism is ideally suited for matching the cardiac output to the venous return. Any sudden, excessive output by one ventricle soon results in a greater venous return to the other ventricle. The consequent increase in diastolic fiber length serves as the stimulus to increase the output of the second ventricle to correspond with that of its mate. For this reason the Frank-Starling mechanism maintains a precise balance between the outputs of the right and left ventricles. Because the two ventricles are arranged in series in a closed circuit, even a small, but maintained, imbalance in the outputs of the two ventricles would be catastrophic.

Frequency-Induced Regulation The effects of

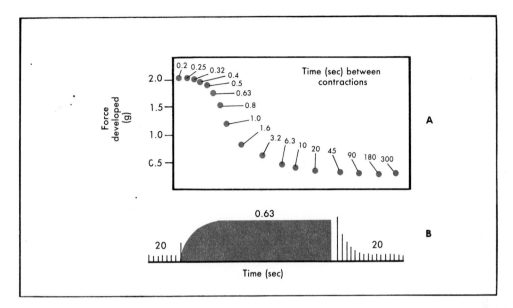

contraction frequency on the force developed in an isometrically contracting cat papillary muscle are shown in Figure 18-15, *B.* Initially the strip of cardiac muscle was stimulated to contract only once every 20 seconds. When the muscle was made to contract once every 0.63 second, the developed force increased progressively over the next several beats. This progressive increase in developed force induced by a change in contraction frequency is known as the *staircase,* or *Treppe, phenomenon.* At the new steady state the developed force was more than five times as great as it was at the lower contraction frequency. A return to the slower rate had the opposite influence on developed force.

The effect of the interval between contractions on the steady-state level of developed force is shown in Figure 18-15, *A,* for a wide range of intervals. As the interval was diminished from 300 seconds down to about 10 to 20 seconds, developed force increased only slightly. However, as the interval was reduced further, to a value of about 0.5 second, force increased sharply. Further reduction of the interval to 0.2 second had little additional effect on developed force.

The progressive rise in developed force as the interval between contractions was suddenly de-creased from 20 to 0.63 second (Figure 18-15, *B*) is ascribable to a gradual rise in intracellular Ca^{++} content. Ca^{++} enters the cell during the action potential plateau (see Chapter 16). When the time between contractions is reduced (i.e., when contraction frequency is increased), the plateau shortens, and therefore less Ca^{++} enters per contraction. However, the influx of Ca^{++} per minute equals the influx per beat times the number of beats per minute. As the frequency of contraction is increased, the increment in the number of beats per minute exceeds the decrement in Ca^{++} influx per beat. Thus the Ca^{++} influx per minute increases. The consequent rise in intracellular Ca^{++} content increases the contractile force, as shown in Figure 18-15.

Postextrasystolic Potentiation   Another influence of the time between beats has been termed postextrasystolic potentiation. When the ventricles contract prematurely, the premature contraction *(extrasystole)* itself is feeble (e.g., beat *A,* in Figure 18-16). However, the next beat *(B),* which usually occurs after a short pause, is very strong. In the intact circulatory system this behavior is partly ascribable to the Frank-Starling mechanism. For most premature beats the time available for ventricular filling is not ade-

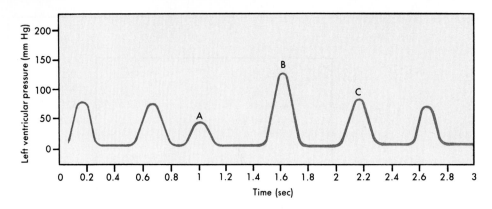

FIGURE 18-16   In an isovolumic canine left ventricle preparation a premature ventricular systole (beat *A*) is typically feeble, whereas the postextrasystolic contraction (beat *B*) is characteristically strong. The enhanced contractility may persist to a diminishing degree over a few beats (e.g., contraction *C*). (Levy MN: Unpublished tracing.)

quate, and this could be mainly responsible for the feeble contraction. Similarly, the augmented filling associated with the subsequent pause could largely explain the vigorous postextrasystolic contraction.

Although the Frank-Starling mechanism is certainly involved in the usual ventricular adaptation to a premature beat, it is **not** the exclusive mechanism. The ventricular pressure curves illustrated in Figure 18-16 were recorded from an isovolumic left ventricle preparation, in which the left ventricle neither fills nor ejects; that is, its volume remains constant throughout the cardiac cycle. Nevertheless, the premature beat *(A)* is feeble, and the succeeding contraction *(B)* is supernormal. The enhanced contractility evident in contraction *B* is an example of postextrasystolic potentiation, and it may persist for one or more additional beats (e.g., contraction *C*). The mechanism responsible for this phenomenon has not yet been fully explained.

Extrinsic Regulation of Contractility

Although the heart possesses effective intrinsic mechanisms of adaptation, a variety of extrinsic mechanisms are also important in regulating myocardial contractility. Under many natural conditions the extrinsic mechanisms dominate the intrinsic mechanisms. The extrinsic regulatory factors may be subdivided into nervous and humoral components.

Nervous Control

Sympathetic influences. Sympathetic neural activity enhances atrial and ventricular contractility. The concentration of norepinephrine in the various

regions of the heart reflects the relative density of the sympathetic innervation to those areas. In the normal heart the norepinephrine concentration in the atria is about three times that in the ventricles. The norepinephrine concentration in the SA and AV nodes is similar to that in the surrounding atrial regions. When the heart is denervated, the tissue concentration of norepinephrine approaches zero.

The alterations in ventricular contraction evoked by electrical stimulation of the left stellate ganglion in a canine isovolumic left ventricle preparation are shown in Figure 18-17. The peak pressure and the maximal rate of pressure rise (dP/dt) during systole are greatly increased. Also, the duration of systole is reduced and the rate of ventricular relaxation (as indicated by the minimal value of dP/dt) is increased during the early phase of diastole.

The shortening of systole and the more rapid ventricular relaxation assist ventricular filling. For a given cardiac cycle length, the abbreviation of systole allows more time for diastole and thus for ventricular filling. In the experiment shown in Figure 18-18, for example, the animal's heart was paced at a constant rapid rate. Sympathetic stimulation *(right panels)* shortened systole, which allowed substantially more time for ventricular filling. These factors gain importance at fast heart rates, which prevail when sympathetic activity is increased.

In the spontaneously beating heart the increase in heart rate induced by an increase in sympathetic neural activity is associated with a substantially greater abridgement of diastole than of systole. The shortening of diastole obviously impedes ventricular filling. However, the concomitant shortening of systole diminishes the extent of this impediment.

Sympathetic nervous activity enhances myocardial performance. Neurally released norepinephrine interacts with β-adrenergic receptors on the cardiac cell membranes (see Chapter 42). This reaction activates adenylate cyclase, which raises the intracellular levels of cyclic adenosine monophosphate (cAMP). Consequently protein kinases are activated and promote the phosphorylation of various proteins within the myocardial cells. Phosphorylation of specific sarcolemmal proteins augments the opening of the calcium channels in the myocardial cell membranes. Thus Ca^{++} influx increases during each action potential plateau, and more Ca^{++} is released from the sarcoplasmic reticulum in response to each cardiac excitation. The contractile strength of the heart is thereby increased. The sympathetically induced acceleration of relaxation may be mediated by the phosphorylation of a specific protein that facilitates the reuptake of cytosolic Ca^{++} by the sarcoplasmic reticulum.

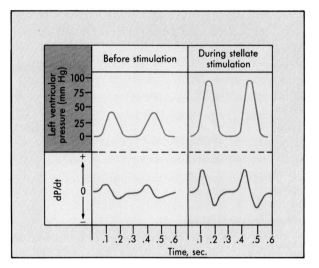

FIGURE 18-17   In an isovolumic canine left ventricle preparation, stimulation of the cardiac sympathetic nerves increases the peak left ventricular pressure and increases the maximal rates of intraventricular pressure rise and fall *(dP/dt)*. (Levy MN: Unpublished tracing.)

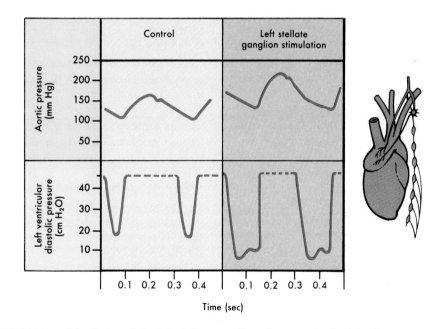

FIGURE 18-18   Stimulation of the left stellate ganglion of an anesthetized dog increases aortic pressure *(top tracing)*. The stroke volume and stroke work increase (not shown), despite a concomitant reduction in left ventricular end-diastolic pressure *(bottom tracing)*. Note also the abridgment of systole, which allows more time for ventricular filling; the heart was paced at a constant rate. In the ventricular pressure tracings the pen excursion is limited *(dashed horizontal lines)* at 45 mm Hg; actual ventricular pressures during systole can be estimated from the aortic pressure tracings. (Redrawn from Mitchell JH et al: Circ Res 8:1100, 1960.)

The overall effect of increased cardiac sympathetic activity on ventricular performance in intact animals can best be appreciated in terms of families of ventricular function curves. When sympathetic activity increases, the ventricular function curves shift progressively to the left. The changes parallel those produced by catecholamine infusions (see Figure 18-14). Thus, for any given left ventricular end-diastolic pressure, ventricular performance improves as the sympathetic nervous activity increases.

During cardiac sympathetic stimulation the increase in performance is usually accompanied by a reduction in left ventricular end-diastolic pressure. An example of the response to stellate ganglion stimulation in a paced heart is shown in Figure 18-18. In this experiment stroke volume and stroke work increased substantially, despite a 7 cm H_2O reduction in the left ventricular end-diastolic pressure. The reason for the reduction in ventricular end-diastolic pressure is explained in Chapter 23.

Parasympathetic influences. The vagus nerves strongly inhibit the cardiac pacemaker, atrial myocardium, and AV conduction tissue. These nerves also depress the ventricular myocardium, but the effects are less pronounced. In the isovolumic left ventricle preparation, vagal stimulation decreases the peak left ventricular pressure, the maximal rate of pressure development (dP/dt), and the maximal rate of pressure decline during diastole. The effects are opposite to those elicited by sympathetic stimulation (see Figure 18-17).

The effects of increased vagal activity on the ventricular myocardium are achieved largely by antagonizing the facilitatory effect of any concurrent sympathetic activity. This antagonism takes place at two levels. At the level of the autonomic nerve endings in the heart, the acetylcholine released from vagal endings inhibits the release of norepinephrine from nearby sympathetic endings. At the level of the cardiac cell membranes, the rise in intracellular cAMP that would ordinarily be produced in response to a given concentration of norepinephrine is attenuated by the acetylcholine released from nearby vagal endings.

These antagonistic interactions between the sympathetic and vagal effects on the ventricular myocardium also take place in other cardiac structures. For example, they probably account for the responses of sinus node pacemaker cells shown in Figure 18-3. In the absence of vagal stimulation, sympathetic stimulation at a frequency of 4 Hz increased heart rate substantially. However, vagal stimulation at a frequency of 8 Hz so attenuated the sympathetic influence that concurrent sympathetic stimulation at a frequency of 4 Hz had no detectable effect on heart rate.

Humoral control

Hormonal control. Various hormones influence cardiac function. The principal hormone secreted by the adrenal medulla is *epinephrine,* although some norepinephrine is also released (see Chapter 42). The rate of secretion of catecholamines by the adrenal medulla is largely regulated by the same mechanisms that control the activity of the sympathetic nervous system. The effects of those catecholamines on the heart are qualitatively similar to those released from the sympathetic nerve endings. However, the concentrations of circulating catecholamines rarely rise sufficiently high to affect cardiac function appreciably.

Thyroid hormones have pronounced effects on cardiac function (see Chapter 40). Cardiac activity is sluggish in patients with inadequate thyroid function *(hypothyroidism);* that is, the heart rate is slow and cardiac output is diminished. The converse is true in patients with overactive thyroid glands *(hyperthyroidism).* Characteristically such patients exhibit tachycardia, high cardiac output, palpitations, and arrhythmias.

Numerous studies on intact animals and humans have demonstrated that thyroid hormones enhance myocardial contractility. The rates of Ca^{++} uptake and adenosine triphosphate (ATP) hydrolysis by the sarcoplasmic reticulum are increased in experimental hyperthyroidism, and the opposite effects occur in hypothyroidism. Thyroid hormones increase protein synthesis in the heart, which leads to cardiac hypertrophy. These hormones also affect the composition of myosin isoenzymes in cardiac muscle. They increase principally those isoenzymes with the greatest ATPase activity, which thereby enhances myocardial contractility.

Insulin has a prominent, direct, positive inotropic effect on the heart of several mammalian species. The effect of insulin is evident even when hypoglycemia is prevented by glucose infusions and when the β-adrenergic receptors are blocked. In fact, the positive inotropic effect of insulin is potentiated by β-adrenergic receptor blockade. The enhancement of contractility cannot be explained satisfactorily by the concomitant augmentation of glucose transport into the myocardial cells.

Glucagon has potent positive inotropic and chronotropic effects on the heart. The endogenous hormone is probably not involved in the normal regulation of the cardiovascular system, but it has been used pharmacologically to treat a variety of cardiac conditions. The effects of glucagon on the heart closely resemble those of the catecholamines, and certain metabolic effects are similar. Both glucagon and catecholamines activate adenylate cyclase to increase the myocardial tissue levels of cAMP. The catecholamines activate adenylate cyclase by interaction with β-adrenergic receptors, but glucagon activates this enzyme through a different mechanism. Nevertheless, the consequent rise in cAMP increases Ca^{++} influx through the Ca^{++} channels in the sarcolemma and facilitates Ca^{++} release and reuptake by the sarcoplasmic reticulum, just as do the catecholamines.

Blood gases. Changes in oxygen tension of the arterial blood (PaO_2) perfusing the brain and the peripheral chemoreceptors affect the heart through nervous system mechanisms (Figure 18-10). These indirect effects of hypoxia are usually prepotent. Moderate degrees of hypoxia characteristically increase heart rate, cardiac output, and myocardial contractility by increasing sympathetic nervous activity. These changes are largely abolished by β-adrenergic receptor blockade.

The PaO_2 of the blood perfusing the myocardium also influences myocardial performance directly. The effect of hypoxia is biphasic; moderate degrees are stimulatory, and more severe degrees are depressant. As shown in Figure 18-19, when the O_2 saturation is reduced to levels below 50% in isolated hearts, the peak left ventricular pressures are less than the control levels. However, with less severe degrees of hypoxia (O_2 saturation > 50%), the peak pressures exceed the control level.

Changes in arterial blood CO_2 tension ($PaCO_2$) may also affect the myocardium directly and indirectly. The indirect, neurally mediated effects produced by increased $PaCO_2$ are similar to those evoked by a decrease in PaO_2 (Figure 18-10).

The direct effects on myocardial performance elicited by changes of $PaCO_2$ in the coronary arterial blood are illustrated in Figure 18-20. In this experiment on an isolated left ventricle preparation, the control $PaCO_2$ was 45 mm Hg (arrow A). Decreasing the $PaCO_2$ to 34 mm Hg (arrow B) was stimulatory, whereas increasing $PaCO_2$ to 86 mm (arrow C) was depressant. In intact animals systemic hypercapnia activates the sympathoadrenal system, which tends to compensate for the direct depressant effect of the increased $PaCO_2$ on the heart.

Neither the $PaCO_2$ nor the blood pH are primary determinants of myocardial behavior. The resultant change in intracellular pH is the critical factor. The

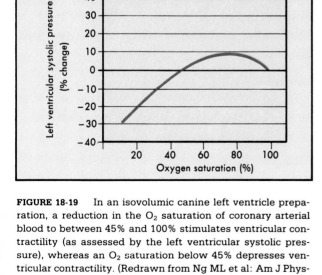

FIGURE 18-19 In an isovolumic canine left ventricle preparation, a reduction in the O_2 saturation of coronary arterial blood to between 45% and 100% stimulates ventricular contractility (as assessed by the left ventricular systolic pressure), whereas an O_2 saturation below 45% depresses ventricular contractility. (Redrawn from Ng ML et al: Am J Physiol 211:43, 1966.)

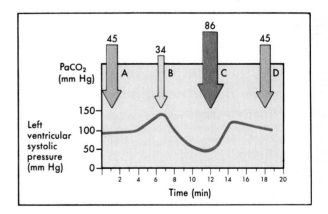

FIGURE 18-20 Decrease in $PaCO_2$ from 45 to 34 mm Hg increases left ventricular systolic pressure (arrow B) in an isovolumic canine left ventricle preparation. A subsequent rise in $PaCO_2$ to 86 mm Hg (arrow C) has the reverse effect. When the $PaCO_2$ is returned to the control level, 45 mm Hg (arrow D), left ventricular systolic pressure returns to its original value (arrow A). (From experiments by Ng ML et al: Am J Physiol 213:115, 1967.)

reduced intracellular pH diminishes the amount of Ca^{++} released from the sarcoplasmic reticulum in response to excitation. The intracellular acidosis also depresses the myofilaments directly. When they are exposed to a given concentration of Ca^{++}, the lower the prevailing pH, the less force the myofibrils develop.

BIBLIOGRAPHY
Journal Articles

Evans DB: Modulation of cAMP: mechanism for positive inotropic effect, J Cardiovasc Pharmacol 8(suppl 9):S22, 1986.

Fleming JW et al: Muscarinic receptor regulation of cardiac adenylate cyclase activity, J Mol Cell Cardiol 19:47, 1987.

Hakumäki MK: Seventy years of the Bainbridge reflex, Acta Physiol Scand 130:177, 1987.

Katona PG et al: Sympathetic and parasympathetic cardiac control in athletes and nonathletes at rest, J Appl Physiol 52:1652, 1982.

Katz AM: Cyclic adenosine monophosphate effects on the myocardium: a man who blows hot and cold with one breath, J Am Coll Cardiol 2:143, 1983.

Kurihara S and Konishi M: Effects of β-adrenoceptor stimulation on intracellular Ca transients and tension in rat ventricular muscle, Pflugers Arch 409:427, 1987.

Lakatta EG: Starling's law of the heart is explained by an intimate interaction of muscle length and myofilament calcium activation, J Am Coll Cardiol 10:1157, 1987.

Löffelholz K and Pappano AJ: The parasympathetic neuroeffector junction of the heart, Pharmacol Rev 37:1, 1985.

Levy MN: Cardiac sympathetic-parasympathetic interactions, Fed Proc 43:2598, 1984.

Mark AL: The Bezold-Jarisch reflex revisited: clinical implications of inhibitory reflexes originating in the heart, J Am Coll Cardiol 1:90, 1983.

Morad M and Cleemann L: Role of Ca^{2+} channel in development of tension in heart muscle, J Mol Cell Cardiol 19:527, 1987.

Schuhmann RE and Hoff HE: Central origin vs. reflex feedback in the respiratory heart rate relationship, Ann Biomed Eng 14:543, 1986.

Seed WA and Walker JM: Relation between beat interval and force of the heartbeat and its clinical implications, Cardiovasc Res 22:303, 1988.

Spyer KM: Neural organization and control of the baroreceptor reflex, Rev Physiol Biochem Pharmacol 88:23, 1981.

Zucker IH: Left ventricular receptors: physiological controllers or pathological curiosities? Basic Res Cardiol 81:539, 1986.

Books and Monographs

Bishop VS et al: Cardiac mechanoreceptors. In Handbook of physiology, section 2: The cardiovascular system—peripheral circulation and organ blood flow, vol III, Bethesda, Md, 1983, American Physiological Society.

Downing SE: Baroreceptor regulation of the heart. In Handbook of physiology, section 2: Cardiovascular system—the heart, vol I, Washington, DC, 1979, American Physiological Society.

Fozzard HA et al, editors: Heart and cardiovascular system: scientific foundations, New York, 1986, Raven Press.

Korner PI: Central nervous control of autonomic cardiovascular function. In Handbook of physiology, section 2: Cardiovascular system—the heart, vol I, Washington, DC, 1979, American Physiological Society.

Kulbertus HE and Franck G, editors: Neurocardiology, Mt Kisco, NY, 1988, Futura Publishing Co.

Levy MN and Martin PJ: Neural control of the heart. In Handbook of physiology, section 2: Cardiovascular system—the heart, vol I, Washington, DC, 1979, American Physiological Society.

Opie L, editor: The heart, Orlando, Fla, 1984, Grune & Stratton, Inc.

Randall WC, editor: Nervous control of cardiovascular function, New York, 1984, Oxford University Press.

Sperelakis N, editor: Physiology and pathophysiology of the heart, ed 2, Boston, 1989, Kluwer Academic Publishers.

19

Hemodynamics

Analysis of the fluid mechanics of the circulatory system is very complicated. The heart is an intermittent pump, and its behavior is regulated by many physical and chemical factors. The blood vessels are branched, distensible conduits of continuously varying dimensions. The blood is a suspension of red and white corpuscles, platelets, and lipid globules, all dispersed in a colloidal solution of proteins. Despite this complexity, however, an understanding of the relevant elementary principles of fluid mechanics provides considerable insight into the physical behavior of the cardiovascular system. Certain basic principles are expounded in this chapter to illuminate the interrelations among vascular geometry, blood flow, and blood pressure.

VELOCITY OF THE BLOODSTREAM

The relationship between the velocity of the bloodstream and the dimensions of the vascular bed is illustrated by the hydraulic system in Figure 19-1. Consider that the conduit is rigid and has a wide section (area $A_1 = 5$ cm^2) and a narrow section (area $A_2 = 1$ cm^2). Also, let an incompressible fluid flow into the wide end of the tube at a rate, $\dot{V}_1$, of 5 cm^3·sec^{-1}. Then the velocity, v_1, of a fluid particle as it passes cross section A_1 would be

$$v_1 = \dot{V}_1/A_1 \qquad (1)$$
$$= 5 \text{ cm}^3 \cdot \text{s}^{-1}/5 \text{ cm}^2$$
$$= 1 \text{ cm} \cdot \text{s}^{-1}$$

Thus a particle of fluid advances a distance (Δl_1) of 1 cm each second. When the fluid enters the narrow section of the tube, the volume of fluid that passes cross section A_2 each second must equal the volume that had passed cross section A_1 each second; that is, $\dot{V}_2 = \dot{V}_1$. The velocity, v_2, in the narrow section would be

$$v_2 = \dot{V}_2/A_2 \qquad (2)$$
$$= 5 \text{ cm}^3 \cdot \text{s}^{-1}/1 \text{ cm}^2$$
$$= 5 \text{ cm} \cdot \text{s}^{-1}$$

Thus each particle of fluid must move past section A_2 five times faster than it did past section A_1. Each particle would have to move a distance (Δl_2) of 5 cm each second.

By the law of conservation of mass, the flow ($\dot{V}_1$) of fluid past A_1 must equal the flow ($\dot{V}_2$) past A_2; that is,

$$\dot{V}_1 = \dot{V}_2 \qquad (3)$$

From equations 1 to 3

$$v_1 A_1 = v_2 A_2 \qquad (4)$$

Therefore

$$v_1/v_2 = A_2/A_1 \qquad (5)$$

Thus, when the caliber of a tube varies with location, the fluid velocities at different sites in the tube are inversely proportional to the cross-sectional areas. This relationship also holds for more complex hydraulic systems, such as the circulatory system. Such systems may be composed of large numbers of conduits, arranged both in series and in parallel.

Note that in Figure 15-2 the velocity decreases progressively as the blood traverses the aorta, its primary and secondary branches, the arterioles, and finally

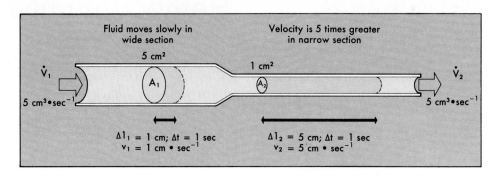

Fluid moves slowly in wide section

Velocity is 5 times greater in narrow section

5 cm²

1 cm²

$\dot{V}_1$

A_1

A_2

$\dot{V}_2$

5 cm³•sec⁻¹

5 cm³•sec⁻¹

$\Delta l_1 = 1$ cm; $\Delta t = 1$ sec
$v_1 = 1$ cm $\cdot$ sec⁻¹

$\Delta l_2 = 5$ cm; $\Delta t = 1$ sec
$v_2 = 5$ cm $\cdot$ sec⁻¹

FIGURE 19-1   In a conduit that contains a wide and a narrow segment, the fluid velocities in the two segments are inversely proportional to the cross-sectional areas of the segments.

the capillaries. As the blood then passes through the venules and continues centrally through the larger veins toward the venae cavae, the velocity increases progressively again. The velocities in the various serial sections of the circulatory system are inversely proportional to the total cross-sectional areas of the respective sections. The total cross-sectional area of all the parallel systemic capillaries greatly exceeds the total cross-sectional area of any other serial section of the systemic vascular bed. Therefore the velocity of the bloodstream in the capillaries is much less than that in any other vascular segment. The very slow movement of the blood through the capillaries allows ample time for exchange of materials between the tissues and the blood.

RELATIONSHIP BETWEEN PRESSURE AND FLOW

Poiseuille's law applies to the flow of fluids through cylindric tubes, but only under special conditions. It applies specifically to the steady, laminar flow of newtonian fluids. The term *steady flow* signifies the absence of variations of flow in time. *Laminar flow* is the type of motion in which the fluid moves as a series of infinitesimally thin layers, with each layer moving at a velocity different from that of its neighboring layers. In the case of laminar flow through a tube, the fluid consists of a series of very thin concentric tubes sliding past one another (see Figure 19-7). A *newtonian fluid* may be considered to be a homogeneous fluid, such as water, in contrast to a suspension, such as blood. Laminar flow and newtonian fluids are described in more detail later.

Effects of Pressure Difference

Pressure is a salient determinant of flow. The pressure, P, in dynes·cm⁻², at a distance h cm below the surface of a liquid is

$$P = h\rho g \qquad \textbf{(6)}$$

where ρ is the density of the liquid in g·cm⁻³, and g is the acceleration of gravity in cm·sec⁻². For convenience, however, pressure is frequently expressed in terms of the height, h, of the column of liquid above an arbitrary reference level.

Consider the tube that connects reservoirs R_1 and R_2 in Figure 19-2. Let reservoir R_1 be filled with liquid to height h_1, and let reservoir R_2 be empty, as in panel A. The outflow pressure, P_o, is therefore equal to the atmospheric pressure, which shall be designated as the zero, or reference, level. The inflow pressure, P_i, is then equal to the same reference level plus the height, h_1, of the column of liquid in reservoir R_1. Under these conditions let the flow, $\dot{V}$, through the tube be 5 ml·sec⁻¹.

If reservoir R_1 is filled to height h_2, which is twice h_1, and reservoir R_2 is again empty (as in panel B), the flow will be twice as great (i.e., 10 ml·sec⁻¹), as it is in panel A. Thus with reservoir R_2 empty, the flow will be directly proportional to the inflow pressure, P_i.

If reservoir R_2 is now allowed to fill to height h_1, and the fluid level in R_1 is maintained at h_2 (as in panel C), the flow will again become 5 ml·sec⁻¹. Thus flow is directly proportional to the difference between inflow and outflow pressures:

$$\dot{V} \propto P_1 - P_o \qquad \textbf{(7)}$$

If the fluid level in R_2 attains the same height as in R_1, flow will cease (panel D).

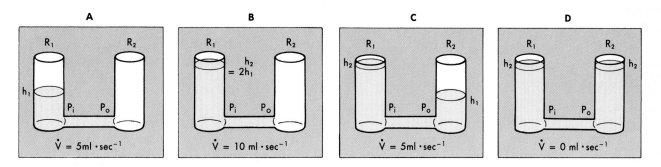

FIGURE 19-2   The flow, $\dot{V}$, of fluid through a tube connecting two reservoirs, R_1 and R_2, is proportional to the difference between the pressure, P_i, at the inflow end and the pressure, P_o, at the outflow end of the tube. **A,** When R_2 is empty, fluid flows from R_1 to R_2 at a rate proportional to the pressure in R_1. **B,** When the fluid level in R_1 is increased twofold, the flow increases proportionately. **C,** Flow from R_1 to R_2 is proportional to the difference between the pressures in R_1 and R_2. **D,** When pressure in R_2 rises to equal the pressure in R_1, flow ceases in the connecting tube.

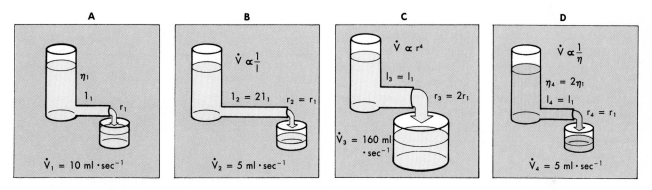

FIGURE 19-3   The flow, $\dot{V}$, of fluid through a tube is inversely proportional to the length, l, and the viscosity, η, and is directly proportional to the fourth power of the radius, r. **A,** Reference condition: for a given pressure, length, radius, and viscosity, let the flow ($\dot{V}_1$) equal 10 ml·s^{-1}. **B,** If tube length doubles, flow decreases by 50%. **C,** If tube radius doubles, flow increases 16-fold. **D,** If viscosity doubles, flow decreases by 50%.

Effects of Tube Dimensions

For any given pressure difference between the two ends of a tube, the flow will depend on the dimensions of the tube. Consider the tube connected to the reservoir in Figure 19-3, *A*. If the tube's length is l_1 and its radius is r_1, the flow $\dot{V}_1$ is observed to be 10 ml·sec^{-1}.

The tube connected to the reservoir in panel *B* has the same radius but is twice as long. Under these conditions the flow $\dot{V}_2$ is found to be 5 ml·sec^{-1}, or only half as great as $\dot{V}_1$. Conversely, for a tube half as long as l_1, the flow would be twice as great as $\dot{V}_1$. In other words, flow is inversely proportional to the length of the tube:

$$\dot{V} \propto 1/l \qquad \textbf{(8)}$$

The length of the tube connected to the reservoir in Figure 19-3, *C*, is the same as l_1, but the radius is twice as great as r_1. Under these conditions the flow $\dot{V}_3$ increases to a value of 160 ml·sec^{-1}, which is 16 times greater than $\dot{V}_1$. The precise measurements of Poiseuille revealed that flow varies directly as the fourth power of the radius:

$$\dot{V} \propto r^4 \qquad \textbf{(9)}$$

Thus, in the previous example, since $r_3 = 2r_1$, $\dot{V}_3$ will be proportional to $(2r_1)^4$, or $16r_1^4$; therefore $\dot{V}_3$ will equal $16\dot{V}_1$.

Effect of Viscosity

Finally, for a given pressure difference across a cylindric tube of given dimensions, the flow will be affected by the nature of the fluid itself. This flow-determining property of fluids is termed *viscosity*, η. Consider that the pressure in reservoir D in Figure 19-3 equals that in reservoir A, and that the tubes connected to reservoirs A and D are identical. However, if the viscosity, η_4, of the fluid in reservoir D is twice the viscosity, η_1, of the fluid in reservoir A, then the flow, $\dot{V}_4$, through the tube in D will be only half the flow, $\dot{V}_1$, through the tube in A. Thus

$$\dot{V} \propto 1/\eta \tag{10}$$

For most homogeneous liquids, such as water or true solutions in water, this inverse proportionality prevails during laminar flow. Such fluids are said to be *newtonian*. For heterogeneous liquids, notably suspensions such as blood, this precise inverse proportionality does not apply. Such fluids are said to be *nonnewtonian*.

Poiseuille's Law

Poiseuille's law takes into account the various factors that influence the laminar flow of a fluid through a tube. It states that for the steady, laminar flow of a newtonian fluid through a cylindric tube, the flow, $\dot{V}$, varies directly as the pressure difference, $P_i - P_o$, and the fourth power of the radius, r, of the tube, and it varies inversely as the length, l, of the tube and the viscosity, η, of the fluid. The full statement of Poiseuille's law is

$$\dot{V} = \frac{\pi(P_i - P_o)r^4}{8\eta l} \tag{11}$$

where $\pi/8$ is the constant of proportionality.

RESISTANCE TO FLOW

In electrical theory, *resistance*, R, is defined as the ratio of voltage drop, E, to current flow, I. By analogy, a hydraulic resistance, R, may be defined as the ratio of pressure drop, $P_i - P_o$, to flow, $\dot{V}$. For the steady, laminar flow of a newtonian fluid through a cylindric tube, the physical components of hydraulic resistance may be appreciated by rearranging Poiseuille's law to yield the hydraulic resistance equation:

$$R = \frac{P_i - P_o}{\dot{V}} = \frac{8\eta l}{\pi r^4} \tag{12}$$

Thus, when Poiseuille's law applies, the resistance to flow depends only on the dimensions (l and r) of the tube and on the viscosity (η) of the fluid.

The principal determinant of the resistance to blood flow through any individual vessel within the circulatory system is its caliber. The resistance to flow through small blood vessels in the cat mesentery has been measured, and the resistance per unit length of vessel (R/l) is plotted against the vessel diameter in Figure 19-4. The resistance is highest in the capillaries (diameter, $7\mu m$), and it diminishes as the vessels increase in diameter on the arterial and

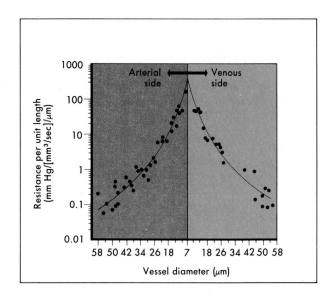

FIGURE 19-4   The resistance per unit length *(R/l)* of individual small blood vessels in the cat mesentery. The capillaries, diameter 7 μm, are denoted by the vertical dashed line. Resistances of the arterioles are plotted to the left and resistances of the venules are plotted to the right of the vertical dashed line. The solid circles represent the actual data. The two curves through the data represent the following regression equations for the arteriole and venule data, respectively: (1) arterioles, $R/l = 1.02 \times 10^6 D^{-4.04}$, and (2) venules, $R/l = 1.07 \times 10^6 D^{-3.94}$. Note that for both types of vessels, the resistance per unit length is inversely proportional to the fourth power (within 1%) of the vessel diameter. (Redrawn from Lipowsky HH et al: Circ Res 43:738, 1978.)

venous sides of the capillaries. The values of R/l were found to be virtually proportional to the fourth power of the diameter for the larger vessels on both sides of the capillaries.

Changes in vascular resistance induced by nervous and humoral mechanisms occur by altering the vessel radius. The principal changes are achieved by alterations in the contractile state of the smooth muscle cells in the vessel wall.

Figure 15-2 shows that among the various types of vessels aligned in series in the circulatory system, the greatest pressure drop occurs across the arterioles. Because the total flow is the same through these serial components of the circulatory system, the pressure drop across any specific component of the vascular system is proportional to its resistance. It follows, therefore, that the greatest resistance to flow resides in the arterioles. The arterioles possess a thick coat of circularly arranged smooth muscle fibers, by means of which the lumen radius may be varied. From the hydraulic resistance equation (equation 12), in which R varies inversely as r^4, it is clear that small changes in radius will alter resistance greatly.

Resistances in Series and in Parallel

In the cardiovascular system the various types of vessels listed along the horizontal axis in Figure 15-2 lie in series with one another. Furthermore, the individual members within each category of vessels are ordinarily arranged in parallel with one another (Figure 15-3). For example, the capillaries throughout the lungs are in parallel with one another, and similarly the capillaries throughout the rest of the body are in most instances in parallel with one another. Notable exceptions are the capillaries in the renal vasculature (where the peritubular capillaries are in series with the glomerular capillaries) and those in the splanchnic vasculature (where the hepatic capillaries are in series with the intestinal capillaries). Formulas for the total hydraulic resistance of components arranged in series and in parallel can be derived in the same manner as for electrical resistances.

Three hydraulic resistances, R_1, R_2, and R_3, are aligned in series in Figure 19-5. The pressure drop across the entire system (i.e., the difference between inflow pressure, P_i, and outflow pressure, P_o) consists of the sum of the pressure drops across each of the individual resistances (Figure 19-5, equation 1). Under steady-state conditions, the flow, $\dot{V}$, through any

given cross section must equal the flow through any other cross section. By dividing each component in equation 1 by $\dot{V}$ (equation 2), it becomes evident from the definition of resistance (i.e., $R = (P_i - P_o)/\dot{V}$) that the total resistance, R_t, of the entire system equals the sum of the individual resistances; that is,

$$R_t = R_1 + R_2 + R_3 \qquad \textbf{(13)}$$

For resistances in parallel (Figure 19-6), all tubes have the same inflow pressures and the same outflow pressures. The total flow, $\dot{V}_t$, through the system equals the sum of the flows through the individual parallel elements (Figure 19-6, equation 1). Because the pressure difference $(P_i - P_o)$ is identical for all parallel elements, each term in equation 1 (Figure 19-6) may be divided by that pressure difference to yield equation 2. From the definition of resistance, equation 3 (Figure 19-6) may be derived. This equation states that the reciprocal of the total resistance, R_t, equals the sum of the reciprocals of the individual resistances; that is,

$$\frac{1}{R_t} = \frac{1}{R_1} + \frac{1}{R_2} + \frac{1}{R_3} \qquad \textbf{(14)}$$

Stated in another way, if we define hydraulic *conductance* as the reciprocal of resistance, it becomes evident that, for tubes in parallel, the total conductance is the sum of the individual conductances.

By considering a few simple illustrations, some of the fundamental properties of parallel hydraulic systems become apparent. For example, if the resistances of the three parallel elements in Figure 19-6 were all equal, then

$$R_1 = R_2 = R_3 \qquad \textbf{(15)}$$

Therefore

$$1/R_t = 3/R_1 \qquad \textbf{(16)}$$

and

$$R_t = R_1/3 \qquad \textbf{(17)}$$

Thus the total resistance is less than any of the individual resistances. Furthermore, for any parallel arrangement, the total resistance must be less than that of any of the individual parallel tubes. For example, consider a system in which a very high-resistance tube is added in parallel to a low-resistance tube. The total resistance must be less than that of the low-resistance component by itself, because the high-resistance component affords an additional pathway, or conductance, for fluid flow.

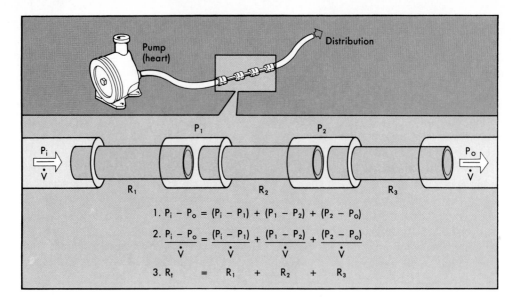

FIGURE 19-5   For resistances $(R_1, R_2,$ and $R_3)$ arranged in series, the total resistance, R_t, equals the sum of the individual resistances.

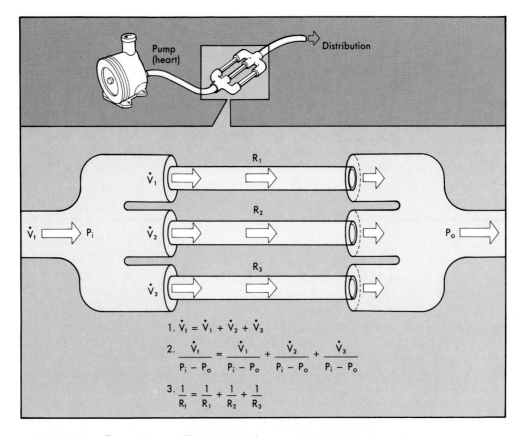

FIGURE 19-6   For resistances $(R_1, R_2,$ and $R_3)$ arranged in parallel, the reciprocal of the total resistance, R_t, equals the sum of the reciprocals of the individual resistances.

LAMINAR AND TURBULENT FLOW

Under certain conditions, the flow of a fluid in a cylindric tube will be *laminar,* as illustrated in Figure 19-7. The thin layer of fluid in contact with the wall of the tube adheres to the wall and thus is motionless. The layer of fluid just central to this external lamina must shear against this motionless layer. Therefore this adjacent layer moves slowly, but with a finite velocity. Similarly, the next more central layer travels still faster. The longitudinal velocity profile is that of a parabola. The velocity of the fluid adjacent to the wall is zero, whereas the velocity at the center of the stream is maximal. The maximal velocity is twice the mean velocity of flow across the entire cross section of the tube. In laminar flow, fluid elements remain in one lamina, or streamline, as the fluid progresses longitudinally along the tube. Flow occurs only in an axial direction, that is, parallel to the axis of the tube. No particles of fluid move in either a radial or a circumferential direction.

Irregular motions of the fluid elements may develop in the flow of fluid through a tube; this irregular flow is called *turbulent* (Figure 19-8). Under such conditions fluid elements do not remain confined to definite laminae, but rapid radial and circumferential mixing occurs, and vortices may develop. More pressure is required to force a given flow of fluid through the same tube when the flow is turbulent than when it is laminar. In turbulent flow the pressure drop is approximately proportional to the square of the flow, whereas in laminar flow the pressure drop is proportional to the first power of the flow. Thus, to produce a given flow, a pump such as the heart must do considerably more work if turbulence develops.

Whether the flow through a tube will be turbulent or laminar may be predicted on the basis of a dimensionless number called *Reynold's number,* N_R, which is defined as follows:

$$N_R = \rho D \bar{v} / \eta \qquad \textbf{(18)}$$

where D is the tube diameter, $\bar{v}$ is the mean velocity, ρ is the fluid density, and η is the fluid viscosity. For

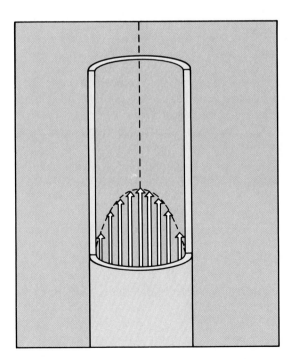

FIGURE 19-7   In laminar flow all elements of the fluid move in streamlines that are parallel to the axis of the tube; movement does not occur in a radial or circumferential direction. The layer of fluid in contact with the wall is motionless; the fluid that moves along the axis of the tube has the maximal velocity.

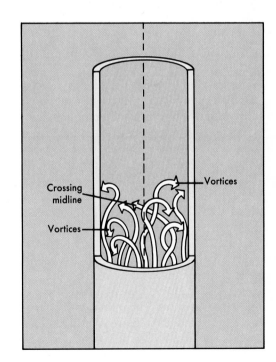

FIGURE 19-8   In turbulent flow the elements of the fluid move irregularly in axial, radial, and circumferential directions. Vortices frequently develop.

N_R <2000, the flow will usually be laminar; for N_R >3000, turbulence will usually prevail. Various flow conditions may develop in the transition range of N_R between 2000 and 3000. Because flow tends to be laminar at low N_R and turbulent at high N_R, it is evident from equation 18 that large diameters, high velocities, and low viscosities predispose to the development of turbulence. In addition to these factors, abrupt variations in tube dimensions or irregularities in the tube walls may produce turbulence.

Turbulence is usually accompanied by vibrations of the fluid and surrounding structures. Some of these vibrations are in the auditory frequency range. Turbulent flow within the cardiovascular system may be detected as a *murmur.*

The factors just cited that predispose to turbulence may account for some murmurs heard clinically. In severe anemia *functional cardiac murmurs* (murmurs not caused by structural abnormalities) are often detectable. Such murmurs are ascribable to the reduced viscosity of the blood (caused by the low red blood cell content) and to the high flow velocities that usually prevail in anemic patients.

RHEOLOGIC PROPERTIES OF BLOOD

The viscosity of a newtonian fluid, such as water, may be determined by measuring the rate of flow of the fluid at a given pressure gradient through a cylindric tube of known length and radius. As long as the fluid flow is laminar, the viscosity may be computed by substituting these values into Poiseuille's equation. The calculated viscosity of a given newtonian fluid at a specified temperature will be constant, regardless of the tube dimensions and flows. However, for a nonnewtonian fluid, the viscosity calculated from Poiseuille's equation may vary considerably when different tube dimensions and flows are used. Therefore, in considering the rheologic properties of a suspension such as blood, the term *viscosity* does not have a unique meaning. The term *apparent viscosity* is frequently applied to the value of viscosity obtained for blood under the particular conditions of measurement.

Rheologically, blood is a suspension, principally of erythrocytes in a relatively homogeneous liquid, the blood plasma. For this reason the apparent viscosity of blood varies as a function of the *hematocrit ratio* (ratio of volume of red blood cells to volume of whole

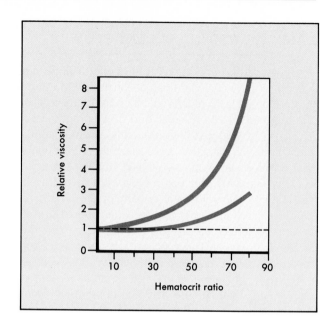

FIGURE 19-9   The viscosity of whole blood, relative to that of plasma, increases at a progressively greater rate as the hematocrit ratio increases. For any given hematocrit ratio, the apparent viscosity of blood is less when measured in a biologic viscometer (e.g., the tissues of an anesthetized dog) than in a conventional capillary tube viscometer. (Redrawn from Levy MN and Share L: Circ Res 1:247, 1953.)

blood). In Figure 19-9 the upper curve represents the ratio of the apparent viscosity of whole blood to that of plasma over a range of hematocrit ratios from 0% to 80%. The data were derived from measurements of flow through a tube 1 mm in diameter. The viscosity of plasma is 1.2 to 1.3 times that of water. Figure 19-9 *(upper curve)* shows that blood, with a normal hematocrit ratio of 45%, has an apparent viscosity 2.4 times that of plasma. In severe anemia blood viscosity is low. With increasing hematocrit ratios the slope of the curve increases progressively; it is especially steep at the upper range of erythrocyte concentrations. If the hematocrit ratio rises to about 70%, which it may in patients with *polycythemia,* the apparent viscosity increases more than twofold, and the resistance to blood flow increases proportionately. The effect of such a change in hematocrit ratio on peripheral resistance may be appreciated when it is recognized that in patients with severe hypertension, the total peripheral resistance rarely increases more than twofold. In hypertension the increase in peripheral resistance is usually achieved by arteriolar vasoconstriction.

For any given hematocrit ratio, the apparent vis-

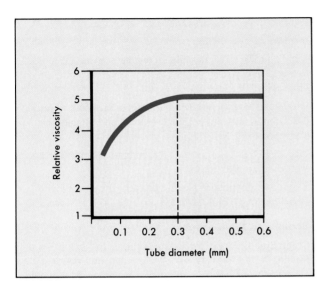

FIGURE 19-10   The viscosity of blood, relative to that of water, increases as a function of tube diameter up to a diameter of about 0.3 mm. (Redrawn from Fahraeus R and Lindqvist T: Am J Physiol 96:562, 1931.)

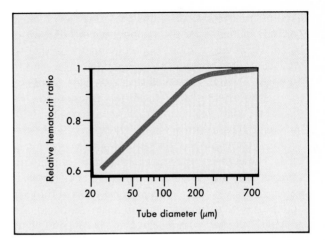

FIGURE 19-11   The "relative hematocrit ratio" of blood flowing from a feed reservoir through capillary tubes of various calibers, as a function of the tube diameter. The relative hematocrit ratio is the ratio of the hematocrit of the blood in the tubes to that of the blood in the feed reservoir. (Redrawn from Barbee JH and Cokelet GR: Microvasc Res 3:6, 1971.)

cosity of blood depends on the dimensions of the tube used to estimate the viscosity. Figure 19-10 demonstrates that the apparent viscosity of blood is not affected appreciably by changes in tube diameter when the diameters exceed 0.3 mm, but the apparent viscosity does diminish progressively as the tube diameter is decreased to values less than about 0.3 mm.

The influence of tube diameter on apparent viscosity is partly ascribable to the difference in the actual composition of the blood as it flows from large tubes into small tubes. The composition changes because in the small tubes the red blood cells tend to accumulate in the faster axial stream, whereas the blood component that flows in the slower marginal layers is largely plasma. Because the red blood cells traverse the tube more quickly than does the plasma, the hematocrit ratio of the blood in the tube is actually less than that of the blood in the reservoir to which the tube is connected (Figure 19-11).

The apparent viscosity of blood diminishes as the shear rate is increased (Figure 19-12), a phenomenon called *shear thinning*. The shear rate is the rate at which one layer of fluid moves with respect to the adjacent layers; the shear rate varies directly with the flow. The greater tendency of the erythrocytes to

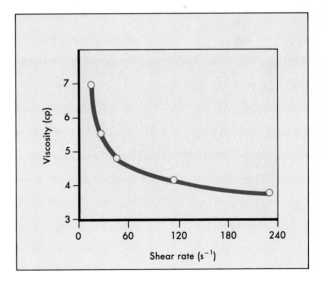

FIGURE 19-12   Decrease in the viscosity of blood (centipoise) at increasing rates of shear. The shear rate refers to the velocity of one layer of fluid relative to that of the adjacent layers and is directionally related to the rate of flow. (Redrawn from Amin TM and Sirs JA: Q J Exp Physiol 70:37, 1985.).

accumulate in the axial laminae at higher flow rates is partly responsible for the nonnewtonian behavior of blood. However, a more important factor is that at very slow rates of shear, the suspended cells tend to aggregate, which increases viscosity. This tendency to aggregate decreases as the flow is augmented. The resultant diminution in apparent viscosity with increasing shear rate is shown in Figure 19-12.

The deformability of the erythrocytes is also a factor in shear thinning, especially at high hematocrit ratios. The mean diameter of human red blood cells is about 7 μm, yet these cells are able to pass through openings with a diameter of only 3 μm. As blood that is densely packed with erythrocytes is caused to flow at progressively greater rates, the erythrocytes become more and more deformed. The greater deformation diminishes the apparent viscosity of the blood. The flexibility of human erythrocytes is enhanced as the concentration of fibrinogen in the plasma increases.

BIBLIOGRAPHY
Journal Articles

Amin TM and Sirs JA: The blood rheology of man and various animal species, J Exp Physiol 70:37, 1985.

Barbee JH and Cokelet GR: The Fahraeus effect, Microvasc Res 3:6, 1971.

Chien S: Role of blood cells in microcirculatory regulation, Microvasc Res 29:129, 1985.

Fahraeus R and Lindqvist T: The viscosity of blood in narrow capillary tubes, Am J Physiol 96:562, 1931.

Goldsmith HL: The microrheology of human blood, Microvasc Res 31:121, 1986.

McMillan DE et al: Rapidly recovered transient flow resistance: a newly discovered property of blood, Am J Physiol 253:H919, 1987.

Sarelius IH and Duling DR: Direct measurement of microvessel hematocrit, red cell flux, velocity, and transit time, Am J Physiol 243:H1018, 1982.

Winter DC and Nerem RM: Turbulence in pulsatile flows, Ann Biomed Eng 12:357, 1984.

Zamir M: The role of shear forces in arterial branching, J Gen Physiol 67:213, 1976.

Books and Monographs

Chien S et al: Blood flow in small tubes. In Renkin EM and Michel CC, editors: Handbook of physiology, section 2: The cardiovascular system—microcirculation, vol IV, Bethesda, MD, 1984, American Physiological Society.

Cokelet, GR et al, editors: Erythrocyte mechanics and blood flow, New York, 1980, Alan R Liss, Inc.

Fung YC: Biodynamics: circulation, New York, 1984, Springer-Verlag New York, Inc.

Milnor WR: Hemodynamics, Baltimore, 1982, Williams & Wilkins.

Noordergraaf A: Circulatory system dynamics, New York, 1979, Academic Press, Inc.

Taylor DEM and Stevens AL, editors: Blood flow: theory and practice, New York, 1983, Academic Press, Inc.

The Arterial System

HYDRAULIC FILTER

The principal function of the systemic and pulmonary arterial systems is to distribute blood to the capillary beds throughout the body. The arterioles, which are the terminal components of the arterial system, regulate the distribution to the various capillary beds. The aorta and pulmonary artery and their major branches constitute a system of conduits that lie between the heart and the arterioles. These conduits have a substantial volume and are very distensible.

An arterial system composed of elastic conduits and high-resistance terminals constitutes a *hydraulic filter,* since it converts the intermittent output of the heart to a steady flow through the capillaries (Figure 20-1). The entire ventricular stroke volume is discharged into the arterial system during ventricular systole, which usually occupies approximately one-third the duration of the cardiac cycle. Most of the stroke volume is pumped during the rapid ejection phase (see Figure 17-10), which constitutes about half of total systole. Part of the energy released by the cardiac contraction is kinetic energy; it is dissipated as forward capillary flow during ventricular systole. The remainder is stored as potential energy, in that much of the stroke volume is retained by the distensible arteries (Figure 20-1, *A*). During diastole the elastic recoil of the arterial walls converts this potential energy into capillary blood flow, which continues throughout diastole (FIgure 20-1, *B*). If the arterial walls were rigid, capillary flow would take place only during ventricular systole (Figure 20-1, *C*), and flow would cease during diastole (Figure 20-1, *D*).

Hydraulic filtering minimizes the workload of the heart. More work is required to pump a given flow intermittently then steadily; the more effective the filtering, the less the work required. A simple example illustrates this point.

Consider first the steady flow of a fluid at a rate of 100 ml·sec^{-1} through a hydraulic system with a resistance of 1 mm Hg·ml^{-1}·sec. This combination of steady flow and resistance would result in a constant pressure of 100 mm Hg, as shown in Figure 20-2, *A*. Neglecting any inertial effect, hydraulic work (W) may be defined as

$$W = \int_{t_1}^{t_2} P\,dV \qquad (1)$$

That is, each small increment of volume, dV, that is pumped at any given moment is multiplied by the pressure, P, existing at that time, and the products are integrated over the time interval of interest, $t^2 - t_1$, to give the total work, W. For steady flow, $W = PV$. In the example in Figure 20-2, *A*, the work done in pumping the fluid for 1 second would be 10,000 mm Hg·ml (or 1.33×10^7 dyne·cm).

Next, consider an intermittent pump that ejects the same volume per second but pumps the entire volume at a steady rate over 0.5 second and then pumps nothing during the next 0.5 second. Thus it pumps at the rate of 200 ml·sec^{-1} for 0.5 second, as shown in Figure 20-2, *B* and *C*. In *B* the conduit is rigid and the fluid is incompressible, but the system has the same resistance as in *A*. During the pumping phase of the cycle (systole) the flow of 200 ml·sec^{-1} through a resistance of 1 mm Hg·ml^{-1}·sec would produce a pressure of 200 mm Hg. During the filling phase of the cycle (diastole) the pressure would be 0

255

Compliant arteries

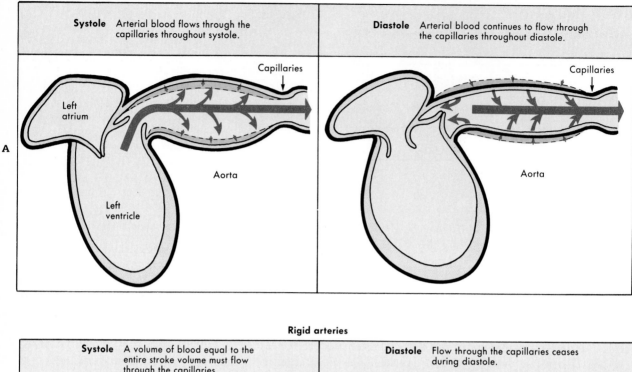

Systole Arterial blood flows through the capillaries throughout systole.	**Diastole** Arterial blood continues to flow through the capillaries throughout diastole.

A

B

Capillaries

Left atrium

Aorta

Left ventricle

Capillaries

Aorta

Rigid arteries

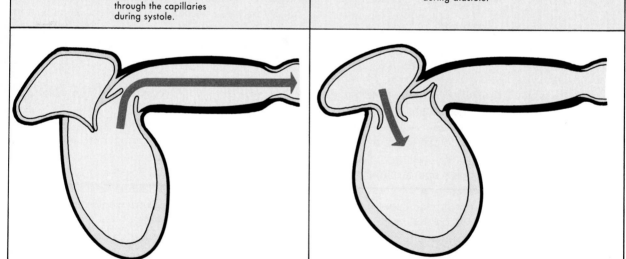

Systole A volume of blood equal to the entire stroke volume must flow through the capillaries during systole.	**Diastole** Flow through the capillaries ceases during diastole.

C

D

FIGURE 20-1   When the arteries are normally compliant, blood flows through the capillaries throughout the cardiac cycle. When the arteries are rigid, blood flows through the capillaries during systole, but flow ceases during diastole. **A,** When the arteries are normally compliant, a substantial fraction of the stroke volume is stored in the arteries during ventricular systole. The arterial walls are stretched. **B,** During ventricular diastole the previously stretched arteries recoil. The volume of blood that is displaced by the recoil furnishes continuous capillary flow throughout diastole. **C,** When the arteries are rigid, virtually none of the stroke volume can be stored in the arteries. **D,** Rigid arteries cannot recoil appreciably during diastole.

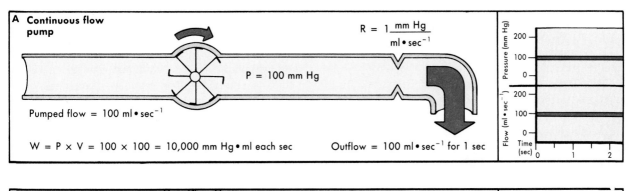

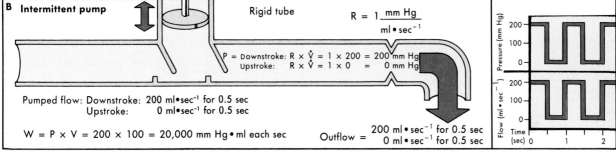

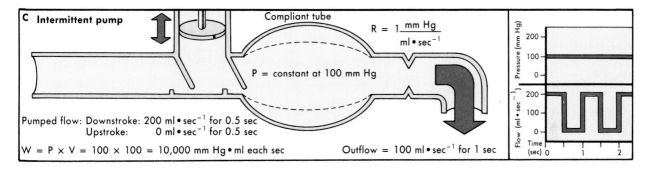

FIGURE 20-2   The relationships between pressure and flow for three hydraulic systems. In each the overall flow is 100 ml·sec^{-1} and the resistance is 1 mm Hg·ml^{-1}·sec. **A,** The flow is steady, and pressure will remain constant regardless of the distensibility of the conduit. **B,** The flow produced by the pump is intermittent; it is steady for half the cycle and ceases for the remainder of the cycle. The conduit is rigid, and therefore the flow produced by the pump during its downstroke must exit through the resistance during the same 0.5 second that elapses during the downstroke. The pump must do twice as much work as the pump in **A. C,** The pump operates as in **B,** but the conduit is infinitely distensible. This results in perfect filtering of the pressure; that is, the pressure is steady, and the flow through the resistance is also steady. The work equals that in **A.**

mm Hg in this rigid system. The work done during systole would be 20,000 mm Hg·ml, which is twice that required in the example shown in Figure 20-2, *A.*

If the system were very distensible, hydraulic filtering would be very effective, and the pressure would remain virtually constant throughout the entire cycle (Figure 20-2, *C*). Of the 100 ml of fluid pumped during

the 0.5 second of systole, only 50 ml would be emitted through the high-resistance outflow end of the system during systole. The remaining 50 ml would be stored by the distensible conduit during systole and would flow out during diastole. Thus the pressure would be virtually constant at 100 mm Hg throughout the cycle. The fluid pumped during systole would be ejected at only half the pressure that prevailed in Figure 20-2, *B,* and therefore the work would be only half as great. With nearly perfect filtering, as in Figure 20-2, *C,* the work would be identical to that for steady flow (Figure 20-2, *A*).

The filtering accomplished by the systemic and pulmonic arterial systems is intermediate between the filtering in the examples in Figure 20-2, *B* and *C.* Under average normal conditions the work increment imposed by intermittent pumping, in excess of that for steady flow, is about 35% for the right ventricle and about 10% for the left ventricle. These fractions change, however, with variations in heart rate, peripheral resistance, and arterial distensibility.

ARTERIAL COMPLIANCE

The elastic properties of the arterial wall may be appreciated by considering first the static pressure–volume relationship for the aorta. To obtain the curves shown in Figure 20-3, aortas were obtained at autopsy from people in different age groups. All branches of the aorta were tied, and successive volumes of liquid were injected into this closed elastic system, just as successive small volumes of water might be intro-

duced into a balloon. After each increment of volume was injected, the internal pressure in the aorta was measured. In Figure 20-3 the curve that relates pressure to volume for the youngest age group (20-24 years) is sigmoidal. The curve is quite linear over most of its extent, but the slope decreases at the upper and lower ends. At any given point the slope (dV/dP) represents the *aortic compliance.* Thus in normal people the aortic compliance is least at very high and low pressures and greatest over the normal range of pressure variations. These compliance changes resemble the familiar changes encountered in inflating a balloon. The greatest difficulty in introducing air into the balloon (i.e., the least compliance) is experienced at the beginning of inflation and again at near-maximal volume, just before rupture of the balloon. At intermediate volumes, the balloon is easy to inflate (i.e., its compliance is large).

It is apparent from Figure 20-3 that the curves become displaced downward and the slopes diminish as a function of advancing age. Thus the compliance decreases with age, which is a manifestation of the progressive changes in the collagen and elastin contents of the arterial walls. The heart cannot eject a given stroke volume into a rigid arterial system as readily as it can into a more compliant system.

DETERMINANTS OF ARTERIAL BLOOD PRESSURE

The factors that determine the arterial blood pressure cannot be evaluated with great precision. The arterial blood pressure is routinely measured in most patients at the physician's office, however, and it provides some useful clues to the patient's cardiovascular status. We therefore take a simplified approach in an attempt to understand the principal determinants of the arterial blood pressure. To accomplish this, we first analyze the determinants of the mean arterial pressure (defined in the following section). The systolic and diastolic arterial pressures then are considered to be the upper and lower limits of periodic oscillations about this mean pressure. Second, we analyze the determinants of these periodic oscillations.

The determinants of the arterial blood pressure are arbitrarily subdivided here into **physical** and **physiologic** factors (Figure 20-4). For the sake of simplicity, the arterial system is assumed to be a static, elastic system. The only two physical factors to be con-

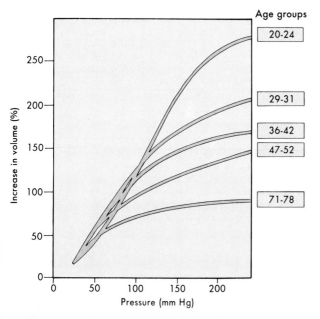

FIGURE 20-3 Pressure-volume relationships for aortas obtained at autopsy from humans in different age groups (denoted by the numbers at the right end of each curve). (Redrawn from Hallock P and Benson IC: J Clin Invest 16:595, 1937.)

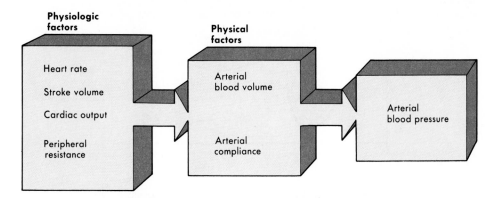

FIGURE 20-4   The arterial blood pressure is determined directly by two major physical factors, the arterial blood volume and the arterial compliance. These physical factors are affected in turn by certain physiologic factors, primarily the heart rate, stroke volume, cardiac output, and peripheral resistance.

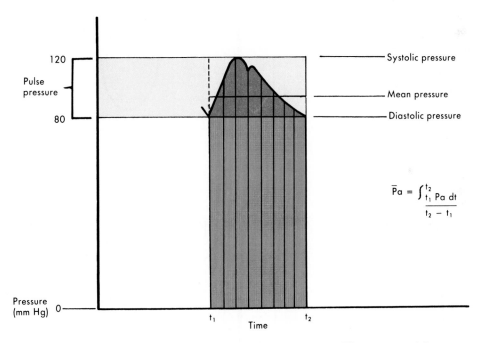

FIGURE 20-5   Arterial systolic, diastolic, pulse, and mean pressures. The mean arterial pressure $(\bar{P}_a)$ represents the area under the arterial pressure curve *(shaded area)* divided by the cardiac cycle duration $(t_2 - t_1)$.

sidered are the *blood volume* within the arterial system and the elastic characteristics *(compliance)* of the system. Several physiologic factors are considered, such as *heart rate, stroke volume, cardiac output,* and *peripheral resistance.* Such physiologic factors are shown to operate through one or both of the physical factors.

Mean Arterial Pressure

The mean arterial pressure is the pressure in the arteries averaged over time. It may be obtained from an arterial pressure tracing by measuring the area under the curve and dividing this area by the appropriate time interval, as shown in Figure 20-5. The

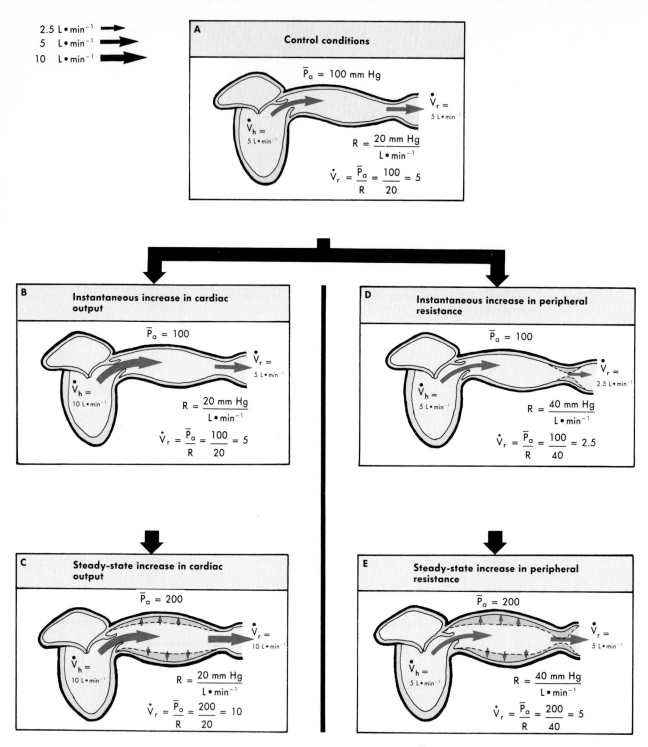

FIGURE 20-6 The relationship of mean arterial blood pressure $(\overline{P}_a)$ to cardiac output $(\dot{V}_h)$, peripheral runoff $(\dot{V}_r)$, and peripheral resistance (R) under control conditions **(A)**, in response to an increase in cardiac output **(B and C)**, and in response to an increase in peripheral resistance **(D and E)**. **A,** Under control conditions $\dot{V}_h = 5$ L·min^{-1}, $\overline{P}_a = 100$ mm Hg, and R = 20 mm Hg·L^{-1}·min. $\dot{V}_r$ must equal $\dot{V}_h$, and therefore the mean blood volume $(\overline{V}_a)$ in the arteries will remain constant from heartbeat to heartbeat. **B,** If $\dot{V}_h$ suddenly increases to 10 L·min^{-1}, $\dot{V}_h$ will initially exceed $\dot{V}_r$, and therefore $\overline{P}_a$ will begin to rise rapidly. **C,** The disparity between $\dot{V}_h$ and $\dot{V}_r$ progressively increases arterial blood volume. The volume continues to increase until $\overline{P}_a$ reaches a level of 200 mm Hg. **D,** If R abruptly increases to 40 mm Hg·L^{-1}·min, $\dot{V}_r$ suddenly decreases and therefore $\dot{V}_h$ exceeds $\dot{V}_r$. Thus $\overline{P}_a$ will rise progressively. **E,** The excess of $\dot{V}_h$ over $\dot{V}_r$ accumulates blood in the arteries. Blood continues to accumulate until $\overline{P}_a$ rises to a level of 200 mm Hg.

mean arterial pressure, $\overline{P}_a$, usually can be estimated satisfactorily from the measured values of the systolic (P_s) and diastolic (P_d) pressures by means of the following formula:

$$\overline{P}_a \cong P_d + \frac{1}{3}(P_s - P_d) \quad \textbf{(2)}$$

The mean pressure depends directly on the mean volume of blood in the arterial system and the compliance of the arterial walls (Figure 20-4). The arterial volume, V_a, in turn depends on the balance between the rate of inflow, $\dot{V}_h$, from the heart into the arteries *(cardiac output)* and the rate of outflow, $\dot{V}_r$, from the arteries through the capillaries *(peripheral runoff)*; expressed mathematically,

$$dV_a/dt = \dot{V}_h - \dot{V}_r \quad \textbf{(3)}$$

This equation is a restatement of the law of conservation of mass. It states that any change in arterial blood volume simply reflects the difference in the rates at which blood enters and leaves the arterial system. If arterial inflow exceeds outflow, arterial volume increases, the arterial walls are stretched more, and pressure rises. The converse occurs when arterial outflow exceeds inflow. Finally, if inflow equals outflow, arterial pressure remains constant.

Cardiac Output The change in pressure in response to an alteration of cardiac output can be better appreciated by considering some simple examples. Under control conditions, let cardiac output be 5 L·min⁻¹ and mean arterial pressure ($\overline{P}_a$) be 100 mm Hg (Figure 20-6, *A*). The definition of *total peripheral resistance is*

$$R = (\overline{P}_a - \overline{P}_{ra})/\dot{V}_r \quad \textbf{(4)}$$

If $\overline{P}_{ra}$ (mean right atrial pressure) is negligible compared with $\overline{P}_a$,

$$R \cong \overline{P}_a/\dot{V}_r \quad \textbf{(5)}$$

Therefore, in the example shown in Figure 20-6, *A*, R is 100/5, or 20 mm Hg·L⁻¹·min.

Now let cardiac output, $\dot{V}_h$, suddenly increase to 10 L·min⁻¹ (Figure 20-6, *B*). Instantaneously $\overline{P}_a$ will be unchanged. Because the outflow, $\dot{V}_r$, from the arteries depends on $\overline{P}_a$ and R, $\dot{V}_r$ also will remain unchanged at first. Therefore $\dot{V}_h$, now 10 L·min⁻¹, will exceed $\dot{V}_r$, still only 5 L·min⁻¹. This will increase the mean arterial blood volume ($\overline{V}_a$). From equation 3, when $\dot{V}_h > \dot{V}_r$, then $d\overline{V}_a/dt > 0$; that is, volume is increasing.

The increase in blood volume in the arteries will raise the blood pressure in the arteries, just as any increase in volume will increase the pressure in a hollow elastic structure, such as a balloon. Blood will continue to accumulate in the arteries until the pressure rises sufficiently high to force through the peripheral resistance an outflow, $\dot{V}_r$, that will be equal to the new, elevated cardiac output, $\dot{V}_h$ (Figure 20-6, *C*). It is evident from equation 5 that $\dot{V}_r$ will not attain a value of 10 L·min⁻¹ until $\overline{P}_a$ reaches a level of 200 mm Hg, as long as R remains constant at 20 mm Hg·L⁻¹·min. Thus, as $\overline{P}_a$ approaches 200, $\dot{V}_r$ will almost equal $\dot{V}_h$, and $\overline{P}_a$ will rise very slowly. When $\dot{V}_h$ is first raised, however, $\dot{V}_h$ greatly exceeds $\dot{V}_r$, and therefore $\overline{P}_a$ will rise sharply. The pressure-time tracing in Figure 20-7 indicates that, regardless of the arterial compliance (C_a), the slope gradually diminishes as pressure rises, to approach a constant pressure asymptotically.

Furthermore, the height to which $\overline{P}_a$ will rise is independent of the compliance of the arterial walls. $\overline{P}_a$ must rise to a level such that $\dot{V}_r = \dot{V}_h$. Rearrangement of equation 4 shows that $\dot{V}_r$ depends only on pressure difference and resistance to flow:

$$\dot{V}_r = (\overline{P}_a - \overline{P}_{ra})/R \quad \textbf{(6)}$$

Thus C_a determines only the **rate** at which the new equilibrium value of $\overline{P}_a$ will be attained, but not the **value** of $\overline{P}_a$, as illustrated in Figure 20-7. When C_a is small (rigid vessels), a relatively slight increment in $\overline{V}_a$ (caused by a transient excess of $\dot{V}_h$ over $\dot{V}_r$) increases $\overline{P}_a$ greatly. Thus $\overline{P}_a$ attains its new equilibrium level quickly. Conversely, when C_a is large, considerable

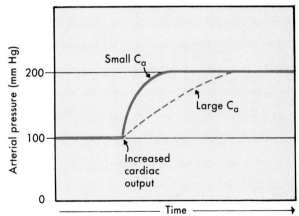

FIGURE 20-7   When cardiac output is suddenly increased, the arterial compliance *(C_a)* determines the rate at which the mean arterial pressure will attain its new, elevated value, but C_a will not determine the magnitude of the new pressure.

volumes can be accommodated with relatively small pressure changes. Therefore the new equilibrium value of $\overline{P}_a$ is reached at a slower rate.

Peripheral Resistance Similar reasoning may now be applied to explain the changes in $\overline{P}_a$ that accompany alterations in peripheral resistance. Let the control conditions be identical with those of the preceding example; that is, $V_h = 5$, $\overline{P}_a = 100$, and $R = 20$ (Figure 20-6, *A*). Then let R suddenly be increased to 40 (Figure 20-6 *D*). Instantaneously $\overline{P}_a$ will be unchanged. With $\overline{P}_a = 100$ and $R = 40$, the peripheral runoff ($\dot{V}_r$) will suddenly decrease to 2.5 $L \cdot min^{-1}$. If the cardiac output, $\dot{V}_h$, remains constant at 5 $L \cdot min^{-1}$, $\dot{V}_h > \dot{V}_r$, and $\overline{V}_a$ will increase; thus $\overline{P}_a$ will rise. $\overline{P}_a$ will continue to rise until it reaches 200 mm Hg (Figure 20-6, *E*). At this level $\dot{V}_r = 200/40 = 5$ $L \cdot min^{-1}$, which equals $\dot{V}_h$. $\overline{P}_a$ will then remain at this new elevated equilibrium level as long as $\dot{V}_h$ and R do not change again.

It is clear, therefore, that the **mean arterial pressure depends only on cardiac output and peripheral resistance** (Figure 20-8). It is immaterial whether the change in cardiac output is accomplished by an alteration of heart rate, of stroke vol-

ume, or of both. Because **cardiac output equals heart rate times stroke volume,** any change in heart rate that is balanced by an inverse change in stroke volume will not alter cardiac output, and therefore $\overline{P}_a$ will not be affected.

Pulse Pressure

The two physical factors that mainly determine the arterial pressure (P_a) at any given time are the *arterial blood volume* (V_a) and the *arterial compliance* (C_a) that prevail at that time (see Figure 20-4). The physiologic factors operate through these physical factors to affect the arterial *pulse pressure* (systolic pressure minus diastolic pressure), just as they did to influence the mean arterial pressure.

Stroke Volume The effect of a change in stroke volume on pulse pressure may be analyzed simply under conditions in which C_a remains constant. C_a is constant over any linear region of a pressure-volume curve (see Figure 20-3). If volume is plotted along the vertical axis and pressure along the horizontal axis, by definition the slope, dV_a/dP_a, is the compliance C_a.

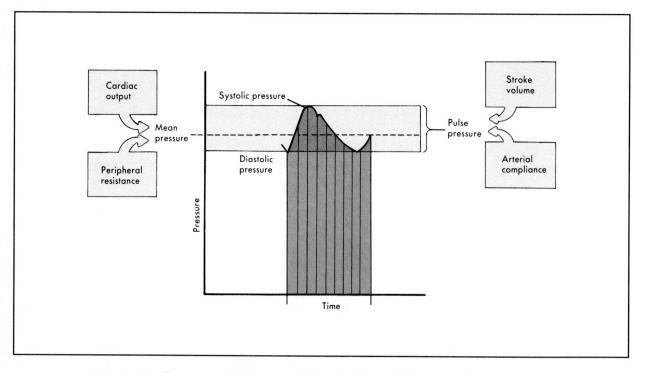

FIGURE 20-8   The mean arterial pressure is determined by cardiac output and peripheral resistance, whereas the pulse pressure is determined mainly by stroke volume and arterial compliance.

To appreciate the effect of a change in stroke volume on the arterial pulse pressure, consider a normal person who initially has a heart rate (HR) of 100 min^{-1}, a stroke volume (SV) of 50 ml, and a peripheral resistance (R) of 20 mm $Hg \cdot L^{-1} \cdot min$. Because cardiac output equals HR times SV, this person's cardiac output is 5 $L \cdot min^{-1}$. As shown in Figure 20-6, A, this person would have a mean arterial pressure ($\bar{P}_a$) of 100 mm Hg. The systolic pressure, P_s, will be greater than 100 mm Hg, and diastolic pressure, P_d, will be less than 100 mm Hg. Let the values of P_s and P_d for this person be 120 and 90 mm Hg, respectively; these values for $\bar{P}_a$, P_s, and P_d satisfy equation 2.

Let us review the hemodynamic events that take place just before and during the rapid ejection phase of systole of a given heartbeat (see Chapter 17). Throughout the ventricular diastole of the preceding heartbeat and during the brief period of isovolumic contraction, the heart has pumped no blood into the arteries, but blood has continuously flowed out of the arteries and through the capillaries. Thus P_a (the pressure as a function of time) would have declined continuously throughout these phases of the cardiac cycle. In an individual with a diastolic pressure of 90 mm Hg, this minimal pressure would have been attained at the end of the isovolumic contraction phase of systole (Figure 17-10).

In healthy people most of SV is ejected during the rapid ejection phase of ventricular systole (see Figure 17-10). Let us assume that 80% of SV (50 ml) in our hypothetic subject is expelled during rapid ejection. Thus during this phase the blood volume (V_h) introduced into the arterial system is 0.80 × 50 ml, or 40 ml (Figure 20-9, A).

Blood exits from the arteries throughout the cardiac cycle. At equilibrium the volume (V_r) that leaves the arteries each cardiac cycle equals SV. The fractional volume that flows out of the arteries during rapid ejection is approximately equal to the fraction of the cardiac cycle duration that is occupied by the rapid ejection phase. Let us assume that during rapid ejection, 16% of SV exits the arteries; that is, the peripheral runoff (V_r) during this phase is 0.16 × 50 ml, or 8 ml.

Thus the volume increment (ΔV) that accumulates in the arteries during the rapid ejection phase of the cardiac cycle equals $V_h - V_r$, or 32 ml. In this person the volume increment causes P_a to rise from the diastolic level of 90 mm Hg to the systolic level of 120 mm Hg. In other words, the volume increment (ΔV_a) of 32 ml produces a pressure increment (ΔP_a) of 30 mm Hg.

Thus the arterial compliance ($C_a = \Delta V / \Delta P$) equals 1.07 ml·mm Hg^{-1}.

Now suppose that the cardiovascular status changes, such that HR decreases to 50 min^{-1} and SV increases to 100 ml. However, let R (20 mm $Hg \cdot L^{-1} \cdot min$) and C_a (1.07 ml·mm Hg^{-1}) remain unchanged (Figure 20-9, B). Because the cardiac output (HR × SV) still equals 5 $L \cdot min^{-1}$ and R is also unchanged, the mean arterial pressure ($\bar{P}_a$) remains 100 mm Hg.

To determine the new pulse pressure, assume that the fraction of SV (100 ml) expelled during rapid ejection is still 80%. Thus V_h during rapid ejection is 0.80 × 100 ml, or 80 ml. Also, assume that 16% of SV exits the arterial system through the capillaries during rapid ejection. Thus the runoff (V_r) during rapid ejection is 0.16 × 100 ml, or 16 ml. Therefore the volume increment (ΔV_a) during rapid ejection is 80 − 16 ml, or 64 ml.

A volume increment of 64 ml in an arterial system with a compliance of 1.07 ml·mm Hg^{-1} would produce a pressure increment (i.e., a pulse pressure) of 60 mm Hg, which is twice the pulse pressure that prevailed under the preceding conditions (Figure 20-9, A). Assuming also that the relation among the mean, systolic, and diastolic pressures satisfies equation 2, this person's systolic and diastolic pressures would equal 140 and 80 mm Hg, respectively. Thus, if cardiac output and C_a remain constant, an increase in SV would raise systolic pressure and lower diastolic pressure but would not affect mean pressure if cardiac output remained constant (Figure 20-9, D).

Arterial Compliance Now let us examine the effect of a change in C_a on the pulse pressure (Figure 20-9, C). Assume that HR, SV, and R are all the same as in Figure 20-9, A. However, consider that the arteries are only half as compliant as previously (i.e., C_a now equals 0.55). Because the cardiac output and peripheral resistance are the same as in the two preceding examples (Figure 20-9, A and B), the $\bar{P}_a$ will still be 100 mm Hg.

Assume also that even though the arteries are much less compliant, the heart can still pump 80% of SV during rapid ejection ($V_h = 40$ ml). In addition, assume that because the less compliant arteries will accommodate a smaller increment in volume, a greater fraction (24%) of SV will run out of the arteries during rapid ejection ($V_r = 12$ ml). In arteries with a compliance of only 0.55 ml·mm Hg^{-1}, the volume increment (40 − 12 = 28 ml) will produce a pressure increment (pulse pressure) of 51 mm Hg. Assuming that

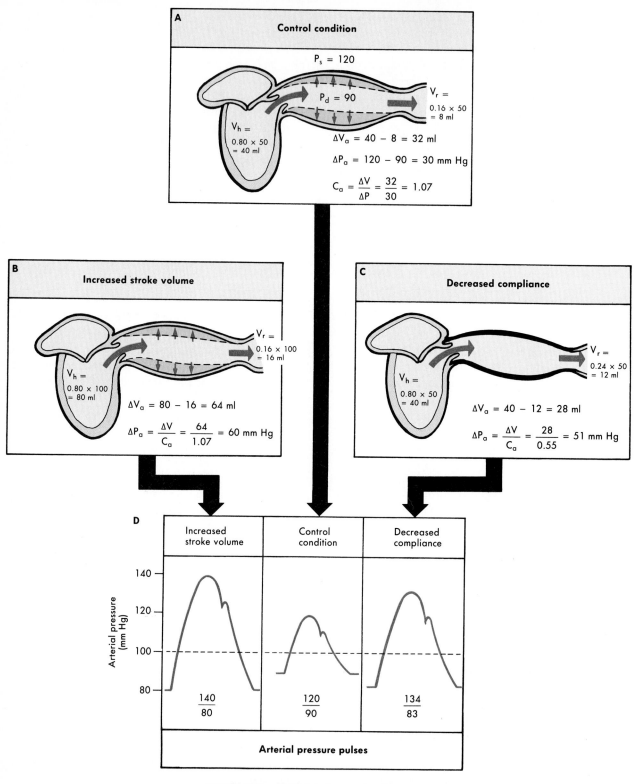

FIGURE 20-9   For legend see opposite page.

FIGURE 20-9 The effects of an increase in stroke volume and of a decrease in arterial compliance on the arterial pulse pressure. **A,** In a hypothetic subject 40 ml of blood is pumped into the arterial system (V_h) each beat during rapid ejection, and 8 ml runs out of the arteries (V_r) during the same interval. Thus a volume increment (ΔV_a) of 32 ml accumulates in the arteries during rapid ejection. This causes a pressure increment (ΔP_a), from the diastolic to the systolic level, of 30 mm Hg; this is the *pulse pressure*. Thus the arterial compliance (C_a) equals 32/30, or 1.07 ml·mm Hg^{-1}. (P_s, Systolic pressure; P_d, diastolic pressure.) **B,** If the heart now pumps 80 ml each beat into the arteries during rapid ejection, and if 16 ml runs out of the arteries during the same period, then ΔV_a is now 64 ml. If C_a is still 1.07, ΔP_a will be 60 mm Hg. **C,** Assume that the heart can still pump 80% of the stroke volume (50 ml) during rapid ejection, despite a decreased C_a. Assume also that, because of the decreased compliance, 24% (rather than 16%) of the stroke volume flows out of the arteries during rapid ejection. The ΔV_a of 28 ml will produce a ΔP_a of 51 mm Hg. **D,** Under conditions that do not alter the mean arterial pressure, an increase in stroke volume *(B)* or a decrease in arterial compliance *(C)* will increase the arterial pulse pressure.

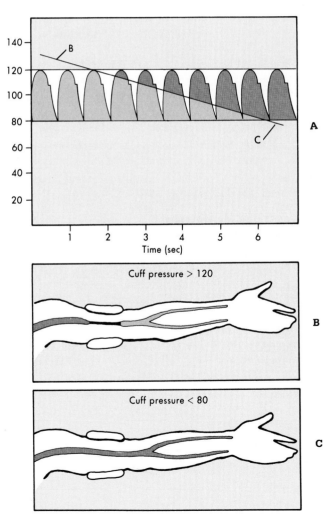

FIGURE 20-10 Measurement of arterial blood pressure with a sphygmomanometer. **A,** Consider that the arterial blood pressure is being measured in a patient whose blood pressure is 120/80 mm Hg. The pressure (represented by the *oblique line*) in a cuff around the patient's arm is allowed to fall from greater than 120 mm Hg (point *B*) to below 80 mm Hg (point *C*) in about 6 seconds. **B,** When the cuff pressure exceeds the systolic arterial pressure (120 mm Hg), no blood progresses through the arterial segment under the cuff, and no sounds can be detected by a stethoscope bell placed on the arm distal to the cuff. **C,** When the cuff pressure falls below the diastolic arterial pressure, arterial flow past the region of the cuff is continuous, and no sounds are audible. When the cuff pressure is between 120 and 80 mm Hg, spurts of blood traverse the artery segment under the cuff with each heartbeat, and the Korotkoff sounds are heard through the stethoscope.

the relation among systolic, diastolic, and mean pressures satisfies equation 2, the systolic and diastolic pressures would be approximately 134 and 83 mm Hg, respectively. Thus the systolic pressure would exceed the control value, the diastolic pressure would be lower than the control value, and the pulse pressure would be substantially greater than the control value (Figure 20-9, *D*).

BLOOD PRESSURE MEASUREMENT IN HUMANS

Needles or catheters may be introduced into peripheral arteries of patients in hospital intensive care units. Arterial blood pressure can then be measured directly by means of strain gauges. Ordinarily, however, the blood pressure is estimated **indirectly** by means of a *sphygmomanometer*. This instrument consists of an inextensible cuff containing an inflatable bag. The cuff is wrapped around the arm so that the inflatable bag lies between the cuff and the skin, directly over the brachial artery. The artery is occluded by inflating the bag, by means of a rubber squeeze bulb, to a pressure in excess of arterial systolic pressure. Pressure is released from the bag at a rate of 2 or 3 mm Hg·sec^{-1} by means of a needle valve in the inflating bulb.

The physician or nurse listens with a stethoscope applied to the skin of the forearm over the brachial artery. While the pressure in the bag exceeds the systolic pressure, the brachial artery is occluded and no sounds are heard. When the inflation pressure falls just below the arterial systolic pressure (*upper horizontal line* in Figure 20-10, *A*), small spurts of blood pass through the artery each time the arterial pressure exceeds the cuff pressure. Consequently, slight tapping sounds *(Korotkoff sounds)* are heard with each heartbeat. The pressure at which the first sound is detected represents the *systolic pressure*. As inflation pressure continues to fall, more blood escapes under the cuff per beat and the sounds become louder. As the inflation pressure approaches the diastolic pressure *(lower horizontal line),* the Korotkoff sounds become muffled. As they fall just below the diastolic level, the sounds disappear; this indicates the *diastolic pressure*. The origin of the Korotkoff sounds is related to the spurt of blood passing under the cuff and meeting a static column of blood; the impact and turbulence generate audible vibrations. Once the inflation pressure is less than the diastolic pressure, flow is continuous in the brachial artery and sounds are no longer heard.

BIBLIOGRAPHY
Journal Articles

Burattini R et al: Total systemic arterial compliance and aortic characteristic impedance in the dog as a function of pressure: a model based study, Comput Biomed Res 20:154, 1987.

Imura T et al: Non-invasive ultrasonic measurement of the elastic properties of the human abdominal aorta, Cardiovasc Res 20:208, 1986.

Kenner T: Arterial blood pressure and its measurement, Basic Res Cardiol 83:107, 1988.

McIlroy MB and Targett RC: Model of the systemic arterial bed showing ventricular-systemic arterial coupling, Am J Physiol 254:H609, 1988.

Nichols WW and Pepine CJ: Left ventricular afterload and aortic input impedance: implications of pulsatile blood flow, Prog Cardiovasc Dis 24:293, 1982.

O'Rourke MF: Vascular impedance in studies of arterial and cardiac function, Physiol Rev 62:570, 1982.

Piene H: Pulmonary arterial impedance and right ventricular function, Physiol Rev 66:606, 1986.

Simon AC et al: An evaluation of large arteries compliance in man, Am J Physiol 237:H550, 1979.

Books and Monographs

Bauer RD and Busse R, editors: Arterial system: dynamics, control theory and regulation, Heidelberg, 1978, Springer-Verlag.

Dobrin PB: Vascular mechanics. In Handbook of physiology, section 2: The cardiovascular system—peripheral circulation and organ blood flow, vol III, Bethesda, Md, 1983, American Physiological Society.

Fung YC: Biodynamics: circulation, Heidelberg, 1984, Springer-Verlag.

Milnor WR: Hemodynamics, Baltimore, 1982, Williams & Wilkins.

Noordergraaf A: Circulatory system dynamics, New York, 1978, Academic Press, Inc.

Taylor DEM and Stevens AL, editors: Blood flow: theory and practice, New York, 1983, Academic Press, Inc.

CHAPTER  21

The Microcirculation and Lymphatics

FUNCTIONAL ANATOMY

The entire circulatory system is geared to supply the body tissues with blood in amounts commensurate with their requirements for oxygen and nutrients. The capillaries, consisting of a single layer of endothelial cells, permit rapid exchange of water and solutes with interstitial fluid. The muscular arterioles, which are the major *resistance vessels,* regulate regional blood flow to the capillary beds, and the venules and veins serve primarily as collecting channels and *storage,* or *capacitance, vessels.*

Arterioles

The arterioles, which range in diameter from about 5 to 100 μm, have a thick smooth muscle layer, a thin adventitial layer, and an endothelial lining (see Figure 15-1). The arterioles give rise directly to the capillaries (5 to 10 μm diameter) or in some tissues to metarterioles (10 to 20 μm diameter), which then give rise to capillaries (Figure 21-1). The metarterioles can serve either as thoroughfare channels to the venules, bypassing the capillary bed, or as conduits to supply the capillary bed. There are often cross connections from arteriole to arteriole and from venule to venule, as well as in the capillary network. Arterioles that give rise directly to capillaries regulate flow through their cognate capillaries by constriction or dilation. The capillaries form an interconnecting network of tubes of different lengths; the average length is 0.5 to 1 mm.

The diameter of the resistance vessels is determined by the balance between the contractile force of the vascular smooth muscle and the distending force produced by the intraluminal pressure. The greater the contractile activity of the vascular smooth muscle of an arteriole, the smaller will be its diameter. At some point, in the case of small arterioles, complete occlusion of the vessel will occur, partly because of infolding of the endothelium. With progressive reduction in the intravascular pressure, vessel diameter decreases, as does tension in the vessel wall (Laplace's law). When perfusion pressure is reduced, a point is reached at which blood flow ceases, even though a positive pressure gradient may still exist. This phenomenon has been referred to as the *critical closing pressure,* and its mechanism is still controversial.

Capillaries

Capillary distribution varies from tissue to tissue. In metabolically active tissues, such as cardiac and skeletal muscle and grandular structures, capillaries are numerous, whereas in less active tissues, such as subcutaneous tissue or cartilage, there are few capillaries. Also, not all capillaries have the same diameter. Because some capillaries have diameters less than that of erythrocytes, the cells must become temporarily deformed to pass through these capillaries. Fortunately, normal red cells are quite flexible and readily change their shape to conform with that of the small capillaries.

Blood flow in the capillaries is not uniform and depends chiefly on the contractile state of the arterioles. The average velocity of blood flow in the capil-

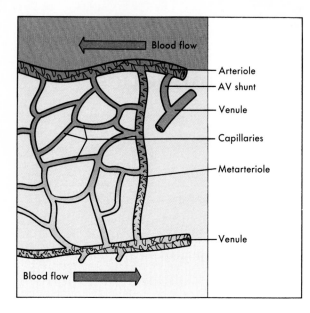

FIGURE 21-1   Schematic drawing of the microcirculation. The circular structures on the arteriole and venule represent smooth muscle fibers, and the branching solid lines represent sympathetic nerve fibers. The arrows indicate the direction of blood flow.

laries is approximately 1 mm·sec^{-1}; however, it can quickly vary from zero to several millimeters per second in the same vessel. The capillary blood flow may vary randomly or it may oscillate rhythmically at different frequencies, as determined by contraction and relaxation *(vasomotion)* of the precapillary vessels. This vasomotion is partly an intrinsic contractile behavior of the vascular smooth muscle and is independent of external influence. Furthermore, changes in transmural pressure (intravascular minus extravascular pressure) affect the contractile state of the precapillary vessels. An increase in transmural pressure, whether produced by an increase in venous pressure or by dilation of arterioles, elicits contraction of the terminal arterioles at the points of origin of the capillaries. Conversely, a decrease in transmural pressure elicits precapillary vessel relaxation. In addition, humoral and possibly neural factors also affect vasomotion.

Although reduced transmural pressure will relax the terminal arterioles, blood flow through the capillaries obviously cannot increase if the reduced intravascular pressure is caused by severe constriction of the parent arterioles or metarterioles. Large arterioles and metarterioles also exhibit vasomotion. However,

in the contraction phase they usually do not completely occlude the lumen of the vessel and arrest blood flow, as may occur with contraction of the terminal arterioles. Thus flow rate may be altered by arteriolar and metarteriolar vasomotion.

Because blood flow through the capillaries provides for exchange of gases and solutes between blood and tissue, it has been termed *nutritional flow,* whereas blood flow that bypasses the capillaries in traveling from the arterial to the venous side of the circulation has been termed *nonnutritional,* or *shunt, flow* (Figure 21-1). In some areas of the body (fingertips) true arteriovenous (AV) shunts exist (see Chapter 24). In many tissues such as muscle, however, evidence of anatomic shunts is lacking.

The true capillaries are devoid of smooth muscle and are therefore incapable of active constriction. Nevertheless, the endothelial cells that form the capillary wall contain actin and myosin and can alter their shape in response to certain chemical stimuli. No evidence shows, however, that changes in endothelial cell shape regulate blood flow through the capillaries. Thus changes in capillary diameter are passive and are caused by alterations in precapillary and postcapillary resistance.

For many years it was believed that capillaries were inert and merely served as barriers to blood cells and large molecules, such as plasma proteins. Recently, however, the metabolic activity of the endothelial cells and the interaction between blood components (particularly platelets) and the endothelial cells have been recognized. For example, stimulation of endothelium with acetylcholine releases a vascular smooth muscle relaxant, *endothelial-derived relaxing factor (EDRF).* In an intact perfused small arterial vessel, acetylcholine produces vasodilation (indirect effect by EDRF release). In a vessel stripped of endothelium, however, acetylcholine elicits vasoconstriction (direct effect of acetylcholine on vascular smooth muscle).

The thin-walled capillaries can withstand high internal pressures without bursting because of their narrow lumen. This can be explained in terms of Laplace's law:

$$T = Pr \qquad \textbf{(1)}$$

where

T = Tension in the vessel wall (dynes/cm)
P = Transmural pressure (dynes/cm^2)
r = Radius of the vessel (cm)

Wall tension is the force per unit length tangential to the vessel wall. This force opposes the distending force (Pr) that tends to pull apart a theoretic longitudinal slit in the vessel (Figure 21-2). Transmural pressure is essentially equal to intraluminal pressure because extravascular pressure is negligible.

At normal aortic and capillary pressures the wall tension of the aorta is about 12,000 times greater than that of the capillary. In a person standing quietly, capillary pressure in the feet may reach 100 mm Hg. Under such conditions capillary wall tension increases to a value that is only one three-thousandth that of the wall tension in the aorta at the same internal pressure. In dilated vessels wall tension increases even when internal pressure remains constant. Under certain circumstances (e.g., aneurysm of the aorta) wall tension may be an important factor in rupture of the vessel.

Capillary Pores

The permeability of the capillary endothelial membrane is not the same in all body tissues. For example, the liver capillaries are very permeable, and albumin

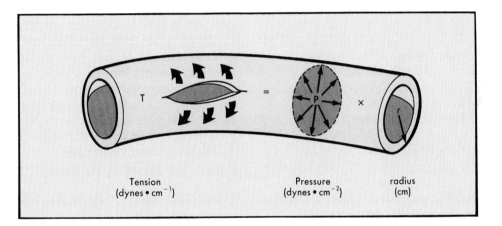

FIGURE 21-2   Diagram of a small blood vessel to illustrate Laplace's law: T = Pr, where P = intraluminal pressure, r = radius of the vessel, and T = wall tension as the force per unit length tangential to the vessel wall, tending to pull apart a theoretic longitudinal slit in the vessel.

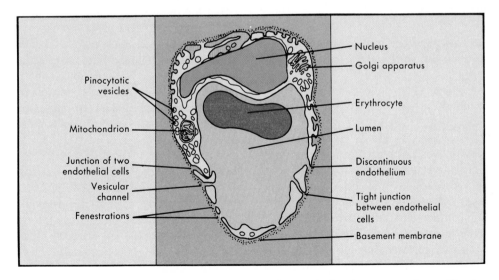

FIGURE 21-3   Diagrammatic sketch of an electron micrograph showing a composite capillary in cross section.

escapes at a rate several-fold greater than from the less permeable muscle capillaries. Also, permeability is not uniform along the whole capillary; the venous ends are more permeable than the arterial ends, and permeability is greatest in the venules. The greater permeability at the venous end of the capillaries and in the venules is caused by the greater number of pores. In cardiac and skeletal muscle many of the clefts (pores) between adjacent endothelial cells are open. These clefts can be visualized by electron microscopy when they are filled with horseradish peroxidase (a small, electron-dense tracer) is applied to the luminal side of the capillaries before fixation of the section. The narrowest point of the cleft is about 4 nm. The clefts are sparse and represent only about 0.02% of the capillary surface area in most tissues. In brain, clefts are absent in the capillaries, and a *blood-brain barrier* to many small molecules exists.

In addition to clefts, some of the more porous capillaries (e.g., in the kidney and the intestine) contain fenestrations 20 to 100 nm wide, whereas others (e.g., in the liver) have a discontinuous endothelium (Figure 21-3). The fenestrations appear to be sealed by a thin diaphragm, but they are quite permeable to horseradish peroxidase and several other tracers. Thus larger molecules can penetrate those capillaries with fenestrations or gaps caused by discontinuous endothelium than can pass through the intercellular clefts of the endothelium.

TRANSCAPILLARY EXCHANGE

Solvent and solute move across the capillary endothelial wall by three processes: diffusion, filtration, and pinocytosis.

Diffusion

Under normal conditions only about 0.06 ml of water per minute moves back and forth across the capillary wall per 100 g of tissue as a result of filtration and absorption. Whereas 300 ml of water per minute per 100 g of tissue do so by diffusion, a 5000-fold difference. Relating filtration and diffusion to blood flow, we find that only about 2% of the plasma passing through the capillaries is filtered. In contrast, the rate that water diffuses across the endothelium is 40 times greater than the rate at which it is delivered to the capillaries by blood flow. The transcapillary exchange

of solutes is also mainly governed by diffusion. **Thus diffusion is the key factor in the exchange of gases, substrates, and waste products between the capillaries and the tissue cells.** However, **net** transfer of fluid across the capillary endothelium is primarily attributable to filtration and absorption.

For small molecules, such as water, NaCl, urea, and glucose, the capillary pores do not restrict diffusion (low *reflection coefficient*). Diffusion proceeds so rapidly that the mean concentration gradient across the capillary endothelium is extremely small. Water passes directly through the endothelial cells and through the capillary pores between endothelial cells. As the size of lipid-insoluble molecules is increased, diffusion through muscle capillaries becomes progressively more restricted. Diffusion of molecules with a molecular weight greater than about 60,000 becomes minimal. With small molecules the only limitation to net movement across the capillary wall is the rate at which blood flow transports the molecules to the capillary; diffusion is said to be *flow limited*.

When transport across the capillary is flow limited, the concentration of a small solute in the blood reaches equilibrium with its concentration in the interstitial fluid near the origin of the capillary from the cognate arteriole (Figure 21-4). A somewhat larger molecule moves farther along the capillary before it reaches an insignificant concentration in the blood, and a still larger molecule cannot pass through the capillary pores. An increase in blood flow extends the detectable concentration of small molecules farther down the capillary and increases the *capillary diffusion capacity* (rate of tissue uptake of the solute).

With large molecules, diffusion across the capillaries becomes the factor that limits exchange *(diffusion limited)*. In other words, capillary permeability to a large molecular solute limits its transport across the capillary wall (Figure 21-4, *A*). The rate of diffusion of small lipid-insoluble molecules is so rapid that the rate-limiting factor in blood-tissue exchange can only become important when long distances exist between capillaries (e.g., tissue edema or very low capillary density) (Figure 21-4, *B*).

In contrast to lipid-insoluble molecules, movement of lipid-soluble molecules across the capillary wall is not limited to capillary pores (only about 0.02% of the capillary surface) because such molecules can pass directly through the lipid membranes of the entire capillary endothelium. Consequently, lipid-soluble molecules move very rapidly between blood and tis-

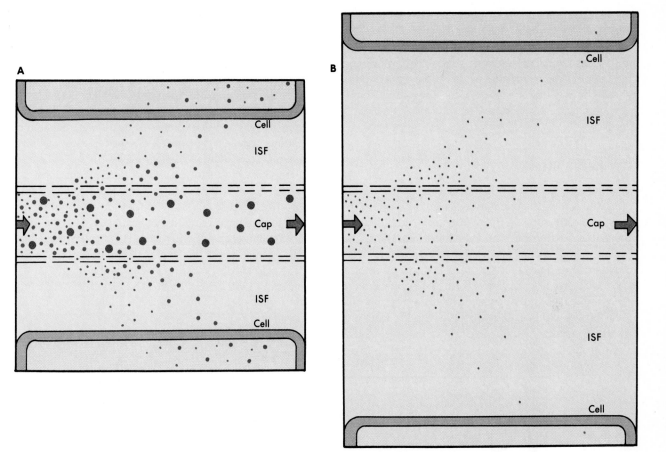

FIGURE 21-4   Flow- and diffusion-limited transport from capillaries *(Cap)* to tissue. **A,** Flow-limited transport. The smallest water-soluble inert tracer particles *(red dots)* reach negligible concentrations after passing only a short distance down the capillary. Larger particles *(blue dots)* with similar properties travel farther along the capillary before reaching insignificant intracapillary concentrations. Both substances cross the interstitial fluid *(ISF)* and reach the parenchymal tissue. Because of their size, more of the smaller particles are taken up by the tissue cells. The largest particles *(purple dots)* cannot penetrate the capillary pores and thus do not escape from the capillary lumen except by pinocytotic vesicle transport. An increase in the volume of blood flow or an increase in capillary density will increase tissue supply for the diffusible solutes. Note that capillary permeability is greater at the venous end of the capillary (also in the venule, not shown) because of the larger number of pores in this region. **B,** Diffusion-limited transport. When the distance between the capillaries and the parenchymal tissue is large, as a result of edema or low capillary density, diffusion becomes a limiting factor in the transport of solutes from capillary to tissue, even at high rates of capillary blood flow.

sue. The *lipid solubility* (oil-to-water partition coefficient) is a good index of the ease of transfer of lipid molecules through the capillary endothelium.

Oxygen and carbon dioxide are both lipid soluble and readily pass through the endothelial cells. Calculations based on (1) the diffusion coefficient for O_2, (2) capillary density and diffusion distances, (3) blood flow, and (4) tissue O_2 consumption indicate that the O_2 supply of normal tissue at rest and during activity

is not limited by diffusion or the number of open capillaries. Measurements of O_2 tension and saturation of blood in the microvessels indicate that in many tissues O_2 saturation at the entrance of the capillaries has already decreased to about 80% as a result of diffusion of O_2 from the arterioles. Such studies also have shown that CO_2 loading and the resultant intravascular shifts in the oxyhemoglobin dissociation curve occur in the precapillary vessels. These findings indicate not only that gas moves to respiring tissue at the precapillary level, but also that O_2 and CO_2 pass directly between adjacent arterioles, venules, and possibly arteries and veins *(countercurrent exchange)*. This exchange of gas represents a diffusional shunt of gas around the capillaries; at low blood flow rates, it may limit the supply of O_2 to the tissue.

Capillary Filtration

The direction and the magnitude of the movement of water across the capillary wall are determined by the algebraic sum of the hydrostatic and osmotic pressures that exist across the membrane. An increase in intracapillary hydrostatic pressure favors movement of fluid from the vessel to the interstitial space. Conversely, an increase in the concentration of osmotically active particles within the vessels favors movement of fluid into the vessels from the interstitial space.

Hydrostatic Forces The hydrostatic pressure (blood pressure) within the capillaries is not constant; it depends on the arterial pressure, the venous pressure, and the precapillary (arteriolar) and postcapillary (venules and small veins) resistances. A gain in arterial or venous pressure increases capillary hydrostatic pressure (P_c), whereas a reduction in each has the opposite effect. An increase in arteriolar resistance reduces capillary pressure, whereas greater venous resistance increases capillary pressure.

Capillary hydrostatic pressure varies from tissue to tissue and even within the same tissue. Average values, obtained from many direct measurements in human skin, are about 32 mm Hg at the arterial end of the capillaries and about 15 mm Hg at the venous end of the capillaries at the level of the heart (Figure 21-5). The hydrostatic pressure in capillaries of the lower extremities will be higher and that of capillaries in the head will be lower when an individual is in the standing position.

Tissue pressure, or more specifically interstitial flu-

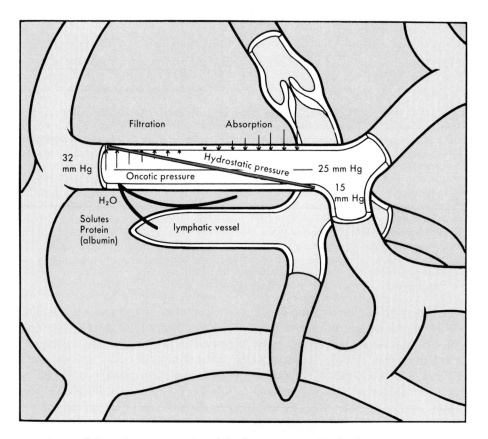

FIGURE 21-5   Schematic representation of the factors responsible for filtration and absorption across the capillary wall and the formation of lymph.

id pressure (P_i) outside the capillaries, opposes capillary filtration. Hydrostatic pressure minus interstitial fluid pressure ($P_c - P_i$) constitutes the driving force for filtration. In the absence of edema, the interstitial fluid pressure is essentially zero.

Osmotic Forces The key factor that restrains fluid loss from the capillaries is the osmotic pressure of the plasma proteins, usually termed the *colloid osmotic pressure,* or *oncotic pressure* (π_p). The total osmotic pressure of plasma is about 6000 mm Hg, whereas the oncotic pressure is only about 25 mm Hg. However, this small oncotic pressure is important in fluid exchange across the capillary wall because the plasma proteins are essentially confined to the intravascular space. The electrolytes that are mainly responsible for the total osmotic pressure of plasma are practically equal in concentration on both sides of the capillary endothelium. The relative permeability of solute to water influences the actual magnitude of the osmotic pressure. The *reflection coefficient* is the relative impediment to the passage of a substance through the capillary membrane. The reflection coefficient of water is 0 and that of albumin (to which the endothelium is essentially impermeable) is 1. Filterable solutes have reflection coefficients between 0 and 1. Also, the same molecule has different reflection coefficients in different tissues.

Of the plasma proteins, *albumin* preponderates in determining oncotic pressure. The average albumin molecule (molecular weight 69,000) is approximately half the size of the average globulin molecule (molecular weight 150,000), and it is present in almost twice the concentration as the globulins (4.0 versus 3.0 $g \cdot dl^{-1}$ of plasma). Albumin also exerts a greater osmotic force than can be accounted for solely on the basis of the number of molecules dissolved in the plasma. Therefore albumin cannot be completely replaced by inert substances of appropriate molecular size, such as dextran. This additional osmotic force becomes disproportionately greater at high concentrations of albumin (as in plasma), and it is weak to absent in dilute solutions of albumin (as in interstitial fluid).

One reason for this behavior of albumin is its negative charge at the normal blood pH and the attraction and retention of cations (principally Na^+) in the vascular compartment (the *Gibbs-Donnan effect*, Chapter 2). Furthermore, albumin binds a small number of chloride ions, which increases its negative charge and thus its ability to retain more sodium ions inside the capillaries. The small increase in electrolyte concentration of the plasma over that of the interstitial fluid produced by the negatively charged albumin enhances its osmotic force to that of an ideal solution containing a solute with a molecular weight of 37,000.

Small amounts of albumin escape from the capillaries and enter the interstitial fluid, where they exert a very small osmotic force (0.1 to 5 mm Hg). This force, π_i, is small because of the low concentration of albumin in the interstitial fluid and because at low albumin concentrations, the osmotic force of albumin simply becomes a function of the number of albumin molecules per unit volume of interstitial fluid.

Balance of Hydrostatic and Osmotic Forces— Starling Hypothesis The relationship between hydrostatic pressure and oncotic pressure and the role of these forces in regulating fluid passage across the capillary endothelium were expounded by Starling in 1896. The *Starling hypothesis* is expressed by the equation

$$\text{Fluid movement} = k[(P_c + \pi_i) - (P_i + \pi_p)] \quad \textbf{(2)}$$

where

P_c = Capillary hydrostatic pressure
P_i = Interstitial fluid hydrostatic pressure
π_p = Plasma oncotic pressure
π_i = Interstitial fluid oncotic pressure
k = Filtration constant for the capillary membrane

Filtration occurs when the algebraic sum is positive, and absorption occurs when the sum is negative.

Classically, filtration has been thought to occur at the arterial end of the capillary and absorption at its venous end because of the gradient of hydrostatic pressure along the capillary. This is true for the idealized capillary, as depicted in Figure 21-5. However, direct observations have revealed that many capillaries filter along their entire length, whereas others only absorb. In some vascular beds (e.g., the renal glomerulus) hydrostatic pressure in the capillary is high enough to result in filtration along the entire length of the capillary. In other vascular beds (e.g., in the intestinal mucosa) the hydrostatic and oncotic forces are such that absorption occurs along the whole capillary.

In the normal steady state, arterial pressure, venous pressure, postcapillary resistance, interstitial fluid hydrostatic and oncotic pressures, and plasma oncotic pressure remain relatively constant, and changes in precapillary resistance determine the movement of fluid across the capillary wall. Because water moves so quickly across the capillary endothelium, the hydrostatic and osmotic forces are nearly in equilibrium along the entire capillary. Thus filtration and absorption normally occur with very small imbalances of pressure across the capillary wall. Only about 2% of the plasma flowing through the vascular system is filtered, and of this about 85% is absorbed in the capillaries and venules. The remainder returns to the vascular system in the lymph, along with the albumin that escapes from the capillaries.

In the lungs the mean capillary hydrostatic pressure is only about 8 mm Hg. Because the plasma oncotic pressure is 25 mm Hg and interstitial fluid oncotic pressure is approximately 15 mm Hg, the net force slightly favors reabsorption. Pulmonary lymph is formed, however, and it consists of fluid that is osmotically drawn out of the capillaries by the small amount of plasma protein that escapes through the capillary endothelium. Only in pathologic conditions, such as left ventricular failure or mitral valve stenosis, does pulmonary capillary hydrostatic pressure exceed plasma oncotic pressure. When this occurs, it may lead to pulmonary edema, a condition that can seriously interfere with gas exchange in the lungs.

Capillary Filtration Coefficient The rate of movement of fluid across the capillary membrane depends not only on the algebraic sum of the hydrostatic and osmotic forces across the endothelium, but also on the area of the capillary wall available for filtration, the distance across the capillary wall, the viscosity of the filtrate, and the filtration constant of the membrane.

Because the thickness of the capillary wall and the viscosity of the filtrate are relatively constant, they can be included in the filtration constant, k. If the area of the capillary membrane is not known, the rate of filtration can be expressed as milliliters per minute per 100 g of tissue per millimeter of mercury pressure difference across the capillary wall.

In any given tissue the filtration coefficient per unit area of capillary surface, and thus capillary permeability, is not changed by certain physiologic conditions, such as arteriolar dilation and capillary distension, nor by such adverse conditions as hypoxia, hypercapnia, or acidosis. Capillary injury (toxins, severe burns) increases capillary permeability greatly (as indicated by an increased filtration coefficient) and significant amounts of fluid and protein leak out of the capillaries into the interstitial space.

Disturbances in Hydrostatic-Osmotic Balance Modest changes in arterial pressure per se may have little effect on filtration, since the change in pressure may be countered by adjustments of the precapillary resistance vessels (*autoregulation,* Chapter 22). However, in a condition such as hemorrhage, in which arterial pressure is severely reduced and venous pressure also decreases because of blood loss, capillary hydrostatic pressure falls. Furthermore, the low blood pressure in hemorrhage decreases the blood flow (and thus O_2 supply) to the tissues, and therefore vasodilator metabolites accumulate and induce relaxation of arterioles. The reduced transmural pressure also induces precapillary vessel relaxation. As a consequence of these several factors, absorption predominates over filtration. This is one of the body's compensatory mechanisms to restore blood volume.

An increase in venous pressure, as occurs in the feet when one changes from the lying to the standing position, would elevate capillary pressure and enhance filtration. However, the increase in transmural pressure causes precapillary vessel closure (*myogenic mechanism,* Chapter 22) so that the capillary filtration coefficient actually decreases. This reduction in capillary surface available for filtration protects against the extravasation of large amounts of fluid into the interstitial space (edema). However, standing associated with elevation of venous pressure (e.g., pregnancy or congestive heart failure) enhances filtration beyond the capacity of the lymphatic system in the legs to remove the capillary filtrate from the interstitial space.

The protein concentration in plasma may also change in different pathologic states and thus may alter the osmotic force and movement of fluid across the capillary membrane. The plasma protein concentration is increased in dehydration (e.g., with water deprivation, prolonged sweating, severe vomiting, and diarrhea), and water moves by osmotic forces from the tissues to the vascular compartment. In contrast, the plasma protein concentration is reduced in nephrosis (a renal disease in which protein is lost in the urine), and edema may occur. When capillaries are injured, as in burns, plasma protein escapes into the interstitial space along with fluid and increases the oncotic pressure of the interstitial fluid. This greater osmotic force outside the capillaries leads to additional fluid loss and possibly to severe dehydration of the patient.

Pinocytosis

Some transfer of substances across the capillary wall can occur in tiny vesicles; this process is called *pinocytosis*. The pinocytotic vesicles, formed by pinching off a section of the surface membrane, can take up substances on one side of the capillary wall, move across the cell, and deposit their contents at the other side. The amount of material that can be transported in this way is much less than that moved by diffusion. However, pinocytosis may move large lipid-insoluble molecules (30 nm) between blood and interstitial fluid. The number of pinocytotic vesicles in endothelium varies with the tissue (muscle > lung > brain) and increases from the arterial to the venous end of the capillary.

LYMPHATICS

The terminal lymphatic vessels consist of a widely distributed closed-end network of highly permeable lymph capillaries that resemble blood capillaries in appearance. However, they generally lack tight junctions between endothelial cells and possess fine filaments that anchor them to the surrounding connective tissue. During skeletal muscle contraction, these fine strands may distort the lymphatic vessel and open spaces between the endothelial cells. This will permit protein, large particles, and cells in the interstitial fluid to enter the lymphatic capillaries. The blood capillary filtrate and the protein and cells that have passed from the intravascular compartment to the interstitial fluid compartment are returned to the circulation by virtue of tissue pressure, facilitated by intermittent skeletal muscle activity, contractions of the lymphatic vessels, and an extensive system of one-way valves. The lymph flows through thin-walled vessels of progressively larger diameter and finally enters the right and left subclavian veins at their junctions with the respective internal jugular veins. Only cartilage, bone, epithelium, and tissues of the central nervous system are devoid of lymphatic vessels.

The volume of fluid that flows through the lymphatic system in 24 hours is about equal to an animal's total plasma volume. The protein returned by the lymphatics to the blood in a day is about one fourth to one half of the circulating plasma proteins. Lymphatic return is the only means whereby protein (mainly albumin) that leaves the vascular compartment can be returned to the blood, since back diffusion into the capillaries is negligible against the large albumin concentration gradient. If the protein was not removed from the interstitial spaces by the lymph vessels, it would accumulate in the interstitial fluid and act as an oncotic force to draw fluid from the blood capillaries and produce severe edema.

In addition to returning fluid and protein to the vascular bed, the lymphatic system filters the lymph at the lymph nodes and removes foreign particles such as bacteria. The largest lymphatic vessel, the *thoracic duct*, drains the lower extremities, returns protein lost through the permeable liver capillaries, and carries substances absorbed from the gastrointestinal tract, principally fat in the form of chylomicrons, to the circulating blood.

Lymph flow varies considerably; it is almost nil in resting skeletal muscle but increases during exercise in proportion to the degree of muscular activity. Lymph flow is increased by any mechanism that enhances the rate of blood capillary filtration (e.g., with increased capillary pressure, increased capillary permeability, or decreased plasma oncotic pressure). If the volume of interstitial fluid exceeds the drainage capacity of the lymphatics or if the lymphatic vessels become blocked, interstitial fluid accumulates, chiefly in the more compliant tissues (e.g., subcutaneous tissue), and clinical edema appears.

BIBLIOGRAPHY
Journal Articles

Boegehold MA and Johnson PC: Response of arteriolar network of skeletal muscle to sympathetic nerve stimulation, Am J Physiol 254:H919, 1988.

Curry FE: Determinants of capillary permeability: a review of mechanisms based on single capillary studies in the frog, Circ Res 59:367, 1986.

Davis MJ: Microvascular control of capillary pressure during increases in local arterial and venous pressure, Am J Physiol 254:H772, 1988.

Duling BR and Klitzman B: Local control of microvascular function: role in tissue oxygen supply, Annu Rev Physiol 42:373, 1980.

Furchgott R: Role of endothelium in responses of vascular smooth muscle, Circ Res 53:557, 1983.

Krogh A: The number and distribution of capillaries in muscles with calculation of the oxygen pressure head necessary for supplying the tissue, J Physiol 52:409, 1919.

Lewis DH, editor: Symposium on lymph circulation, Acta Physiol Scand Suppl 463:9, 1979.

Rubanyi GM: Endothelium-dependent pressure-induced contraction of isolated canine carotid arteries, Am J Physiol 255:H783, 1988.

Starling EH: On the absorption of fluids from the connective tissue spaces, J Physiol 19:312, 1896.

Books and Monographs

Bert JL and Pearce RH: The interstitium and microvascular exchange. In Handbook of physiology, section 2: The cardiovascular system—microcirculation, vol IV, Bethesda, Md, 1984, American Physiological Society.

Crone C and Levitt DG: Capillary permeability to small solutes. In Handbook of physiology, section 2: The cardiovascular system—microcirculation, vol IV, Bethesda, Md, 1984, American Physiological Society.

Hudlicka O: Development of microcirculation: capillary growth and adaptation. In Handbook of physiology, section 2: The cardiovascular system—microcirculation, vol IV, Bethesda, Md, 1984, American Physiological Society.

Johnston MG, editor: Experimental biology of the lymphatic circulation, Amsterdam, 1985, Elsevier.

Michel CC: Fluid movements through capillary walls. In Handbook of physiology, section 2: The cardiovascular system—microcirculation, vol IV, Bethesda, Md, 1984, American Physiological Society.

Mortillaro NA: Physiology and pharmacology of the microcirculation, vol I, New York, 1983, Academic Press, Inc.

Renkin EM: Control of microcirculation and blood-tissue exchange. In Handbook of physiology, section 2: The cardiovascular system—microcirculation, vol IV, Bethesda, Md, 1984, American Physiological Society.

Simionescu M and Simionescu N: Ultrastructure of the microvascular wall: functional correlations. In Handbook of physiology, section 2: The cardiovascular system—microcirculation, vol IV, Bethesda, Md, 1984, American Physiological Society.

Taylor AE and Granger DN: Exchange of macromolecules across the microcirculation. In Handbook of physiology, section 2: The cardiovascular system—microcirculation, vol IV, Bethesda, Md, 1984, American Physiological Society.

Wiedeman MP: Architecture. In Handbook of physiology, section 2: The cardiovascular system—microcirculation, vol IV, Bethesda, Md, 1984, American Physiological Society.

Zweifach BW and Lipowsky HH: Pressure-flow relations in blood and lymph microcirculation. In Handbook of physiology, section 2: The cardiovascular system—microcirculation, vol IV, Bethesda, Md, 1984, American Physiological Society.

The Peripheral Circulation and Its Control

The peripheral circulation is essentially under dual control: **centrally by the nervous system, and locally in the tissues** by the conditions in the immediate vicinity of the blood vessels. The relative importance of these two control mechanisms is not the same in all tissues. In some areas of the body, such as the skin and the splanchnic regions, neural regulation of blood flow predominates, whereas in others, such as the heart and brain, local factors are dominant.

The vessels that regulate the blood flow throughout the body are called the *resistance vessels* (arterioles). These vessels offer the greatest resistance to flow of blood pumped to the tissues by the heart and thereby are important in the maintenance of arterial blood pressure. Smooth muscle fibers are the main component of the walls of the resistance vessels (see Figure 15-1). Therefore the vessel lumen can be varied from complete obliteration by strong contraction of the smooth muscle, with infolding of the endothelial lining, to maximal dilation by full relaxation of the smooth muscle. Some resistance vessels are closed at any given time by partial contraction (or tone) of the arteriolar smooth muscle. If all the resistance vessels in the body dilated simultaneously, blood pressure would fall precipitously.

VASCULAR SMOOTH MUSCLE

Vascular smooth muscle is responsible for the control of total peripheral resistance, arterial and venous tone, and the distribution of blood flow throughout the body. The smooth muscle cells are small, mononucleate, and spindle shaped. They are generally arranged in helical or circular layers around the larger blood vessels and in a single circular layer around the arterioles (see Chapter 13).

INTRINSIC OR LOCAL CONTROL OF PERIPHERAL BLOOD FLOW

In certain tissues the blood flow is adjusted to the existing metabolic activity of the tissue. Furthermore, at constant levels of tissue metabolism, imposed changes in the perfusion pressure are met with vascular resistance changes that maintain a constant blood flow. This mechanism is commonly referred to as *autoregulation of blood flow* and is illustrated graphically in Figure 22-1. In the skeletal muscle preparation from which these data were gathered, the muscle was completely isolated from the rest of the animal and was in a resting state. The pressure was abruptly increased or decreased from a control pressure of 100 mm Hg. The blood flows observed immediately after changing the perfusion pressure are represented by the black curve. Maintenance of the pressure at each new level was followed within 60 seconds by a return of flow to or toward the control levels; the red curve represents these steady-state flows. Over the pressure range of 20 to 120 mm Hg, the steady-state flow is relatively constant. Calculation of resistance (pressure/flow) across the vascular bed during steady-state conditions indicates that with elevation of perfusion pressure, the resistance vessels constricted, whereas with reduction of perfusion pressure, dilation occurred.

The mechanism responsible for this constancy of blood flow in the presence of an altered perfusion

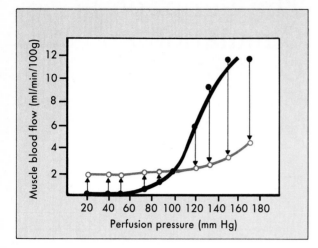

FIGURE 22-1   Pressure-flow relationship in the skeletal muscle vascular bed of the dog. The black curve represents the flows obtained immediately after abrupt changes in perfusion pressure from the control level (point where lines cross). The red curve represents the steady-state flows obtained at the new perfusion pressure. (Redrawn from Jones RD and Berne RM: Circ Res 14:126, 1964. By permission of the American Heart Association, Inc.)

pressure is not known. However, two explanations have been suggested, the myogenic hypothesis and the metabolic hypothesis.

The *myogenic hypothesis* states that the vascular smooth muscle contracts in response to stretch and relaxes with a reduction in tension. An abrupt increase in perfusion initially distends the blood vessels. This passive vascular distension is followed by contraction of the smooth muscles of the resistance vessels and a return of flow to the previous control level.

Because blood pressure is reflexly maintained at a fairly constant level under normal conditions, we would expect the operation of a myogenic mechanism to be minimized. However, when one changes from a lying to a standing position, a large increase in transmural pressure occurs in the vessels of the lower extremities. The precapillary vessels constrict in response to this imposed stretch. The constriction diminishes capillary filtration until the increase in plasma oncotic pressure and the increase in interstitial fluid pressure balance the elevated capillary hydrostatic pressure associated with the vertical position. If arteriolar resistance did not increase with standing, the hydrostatic pressure in the lower parts of the legs would reach such high levels that large

volumes of fluid would pass from the capillaries into the interstitial fluid compartment and produce edema.

According to the *metabolic hypothesis*, the blood flow is governed by the metabolic activity of the tissue. Any intervention that results in an oxygen supply that is inadequate for the tissue requirements releases vasodilator metabolites from the tissue and dilates the resistance vessels. When the metabolic rate of the tissue increases or the oxygen delivery to the tissue decreases, more vasodilator substance is formed. If perfusion pressure is constant, a decrease in metabolic activity will decrease the concentration of the vasodilator in the tissue. Similarly, if metabolic activity is constant, an increase in perfusion pressure will decrease the tissue concentration of the vasodilator agent. A decrease in metabolite production or an increase in washout or inactivation of the metabolite will increase the precapillary resistance. An attractive feature of the metabolic hypothesis is that in most tissues, blood flow closely parallels metabolic activity. Thus, although blood pressure is kept fairly constant, metabolic activity and blood flow in the different tissues vary together under physiologic conditions.

Many substances have been proposed as mediators of metabolic vasodilation. Some of the dilators suggested initially were lactic acid, carbon dioxide, and hydrogen ions. However, the decrease in vascular resistance induced by supernormal concentrations of these dilator agents could not account for the dilation observed under physiologic conditions of increased metabolic activity. However, few data support the concept that oxygen tension (PO_2) per se serves as a mediator of metabolically induced vasodilation. Experimental observations are more compatible with the release of a vasodilator metabolite from the tissue than with a direct effect of PO_2 on the vascular smooth muscle.

Potassium ions, inorganic phosphate, and interstitial fluid osmolarity can also induce vasodilation, and they have been proposed as factors that contribute to active hyperemia (increased blood flow caused by enhanced tissue activity). However, significant increases of phosphate concentration and osmolarity are not consistently observed during muscle contraction. Furthermore, the changes in these factors that occur with the onset of skeletal muscle contraction or an increase in cardiac activity are not sustained, even though arteriolar dilation persists throughout the period of enhanced muscle activity. Therefore some other agent must mediate the vasodilation associated with the increased tissue metabolism.

Recent evidence indicates that adenosine, which is involved in the regulation of coronary blood flow, may also participate in the control of the resistance vessels in skeletal muscle. Some of the prostaglandins also may be important vasodilator mediators in certain vascular beds.

Metabolic control of vascular resistance by the release of a vasodilator is predicated on the existence of *basal tone,* which is the partial contraction (tonic activity) of vascular smooth muscle. In contrast to tone in skeletal muscle, basal tone is independent of the nervous system, and the factor responsible is not known.

If arterial inflow to a vascular bed is stopped for a few seconds to several minutes, the blood flow immediately after release of the occlusion *(reactive hyperemia)* exceeds the flow before occlusion, and it returns only gradually to the control level. This is illustrated in Figure 22-2, where blood flow to the leg was stopped by clamping the femoral artery for 15, 30, and 60 seconds. Release of the 60-second occlusion resulted in a peak blood flow 70% greater than the control flow, with a return to control flow within about 110 seconds. When this same experiment is done in humans by inflating a blood pressure cuff on the

upper arm, dilation of the resistance vessels of the hand and forearm, immediately after release of the cuff, is evident from the bright red color of the skin and the fullness of the veins. Within limits the peak flow and particularly the duration of the reactive hyperemia are proportional to the duration of the occlusion (Figure 22-2). If the arm is exercised during the occlusion period, reactive hyperemia is increased. These observations and the close relationship that exists between metabolic activity and blood flow in the unoccluded limb are consonant with the hypothesis that a metabolic mechanism regulates tissue blood flow.

EXTRINSIC CONTROL OF PERIPHERAL BLOOD FLOW

Neural Sympathetic Vasoconstriction

Several regions in the medulla oblongata influence cardiovascular activity (Chapter 18). Some of the effects of stimulation of the dorsal lateral medulla are vasoconstriction, cardiac acceleration, and enhanced myocardial contractility. Caudal and ventromedial to the pressor region is a zone that decreases blood pressure on stimulation. This depressor area exerts its effect by direct inhibition of spinal neurons and by inhibition of the medullary pressor region. However, the precise mechanism of its depressor actions is still unknown. These areas do not comprise a center in an anatomic sense because no discrete group of cells is discernible. However, they do constitute a center in a physiologic sense, in that stimulation of the pressor region produces the responses mentioned previously.

From the vasoconstrictor regions, fibers descend in the spinal cord and synapse at different levels of the thoracolumbar region (T1 to L2 or L3). Fibers from the intermediolateral gray matter of the cord emerge with the ventral roots but leave the motor fibers to join the paravertebral sympathetic chains through the white communicating branches. These preganglionic white (myelinated) fibers may pass up or down the sympathetic chains to synapse in the various ganglia within the chains or in certain outlying ganglia. Postganglionic gray branches (unmyelinated) then join the corresponding segmental spinal nerves and accompany them to the periphery to innervate the arteries and veins. Postganglionic sympathetic fibers from the various ganglia join the large arteries and

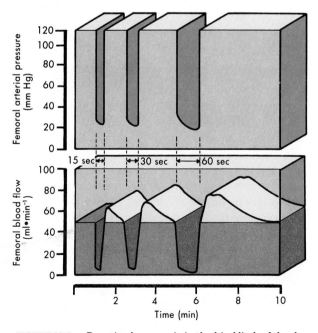

FIGURE 22-2   Reactive hyperemia in the hind limb of the dog after 1-, 30-, and 60-second occlusions of the femoral artery.

accompany them as an investing network of fibers to the resistance (arterioles) and capacitance (veins) vessels.

The vasoconstrictor regions are tonically active. Reflexes or humoral stimuli that enhance this activity increase the frequency of impulses reaching the terminal branches of the vessels. A constrictor neurohumor (norepinephrine) is released from the postganglionic nerve endings and elicits constriction (α-adrenergic effect) of the resistance vessels. Inhibition of the vasoconstrictor areas diminishes the frequency of impulses in the efferent nerve fibers, resulting in vasodilation. In this manner neural regulation of the peripheral circulation is accomplished primarily by altering the number of impulses passing down the vasoconstrictor fibers of the sympathetic nerves to the blood vessels. The tonic activity of the vasomotor regions may vary rhythmically, which is manifested as oscillations of arterial pressure. Some oscillations occur at the frequency of respiration *(Traube-Hering waves)* and are caused by an increase in sympathetic impulses to the resistance vessels coincident with inspiration. Other oscillations *(Mayer waves)* occur at a lower frequency than respiration.

Sympathetic Constrictor Influence on Resistance and Capacitance Vessels

The vasoconstrictor fibers of the sympathetic nervous system supply the arteries, arterioles, and veins, but the neural influence on the large vessels is far less important functionally than it is on the microcirculation. Capacitance vessels are more responsive to sympathetic nerve stimulation than are resistance vessels; they are maximally constricted at a lower frequency of stimulation than are the resistance vessels. However, capacitance vessels do not possess β-adrenergic receptors, nor do they respond to vasodilator metabolites. Norepinephrine is the neurotransmitter released at the sympathetic nerve terminals at the blood vessels. Many factors, such as circulating hormones and particularly locally released substances, modify the liberation of norepinephrine from the nerve terminals.

At basal tone approximately one third of the blood volume of a tissue can be mobilized on stimulation of the sympathetic nerves at physiologic frequencies. The basal tone is very low in capacitance vessels, and only small increases in volume are obtained with maximal doses of the vasodilator, acetylcholine.

Therefore at basal tone the tissue blood volume is close to its maximal value.

Blood is mobilized from capacitance vessels in response to physiologic stimuli. In exercise, activation of the sympathetic nerve fibers constricts veins and thus augments the cardiac filling pressure. Also, in arterial hypotension (as in hemorrhage) the capacitance vessels constrict and thereby aid in overcoming the associated decrease in central venous pressure. In addition, the resistance vessels constrict in shock and thereby assist in the restoration of arterial pressure. Furthermore, extravascular fluid is mobilized by a greater reabsorption of fluid into the capillaries in response to the lowered capillary hydrostatic pressure.

Parasympathetic Neural Influence

The efferent fibers of the cranial division of the parasympathetic nervous system supply blood vessels of the head and viscera, whereas fibers of the sacral division supply blood vessels of the genitalia, bladder, and large bowel. Skeletal muscle and skin do not receive parasympathetic innervation. Because only a small proportion of the resistance vessels of the body receives parasympathetic fibers, the effect of these cholinergic fibers on total vascular resistance is small.

Humoral Factors

Epinephrine and norepinephrine exert a profound effect on the peripheral blood vessels. In skeletal muscle epinephrine in low concentrations dilates resistance vessels (β-adrenergic effect) and in high concentrations constricts them (α-adrenergic effect). In skin only vasoconstriction is obtained with epinephrine, whereas in all vascular beds the main effect of norepinephrine is vasoconstriction. When stimulated, the adrenal gland can release epinephrine and norepinephrine into the systemic circulation. Under physiologic conditions, however, the effect of catecholamine release from the adrenal medulla is less important than norepinephrine release from the sympathetic nerves.

Vascular Reflexes

Areas of the medulla that mediate sympathetic and vagal effects are under the influence of neural impulses (arising in the baroreceptors, chemoreceptors, hypothalamus, cerebral cortex, and skin) and of local carbon dioxide and oxygen concentrations.

Baroreceptors The baroreceptors (or *pressoreceptors*) are stretch receptors located in the carotid sinuses (slightly widened areas of the internal carotid arteries at their points of origin from the common carotid arteries) and in the aortic arch (Figures 22-3 and 22-4). Impulses arising in the carotid sinus travel up afferent fibers in the sinus nerve, which is a branch of the glossopharnygeal nerve. These fibers travel with the latter to the nucleus of the tractus solitarius (NTS) in the medulla. The NTS is the site of central projection of the chemoreceptors and baroreceptors. Stimulation of the NTS inhibits sympathetic nerve impulses to the peripheral blood vessels *(depressor effect),* whereas lesions of the NTS produce vasoconstriction *(pressor effect).* Impulses arising in the baroreceptors of the aortic arch reach the NTS via afferent fibers in the vagus nerves.

The baroreceptor nerve terminals in the walls of the carotid sinus and aortic arch respond to the stretch and deformation of the vessel induced by the arterial pressure. The frequency of firing is enhanced by an increase in blood pressure and diminished by a reduction in blood pressure. An increase in impulse frequency inhibits the vasoconstrictor regions and results in peripheral vasodilation and a lowering of blood pressure. Contributing to a lowering of the blood pressure is a bradycardia brought about by stimulation of the vagal regions. The carotid sinus baroreceptors are more sensitive to pressure than are the aortic baroreceptors. However, with pulsatile changes in blood pressure the two sets of baroreceptors respond similarly.

The carotid sinus with the sinus nerve intact can be isolated from the rest of the circulation and artificially perfused. Under these conditions, changes in the pressure within the carotid sinus elicit reciprocal changes in the blood pressure of the experimental animal. The receptors in the walls of the carotid sinus show some adaptation, and therefore they are more responsive to constantly changing pressures than to sustained constant pressures; this is illustrated in Figure 22-5. At normal levels of blood pressure a barrage of impulses from a single fiber of the sinus nerve

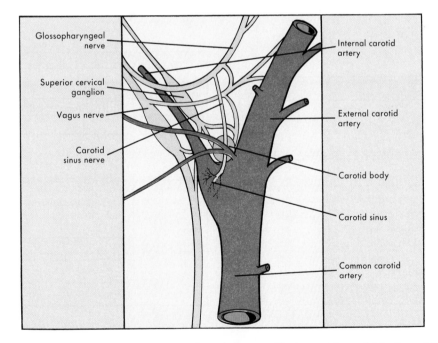

FIGURE 22-3 Diagrammatic representation of the carotid sinus and carotid body and their innervation in the dog. (Redrawn from Adams WE: The comparative morphology of the carotid body and carotid sinus, Springfield, Ill, 1958, Charles C Thomas, Publisher.)

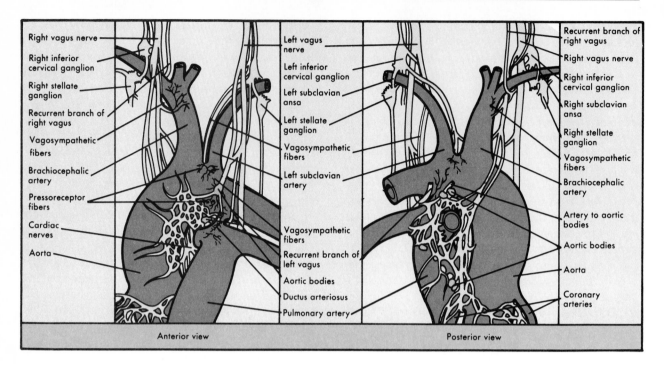

FIGURE 22-4   Anterior view and posterior view of the aortic arch showing the innervation of the aortic bodies and pressoreceptors in the dog. (Modified from Nonidez JF: Anat Rec 69:299, 1937.)

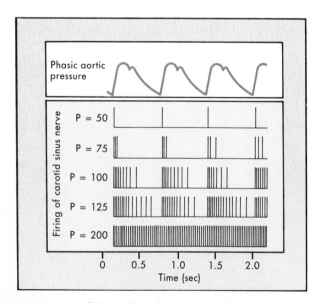

FIGURE 22-5   Relationship of phasic aortic blood pressure in the firing of a single afferent nerve fiber from the carotid sinus at different levels of mean arterial pressure.

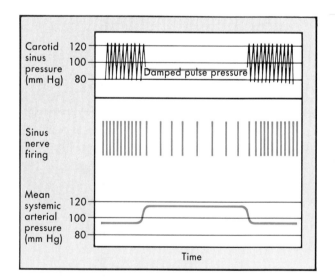

FIGURE 22-6   Effect of reducing pulse pressure in the vascularly isolated perfused carotid sinuses *(top record)* on impulses recorded from a fiber of a sinus nerve *(middle record)* and on mean systemic arterial pressure *(bottom record)*. Mean pressure in the carotid sinuses *(red line, top record)* is held constant when pulse pressure is damped.

is initiated in early systole by the pressure rise; only a few spikes are observed during late systole and early diastole. At lower pressures these phasic changes are even more evident, but the overall frequency of discharge is reduced. The blood pressure threshold for eliciting sinus nerve impulses is about 50 mm Hg, and a maximal sustained firing is reached at approximately 200 mm Hg.

Because the baroreceptors do adapt, their response at any level of mean arterial pressure is greater with a large than with a small pulse pressure. This is illustrated in Figure 22-6, which shows the effects of damping pulsations in the carotid sinus on the frequency of firing in a sinus nerve fiber and on the systemic arterial pressure. When the pulse pressure in the carotid sinuses is reduced but mean pressure remains constant, the rate of electrical impulses recorded from a sinus nerve fiber decreases and the systemic arterial pressure increases. Restoration of the pulse pressure in the carotid sinus restores the frequency of sinus nerve discharge and systemic arterial pressure to control levels.

The resistance increases that occur in the peripheral vascular beds in response to a reduced pressure in the carotid sinus vary from one vascular bed to another and thereby redistribute blood flow. For example, the resistance changes elicited in the dog by altering carotid sinus pressure are greatest in the femoral vessels, less in the renal, and least in the mesenteric and celiac vessels. Furthermore, the sensitivity of the carotid sinus reflex can be altered. For example, in hypertension, when the carotid sinus becomes stiffer and less deformable as a result of the high intraarterial pressure, baroreceptor sensitivity decreases.

The **baroreceptors play a key role in short-term adjustments of blood pressure,** when relatively abrupt changes in blood volume, cardiac output, or peripheral resistance (as in exercise) occur. However, **long-term control of blood pressure (i.e., over days or weeks) is determined by the individual's fluid balance,** namely, the balance between fluid intake and fluid output and the resulting blood volume. (At constant peripheral resistance, an increase in blood volume increases blood pressure by augmenting cardiac output. See Chapter 23.) By far **the single most important organ in the control of body fluid volume, and thus blood pressure, is the kidney.** With overhydration excessive fluid intake is excreted, whereas with dehydration urine output is reduced.

Cardiopulmonary Baroreceptors In addition to the carotid sinus and aortic baroreceptors, cardiopulmonary receptors also exist; both types of receptors are necessary for the full expression of blood pressure regulation. The cardiopulmonary receptors have vagal and sympathetic afferent and efferent nerves, are tonically active, and can alter peripheral resistance with changes in intracardiac, venous, or pulmonary vascular pressures.

Peripheral Chemoreceptors The *peripheral chemoreceptors* consist of small, highly vascular bodies in the regions of the aortic arch and of the carotid artery bifurcations, just medial to the carotid sinuses (see Figures 22-3 and 22-4). The chemoreceptors are sensitive to changes in the oxygen tension (PO_2), carbon dioxide tension (PCO_2) and pH of the blood. Although they are mainly concerned with the regulation of respiration (Chapter 28), the peripheral chemoreceptors reflexly influence the circulatory system to a minor degree. A reduction in arterial blood oxygen tension (PaO_2) stimulates the chemoreceptors. The resultant increase in the frequency of impulses in the afferent nerve fibers from the carotid and aortic bodies stimulates the vasoconstrictor regions; this action increases tone in the resistance and capacitance vessels. The chemoreceptors are also stimulated by increased arterial blood carbon dioxide tension ($PaCO_2$) and by reduced pH. However, the reflex effect is much less than the direct effect of hypercapnia and hydrogen ions on the vasomotor regions in the medulla. When hypoxia and hypercapnia occur at the same time, the stimulation of the chemoreceptors is greater than the sum of the two stimuli when they act alone. Stimulation of the chemoreceptors simultaneously with a reduction of pressure in the baroreceptors potentiates the vasoconstrictor response of hypertension on the peripheral vessels. However, when the baroreceptors and chemoreceptors are both stimulated (e.g., high carotid sinus pressure and low PaO_2), the effects of the baroreceptors predominate.

Chemoreceptors with sympathetic afferent fibers exist in the heart. These cardiac chemoreceptors are activated by ischemia, and they transmit the precordial pain (angina pectoris) associated with an inadequate blood supply to the myocardium.

Hypothalamus Optimal function of the cardiovascular reflexes requires the integrity of pontine and hypothalamic structures. Furthermore, these structures are responsible for behavioral and emotional control of the cardiovascular system. Stimulation of the anterior hypothalamus decreases blood pressure and heart rate, whereas stimulation of the posterolateral region of the hypothalamus increases blood pressure and heart rate. The hypothalamus also contains a temperature-regulating center that affects the skin vessels. Cooling the skin or the blood perfusing the hypothalamus results in constriction of the skin vessels and heat conservation, whereas warm stimuli have the opposite effects.

Cerebrum The cerebral cortex can also affect the blood flow distribution in the body. Stimulation of the motor and premotor areas can affect blood pressure; usually a pressor response is obtained. However, vasodilation and depressor responses may be evoked (e.g., blushing or fainting) in response to an emotional stimulus.

Skin and Viscera Painful stimuli can elicit either pressor or depressor responses, depending on the magnitude and location of the stimulus. Distension of the viscera often decreases blood pressure, whereas painful stimuli on the body surface usually raise blood pressure.

Pulmonary Reflexes

Inflation of the lungs reflexly dilates systemic vessels and decreases arterial blood pressure. Conversely, collapse of the lungs constricts systemic vessels. Afferent fibers that mediate this reflex are carried in the vagus nerves. Stimulation of the pulmonary stretch receptors inhibits the vasomotor areas. The magnitude of the depressor response to lung inflation is directly related to the degree of inflation and to the existing level of vasoconstrictor tone.

Chemosensitive Regions of the Medulla

Increases of $PaCO_2$ stimulate the medullary vasoconstrictor regions and thereby increase peripheral resistance. Reduction in $PaCO_2$ below normal levels (as with hyperventilation) decreases the tonic activity in these areas and thus decreases peripheral resistance. The chemosensitive regions are also affected by changes in pH. A lowering of blood pH stimulates and a rise in blood pH inhibits these areas.

Oxygen tension has little direct effect on the vasomotor region. The reflex effect of hypoxia is mainly mediated by the carotid and aortic chemoreceptors. Moderate reduction of PaO_2 stimulates the vasomotor region, but severe reduction depresses vasomotor activity, just as very low O_2 tensions depress other areas of the brain.

Cerebral ischemia results in severe peripheral vasoconstriction. The stimulation is probably caused by a local accumulation of carbon dioxide and a reduction of oxygen in the brain. With prolonged severe ischemia, extreme depression of cerebral function eventually supervenes and blood pressure falls.

BALANCE BETWEEN EXTRINSIC AND INTRINSIC FACTORS IN REGULATION OF PERIPHERAL BLOOD FLOW

Dual control of the peripheral vessels by intrinsic and extrinsic mechanisms constitutes a complex system of vascular regulation. This system enables the body to direct blood flow to areas where it is in greater need and to divert it away from areas where it is in lesser need. In some tissues the relative potency of extrinsic and intrinsic mechanisms is constant. However, in other tissues the ratio is changeable, depending on the state of activity of that tissue.

In the brain and the heart, which are vital structures with very limited tolerance for a reduced blood supply, intrinsic flow-regulating mechanisms are dominant. For instance, massive discharge of the medullary vasoconstrictor region (which might occur in severe, acute hemorrhage) has negligible effects on the cerebral and cardiac resistance vessels but greatly constricts the skin, renal, and splanchnic blood vessels.

In the skin the extrinsic vascular control is dominant. Not only do the cutaneous vessels participate strongly in a general vasoconstrictor discharge, but they also respond selectively through the hypothalamic pathways that subserve body temperature regulation. However, intrinsic control can be demonstrated by local changes of skin temperature that can modify or override the central influence on the resistance and capacitance vessels.

In skeletal muscle the changing balance between extrinsic and intrinsic mechanisms can be clearly seen. In resting skeletal muscle neural control (vasoconstrictor tone) is dominant, as can be demonstrated by the large increment in blood flow that occurs immediately after sectioning the sympathetic nerves to the tissue. Just before and at the start of running, blood flow increases in the leg muscles. After the onset of exercise the intrinsic flow-regulating mechanism assumes control. Because of the local increase in metabolites, vasodilation occurs in the active muscles. Vasoconstriction occurs in the inactive tissues

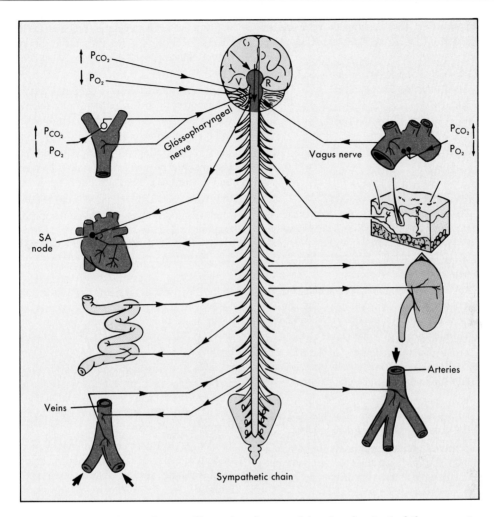

FIGURE 22-7   Schematic diagram illustrating the neural input and output of the vasomotor region *(VR). SA node;* Sinoatrial node.

as a manifestation of the general sympathetic discharge, but constrictor impulses reaching the resistance vessels of the active muscles are overridden by the local metabolic effect. Operation of this dual-control mechanism thus provides increased blood where it is required and shunts it away from the inactive areas. Similar effects may be achieved by an increase in $PaCO_2$. Normally the hyperventilation associated with exercise keeps $PaCO_2$ at normal levels. However, if $PaCO_2$ increased, a generalized vasoconstriction would occur because of stimulation of the medullary vasoconstrictor region by CO_2. In the active muscles, where the CO_2 concentration is highest, the smooth muscle of the arterioles relaxes in response to the local PCO_2. Factors that affect and that are affected by the medullary vasomotor region are summarized in Figure 22-7.

BIBLIOGRAPHY
Journal Articles

Berne RM et al: Adenosine in the local regulation of blood flow: a brief overview, Fed Proc 42:3136, 1983.

Bevan JA and Brayden JE: Nonadrenergic neural vasodilator mechanisms, Circ Res 60:309, 1987.

Donald DE and Shepherd JT: Autonomic regulation of the peripheral circulation, Annu Rev Physiol 42:429, 1980.

Duling BR and Damon DH: An examination of the measurement of flow heterogeneity in striated muscle, Circ Res 60:1, 1987.

Johnson PC: Autoregulation of blood flow, Circ Res 59:483, 1986.

Laughlin MH: Skeletal muscle blood flow capacity: role of muscle pump in exercise hyperemia, Am J Physiol 253:H993, 1987.

Meininger GA et al: Myogenic vasoregulation overrides local metabolic control in resting rat skeletal muscle, Circ Res 60:861, 1987.

Metting PJ et al: Evaluation of whole body autoregulation in conscious dogs, Am J Physiol 255:H44, 1988.

O'Leary DS and Scher AM: Arterial pressure control after chronic carotid sinus denervation, Am J Physiol 255:H910, 1988.

Shepherd JT: Reflex control of arterial blood pressure, Cardiovasc Res 16:357, 1982.

Books and Monographs

Abboud FM and Thames MD: Interaction of cardiovascular reflexes in circulatory control. In Handbook of physiology, section 2: The cardiovascular system—peripheral circulation and organ blood flow, vol III, Bethesda, Md, 1983, American Physiological Society.

Bevan JA et al: Adrenergic regulation of vascular smooth muscle. In Handbook of physiology, section 2: The cardiovascular system—vascular smooth muscle, vol II, Bethesda, Md, 1980, American Physiological Society.

Bishop VS et al: Cardiac mechanoreceptors. In Handbook of physiology, section 2: The cardiovascular system, vol III, Bethesda, Md, 1983, American Physiological Society.

Brown, AM: Cardiac reflexes. In Handbook of physiology, section 2: The cardiovascular system—the heart, vol I, Bethesda, Md, 1979, American Physiological Society.

Eyzaguirre C et al: Arterial chemoreceptors. In Handbook of physiology, section 2: The cardiovascular system—peripheral circulation and organ blood flow, vol III, Bethesda, Md, 1983, American Physiological Society.

Johnson PC: The myogenic response. In Handbook of physiology, section 2: The cardiovascular system—vascular smooth muscle, vol II, Bethesda, Md, 1980, American Physiological Society.

Korner PI: Central nervous control of autonomic cardiovascular function. In Handbook of physiology, section 2: The cardiovascular system—the heart, vol I, Bethesda, Md, 1979, American Physiological Society.

Mancia G and Mark AL: Arterial baroreflexes in humans. In Handbook of physiology, section 2: The cardiovascular system—peripheral circulation and organ blood flow, vol III, Bethesda, Md, 1983, American Physiological Society.

Mark AL and Mancia G: Cardiopulmonary baroreflexes in humans. In Handbook of physiology, section 2: The cardiovascular system—peripheral circulation and organ blood flow, vol III, Bethesda, Md, 1983, American Physiological Society.

Rothe CF: Venous system: physiology of the capacitance vessels. In Handbook of physiology, section 2: The cardiovascular system—peripheral circulation and organ blood flow, vol III, Bethesda, Md, 1983, American Physiological Society.

Sagawa K: Baroreflex control of systemic arterial pressure and vascular bed. In Handbook of physiology, section 2: The cardiovascular system—peripheral circulation and organ blood flow, vol III, Bethesda, Md, 1983, American Physiological Society.

Sparks HV Jr: Effect of local metabolic factors on vascular smooth muscle. In Handbook of physiology, section 2: The cardiovascular system—vascular smooth muscle, vol II, Bethesda, Md, 1980, American Physiological Society.

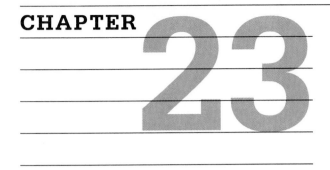

Control of Cardiac Output: Coupling of the Heart and Blood Vessels

CONTROLLING FACTORS

The following four factors (Figure 23-1) control cardiac output: *heart rate, myocardial contractility, preload,* and *afterload.* Heart rate and myocardial contractility are strictly *cardiac factors.* They are characteristics of the cardiac tissues, although they are modulated by various neural and humoral mechanisms.

Preload and afterload, however, depend on the characteristics of both the heart and the vascular system. On the one hand, preload and afterload are important determinants of cardiac output. On the other hand, preload and afterload are themselves determined by the cardiac output and by certain vascular characteristics as well. Preload and afterload may be designated *coupling factors* because they constitute a functional coupling between the heart and blood vessels. The heart pumps the blood around the vascular system. The rate at which it pumps that blood is an important determinant of the preload and afterload. Concomitantly, the vascular characteristics also determine the preload and afterload and thus regulate the quantity of blood that the heart will pump around the circuit per unit time. To understand the regulation of cardiac output, therefore, it is important to appreciate the nature of the coupling between the heart and the vascular system.

Guyton and his colleagues have developed a graphic technique that we shall use in modified form to analyze the interactions between the cardiac and vascular components of the circulatory system. This analysis involves two independent functional relationships between the *cardiac output* and the *central venous pressure* (i.e., the pressure in the right atrium and thoracic venae cavae).

The curve defining one of these relationships will be called the *cardiac function curve.* It is an expression of the well-known Frank-Starling relationship (see Chapter 18) and reflects that the cardiac output depends partly on the preload (i.e., the central venous, or right atrial, pressure). The cardiac function curve is a characteristic of the heart itself; it has been studied in hearts completely isolated from the rest of the circulatory system. This curve indicates that a rise in right atrial pressure ordinarily increases cardiac output; that is, cardiac output varies **directly** with right atrial pressure.

The second functional relationship between the central venous pressure and the cardiac output is defined by a second curve that we call the *vascular function curve.* This relationship depends only on certain characteristics of the vascular system: the *peripheral resistance, arterial* and *venous capacitances,* and *blood volume.* The vascular function curve is entirely independent of the characteristics of the heart; it can be derived even if the heart were replaced by a mechanical pump. The vascular function curve shows that if the rate at which the blood is pumped around the body is increased, the central venous pressure decreases; that is, the central venous pressure varies **inversely** with the cardiac output.

Thus the cardiac function curve illustrates that cardiac output varies directly with central venous pressure, whereas the vascular function curve indicates that central venous pressure varies inversely

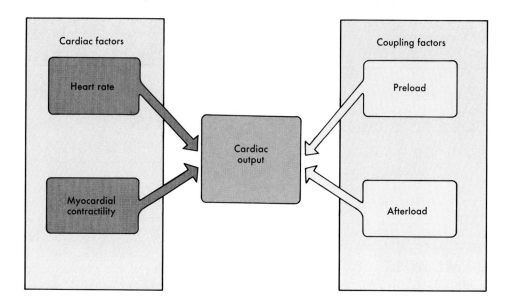

FIGURE 23-1   The four determinants of cardiac output.

with cardiac output. On first exposure to this pair of curves, one often thinks that these assertions are contradictory. The explanations that follow are intended to avert this misconception. This chapter shows that the pair of graphs, one with a direct and one with an inverse relationship between the same two variables, are not at all contradictory. We explain that the combination of a direct and an inverse relationship are essential for the operation of the cardiovascular system to be stable.

We first examine the vascular function curve and explain why central venous pressure varies inversely with the cardiac output.

VASCULAR FUNCTION CURVE

The vascular function curve defines the change in central venous pressure that occurs as a consequence of a change in cardiac output; that is, central venous pressure is the *dependent variable* (or response), and cardiac output is the *independent variable* (or stimulus).

The simplified model of the circulation illustrated in Figure 23-2 helps explain how the cardiac output determines the level of the central venous pressure.

The essential components of the cardiovascular system have been combined into four elements. The right and left sides of the heart and the pulmonary vascular bed are considered simply to be a pump-oxygenator, much as that employed during open-heart surgery. In Figure 23-2, it is simply called a *pump*. The high-resistance microcirculation is designated the *peripheral resistance*. Finally, the entire compliance of the system is subdivided into two components, the total *arterial compliance, C_a,* and the total *venous compliance, C_v.* As defined in Chapter 20, C is the increment of volume (dV) accommodated per unit change of pressure (dP); that is,

$$C = dV/dP \qquad \textbf{(1)}$$

The venous compliance normally is about 20 times as great as the arterial compliance. In the following example the ratio of C_v to C_a is set at 19:1 to simplify certain calculations. Thus, if it were necessary to add x ml of blood to the arterial system to increment the arterial pressure by 1 mm Hg, then it would be necessary to add 19 x ml of blood to the venous system to raise venous pressure by the same amount.

Our model illustrates very simply why the central venous pressure varies inversely with the cardiac output. For this example, let us first endow our model

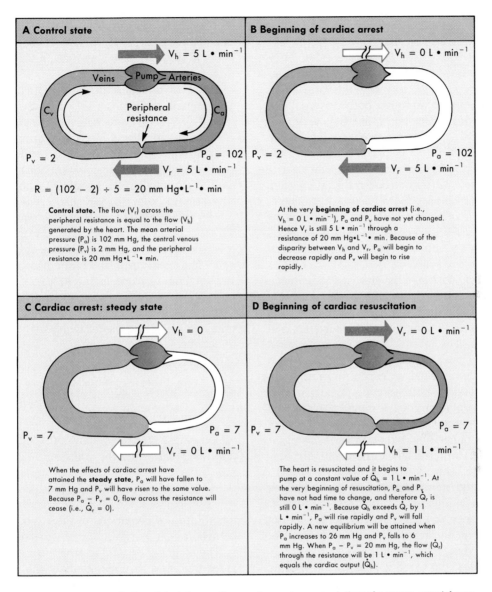

A Control state

Veins   Pump   Arteries

$V_h = 5$ L • min^{-1}

Peripheral resistance

C_v C_a

$P_v = 2$ $P_a = 102$

$V_r = 5$ L • min^{-1}

$R = (102 - 2) \div 5 = 20$ mm Hg•L^{-1}• min

Control state. The flow (V_r) across the peripheral resistance is equal to the flow (V_h) generated by the heart. The mean arterial pressure (P_a) is 102 mm Hg, the central venous pressure (P_v) is 2 mm Hg, and the peripheral resistance is 20 mm Hg•L^{-1}• min.

B Beginning of cardiac arrest

$V_h = 0$ L • min^{-1}

$P_v = 2$ $P_a = 102$

$V_r = 5$ L • min^{-1}

At the very **beginning of cardiac arrest** (i.e., $V_h = 0$ • min^{-1}), P_a and P_v have not yet changed. Hence V_r is still 5 • min^{-1} through a resistance of 20 mm Hg•L^{-1}• min. Because of the disparity between V_h and V_r, P_a will begin to decrease rapidly and P_v will begin to rise rapidly.

C Cardiac arrest: steady state

$V_h = 0$

$P_v = 7$ $P_a = 7$

$V_r = 0$ L • min^{-1}

When the effects of cardiac arrest have attained the **steady state**, P_a will have fallen to 7 mm Hg and P_v will have risen to the same value. Because $P_a - P_v = 0$, flow across the resistance will cease (i.e., $\dot{Q}_r = 0$).

D Beginning of cardiac resuscitation

$V_r = 0$ L • min^{-1}

$P_v = 7$ $P_a = 7$

$V_h = 1$ L • min^{-1}

The heart is resuscitated and it begins to pump at a constant value of $\dot{Q}_h = 1$ L • min^{-1}. At the very beginning of resuscitation, P_a and P_v have not had time to change, and therefore $\dot{Q}_r$ is still 0 L • min^{-1}. Because $\dot{Q}_h$ exceeds $\dot{Q}_r$ by 1 L • min^{-1}, P_a will rise rapidly and P_v will fall rapidly. A new equilibrium will be attained when P_a increases to 26 mm Hg and P_v falls to 6 mm Hg. When $P_a - P_v = 20$ mm Hg, the flow ($\dot{Q}_r$) through the resistance will be 1 L • min^{-1}, which equals the cardiac output ($\dot{Q}_h$).

FIGURE 23-2   Simplified model of the cardiovascular system, consisting of a pump, arterial compliance (C_a), peripheral resistance, and venous compliance (C_v). **A,** Control state. The flow $(\dot{V}_r)$ across the peripheral resistance equals the flow $(\dot{V}_h)$ generated by the heart. The mean arterial pressure (P_a) is 102 mm Hg, the central venous pressure (P_v) is 2 mm Hg, and the peripheral resistance (R) is 20 mm Hg·L^{-1}·min. **B,** At the very beginning of cardiac arrest (i.e., $\dot{V}_h = 0$ L·min^{-1}), P_a and P_v have not yet changed, and thus $\dot{V}_r$ is still 5 L·min^{-1}. Because of the disparity between $\dot{V}_h$ and $\dot{V}_r$, P_a will begin to decrease rapidly and P_v will begin to rise rapidly. **C,** When the effects of cardiac arrest have attained equilibrium, P_a will have fallen to 7 mm Hg and P_v will have risen to the same value. **D,** The heart is restarted, and it begins to pump at a constant value of $\dot{V}_h = 1$ L·min^{-1}. Because $\dot{V}_h$ exceeds $\dot{V}_r$ by 1 L·min^{-1}, P_a will rise rapidly and P_v will fall rapidly. A new equilibrium will be attained when P_a increases to 26 mm Hg and P_v falls to 6 mm Hg.

with characteristics that resemble those of a normal, resting, adult person (Figure 23-2, *A*). Let the flow, $\dot{V}_h$, generated by the heart (i.e., the cardiac output) be 5 L·min^{-1}, the mean arterial pressure, P_a, be 102 mm Hg, and the central venous pressure, P_v, be 2 mm Hg. The peripheral resistance, R, is the ratio of pressure difference ($P_a - P_v$) to flow ($\dot{V}_h$); this ratio equals 20 mm Hg·L^{-1}·min. An arteriovenous pressure difference of 100 mm Hg is sufficient to force a flow ($\dot{V}_r$) of 5 L·min^{-1} through a peripheral resistance of 20 mm Hg·L^{-1}·min; this flow ($\dot{V}_r$) is precisely equal to the flow ($\dot{V}_h$) generated by the heart. From heartbeat to heartbeat, the volume (V_a) of blood in the arteries and the volume (V_v) of blood in the veins remain constant, since the volume of blood transferred from the veins to the arteries by the heart equals the volume of blood that flows from the arteries through the resistance vessels and into the veins.

Figure 23-2, *B*, illustrates the status of the circulation at the very beginning of an episode of cardiac arrest (i.e., $\dot{V}_h = 0$ L·min^{-1}). Initially the volumes of blood in the arteries (V_a) and veins (V_v) have not had time to change. The arterial and venous pressures depend on V_a and V_v, respectively. Therefore these pressures are identical to the respective pressure in Figure 23-2, *A* (i.e., $P_a = 102$ and $P_v = 2$). The arteriovenous pressure gradient of 100 mm Hg will force a flow of 5 L·min^{-1} through the peripheral resistance of 20 mm Hg·L^{-1}·min. Thus, although cardiac output now equals 0 L·min^{-1}, the flow through the microcirculation equals 5 L·min^{-1}. In other words, the potential energy stored in the arteries by the previous pumping action of the heart causes blood to be transferred from arteries to veins, initially at the control rate, even though the heart can no longer transfer blood from the veins into the arteries.

As time passes, the blood volume in the arteries progressively decreases and the blood volume in the veins progressively increases. Because the vessels are elastic structures, the arterial pressure falls progressively and the venous pressure rises progressively. This process will continue until the arterial and venous pressures become equal (Figure 23-2, *C*). Once this condition is reached, the flow ($\dot{V}_r$) from arteries to veins through the resistance vessels will be zero, as is the cardiac output ($\dot{V}_h$).

At zero flow equilibrium (Figure 23-2, *C*), the pressure attained in the arteries and veins depends on the relative compliances of these vessels. If the arterial (C_a) and venous (C_v) compliances had been equal, the

decline in P_a would have been equal to the rise in P_v, since the decrement in arterial volume equals the increment in venous volume (principle of conservation of mass). P_a and P_v would have both attained the average of P_a and P_v in Figure 23-2, *A* and *B* (i.e., $P_a = P_v = (102 + 2)/2 = 52$ mm Hg).

However, the veins are much more compliant than the arteries; the ratio is approximately equal to the ratio ($C_v/C_a = 19$) that we have assumed for our model. Thus the transfer of blood from arteries to veins at equilibrium would induce a fall in arterial pressure 19 times as great as the concomitant rise in venous pressure. As Figure 23-2, *C*, shows, P_v would increase by 5 mm Hg (to 7 mm Hg), whereas P_a would fall by 19 × 5 = 95 mm Hg (to 7 mm Hg). This equilibrium pressure that exists in the absence of flow is often referred to as the *mean circulatory pressure,* or the *static pressure.* The pressure in the static system reflects the total volume of blood in the system and the overall compliance of the system.

The algebraic solution of the previous problem was obtained by solving the following set of simultaneous equations:

$$C_v = 19\ C_a \tag{2}$$

as stated above,

$$\Delta V_v = -\Delta V_a \tag{3}$$

which states that the volume of blood lost from the arteries is gained by the veins, and

$$C_v = \Delta V_v/\Delta P_v$$
$$C_a = \Delta V_a/\Delta P_a \tag{4}$$

which are the definitions of the venous and arterial compliances.

The example of cardiac arrest in Figure 23-2 provides the basis for understanding the vascular function curves. Two important points on the curve have already been derived, as shown in Figure 23-3. One point *(A)* represents the normal status (depicted in Figure 23-2, *A*). Thus, when cardiac output was 5 L·min^{-1}, the central venous pressure was 2 mm Hg. Then, when flow stopped (cardiac output was zero), the central venous pressure became 7 mm Hg, at equilibrium; this pressure is the mean circulatory pressure, P_{mc}.

The inverse relation between central venous pressure and cardiac output simply expresses that when cardiac output is suddenly decreased, the rate at which blood flows from arteries to veins through the

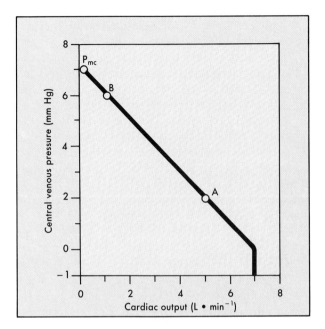

FIGURE 23-3   Changes in central venous pressure produced by changes in cardiac output. The mean circulatory pressure (or static pressure), P_{mc}, is the equilibrium pressure throughout the cardiovascular system when cardiac output is 0. Points B and A represent the values of venous pressure at cardiac outputs of 1 and 5 L·min^{-1}, respectively.

capillaries is temporarily greater than the rate at which the heart pumps it from the veins back into the arteries. During that transient period a net volume of blood is translocated from arteries to veins, and thus arterial pressure falls and venous pressure rises.

An example of a sudden increase in cardiac output will illustrate how a third point (B in Figure 23-3) on the vascular function curve is derived. Consider that the arrested heart is suddenly restarted and immediately begins pumping blood from the veins into the arteries at a rate of 1 L·min^{-1} (Figure 23-2, D). When the heart just begins to beat, the arteriovenous pressure gradient is zero, and thus no blood is being transferred from the arteries through the capillaries and into the veins. Therefore, when beating has just resumed, blood is being depleted from the veins at the rate of 1 L·min^{-1}, and the arterial volume is being repleted at this same rate. Thus venous pressure begins to fall, and arterial pressure begins to rise. Because of the difference in compliances, arterial

pressure will rise 19 times more rapidly than venous pressure will fall.

The resultant pressure gradient will cause blood to flow through the resistance. If the heart maintains a constant output of 1 L·min^{-1}, the arterial pressure will continue to rise and the venous pressure will continue to fall until the pressure gradient becomes 20 mm Hg. This gradient will force a flow of 1 L·min^{-1} through a resistance of 20 mm Hg·L^{-1}·min. This gradient will be achieved by a 19 mm Hg rise (to 26 mm Hg) in arterial pressure and a 1 mm Hg fall (to 6 mm Hg) in venous pressure. This equilibrium value of $P_v = 6$ mm Hg for a cardiac output of 1 L·min^{-1} also appears (point B) on the vascular function curve of Figure 23-3. It reflects a net transfer of blood from the venous to the arterial side of the circuit and a consequent reduction of the venous pressure.

The reduction of venous pressure that can be achieved by an increase in cardiac output is limited. At some critical maximal value of cardiac output, sufficient fluid will be translocated from the venous to the arterial side of the circuit to reduce the venous pressure below the ambient pressure. In a system of very distensible vessels, such as the venous system, the vessels will be collapsed by the greater external pressure. This venous collapse constitutes an impediment to venous return to the heart. Thus, it will limit the maximal value of cardiac output, regardless of the capabilities of the pump. Note that, in Figure 23-3, cardiac output remains constant as venous pressure decreases below zero.

Blood Volume

The vascular function curve is affected by changes in total blood volume. During circulatory standstill (zero cardiac output), the mean circulatory pressure depends only on vascular compliance and blood volume, as stated previously. Thus for a given vascular compliance P_{mc} will increase when the blood volume is expanded (*hypervolemia*) and will decrease when the blood volume is diminished (*hypovolemia*). In the family of vascular function curves shown in Figure 23-4, for example, consider that blood was either transfused into the static system until P_{mc} reached 9 mm Hg at equilibrium (*top curve*), or blood was withdrawn from the static system until P_{mc} reached 5 mm Hg at equilibrium (*bottom curve*).

Note also that the various vascular function curves in Figure 23-4 are all parallel to each other. To under-

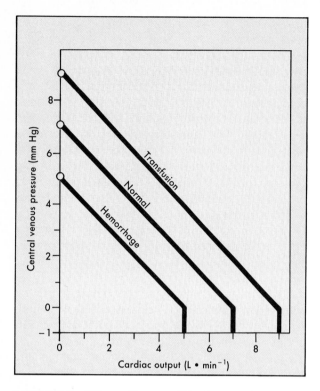

FIGURE 23-4   Effects of increased blood volume *(transfusion curve)* and of decreased blood volume *(hemorrhage curve)* on the vascular function curve. Similar shifts in the vascular function curve are produced by increases and decreases, respectively, in venomotor tone.

stand why the curves are parallel, consider the example of hypervolemia, in which the mean circulatory pressure had been raised to 9 mm Hg (as in the *top curve,* Figure 23-4). When the system is static, arterial and venous pressures would both be 9 mm Hg. If cardiac output were then increased suddenly to 1 $L \cdot min^{-1}$ (as in Figure 23-2, *D*), and if the peripheral resistance were still 20 mm $Hg \cdot L^{-1} \cdot min$, an arteriovenous pressure gradient of 20 mm Hg would still be necessary for 1 $L \cdot min^{-1}$ to flow through the resistance vessels. This does not differ from the example for normovolemia. Assuming the same C_v/C_a ratio of 19:1, the pressure gradient would be achieved by a 1 mm Hg decline in P_v and a 19 mm Hg rise in P_a.

Therefore a change in cardiac output from 0 to 1 $L \cdot min^{-1}$ would evoke the same 1 mm Hg reduction in P_v irrespective of the blood volume, as long as the C_v/C_a ratio and the peripheral resistance remain constant. The slope of the vascular function curve is, by definition, the change in P_v per unit change in cardiac output. Because the change in P_v induced by a unit change in cardiac output is not affected by the blood volume, the vascular function curves that represent different blood volumes are parallel to each other, as shown in Figure 23-4.

Figure 23-4 also shows that the cardiac output at which P_v becomes zero varies directly with the blood volume. Therefore the maximal value that cardiac output can attain becomes progressively more limited as the total blood volume is reduced. However, the pressure at which the veins collapse (sharp change in slope of the vascular function curve) is not altered appreciably by changes in blood volume. This pressure depends only on the ambient pressure. When P_v falls below the ambient pressure, the veins collapse and limit venous return to the heart.

Venomotor Tone

The effects of changes in venomotor tone on the vascular function curve closely resemble those for changes in blood volume. In Figure 23-4, for example, the transfusion curve could just as well represent increased venomotor tone, whereas the hemorrhage curve could represent decreased tone. A given increment in P_v could be achieved as readily by constriction of the smooth muscle in the venous walls as by an increase in the volume of blood in the veins.

The extent of venoconstriction is considerably greater in certain regions of the body than in others. In effect, vascular beds that undergo appreciable venoconstriction constitute blood reservoirs. The vascular bed of the skin is one of the major blood reservoirs in humans. During blood loss profound subcutaneous venoconstriction occurs, giving rise to the characteristic pale appearance of the skin. The resultant redistribution of blood away from the skin liberates several hundred milliliters of blood to be perfused through more vital regions. The vascular beds of the liver, lungs, and spleen are important blood reservoirs.

Peripheral Resistance

The modification of the vascular function curve introduced by changes in peripheral resistance is shown in Figure 23-5. The arterioles contain only about 3% of the total blood volume (see Figure 15-2). Thus, changes in the contractile state of these vessels

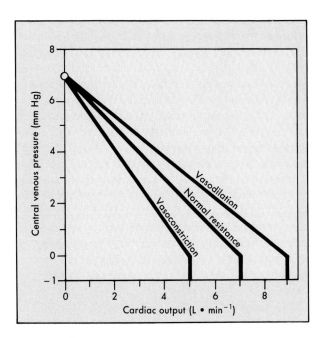

FIGURE 23-5   Effects of arteriolar vasodilation and vasoconstriction on the vascular function curve.

do not significantly alter P_{mc}. Therefore the family of vascular function curves that represent various peripheral resistances converges at a common point (P_{mc}) on the Y axis.

At any given cardiac output, venous pressure varies inversely with the peripheral resistance, all other factors remaining constant. Arteriolar constriction sufficient to double the peripheral resistance will cause a twofold rise in arterial pressure. In the example shown in Figure 23-2, *D*, a change in the cardiac output from 0 to 1 L·min^{-1} raised arterial pressure from 7 to 26 mm Hg, an increment of 19 mm Hg. If peripheral resistance had been twice as great, the same change in cardiac output would have evoked twice as great an increment in arterial pressure (Chapter 20).

To achieve a twofold increase in arterial pressure, twice as great an increment in blood volume would be required on the arterial side of the circulation, assuming a constant arterial compliance. Given a constant total blood volume, this larger arterial volume signifies a corresponding reduction in venous volume.

Thus the decrement in venous volume would be twice as great when the peripheral resistance is doubled. With a constant venous compliance a twofold reduction in venous volume would be reflected by a twofold decline in venous pressure. Therefore in Figure 23-2, *D*, an increase in cardiac output to 1 L·min^{-1} would have caused a 2 mm Hg decrement in venous pressure. Similarly, greater increases in cardiac output would have evoked proportionately greater decrements in venous pressure when peripheral resistance is increased than when resistance is normal.

Increases in peripheral resistance produce a clockwise rotation of the vascular function curves about a common intercept on the venous pressure axis, since an increase in peripheral resistance tends to decrease P_v but does not affect P_{mc} (Figure 23-5). Conversely, arteriolar vasodilation produces a counterclockwise rotation about the same venous pressure axis intercept. A higher maximal cardiac output is attainable when the arterioles are dilated that when they are normal or constricted (Figure 23-5).

COUPLING BETWEEN THE HEART AND THE VASCULATURE

The central venous pressure constitutes the filling pressure (essentially the preload) for the right ventricle. In accordance with the Frank-Starling mechanism (see Chapters 17 and 18), the central venous pressure is a cardinal determinant of the cardiac output. Ordinarily cardiac output varies directly with the central venous pressure; that is, over a wide range of venous pressures, a rise in venous pressure increases the cardiac output. In the following discussion, graphs of cardiac output as a function of venous pressure are called *cardiac function curves*. Alterations in *myocardial contractility* are represented by shifts in these curves.

To appreciate the coupling between the heart and the blood vessels, we examine the interrelations between the cardiac function curve and the vascular function curve (Figure 23-6). Both curves reflect the relations between cardiac output and central venous pressure. As just stated, the cardiac function curve expresses how cardiac output varies in response to a change in venous pressure. Thus, cardiac output here is the dependent variable (or response) and venous pressure is the independent variable (or stimulus). By convention, the dependent variable is scaled along

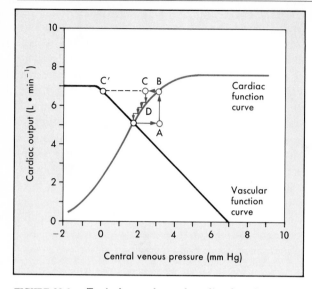

FIGURE 23-6 Typical vascular and cardiac function curves plotted on the same coordinate axes. Note that to plot both curves on the same graph, it is necessary to switch the X and Y axes for the vascular function curve (as compared to the arrangement of axes in Figures 23-3 to 23-5). The coordinates of the equilibrium point, at the intersection of the cardiac and vascular function curves, represent the stable values of cardiac output and central venous pressure at which the system tends to operate. Any perturbation (e.g., when venous pressure is suddenly increased to point *A*) institutes a sequence of changes in cardiac output and venous pressure such that these variables are returned to their equilibrium values.

the Y axis and the independent variable is scaled along the X axis. Note that in Figure 23-6, the assignment of X and Y axes is **conventional** for the cardiac function curve.

Conversely, the vascular function curve reflects how central venous pressure is affected by a change in cardiac output. For the vascular function curve venous pressure is the dependent variable (or response) and cardiac output is the independent variable (or stimulus). By convention venous pressure should be scaled along the Y axis and cardiac output along the X axis. Note that this convention was observed for the vascular function curves displayed in Figures 23-3 to 23-5.

However, to include the vascular function curve on the same set of coordinate axes with the cardiac function curve (Figure 23-6), it was necessary to violate the plotting convention for one of these curves. **We have arbitrarily violated the convention for the vascular function curve.** Note that the vascular function curve in Figure 23-6 reflects how the central venous pressure (scaled along the X axis) varies in

response to a change of cardiac output (scaled along the Y axis).

Simultaneous examination of the two curves, one that characterizes the heart and the other that characterizes the vessels, provides some insight about the coupling between the heart and the vessels. Theoretically the heart can operate at all combinations of venous pressure and cardiac output that fall on the appropriate cardiac function curve. Similarly the vascular system can operate at all combinations of venous pressure and cardiac output that fall on the appropriate vascular function curve. At equilibrium, therefore, the entire cardiovascular system (i.e., the combination of heart and vessels) must operate at the point of intersection of these two curves. Only at this point of intersection will the prevailing venous pressure evoke the cardiac output defined by the cardiac function curve; simultaneously, only at this point of intersection will the prevailing cardiac output evoke the venous pressure defined by the vascular function curve.

The tendency for the cardiovascular system to operate about such an equilibrium point may best be illustrated by examining its response to a sudden perturbation. Consider the changes elicited by a sudden rise in venous pressure from the equilibrium point to point *A* in Figure 23-6. Such a change might be induced by the rapid injection, during ventricular systole, of a given volume of blood on the venous side of the circuit, accompanied by the rapid withdrawal of an equal volume from the arterial side; the total blood volume would remain constant.

As defined by the cardiac function curve, this elevated venous pressure would increase cardiac output (from point *A* to *B* in Figure 23-6) during the very next ventricular systole. The increased cardiac output, in turn, would result in the net transfer of blood from the venous to the arterial side of the circuit, with a consequent reduction in venous pressure.

In one heartbeat, the reduction in venous pressure would be small (from point *B* to *C* in Figure 23-6), since the heart would transfer only a tiny fraction of the total venous blood volume over to the arterial side. Because of this reduction in venous pressure, the cardiac output during the next beat diminished (point *C* to *D*) by an amount dictated by the cardiac function curve. Because point *D* is still above the intersection point, the heart will pump blood from the veins to the arteries at a rate greater than the blood will flow across the peripheral resistance from arteries to veins. Thus central venous pressure will continue to fall. This process will continue in ever-diminishing steps until the point of intersection is reached. Only one specific combination of cardiac output and venous pressure (denoted by the coordinates of the point of intersection) will satisfy simultaneously the requirements of the cardiac and vascular function curves.

Enhanced Myocardial Contractility

Graphs of cardiac and vascular function curves help explain the effects of alterations in ventricular contractility. In Figure 23-7 the lower cardiac function curve represents the control contractility state, whereas the upper curve reflects an enhanced contractility (e.g., as might be achieved by stimulation of the cardiac sympathetic nerves). This pair of curves is analogous to the family of ventricular function curves shown in Figure 18-14. If the technique used to enhance contractility in Figure 23-7 influenced only the heart, the vascular function curve would not be affected, and therefore one vascular function curve would suffice.

During the control state the equilibrium values for cardiac output and venous pressure in Figure 23-7 are designated by point A. With the onset of cardiac sympathetic nerve stimulation (assuming the effects to be instantaneous and constant), the prevailing level of P_v would abruptly raise cardiac output to point B because of the enhanced contractility. However, this high cardiac output would increase the net transfer of blood from the venous to the arterial side of the circuit, and consequently venous pressure will begin to fall (to point C). Cardiac output will continue to fall until it reaches a new equilibrium point (D), which is located at the intersection of the vascular function curve with the new cardiac function curve. The new equilibrium point (D) lies above and to the left of the control equilibrium point (A). This shift reveals that sympathetic stimulation increases cardiac output, despite the diminution of the ventricular filling pressure (i.e., central venous pressure). Such a change accurately describes the true response. In the experiment shown in Figure 23-8, stimulation of the left stellate ganglion in an anesthetized dog increased cardiac output but decreased right and left atrial pressures (P_{RA} and P_{LA}).

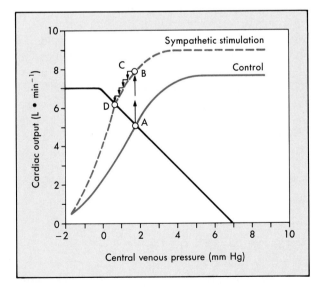

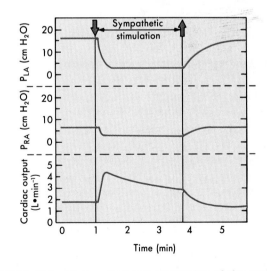

FIGURE 23-7 Enhancement of myocardial contractility, as accomplished by cardiac sympathetic nerve stimulation, causes the equilibrium values of cardiac output and central venous pressure to shift from the intersection (A) of the control vascular and cardiac function curves (continuous lines) to the intersection (D) of the same vascular function curve with the cardiac function curve (dashed line) that represents enhanced myocardial contractility.

FIGURE 23-8 During electrical stimulation of the cardiac sympathetic nerves, aortic blood flow increased while pressures in the left atrium (P_{LA}) and right atrium (P_{RA}) diminished. These data support the graphic analysis shown in Figure 23-7, in which, at equilibrium, cardiac output is predicted to increase and venous pressure to decrease during stimulation of the cardiac sympathetic nerves. (Redrawn from Sarnoff SJ et al: Circ Res 8:1108, 1960.)

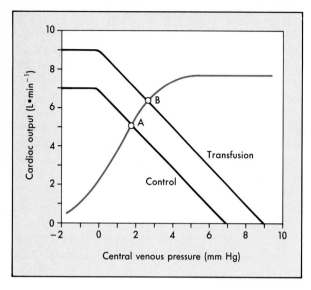

FIGURE 23-9   After a blood transfusion the vascular function curve is shifted to the right *(dashed curve)*. Therefore cardiac output and venous pressure are both increased, as denoted by the translocation of the equilibrium point from *A* to *B*.

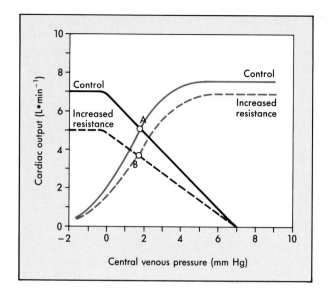

FIGURE 23-10   An increase in peripheral resistance shifts the cardiac and the vascular function curves downward. At equilibrium the cardiac output is less *(B)* when the peripheral resistance is high than when it is normal *(A)*.

Blood Volume

Changes in blood volume do not directly affect the cardiac function curve, but they do influence the vascular function curve in the manner shown in Figure 23-4. Therefore, to understand the circulatory alterations evoked by a given change in blood volume, the appropriate cardiac function curve must be plotted along with the vascular function curves that represent the control and experimental states.

Figure 23-9 illustrates the response to a blood transfusion. Equilibrium point *B*, which denotes the values for cardiac output and venous pressure after transfusion, lies above and to the right of the control equilibrium point *A*. Thus transfusion increases both cardiac output and venous pressure. Hemorrhage has the opposite effect. Pure increases or decreases in venomotor tone elicit responses that are analogous to those evoked by augmentations or reductions, respectively, of the total blood volume, for reasons already discussed.

Peripheral Resistance

Predictions concerning the effects of changes in peripheral resistance on cardiac output and central venous pressure are more complex because the car-

diac and vascular function curves both shift with changes in peripheral resistance (Figure 23-10). The vascular function curve is displaced downward by an increase in resistance, but the control curve and the curve for increased resistance both converge to the same venous pressure axis intercept, as explained previously (see Figure 23-5). The cardiac function curve is also shifted downward because at any given central venous pressure, the heart pumps less blood when the resistance (afterload) is increased. Because both curves are displaced downward, the new equilibrium point, *B*, will fall below the control point, *A*. Note that the vascular function curves in Figure 23-10 differ from those in Figure 23-5 only because the X and Y axes for the vascular function curves have been switched.

ROLE OF HEART RATE

Cardiac output is the product of stroke volume and heart rate. The preceding analysis of the control of cardiac output was restricted to the control of stroke volume, and the role of heart rate was ignored.

Analysis of the effect of changes in heart rate on cardiac output is complex because a change in heart

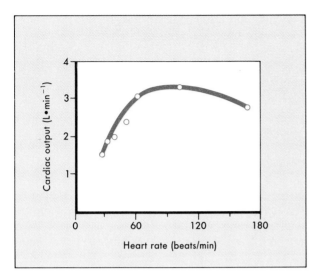

FIGURE 23-11   The changes in cardiac output induced by changing the rate of ventricular pacing in a dog with complete heart block. (Redrawn from Miller DE et al: Effect of ventricular rate on the cardiac output in the dog with chronic heart block, Circ Res 10:658, 1962. By permission of the American Heart Association, Inc.)

rate will alter the other factors (preload, afterload, contractility) that determine stroke volume (Figure 23-1). An increase in heart rate, for example, decreases the duration of diastole. Thus ventricular filling is abridged; that is, preload is reduced. If the proposed increase in heart rate did alter cardiac output, the arterial pressure would change; that is, afterload would be altered. Finally, the rise in heart rate would increase the net influx of Ca^{++} into the cardiac myocytes, and this would enhance myocardial contractility (see Chapters 17 and 18).

Heart rate has been varied by artificial pacing in experimental animals and in humans. The effects on cardiac output have usually resembled the experimental results shown in Figure 23-11. In that experiment on a dog with atrioventricular block, contraction frequency was varied by ventricular pacing. As the ventricular frequency was increased from 30 to 60 min^{-1}, the cardiac output increased substantially. Presumably, at the lower frequencies within this range, the greater filling per cardiac cycle does not increase stroke volume sufficiently to compensate for the decreased number of contractions per minute.

Over the frequency range from 60 to 170 min^{-1}, however, cardiac output did not change much with changes in frequency (Figure 23-11). Thus, as the

pacing frequency was increased, the stroke volume decreased proportionately. The diminished time for filling at the higher frequencies partly accounts for the observed proportionality between heart rate increment and stroke volume decrement. Also, vascular autoregulation tends to hold tissue blood flow constant (see Chapter 22). This adaptation would induce changes in preload and afterload that would tend to maintain a nearly constant cardiac output.

ANCILLARY FACTORS THAT AFFECT THE VENOUS SYSTEM AND CARDIAC OUTPUT

In the previous discussion we oversimplified the interrelations between central venous pressure and cardiac output. We attempted to explain the effects elicited by changes in just one factor. However, because many feedback control mechanisms regulate the cardiovascular system, an isolated change in a single variable rarely occurs. A change in blood volume, for example, reflexly alters cardiac function, peripheral resistance, and venomotor tone. Several auxiliary factors also regulate cardiac output.

Gravity

Gravitational forces may affect cardiac output profoundly. Gravitational effects on cardiac output are exaggerated in airplane pilots during pullouts from dives. The centrifugal force in the footward direction may be several times greater than the force of gravity. Such individuals characteristically black out during the maneuver, as blood is drained from the cephalic regions and pooled in veins in the lower parts of the body.

Among soldiers standing at attention for long periods, particularly in hot weather, some individuals may faint because their cardiac outputs decrease. The explanation for the reduction in cardiac output under such conditions is often specious. It is argued that when an individual is standing, the forces of gravity impede venous return from the dependent regions of the body. This statement is only partly correct, however, because it ignores the gravitational counterforce on the arterial side of the same circuit, which by itself would tend to promote flow.

The reason that gravitational forces can affect cardiac output resides in the distensibility (or compliance) of the blood vessels. When a person stands, the blood vessels below the level of the heart will be distended by the gravitational forces acting on the columns of blood in the vessels. The distension will be more prominent on the venous than on the arterial side of the circuit because the venous compliance is so much greater than the arterial compliance. Such venous distension is readily observed on the backs of the hands when the arms are allowed to hang below the level of the heart. The hemodynamic effects of such venous distension (venous pooling) resemble those caused by the loss of an equivalent volume of blood from the body. When a person shifts from a supine position to a relaxed standing position, 300 to 800 ml of blood are pooled in the legs. This may reduce cardiac output by about 2 L·min^{-1}.

The compensatory adjustments to the erect position are similar to the adjustments to blood loss. For example, venous pooling and other gravitational effects tend to lower the pressure in the regions of the arterial baroreceptors. The resultant diminution in baroreceptor excitation reflexly speeds the heart, strengthens the cardiac contraction, and constricts the arterioles and veins. The baroreceptor reflex has a greater effect on the resistance vessels (arterioles) than on the capacitance vessels (veins). Warm ambient temperatures tend to interfere with the compensatory vasomotor reactions, and the absence of muscular activity exaggerates the gravitational effects, as explained in the following section.

Many of the drugs used to treat hypertension also interfere with the reflex adaptation to standing. Similarly, astronauts exposed to weightlessness lose their adaptations after a few days, and they experience difficulties when they first return to a normal gravitational field. When individuals with impaired reflex adaptations stand, their blood pressures may fall dramatically. This response is called orthostatic hypotension, which may cause lightheadedness or fainting.

Muscular Activity and Venous Valves

When an individual stands but is quiet and relaxed, the pressure rises in the veins below the heart. The venous pressure in the legs increases gradually and does not reach an equilibrium value until almost 1 minute after standing. The slowness of the rise in venous pressure is attributable to the venous valves, which permit flow only toward the heart. When a person stands, the valves prevent blood in the

veins from actually falling toward the feet. Thus the column of venous blood is supported at numerous levels by these valves; temporarily the venous column consists of many separate segments. However, blood continues to enter the column from many venules and small tributary veins, and pressure continues to rise. As soon as the pressure in one segment exceeds that in the segment just above it, the valve is forced open. Ultimately all the valves are open, and the column is continuous.

Precise measurement reveals that the final level of venous pressure in the feet during quiet standing is only slightly greater than that in a static column of blood extending from the right atrium to the feet. This indicates that the pressure drop caused by flow from foot veins to the right atrium is very small; thus the resistance to flow is also small. This very low venous resistance justifies our combining the veins together as a common venous compliance in the model illustrated in Figure 23-2.

When an individual who has been standing quietly begins to walk, the venous pressure in the legs and feet decreases appreciably. Because of the intermittent venous compression produced by the contracting muscles and because of the presence of the venous valves, blood is forced from the veins toward the heart. Therefore muscular contraction lowers the mean venous pressure in the legs and serves as an auxiliary pump. Furthermore, it prevents venous pooling and lowers capillary hydrostatic pressure, thereby reducing the tendency for edema fluid to collect in the feet during standing.

Respiratory Effects

The normal, periodic activity of the respiratory muscles causes rhythmic variations in vena caval flow, and it constitutes an auxiliary pump to promote venous return of blood to the heart. Coughing, straining at stool, and other activities that require the respiratory muscles may affect cardiac output substantially.

The changes in blood flow in the superior vena cava during the respiratory cycle of an anesthetized dog are shown in Figure 23-12. During respiration the changes in intrathoracic pressure are transmitted to the lumina of the thoracic blood vessels. The reduction in central venous pressure during inspiration increases the pressure gradient between extrathoracic and intrathoracic veins. The consequent accelera-

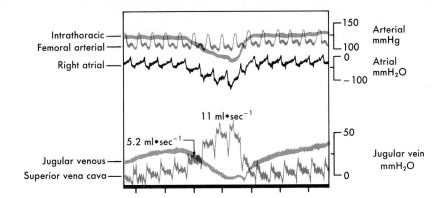

FIGURE 23-12 During inspiration in an anesthetized dog intrathoracic *(ITP)*, right atrial *(RAP)*, and jugular venous *(JVP)* pressures decrease, and flow in the superior vena cava *(SVCF)* increases (from 5.2 to 11 ml·sec⁻¹). All pressures are in mm H₂O, except for femoral arterial pressure *(FAP)*, which is in mm Hg. (Modified from Brecher GA: Venous return, New York, 1956, Grune & Stratton, Inc.)

tion of venous return to the right atrium is displayed in Figure 23-12 as an increase in superior vena caval blood flow from 5.2 ml·sec^{-1} during expiration to 11 ml·sec^{-1} during inspiration.

During expiration, flow into the central veins decelerates. However, the mean rate of venous return during normal respiration exceeds the flow rate in the temporary absence of respiration. Therefore normal inspiration apparently facilitates venous return more than normal expiration impedes it. This must be attributable partly to the valves in the veins of the extremities and neck. These valves prevent any reversal of flow during expiration. Thus the respiratory muscles and venous valves constitute an auxiliary pump for venous return.

BIBLIOGRAPHY
Journal Articles

Bromberger-Barnea B: Mechanical effects of inspiration on heart functions: a review, Fed Proc 40:2172, 1981.

Cassidy SS et al: Geometric left-ventricular responses to interactions between the lung and left ventricle: positive pressure breathing, Ann Biomed Eng 15:285, 1987.

Freeman GL et al: Influence of heart rate on left ventricular performance in conscious dogs, Circ Res, 61:455, 1987.

Greenway CV and Lautt WW: Blood volume, the venous system, preload, and cardiac output, Can J Physiol Pharmacol 64:383, 1986.

Levy MN: The cardiac and vascular factors that determine systemic blood flow, Circ Res 44:739, 1979.

Melbin J et al: Coherence of cardiac output with rate changes, Am J Physiol 243:H499, 1982.

Rothe CF: Reflex control of veins and vascular capacitance, Physiol Rev 63:1281, 1983.

Rothe CF: Physiology of venous return: an unappreciated boost to the heart, Arch Intern Med 146:977, 1986.

Sagawa K: Closed-loop physiological control of the heart, Ann Biomed Eng 8:415, 1980.

Tözeren A and Chien S: Modeling of time-variant coupling between left ventricle and aorta in cardiac cycle, Am J Physiol 249:H560, 1985.

Books and Monographs

Green JF: Determinants of systemic blood flow. International review of physiology. III. Cardiovascular physiology, vol 18, Baltimore, 1979, University Park Press.

Guyton AC et al: Circulatory physiology: cardiac output and its regulation, ed 2, Philadelphia, 1973, WB Saunders Co.

Rothe CF: Venous system: physiology of the capacitance vessels. In Handbook of physiology, section 2: The cardiovascular system—peripheral circulation and organ blood flow, vol III, Bethesda, Md, 1983, American Physiological Society.

Shepherd JT and Vanhoutte PM: Veins and their control, Philadelphia, 1975, WB Saunders Co.

Yin FCP, editor: Ventricular/vascular coupling, New York, 1987, Springer-Verlag.

CUTANEOUS CIRCULATION

The oxygen and nutrient requirements of the skin are relatively small. In contrast to most other body tissues, the supply of these essential materials is not the chief governing factor in the regulation of cutaneous blood flow. The primary function of the cutaneous circulation is maintenance of a constant body temperature. Consequently blood flow to the skin fluctuates widely, depending on the need for loss or conservation of body heat. Mechanisms responsible for alterations in skin blood flow are mainly activated by changes in ambient and internal body temperatures.

Regulation of Skin Blood Flow

There are essentially two types of resistance vessels in skin: arterioles and arteriovenous (AV) anastomoses. The arterioles are similar to those found elsewhere in the body. AV anastomoses shunt blood from arterioles to venules and venous plexuses; thus they bypass the capillary bed. They are found mainly in the fingertips, palms of the hands, toes, soles of the feet, ears, nose, and lips. AV anastomoses differ morphologically from the arterioles in that they are either short and straight or long and coiled vessels about 20 to 40 μm in lumen diameter, with thick muscular walls richly supplied with nerve fibers. These vessels are almost exclusively under sympathetic neural control, and they dilate maximally when their nerve supply is interrupted. Conversely, reflex stimulation of the sympathetic fibers to these vessels may constrict

them to the point of complete obliteration of the vascular lumen. Although AV anastomoses do not exhibit basal tone (tonic activity of the vascular smooth muscle independent of innervation), they are highly sensitive to vasoconstrictor agents such as epinephrine and norepinephrine. Furthermore, AV anastomoses are not under metabolic control, and they fail to show reactive hyperemia or autoregulation of blood flow. Thus the regulation of blood flow through these anastomotic channels is governed principally by the nervous system in response to reflex activation by temperature receptors or by higher centers of the central nervous system.

The bulk of the skin resistance vessels exhibits some basal tone, and vascular resistance in the skin is under the dual control of the sympathetic nervous system and local regulatory factors, in much the same manner as are other vascular beds. In the skin, however, neural control is more important than local factors. Stimulation of sympathetic nerve fibers to skin blood vessels (arteries and veins, as well as arterioles) induces vasoconstriction, and severance of the sympathetic nerves induces vasodilation. With chronic denervation of the cutaneous blood vessels, the degree of tone that existed before denervation is gradually regained over several weeks. This is accomplished by an enhancement of basal tone that compensates for the degree of tone previously contributed by sympathetic nerve fiber activity. Epinephrine and norepinephrine elicit only vasoconstriction in cutaneous vessels.

Parasympathetic vasodilator nerve fibers do not supply the cutaneous blood vessels. However, stimulation of the sweat glands, which are innervated by cholinergic fibers of the sympathetic nervous system, results in dilation of the skin resistance vessels. Sweat contains an enzyme that acts on a protein moiety in the tissue fluid to release *bradykinin*, a polypeptide with potent vasodilator properties. Bradykinin formed in the tissue can act locally to dilate the arterioles and increase blood flow to the skin. The skin vessels of certain body regions, particularly the head, neck, shoulders, and upper chest, are under the influence of the higher centers of the central nervous system. Blushing, caused by embarrassment or anger, and blanching, caused by fear or anxiety, are examples of cerebral inhibition and stimulation, respectively, of the sympathetic nerve fibers to the affected regions.

In contrast to AV anastomoses in the skin, the cutaneous resistance vessels show autoregulation of blood flow and reactive hyperemia. If the arterial inflow to a limb is stopped by inflating a blood pressure cuff briefly, the skin reddens greatly below the point of vascular occlusion when the cuff is deflated. This increased cutaneous blood flow (reactive hyperemia) is also manifested by the distension of the superficial veins in the erythematous extremity.

Ambient and Body Temperature Because the primary function of the skin is to preserve the internal milieu and protect it from adverse changes in the environment, and because the ambient temperature is one of the most important external variables with which the body must contend, it is not surprising that the vasculature of the skin is chiefly influenced by environmental temperature. Exposure to cold elicits a generalized cutaneous vasoconstriction that is most pronounced in the hands and feet. This response is chiefly mediated by the nervous system, since arrest of the circulation to a hand with a pressure cuff and immersion of that hand in cold water results in vasoconstriction in the skin of the other extremities that are exposed to room temperature. When the circulation to the chilled hand is not occluded, the reflex vasoconstriction is caused partly by the cooled blood returning to the general circulation and stimulating the temperature-regulating center in the anterior hypothalamus. Direct application of cold to this region of the brain produces cutaneous vasoconstriction.

The skin vessels of the cooled hand also respond directly to cold. Moderate cooling or exposure of the hand for brief periods to severe cold ($0°$ to $15° C$) constricts the resistance and capacitance vessels, including AV anastomoses. However, prolonged exposure of the hand to severe cold has a secondary vasodilator effect. Prompt vasoconstriction and severe pain are elicited by immersion of the hand in water near $0° C$, but they are soon followed by dilation of the skin vessels, reddening of the immersed part, and alleviation of the pain. With continued immersion of the hand, alternating periods of constriction and dilation occur, but the skin termperature rarely drops as low as it did during the initial vasoconstriction. Prolonged severe cold damages the tissue. The rosy faces of people outdoors in the cold are examples of cold vasodilation. However, the blood flow through the skin of the face may be very low, despite the flushed appearance. The red color of the slowly flowing blood is largely the result of the reduced oxygen uptake by the cold skin and the cold-induced shift to the left of the oxyhemoglobin dissociation curve.

Direct application of heat produces not only local vasodilation of resistance and capacitance vessels and AV anastomoses, but also reflex dilation in other parts of the body. The local effect is independent of the vascular nerve supply, whereas the reflex vasodilation is a combination of anterior hypothalamic stimulation by the returning warmed blood and of stimulation of receptors in the heated part.

The proximity of the major arteries and veins to each other permits considerable heat exchange (countercurrent) between artery and vein. Cold blood that flows in veins from a cooled hand toward the heart takes up heat from adjacent arteries; this warms the venous blood and cools the arterial blood. Heat exchange occurs in the opposite direction when the extremity is exposed to heat. Thus heat conservation is enhanced during exposure of the extremities to cold, and heat gain is minimized during exposure of the extremities to warmth.

Skin Color and Special Reactions of the Skin Vessels

The color of the skin is determined in large part by pigment. In all but very dark skin, however, the pallor or ruddiness is primarily a function of the amount of blood in the skin. With little blood in the venous plexus the skin appears pale, whereas with larger quantities of blood in the venous plexus the skin has more color. Whether this color is bright red, blue, or some intermediate shade is determined by the degree of oxygenation of the blood in the subcutaneous vessels.

For example, a combination of vasoconstriction and reduced hemoglobin can produce an ashen gray color of the skin, whereas a combination of venous engorgement and reduced hemoglobin can result in a dark purple hue. Skin color provides little information about the rate of cutaneous blood flow. Rapid blood flow and pale skin may coexist when the AV anastomoses are open, and slow blood flow and red skin may coexist when the extremity is exposed to cold.

White Reaction and Triple Response If the skin of the forearm of many individuals is lightly stroked with a blunt instrument, a white line appears at the site of the stroke within 20 seconds. The blanching becomes maximal in about 30 to 40 seconds and then gradually disappears within 3 to 5 minutes. This response is known as a *white reaction* and is attributable to microvascular constriction induced by the mechanical stimulation.

If the skin is stroked more strongly with a sharp pointed instrument, a *triple response* is elicited. Within 3 to 15 seconds a thin red line appears at the site of the stroke. It is followed in about 15 to 30 seconds by a red blush, or flare, extending out 1 to 2 cm from either side of the red line. This in turn is followed in 3 to 5 minutes by an elevation of the skin along the red line, with gradual fading of the red line as the elevation, a *wheal*, becomes more prominent. Mechanical stimulation causes the red line through dilation of the vessels. The flare, however, is the result of dilation of neighboring arterioles by an axon reflex that originates at the site of mechanical stimulation. In an *axon reflex* the nerve impulse travels centripetally in the cutaneous sensory nerve fiber and then antidromically down the small branches of the afferent nerve to adjacent arterioles to elicit vasodilation (Figure 24-1). The flare is not affected by acute section or anesthetic block of the sensory nerve central to the point of branching, whereas it is abolished when the nerve degenerates after section. The wheal is caused by increased capillary permeability induced by the trauma. Fluid that contains protein leaks out of the capillaries locally and produces edema at the site

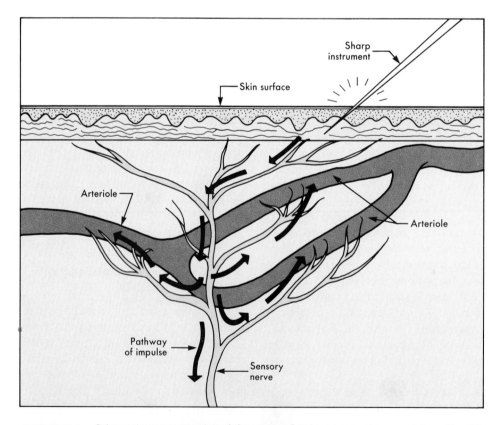

FIGURE 24-1   Schematic representation of the axon reflex in response to a scratch on the skin surface with a sharp instrument. Arrows indicate the pathways of impulses in a sensory nerve from the site of stimulation to adjacent arterioles to produce local vasodilation (flare).

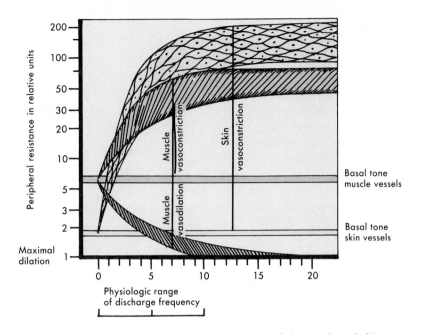

FIGURE 24-2 Basal tone and the range of the resistance vessels in muscle and skin to sympathetic nerve stimulation. Peripheral resistance is plotted on a logarithmic scale. (Redrawn from Celander O and Folkow B: Acta Physiol Scand 29:241, 1953.)

of injury. Because the triple response can be elicited by an intradermal injection of histamine, it has been attributed to histamine or a histamine-like substance (H-substance).

SKELETAL MUSCLE CIRCULATION

The blood flow to skeletal muscle varies directly with the contractile activity of the tissue and the type of muscle. Blood flow and capillary density in red (slow-twitch, high-oxidative) muscle are greater than in white (fast-twitch, low-oxidative) muscle. In resting muscle the arterioles exhibit asynchronous intermittent contractions and relaxations. Thus, at any given moment, a large percentage of the capillary bed is not perfused. Consequently total blood flow through quiescent skeletal muscle is low (1.4 to 4.5 ml·min^{-1}·100 g^{-1}). During exercise the resistance vessels relax and the muscle blood flow may increase manyfold (up to 15 to 20 times the resting level); the magnitude of the increase depends largely on the severity of the exercise.

Regulation of Skeletal Muscle Blood Flow

Control of muscle circulation is achieved by neural and local factors; the relative contribution of these

factors is dictated by muscle activity. At rest neural and myogenic regulations are predominant, whereas during exercise metabolic control supervenes. As with all tissues, physical factors such as arterial pressure, tissue pressure, and blood viscosity influence blood flow to muscle. However, another physical factor plays a role during exercise—the squeezing effect of the active muscle on the vessels. With intermittent contractions inflow is restricted and venous outflow is enhanced during each brief contraction. The presence of the venous valves prevents backflow of blood in the veins between contractions, thereby aiding in the forward propulsion of the blood. With strong sustained contractions, the vascular bed can be compressed to the point at which blood flow actually ceases temporarily.

Neural Factors Although the resistance vessels of muscle have a high basal tone, they also display tone attributable to continuous low-frequency activity in the sympathetic vasoconstrictor nerve fibers.

The tonic activity of the sympathetic nerves is greatly influenced by the baroreceptor reflex. An increase in carotid sinus pressure dilates the vascular bed of the muscle, and a decrease in carotid sinus pressure elicits vasoconstriction. Because muscle is the major body component on the basis of mass and thereby represents the largest vascular bed, the participation of its resistance vessels in vascular reflexes is important in maintaining a constant arterial blood pressure.

A comparison of the vasoconstrictor and vasodilator effects of the sympathetic nerves to blood vessels of muscle and skin is summarized in Figure 24-2. Note the lower basal tone of the skin vessels, their greater constrictor response, and the absence of active cutaneous vasodilation.

Local Factors It already has been stressed that neural regulation of muscle blood flow is superseded by metabolic regulation (see Chapter 22) when the muscle changes from the resting to the contracting state. However, local control also occurs in innervated resting skeletal muscle when the vasomotor nerves are not active. Thus autoregulation can be observed in innervated as well as denervated muscle, and in both conditions it is characterized by a low venous blood oxygen saturation.

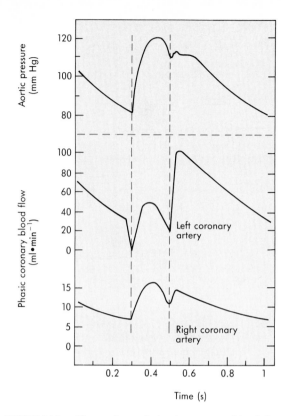

FIGURE 24-3 Comparison of phasic coronary blood flow in the left and right coronary arteries.

CORONARY CIRCULATION

Factors That Influence Coronary Blood Flow

Physical Factors The main factor responsible for perfusion of the myocardium is the aortic pressure, which is generated by the heart itself. Changes in aortic pressure generally shift coronary blood flow in the same direction. However, alterations of cardiac work, produced by an increase or decrease in aortic pressure, have a considerable effect on coronary resistance. Increased metabolic activity of the heart decreases coronary resistance, and a reduction in cardiac metabolism increases coronary resistance. Under normal conditions blood pressure is kept within relatively narrow limits by the baroreceptor reflex. Therefore changes in coronary blood flow are primarily caused by caliber changes of the coronary resistance vessels in response to metabolic demands of the heart. When the role of myocardial metabolism is unchanged and coronary perfusion pressure is raised or lowered, coronary blood flow remains relatively constant (autoregulation of blood flow).

In addition to providing the head of pressure to drive blood through the coronary vessels, the heart also influences its blood supply by the squeezing effect of the contracting myocardium on the blood vessels that course through it (extravascular com-

pression or extracoronary resistance). This force is so great during early ventricular systole that blood flow in a large coronary artery supplying the left ventricle is briefly reversed. Maximal left coronary inflow occurs in early diastole, when the ventricles have relaxed and extravascular compression of the coronary vessels is virtually absent. This flow pattern is seen in the phasic coronary flow curve for the left coronary artery (Figure 24-3). After an initial reversal in early systole, left coronary blood flow parallels the aortic pressure until early diastole, when it rises abruptly and then declines slowly as aortic pressure falls during the remainder of diastole.

Flow in the right coronary artery shows a similar pattern (Figure 24-3), but because of the lower pressure developed by the thin right ventricle during systole, blood flow does not reverse in early systole. Systolic blood flow constitutes a much greater proportion of total coronary inflow than it does in the left coronary artery.

Tachycardia and bradycardia have dual effects on coronary blood flow. A change in heart rate is accomplished chiefly by shortening or lengthening diastole. During tachycardia the proportion of time spent in systole, and consequently the period of restricted inflow, increases. However, this mechanical reduction in mean coronary flow is overridden by the coronary dilation associated with the increased metabolic activity of the more rapidly beating heart. During bradycardia the opposite is true; restriction of coronary inflow is less (more time is diastole) but so are the metabolic (O_2) requirements of the myocardium.

Neural and Neurohumoral Factors The primary effect of stimulation of the sympathetic nerves to the coronary vessels is vasoconstriction. However, the observed effect is a great increase in coronary blood flow. The increase in flow is associated with cardiac acceleration and a more forceful systole. The stronger myocardial contractions and the tachycardia (with the consequent greater proportion of time spent in systole) tend to restrict coronary flow. However, the increase in myocardial metabolic activity, as evidenced by the rate and contractility changes, tends to dilate the coronary resistance vessels. The increase in coronary blood flow observed with cardiac sympathetic nerve stimulation is the algebraic sum of these factors.

α-Receptors (constrictors) and β-receptors (dilators) are present on the coronary vessels. Furthermore, the coronary resistance vessels participate in the baroreceptor and chemoreceptor reflexes, and the sympathetic constrictor tone of the coronary arterioles can be reflexly modulated. Nevertheless, coronary resistance is predominantly under local nonneural control.

Vagus nerve stimulation has little direct effect on the caliber of the coronary arterioles. Small increments in coronary blood flow can be induced by stimulation of the peripheral ends of the vagi, and activation of the carotid and aortic chemoreceptors can decrease coronary resistance slightly via the vagus nerves to the heart.

Metabolic Factors One of the most striking characteristics of the coronary circulation is the close parallelism between the level of myocardial metabolic activity and the magnitude of the coronary blood flow. This relationship is also found in the denervated heart or the completely isolated heart.

The link between cardiac metabolic rate and coronary blood flow remains unsettled. Numerous agents, generally referred to as metabolites, have been suggested as mediators of the vasodilation observed with increased cardiac work. Accumulation of vasoactive metabolites also may be responsible for reactive hyperemia, since the duration of coronary flow after release of the briefly occluded vessel is, within certain limits, proportional to the duration of occlusion. Among the substances implicated are carbon dioxide, oxygen (reduced O_2 tension), lactic acid, hydrogen ions, histamine, potassium ions, increased osmolarity, polypeptides, prostaglandins, and adenosine. Of these agents, adenosine comes closest to satisfying the criteria for the physiologic mediator.

Cardiac Oxygen Consumption, Work, and Efficiency

The volume of O_2 consumed by the heart is determined by the amount and the type of activity the heart performs. Under basal conditions myocardial O_2 consumption is about 8 to 10 ml$\cdot$min$^{-1}\cdot$100 g^{-1} of heart. It can increase several-fold with exercise and decrease moderately under such conditions as hypotension and hypothermia. The cardiac venous blood normally has a low O_2 content (about 5 ml$\cdot$dl^{-1}), and the myocardium can therefore receive little additional O_2 by further O_2 extraction from the coronary blood.

Left ventricular work per beat (stroke work) is generally considered to be equal to the product of the stroke volume and the mean aortic pressure against which the blood is ejected by the left ventricle. At resting levels of cardiac output, the kinetic energy component is negligible (see Chapter 19). However, at high cardiac outputs, as in severe exercise, the kinetic component can account for up to 50% of total cardiac work. One can simultaneously halve the aortic pressure and double the cardiac output, or vice versa, and still arrive at the same value for cardiac work. However, the O_2 requirements are greater for any given amount of cardiac work when a major fraction is pressure work as opposed to volume work. Pumping an increase in cardiac output at a constant aortic pressure *(volume work)* is accomplished with a small increase in left ventricular O_2 consumption. Conversely, pumping against an increased arterial pressure at a constant cardiac output *(pressure work)* is accompanied by a large increment in myocardial O_2 consumption. The greater energy demand of pressure work over volume work is clinically important. For example, in aortic stenosis (a narrow aortic valve), left ventricular O_2 consumption is increased because of the high intraventricular pressures developed during systole, whereas coronary perfusion pressure is normal or reduced because of the pressure drop across the narrowed orifice of the diseased aortic valve.

As with an engine, the efficiency of the heart is the ratio of the work accomplished to the total energy utilized. Assuming an average O_2 consumption of 9 ml·min^{-1}·100 g^{-1} for the two ventricles, a 300 g heart consumes 27 ml O_2·min^{-1}, which is equivalent to 130 small calories at a respiratory quotient of 0.82. Together the two ventricles do about 8 kg·m of work/min, which is equivalent to 18.7 small calories. Therefore the gross efficiency is 14%:

$$\frac{18.7}{130} \times 100 = 14\% \tag{1}$$

With exercise, efficiency improves because mean blood pressure changes little, whereas cardiac output and work increase considerably without a proportional increase in myocardial O_2 consumption. The energy that is expended in cardiac metabolism and that does not contribute to the propulsion of blood through the body appears in the form of heat. The energy of the flowing blood is also dissipated as heat, chiefly in passage through the arterioles.

Substrate Utilization

The heart is versatile in its use of substrates. Within certain limits the uptake of a particular substrate is directly proportional to its arterial concentration. The utilization of one substrate is also influenced by the presence or absence of other substrates. For example, the addition of lactate to the blood perfusing a heart metabolizing glucose leads to a reduction in glucose uptake, and vice versa. At normal blood concentrations, glucose and lactate are consumed at about equal rates, whereas pyruvate uptake is very low because its arterial concentration is small. For glucose the threshold concentration is about 4 mM; below this blood level, no myocardial glucose uptake occurs.

Of the total cardiac O_2 consumption, only about 40% can be accounted for by the oxidation of carbohydrate. Thus the heart derives the major part of its energy from oxidation of noncarbohydrate sources. The chief noncarbohydrate fuels used by the heart are esterified and nonesterified fatty acids, which account for about 60% of myocardial O_2 consumption in the postabsorptive state. The various fatty acids have different thresholds for myocardial uptake, but are generally used in direct proportion to their arterial

concentration. Ketone bodies, especially acetoacetate, are readily oxidized by the heart and contribute a major source of energy in diabetic acidosis. As is true of carbohydrate substrates, utilization of specific noncarbohydrate substrate is influenced by the presence of other substrates, both noncarbohydrate and carbohydrate. Therefore, within certain limits, the heart preferentially uses that substrate available in the largest concentration.

CEREBRAL CIRCULATION

Blood reaches the brain through the internal carotid and vertebral arteries. The latter join to form the basilar artery, which, in conjunction with branches of the internal carotid arteries, forms the *circle of Willis*. A unique feature of the cerebral circulation is that it all lies within a rigid structure, the cranium. Because intracranial contents are incompressible, any increase in arterial inflow, as occurs with arteriolar dilation, must be associated with a comparable increase in venous outflow. The volume of blood and of extravascular fluid can vary considerably in most tissues. In brain the volume of blood and extravascular fluid is relatively constant; changes in either of these fluid volumes must be accompanied by a reciprocal change in the other. In contrast to most other organs, the total cerebral blood flow is held within a relatively narrow range; in humans it averages 55 ml·min^{-1}·100 g^{-1} of brain.

Regulation of Cerebral Blood Flow

Of the various body tissues, the brain is the least tolerant of ischemia. Interruption of cerebral blood flow for as little as 5 seconds results in loss of consciousness. Ischemia lasting just a few minutes results in irreversible tissue damage. Fortunately regulation of the cerebral circulation is mainly under direction of the brain itself. Local regulatory mechanisms and reflexes that originate in the brain tend to maintain cerebral circulation relatively constant. This constancy prevails even in the presence of possible adverse extrinsic effects, such as sympathetic vasomotor nerve activity, circulating humoral vasoactive agents, and changes in arterial blood pressure. Under certain conditions the brain also regulates its blood flow by initiating changes in systemic blood pressure. For example, elevation of intracranial pressure results in an increase in systemic blood pressure. This response, called *Cushing's phenomenon,* is apparently caused by ischemic stimulation of vasomotor regions of the medulla. It aids in maintaining cerebral blood flow in such conditions as expanding intracranial tumors.

Neural Factors The cerebral vessels are innervated by the cervical sympathetic nerve fibers that accompany the internal carotid and vertebral arteries into the cranial cavity. Relative to the control of other vascular beds, the sympathetic control of the cerebral vessels is weak, and the contractile state of the cerebrovascular smooth muscle depends mainly on local metabolic factors. There are no known sympathetic vasodilator nerves to the cerebral vessels. However, the vessels do receive parasympathetic fibers from the facial nerve, which produce a slight vasodilation on stimulation.

Local Factors Generally, total cerebral blood flow is constant. However, regional cortical blood flow varies with and is tightly coupled with regional metabolic activity. For example, movement of one hand results in increased blood flow only in the hand area of the contralateral sensory-motor and premotor cortex. The mediator of the link between cerebral metabolism and blood flow has not been established, but the three principal candidates are pH, potassium, and adenosine.

It is well known that the cerebral vessels are very sensitive to carbon dioxide tension (Pco_2). Increases in arterial blood CO_2 tension ($Paco_2$) elicit extreme cerebral vasodilation; inhalation of 7% CO_2 results in a twofold increment in cerebral blood flow. By the same token decreases in $Paco_2$, which may be elicited by hyperventilation, decrease the cerebral blood flow. Carbon dioxide changes arteriolar resistance by altering the perivascular pH and probably the intracellular pH of the vascular smooth muscle. By independently changing the Pco_2 and the bicarbonate concentration, it has been demonstrated that pial vessel diameter (and presumably blood flow) and pH are inversely related, regardless of the level of the Pco_2.

The cerebral circulation shows reactive hyperemia and excellent autoregulation between pressures of about 60 and 160 mm Hg. Mean arterial pressures less than 60 mm Hg result in reduced cerebral blood flow and syncope, whereas mean pressures greater than 160 may increase the permeability of the blood-brain barrier and cause cerebral edema.

SPLANCHNIC CIRCULATION

The splanchnic circulation consists of the blood supply to the gastrointestinal tract, liver, spleen, and pancreas. The most noteworthy feature of the splanchnic circulation is that two large capillary beds are partly in series with one another. The small splanchnic arterial branches supply the capillary beds in the gastrointestinal tract, spleen, and pancreas. From these capillary beds the venous blood ultimately flows into the portal vein, which normally provides most of the blood supply to the liver. However, the hepatic artery also supplies blood to the liver.

Intestinal Circulation

Neural Regulation The neural control of the mesenteric circulation is almost exclusively sympathetic. Increased sympathetic activity constricts the mesenteric arterioles and capacitance vessels. These responses are mediated by α-receptors, which are prepotent in the mesenteric circulation; however, β-receptors are also present. Infusion of a β-receptor agonist, such as isoproterenol, causes vasodilation.

Autoregulation Autoregulation in the intestinal circulation is not as well developed as it is in certain other vascular beds, such as those in the brain and kidney. The principal mechanism responsible for autoregulation is metabolic, although a myogenic mechanism probably also participates.

Functional Hyperemia Food ingestion increases intestinal blood flow. The secretion of the gastroinestinal hormones, gastrin and cholecystokinin, augment intestinal blood flow. The absorption of food also increases intestinal blood flow; the principal mediators of mesenteric hyperemia are glucose and fatty acids.

Hepatic Circulation

Regulation of Flow Blood flow in the portal venous and hepatic arterial systems vary reciprocally. When blood flow is curtailed in one system, the flow increases in the other system. However, the resultant increase in flow in one system usually does not fully compensate for the initiating reduction in flow in the other system.

The portal venous system is not autoregulating. As the portal venous pressure and flow are raised, resistance either remains constant or decreases. However, the hepatic arterial system is autoregulating.

The sympathetic nerves constrict the presinusoidal resistance vessels in the portal venous and hepatic arterial systems. However, neural effects on the capacitance vessels are more important. The effects are mediated mainly by α-receptors.

Capacitance Vessels The liver contains about 15% of the total blood volume of the body. Under appropriate conditions, such as in response to hemorrhage, about half of the hepatic blood volume can be rapidly expelled. Thus the liver constitutes an important blood reservoir in humans.

FETAL CIRCULATION

The circulation of the fetus differs from that of the postnatal infant. The fetal lungs are functionally inactive, and the fetus depends completely on the placenta for oxygen and nutrient supply. Oxygenated fetal blood from the placenta passes through the umbilical vein to the liver. A major fraction passes through the liver, and a small fraction bypasses the liver to the inferior vena cava through the ductus venosus (Figure 24-4). In the inferior vena cava blood from the ductus venosus joins blood returning from the lower trunk and extremities, and this combined stream is in turn joined by blood from the liver through the hepatic veins. The streams of blood tend to maintain their identity in the inferior vena cava, but they are divided into two streams of unequal size by the edge of the interatrial septum *(crista dividens)*. The larger stream, which is mainly blood from the umbilical vein, is shunted to the left atrium through the foreamen ovale, which lies between the inferior vena cava and the left atrium (*inset,* Figure 24-4). The other stream passes into the right atrium, where it is joined by superior vena caval blood returning from the upper parts of the body and by blood from the myocardium.

In contrast to the adult, in whom the right and left ventricles pump in series, the ventricles in the fetus operate essentially in parallel. Because of the large pulmonary resistance, less than one third of the right ventricular output goes through the lungs. The remainder passes through the ductus arteriosus from the pulmonary artery to the aorta at a point distal to the origins of the arteries to the head and upper extremities. Flow from the pulmonary artery to the aorta occurs because the pulmonary artery pressure is about 5 mm Hg higher than the aortic pressure in the fetus. The large volume of blood coming through the foramen ovale into the left atrium is joined by blood returning from the lungs and is pumped out by the left ventricle into the aorta. About one third of the aortic blood goes to the head, upper thorax, and arms, and the remaining two thirds goes to the rest of the body and the placenta. The amount of blood pumped by the left ventricle is about 20% greater than that pumped by the right ventricle. The major fraction of the blood that passes down the descending aorta flows by way of the two umbilical arteries to the placenta.

In Figure 24-4 the distribution of fetal blood flow is given as a percentage of the combined right and left ventricular outputs. Note that more than half of the combined cardiac output is returned directly to the placenta without passing through any capillary bed. Also indicated in Figure 24-4 are the O_2 saturations of the blood (numbers in parentheses) at various points of the fetal circulation. Fetal blood leaving the placenta is 80% saturated, but the saturation of the blood passing through the foramen ovale is reduced to 67% by mixing with desaturated blood returning from the lower part of the body and the liver. Addition of the desaturated blood from the lungs reduces the O_2 saturation of left ventricular blood to 62%, which is the level of saturation of the blood reaching the head and upper extremities.

The blood in the right ventricle, a mixture of desaturated superior vena caval blood, coronary venous blood, and inferior vena caval blood, is only 52% saturated with O_2. When the major portion of this blood traverses the ductus arteriosus and joins that pumped out by the left ventricle, the resultant O_2 saturation of blood traveling to the lower part of the body and back to the placenta is 58% saturated. Thus it is apparent that the tissues receiving blood of the highest O_2 saturation are the liver, heart, and upper parts of the body, including the head.

At the placenta the chorionic villi dip into the maternal sinuses, and O_2, CO_2, nutrients, and metabolic waste products exchange across the membranes. The barrier to exchange is large, and the equilibrium of O_2 tension (P_{O_2}) between the two circulations is not reached at normal rates of blood flow. Therefore, P_{O_2} of the fetal blood leaving the placenta is very low. If fetal hemoglobin did not have a greater affinity for O_2 than adult hemoglobin, the fetus would not receive an adequate O_2 supply. The fetal oxyhemoglobin dissociation curve is shifted to the left. Therefore, at equal O_2 pressures, fetal blood carries significantly more O_2 than does maternal blood. If the

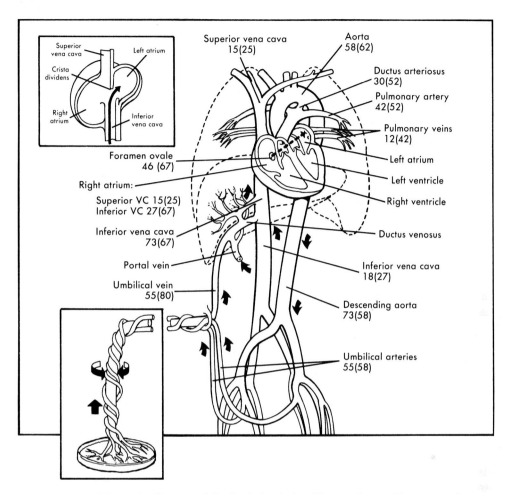

FIGURE 24-4   Schematic diagram of the fetal circulation. The numbers without parentheses represent the distribution of cardiac output as a percentage of the sum of the right and left ventricular outputs, and the numbers within parentheses represent the percentage of O_2 saturation of the blood flowing in the indicated blood vessel. The insert at the upper left illustrates the direction of flow of a major portion of the inferior vena cava blood through the foramen ovale to the left atrium. (Values for percentage distribution of blood flow and O_2 saturations are from Dawes GS et al: J Physiol 126:563, 1954.)

mother is subjected to hypoxia, the reduced blood PO_2 is reflected in the fetus by tachycardia and an increase in blood flow through the umbilical vessels. If the hypoxia persists or if flow through the umbilical vessels is impaired, fetal distress occurs; it is first manifested as bradycardia. In early fetal life the high cardiac glycogen levels that prevail may protect the heart from acute periods of hypoxia. The glycogen levels gradually decrease to adult levels by term.

Circulatory Changes That Occur at Birth

The umbilical vessels have thick muscular walls that are very reactive to trauma, tension, sympatho-

mimetic amines, bradykinin, angiotensin, and changes in PO_2. In animals in which the umbilical cord is not tied, hemorrhage of the newborn is prevented because these large vessels constrict in response to one or more of the stimuli just listed. Closure of the umbilical vessels increases total peripheral resistance and blood pressure. When blood flow through the umbilical vein ceases, the ductus venosus, a thick-walled vessel with a muscular sphincter, closes. The factor initiating closure of the ductus venosus is still unknown. The asphyxia, which starts with constriction or clamping of the umbilical vessels, plus the cooling of the body activate the respiratory center of the newborn infant. After the lungs fill with air, pulmonary vascular resistance decreases to about one tenth of the value that existed before lung expansion. This resistance change is not caused by the presence of O_2 in the lungs because the change is just as great if the lungs are filled with nitrogen.

The left atrial pressure is raised above that in the inferior vena cava and right atrium by (1) the decrease in pulmonary resistance, with the resulting large flow of blood through the lungs to the left atrium; (2) the reduction of flow to the right atrium caused by occlusion of the umbilical vein; and (3) the increased resistance to left ventricular output produced by occlusion of the umbilical arteries. This reversal of the pressure gradient across the atria abruptly closes the valve over the foramen ovale, and the septal leaflets fuse over several days.

With the decrease in pulmonary vascular resistance, the pressure in the pulmonary artery falls to about half its previous level (to about 35 mm Hg). This change in pressure, coupled with a slight increase in aortic pressure, reverses the flow of blood through the ductus arteriosus. However, within several minutes the large ductus arteriosus begins to constrict, producing turbulent flow, which manifests as a murmur in the newborn. Constriction of the ductus arteriosus is progressive and usually is complete within 1 to 2 days after birth. Closure of the ductus arteriosus appears to be initiated by the high PaO_2 of the arterial blood passing through it; pulmonary ventilation with O_2 closes the ductus, whereas ventilation with air low in O_2 opens this shunt vessel. Whether O_2 acts directly on the ductus or mediates the release of a vasoconstrictor substance is not known.

At birth the walls of the two ventricles are about equally thick, or the right ventricle is slightly thicker. Also, in the newborn, the pulmonary arterioles are thick, which is partly responsible for the high pulmonary vascular resistance of the fetus. After birth the thickness of the walls of the right ventricle diminishes, as does the muscle layer of the pulmonary arterioles; the left ventricular walls become thicker. These changes are progressive for several weeks after birth.

The foramen ovale or ductus arteriosus occasionally fails to close after birth. This failure leads to two of the more common congenital cardiac abnormalities that are amenable to surgical correction.

BIBLIOGRAPHY
Journal articles

Abboud FM, editor: Regulation of the cerebral circulation, Fed Proc 40:2296, 1981.

Berne RM: Role of adenosine in the regulation of coronary blood flow, Circ Res 47:807, 1980.

Berne RM et al: The local regulation of cerebral blood flow, Prog Cardiovasc Dis 24:243, 1981.

Chou CC: Contribution of splanchnic circulation to overall cardiovascular and metabolic homeostasis, Fed Proc 42:1656, 1983.

Feigl EO: Coronary physiology, Physiol Rev 63:1, 1983.

Granger DN and Kvietys PR: The splanchnic circulation: intrinsic regulation, Annu Rev Physiol 43:409, 1981.

Greenway CV: The role of the splanchnic venous system in overall cardiovascular homeostasis, Fed Proc 42:1678, 1983.

Heymann MA et al: Factors affecting changes in the neonatal systemic circulation, Annu Rev Physiol 43:371, 1981.

Klocke FJ et al: Coronary pressure-flow relationships—controversial issues and probable implications, Circ Res 56:310, 1985.

Kontos HA: Regulation of the cerebral circulation, Annu Rev Physiol 43:397, 1981.

Olsson RA: Local factors regulating cardiac and skeletal muscle blood flow, Annu Rev Physiol 43:385, 1981.

Sparks HV Jr and Bardenheuer H: Regulation of adenosine formation by the heart, Circ Res 58:193, 1986.

Wearn JT et al: The nature of the vascular communications between the coronary arteries and the chambers of the heart, Am Heart J 9:143, 1933.

Books and Monographs

Berne RM and Rubio R: Coronary circulation. In Handbook of physiology, section 2: The cardiovascular system—the heart, vol I, Bethesda, Md, 1979, American Physiological Society.

Berne RM et al: Metabolic regulation of cerebral blood flow. In Mechanisms of vasodilation—second symposium, New York, 1981, Raven Press.

Donald DE: Splanchnic circulation. In Handbook of physiology, section 2: The cardiovascular system—peripheral circulation and organ blood flow, vol III, Bethesda, Md, 1983, American Physiological Society.

Faber JJ and Thornburg KL: Placental physiology, New York, 1983, Raven Press.

Gregg DE: Coronary circulation in health and disease, Philadelphia, 1950, Lea & Febiger.

Heistad DD and Kontos HA: Cerebral circulation. In Handbook of physiology, section 2: The cardiovascular system—peripheral circulation and organ blood flow, vol III, Bethesda, Md, 1983, American Physiological Society.

Hellon R: Thermoreceptors. In Handbook of physiology, section 2: The cardiovascular system—peripheral circulation and organ blood flow, vol III, Bethesda, Md, 1983, American Physiological Society.

Lautt WW, editor: Hepatic circulation in health and disease, New York, 1981, Raven Press.

Lewis T: Blood vessels of the human skin and their responses, London, 1927, Shaw & Son, Ltd.

Marcus ML: The coronary circulation in health and disease, New York, 1983, McGraw-Hill Book Co.

Morgan HE et al: Protein metabolism of the heart. In Handbook of physiology, section 2: The cardiovascular system—the heart, vol I, Bethesda, Md, 1979, American Physiological Society.

Mott JC and Walker DW: Neural and endocrine regulation of circulation in the fetus and newborn. In Handbook of physiology, section 2: The cardiovascular system—peripheral circulation and organ blood flow, vol III, Bethesda, Md, 1983, American Physiological Society.

Owman C and Hardebo JE, editors: Neural regulation of brain circulation, Amsterdam, 1985, Elsevier.

Randle PJ and Tubbs PK: Carbohydrate and fatty acid metabolism. In Handbook of physiology, section 2: The cardiovascular system—the heart, vol I, Bethesda, Md, 1979, American Physiological Society.

Roddie EC: Circulation to skin and adipose tissue. In Handbook of physiology, section 2: The cardiovascular system—peripheral circulation and organ blood flow, vol III, Bethesda, Md, 1983, American Physiological Society.

Shepherd AP and Granger DN: Physiology of the intestinal circulation, New York, 1984, Raven Press.

Shepherd JT: Circulation to skeletal muscle. In Handbook of physiology, section 2: The cardiovascular system—peripheral circulation and organ blood flow, vol III, Bethesda, Md, 1983, American Physiological Society.

Sparks HV Jr et al: Control of the coronary circulation. In Sperelakis N, editor: Physiology and pathophysiology of the heart, ed 2, Boston, 1989, Kluwer Academic Publishers.

RESPIRATORY SYSTEM

ROBERT M. BERNE
MATTHEW N. LEVY

PART

Mechanics of Respiration

FUNCTIONAL ANATOMY

The major function of the respiratory system is to ventilate the lungs, thereby adding oxygen (O_2) and removing carbon dioxide (CO_2) from the circulating blood. The respiratory system responds to neural and chemical stimuli and thereby adjusts the rate of ventilation to the metabolic needs of the body, as outlined in Figure 25-1 and described in detail in Chapter 28.

In a resting subject inspiration is an active process produced by contractions of the diaphragm and the external intercostal muscles, whereas expiration occurs passively by elastic recoil of the lungs and the chest wall. Thus normal respiration at rest requires little effort. However, with exercise, hypoxia, or partial airway obstruction, inspiration requires greater effort, and expiration is assisted by active contraction of the muscles of expiration (internal intercostal, anterior abdominal muscles). Under resting conditions expiration is followed by a brief pause before the next inspiration occurs. During this pause the muscles of respiration are relaxed, and forces acting on the lung and chest wall are in equilibrium; air does not move into or out of the lungs. The opposing pulmonary forces at the end of a normal expiration are balanced. They consist of the inwardly directed elastic recoil force of the lungs and the outwardly directed elastic force of the chest wall.

The lungs fill the chest cavity and are separated from the internal chest wall by a thin layer of fluid that lies between the visceral pleura, which invests the lungs, and the parietal pleura, which lines the internal chest wall. The lungs are elastic and are stretched even at the end of a maximal expiration. They would collapse without the adherence between the visceral and parietal pleura. A potential space exists between the two pleural surfaces, and if a small amount of air is introduced into this space, a negative pressure (relative to atmospheric pressure) of 5 cm of H_2O (equal to 3.68 mm Hg) is observed. This negative pressure represents the elastic recoil force of the lungs. In contrast, the pressure within the entire respiratory tree equals atmospheric pressure when there is no airflow and the airways are open (Figure 25-2).

With expansion of the lungs by contraction of the muscles of inspiration, pleural and alveolar pressures are reduced and air enters the lungs. When the inspiratory effort ceases, pulmonary air inflow ceases. If the airways remain open, alveolar pressure rapidly equilibrates with atmospheric pressure. With relaxation of the inspiratory muscles, the recoil of the chest wall raises pleural and alveolar pressures. The increase in alveolar pressure above that of atmospheric pressure causes air to flow, first rapidly and then more slowly, out of the lungs. As expiration ends, pleural pressure returns to its resting level of -5 cm H_2O, and alveolar pressure decreases to the atmospheric level. The temporal relationships among alveolar pressure, pleural pressure, velocity of airflow, and lung volume changes in a single respiratory cycle in a person at rest are depicted in Figure 25-3.

The inspired air passes through the pharynx, larynx, trachea, and bronchi, which repeatedly divide dichotomously in an irregular pattern into smaller, narrower branches. The bronchi divide into terminal bronchioles and ultimately into the alveolar ducts and

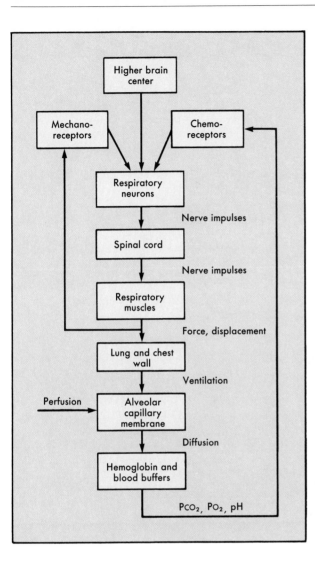

FIGURE 25-1 Overview of the factors that regulate respiration.

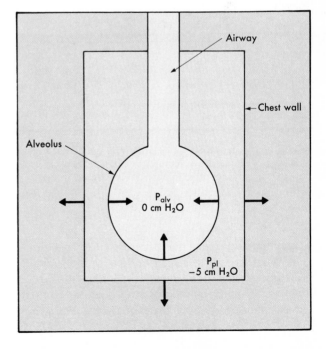

FIGURE 25-2   An airway and an alveolus (representing the tracheobronchial tree and lungs) are enclosed within the chest wall. The pleural pressure (P_{pl}) is the pressure in the potential space between the lungs and chest wall. The alveolar pressure (P_{alv}) is the pressure within the alveoli. The transpulmonary pressure is the difference between P_{alv} and P_{pl}. The pressure difference across the chest wall is the difference between the pleural pressure and the pressure at the body surface. At the end of a normal exhalation the elastic recoil forces of the lung are directed inward, whereas the elastic recoil forces of the chest wall are directed outward.

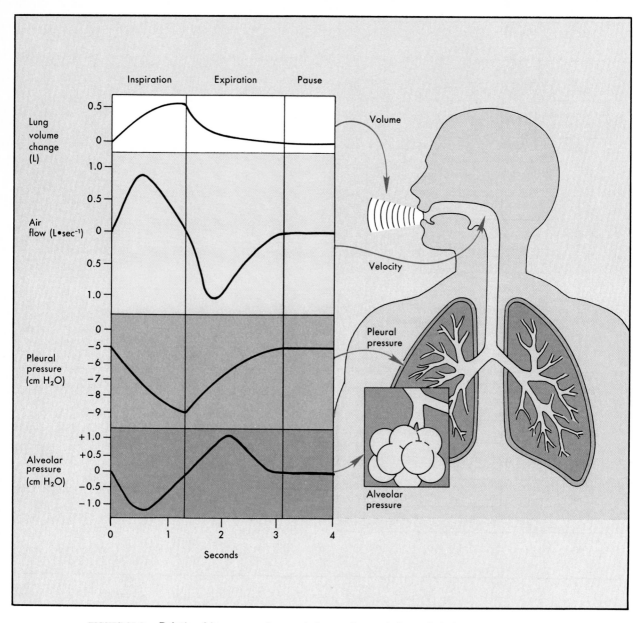

FIGURE 25-3   Relationships among changes in lung volume, air flow, pleural pressure, and alveolar pressure during a single respiratory cycle in a normal resting person. Note that changes are most rapid in all variables during the early phases of inspiration and expiration.

sacs, where gas exchange occurs (Figure 25-4, *A*).

Because the cross-sectional area of the airways progressively increases toward the alveoli, the velocity of the inflowing air progressively decreases until it moves almost solely by diffusion in the smallest bronchioles and alveoli. Diffusion of gases between blood and alveoli occurs at the alveolar-capillary membrane, which consists of the alveolar epithelium and the capillary endothelium separated by their respective basement membranes (Figure 25-4, *B*).

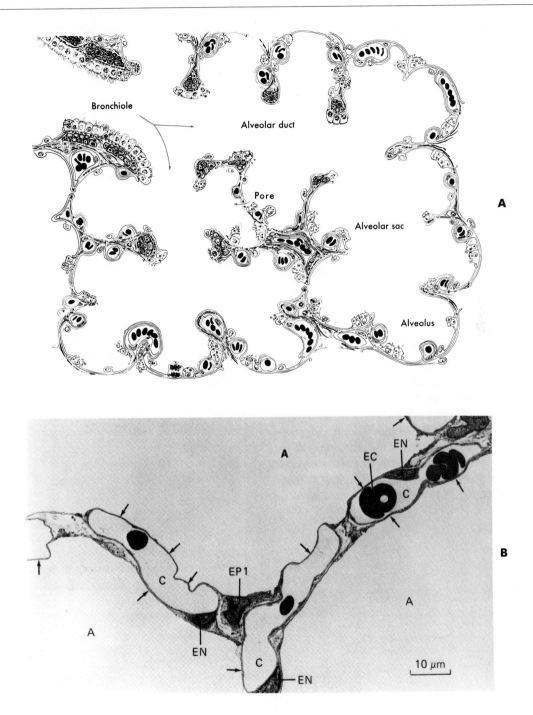

FIGURE 25-4 **A,** Diagram of the respiratory subdivisions in the lung. Smooth muscle (dark cells) ends in alveolar ducts. **B,** Histologic section of alveolar septa from human lung. Note that the combined alveolar and capillary membranes (denoted by the *arrows*) are very thin. Abbreviations: *A,* Alveolus; *C,* capillary; *EC,* erythrocyte; *EN,* endothelial cell; *EPl,* epithelial cell, type 1. (From Weibel ER: In Fishman AP, editor: Pulmonary diseases and disorders, ed 1, New York, 1980, McGraw-Hill Book Co.)

RESPIRATORY MUSCLES

The muscles used for inspiration are the diaphragm, the parasternal muscles, the external intercostal muscles, and the scalene and sternocleidomastoid muscles. At rest, expiration is passive, but when it becomes active (as in exercise), the internal intercostal and the abdominal muscles are used.

The Diaphragm

The main muscle for inspiration is the diaphragm, a thin dome-shaped muscle that is innervated by the phrenic nerves and separates the chest and abdominal cavities. The descent of the diaphragm during inspiration increases the vertical, anteroposterior, and lateral dimensions of the thoracic cage and compresses the abdominal contents (Figure 25-5).

External Intercostal Muscles

The external intercostal muscles originate from the lower edges of ribs 1 through 11 and pass caudally and anteriorly to insert on the upper edge of the adjacent rib. They are innervated by the intercostal nerves. Contraction of the external intercostal muscles during inspiration raises the ribs. Elevation of the upper ribs enlarges the chest anteroposteriorly, whereas elevation of the lower ribs enlarges it transversely. In addition, the contracted intercostal muscles stabilize the chest and prevent retraction of the intercostal spaces during inspiration, when pleural pressure becomes more negative relative to atmospheric pressure. Although the intercostal muscles play a minor role in inspiration under normal resting conditions, they become important in exercise or when the diaphragm is paralyzed.

Accessory Muscles of Inspiration

During exercise or when airways are obstructed, the scalene and sternocleidomastoid muscles are recruited in inspiration. These muscles raise and expand the upper chest.

Muscles of Expiration

Contraction of the abdominal muscles is mainly responsible for active expiration. They increase abdominal pressure by pulling down the lower ribs and sternum. This action forces the diaphragm upward, thereby decreasing the volume of air in the lungs.

The internal intercostal muscles run downward and posteriorly from the lower edge of one rib to the upper edge of the next rib. They assist expiration by pulling the ribs downward and inward (posteriorly). These muscles also stabilize the rib cage and prevent bulging of the intercostal spaces during forced expiration (e.g., straining and coughing).

OXYGEN COST OF BREATHING

Breathing requires the expenditure of energy, and the amount of this energy can be estimated from measurements of the quantity of oxygen used by the muscles of respiration. Normally the oxygen cost of breathing is about 5% of the total oxygen consumption of the body. However, it can increase manyfold with voluntary hyperventilation, with severe exercise, or with such abnormal conditions as hypercapnia, airway obstruction, or increased stiffness of the chest walls or lungs.

LUNG VOLUMES

The *total lung capacity (TLC)* is the lung volume at the end of a maximal inspiration. This TLC is divided into several compartments (Figure 25-6). The volume

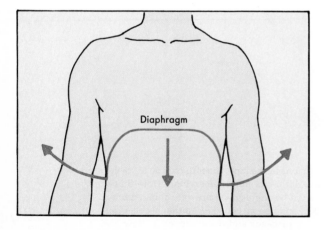

FIGURE 25-5   Contraction of the diaphragm during inspiration increases the vertical, anteroposterior, and lateral dimensions of the thorax.

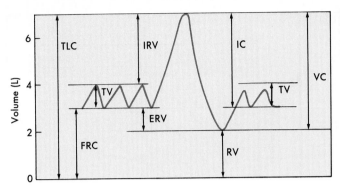

FIGURE 25-6   Lung volume and subdivisions. *TLC*, Total lung capacity; *FRC*, functional residual capacity; *IRV*, inspiratory reserve volume; *ERV*, expiratory reserve volume; *RV*, residual volume; *IC*, inspiratory capacity; *TV*, tidal volume; *VC*, vital capacity.

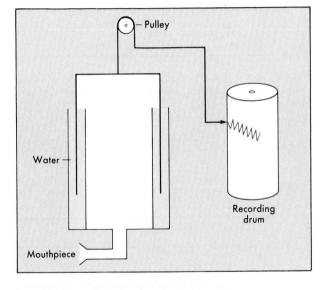

FIGURE 25-7   Diagram of a spirometer.

of air inhaled and exhaled during normal quiet respiration is the *tidal volume (TV)*. The difference in lung volume between the end of normal inspiration and maximal inspiration is called the *inspiratory reserve volume (IRV)*, whereas the lung volume between maximal inspiration and the end of normal expiration (TV plus IRV) is called the *inspiratory capacity (IC)*. The lung volume between the end of normal expiration and maximal expiration is termed the *expiratory reserve volume (ERV)*, and the volume of air that remains in the lung at the end of maximal expiration is called the *residual volume (RV)*. *Functional residual capacity (FRC)* is the sum of ERV and RV, and *vital capacity (VC)* is the amount of air that can be expired from maximal inspiration to maximal expiration (Figure 25-6). The volume of the respiratory tree that conducts air to and from the alveoli is termed the *ana-*

tomic dead space because no gas exchange between lungs and blood occurs there.

The IRV, TV, and ERV can be measured with a spirometer (Figure 25-7). The RV is determined by first measuring FRC by one of several techniques (gas dilution or whole body plethysmography) and then subtracting the ERV obtained by spirometry from the FRC.

ELASTIC PROPERTIES OF THE LUNG

Volume-Pressure Relationships

The elastic properties of the lungs (lung compliance) can be determined by measurements of the pressure-volume relationships during the course of expiration from total lung capacity (Figure 25-8). Airflow is temporarily arrested at successive lung volumes so that measurements can be made under static conditions. At each lung volume with the airways open, *transpulmonary pressure* is determined from the difference between alveolar and pleural pressures. Pleural pressure is determined indirectly by measuring the pressure in the esophagus with a loose, thin-walled balloon attached to a fine catheter. When airflow along the tracheobronchial tree ceases, alveolar pressure equals atmospheric pressure.

At lung volumes near FRC, lung compliance is relatively great. However, it decreases progressively as the lungs are inflated toward TLC. Near TLC, large increments in transpulmonary pressure produce only small increments in lung volume (flat portion of the pressure-volume curve in Figure 25-8). The nonelastic collagen fibers in the lung are mainly responsible for the decreased compliance at the higher lung volumes.

Lung compliance can be seriously affected by dis-

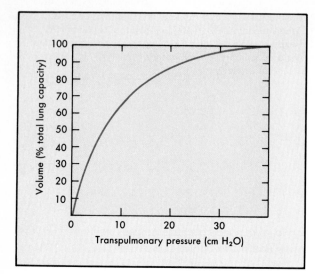

FIGURE 25-8   Relationship between lung volume and transpulmonary pressure (difference between alveolar and pleural pressures).

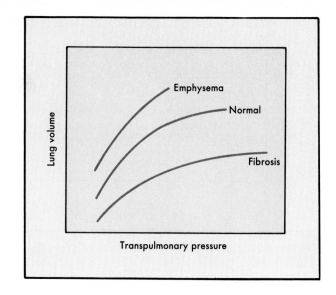

FIGURE 25-9   Volume-pressure relationships in a normal subject and in patients with emphysema and pulmonary fibrosis.

ease. In *emphysema* lung tissue is damaged and alveolar walls are destroyed, which greatly increases lung compliance. Thus small changes in transpulmonary pressure evoke large changes in lung volume in emphysema (Figure 25-9). Conversely, *pulmonary fibrosis,* characterized by an increase in inelastic fibrous tissue in the lungs, greatly decreases lung compliance. Thus large changes in transpulmonary pressure in patients with pulmonary fibrosis produce small changes in lung volume (Figure 25-9). If the chest wall is opened (e.g., chest wound or surgery) or a communication develops between a bronchus and the pleural space, air enters the pleural space and the lungs collapse. This condition is termed a *pneumothorax.*

Surface Tension in Alveoli

The alveoli are lined with a thin film of liquid, and at the air-liquid interface strong intermolecular forces in the liquid cause the area of the lining to shrink. These forces contribute to the elastic recoil of the lung. The effects of surface forces on lung compliance become evident when the pressure-volume relationships of air-filled and saline-filled lungs are compared (Figure 25-10). Filling the lung with saline eliminates the air-liquid interface and abolishes the surface forces without affecting the elasticity of the pulmo-

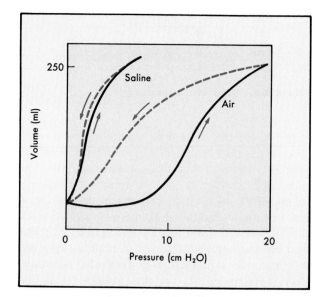

FIGURE 25-10   Pressure-volume relationships of an air-filled and saline-filled lung. In the saline-filled lung an air-liquid interface is absent, and surface forces are abolished. The compliance of the saline-filled lung is considerably greater than that of the air-filled lung. Hysteresis, which is considerable in the air-filled lung, is greatly reduced in the saline-filled lung.

nary tissues. At any given volume, the transpulmo-nary pressure of the liquid-distended lung is only about half that of the air-inflated lung. Thus, when the air-liquid interface is removed, the lung becomes more compliant. The contribution of surface forces to the elastic recoil of the lung is greater at low than at high lung volumes.

The most important component of the liquid film lining the alveolar walls is *surfactant.* It is produced by type II granular pneumocytes, and its major constituent is dipalmitoyl lecithin, a phospho-lipid with detergent-like properties.

The surface tension is greater in small spheres, such as alveoli, than in large spheres. Therefore, with-out surfactant, small alveoli would tend to empty into adjacent large alveoli because of the surface tension–induced pressure difference (Figure 25-11, *A*). This would result in the collapse of the small alveoli (Figure 25-11, *B*). Surfactant reduces the surface tension in the alveoli, particularly in small alveoli, so that the surface tension decreases as alveoli shrink. Thus the alveolar pressures of parallel large and small alveoli are virtually the same, and the tendency of alveoli to collapse at low gas volumes is abrogated by the pres-ence of surfactant.

By reducing surface tension, surfactant makes the lung more compliant and thereby decreases the effort needed to inflate the lung. Furthermore, if alveolar surface tension was not reduced by surfactant, the high surface tension of the alveolar fluid film would cause plasma transudate to enter the alveoli from the pulmonary capillaries; this would produce pulmonary edema and seriously interfere with gas exchange.

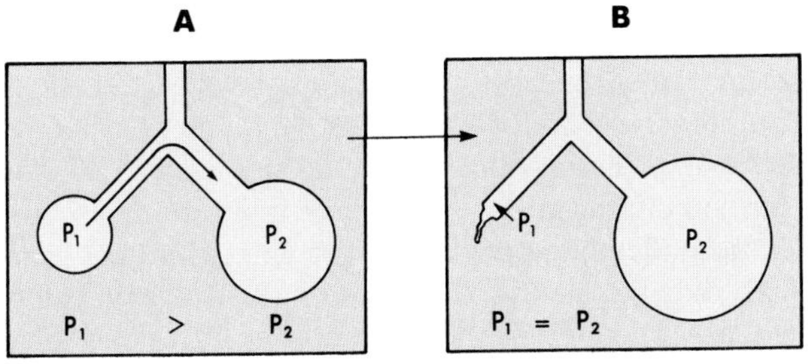

FIGURE 25-11   Communicating alveoli of different sizes. If the surface tension in both alveoli were the same **(A)**, the transpulmonary pressure of the smaller alveolus (P_1) would exceed that of the larger alveolus (P_2). The smaller alveolus would collapse **(B)** and empty into the larger alveolus.

ELASTIC PROPERTIES OF THE CHEST WALL

When the chest wall is at FRC, its elastic recoil is directed outward and assists inspiration. If the elastic recoil of the relaxed chest wall were unopposed by the recoil of the lungs, the chest would expand to about 70% of TLC. With further expansion of the thorax, the elastic recoil of the chest wall (as with the lung) is directed inward.

As lung volume is progressively reduced below 70% of TLC, the outward-directed recoil force of the chest wall increases and its compliance decreases. The increased stiffness of the chest wall at low lung volumes is a major determinant of RV.

ELASTIC PROPERTIES OF THE LUNG–CHEST WALL SYSTEM

The pressures exerted by the passive recoil of the lung and the passive recoil of the chest wall together yield the overall recoil pressure of the respiratory sys-tem. When the thorax is at FRC, the elastic recoil pressures of the lung and chest wall are equal in mag-nitude but opposite in direction; the algebraic sum of the two recoil pressures is zero. Thus the FRC repre-sents the equilibrium or relaxed position of the respi-ratory system. Lung volumes greater than the FRC can be maintained only by the action of the muscles of inspiration. Lung volumes less than the FRC can be maintained only by the action of the muscles of expi-ration.

RESISTANCE TO AIRFLOW

Total resistance of the ventilatory system consists of flow resistance of the airways and frictional resis-tance to the displacement of lung and chest wall tis-sues during breathing. Tissue resistance normally comprises only about 10% of the total resistance, but it may increase considerably with diseases of the lung parenchyma.

Anatomic Airway Resistance

The progressive increase in total cross-sectional area from the upper airways to alveoli results in a corresponding reduction in resistance to airflow. Most of the resistance to airflow occurs in the upper respiratory tract. The resistance of the nasal passages is very high; it may comprise as much as 50% of the total airway resistance during nose breathing. The mouth, pharynx, larynx, and trachea account for approximately 20% to 30% of airway resistance during quiet mouth breathing. However, this fraction may increase to about 50% at increased levels of ventilation, as occurs during exercise.

Transmural Airway Pressure Difference

The airways are elastic and can be compressed or distended. The caliber of the airways depends on the difference between the pressures inside (P_i) and outside (P_o) the airways, that is, the *transmural airway pressure difference*. Since P_o is approximately equal to pleural pressure (P_{pl}), the transmural pressure difference of the airways is essentially the same as transpulmonary pressure. In short, the terms are interchangeable.

With inspiration P_{pl} becomes more negative and airway calibers enlarge. Furthermore, the airways are essentially tethered to the parenchymal lung tissue, and the expanding lung tissue pulls outward on the airways. The increased cross section of the airways decreases their resistance to airflow. With forced expiration P_{pl} exceeds airway pressure (transmural pressure difference is reversed), which narrows the airways and increases their resistance to airflow.

The extent to which the transmural pressure difference changes airway caliber depends on the compliance of the trachea and bronchi. The trachea is almost completely encircled by cartilaginous rings, which prevent collapse when the pressure surrounding the trachea exceeds the intraluminal pressure. The bronchi are less well supported by incomplete cartilaginous rings and cartilaginous plates, and the bronchioles lack any cartilaginous support. Thus the bronchi and particularly the bronchioles may narrow considerably when the extraluminal pressure exceeds the intraluminal pressure.

Airway Contraction and Relaxation

The smooth muscle that encircles the airways can change the caliber of the small bronchi and bronchioles and thus can alter airflow resistance. Airway smooth muscle reacts to autonomic nervous activity, circulating hormones, inhaled particles, and chemicals released by cells located near the tracheobronchial tree (see Chapter 28).

Air Turbulence

From the trachea to the alveoli the airway progressively enlarges, and thus velocity of airflow diminishes. At the lower velocities in the small bronchi, airflow is laminar and therefore engenders little resistance. However, in the trachea, larger bronchi, and at branch points of bronchi, airflow is turbulent, which increases airway resistance.

ALVEOLAR VENTILATION

As discussed earlier, pulmonary ventilation depends on the action of the respiratory muscles, the physical characteristics of the chest wall and lung, and the effect of surfactant on the surface tension of the fluid film lining the alveoli. Ventilation of the alveoli is also influenced by gravity, and distribution of inspired air depends on regional differences in compliance and resistance within the lungs.

Effects of Gravity

In the upright human being P_{pl} is several centimeters of H_2O higher at the base of the lung than at the apex. The average P_{pl} is -5 cm H_2O, but at the apex it may be -10 cm H_2O, and at the base, -2.5 cm H_2O. The reason for this P_{pl} gradient from lung apex to base is the pull of gravity and the weight of the lung. Gravity also influences the distribution of pulmonary blood flow, as described in Chapter 26 (see Figure 26-4).

The consequence of a gravitational P_{pl} gradient is that alveoli at the apex are subjected to a greater distending force (transmural pressure difference) than are those at the base of the lung. However, the effect of this force on alveolar size depends on the magnitude of the lung volume. With the lungs near TLC (flat portion of pressure-volume curve in Figure 25-8), the P_{pl} is sufficiently negative at the apices and bases to fully expand the alveoli (Figure 25-12). At a lower lung volume (e.g., FRC) the P_{pl}, particularly at the lung base, is less negative than when the lung is at TLC. Therefore, at FRC, the alveoli at the lung base are smaller than those at the lung apex. Also, the apical alveoli at FRC are smaller than those at the apex **and**

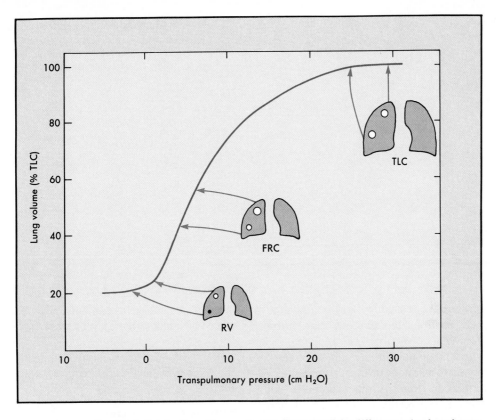

FIGURE 25-12   Regional distribution of lung volumes. Because of the differences in pleural pressure from the apex to the base of the lung, the transpulmonary pressures of the alveoli at the lung apex will differ from those of alveoli at the lung base. At a given lung volume, alveoli at the apex and at the base will fall on different locations of the same pressure-volume curve.

at the base when the lung is at TLC (Figure 25-12). A decrease in lung volume to RV increases P_{pl} at the apex to a less negative value than that prevailing with lung volume at FRC, and it increases P_{pl} at the base to a positive value. This decreases further the size of alveoli at the apex and collapses alveoli at the base of the lung (Figure 25-12).

Because the change in the size of the alveoli at the lung base is greater than that of the alveoli at the apex, the movement of air in and out of the alveoli is also greater at the lung base than at the apex. **Thus, despite alveoli at the base being smaller than those at the apex, ventilation of alveoli is greater at the lung base.**

Distribution of Ventilation

The distribution of inspired air throughout the lung is determined by the compliance of the lung tissue and the resistance of the branching airways. If lung compliance and airway resistance were the same in all the lobes and lobules in the lung, ventilation would be evenly distributed in the lung, and lung volume would reach a plateau by the end of inspiration, even at rapid respiratory rates (*curve 1,* Figure 25-13). If resistance in the bronchus to a unit (lobe or lobule) of the lung were normal but if compliance were decreased, airflow into the unit in early inspiration would be rapid but would reach a plateau before the end of inspiration (*curve 2,* Figure 25-13). Also, the volume of inspired air would be less than that in the more compliant unit. However, if compliance were normal in a lung unit but resistance to flow through the bronchus to that unit were increased, air inflow during inspiration would be slow, and normal inspiratory volume would not be reached by the end of inspiration (*curve 3,* Figure 25-13). At high respiratory frequencies the inspiratory airflow to a unit with increased airway resistance would be seriously impaired. To achieve uniform ventilation throughout

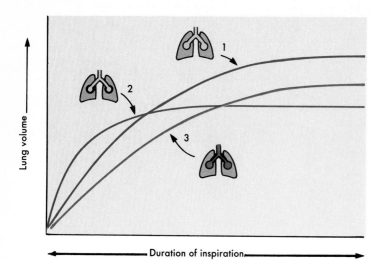

FIGURE 25-13   Comparison of volume changes during inspiration of a normal respiratory unit *(1)*, a low compliance unit *(2)*, and a unit *(3)* with high airway resistance in parallel.

the lung, the product of resistance and compliance must be about the same for the different units in the lung.

The existence of pulmonary units of different compliances and airway resistances results in uneven distribution of ventilation. Mucous plugs or bronchial lesions (e.g., cancer) can increase airway resistance in parts of the lung. Pulmonary fibrosis, venous congestion, or pulmonary edema can decrease the compliance in affected regions of the lung, particularly in the lung bases.

BIBLIOGRAPHY
Journal Articles

Chang HK and El Masry OA: A model study of flow dynamics in human central airways. I. Axial velocity profiles, Respir Physiol 49:75, 1982.

Engel LA and Paiva M: Analyses of sequential filling and emptying of the lung, Respir Physiol 45:309, 1981.

Haber PS et al: Alveolar size as a determinant of pulmonary distensibility in mammalian lungs, J Appl Physiol 54:837, 1983.

Isabey D and Chang HK: Steady and unsteady pressure-flow relationships in central airways, J Appl Physiol 51:1338, 1981.

Macklem PT: Airway obstruction and collateral ventilation, Physiol Rev 51:365, 1971.

Sharp JT et al: Relative contributions of rib cage and abdomen to breathing in normal subjects, J Appl Physiol 39:608, 1975.

Wilson TA: Relations among recoil pressure, surface area and surface tension in the lung, J Appl Physiol 50:921, 1981.

Books and Monographs

Chang HK and Paiva M: Respiratory physiology—an analytical approach, New York, 1989, Marcel Dekker, Inc.

Clements JA and King RJ: Composition of the surface active material. In Crystal RG, editor: The biochemical basis of pulmonary function, New York, 1976, Marcel Dekker, Inc.

Macklem PT and Mead J, editors: Handbook of physiology, section 3: Respiratory system—mechanics of breathing, vol III, Bethesda, Md, 1986, American Physiological Society.

Martin DE and Youtsey JW: Respiratory anatomy and physiology, St Louis, 1988, The CV Mosby Co.

Murray JF: The normal lung, Philadelphia, 1988, WB Saunders Co.

Slonim NB and Hamilton LH: Respiratory physiology, ed 5, St Louis, 1987, The CV Mosby Co.

Taylor AE et al: Clinical respiratory physiology, Philadelphia, 1989, WB Saunders Co.

West JB: Respiratory physiology—the essentials, ed 3, Baltimore, 1985, Williams & Wilkins.

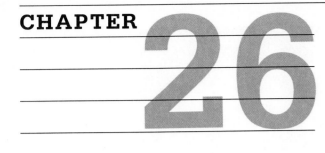

CHAPTER 26

Pulmonary Circulation

The lungs are supplied with blood from the pulmonary and systemic circuits. The *pulmonary vasculature* ordinarily carries about 99% of the blood flow to the lungs, and its function is to exchange oxygen (O_2) and carbon dioxide (CO_2) between the blood and alveoli. The *bronchial vasculature* is actually a component of the systemic circulation and normally carries only about 1% of the blood flow to the lungs. Its function is to provide blood circulation to the tracheobronchial tree.

Pulmonary Vasculature

The pulmonary vascular system is a low-resistance network of highly distensible vessels. The main pulmonary artery is much shorter than the aorta. The walls of the pulmonary artery and its major branches are much thinner than the walls of the aorta, and they contain less smooth muscle and elastin. Similarly, the pulmonary arterioles are thin and contain much less smooth muscle than do systemic arterioles. Consequently, the pulmonary arterioles cannot constrict as effectively as systemic arterioles. The walls of the pulmonary venules and veins are also very thin and possess little smooth muscle.

Pulmonary capillary networks differ greatly from systemic capillary networks. The systemic capillaries constitute a network of interconnecting tubular vessels (see Chapter 21). However, the pulmonary capillaries are aligned such that the blood flows in thin sheets between adjacent alveoli (Figure 26-1). This arrangement exposes the capillary blood optimally to the alveolar gases. Only thin layers of vascular and alveolar endothelium separate the blood and alveolar gas (Figure 26-1; see also Figure 25-4). The total surface area for exchange between alveoli and blood is about 50 to 70 m^2 in adult humans.

The thickness of the sheets of blood between adjacent alveoli depends on the intravascular and intraalveolar pressures (Figure 26-2). Ordinarily the width of an interalveolar sheet of blood is about equal to the diameter of a red blood cell (Figure 26-1). During pulmonary vascular congestion, which occurs when the left atrial pressure is elevated, the width of the sheet may increase several-fold. Conversely, when the local alveolar pressure exceeds the adjacent capillary pressure, the capillaries may collapse and blood will not flow to those alveolar capillaries. Gravitational factors participate in this phenomenon, particularly with respect to the regional distribution of blood flow to the lungs (see Figure 26-4).

Bronchial Vasculature

The bronchial arteries are branches of the thoracic aorta. These arteries and their branches have the structural characteristics of systemic arteries; that is, they have much thicker walls and more smooth muscle than do pulmonary arteries of equivalent caliber.

The bronchial veins drain partly into the pulmonary venous system and partly into the azygos veins, which are components of the systemic venous system. As noted, the bronchial blood flow normally constitutes about 1% of the cardiac output. Therefore the fraction of the bronchial blood flow that drains into the pulmonary veins and then into the left atrium constitutes at most 1% of the venous return to the heart.

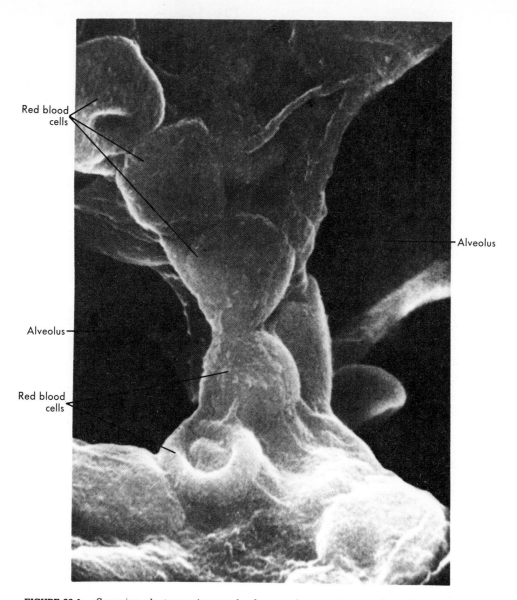

Red blood cells

Red blood cells

Alveolus

Alveolus

FIGURE 26-1   Scanning electron micrograph of mouse lung to show an interalveolar septum. Note that the membranes that separate an alveolus from a capillary are so thin that the shapes of the erythrocytes in the capillary can easily be discerned. (From Greenwood MF and Holland P: Lab Invest 27:296, 1972.)

This small quantity of bronchial venous blood, plus a small amount of coronary venous blood that drains directly into the left atrium or left ventricle, contaminates the pulmonary venous blood, which would otherwise be fully saturated with O_2. Thus, the aortic blood is very slightly desaturated. This small quantity of venous drainage directly into the left side of the heart also accounts for, even under true equilibrium conditions, the output of the left ventricle slightly exceeding that of the right ventricle. In certain pathologic states the bronchial circulation may become substantial, and the admixture of blood between the systemic and pulmonary circuits may be appreciable. In patients with such disturbances the O_2 saturation of their arterial blood may be abnormally low.

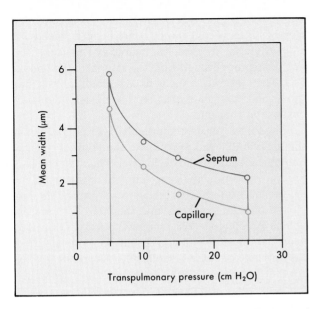

FIGURE 26-2   Mean widths of pulmonary capillaries and interalveolar septa as functions of the transpulmonary pressure in rat lungs. (Modified from Mazzone RW et al: J Appl Physiol 45:325, 1978.)

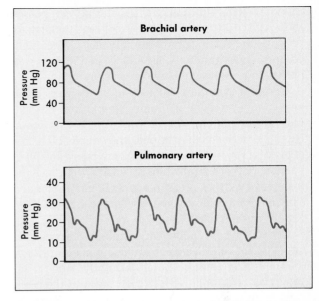

FIGURE 26-3   Pressures recorded in the brachial and pulmonary arteries of a normal human subject. (Modified from Harris P and Heath D: Human pulmonary circulation, Edinburgh, 1962, E and S Livingstone Ltd.)

PULMONARY HEMODYNAMICS

Pressures in the Pulmonary Circulation

In normal individuals the systolic pressure in the pulmonary artery is about 25 or 30 mm Hg, the diastolic pressure is approximately 10 mm Hg, and the mean pressure is about 15 mm Hg (Figure 26-3). These pressures are much lower than those in the aorta because the pulmonary vascular resistance is only about one-tenth the resistance of the systemic vascular bed. The mean pressure in the left atrium is normally about 5 mm Hg, and thus the total pulmonary arteriovenous pressure gradient is only about 10 mm Hg. The mean pressure in the pulmonary capillaries lies between the pulmonary arterial and pulmonary venous pressures, but somewhat closer to the latter.

Pulmonary Blood Flow

Because pressure is low in the pulmonary blood vessels, which also are very distensible, the regional distribution of blood flow in the lungs is affected by gravity. Three distinct flow patterns may be found at different hydrostatic levels in the lung, as illustrated

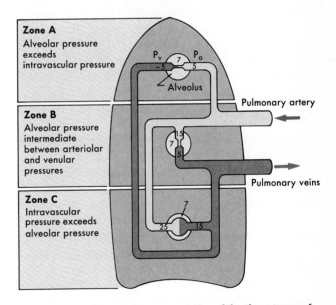

FIGURE 26-4   Schematic representation of the three types of flow patterns that exist in the pulmonary circulation. In *zone A* alveolar pressure exceeds intravascular pressures. Pulmonary capillaries in this zone will not be perfused. In *zone B* alveolar pressure lies between pulmonary arterial and venous pressures. Pulmonary capillaries will flutter between the open and closed states. In *zone C* intravascular pressures exceed alveolar pressure. The pulmonary capillaries are always open, and the flow resistances in individual vessels vary with the hydrostatic pressure in the vessel.

in Figure 26-4. The flow patterns, in turn, depend on the relative pressures in the alveoli and pulmonary vasculature. The regions of the lung may be subdivided into three zones (A, B, and C in Figure 26-4), depending on these relative pressures.

Consider that the pulmonary artery delivers blood at a steady pressure of 15 mm Hg and that the pulmonary venous pressure remains constant at 5 mm Hg. In those pulmonary arterial and venous branches (zone C in Figure 26-4) that are 13 cm below the main pulmonary vessels, gravitational effects will cause the respective vascular pressures to be 10 mm Hg (equivalent of 13 cm of blood) greater than those in the main vessels at the pulmonary hilum.

Conversely, in pulmonary arterial and venous branches that are 13 cm above (zone A in Figure 26-4) the main hilar vessels, the arterial and venous pressures will be 10 mm Hg less than those in the main vessels. At the same hydrostatic level (zone B) as the main vessels, the arterial and venous pressures in the branches will be approximately equal to those in the main vessels.

For illustrative purposes, assume that the alveolar pressure equals 7 mm Hg in all alveoli. Such an alveolar pressure might exist in an individual receiving positive-pressure respiration. In zone A the alveolar pressure would exceed the local arterial and venous pressures (Figure 26-4). The pulmonary capillary pressures lie between those of the arteries and veins, and therefore alveolar pressure would also exceed capillary pressure. Capillaries lying between adjacent alveoli would collapse, and therefore no gas would exchange between those alveoli and the blood.

In a normal person not attached to an artificial ventilator, the mean pressure in the alveoli, averaged over the respiratory cycle, is equal to the atmospheric pressure. Therefore the conditions depicted in zone A do not ordinarily prevail in any region of the lungs. In hypovolemic states, however, the mean pulmonary artery pressure is often very low. Thus vascular pressures in the lung apices might be subatmospheric. The atmospheric pressure in the alveoli would then compress the apical capillaries, and therefore almost no blood would flow in the pulmonary vessels to the apices. However, the bronchial circulation, which operates at much higher pressures, would be unaffected.

In zone B the alveolar pressure lies between the local arterial and venous pressures (Figure 26-4). Again, if alveolar pressure equals 7 mm Hg, a capillary in that region will flutter between the open and closed states. When the capillary is open, blood will flow through it, and the capillary pressure will decrease progressively from the arterial to the venous end. The pressure at the venous end will be less than the alveolar pressure, causing the capillary to quickly collapse at that end.

When flow ceases, the pressures will equalize in all the vessels that are proximal to the point of collapse and at the same hydrostatic level. Thus the capillary pressure will quickly rise to equal the pressure in the local small arteries. This pressure will exceed the prevailing alveolar pressure, and therefore the capillary will be forced open. With the restitution of flow, however, the pressure will drop again along the length of the capillary because viscous flow dissipates energy. As the pressure at the venous end drops below the ambient alveolar pressure, the capillary will again close.

The critical pressure gradient for flow in zone B is the arterioalveolar pressure difference. The resistance to flow is **not** proportional to the arteriovenous pressure difference, as it is for most of the other vessels in the body. As long as the venous pressure is less than the alveolar pressure, the venous pressure has no influence on the flow. Such a flow condition is called a waterfall effect, since the height of a waterfall has no influence on the rate of fluid flow over the edge.

In zone C the arterial and venous pressures both exceed the alveolar pressure (Figure 26-4). Thus the pressure everywhere along the capillary exceeds the alveolar pressure, and the capillary remains permanently open. In this zone the flow is determined by the arteriovenous pressure gradient, and the resistance may be calculated by the conventional formula.

The large and small pulmonary vessels, including the capillaries (see Figure 26-2), are very distensible, as already noted. The pressure difference that determines the caliber of a distensible tube is the transmural pressure; that is, the difference between the internal and external pressures. In an upright individual the intravascular pressures in the lungs increase from apex to base, but the alveolar pressures do not. The transmural pressures increase accordingly, and thus the diameters of the pulmonary vessels increase from apex to base. Because resistance to flow varies inversely with vessel caliber, resistance decreases and flow increases in zone C (Figure 26-4) in the apex-to-base direction. Such predicted changes in flow have been verified in humans.

REGULATION OF THE PULMONARY CIRCULATION

The pulmonary and systemic blood flows are virtually equal in normal people. Therefore the various cardiac and vascular factors that determine cardiac output in general also determine the total pulmonary blood flow. These factors are discussed in Chapter 23.

The autonomic nervous system innervates the pulmonary blood vessels. Although the small pulmonary vessels contain little smooth muscle, slight changes in smooth muscle tone may alter vascular resistance substantially because the pressures in the pulmonary vessels are so low.

The pulmonary arterioles are innervated by parasympathetic (vagus) and sympathetic nerves. Efferent vagal stimulation dilates the pulmonary vessels. The vasodilation is ascribable to the action of acetylcholine, which is released from the vagal endings and acts on cholinergic receptors in the pulmonary vessels.

Sympathetic nerve fibers in the pulmonary vessels release norepinephrine, which can interact with α- and β-adrenergic receptors on the membranes of vascular smooth muscle cells. Interaction of norepinephrine with the α-receptors causes vasoconstriction, whereas interaction with the β-receptors induces vasodilation. The α-receptors predominate in the pulmonary circulation, and therefore sympathetic nerve stimulation constricts the pulmonary blood vessels (Figure 26-5). However, if the α-receptors are blocked with a drug, such as phenoxybenzamine, sympathetic nerve stimulation may then dilate the vessels.

Baroreceptor stimulation can dilate pulmonary resistance vessels reflexly, just as it does to systemic

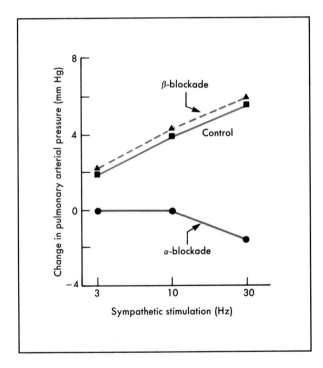

FIGURE 26-5 Effects of sympathetic nerve stimulation on pressure in a pulmonary artery in an anesthetized cat. The pressure changes reflect concordant changes in pulmonary vascular resistance. Raising the frequency of sympathetic stimulation induced progressively greater vasoconstriction. This was mediated predominantly by α-receptors; phenoxybenzamine, an α-receptor blocker, either abolished or reversed the effects of sympathetic stimulation. β-Receptor blockade had a negligible effect on the response. (Modified from Hyman AL et al: In Said SI, editor: Pulmonary circulation and acute lung injury, Mt Kisco, NY, 1985, Futura Publishing Co.)

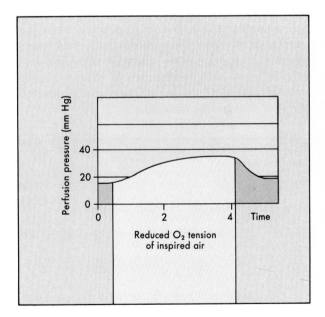

FIGURE 26-6 The effect of hypoxia on vascular resistance of an isolated rat lung. The lung was perfused with blood at a constant flow. When the O_2 tension of the inspired air was reduced, the pulmonary resistance vessels constricted, as indicated by the substantial rise in perfusion pressure. (Modified from Grover RF et al: In Handbook of physiology, section 2: The cardiovascular system—peripheral circulation and organ blood flow, vol III, Bethesda, Md, 1983, American Physiological Society.)

arterioles. Conversely, peripheral chemoreceptor stimulation constricts the pulmonary vessels. However, the physiologic importance of such reflex regulation has not yet been established.

Hypoxia is the most important influence on pulmonary vasomotor tone. Both acute and chronic hypoxia increase pulmonary vascular resistance (Figure 26-6). Hypoxia will constrict pulmonary vessels, regardless of whether the hypoxia originates in the airways (e.g., impaired alveolar ventilation) or in the bloodstream (e.g., decreased O_2 tension, Po_2, in the pulmonary arterial blood). The regional constriction of arterioles in response to local reductions in alveolar Po_2 helps maintain an optimal ventilation-perfusion ratio (see Figure 27-6). The Po_2 level in poorly ventilated alveoli will approach the Po_2 level in the pulmonary arterial blood. Blood flowing by such poorly ventilated alveoli will not be well oxygenated, and therefore the Po_2 of the blood returning to the left atrium will decrease. Arteriolar vasoconstriction, induced by the hypoxia, reduces the blood flow to such poorly ventilated alveoli. Therefore, it minimizes the contamination of the pulmonary venous blood with poorly oxygenated blood. In effect this mechanism shunts the pulmonary blood flow from the poorly ventilated regions to the better ventilated regions of the lungs, thereby improving the O_2 saturation of the systemic arterial blood. The mechanism by which hypoxia causes pulmonary vasoconstriction is still obscure, despite considerable investigation.

BIBLIOGRAPHY
Journal Articles

Baile EM, et al: Measurement of regional bronchial arterial blood flow and bronchovascular resistance in dogs, J Appl Physiol 53:1044, 1982.

Bhattacharya J et al: Factors affecting lung microvascular pressure, Ann NY Acad Sci 384:107, 1982.

Dawson CA et al: Pulmonary microcirculatory hemodynamics, Ann NY Acad Sci 384:90, 1982.

Gil J: Organization of microcirculation in the lung, Annu Rev Physiol 42:177, 1980.

Lucas CL: Fluid mechanics of the pulmonary circulation, CRC Crit Rev Biomed Eng 10:317, 1984.

Marshall C and Marshall B: Site and sensitivity for stimulation of hypoxic pulmonary vasoconstriction, J Appl Physiol 55:711, 1983.

McMurtry IF et al: Studies of the mechanism of hypoxic pulmonary vasoconstriction, Adv Shock Res 8:21, 1982.

Modell HI et al: Functional aspects of canine bronchial-pulmonary vascular communications, J Appl Physiol 50:1045, 1981.

Nandiwada PA et al: Pulmonary vasodilator responses to vagal stimulation and acetylcholine in the cat, Circ Res 53:86, 1983.

Orchard CH et al: The relationship between hypoxic pulmonary vasoconstriction and arterial tension in the intact dog, J Physiol (Lond) 338:64, 1983.

Shirai M et al: Effects of regional alveolar hypoxia and hypercapnia on microcirculation in small pulmonary vessels in cats, J Appl Physiol 61:440, 1986.

Zasslow MA et al: Hypoxic pulmonary vasoconstriction and the size of hypoxic compartment, J Appl Physiol 53:626, 1982.

Books and Monographs

Cumming G and Bonsignore G, editors: Pulmonary circulation in health and disease, New York, 1980, Plenum Press.

Fishman AP: Pulmonary circulation. In Handbook of physiology, section 3: Respiratory system, vol I, Bethesda, Md, 1985, American Physiological Society.

Fishman AP: Normal pulmonary circulation. In Fishman AP, editor: Pulmonary diseases and disorders, ed 2, vol 2, New York, 1988, McGraw-Hill Book Co.

Fishman AP and Renkin EM, editors: Pulmonary edema, Bethesda, Md, 1979, American Physiological Society.

Grover RF et al: Pulmonary circulation. In Handbook of physiology, section 2: The cardiovascular system—peripheral circulation and organ blood flow, vol III, Bethesda, Md, 1983, American Physiological Society.

Parker JC et al: Pulmonary transcapillary exchange and pulmonary edema. In Guyton AC and Young DB, editors: International review of physiology: cardiovascular physiology III, vol 18, Baltimore, 1979, University Park Press.

Said SI: Pulmonary circulation and acute lung injury, Mt Kisco, NY 1985, Futura Publishing Co.

Will JA et al, editors: Pulmonary circulation in health and disease, Orlando, Fla, 1987, Academic Press.

Gas Transport

The exchange of gases between the body and the environment occurs mainly at the interface between the pulmonary capillaries and the alveoli. The important gases exchanged are oxygen and carbon dioxide.

OXYGEN TRANSPORT

Oxygen (O_2) is carried by the blood partly in physical solution in water but primarily in chemical combination with hemoglobin.

Physical Solution

If a sample of blood is allowed to equilibrate with a mixture of gases containing O_2, the concentration of O_2 in the blood in physical solution will be proportional to the partial pressure of O_2 in the gas phase. In accordance with *Henry's law*,

$$C_{O_2} = k_s P_{O_2} \tag{1}$$

where C_{O_2} is the concentration of O_2 in physical solution at equilibrium, P_{O_2} is the partial pressure of O_2 in the gas phase, and k_s is the solubility coefficient. The value of k_s for O_2 in water at 37° C is 0.003 ml·dl^{-1}·mm Hg^{-1}. This value signifies that for each 1 mm Hg increment in P_{O_2}, an additional 0.003 ml of O_2 (standard conditions) will be physically dissolved in each deciliter of water.

Thus, if pulmonary capillary blood equilibrates with alveolar gas, and if the alveolar gas has a P_{O_2} of 100 mm Hg, the concentration of O_2 in physical solution in the pulmonary venous blood will be 0.3 ml·dl^{-1}. Ordinarily the total concentration of O_2 in pulmonary venous blood is about 20 ml·dl^{-1}. Therefore only a tiny fraction of the O_2 in pulmonary venous blood is usually in physical solution, and most of the O_2 is combined with hemoglobin. Even if a person breathes pure O_2 and increases alveolar P_{O_2} to 700 mm Hg, the concentration of physically dissolved O_2 will increase to just 2.1 ml·dl^{-1}. Only if a subject breathed pure O_2 in a hyperbaric chamber at a pressure of 3 or more atmospheres could a quantity of O_2 be dissolved in blood that would be adequate to meet the body's metabolic needs. However, the resultant high blood P_{O_2}, if sustained, would be very injurious, a condition known as *oxygen toxicity.*

Combination of Oxygen with Hemoglobin

As stated, almost all the oxygen carried in the blood is combined with hemoglobin. This protein has a molecular weight of 67,000 daltons. Hemoglobin exists within the red blood cells and imparts the red color to the blood. Normally the concentration of hemoglobin in the blood is about 15 g·dl^{-1}. Each gram of hemoglobin can combine with about 1.36 ml of O_2. Thus, the hemoglobin in each deciliter of blood can combine with approximately 20 ml of O_2. Therefore, when the hemoglobin in normal blood is fully *saturated* with O_2, the blood contains about 20 ml·dl^{-1}

Oxyhemoglobin Dissociation Curve The quantity of O_2 physically dissolved in blood is directly proportional to the P_{O_2} in the gas phase with which the blood has equilibrated, as already stated. Thus the relation between C_{O_2} and P_{O_2} is linear. In contrast, the relation between the quantity of O_2 that

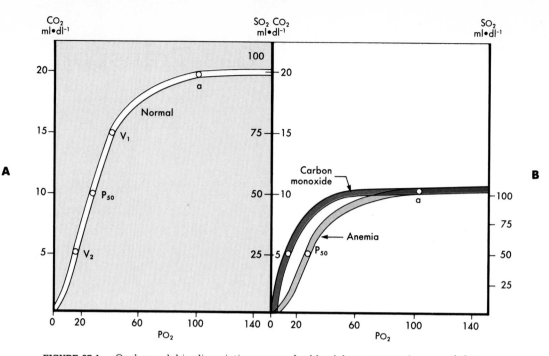

FIGURE 27-1 Oxyhemoglobin dissociation curves for blood from a normal person (**A**), for an anemic person (**B**, *red curve*), and for a person who had been exposed to carbon monoxide (**B**, *blue curve*). **A**, The oxygen content (CO_2) and oxygen saturation (SO_2) as a function of the partial pressure of oxygen (PO_2) for normal blood. Point a represents arterial blood. Because a falls on the flat portion of the curve, relatively large changes in alveolar PO_2 (which ordinarily determines the value of PO_2 in the systemic arterial blood) will have very little effect on the CO_2 and SO_2. Point v_1 represents mixed systemic venous blood in a resting subject. Point v_2 represents the mixed systemic venous blood in a normal subject during strenuous exercise. Because v_1 and v_2 lie on the steep portion of the curve, a relatively small change in tissue PO_2 leads to the dissociation of a relatively large amount of O_2 from hemoglobin, and consequently facilitates the transport of O_2 to the more actively metabolizing tissues (i.e., to the tissues with a lower PO_2). P_{50} denotes the value of the PO_2 at which the hemoglobin is half saturated with O_2. **B**, The blue curve is the oxyhemoglobin dissociation curve for an anemic subject whose blood hemoglobin concentration is half normal. Consequently, the O_2 content of the systemic arterial blood is only half the normal value when the hemoglobin is fully saturated with O_2. Otherwise, the curve resembles that for normal blood (**A**). The red curve is the oxyhemoglobin dissociation curve for an individual who has a normal blood hemoglobin concentration but who was exposed to a level of carbon monoxide sufficient to combine with half the hemoglobin. Thus only half the hemoglobin is available for combination with O_2, which accounts for the CO_2 of only 10 ml · dl^{-1} when the blood is 100% saturated. Carbon monoxide increases the affinity of hemoglobin for O_2, as reflected by (1) the greater values of CO_2 and SO_2 in the carbon monoxide curve than in the anemia curve when the PO_2 is below 80 mm Hg and (2) the lower P_{50} value in the carbon monoxide curve than in the anemia curve.

combines with hemoglobin and the PO_2 is highly non-linear, as shown in Figure 27-1.

Gradients in the **partial pressure** of O_2 (PO_2) indicate the **direction** of net diffusion of O_2, whereas gradients in the **concentration** of O_2 (CO_2) indicate the **amount** of O_2 potentially available for diffusion. With regard to gradients in PO_2, oxygen tends to diffuse from a region of higher PO_2 to a region of lower PO_2, regardless of the concomitant gradient in CO_2. The O_2 exchanges that occur at the arterial end of a

pulmonary capillary illustrate this principle (Figure 27-2).

Consider that the venous blood returning to the right atrium from the peripheral tissues has a PO_2 of 40 mm Hg and a CO_2 of 15 ml·dl^{-1}, as indicated by point v_1 on the oxyhemoglobin dissociation curve (Figure 27-1, *A*). As the blood is pumped by the right ventricle into the pulmonary circulation and approaches the arterial end of an alveolar capillary, the plasma PO_2 tends to equilibrate quickly with the PO_2

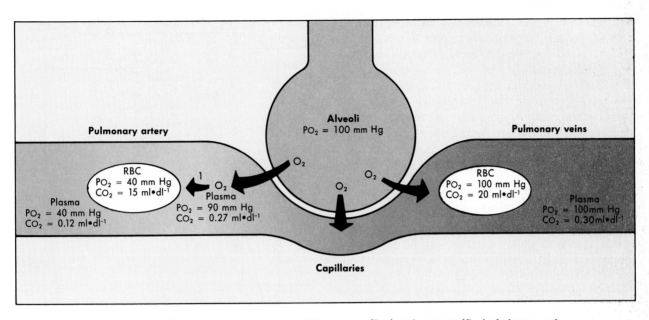

FIGURE 27-2   The changes in oxygen partial pressure (Po_2) and content (Co_2) of plasma and whole blood as the systemic mixed venous blood enters the pulmonary artery, passes through a pulmonary capillary, and then flows into a pulmonary vein. Note that in the red blood cells (RBC), the Co_2 is expressed as $ml \cdot dl^{-1}$ of *whole blood* rather than as $ml \cdot dl^{-1}$ of erythrocytes.

in the neighboring alveoli. However, the equilibration of the plasma Po_2 with the erythrocyte Po_2 is slower. Therefore, as blood enters the pulmonary capillary, O_2 diffuses from the plasma into the red blood cell (*arrow 1*, Figure 27-2). O_2 diffuses in this direction because the Po_2 is higher in the plasma (90 mm Hg) than in the red cell (40 mm Hg), even though the CO_2 in the red cell (15 $ml \cdot dl^{-1}$) is much higher than it is in the plasma (0.27 $ml \cdot dl^{-1}$). In a resting subject about one fourth of the O_2 is extracted from the blood as it passes through the systemic capillaries. Thus the CO_2 in the mixed systemic venous blood (and thus also in the pulmonary arterial blood) averages about 15 $ml \cdot dl^{-1}$, as shown in Figure 27-2.

The sigmoidal shape of the oxyhemoglobin dissociation curve (Figure 27-1, *A*) indicates certain important features of the transport of O_2 by the blood. The flatness of the upper region of the curve emphasizes that once hemoglobin is fully saturated, large changes in Po_2 are associated with very small changes in CO_2. Thus in normal subjects any increases in alveolar Po_2, produced by breathing O_2-enriched gas mixtures, will not greatly affect the CO_2 in the blood. Once hemoglobin is fully saturated, large increases in Po_2 will only increase the concentration

of physically dissolved O_2. As stated, O_2 is not very soluble in water. Therefore, the total CO_2 of a given volume of blood will not increase much unless Po_2 is raised to unusually high levels.

The flatness of the curve over the Po_2 ranging from about 75 to 100 mm Hg (Figure 27-1, *A*) also indicates that the systemic arterial blood will be almost fully saturated even when alveolar Po_2 is somewhat reduced. Such reductions in alveolar Po_2 may occur when an individual ascends to moderately high altitudes or hypoventilates. This characteristic of full saturation of the blood with O_2, even though alveolar Po_2 may be moderately reduced, constitutes a safety factor that tends to ensure the normal delivery of O_2 to the tissues.

The steepness of the oxyhemoglobin dissociation curve at Po_2 values less than about 70 mm Hg is a factor that ensures ample unloading of O_2 from the systemic capillary blood to the tissues; this factor also ensures an ample uptake of O_2 by the blood in the pulmonary capillaries. As stated, mixed systemic venous blood in a resting subject has a Po_2 of approximately 40 mm Hg and a CO_2 of about 15 $ml \cdot dl^{-1}$ (point v_1, Figure 27-1, *A*). It indicates that the Po_2 of the mixed venous blood has approached equilibrium with a mean PO_2 in the tissues of about 40 mm Hg. Let us assume that this PO_2 is also the PO_2 of the venous blood draining the resting skeletal muscles. If the subject then begins to exercise, for example, the PO_2 in the contracting muscles will decrease. If the PO_2 in the muscle tissues decreases to the value denoted by point v_2 (Figure 27-1, *A*), the blood in the tissue capillaries will now decrease its CO_2 to 5 $ml·dl^{-1}$, rather than to 15 $ml·dl^{-1}$. That is, each deciliter of blood will yield an additional 10 ml O_2 to the exercising muscle in response to a relatively small additional decrease in the tissue PO_2.

Effects of Anemia The shape of the oxyhemoglobin dissociation curve in patients with anemia is the same as it is in normal persons, except the scaling factor is different. For example, in an anemic patient whose hemoglobin content and red blood cell count are 50% of normal, equilibration of the blood with a gas mixture that has PO_2 of 100 mm Hg will fully saturate the hemoglobin; that is, the O_2 saturation (SO_2) of the blood will be 100%. However, the CO_2 in the blood will be only 10 $ml·dl^{-1}$ (Figure 27-1, *B*), rather than 20 $ml·dl^{-1}$, as in normal blood (Figure 27-1, *A*). Similarly, at any given PO_2, the CO_2 in the anemic blood will be just half that in the normal blood.

Two types of compensatory mechanisms would allow a normal quantity of O_2 per minute to be delivered to the tissues of anemic subjects. First, because the tissue PO_2 is lower in anemic subjects than in normal subjects, a greater fraction of CO_2 of the blood can be extracted by the tissues from each unit volume of blood that flows through the tissue capillaries of the anemic subject. Second, more blood flows through the tissues per unit time (i.e., cardiac output is greater) in anemic than in normal individuals. Thus, even though less O_2 is carried by each milliliter of blood in anemic patients, more milliliters per minute of blood perfuse their tissues than perfuse those of normal subjects.

Altered Affinity between Hemoglobin and Oxygen

Normal factors. Several factors normally affect the affinity between hemoglobin and O_2 and thereby alter the oxyhemoglobin dissociation curve (Figure 27-3). Carbon dioxide (CO_2) is the most important of these factors. As the partial pressure of CO_2 (PCO_2) is raised, the curve shifts to the right; reductions in PCO_2 have the opposite effect. Such a shift in the oxyhemoglobin dissociation curve, caused by a change in PCO_2, is called the *Bohr effect*.

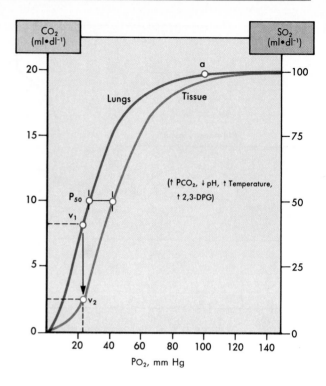

FIGURE 27-3   The oxyhemoglobin dissociation curves for blood as it passes through the pulmonary capillaries *(left curve)* and the systemic capillaries *(right curve)*. Increases in PCO_2, temperature, H^+ concentration, and 2,3-DPG concentration shift the curve to the right. More O_2 is delivered to the tissues because the PCO_2, H^+ concentration, and temperature are elevated in the tissues. These factors alter the affinity of hemoglobin for O_2, and this is attended by a shift in the O_2 dissociation curve *(right curve)*. Consequently, for every 100 ml (deciliter) of blood that perfuses the tissues, an additional 6 ml of O_2 are unloaded from the blood to the tissues, as denoted by the difference between the Y coordinate values of points v_1 and v_2.

Reductions in the pH of the blood shift the oxyhemoglobin dissociation curve to the right, whereas increases in pH shift the curve to the left. The shifts caused by changes in PCO_2 are ascribable largely to the associated alterations in pH. Finally, increases in the temperature of the blood shift the curve to the right, and reductions in temperature have the opposite effect.

The magnitude of a shift in the oxyhemoglobin dissociation curve is often expressed in terms of the P_{50}, which reflects the PO_2 at which the blood would be 50% saturated (Figures 27-1 and 27-3). An increase in P_{50} denotes a diminished affinity between hemoglobin and O_2; a decrease in P_{50} denotes an increased affinity.

When a skeletal muscle actively contracts, it generates more CO_2 and hydrogen ions and its temperature rises. The resultant changes in local PCO_2, pH, and temperature decrease the affinity between hemoglobin and O_2, thereby shifting the oxyhemoglobin dissociation curve to the right.

These changes in affinity promote the delivery of more O_2 to the active tissue, as shown in Figure 27-3. Consider that the left curve represents the oxyhemoglobin dissociation curve for blood at the PCO_2, pH, and temperature that prevailed at the venous ends of the pulmonary capillaries. Consider also that the PO_2 in the actively metabolizing muscle was 22 mm Hg, and that the PO_2 in the capillary blood in the muscle reached the same value. If the PCO_2, pH, and temperature of the blood in the muscle capillaries would remain the same as their respective values in the pulmonary capillary blood, then 12 ml O_2 would be delivered to the muscle from each 100 ml (deciliter) of blood flowing through the muscle. The blood oxygen content (CO_2) would decrease from 20 to 8 ml·dl^{-1} (point a to point v_1, Figure 27-3).

However, the PCO_2, H^+ concentration, and temperature are all elevated in the active muscle. These characteristics of the blood tend to equilibrate with the elevated PCO_2, H^+ concentration, and temperature in the tissue when the blood arrives in the muscle capillaries. Thus, the affinity between hemoglobin and O_2 will be diminished, and the oxyhemoglobin dissociation curve will be shifted to the right. Consequently, more O_2 will be delivered to the tissue from each deciliter of blood flowing through the muscle capillaries. Note in Figure 27-3 that if the capillary blood equilibrates at a PO_2 level of 22 mm Hg, approximately 6 ml more O_2 will be delivered to the tissues by each deciliter of blood because of the shift in the oxyhemoglobin dissociation curve. This disparity in O_2 delivery can be appreciated by comparing point v_2 with point v_1 in Figure 27-3.

This comparison also illustrates the different affinity of hemoglobin for O_2 in pulmonary capillary blood from that in muscle capillary blood. The blood at both sites has the same PO_2, namely 22 mm Hg. The hemoglobin in 1 dl of pulmonary capillary blood is able to combine with 8 ml O_2 (v_1). However, the hemoglobin in 1 dl of muscle capillary blood can combine with only 2 ml O_2 (v_2). This disparity denotes that the hemoglobin in the muscle capillary blood has a lower affinity for O_2 than does the hemoglobin in the pulmonary capillary blood.

The affinity of hemoglobin for O_2 is also affected by organic polyphosphates, which regulate O_2 transport by the blood. The most important of these compounds is *2, 3-diphosphoglycerate (2,3-DPG)*, which is a byproduct of anaerobic glycolysis. It is present in especially high concentrations in red blood cells because of their content of 2,3-DPG mutase. The affinity of hemoglobin for O_2 diminishes as the concentration of 2,3-DPG increases in the red blood cells. Such increases in 2,3-DPG concentration shift the oxyhemoglobin dissociation curve to the right (Figure 27-3), whereas reductions in its concentration shift the curve to the left.

The affinity of hemoglobin for O_2 caused by the 2, 3-DPG normally in the red blood cell promotes O_2 delivery to the tissues, in much the same way as do elevations in PCO_2, H^+ concentration, and temperature (Figure 27-3). At high values of PO_2 (> 80 mm Hg), 2,3-DPG has little effect on affinity. Therefore, the O_2-carrying capacity of pulmonary venous (and systemic arterial) blood is not influenced by the red cell content of 2, 3-DPG. When the blood enters the capillaries in the peripheral tissues, the affinity of hemoglobin for O_2 is diminished by the 2,3-DPG normally present in the red blood cells, and thus more oxygen (e.g., difference between v_1 and v_2 in Figure 27-3) is delivered to the tissues because of the diminished affinity.

Abnormal factors. Carbon monoxide (CO) is a toxic gas that profoundly alters the ability of blood to deliver O_2 to the tissues. CO affects O_2 transport in two ways: (1) it prevents some of the hemoglobin from combining with O_2, and (2) for that fraction of the hemoglobin that does combine with O_2 rather than CO, the presence of CO greatly increases the affinity between hemoglobin and O_2. This latter effect interferes with the release of O_2 from the blood to the tissues.

Carbon monoxide has an affinity for hemoglobin that is about 200 times greater than the affinity of O_2 for hemoglobin. Thus, when a subject inhales CO, even in low concentrations, it combines avidly with hemoglobin. CO combines at the same locus in the hemoglobin molecule at which O_2 combines. Therefore CO interferes with the ability of O_2 to combine with hemoglobin, and thus it decreases the O_2 content of the blood.

Because the affinity of hemoglobin for CO is 200 times greater than its affinity for O_2, exposure of blood to a gas mixture with a PO_2 of 100 mm Hg and a PCO

of 0.05 mm Hg will cause half the hemoglobin to be in the form of oxyhemoglobin and half in the form of *carboxyhemoglobin*. Thus, when P_{O_2} is 100 mm Hg, the O_2 content (C_{O_2}) of the blood will be only half normal (point *a*, Figure 27-1, *B*). These values are equivalent to those in an anemic subject who had not been exposed to CO, but whose hemoglobin concentration and red blood cell counts are half the normal values. The comparison is shown in Figure 27-1, *B*. Note that in either case, the C_{O_2} at a P_{O_2} of 100 mm Hg (point *a*) is only 10 ml·dl^{-1}, just half that in blood from a normal subject (point *a*, Figure 27-1, *A*).

The secondary effect of CO, to increase the affinity between O_2 and hemoglobin, is evident in the lower ranges of P_{O_2} in Figure 27-1, *B*. Note that when the P_{O_2} is less than 80 mm Hg, the carboxyhemoglobin-containing blood actually has a substantially greater C_{O_2} than does the blood from the anemic subject who had not been exposed to CO.

The deleterious effect of this CO-induced change in affinity for O_2 is that the P_{O_2} in the tissues must drop to dangerously low levels after exposure to CO before O_2 will be transferred from the capillary blood to the tissues. Figure 27-1, *B*, shows that in subjects exposed to CO, the tissue P_{O_2} must drop to about 10 mm Hg *(P_{50} of left curve)* before each mililiter of blood in the tissue capillaries would yield half its O_2 to the tissues. Conversely, in the anemic subject, the tissue P_{O_2} would have to drop to only about 25 mm Hg *(P_{50} of right curve)* before each mililiter of blood in the tissue capillaries would yield half its O_2 to the tissues.

CARBON DIOXIDE TRANSPORT

Carbon dioxide is produced in the body as a consequence of cell metabolism. The CO_2 tension (P_{CO_2}) in the tissues is higher than that in the capillary blood perfusing the tissues, and therefore CO_2 tends to diffuse from tissue to capillary blood. CO_2 is carried in the blood in three principal forms: in physical solution, as bicarbonate ions, and as carbamino compounds.

Physical Solution

CO_2 is about 20 times more soluble in water than O_2. The solubility coefficient (k_s) for CO_2 is about 0.06 ml·dl^{-1}·mm Hg^{-1}. Therefore a significant fraction (about 5%) of the CO_2 carried by the blood exists as physically dissolved CO_2. The CO_2 diffuses readily across the capillary endothelium into the plasma and across the erythrocyte cell membrane into the cell interior (Figure 27-4).

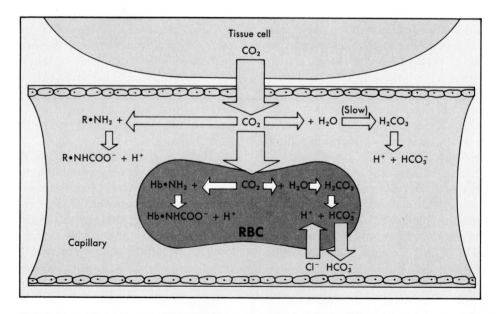

FIGURE 27-4   The role of carbamino compounds in the transport of CO_2 by the capillary blood away from the tissues.

Bicarbonate

In the plasma and in the erythrocyte, the dissolved CO_2 combines with water to form carbonic acid (H_2CO_3), which in turn dissociates to form H and bicarbonate (HCO_3) ions:

$$CO_2 + H_2O \rightleftharpoons H_2CO_3 \rightleftharpoons H^+ + HCO_3^- \qquad (2)$$

Bicarbonate is the principal form by which CO_2 is carried in the blood.

The formation of H_2CO_3 proceeds very slowly in the plasma because the enzyme *carbonic anhydrase* is not present in the plasma. This enzyme is present in the red cell, however, and therefore H_2CO_3 is formed at a much more rapid rate (more than 10,000 times more rapidly) in the red cell than in the plasma. Because of these differences in reaction rates, the HCO_3^- concentration rises much more rapidly in the erythrocytes than in the plasma of the tissue capillaries.

The red blood cell membrane is very permeable to HCO_3^- and other anions, and therefore HCO_3^- diffuses rapidly down the concentration gradient from erythrocyte to plasma. The cell membrane is not very permeable to cations, such as K^+, Na^+, or H^+. Therefore the diffusion of HCO_3^- from cell to plasma tends to depolarize the red blood cell membrane. This ionic imbalance attracts anions into the red blood cells from the plasma. The most abundant of these anions in the plasma is Cl^-, and therefore Cl^- is the principal anion that moves into the cell to counterbalance the outward flux of HCO_3^-. This exchange of Cl^- for HCO_3^- in tissue capillaries, as well as the reverse exchange that takes place in pulmonary capillaries, is termed the *chloride shift*.

Because the red blood cell membrane is very permeable to HCO_3^-, the concentration of HCO_3^- is essentially the same in the red blood cell and in the plasma. However, the fractional volume of plasma is greater than the fractional volume of erythrocytes in blood; the ratio is about 55:45. Thus, more of the CO_2 that is carried as HCO_3^- is carried in the plasma than in the red blood cells, even though the chemical reactions that generate the HCO_3^- occur in the red blood cells.

Carbamino Compounds

Dissolved CO_2 can react with free amino groups of proteins to form carbamino compounds:

$$R \cdot NH_2 + CO_2 \rightleftharpoons R \cdot NHCOO^- + H^+ \qquad (3)$$

The CO_2 dissolved in the plasma forms carbamino compounds with the various plasma proteins (Figure 27-4). The H^+ formed from the dissociation of these carbamino compounds is buffered by the plasma proteins and other buffer systems in the plasma.

The CO_2 that diffuses into the red blood cells reacts with hemoglobin to form *carbaminohemoglobin* (Figure 27-4). The affinity of reduced hemoglobin for CO_2 is greater than that of oxyhemoglobin. Thus the formation of carbaminohemoglobin in the tissue capillaries is enhanced as O_2 dissociates from oxyhemoglobin in the tissue capillaries. Also, the H^+ formed by the dissociation of carbaminohemoglobin (Figure 27-4) is largely neutralized by buffering groups on the hemoglobin molecule itself. Proteins are effective buffers, and hemoglobin is the most abundant protein in the red blood cell.

Carbon Dioxide Dissociation Curve

The CO_2 content of the blood varies with PCO_2 in much the same way that the O_2 content varies with PO_2. CO_2 is carried in the blood partly in association with hemoglobin but mainly as bicarbonate. In normal arterial blood the relative contents of dissolved CO_2 to carbamino CO_2 to bicarbonate CO_2 are about 0.3:1.0:4.3. Because bicarbonate is the principal carrier of CO_2, and because increases in PCO_2 augment bicarbonate formation without limit, the CO_2 dissociation curve (Figure 27-5) has no plateau. This contrasts with the oxyhemoglobin dissociation curve (see Figure 27-1), which has a prominent plateau.

The CO_2 dissociation curve is affected by the level of PO_2 (Figure 27-5), just as oxyhemoglobin dissociation curve is affected by the level of PCO_2 (see Figure 27-3). Note that in Figure 27-5, when the blood has been equilibrated with a PO_2 of 40 mm Hg *(upper curve)*, the CO_2 dissociation curve lies above the CO_2 dissociation curve obtained with blood that had been equilibrated with a PO_2 of 100 mm Hg *(lower curve)*. Therefore, for any given PCO_2, the blood will hold more CO_2 when the PO_2 has been diminished. This phenomenon is known as the *Haldane effect;* it is analogous to the Bohr effect, which modulates the oxyhemoglobin dissociation curve (see Figure 27-3). The Haldane effect reflects the tendency for an increase in PO_2 to diminish the affinity of hemoglobin for CO_2.

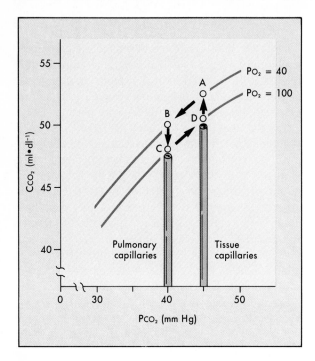

FIGURE 27-5   The CO_2 dissociation curves for systemic arterial ($Po_2 = 100$ mm Hg) and mixed venous ($Po_2 = 40$ mm Hg) blood. In the *pulmonary capillaries* the Po_2 of the blood increases from 40 to 100 mm Hg. The higher Po_2 alters the affinity of hemoglobin for CO_2, which is accompanied by a shift in the CO_2 dissociation curve. Consequently the capillary blood is able to hold less CO_2 (point *B* to *C*), and thus more CO_2 passes from blood to alveolus. Conversely, in the *tissue capillaries* the Po_2 of the blood decreases from 100 to 40 mm Hg. Because of the resulting change in affinity of hemoglobin for CO_2 (curve shift), more CO_2 (from point *D* to *A*) can be accommodated by the blood in the tissue capillaries and be carried back to the pulmonary circulation.

The Haldane effect enhances CO_2 transport away from the metabolizing tissue and promotes the elimination of CO_2 by the lungs, as shown in Figure 27-5. Assume the Pco_2 of the capillary blood in the tissues is 45 mm Hg, and the CO_2 content (Cco_2) of that blood is 53 ml·dl^{-1}, as represented by point *A* in Figure 27-5. When the blood finally reaches the pulmonary capillaries and equilibrates with an alveolar Pco_2 of 40 mm Hg, the Cco_2 of that blood might be expected to diminish to 50 ml·dl^{-1} (point *B*); that is, 3 ml CO_2 would be transferred to the alveoli from each deciliter of capillary blood, assuming no change in the affinity of hemoglobin for CO_2. However, in the pulmonary capillaries the influx of O_2 into the blood from the alveoli reduces the affinity of hemoglobin for CO_2, thereby enhancing the efflux of CO_2 into the alveoli.

Consequently the Cco_2 of the blood decreases to 48 ml·dl^{-1} (point *C*).

Similarly, the Haldane effect improves the ability of the blood in the tissue capillaries to take up CO_2 and transport it away from the metabolizing tissues. As the blood travels away from the pulmonary capillaries (point *C*, Figure 27-5) and arrives at the tissue capillaries (point *D*), the higher Pco_2 in the tissues might be expected to cause each deciliter of capillary blood to take on an additional 3 ml CO_2, assuming no change in the affinity of hemoglobin for CO_2. However, the reduction of Po_2 to 40 mm Hg in the tissue capillary blood enables each deciliter of blood to pick up an additional 2 ml of CO_2 (from point *D* to point *A*), since the reduced Po_2 increases the affinity of hemoglobin for CO_2.

Role of Carbon Dioxide in Acid-Base Balance

The equilibrium between the concentrations of CO_2, H_2CO_3, and HCO_3^- in the blood is very important in the regulation of the acid-base balance in the body. The role of these compounds as determinants of blood pH is defined by the Henderson-Hasselbalch equation (see Chapter 34). Changes in pulmonary ventilation can rapidly alter the pH of the pulmonary capillary blood (and thus of the systemic arterial blood) by changing the concentration of dissolved CO_2, and therefore of H_2CO_3, in the blood.

When the systemic arterial blood pH diminishes, as it does in *metabolic acidosis* (see Table 34-2), the peripheral chemoreceptors are stimulated and pulmonary ventilation increases (see Chapter 28). The arterial blood Pco_2 is thereby reduced, and thus the blood pH rises toward normal. Conversely, when the arterial blood pH is abnormally elevated, as it is in *metabolic alkalosis*, the excitation of the peripheral chemoreceptors is diminished and pulmonary ventilation decreases. This raises the arterial blood Pco_2, which tends to lower the blood pH toward normal.

If pulmonary ventilation is impaired, as it is in certain types of lung disease, alveolar Pco_2 is abnormally elevated. Consequently the Pco_2 of the systemic arterial blood is also increased, and thus the pH of the blood is diminished. Such a reduction in blood pH induced by an impairment of alveolar ventilation is known as *respiratory acidosis*. The kidneys tend to compensate for this derangement by conserving HCO_3^- (see Table 34-2).

An increase in pulmonary ventilation sometimes

occurs during states of emotional excitement. The resultant reduction in alveolar P_{CO_2} raises the pH of the arterial blood, a condition known as *respiratory alkalosis*. In a sustained state of respiratory alkalosis the kidneys compensate by excreting HCO_3^- (see Table 34-2).

REGIONAL VARIATIONS IN ALVEOLAR VENTILATION AND PERFUSION

Gravity affects the regional distribution of pulmonary blood flow and alveolar ventilation. When a person is in the upright position, blood flow per gram of lung tissue varies with location in the lung; the flow becomes progressively greater in the apex-to-base direction (see Chapter 26). Similarly, alveolar ventilation per gram of lung tissue increases in the apex-to-base direction (see Chapter 25); the relative effect of location however, is not nearly so great for ventila-

tion as for blood flow. Thus the ratio of alveolar ventilation ($\dot{V}_A$) to blood flow ($\dot{Q}$) tends to decrease in the direction from lung apex to base.* Regional variations in the *ventilation/perfusion ($\dot{V}_A/\dot{Q}$) ratio* are much more pronounced in patients with various types of lung disease than in healthy people.

Normal lung function depends on a proper matching of ventilation and perfusion. An abnormal $\dot{V}_A/\dot{Q}$ ratio is actually the most common cause of inadequate oxygenation of the arterial blood in patients with lung disease. The example shown in Figure 27-6 will illustrate the importance of proper matching between $\dot{V}_A$ and $\dot{Q}$. In healthy resting adults the average rate of alveolar ventilation is about 4 L·min^{-1}, and the average pulmonary blood flow is about 5 L·min^{-1}. Thus the $\dot{V}_A/\dot{Q}$ ratio normally is about 0.8. In Figure

---

*In respiratory physiology, "$\dot{Q}$" is used to represent blood flow.

A

Normal $\dot{V}_A /\dot{Q}$ = 0.8—both lungs

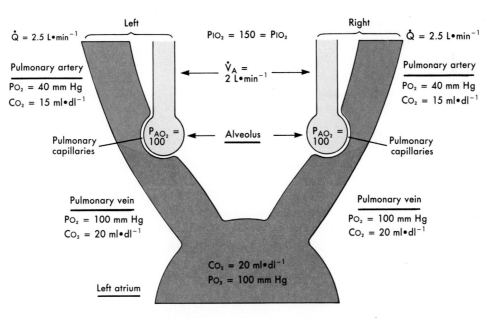

FIGURE 27-6   The effects of disparities in the ratio of alveolar ventilation $(\dot{V}_A)$ to alveolar blood perfusion $(\dot{Q})$ on the O_2 content (C_{O_2}) and partial pressure (P_{O_2}) of left atrial (and systemic arterial) blood. **A,** $\dot{V}_A$ of the left and right lungs is 2 L · min^{-1}, and $\dot{Q}$ to the left and right lungs is 2.5 L · min^{-1}. The pulmonary venous blood from both lungs is fully saturated with O_2, and therefore the blood in the left atrium is fully saturated with O_2. P_{IO_2}, O_2 tension of inspired air; P_{AO_2}, alveolar O_2 tension.   ***Continued.***

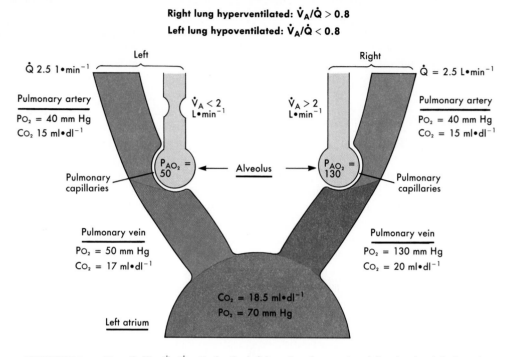

B

Right lung hyperventilated: $\dot{V}_A/\dot{Q} > 0.8$

Left lung hypoventilated: $\dot{V}_A/\dot{Q} < 0.8$

Left   Right

$\dot{Q}$ 2.5 1•min^{-1} $\dot{Q}$ = 2.5 L•min^{-1}

Pulmonary artery $\dot{V}_A < 2$ $\dot{V}_A > 2$ Pulmonary artery

Po_2 = 40 mm Hg   L•min^{-1} L•min^{-1} Po_2 = 40 mm Hg

Co_2 15 ml•dl^{-1} Co_2 = 15 ml•dl^{-1}

P_{AO_2} = 50   ← Alveolus → P_{AO_2} = 130

Pulmonary capillaries   Pulmonary capillaries

Pulmonary vein   Pulmonary vein

Po_2 = 50 mm Hg Po_2 = 130 mm Hg

Co_2 = 17 ml•dl^{-1} Co_2 = 20 ml•dl^{-1}

Co_2 = 18.5 ml·dl^{-1}

Po_2 = 70 mm Hg

Left atrium

FIGURE 27-6, cont'd.   B, The $\dot{V}_A/\dot{Q}$ ratio for the left lung is subnormal and that for the right lung is supernormal. The Po_2 and Co_2 of the blood in the left pulmonary veins are subnormal because the alveoli in the left lung are hypoventilated. The Po_2 of the blood in the right pulmonary veins is elevated (130 mm Hg) because the alveoli in the right lung are hyperventilated. However, the Co_2 of the blood in the right pulmonary veins is not above normal (20 ml · dl^{-1}) because the hemoglobin is fully saturated with O_2 when Po_2 exceeds about 90 mm Hg. Equal volumes of desaturated blood from the left lung and fully saturated blood from the right lung are admixed in the left atrium. Consequently the blood in the left atrium and systemic arteries is not fully saturated with O_2; the Co_2 is 18.5, rather than 20.0, ml · dl^{-1}.

27-6, *A,* half the alveolar ventilation (2.0 L·min^{-1}) and half the pulmonary blood flow (2.5 L·min^{-1}) are assigned to each lung, and thus the $\dot{V}_A/\dot{Q}$ ratio for each lung is 0.8. In the right and left pulmonary veins the hemoglobin is fully saturated with O_2; that is, the O_2 content (Co_2) is 20 ml·dl^{-1}. Therefore, as the venous blood from both lungs admixes in the left atrium, the Co_2 of the blood in the left atrium (and thus in the systemic arteries) will also be 20 ml·dl^{-1}.

The effect of disparate $\dot{V}_A/\dot{Q}$ ratios on the Co_2 of the systemic arterial blood is illustrated in Figure 27-6, *B.* In this example the right lung is hyperventilated, the left lung is hypoventilated, and the blood flows to both lungs are normal and equal. Therefore the $\dot{V}_A/\dot{Q}$ ratio for the right lung is greater than normal, and that for the left lung is less than normal. The alveolar Po_2 (P_{AO_2}) is subnormal (50 mm Hg) for the hypoventi-

lated left lung and supernormal (130 mm Hg) for the hyperventilated right lung. Therefore the Po_2 of the blood in the left and right pulmonary veins will be 50 and 130 mm Hg, respectively.

Equilibration of the blood with a Po_2 of 50 mm Hg will not fully saturate the hemoglobin (see Figure 27-1, *A*). Thus the Co_2 of venous blood draining the left lung will be 17 ml·dl^{-1} (Figure 27-6, *B*) rather than 20 ml·dl^{-1}. However, the hyperventilation of the right lung does not achieve a greater than normal Co_2 of the blood in the right pulmonary veins. The plateau region of the oxyhemoglobin dissociation curve indicates that the hemoglobin is fully saturated with O_2 when Po_2 exceeds about 90 mm Hg. Therefore, even though the Po_2 of the blood in the right pulmonary veins is supernormal (130 mm Hg), the Co_2 is 20 ml·dl^{-1}, which is not greater than normal.

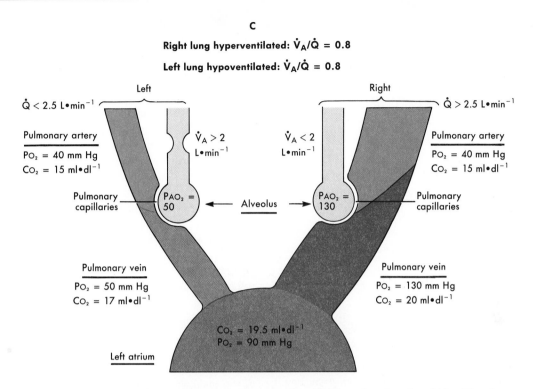

C

Right lung hyperventilated: $\dot{V}_A/\dot{Q} = 0.8$

Left lung hypoventilated: $\dot{V}_A/\dot{Q} = 0.8$

Left   Right

$\dot{Q} < 2.5$ L•min^{-1} $\dot{Q} > 2.5$ L•min^{-1}

Pulmonary artery $\dot{V}_A > 2$ $\dot{V}_A < 2$ Pulmonary artery

$PO_2 = 40$ mm Hg   L•min^{-1} L•min^{-1} $PO_2 = 40$ mm Hg

$CO_2 = 15$ ml•dl^{-1} $CO_2 = 15$ ml•dl^{-1}

Pulmonary $PAO_2 =$ $PAO_2 =$ Pulmonary

capillaries   50   ◄— Alveolus —►   130   capillaries

Pulmonary vein   Pulmonary vein

$PO_2 = 50$ mm Hg $PO_2 = 130$ mm Hg

$CO_2 = 17$ ml•dl^{-1} $CO_2 = 20$ ml•dl^{-1}

$CO_2 = 19.5$ ml•dl^{-1}

$PO_2 = 90$ mm Hg

Left atrium

FIGURE 27-6, cont'd.   C, Blood flow to the hypoventilated left lung is reduced, and blood flow to the hyperventilated right lung is increased, such that the $\dot{V}_A/\dot{Q}$ ratios for both lungs equal 0.8. The volume of fully saturated blood that enters the left atrium from the right lung is substantially greater than the volume of desaturated blood that enters the left atrium from the left lung. Therefore the CO_2 of the systemic arterial blood is closer to full saturation than under conditions in which the $\dot{V}_A/\dot{Q}$ ratios were disparate **(B).**

Because the blood flows to the left and right lungs are equal in Figure 26-7, *B*, the CO_2 of the blood in the left atrium (and systemic arteries) will be the average (18.5 ml·dl^{-1}) of the contents in the right and left pulmonary veins, and its PO_2 will be only 70 mm Hg. Thus, because hemoglobin is fully saturated with O_2 when PO_2 exceeds about 90 mm Hg, hyperventilation of some alveoli cannot compensate adequately for the effects of hypoventilation of other alveoli.

The effects of disparities in $\dot{V}_A/\dot{Q}$ ratios on the CO_2 content of systemic arterial blood are not as pronounced as on the O_2 content, since the CO_2 dissociation curve (see Figure 27-5) does not have the plateau that characterizes the oxyhemoglobin dissociation curve. Thus, although hypoventilation of some alveoli will tend to raise the CO_2 content of the systemic arterial blood, hyperventilation of other alveoli will have the opposite effect, and the two effects will tend to counterbalance each other.

The body possesses mechanisms that can compensate effectively for the reductions in O_2 content in the systemic arterial blood caused by moderate mismatches between ventilation and perfusion. For example, if ventilation is suddenly diminished to a group of alveoli, the PO_2 in those alveoli will decrease, as illustrated in Figure 27-6, *B*. The reduction in alveolar PO_2 will initiate vasoconstriction in the arterioles supplying those alveoli (see Figure 26-6). The consequent reduction in perfusion will partly compensate for the reduction in ventilation and will tend to restore the $\dot{V}_A/\dot{Q}$ ratio to normal.

Figure 27-6, *C*, shows that if the regional pulmonary blood flows adjust so that the $\dot{V}_A/\dot{Q}$ ratios to the hypoventilated and hyperventilated alveoli are equal and normal, the fraction of pulmonary venous blood from poorly ventilated alveoli entering the left atrium will be substantially less than the fraction from well-ventilated alveoli. Thus the O_2 content of left atrial and systemic arterial blood will be closer to normal when the $\dot{V}_A/\dot{Q}$ ratios are equal (Figure 27-6, *C*) than when they are disparate *(B)*.

BIBLIOGRAPHY
Journal Articles

Bidani A: Velocity of CO_2 exchanges in the lungs, Annu Rev Physiol 50:639, 1988.

Crandall ED and Bidani A: Effects of red blood cell HCO_3^-/Cl^- exchange kinetics on lung CO_2 transfer: theory, J Appl Physiol 50:265, 1981.

Grant BJB: Influence of Bohr-Haldane effect on steady-state gas exchange, J Appl Physiol 52:1330, 1982.

Henry RP et al: Rat carbonic anhydrase: activity, localization and isozymes, J Appl Physiol 60:638, 1986.

Jennings ML: Kinetics and mechanism of anion transport in red blood cells, Annu Rev Physiol 47:519, 1985.

Klocke RA: Velocity of CO_2 exchange in blood, Annu Rev Physiol 50:625, 1988.

Lönnerholm G: Pulmonary carbonic anhydrase in the human, monkey, and rat, J Appl Physiol 52:352, 1982.

Schuster KD: Diffusion limitation and limitation by chemical reactions during alveolar-capillary transfer of oxygen-labeled CO_2, Respir Physiol 67:13, 1987.

Sylvester JT et al: Components of alveolar-arterial O_2 gradient during rest and exercise at sea level and high altitude, J Appl Physiol 50:1129, 1981.

Thews G: Theoretical analysis of the pulmonary gas exchange at rest and during exercise, Int J Sports Med 5:113, 1984.

Werlen C et al: Alveolar-arterial equilibration in the lung of sheep, Respir Physiol 55:205, 1984.

Books and Monographs

Baumann R: Interaction between hemoglobins, CO_2 and anions. In Bauer C et al, editors: Biophysics and physiology of carbon dioxide, New York, 1980, Springer-Verlag.

Farhi LE and Tenney SM, editors: Handbook of physiology, section 3: Respiratory system, vol IV: Gas exchange, Bethesda, Md, 1987, American Physiological Society.

Fishman AP, editor: Pulmonary diseases and disorders, ed 2, New York, 1988, McGraw-Hill Book Co.

Klocke RA: Kinetics of pulmonary gas exchange. In West JB, editor: Pulmonary gas exchange, vol I: Ventilation, blood flow, and diffusion, New York, 1980, Academic Press.

Knoche W: Chemical reactions of CO_2 in water. In Bauer C et al, editors: Biophysics and physiology of carbon dioxide, New York, 1980, Springer-Verlag.

Weibel ER: The pathway for oxygen: structure and function in the mammalian respiratory system, Cambridge, 1984, Harvard University Press.

West JB, editor: Pulmonary gas exchange, vol I: Ventilation, blood flow, and diffusion, New York, 1980, Academic Press.

28

Control of Pulmonary Ventilation

NEURAL CONTROL

Centers in the Brain

Normal respiration depends on an intact innervation of the muscles of respiration. The activity of the respiratory muscles is integrated by neuronal centers in the brainstem. These centers in turn are regulated by higher and lower regions of the central nervous system.

Respiration is affected by reflexes that originate from mechanoreceptors in the lungs, respiratory muscles, chest wall, carotid sinuses, and aortic arch. Respiration is also affected by certain chemical changes in the body, notably by changes in the partial pressures of carbon dioxide (PCO_2) and oxygen (PO_2) and in the concentration of H^+ (expressed as pH) in the arterial blood. *Central chemoreceptors* in the medulla oblongata and *peripheral chemoreceptors* in the carotid and aortic bodies detect changes in these chemical constituents.

The arterial PCO_2, PO_2, and pH are held remarkably constant in normal subjects. This fact documents the effectiveness of the neural mechanisms that regulate respiration. The mechanoreceptors are primarily responsible for breath-by-breath adjustments in respiratory activity, whereas the chemoreceptors are mainly responsible for blood gas homeostasis.

Medullary Centers The primary neurons responsible for respiratory activity are located in the reticular formation of the medulla oblongata. Destruction of these neurons or transection through the caudal end of the medulla (*level 4,* Figure 28-1) terminates spon-

taneous respiratory activity; that is, *apnea* ensues (Figure 28-2).

The medullary respiratory neurons occur in two anatomically separate groups, one dorsal and the other ventral (Figure 28-1). The *dorsal respiratory group (DRG)* is a circumscribed group of neurons located dorsomedially in the medulla, near the nucleus of the solitary tract and the dorsal motor nucleus of the vagus. Neurons in the dorsal respiratory group fire mainly during **inspiration;** their discharges increase progressively throughout inspiration. These neurons are responsible for the contraction of the inspiratory muscles (diaphragm and intercostal muscles).

The *ventral respiratory group (VRG)* of neurons extends throughout almost the entire length of the medulla (Figure 28-1). This group of neurons includes the nucleus ambiguus and nucleus retroambiguus. Some cells in the ventral respiratory group fire during **inspiration,** whereas others fire during **expiration.** The precise interrelations among the various functional types of neurons in the dorsal and ventral groups remain to be established.

The respiratory centers in the medulla can bring about rhythmic respiration, even when the brainstem is transected at the rostral end of the medulla (*level 3,* Figure 28-1). The respiratory activity consists of irregular gasps, and the pattern is not affected much by vagal transection (Figure 28-2).

Pontine Centers Centers in the pons modulate the activity of the medullary respiratory centers. Although regions of the brain above the pons can modify respiration substantially, the respiratory activity is very regular (Figure 28-2) when a transection is

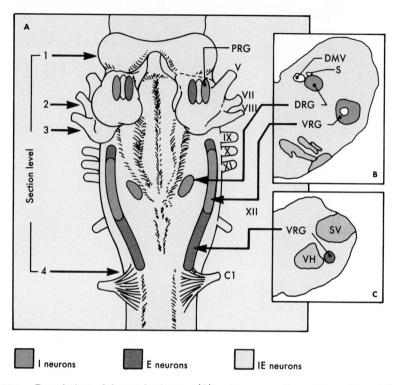

FIGURE 28-1   Dorsal view of the cat brainstem **(A)** and cross sections at the midmedullary **(B)** and caudal medullary **(C)** levels to illustrate locations of the dorsal *(DRG)*, ventral *(VRG)*, and pontine *(PRG)* groups of respiratory neurons. The three types of shading indicate the locations of neurons that fire during inspiration only *(I)*, during expiration only *(E)*, and during both phases *(IE)* of respiration. The Roman numerals along the right side of **A** denote the points of exit of the corresponding cranial nerves. The Arabic numerals along the left side of **A** denote the levels of certain experimental transections that evoke changes in the pattern of respiration; these changes are shown in Figure 28-2. *C1,* first cervical nerve root; *DMV,* dorsal motor nucleus of the vagus; *S,* nucleus of the solitary tract; *SV,* spinal trigeminal nucleus; and *VH,* ventral horn. (Modified from Feldman JL: In Mountcastle VB and Bloom FE, editors: Handbook of physiology, section 1: Nervous system, vol IV, Bethesda, Md, 1986, American Physiological Society.)

made at the rostral end of the pons *(level 1,* Figure 28-1). Thus centers in the pons and medulla, acting in concert, can generate a regular respiratory rhythm.

The *pneumotaxic center,* or *pontine respiratory group (PRG,* Figure 28-1), is located in the rostral half of the pons. These neurons **inhibit inspiratory activity.** Electrical stimulation of this center terminates inspiration early and thereby decreases tidal volume. Conversely, destruction of the pneumotaxic center, or a midpontine transection *(level 2,* Figure 28-1) just caudal to it, exaggerates inspiration and thereby increases the tidal volume (Figure 28-2).

If the midpontine transection *(level 2)* is combined with bilateral vagotomy in an anesthetized animal, inspiration will be much deeper and much more prolonged (Figure 28-2). This respiratory pattern is called *apneusis.*

Higher Brain Centers Higher brain centers frequently modulate the activity of the more primitive controlling centers in the medulla and pons. These higher centers allow the rate and depth of respiration to be controlled voluntarily. The pattern of respiration is greatly altered by various ordinary activities, such as speaking, laughing, crying, eating, defecating,

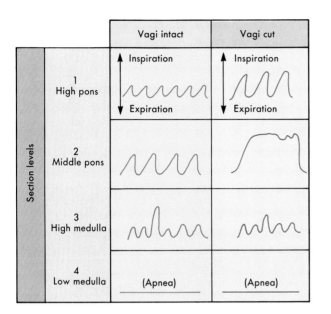

Section levels		Vagi intact	Vagi cut
	1 High pons	Inspiration [curve] Expiration	Inspiration [curve] Expiration
	2 Middle pons	[curve]	[curve]
	3 High medulla	[curve]	[curve]
	4 Low medulla	(Apnea)	(Apnea)

FIGURE 28-2   The respiratory patterns produced by various transections of the brainstem in anesthetized animals. The levels of transection are denoted by the numbers along the left side of the brainstem in Figure 28-1, **A.** The vagi were either intact or transected. When the brainstem was transected at the rostral end of the pons *(level 1)*, the respiratory pattern was regular. However, after the vagi had been cut, the respiration became slower and deeper. When the transection was midpontine *(level 2,* below the pontine respiratory group, PRG), respiration was regular but inspiration was deep. After vagotomy, inspiration was even deeper and very prolonged (apneusis). When the brainstem was transected between the medulla and pons *(level 3),* an irregular, gasping type of respiration was observed. The pattern of respiration was not appreciably influenced by vagal transection. When the brainstem was transected at the caudal end of the medulla *(level 4),* respiration ceased *(apnea).* (Modified from Mitchell RA and Berger AJ: Am Rev Respir Dis 111:206, 1975.)

coughing, and sneezing. Adaptations to changes in environmental temperature may involve pronounced changes in the pattern of respiration. Panting, for example, is a major mechanism in dogs for adapting to high environmental temperatures.

Pulmonary Reflexes

Reflexes that originate from receptors in the tracheobronchial tree and lungs subserve two principal functions: (1) they influence the normal pattern of breathing, and (2) they protect the respiratory system

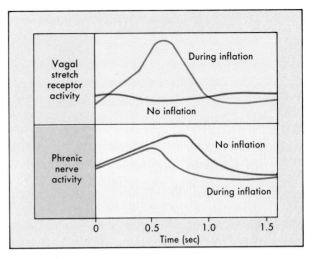

FIGURE 28-3   The effects of lung inflation on inspiratory effort, as reflected by efferent phrenic nerve activity. The lower tracings show that when the lung was not inflated, inspiratory effort was substantially longer and more intense than it was when the lungs were inflated. (See text.) (Modified from Euler C von: Fed Proc 36:2375, 1977.)

from harmful stimuli. The vagus nerves carry most of the afferent fibers for both types of reflexes.

Pulmonary Stretch Receptors   Stretch receptors are located throughout the pulmonary airways. They adapt very slowly to a sustained stretch and provide information about lung volume to the brain. Hering and Breuer showed more than a century ago that an experimentally induced lung inflation could reflexly decrease tidal volume and respiratory frequency in anesthetized animals. This reflex is often referred to as the *Hering-Breuer inflation reflex.*

The experiment illustrated in Figure 28-3 was carried out in an anesthetized animal in which a curare-like drug had been injected to abolish respiratory muscle contractions. Artificial respiration was synchronized with the efferent neural activity *(bottom tracing)* recorded from a phrenic nerve, which innervates the diaphragm. The afferent pulmonary stretch receptor activity associated with lung inflation is also shown *(top tracing).*

Periodically the respirator could be stopped briefly to eliminate lung inflation during one efferent phrenic discharge. The consequent absence of lung inflation is denoted by the flatness of the stretch receptor tracing (Figure 28-3). Note that when the lung was not inflated, the inspiratory activity (denoted by the phrenic neurogram) terminated much later and reached a higher level than when the lung had been inflated by the respirator. Thus, when the pulmonary stretch receptors are excited by lung inflation, efferent respiratory activity is less intense and less prolonged. Therefore, if the respiratory muscles had not been paralyzed, tidal volume would have been diminished and the duration of inspiration abridged by the reflex initiated by lung inflation.

In accordance with these results, the inspiratory excursions recorded in animals with a brainstem transection at level 1 are substantially smaller and more frequent when the vagi are intact than after the vagi have been cut (see Figures 28-1 and 28-2). Similarly, after brainstem transection at level 2, the depth and duration of inspiration are much greater after bilateral vagotomy; the breathing pattern is decidedly apneustic. Thus, when the vagi are intact, afferent information from the pulmonary stretch receptors attenuates the apneusis that would otherwise be induced by midpontine transections.

In anesthetized intact animals bilateral vagotomy increases the depth and decreases the frequency of inspiration, much as it does in animals with high pontine transections (*level 1,* Figure 28-2). In unanesthetized intact animals, however, the effects of bilateral vagotomy on the breathing pattern is much less pronounced. Thus the contribution of the Hering-Breuer inflation reflex to the regulation of respiration in normal, unanesthetized animals and humans is questionable. However, it may protect against excessive lung inflations.

Other reflexes ascribable to the pulmonary stretch receptors have also been demonstrated in anesthetized animals. For example, rapid deflation of the lungs reflexly induces an earlier and deeper inspiration; this reflex has been called the *Hering-Breuer deflation reflex.* Again, its role in the regulation of normal respiratory activity is questionable.

Irritant and Juxtacapillary Receptors The irritant and juxtacapillary receptors rapidly adapt to sustained stimuli. The irritant receptors are distributed in the epithelial linings of the large and small airways, whereas the juxtacapillary receptors lie in the lung parenchyma close to the alveolar capillaries. Their afferent neural fibers travel to the brain in the vagus nerve trunks.

The *irritant receptors* respond to noxious materials, such as cigarette smoke, sulfur dioxide, and ammonia, in the inspired air. Stimulation of irritant receptors in the nose, pharynx, larynx, and trachea can induce sneezing, coughing, and laryngeal spasm. The irritant receptors also respond to certain mediators, such as histamine, that are released in the tissues during allergic reactions.

Stimulation of the irritant receptors reflexly constricts the airways and induces a shallow, rapid respiratory pattern. This reflex minimizes the entry of noxious materials into the alveoli. Both divisions of the autonomic nervous system carry the efferent neural impulses to the airways. Efferent vagal fibers release acetylcholine, which interacts with muscarinic receptors in the smooth muscle cell membranes and constricts the airways. Efferent sympathetic fibers release norepinephrine, which interacts with β-adrenergic receptors in the smooth muscle cell membranes and relaxes the airways.

The *juxtacapillary receptors,* more commonly called *J receptors,* respond to certain anesthetic gases and to certain mediators, such as histamine. They also respond to engorgement of the pulmonary capillaries and to interstitial edema. Stimulation of the J receptors reflexly induces rapid, shallow breathing. This respiratory pattern is often observed in patients with heart failure, pneumonia, or pulmonary embolism. The capillary engorgement and interstitial edema that occur in these conditions undoubtedly stimulate J receptors, which probably contributes to the characteristic breathing pattern.

Reflexes from Muscles and Joints

The muscle spindles in the diaphragm, intercostal muscles, and other muscles of respiration sense the degree of stretch of those muscles and reflexly regulate their extent of contraction. Afferent impulses from the muscles, joints, and tendons in the limbs also influence respiration, particularly during muscular exercise.

CHEMICAL CONTROL

Changes in the arterial concentrations of CO_2, O_2, and H^+ are sensed by the central and peripheral chemoreceptors.

Central Chemoreceptors

Of the various constituents of the arterial blood that regulate respiration, CO_2 is by far the most

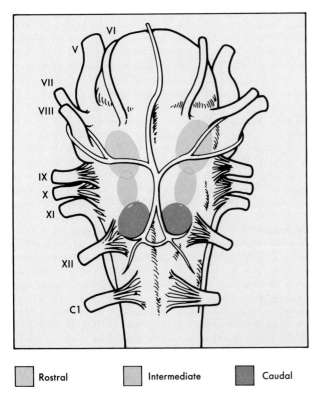

Rostral	Intermediate	Caudal

FIGURE 28-4   Ventral view of the cat brainstem to show the location of the central chemoreceptors, which have been identified in rostral, intermediate, and caudal areas. The chemosensitive cells are believed to lie very close (less than 0.5 mm) to the ventral surface of the medulla. (Modified from Feldman JL: In Mountcastle VB and Bloom FE, editors: Handbook of physiology, section 1: Nervous system, vol IV, Bethesda, Md, 1986, American Physiological Society.)

important. Any rise in arterial P_{CO_2} (Pa_{CO_2}) promptly increases ventilation, which tends to restore Pa_{CO_2} to the normal level. An increase in the Pa_{CO_2} tends to increase the blood pH; that is, it tends to produce an *acidosis*. Therefore the chemoreceptor-induced ventilatory response to the elevation of Pa_{CO_2} constitutes an important component of the body's capacity to regulate its *acid-base balance* (see Chapters 27 and 34).

Approximately 80% of this regulation of the Pa_{CO_2} is accomplished by the central chemoreceptors; the other 20%, by the peripheral chemoreceptors. However, the peripheral chemoreceptors respond much more **rapidly** to a sudden change in Pa_{CO_2} than do the central chemoreceptors.

The location of the central chemoreceptors has not been defined precisely. They appear to be distributed over a fairly extensive portion of the medulla (Figure 28-4), within 0.5 mm of its ventral surface. Rostral, intermediate, and caudal chemosensitive areas have been detected. The central chemoreceptors are distinct from the medullary respiratory centers, but they do communicate with the respiratory centers, especially the dorsal respiratory group, via axonal connections (Figure 28-5).

Ionic constituents of the blood, such as H^+ and bicarbonate (HCO_3^-), cannot influence the central chemoreceptors because these charged substances cannot diffuse through the *blood-brain barrier* into the medullary extracellular spaces. However, CO_2, which has no charge, freely diffuses out of the cerebral capillaries and into the extracellular fluid (Figure 28-5). Thus the CO_2 in the interstitial fluid in contact with the central chemoreceptors can stimulate them. Alternatively, the CO_2 in the extracellular fluid of the brain can diffuse into the cerebrospinal fluid (Figure 28-5). Because the central chemoreceptors are so near the medullary surface, they can also be influenced by the P_{CO_2} in the cerebrospinal fluid.

Peripheral Chemoreceptors

The peripheral chemoreceptors are located in the carotid and aortic bodies. The *carotid bodies* reside at the bifurcations of the common carotid arteries, very close to the carotid sinuses (see Chapter 22). Afferent fibers from the carotid chemoreceptors accompany the afferent fibers from the carotid baroreceptors in the carotid sinus nerve, which is a branch of the glossopharyngeal (IX cranial) nerve. The *aortic bodies* are small clusters of chemosensitive cells that lie adjacent to the aortic arch. Afferent fibers from the aortic bodies travel centrally with the aortic baroreceptor fibers in the vagus nerve trunks.

The peripheral chemoreceptors are sensitive to changes in the P_{CO_2}, P_{O_2}, and pH of the arterial blood. The firing frequency of the peripheral chemoreceptors increases in response to elevations of Pa_{CO_2} (Figure 28-6) and to reductions of arterial pH and Pa_{O_2} (Figure 28-7). The carotid body chemoreceptors are more sensitive to changes in blood gas composition than are the aortic arch chemoreceptors.

The responsiveness of the peripheral chemoreceptors to changes in arterial pH constitutes another of the body's defenses against acid-base disturbances. Acidosis, for example, is associated with a reduction of the blood pH. By stimulating the peripheral chemo-

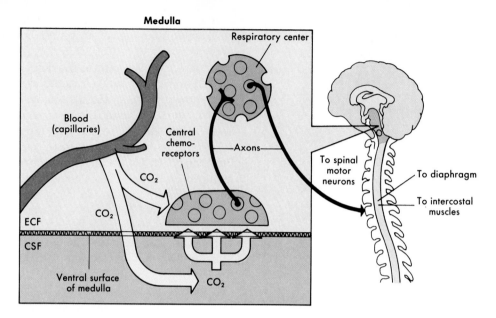

FIGURE 28-5   The role of the central chemoreceptors in the chemical control of respiration. CO_2 diffuses out of the capillaries in the medulla oblongata and into the extracellular fluid *(ECF)* of the medullary tissues. Ultimately the CO_2 released from the medullary capillaries also diffuses into the cerebrospinal fluid *(CSF)*. Cells sensitive to CO_2 (the *central chemoreceptors*) are located very near the ventral surface of the medulla, within 0.5 mm of the CSF. When these cells are stimulated by the CO_2 in the ECF and CSF, their firing frequency increases. This excitation is conveyed to the respiratory centers, mainly the dorsal respiratory group, by way of axonal connections. The neurons in the respiratory centers then transmit this excitation to the respiratory muscle motor neurons in the spinal cord.

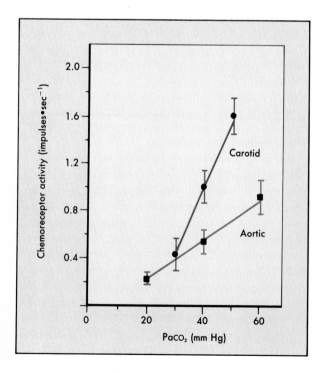

FIGURE 28-6   The firing frequencies (impulses · sec^{-1}) of aortic and carotid chemoreceptors as functions of the arterial Pco_2 *(Paco_2)* in anesthetized cats. Each point represents the mean value obtained from 15 chemoreceptors. The Pao_2 was held at 400 mm Hg so that the chemoreceptors would not be responding to hypoxia. (From Lahiri S et al: J Appl Physiol 51:55, 1981.)

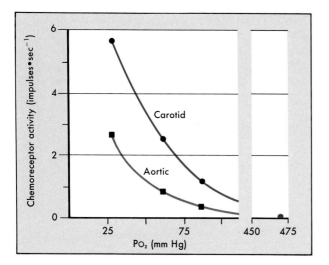

FIGURE 28-7   The firing frequencies (impulses · sec^{-1}) of aortic and carotid chemoreceptors as functions of the arterial P_{O_2} *(PaO$_2$)* in an anesthetized cat. The Pa_{CO_2} was held constant at 29 mm Hg so that the chemoreceptors would not be responding to hypercapnia. (From Lahiri S et al: J Appl Physiol 51:55, 1981.)

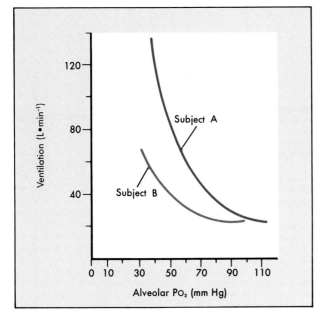

FIGURE 28-8   The effect of changes in alveolar P_{O_2} (and consequently in Pa_{O_2}) on pulmonary ventilation in two normal human subjects. Each subject rebreathed from a large bag. The P_{O_2} of the gas mixture decreased progressively, but its P_{CO_2} was held constant experimentally. (Modified from Rebuck AS and Slutsky AS: Am Rev Respir Dis, 109:345, 1974.)

receptors, ventilation increases and blood P_{CO_2} declines. This tends to restore a normal blood pH (Chapter 34).

The ventilatory response to peripheral chemoreceptor stimulation constitutes the body's principal compensation to *hypoxia*, or reductions in Pa_{O_2}. When a normal subject breathes air with a low P_{O_2} (e.g., ascending to a high altitude), the resultant decrease in Pa_{O_2} elicits a pronounced hyperventilation (Figure 28-8). The increased ventilation is ascribable to stimulation of the peripheral chemoreceptors, and it tends to restore the Pa_{O_2} toward normal. In experimental animals in which the carotid and aortic bodies have been removed, a reduction in the Pa_{O_2} **depresses** the neurons in the medullary respiratory centers and actually **diminishes** pulmonary ventilation.

The carotid and aortic bodies have extremely high metabolic rates and blood flows per unit mass of tissue. The blood flows are actually so great that the O_2 content of the venous blood draining the peripheral chemoreceptors is not much less than the arterial O_2 content. Because of the high blood flow, the peripheral chemoreceptors are relatively insensitive to changes in the O_2 **content** of the arterial blood, but they are sensitive to the **partial pressure** of O_2 in that blood.

BIBLIOGRAPHY
Journal Articles

Berger AJ: Properties of medullary respiratory neurons, Fed Proc 40:2378, 1981.

Cohen MI: Neurogenesis of respiratory rhythm in the mammal, Physiol Rev 59:1105, 1979.

Euler C von: On the central pattern generator for the basic breathing rhythmicity, J Appl Physiol 55:1647, 1983.

Kalia MP: Organization of central control of airways, Annu Rev Physiol 49:595, 1987.

Lahiri S et al: Comparison of aortic and carotid chemoreceptor responses to hypercapnia and hypoxia, J Appl Physiol 51:55, 1981.

Laitinen LA and Laitinen A: Innervation of airway smooth muscle, Am Rev Respir Dis 136:S38, 1987.

Mitchell RA and Berger AJ: Neural regulation of respiration, Am Rev Respir Dis 111:206, 1975.

Richter DW: Generation and maintenance of the respiratory rhythm, J Exp Biol 100:93, 1982.

Sant'Ambrogio G: Nervous receptors of the tracheobronchial tree, Annu Rev Physiol 49:611, 1987.

Schlaefke ME: Central chemosensitivity: a respiratory drive, Rev Physiol Biochem Pharmacol 90:171, 1981.

Books and Monographs

Cherniack NS and Widdicombe JG editors: The respiratory system. In Handbook of physiology, section 3: The control of breathing, part 1, vol II, Bethesda, Md, 1986, American Physiological Society.

Euler C von and Lagercrantz H, editors: Neurobiology of the control of breathing, New York, 1987, Raven Press.

Feldman JL: Neurophysiology of breathing in mammals. In Mountcastle VB and Bloom FE, editors: Handbook of physiology, section 1: Nervous system, vol IV: Intrinsic regulatory systems of the brain, Bethesda, Md, 1986, American Physiological Society.

GASTROINTESTINAL SYSTEM

HOWARD C. KUTCHAI

PART VI

Motility

INTRODUCTION TO THE GASTROINTESTINAL SYSTEM

Components

The gastrointestinal (GI) system consists of the *gastrointestinal tract* and certain *associated glandular organs* that produce secretions that function in the GI tract. The major subdivisions of the GI tract are the mouth, pharynx, esophagus, stomach, duodenum, jejunum and ileum (small intestine), colon, rectum, and anus. Associated glandular organs include salivary glands, liver, gallbladder, and pancreas.

Functions

The major physiologic functions of the GI system are to digest foodstuffs and absorb nutrient molecules into the bloodstream. The activities by which the GI system carries out these functions may be subdivided into motility, secretion, digestion, and absorption. *Motility* refers to movements of the GI tract that mix and circulate its contents and propel these along its length. Usually net propulsion occurs in the *orthograde* direction, that is, away from the mouth and toward the anus. *Retrograde* propulsion does occur, however; vomiting is a notable example. *Secretion* refers to the processes by which the glands associated with the GI tract secrete water and substances into the tract. *Digestion* is defined as the processes by which large ingested molecules are chemically degraded to produce smaller molecules that can be absorbed across the wall of the GI tract. *Absorption*

refers to the processes by which nutrient molecules are absorbed by the GI tract and enter the bloodstream.

STRUCTURE-FUNCTION RELATIONSHIPS

Layers of the Gastrointestinal Tract Wall

The structure of the GI tract varies greatly from region to region, but common features exist in the overall organization of the tissue. Figure 29-1 depicts the general layered structure of the GI tract wall.

The *mucosa* consists of an epithelium, the lamina propria, and the muscularis mucosae. The epithelium is a single layer of specialized cells that lines the lumen of the GI tract. The nature of the epithelium varies greatly from one part of the digestive tract to another. The *lamina propria* consists largely of loose connective tissue containing collagen and elastin fibrils. The lamina propria is rich in several types of glands and contains lymph nodules and capillaries. The *muscularis muscosae* is the thin, innermost layer of intestinal smooth muscle. Contractions of the muscularis mucosae throw the mucosa into folds and ridges.

The *submucosa* consists largely of loose connective tissue with collagen and elastin fibrils. In some regions submucosal glands are present. The larger nerve trunks and blood vessels of the intestinal wall travel in the submucosa.

The *muscularis externa* characteristically consists of two substantial layers of smooth muscle cells: an inner circular layer and an outer longitudinal layer.

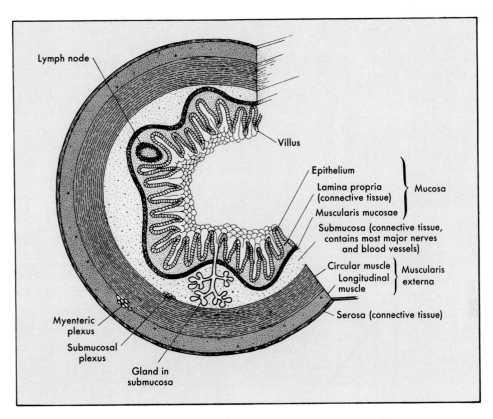

FIGURE 29-1   The general organization of the layers of the GI tract.

Contractions of the muscularis externa mix and circulate the contents of the lumen and propel them along the GI tract.

The wall of the GI tract contains many neurons that are highly interconnected. A dense network of nerve cells in the submucosa is the *submucosal plexus* (Meissner's plexus). The prominent *myenteric plexus* (Auerbach's plexus) is located between the circular and longitudinal smooth muscle layers. The submucosal and myenteric plexuses *(intramural plexuses),* together with the other neurons of the GI tract, constitute the *enteric nervous system,* which helps to integrate the motor and secretory activities of the GI system. If the sympathetic and parasympathetic nerves to the gut are cut, many motor and secretory activities continue because these processes are controlled by the enteric nervous system.

The *serosa,* or adventitia, is the outmost layer and consists mainly of connective tissue covered with a layer of squamous mesothelial cells.

Innervation

Sympathetic Innervation Sympathetic innervation of the GI tract is primarily via postganglionic adrenergic fibers whose cell bodies are in prevertebral and paravertebral plexuses (Figure 29-2). The celiac, superior and inferior mesenteric, and hypogastric plexuses provide postganglionic sympathetic innervation to various segments of the GI tract. Most of the sympathetic fibers terminate in the submucosal and myenteric plexuses. Activation of the sympathetic nerves usually inhibits the motor and secretory activities of the GI system. Some sympathetic fibers innervate blood vessels of the GI tract, causing vasoconstriction. Other sympathetic fibers innervate glandular structures in the wall of the gut.

Relatively few of the sympathetic fibers terminate in the muscularis externa. Stimulation of the sympathetic input to the GI tract inhibits motor activity of the muscularis externa but stimulates contraction of the muscularis mucosae and certain sphincters. The

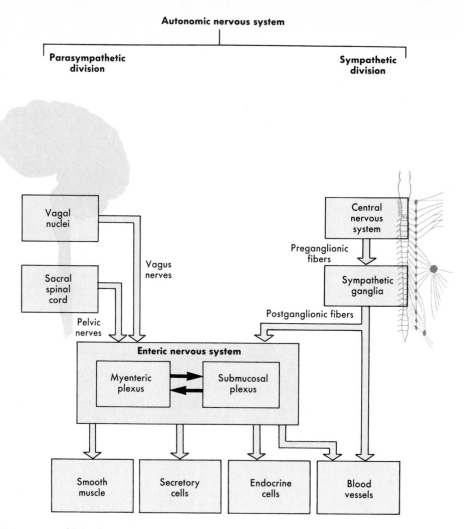

FIGURE 29-2 Major features of the autonomic innervation of the GI tract. In most cases the autonomic nerves influence the functions of the GI tract by modulating the activities of neurons of the enteric nervous system. (Redrawn from Costa M and Furness JB: Br Med Bull 38:247, 1982.)

inhibitory effect of the sympathetic nerves on the muscularis externa is not a direct action on the smooth muscle cells, since few sympathetic nerve endings are in the muscularis externa. Rather, the sympathetic nerves act to influence neural circuits in the intrinsic plexuses that provide input to the smooth muscle cells. This effect may be reinforced by the action of the sympathetic nerves in reducing blood flow to the muscularis externa. Other fibers that travel with the sympathetic nerves may be cholinergic; still others release neurotransmitters that remain to be identified.

Parasympathetic Innervation Parasympathetic innervation of the GI tract down to the level of the transverse colon is provided by branches of the vagus

nerves (Figure 29-2). The remainder of the colon, the rectum, and the anus receive parasympathetic fibers from the pelvic nerves. These parasympathetic fibers are preganglionic and predominantly cholinergic. Other fibers that travel in the vagus and its branches have other transmitters that in most cases have not been identified. The parasympthetic fibers terminate predominantly on the ganglion cells in the intramural plexuses. The ganglion cells then directly innervate the smooth muscle and secretory cells of the GI tract. Stimulation of parasympathetic nerves usually stimulates the motor and secretory activity of the gut.

Enteric Nervous System The myenteric and submucosal plexuses are the most well-defined plexuses in the wall of the GI tract (Figure 29-3). The plexuses are networks of nerve fibers and ganglion cell bodies. Some of the incoming axons are preganglionic, whereas others are postganglionic sympathetic fibers. Interneurons in the plexuses connect afferent sensory fibers with efferent neurons to smooth muscle and secretory cells to form reflex arcs that are wholly within the GI tract wall. Consequently the myenteric and submucosal plexuses can control much coordinated activity in the absence of extrinsic innervation of the GI tract. Axons of plexus neurons innervate gland cells in the mucosa and submucosa, smooth muscle cells in the muscularis externa and muscularis mucosae, and intramural endocrine and exocrine cells.

Afferent Fibers Afferent fibers in the gut provide the afferent limbs of both local and central reflex arcs (Figure 29-4). Chemoreceptor and mechanoreceptor endings are present in the mucosa and muscularis

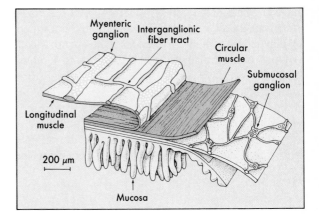

FIGURE 29-3   Enteric neurons of the submucosal and myenteric plexuses in the wall of the GI tract. The plexuses consist of ganglia interconnected by fiber tracts. (Redrawn from Wood JD: In Johnson RL, editor: Physiology of the gastrointestinal tract, ed 2, New York, 1987, Raven Press.)

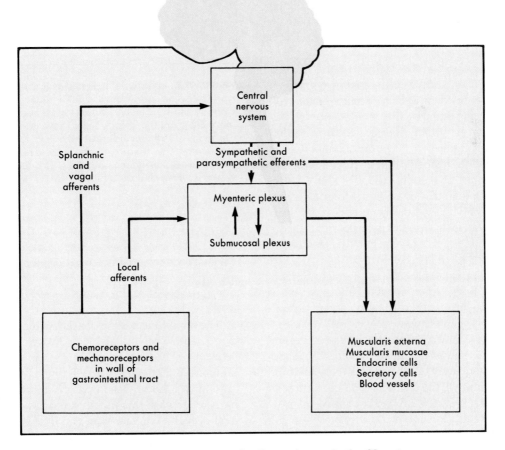

FIGURE 29-4   Local and central reflex pathways in the GI system.

externa. The cell bodies of certain other sensory receptors are located in the myenteric and submucosal plexuses. The axons of some of these receptor cells synapse with other cells in the plexuses to mediate local reflex activity. Other such receptors send their axons back to the central nervous system. The cell bodies of some of the sensory neurons are located more centrally. Many sensory afferent fibers arise from the GI tract. The complex afferent and efferent innervation allows for fine control of secretory and motor activities.

Electrophysiology of Gastrointestinal Smooth Muscle Cells

Resting Membrane Potential. The resting membrane potential of GI smooth muscle cells ranges from approximately −40 to −80 mV. As discussed in Chapter 2, the relative membrane conductances of K^+, Na^+, and Cl^- are important in determining the resting membrane potential. Compared with skeletal muscle cells, GI smooth muscle cells have a higher ratio of Na^+ to K^+ conductance. This contributes to the somewhat lower resting membrane potential of GI smooth muscle.

The electrogenic Na^+, K^+ pump contributes significantly to the resting membrane potential in smooth muscle. Because 3 Na^+ are extruded for every 2 K^+ taken up, the pump produces a net outward flow of positive charge, which contributes to the membrane potential. If the Na^+, K^+ pump of the teniae coli is inhibited with ouabain, the resting membrane potential changes from about −60 to −40 mV. When ouabain is washed out, the resting potential returns to almost −60 mV. Based on this observation, about one third of the resting membrane potential in this type of GI smooth muscle results from the electrogenicity of the Na^+,K^+-ATPase.

In most other excitable tissues the resting membrane potential is rather constant in time. In GI smooth muscle the resting membrane potential characteristically varies in time.

Slow waves are oscillations of the resting membrane potential and are characteristic of GI smooth muscle. The frequency of the oscillations varies from about 3 per minute in the stomach to 12 per minute in the duodenum. Figure 29-5 shows sinusoidal slow waves in the rabbit jejunum. Slow waves also are referred to as the *basic electrical rhythm*.

In each segment of the GI tract the basic electrical

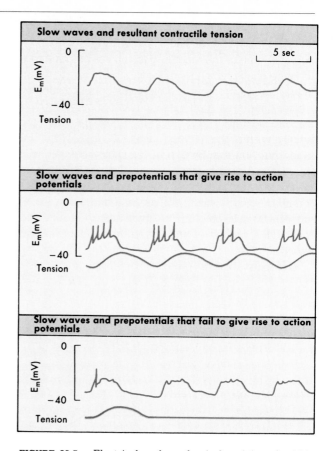

FIGURE 29-5   Electrical and mechanical activity of rabbit small intestinal, smooth muscle. The upper tracing in each panel is the transmembrane electrical potential difference (millivolts), and the lower tracing shows contractile tension. The range of different electrical behaviors in a single tissue is noteworthy. The tissue contracts only in response to action potentials. E_m, membrane potential. (Modified from Bortoff A: Am J Physiol 201:203, 1961.)

rhythm of some cells is faster than that of the other cells. The faster cells serve as *pacemakers* for the slow waves in the region. Because the smooth muscle cells are well coupled electrically, the wave of depolarization (or repolarization) spreads throughout that segment of the gut.

The amplitude and, to a lesser extent, the frequency of the slow waves can be modulated by the activity of intrinsic and extrinsic nerves and by circulating hormones. In general, circulating epinephrine and norepinephrine released from sympathetic nerve terminals decrease the amplitude of the slow waves or abolish them, whereas stimulation of parasympathetic nerves increases the size of the slow waves.

Depending on the excitability of the smooth muscle cells, the slow waves may induce action potentials. If the peak of the slow wave is above threshold for the cells to fire action potentials, one or more action potentials may be triggered during the peak of the slow wave (Figure 29-5). Typically the slow waves themselves do not elicit contractions of the smooth muscle layers. Contraction is evoked by the action potentials that are intermittently triggered near the peaks of the slow waves. The greater the number of action potentials that occur at the peak of a slow wave, the more intense is the contraction of the smooth muscle.

Action potentials in GI smooth muscle are more prolonged (10 to 20 msec) than those of skeletal muscle and have little or no overshoot. The rising phase of the action potential is caused by ion flow through channels that conduct both Ca^{++} and Na^{+} and are relatively slow to open. The Ca^{++} that enters the cell during the action potential plays a significant role in initiating contraction. Repolarization is aided by a delayed increase in K^{+} conductance.

When the membrane potential of GI smooth muscle reaches threshold, a train of action potentials (1 to 10 per second) occurs (Figure 29-5). The extent of depolarization of the cells and the frequency of action potentials are enhanced by excitatory hormones and compounds liberated from nerve endings. Inhibitory hormones and neuroeffector substances hyperpolarize the smooth muscle cells and may abolish action potential spikes.

Electrical Coupling Between Smooth Muscle Cells

Neighboring cells are said to be *well coupled* electrically if a perturbation of the membrane potential of one cell spreads rapidly, and with little decrement, to the other cell. The smooth muscle cells of the muscularis externa are well coupled.

The smooth muscle cells of the circular layer are better coupled than those of the longitudinal layer. The cells of the circular layer are joined by frequent low-resistance gap junctions that allow the spread of electrical current from one cell to another.

Because the electrical resistance of membranes is much higher than the resistance of cytoplasm, the resistance along the long axis of smooth muscle cells is less than that in the transverse direction. Thus an electrical depolarization in the *circular muscle* is readily conducted circumferentially around the gut

wall and quickly depolarizes the ring of circular muscle that includes the original site of depolarization. The ring of depolarization then spreads more slowly along the long axis of the gut. The *longitudinal smooth muscle* cells of the muscularis externa are also well coupled electrically. Therefore an electrical disturbance is rapidly propagated along the long axis of the gut in the longitudinal layer.

Contractility of Intestinal Smooth Muscle: Relationship Between Membrane Potential and Tension

Gastrointestinal smooth muscle cells fire action potentials (Figure 29-5), and the cells contract in response to these potentials. The action potentials usually occur in bursts at the peaks of the slow waves and cause phasic contractions that are superimposed on the baseline level of contraction. Because smooth muscle cells contract rather slowly (about one-tenth as fast as skeletal muscle), the individual contractions caused by each action potential in a burst are not visible as distinct twitches; rather, they sum temporally to produce a smoothly increasing level of tension. The increase in tension in response to a burst of action potentials is proportional to the number of action potentials in the burst. Gastric smooth muscle is exceptional in that it contracts in response to the depolarizing phase of the slow waves, even in the absence of action potentials.

Between bursts of action potentials the tension developed by GI smooth muscle falls, but not to zero. This nonzero resting, or baseline, tension developed by the smooth muscle is called *tone*. The tone of GI smooth muscle may be altered by neuroeffectors, hormones, and drugs.

INTEGRATION AND CONTROL OF GASTROINTESTINAL MOTOR ACTIVITIES

Control of the contractile activities of GI smooth muscle involves the central nervous system, the intrinsic plexuses of the gut, humoral factors, and electrical coupling among the smooth muscle cells. The GI tract displays much intrinsic control: the intramural plexuses can mediate control and integration of much of the gut's contractile behavior without intervention by the central nervous system. The autonomic nervous system typically modulates the patterns of muscular and secretory activity that are controlled more directly by the enteric nervous system.

Neuromuscular Interactions

Neuromuscular interactions in the GI tract do not involve true neuromuscular junctions with specialization of the postjunctional membrane, as occurs at the motor endplate. The neurons of the intramural plexuses send axons to the smooth muscle layers, and each axon branches extensively to innervate many smooth muscle cells.

The circular smooth muscle layer of the muscularis externa is heavily innervated by nerve terminals that form close association with the plasma membranes of the smooth muscle cells. The predominant innervation is by fibers that *inhibit* the electrical and contractile activity of the smooth muscle cells.

The longitudinal smooth muscle cells are much less richly innervated by the neurons of the intrinsic plexuses, and the neuromuscular contacts are not so intimate. In contrast to the innervation of the circular layer, the longitudinal layer appears to be totally devoid of inhibitory nerve endings. The sparse excitatory nerve fibers are predominantly cholinergic.

Thus neural activity of the plexus neurons inhibits the circular muscle but stimulates the longitudinal muscle. This is physiologically relevant because the two layers are antagonistic in the sense that contraction of one layer opposes contraction of the other; contraction of the circular layer makes the gut longer and thinner, whereas contraction of the longitudinal layer makes the gut shorter and wider.

Enteric Nervous System

The plexuses of the GI tract wall function as a semiautonomous nervous system that controls the motor and secretory activities of the digestive system. The enteric nervous system of the large and small intestines alone contains about 10^8 neurons, about as many neurons as in the spinal cord. Figure 29-3 depicts the myenteric and submucosal plexuses and their locations in the wall of the intestine. Both plexuses consist of ganglia that are interconnected by tracts of fine, unmyelinated nerve fibers. The neurons in the ganglia include sensory neurons, with their sensory endings in GI tract wall. Neurons sensitive to mechanical deformation, to particular chemical stimuli, and to temperature have been identified. Some of

Neurotransmitters or Neuromodulatory Substances in the Enteric Nervous System	
Acetylcholine	Somatostatin
Norepinephrine	Vasoactive intestinal
5-Hydroxytryptamine	polypeptide (VIP)
(5-HT)	Enkephalin
Purine nucleotides	Substance P
Dopamine	Bombesin
Neurotensin	γ-Aminobutyric acid
Cholecystokinin (CCK)	(GABA)
Glycine	Gastrin
Motilin	Histamine
Angiotensin	Thyrotropin-releasing
Secretin	hormone
Galanin	Gastrin-releasing peptide (GRP)
Neuropeptide y	Prostaglandins
	Peptide, histidine, isoleucine (PHI)

Modified from Wood JD: Physiology of the enteric nervous system. In Johnson RL, editor: Physiology of the gastrointestinal tract, New York, 1987, Raven Press, p. 69.

the neurons in the enteric ganglia are effector neurons that send axons to smooth muscle cells of the circular or longitudinal layers, to secretory cells of the GI tract, or to GI blood vessels. Many of the neurons in the enteric ganglia are interneurons. They are part of a network of neurons that integrates the sensory input to the ganglia and formulates the output of the effector neurons.

Neuromodulatory Substances The number of neuromodulatory substances present in the GI tract wall is similar to that of such compounds in the brain; most of the known neuromodulators are present in both gut and brain. The box lists some of the neuroactive substances present in the GI tract. The functions of some of these neuroactive substances in the central nervous system are discussed in Chapter 4. Table 29-1 lists the densities of enteric neurons that contain particular candidate transmitters and neuromodulators.

Having described the basic properties of GI smooth muscle and its innervation by intrinsic and extrinsic neuronal circuits, we now discuss the mechanical activities that occur in the different regions of the GI tract.

Table 29-1 Number and Type of Neurons in 1 cm Length of Guinea Pig Small Intestine

	Myenteric Plexus	Submucous Plexus
Total number	10,000	7200
Substance P	350	820
VIP	240	3060
Somatostatin	470	1260
Enkephalin	2450	0
5-HT	200	0
Amine handling	50	850

From Costa M and Furness JB: Br Med Bull 38:247, 1982.

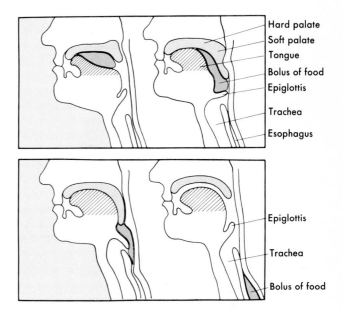

FIGURE 29-6 Major events involved in the swallowing reflex.

CHEWING (MASTICATION)

Chewing can be carried out voluntarily, but it is more frequently a reflex behavior. Chewing lubricates the food by mixing it with salivary mucus, mixes starch-containing food with the salivary amylase, and subdivides the food so that it can be mixed more readily with the digestive secretions of the stomach and duodenum.

SWALLOWING

Swallowing can be initiated voluntarily, but thereafter it is almost entirely under reflex control. The swallowing reflex is a rigidly ordered sequence of events that propels food from the mouth to the stomach. At the same time it inhibits respiration and prevents the entrance of food into the trachea (Figure 29-6). The afferent limb of the swallowing reflex begins with touch receptors, most notably those near the opening of the pharynx. Sensory impulses from these receptors are transmitted to certain areas in the medulla. The central integrating areas for swallowing lie in the medulla and lower pons; they are collectively called the *swallowing center*. Motor impulses travel from the swallowing center to the musculature of the pharynx and upper esophagus via various cranial nerves.

The oral, or voluntary, phase of swallowing is initiated by separating a bolus of food from the mass in the mouth with the tip of the tongue. The bolus to be swallowed is moved upward and backward in the mouth by pressing first the tip of the tongue and later the more posterior portions of the tongue as well against the hard palate. This forces the bolus into the pharynx, where it stimulates the tactile receptors that initiate the swallowing reflex.

The pharyngeal phase of swallowing involves the following sequence of events, which occur in less than 1 second:

1. The soft palate is pulled upward, and the palatopharyngeal folds move inward toward one another. This prevents reflux of food into the nasopharynx and provides a narrow passage through which food moves into the pharynx.
2. The vocal cords are pulled together. The larynx is moved upward against the epiglottis. These actions prevent food from entering the trachea.
3. The upper esophageal sphincter relaxes to receive the bolus of food. Then the pharyngeal superior constrictor muscles contract strongly to force the bolus deeply into the pharynx.
4. A peristaltic wave is initiated with contraction of the pharyngeal superior constrictor muscles, and moves toward the esophagus. This forces the bolus of food through the relaxed upper esophageal sphincter.

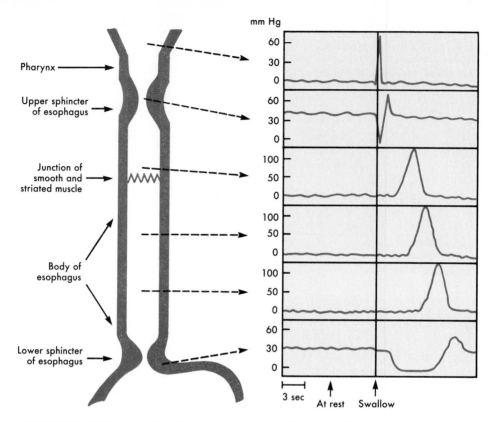

FIGURE 29-7   Pressures in the pharynx, esophagus, and esophageal sphincters during swallowing. Note the reflex relaxation of the upper and lower esophageal sphincters and the timing of the relaxation. (Redrawn from Christensen JL: In Christensen J and Wingate DL, editors: A guide to gastrointestinal motility, Bristol, UK, 1983, John Wright and Sons.)

During the pharyngeal stage of swallowing, respiration is reflexly inhibited.

The esophageal phase of swallowing is controlled mainly by the swallowing center. After the bolus of food passes the upper esophageal sphincter, the sphincter reflexly constricts. A peristaltic wave then begins just below the upper esophageal sphincter and traverses the entire esophagus in about 10 seconds (Figure 29-7). This initial wave of peristalsis, called *primary peristalsis,* is controlled by the swallowing center. The peristaltic wave travels down the esophagus at 3 to 5 cm/sec. If the primary peristalsis is insufficient to clear the esophagus of food, distension of the esophagus initiates another peristaltic wave, which begins above the site of distension and moves downward. This latter type of peristalsis, termed *secondary peristalsis,* is partly mediated by the enteric nervous system; it occurs, although more weakly, even after the extrinsic nerves to the esoph-

agus are cut. Input from esophageal sensory fibers to the central and enteric nervous systems modulates esophageal peristalsis.

ESOPHAGEAL FUNCTION

After food is swallowed, the esophagus functions as a conduit to move the food from the pharynx to the stomach. It is important to prevent air from entering at the upper end of the esophagus and to keep corrosive gastric contents from refluxing back into the esophagus at its lower end. Reflux is particularly problematic because the pressure in the thoracic esophagus is close to intrathoracic pressure, which is less than atmospheric pressure and thus less than intraabdominal pressure.

In the upper third of the esophagus both the inner circular and the outer longitudinal muscle layers are striated. In the lower third, the muscle layers are composed entirely of smooth muscle cells. In the middle third, skeletal and smooth muscles coexist, with a gradient from all skeletal above to all smooth below.

The esophageal musculature, both striated and smooth, is mainly innervated by branches of the vagus nerve. Somatic motor fibers of the vagus form motor endplates on striated muscle fibers. Visceral motor nerves are preganglionic parasympathetic fibers that synapse primarily on the nerve cells of the myenteric plexus. Neurons of the myenteric plexus directly innervate the smooth muscle cells of the esophagus and communicate with one another. The neural circuits that control the esophagus are schematized in Figure 29-8.

The upper and lower ends of the esophagus function as sphincters to prevent the entry of air and gastric contents, respectively, into the esophagus. The sphincters are known as the upper and lower esophageal sphincters. The pressure in the upper esophageal sphincter is about 40 mm Hg at rest (see Figure 29-7). The lower 1 to 2 cm of the esophagus functions as a sphincter. In normal individuals the pressure at the lower esophageal sphincter is always greater than that in the stomach.

The lower esophageal sphincter opens when a wave of esophageal peristalsis begins. The opening is vagally mediated. In the absence of esophageal peristalsis the sphincter must remain tightly closed to prevent reflux of the gastric contents, which would cause esophagitis and the sensation of heartburn.

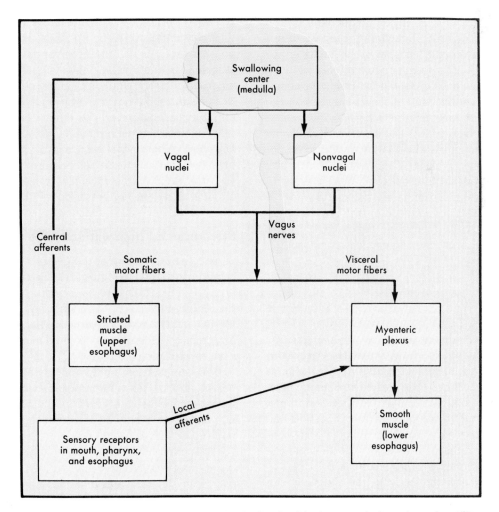

FIGURE 29-8   Local and central neural circuits involved in the control of esophageal motility.

Control of the Lower Esophageal Sphincter

Control of the Tone The resting pressure in the lower esophageal sphincter is about 30 mm Hg. The tonic contraction of the circular musculature of the sphincter is regulated by nerves, both intrinsic and extrinsic, and by hormones and neuromodulators. A significant fraction of basal tone in this sphincter is mediated by vagal cholinergic nerves. Stimulating sympathetic nerves to the sphincter also causes contraction. Significant sphincter tone and regulation of that tone persist when the extrinsic nerves to the sphincter are sectioned. Thus the enteric nervous system helps regulate the function of the lower esophageal sphincter. Several neuroactive peptides can influence the tone of the lower esophageal sphincter, but their physiologic roles remain to be determined.

Relaxation The intrinsic and extrinsic innervation of the lower esophageal sphincter is both excitatory and inhibitory. A major component of the sphincter's relaxation that occurs in response to primary peristalsis in the esophagus is mediated by vagal fibers inhibitory to the circulatory muscle.

In some individuals the sphincter fails to relax sufficiently during swallowing to allow food to enter the stomach, a condition known as *achalasia*. Therapy for achalsia may involve surgically weakening the lower esophageal sphincter. Individuals with *diffuse esophageal spasm* have prolonged and painful contraction of the lower part of the esophagus after swallowing instead of the normal esophageal peristaltic wave. In individuals with *incompetence* of the lower esophageal sphincter, gastric juice can reflux back up into the lower esophagus, causing erosion of the esophageal mucosa.

GASTRIC MOTILITY

The major functions of gastric motility are (1) to allow the stomach to serve as a reservoir for the large volume of food that may be ingested at a single meal, (2) to fragment food into smaller particles and to mix food with gastric secretions so that digestion can begin, and (3) to empty gastric contents into the duodenum at a controlled rate. Figure 29-9 shows the major anatomic divisions of the stomach.

The *fundus* and the *body* of the stomach can accommodate volume increases as large as 1.5 L without a great increase in intragastric pressure.

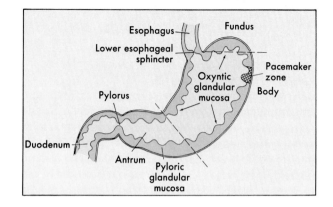

FIGURE 29-9   The major anatomic subdivisions of the stomach.

Contractions of the fundus and body are normally weak, so that much of the gastric contents remains relatively unmixed for long periods. Thus the fundus and the body serve the reservoir functions of the stomach. In the *antrum,* however, contractions are vigorous and thoroughly mix antral chyme with gastric juice and subdivide food into smaller particles. The antral contractions serve to empty the gastric contents in small squirts into the duodenal bulb. The rate of gastric emptying is adjusted by several mechanisms so that chyme is not delivered to the duodenum too rapidly. The physiologic mechanisms that underlie this behavior are discussed later.

Structure and Innervation of the Stomach

The basic structure of the gastric wall follows the scheme presented in Figure 29-1. The circular muscle layer of muscularis externa is more prominent than the longitudinal layer. The muscularis externa of the fundus and the body is relatively thin, but that of the antrum is considerably thicker and increases in thickness toward the pylorus.

The stomach is richly innervated by extrinsic nerves and by the neurons of the submucosal and myenteric plexuses. Axons from the cells of the intramural plexuses innervate smooth muscle and secretory cells.

Parasympathetic innervation comes via the vagus nerves and sympathetic innervation from the celiac plexus. In general, parasympathetic nerves stimulate gastric smooth muscle motility and gastric secretions, whereas sympathetic activity inhibits these functions. Numerous sensory afferent fibers leave the stomach in the vagus nerves, and some travel with sympathetic nerves. Other fibers are the afferent links of intrinsic reflex arcs via the intramural plexuses of the stomach. Some of these afferent fibers relay information about intragastric pressure, gastric distension, intragastric pH, or pain.

Responses to Gastric Filling

When a wave of esophageal peristalsis begins the lower esophageal sphincter reflexly relaxes. This is followed by *receptive relaxation* of the fundus and body of the stomach. The stomach also will relax if it is directly filled with gas or liquid. The nerve fibers in the vagi are a major efferent pathway for reflex relaxation of the stomach. The vagal fibers that mediate this response have *vasoactive intestinal polypeptide (VIP)* as their transmitter. When the stomach relaxes in response to filling, the sensory afferent fibers that report gastric distension constitute the afferent limb of the response. The smooth muscle of the fundus and body of the stomach have a lower resistance to stretch than does the smooth muscle of the antrum. This property enhances the ability of the fundus and body to accommodate a large increase in volume with little increase in intragastric pressure.

Mixing and Emptying of Gastric Contents

The muscle layers in the fundus and body are thin; weak contractions characterize these parts of the stomach. As a result, the contents of the fundus and the body tend to form layers that are based on their density. The gastric contents may remain unmixed for as long as 1 hour after eating. Fats tend to form an oily layer on top of the other gastric contents. Consequently, fats are emptied later than other gastric contents. Liquids can flow around the mass of food contained in the body of the stomach and are emptied more rapidly into the duodenum. Large or indigestible particles are retained in the stomach for a longer period (Figure 29-10).

Gastric contractions usually begin in the middle of the body of the stomach and travel toward the pylorus. The contractions increase in force and velocity as they approach the gastroduodenal junction. As a result, the major mixing activity occurs in the antrum, the contents of which are mixed rapidly and thoroughly with gastric secretions. Immediately after eat-

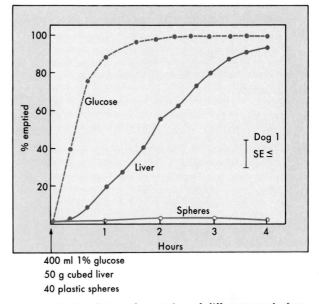

400 ml 1% glucose
50 g cubed liver
40 plastic spheres

FIGURE 29-10   Rates of emptying of different meals from dog stomach. A solution (1% glucose) is emptied faster than a digestible solid (cubed liver). An indigestible solid (7 mm plastic spheres) remains in the stomach under these conditions. (Redrawn from Hinder RA and Kelly KA: Am J Physiol 233:E335, 1977.)

ing, the antral contractions are relatively weak, but as digestion proceeds, they become more forceful. Typically the frequency of gastric contraction is about three per minute.

Fed Versus Fasted Animals After an animal eats, regular contractions of the antrum occur at the frequency of the gastric slow waves. As discussed later, the rate of gastric emptying is regulated by feedback mechanisms that diminish the force of antral contractions. Contractions of the antrum after ingestion of food vary from moderately forceful to weak.

In a fasted animal a different pattern of antral contractions occurs. The antrum is quiescent for 1 to 2 hours; then a short period of intense electrical and motor activity lasts for 10 to 20 minutes. This activity is characterized by strong contractions of the antrum with a relaxed pylorus. During this period even large chunks of material that remain from the previous meal are emptied from the stomach. The period of intense contractions is followed by another 1 to 2 hours of quiescence. This cycle of contractions in the stomach is part of a pattern of contractile activity that periodically sweeps from the stomach to the terminal ileum during fasting. This cyclic contractile activity is known as the *migrating myoelectric complex (MMC)* and is discussed later.

Electrical Activity that Underlies Gastric Contractions

The gastric peristaltic waves occur at the frequency of the gastric slow waves that are generated by a *pacemaker zone* (see Figure 29-9) near the middle of the body of the stomach. These waves are conducted toward the pylorus. In humans the frequency of slow waves is about three per minute.

The gastric slow wave is triphasic (Figure 29-11). Its shape resembles the action potentials in cardiac muscle. However, the gastric slow wave lasts about 10 times longer than the cardiac action potential and does not overshoot. The initial, rapid depolarizing phase of the slow wave is probably caused by the entry of Ca^{++} via voltage-gated Ca^{++} channels. The plateau of the slow wave is probably caused by the entry of Ca^{++} and Na^+ through slower voltage-gated channels. Repolarization of the membrane is associated with a delayed increase in K^+ conductance

Gastric smooth muscle contracts when the depolarization during the slow wave exceeds the threshold for contraction (Figure 29-11). The greater the extent of depolarization and the longer the cell remains depolarized above the threshold, the greater the force of contraction. In this respect, gastric smooth muscle differs from intestinal smooth muscle, which contracts only in response to action potential spikes that are generated during the plateau of the slow wave (see Figure 29-5). In the gastric antrum, action potential spikes may occur during the plateau phase; when action potentials occur, the resulting contraction is stronger than in the absence of action potentials. Acetylcholine and the hormone gastrin stimulate gastric contractility by increasing the amplitude and duration of the plateau phase of the gastric slow wave. Norepinephrine has the opposite effect.

Gastroduodenal Junction

The pylorus separates the gastric antrum from the first part of the duodenum, the duodenal bulb (or cap). It is debatable whether the pylorus is a true anatomic sphincter, but it functions physiologically as a sphincter. The circular smooth muscle of the pylorus forms two ringlike thickenings that are followed by a connective tissue ring separating pylorus from duodenum. The mucosa, submucosa, and muscle layers of the pylorus and duodenal bulb are separate, with the exception of a few longitudinal muscle fibers that cross over the junction. The myenteric plexuses of the pylorus and duodenal bulb are continuous.

The duodenum has a basic electrical rhythm of 10 to 12 per minute, far faster than the 3 per minute of the stomach. The duodenal bulb is influenced by the basic electrical rhythms of both the stomach and the postbulbar duodenum. It thus contracts somewhat irregularly. However, the antrum and duodenum are coordinated; when the antrum contracts, the duodenal bulb is relaxed.

The essential functions of the gastroduodenal junction are (1) to allow the carefully regulated emptying of gastric contents at a rate consistent with the ability of the duodenum to process the chyme and (2) to prevent regurgitation of duodenal contents back into the stomach. The gastric mucosa is highly resistant to acid but may be damaged by bile. The duodenal mucosa has the opposite properties. Thus too rapid gastric emptying may lead to duodenal ulcers, whereas regurgitation of duodenal contents may contribute to gastric ulcers.

The pylorus is densely innervated by both vagal and sympathetic nerve fibers. Sympathetic fibers increase the constriction of the sphincter. Vagal

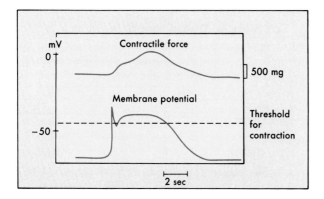

FIGURE 29-11   Relationship between contraction of smooth muscle of dog stomach *(upper tracing)* and intracellularly recorded slow wave *(lower tracing)*. Note the triphasic shape of the slow wave in gastric smooth muscle. Contraction occurs when the depolarizing phase of the slow wave exceeds the threshold for contraction, even though there are no action potential spikes on the plateau of the slow wave. (Redrawn from Szurszewski J: Electrical basis for gastrointestinal motility. In Johnson LR, editor, Physiology of the gastrointestinal tract, New York, 1981, Raven Press, pp. 1435–1466.)

fibers are both excitatory and inhibitory to pyloric smooth muscle. Cholinergic vagal fibers stimulate constriction of the sphincter. Inhibitory vagal fibers release another transmitter, probably VIP, that relaxes the sphincter.

The hormones cholecystokinin, gastrin, gastric inhibitory peptide, and secretin all elicit constriction of the pyloric sphincter.

Regulation of Gastric Emptying Rate

The emptying of gastric contents is regulated by both neural and humoral mechanisms. The duodenal and jejunal mucosa have receptors that sense acidity, osmotic pressure, and fat content (Figure 29-12).

The presence of fatty acids or monoglycerides (products of fat digestion) in the duodenum dramatically decreases the rate of gastric emptying. The chyme that leaves the stomach is usually hypertonic,

and it becomes more hypertonic because of the action of the digestive enzymes in the duodenum. Gastric emptying is retarded by hypertonic solutions in the duodenum, pH less than 3.5 in the duodenum, and the presence of amino acids and peptides in the duodenum. As a result of these mechanisms:

1. Fat is not emptied into the duodenum at a rate greater than that at which it can be emulsified by the bile acids and lecithin of the bile.
2. Acid is not dumped into the duodenum more rapidly than it can be neutralized by pancreatic and duodenal secretions and by other mechanisms.
3. The rates at which the other components of chyme are presented to the small intestine usually correspond to the rates at which the small intestine can process those components.

Mechanisms The slowing of gastric emptying in response to fatty acids, low pH, or hypertonicity of

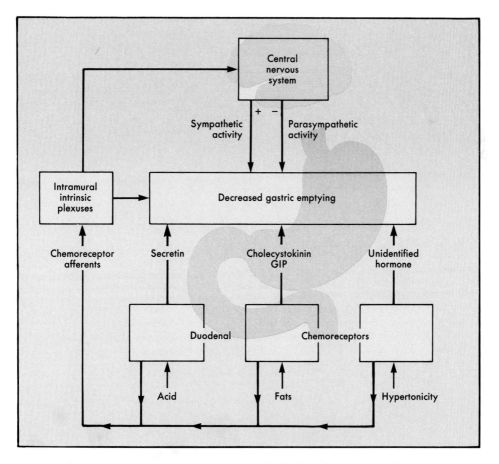

FIGURE 29-12   Neural and hormonal inhibition of gastric emptying.

duodenal contents is mediated by neural and hormonal mechanisms.

1. *Acid in the duodenum.* In response to acid in the duodenum, the force of gastric contractions promptly decreases and duodenal motility increases. This response has neural and hormonal components. The presence of acid in the duodenum releases *secretin,* which diminishes the rate of gastric emptying by inhibiting antral contractions and stimulating contraction of the pyloric sphincter. Secretin also stimulates the output of the bicarbonate-rich secretions of the pancreas and liver, which buffer the duodenal pH. (See Figure 29-12.)

2. *Fat-digestion products.* The presence of fat-digestion products in the duodenum and jejunum decreases the rate of gastric emptying. This response results mainly from release of *cholecystokinin* from the duodenum and jejunum. Cholecystokinin decreases the rate of gastric emptying. The presence of fatty acids in the duodenum and jejunum releases another hormone, *gastric inhibitory peptide,* that also decreases the rate of gastric emptying.

3. *Osmotic pressure of duodenal contents.* Hyperosmotic solutions in the duodenum and jejunum slow the rate of gastric emptying. Osmoreceptors in the mucosa or submucosa of the duodenum and jejunum may initiate this response, which has both neural and hormonal components. Hypertonic solutions in the duodenum also release an unidentified hormone that diminishes the rate of gastric emptying.

4. *Peptides and amino acids in the duodenum.* Peptides and amino acids release *gastrin* from G cells

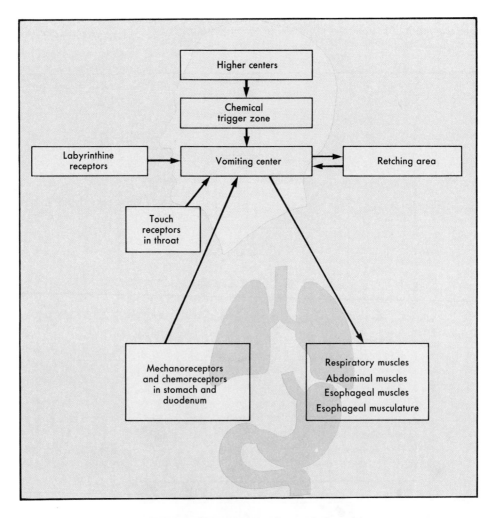

FIGURE 29-13   Some aspects of control of vomiting.

located in the antrum of the stomach and the duodenum. Gastrin increases the strength of antral contractions and increases constriction of the pyloric sphincter; the net effect diminishes the rate of gastric emptying.

Vomiting

Vomiting is the expulsion of gastric (and sometimes duodenal) contents from the GI tract via the mouth. Vomiting often is preceded by a feeling of nausea, a rapid or irregular heartbeat, dizziness, sweating, pallor, and dilation of the pupils. Vomiting usually is also preceded by *retching,* in which gastric contents are forced up into the esophagus but do not enter the pharynx.

Vomiting is a reflex behavior controlled and coordinated by a *vomiting center* in the medulla oblongata (Figure 29-13). Many areas in the body have receptors that provide afferent input to the vomiting center. Distension of the stomach and duodenum is a strong stimulus that elicits vomiting. Tickling the back of the throat, painful injury to the genitourinary system, dizziness, and certain other stimuli can bring about vomiting.

Certain chemicals, called *emetics,* can elicit vomiting. Some emetics do this by stimulating receptors in the stomach or more often in the duodenum. The widely used emetic *ipecac* stimulates duodenal receptors. Certain other emetics act at the level of the central nervous system on receptors in the floor of the fourth ventricle, in an area known as the *chemoreceptor trigger zone.* The chemoreceptor trigger zone lies on the blood side of the blood-brain barrier and thus can be reached by most blood-borne substances.

When the vomiting reflex is initiated, the sequence of events is the same regardless of the stimulus that initiates the reflex. Early events in the vomiting reflex include a wave of reverse peristalsis that sweeps from the middle of the small intestine to the duodenum. The pyloric sphincter and the stomach relax to receive intestinal contents. Then a forced inspiration occurs against a closed glottis. This decreases intrathoracic pressure, whereas the lowering of the diaphragm increases intraabdominal pressure. Then a forceful contraction of abdominal muscles sharply elevates intraabdominal pressure, driving gastric contents into the esophagus. The lower esophageal sphincter relaxes reflexly to receive gastric contents, and the pylorus and antrum contract reflexly. When a person retches, the upper esophageal sphincter remains closed, preventing vomiting. When the respiratory and abdominal muscles relax, the esophagus is emptied by secondary peristalsis into the stomach. Often a series of stronger and stronger retches precedes vomiting.

When a person vomits, the rapid propulsion of gastric contents into the esophagus is accompanied by a reflex relaxation of the upper esophageal sphincter. Vomitus is projected into the pharynx and mouth. Entry of vomitus into the trachea is prevented by approximation of the vocal chords, closure of the glottis, and inhibition of inspiration.

MOTILITY OF THE SMALL INTESTINE

The small intestine, the longest segment of the GI system, accounts for about three fourths of the length of the human GI tract. The small intestine is about 5 m in length, and chyme typically takes 2 to 4 hours to traverse it. The first 5% or so of the small intestine is the *duodenum,* which has no mesentery and can be distinguished from the rest of the small intestine histologically. The remaining small intestine is divided into the *jejunum* and the *ileum.* The jejunum is more proximal and occupies about 40% of the length of the small bowel. The ileum is the distal part of the small intestine and accounts for its remaining length.

The small intestine, particularly the duodenum and jejunum, is the site of most digestion and absorption. The movements of the small intestine mix chyme with digestive secretions, bring fresh chyme into contact with the absorptive surface of the microvilli, and propel chyme toward the colon.

The most frequent type of movement of the small intestine is termed *segmentation.* Segmentation (Figure 29-4) is characterized by closely spaced contractions of the circular muscle layer. The contractions divide the small intestine into small neighboring segments. In rhythmic segmentation the sites of the circular contractions alternate, so that a given segment of gut contracts and then relaxes. Segmentation effectively mixes chyme with digestive secretions and brings fresh chyme into contact with the mucosal surface.

Peristalsis is the progressive contraction of successive sections of circular smooth muscle. The contractions move along the GI tract in an orthograde direction. Short peristaltic waves do occur in the small intestine, but they usually involve only a small length of intestine. As in other parts of the digestive tract, the slow waves of the smooth muscle cells determine the timing of intestinal contractions.

Electrical Activity of Small Intestinal Smooth Muscle

Regular slow waves occur all along the small intestine. The frequency is highest (11 to 13 per minute in humans) in the duodenum and declines along the length of the small bowel (to a minimum 8 or 9 per minute in humans in the terminal part of the ileum). As in the stomach, the slow waves may or may not be accompanied by bursts of action potential spikes during the depolarizing part of the slow waves. In contrast to the behavior of gastric smooth muscle, the action potentials, but not the slow waves, elicit the

contractions of the smooth muscle that cause the major mixing and propulsive movements of the small intestine. Because the action potential bursts occur near the peaks of the slow waves, the slow wave frequency determines the maximum possible frequency of intestinal movements. Action potential bursts are localized to short segments of the intestine. Thus they elicit the highly localized contractions of the circular smooth muscle that cause segmentation.

The basic electrical rhythm of the small intestine is independent of the extrinsic innervation; the extrinsically denervated intestine can carry out segmentation and short peristaltic movements. The frequency of the action potential spike bursts that elicit contractions depends on the excitability of the smooth muscle cells of the small intestine. The excitability of the smooth muscle is influenced by circulating hormones, the autonomic nervous system, and enteric neurons. Excitability is enhanced by parasympathet-

A

B

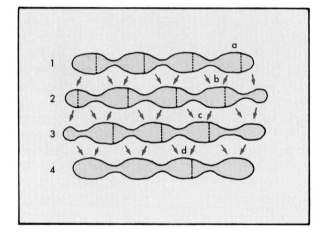

FIGURE 29-14 **A,** X-ray view showing the stomach and small intestine filled with barium contrast medium in a normal individual. Note that segmentation of the small intestine divides its contents into ovoid segments. **B,** The sequence of segmental contractions in a portion of cat's small intestine. Lines *1* through *4* indicate successive patterns in time. The dotted lines indicate where contractions will occur next. The arrows show the direction of chyme movement. (**A** from Gardner EM et al: Anatomy, a regional study of human structure, ed 4, Philadelphia, 1975, WB Saunders Co. **B** redrawn from Cannon WB: Am J Physiol 6:251, 1902.)

ic nerves and is inhibited by sympathetic nerves, both acting primarily in the intramural plexuses. Even though much of the direct control of intestinal motility resides in the intramural plexuses, the parasympathetic and sympathetic innervation of the small intestine is important in modulating contractile activity. The extrinsic neural circuits are essential for certain long-range intestinal reflexes discussed later.

Contractile Behavior of the Small Intestine

Contractions of the duodenal bulb mix chyme with pancreatic and biliary secretions, and they propel the chyme along the duodenum. Contractions of the duodenal bulb typically follow contractions of the gastric antrum. This helps prevent regurgitation of duodenal contents back into the stomach.

Segmentation is the most frequent type of movement by the small intestine (Figure 29-14). Segmen-

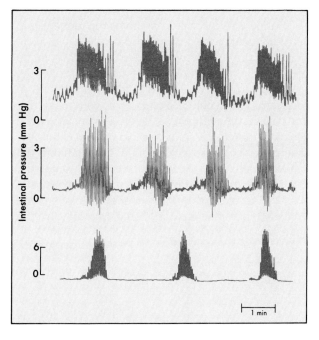

FIGURE 29-15   The minute rhythm recorded from the jejunums of three different ferrets in the fed state. Similar minute rhythms occur in the human jejunum. (From Collman PI et al: J Physiol [Lond] 345:65, 1983.)

tation occurs at a frequency similar to that of the small intestinal slow waves: about 11 or 12 contractions per minute in the duodenum and 8 or 9 contractions per minute in the ileum. A group of sequential contractions tends to be followed by a period of rest.

In the jejunum contractile activity usually occurs in bursts, separated by an interval in which contractions are weak or absent. Because the bursts are approximately 1 minute apart, the pattern has been termed the *minute rhythm* of the jejunum (Figure 29-15).

Short-range peristalsis also occurs in the small intestine, although much less frequently than segmentation. A peristaltic wave rarely traverses more than 10 cm of the small intestine. The relatively low rate of net propulsion of chyme in the small intestine allows time for digestion and absorption.

Intestinal Reflexes

Intestinal reflexes can occur along a considerable length of the GI tract. These depend on the function of both intrinsic and extrinsic nerves.

When a bolus of material is placed in the small intestine, the intestine may contract behind the bolus and relax ahead of it, a response known as the *law of the intestine*. This may propel the bolus in an orthograde direction, similar to a peristaltic wave.

Overdistension of one segment of the intestine relaxes the smooth muscle in the rest of the intestine. This response is known as the *intestinointestinal reflex*.

The stomach and the terminal part of the ileum interact reflexly. Elevated secretory and motor functions of the stomach increase the motility of the terminal part of the ileum and accelerate the movement of material through the ileocecal sphincter. This response is called the *gastroileal reflex*. Increased motor activity of the colon is stimulated by gastric distension, a response known as the *gastrocolic reflex*.

Migrating Myoelectric Complex

The contractile behavior of the small intestine just discussed is characteristic of the period after ingestion of a meal. In a *fasted* individual or some hours after the processing of the previous meal, small intestinal motility follows a different pattern characterized by bursts of intense electrical and contractile activity

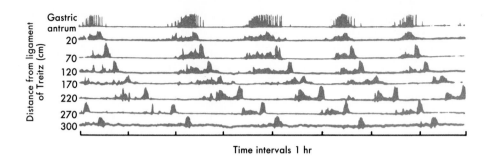

FIGURE 29-16   The occurrence of the migrating myoelectric complex in the stomach and small intestine of a fasting dog. (From Itoh Z and Sekiguchi, T: Interdigestive motor activity in health and disease, Scand J Gastroenterol Suppl 82:121, 1983.)

separated by longer quiescent periods. This pattern is propagated from the stomach to the terminal ileum (Figure 29-16) and is known as the *migrating myoelectric complex (MMC)*. The MMC in the stomach is discussed earlier in this chapter.

The MMC repeats every 75 to 90 minutes in humans and occurs in four phases. Phase 1, the quiescent phase, is characterized by slow waves with very few action potentials and very few contractions. In phase 2 irregular action potentials and contractions occur and gradually increase in intensity and frequency. Phase 3 is a period of intense electrical and contractile activity lasting from 3 to 6 minutes. During phase 4 the electrical and contractile activity rapidly decline. Phase 4 varies in duration and blends almost imperceptibly into phase 1. About the time that one MMC reaches the distal ileum, the next MMC begins in the stomach.

The contractions of phase 3 of the MMC, both in the stomach and in the small intestine, are more vigorous and more propulsive than the contractions that occur in the fed individual. Phase 3 contractions sweep the small bowel clean and empty its contents into the cecum. Thus the MMC has been termed the "housekeeper of the small intestine." The MMC inhibits the migration of colonic bacteria into the terminal ileum.

Contractile Activity of the Muscularis Mucosae

Sections of the muscularis mucosae contract irregularly at a rate of about three contractions per minute. These contractions alter the pattern of ridges and folds of the mucosa, mix the luminal contents, and bring different parts of the mucosal surface into contact with freshly mixed chyme. The villi contract irregularly, especially in the proximal part of the small intestine. This helps to empty the central lacteals of the villi and increases intestinal lymph flow.

Emptying the Ileum

The *ileocecal sphincter* separates the terminal end of the ileum from the cecum, the first part of the colon. Normally the sphincter is closed, but short-range peristalsis in the terminal part of the ileum relaxes the sphincter and allows a small amount of chyme to squirt into the cecum. Distension of the terminal ileum relaxes the sphincter reflexly. Distension of the cecum contracts the sphincter and prevents additional emptying of the ileum. Under normal conditions the ileocecal sphincter allows ileal chyme to enter the colon at a slow enough rate so that the colon can absorb most of the salts and water of the chyme. The ileocecal sphincter is coordinated primarily by the neurons of the intramural plexuses.

MOTILITY OF THE COLON

The colon receives 500 to 1500 ml of chyme per day from the ileum. Most of the salts and water that enter the colon are absorbed; the feces normally contain only about 50 to 100 ml of water each day. Colonic contractions mix the chyme and circulate it across the mucosal surface of the colon. As the chyme becomes semisolid, this mixing resembles a kneading process. The progress of colonic contents is slow, about 5 to 10 cm/hr at most. One to three times daily a wave of contraction, called a *mass movement,* occurs. A mass movement resembles a peristaltic wave in which the contracted segments remain contracted for some time. Mass movements push the contents of a significant length of colon in an orthograde direction.

Structure and Innervation of the Large Intestine

As shown in Figure 29-17, the major subdivisions of the large intestine are the cecum, the ascending colon, the transverse colon, the descending colon, the sigmoid colon, the rectum, and the anal canal. The structure of the wall of the large bowel follows the general plan presented earlier in this chapter, but the longitudinal muscle layer of the muscularis externa is concentrated into three bands called the teniae coli. In between the teniae coli the longitudinal layer is thin. The longitudinal muscle of the rectum and anal canal is substantial and continuous.

The extrinsic innervation of the large intestine is predominantly autonomic. Parasympathetic innervation of the cecum and the ascending and transverse colon is via branches of the vagus nerve; that of the descending and sigmoid colon, the rectum, and the anal canal is via the pelvic nerves from the sacral spinal cord. The parasympathetic fibers end primarily on neurons of the intramural plexuses. Sympathetic fibers innervate the proximal part of the large intestine via the superior mesenteric plexus, the distal part of the large intestine via the inferior mesenteric and superior hypogastric plexuses, and the rectum and anal canal via the inferior hypogastric plexus.

Stimulation of the sympathetic nerves stops colonic movements. Vagal stimulation causes segmental contractions of the proximal part of the colon. Stimulation of the pelvic nerves brings about expulsive movements of the distal colon and sustained contraction of some segments.

The anal canal usually is kept closed by the internal and external sphincters. The internal anal sphincter is a thickening of the circular smooth muscle of the anal canal. The external anal sphincter is more distal, and it consists entirely of striated muscle. The external anal sphincter is innervated by somatic motor fibers via the pudendal nerves, which allow it to be controlled both reflexly and voluntarily.

Motility of the Cecum and Proximal Part of the Colon

Most contractions of the cecum and proximal part of the large bowel are segmental and are more effective at mixing and circulating the contents than at propelling them. The mixing action facilitates absorption of salts and water by the mucosal epithelium.

Localized segmental contractions divide the colon into neighboring ovoid segments, called *haustra* (Figure 29-18). Thus, segmentation in the colon is known as *haustration.* The most dramatic difference between haustration and the segmentation that occurs in the small intestine is the regularity of the segments (haustra) produced by haustration and the large length of bowel involved in haustration at one time. Segmental thickenings of the circular smooth muscle are probably an anatomic component of haustration. The pattern of haustra is not fixed, however; it fluctuates as segments of the large bowel relax and contract, which results in back-and-forth mixing of luminal contents.

In the proximal colon "antipropulsive" patterns sometimes occur. Reverse peristalsis and segmental propulsion toward the cecum both take place. These antipropulsive movements retain chyme in the proximal colon and may thus facilitate the absorption of salts and water in this section of the colon.

The net rate of chyme flow in the proximal part of the colon is only about 5 cm/hr in a fasting individual. Colonic propulsive motility increases after eating, and the net rate of flow increases to about 10 cm/hr.

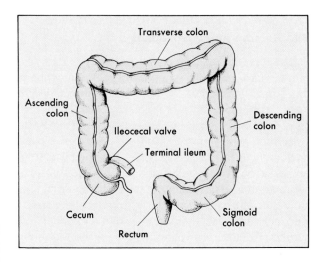

FIGURE 29-17   Major anatomic subdivisions of the colon.

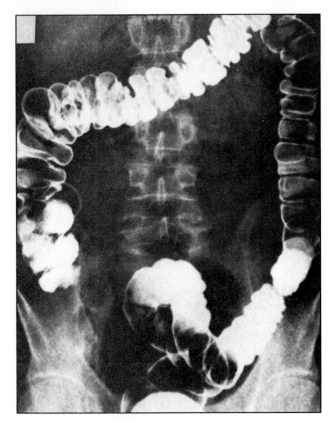

FIGURE 29-18   X-ray image showing a prominent haustral pattern in the colon of a normal individual. (Reproduced with permission from Keats TE: An atlas of normal roentgen variants, ed 2. Copyright © 1979 by Year Book Medical Publishers, Inc, Chicago.)

Motility of the Distal Part of the Colon

Normally the distal part of the colon is filled with semisolid feces by a mass movement. Segmental contractions knead the feces, facilitating absorption of remaining salts and water. About one to three times daily, mass movements occur and sweep the feces toward the rectum.

Control of Colonic Movements

As in other segments of the GI tract, the intramural plexuses control the contractile behavior of the colon, and the extrinsic autonomic nerves to the colon are modulatory. The defecation reflex, discussed later, is an exception to this rule: it requires the function of the spinal cord via the pelvic nerves. Enteric neurons are both excitatory and inhibitory to colonic smooth

muscle. The net effect is inhibitory because blocking enteric neurons with drugs causes sustained contractions of the circular muscle of the colon. In *Hirschsprung's disease,* in which enteric neurons are congenitally absent from part of the colon, tonic contractions of the colonic circular muscle can cause obstruction of the colon.

Electrophysiology of the Colon   The colons of some mammals, including dogs and cats, show continuous slow wave activity at frequencies of 4 to 6 per minute. The colons of other mammals, including humans, show intermittent, irregular slow waves. As in the small intestine, contractions are associated with bursts of action potentials that occur near the crests of the slow waves. The action potentials usually occur in short bursts of spiking and do not propagate very far from the site of origin. They are associated with weak, highly localized contractions of the circular muscle (type 1 contractions). Longer bursts of action potentials occur less frequently. They are propagated for greater distances up and down the colon and are associated with more vigorous contractions (type 2 contractions) of longer segments of the colon. A pattern of intense spiking activity that spreads in the orthograde direction is the electrophysiologic correlate of a mass movement.

Reflex Control   Distension of one part of the colon reflexly relaxes other parts of the colon. This *colonocolonic reflex* is mediated partly by the sympathetic fibers that supply the colon. The motility of proximal and distal colon and the frequency of mass movements increase reflexly via the *gastrocolic reflex* after a meal enters the stomach.

The Rectum and Anal Canal

The rectum is usually empty, or nearly so. The rectum is more active in segmental contractions than is the sigmoid colon, so that the rectal contents tend to move retrogradely into the sigmoid colon. The anal canal is tightly closed by the anal sphincters. Before defecation the rectum is filled as a result of a mass movement in the sigmoid colon. Filling the rectum brings about reflex relaxation of the internal anal sphincter and reflex constriction of the external anal sphincter and causes the urge to defecate. Persons who lack functional motor nerves to the external anal sphincter defecate involuntarily when the rectum is filled. The reflex reactions of the sphincters to rectal distension are transient. If defecation is postponed, the sphincters regain their normal tone, and the urge to defecate temporarily subsides.

Defecation

When an individual feels the circumstances are appropriate, he or she voluntarily relaxes the external anal sphincter to allow defecation to proceed. Defecation is a complex behavior involving both reflex and voluntary actions. The integrating center for the reflex actions is in the sacral spinal cord and is modulated by higher centers. The efferent pathways are cholinergic parasympathetic fibers in the pelvic nerves. The sympathetic nervous system does not play a significant role in normal defecation.

Voluntary actions are important in defecation. The external anal sphincter is voluntarily held in the relaxed state. Intraabdominal pressure is elevated to aid in expulsion of feces. Evacuation is normally preceded by a deep breath, which moves the diaphragm downward. The glottis is then closed, and contractions of the respiratory muscles on full lungs elevates both the intrathoracic and the intraabdominal pressure. Contractions of the muscles of the abdominal wall further increase intraabdominal pressure, which may be as great as 200 cm H_2O. This helps to force feces through the relaxed sphincters. The muscles of the pelvic floor are relaxed to allow the floor to drop. This helps to straighten out the rectum and prevent rectal prolapse.

BIBLIOGRAPHY
Journal Article

Wood JD: Enteric neurophysiology, Am J Physiol 247:G585, 1984.

Books and Monographs

Christensen J and Wingate DL, editors: A guide to gastrointestinal motility, Bristol, UK, 1983, John Wright and Sons.

Davenport HW: Physiology of the digestive tract, ed 5, Chicago, 1985, Year Book Medical Publishers, Inc.

Gregory RA, editor: Regulatory peptides of gut and brain, Br Med Bull 38(3):319, 1982.

Grundy D: Gastrointestinal motility: the integration of physiological mechanisms, Lancaster, UK, 1985, MTP Press.

Johnson LR, editor: Physiology of the gastrointestinal tract, ed 2, vols 1 and New York, 1987, Raven Press.

Secretion

REGULATION OF SECRETION—GENERAL ASPECTS

This chapter deals with the glandular secretion of fluids and compounds that have important functions in the digestive tract. In particular, the secretions of salivary glands, gastric glands, the exocrine pancreas, and the liver are considered. In each case the nature of the secretions and their functions in digestion are discussed, and the regulation of the secretory processes is emphasized.

The secretions just mentioned are elicited by the action of specific effector substances on the secretory cells. These substances may be classified as neurocrine, endocrine, or paracrine. *Neurocrine* substances are released from the endings of neurons that innervate the secretory cells. *Endocrine* modulators are produced by specific cells located some distance from their target cell, and they reach the target cell via the circulation. *Paracrine* regulatory substances are released near the target secretory cell and reach the target cell by diffusion.

A substance that stimulates a particular cell to secrete is called a *secretagogue*. Many secretagogues exist, but there are few cellular mechanisms of action of secretagogues. Most secretagogues increase the intracellular level of cyclic adenosine monophosphate (cAMP) or increase the turnover of certain inositol-containing phospholipids in the plasma membrane of the secretory cell (Figures 30-1 and 30-2). The phospholipid turnover may result in an increase in the concentration of calcium ions in the cytosol. Ca^{++} and cAMP, called *second messengers*, initiate chains of

events that culminate in increased secretion (or in other responses in nonsecretory cells).

SECRETION OF SALIVA

In humans the salivary glands produce about 1 L of saliva each day. Saliva lubricates food for greater ease of swallowing and also facilitates speaking. In people lacking functional salivary glands, *xerostomia* (dry mouth), dental caries, and infections of the buccal mucosa are prevalent.

Functions of Saliva

Mucins, glycoproteins produced by the submaxillary and sublingual glands, lubricate food so that it may be more readily swallowed. The major digestive function of saliva results from the action of *salivary amylase* on starch. Salivary amylase is an enzyme that has the same specificity as the α-amylase of pancreatic juice, and it reduces starch to oligosaccharide molecules. The pH optimum of salivary amylase is about 7, but it has activity between pH 4 and 11. Amylase action continues in the mass of food in the stomach and is terminated only when the contents of the antrum are mixed with enough gastric acid to lower the pH to less than 4. More than half the starch in a well-chewed meal may be reduced to small oligosaccharides by action of salivary amylase. However, because of the large capacity of the pancreatic α-amylase to digest starch in the small intestine, the absence of salivary amylase causes no malabsorption of starch.

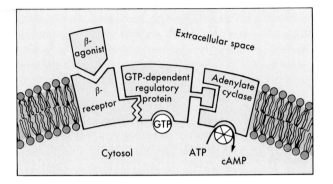

FIGURE 30-1   The mechanism whereby β-adrenergic agonists increase the intracellular concentration of cyclic adenosine monophosphate *(cAMP)*. *GTP,* Guanosine triphosphate; *ATP,* adenosine triphosphate.

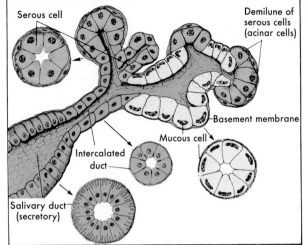

FIGURE 30-3   The structure of the human submandibular gland, as seen with the light microscope. (Redrawn from Braus H: Anatomie des Menschen, Berlin, 1934, Julius Springer.)

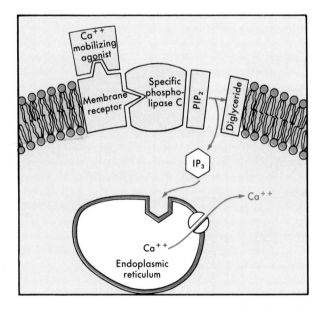

FIGURE 30-2   The mechanisms whereby agonists that stimulate hydrolysis of inositol phosphatides lead to an increase in the intracellular level of free Ca^{++}. *IP_3,* Inositol-(1,3,5)-trisphosphate.

Major Salivary Glands and Their Structure

In humans the parotid glands, the largest glands, are entirely serous glands. Their watery secretion lacks mucins. The submaxillary and sublingual glands are mixed mucous and serous glands, and they secrete a more viscous saliva containing mucins. Many smaller salivary glands are present in the oral cavity. The microscopic structure of mixed salivary glands is depicted in Figure 30-3. The salivary glands resemble the exocrine pancreas in many respects. The serous acinar cells, located in the end-pieces, have apical zymogen granules that contain salivary amylase and perhaps certain other salivary proteins as well (Figure 30-4). Mucous acinar cells secrete glycoprotein mucins into the saliva. Intercalated ducts drain the acinar fluid into somewhat larger ducts, the striated ducts, which empty into still larger excretory ducts. A single large duct brings the secretions of each major gland into the mouth.

The current concept is that a primary secretion is elaborated in the secretory end-pieces and that the cells that line the ducts modify the primary secretion.

Metabolism and Blood Flow of Salivary Glands

For their size the salivary glands produce a prodigious flow of saliva: the maximal rate in humans is about 1 ml/min/g of gland. Salivary glands have a high rate of metabolism and a high blood flow; both are proportional to the rate of saliva formation. The blood flow to maximally secreting salivary glands is approximately 10 times that of an equal mass of actively contracting skeletal muscle. Stimulation of

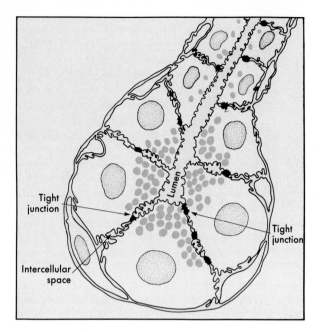

FIGURE 30-4 A schematic representation of the cellular morphology of a secretory end-piece of a serous salivary gland. Secretory canaliculi drain into the acinar lumen. Tight junctions separate the secretory canaliculi from the lateral intercellular spaces. Purple circles represent zymogen granules. (Redrawn from Young JA and van Lennep LW: Morphology of salivary glands, London, 1978, Academic Press.)

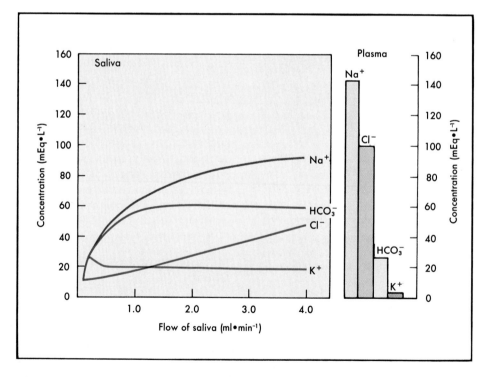

FIGURE 30-5 Average composition of the parotid saliva as a function of the salivary flow rate. (Redrawn from Thaysen JH et al: Am J Physiol 178:155, 1954).

the parasympathetic nerves to salivary glands increases blood flow by dilation of the vasculature of the glands. *Vasoactive intestinal polypeptide (VIP)* is released with acetylcholine from parasympathetic nerve terminals in the salivary glands, and VIP may contribute to vasodilation during secretory activity.

Ionic Composition of Saliva and Secretion of Water and Electrolytes

In humans saliva is always hypotonic to plasma. As shown in Figure 30-5, salivary concentrations of Na^+ and Cl^- are less than those of plasma. The tonicity of saliva and its ionic composition vary from species to species and from one salivary gland to another in the same species. The greater the secretory flow rate, the higher is the tonicity of the saliva; at maximal flow rates the tonicity of saliva in humans is about 70% of that of plasma. The pH of saliva from resting glands is slightly acidic. During active secretion, however, the saliva becomes basic, with pH approaching 8. The increase in pH with secretory flow rate is partly caused by the increase in salivary bicarbonate (HCO_3^-) concentration. The concentration of K^+ in saliva is almost independent of salivary flow rate over a wide range of flows. However, at very low salivary flow rates the concentration of K^+ in saliva increases steeply with decreasing flow rate.

Available evidence is consistent with a *two-stage model* of salivary secretion (Figure 30-6). The two-stage model postulates the following:

1. The end-pieces, perhaps with the participation of intercalated ducts, produce an isotonic primary secretion. The amylase concentration and the rate of fluid secretion vary with the level and type of stimulation. However, the electrolyte composition of the secretion is fairly constant, and the levels of Na^+, K^+, and Cl^- are close to plasma levels.

2. The execretory ducts, and probably the striated ducts as well, modify the primary secretion by extracting Na^+ and Cl^- from and adding K^+ and HCO_3^- to the saliva. The ducts do not add to the volume of saliva.

As saliva flows down the ducts, it becomes progressively more hypotonic. Thus the ducts remove more ions from saliva than they contribute to it. The faster the flow rate of the saliva down the striated and excretory ducts, the closer to isotonicity is the saliva.

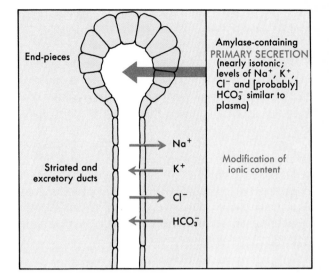

FIGURE 30-6   Schematic representation of the two-stage model of salivary secretion.

Secretion of Salivary Amylase

In their apical cytoplasm, serous acinar cells have zymogen granules (see Figure 30-4) that contain salivary amylase. When the gland is stimulated to secrete, the zymogen granules fuse with the plasma membrane and release their contents into the lumen of an acinus by exocytosis.

Neural Control of Salivary Gland Function

The primary physiologic control of the salivary glands is by the autonomic nervous system. Stimulation of either sympathetic or parasympathetic nerves to the salivary glands stimulates salivary secretion, but the effects of the parasympathetic nerves are stronger and more long lasting. Interruption of the sympathetic nerves causes no major defect in the function of the salivary glands. If the parasympathetic supply is interrupted, the salivary glands atrophy. This response suggests that the essential physiologic control is by way of the parasympathetic nervous system.

Sympathetic fibers to the salivary glands come from the superior cervical ganglion. Preganglionic parasympathetic fibers come via branches of the facial and glossopharyngeal nerves (cranial nerves VII and IX, respectively), and they synapse with postganglionic neurons in or near the salivary glands. Both acinar cells and ducts are supplied with parasympathetic nerve endings.

Parasympathetic stimulation increases the synthesis and secretion of salivary amylase and mucins, enhances the transport activities of the ductular epithelium, greatly increases blood flow to the glands, and stimulates glandular metabolism and growth.

Sympathetic stimulation and circulating catecholamines stimulate the secretion of saliva that is rich in amylase, K^+, and HCO_3^-. Sympathetic stimulation constricts blood vessels, with consequent reductions in salivary gland blood flow. The increase in salivary secretion that results from stimulation of sympathetic nerves is transient.

Cellular Mechanisms of Salivary Secretion

Duct System Current knowledge of cellular mechanisms that control ion transport activities in the striated and excretory ducts of salivary glands is sparse. The ducts respond to both cholinergic and adrenergic agonists by increasing their rates of secretion of K^+ and HCO_3^-.

Acinar Cells The neuroeffector substances that stimulate acinar cell secretions act primarily by one of the two general mechanisms outlined at the beginning of this chapter: by elevating intracellular cAMP or by increasing the level of Ca^{++} in the cytosol (Figure 30-7).

Acetylcholine, norepinephrine, substance P, and VIP are released in salivary glands by specific nerve

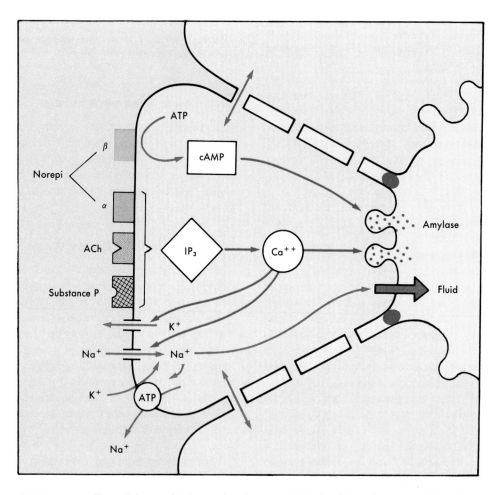

FIGURE 30-7   The cellular mechanisms whereby norepinephrine *(Norepi)*, acetylcholine *(ACh)*, and substance P evoke salivary secretion. (Modified from Peterson OH. In Johnson RL, editor: *Physiology of the gastrointestinal tract*, New York, 1981, Raven Press.)

terminals. Each of these neuroeffectors may increase the secretion of salivary amylase and the flow of saliva.

Acinar cells have both α- and β-adrenergic receptors, and norepinephrine binds to both classes of receptors. Activation of β-receptors in salivary glands elevates intracellular cAMP. VIP also increases cAMP. By contrast, acetylcholine, substance P, and norepinephrine—acting on α-receptors—increase intracellular Ca^{++}.

In general, effectors that increase cellular cAMP elicit a primary secretion that is richer in amylase than the secretion evoked by agents that increase intracellular Ca^{++}. Substances that increase intracellular Ca^{++} produce a greater volume of acinar cell secretion than agonists that increase intracellular cAMP.

GASTRIC SECRETIONS

Structure of the Gastric Mucosa

The surface of the gastric mucosa (Figure 30-8) is covered by columnar epithelial cells that secrete mucus and an alkaline fluid that protects the epithelium from mechanical injury and gastric acid. The surface is studded with gastric pits; each pit is the opening of a duct into which one or more gastric glands empty (Figure 30-8, A). The gastric pits are so numerous that they account for a significant fraction of the total surface area.

The gastric mucosa can be divided into three regions, based on the structures of the glands present. The region just below the lower esophageal sphincter is called the cardiac glandular region. In humans the cardiac glandular region is, at most, a few centimeters wide. The glands of this region are tortuous and contain primarily mucus-secreting cells. The remainder of the gastric mucosa is divided into the oxyntic (acid-secreting) glandular region, above the notch, and the pyloric glandular region, below the notch.

The structure of a gastric gland from the oxyntic glandular region is illustrated in Figure 30-8, B. The surface epithelial cells extend slightly into the duct opening. In the narrow neck of the gland are the mucous neck cells, which secrete mucus. Deeper in the gland are parietal or oxyntic cells, which secrete HCl and intrinsic factor, and chief or peptic cells, which secrete pepsinogen. Oxyntic cells are particularly numerous in glands in the fundus.

The glands of the pyloric glandular region contain few oxyntic and peptic cells; mucus-secreting cells predominate there. The pyloric glands also contain G cells, which secrete gastrin.

Surface epithelial cells are exfoliated into the lumen at a considerable rate during normal gastric function. They are replaced by mucous neck cells, which differentiate into columnar epithelial cells and migrate up out of the necks of the glands. The capacity of the stomach to repair damage to its epithelial surface in this way is remarkable.

Gastric Acid Secretion

The fluid secreted into the stomach is called gastric juice. Gastric juice is a mixture of the secretions of the surface epithelial cells and the secretion of gastric glands. Among the important components of gastric juice are salts, water, HCl, pepsins, intrinsic factor, and mucus. Secretion of all these components increases after a meal.

Ionic Composition of Gastric Juice

The ionic composition of gastric juice depends on the rate of secretion. Figure 30-9 shows that the higher the secretory rate, the higher the concentration of hydrogen ion. At lower secretory rates $[H^+]$ diminishes, and $[Na^+]$ increases. $[K^+]$ in gastric juice is always higher than in plasma, and consequently prolonged vomiting may lead to hypokalemia. At all rates of secretion, Cl^- is the major anion of gastric juice. At high rates of secretion the composition of gastric juice resembles that of an isotonic solution of HCl. At low rates of secretion gastric juice is hypotonic to plasma. Gastric HCl converts pepsinogen to pepsin, provides an acid pH at which pepsin is active, and kills most ingested microorganisms.

Rate of Secretion of Gastric Acid

The rate of gastric acid secretion varies considerably among individuals, partly because of variations in the number of parietal cells. Basal (unstimulated) rates of gastric acid production typically range from about 1 to 5 mEq/hr in humans. On maximal stimu-

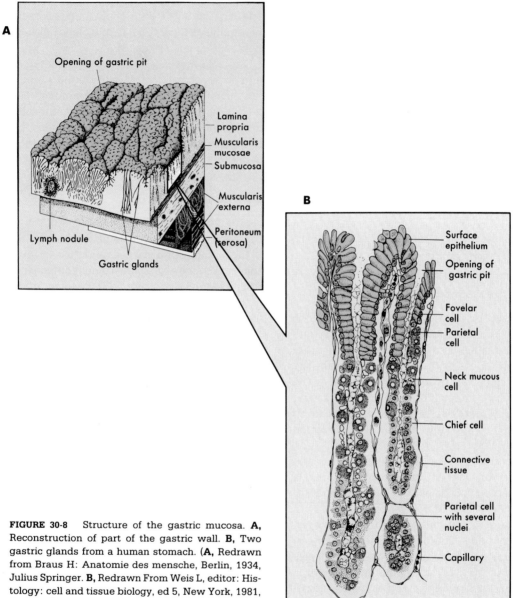

FIGURE 30-8   Structure of the gastric mucosa. **A,** Reconstruction of part of the gastric wall. **B,** Two gastric glands from a human stomach. (**A,** Redrawn from Braus H: Anatomie des mensche, Berlin, 1934, Julius Springer. **B,** Redrawn From Weis L, editor: Histology: cell and tissue biology, ed 5, New York, 1981, Elsevier.)

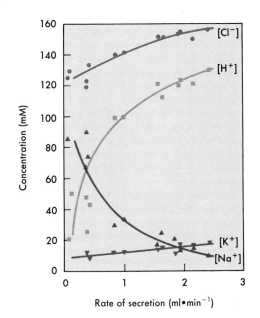

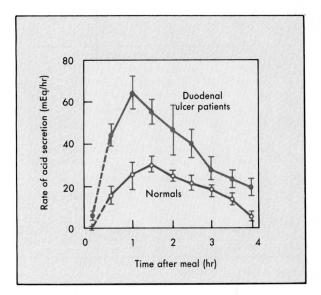

FIGURE 30-9 Concentrations of ions in gastric juice as a function of the rate of secretion in a normal young person. (Adapted from Nordgren B: Acta Physiol Scand 58[suppl 202]:1, 1963.)

FIGURE 30-10 Rate of gastric acid secretion after a meal in six normal subjects and seven patients with duodenal ulcers. (Reproduced from Fordtran JS and Walsh JH: the Journal of Clinical Investigation, 1973, vol 52, p 645. By copyright permission of the American Society for Clinical Investigation.)

lation, HCl production rises to 6 to 40 mEq/hr. On average, patients with gastric ulcers secrete less HCl but those with duodenal ulcers secrete more HCl than do normal individuals (Figure 30-10).

Morphologic Changes Accompanying Gastric Acid Secretion

Parietal cells have a distinctive ultrastructure (Figure 30-11) and an elaborate system of branching *secretory canaliculi,* which course through the cytoplasm and are connected by a common outlet to the cell's luminal surface. Microvilli line the surfaces of the canaliculi. In addition, the cytoplasm of the parietal cells contains extensive tubules and vesicles— the *tubulovesicular system.*

When parietal cells are stimulated to secrete, their morphology changes greatly (Figure 30-11). Tubules and vesicles of the tubulovesicular system fuse with the plasma membrane of the secretory canaliculi, greatly diminishing the content of tubulovesicles and greatly increasing the surface area of the secretory canaliculi. Because the tubulovesicles contain the HCl secretory apparatus, the extensive membrane fusion that occurs on stimulation greatly increases

the number of HCl pumping sites available at the surface of the secretory canaliculi.

Cellular Mechanism of Gastric Acid Secretion

At maximal rates of secretion H^+ is pumped against a concentration gradient that is more than one million to one. Cl^- also enters the gastric lumen against a large electrochemical potential difference. Thus energy is required for transport of both H^+ and Cl^-.

A current model of the mechanism of gastric acid transport is shown in Figure 30-12. The apical membrane of the parietal cell (the membrane facing the secretory canaliculus) contains an H^+, K^+-activated ATPase, which exchanges H^+ for K^+. This ATPase appears to be the primary H^+ pump. Both H^+ and K^+ are pumped against their electrochemical potential gradients.

When H^+ is pumped out of the parietal cell, an excess of HCO_3^- is left behind. HCO_3^- flows down its electrochemical gradient across the basolateral plasma membrane. The protein that mediates HCO_3^- efflux transports Cl^- in the opposite direction. Cl^- moves against its electrochemical potential gradient

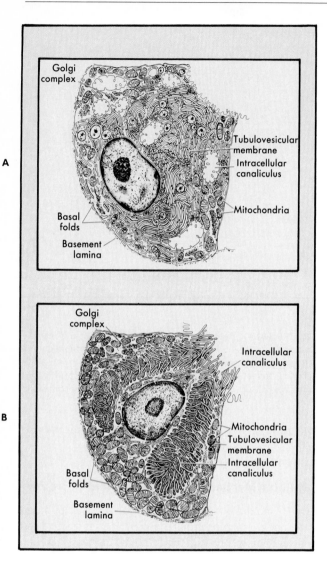

A, Drawing of a resting parietal cell with cytoplasm full of tubulovesicles and an internalized intracellular canaliculus. **B,** An acid-secreting parietal cell. Tubulovesicles have fused with the membrane of the intracellular canaliculus, which is now open to the lumen of the gland and lined with abundant, long microvilli. (Redrawn after Ito S. In Johnson RL, editor: Physiology of the gastrointestinal tract, New York, 1981, Raven Press.)

Labels in figure (top, A): Golgi complex; Tubulovesicular membrane; Intracellular canaliculus; Mitochondria; Basal folds; Basement lamina

Labels in figure (bottom, B): Golgi complex; Intracellular canaliculus; Mitochondria; Tubulovesicular membrane; Intracellular canaliculus; Basal folds; Basement lamina

FIGURE 30-11 (caption as above)

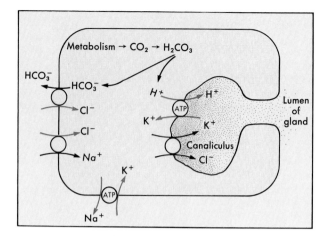

FIGURE 30-12   Hypothetical scheme for HCl secretion by the parietal cell. Ion fluxes against an electrochemical potential gradient are shown in red.

Labels in figure: Metabolism → CO_2 → H_2CO_3; HCO_3^-; HCO_3^-; Cl^-; Cl^-; Na^+; K^+; Na^+; H^+; ATP; H^+; K^+; K^+; Canaliculus; Cl^-; Lumen of gland; ATP

plasma membrane enables Na^+ to enter the cell down its electrochemical potential gradient, and this provides energy to bring Cl^- into the cell against its electrochemical potential gradient. The maintenance of the Na^+ gradient depends on the action of the Na^+, K^+-ATPase in the basolateral membrane. As a result of the action of the Cl^-/HCO_3^- counter-transporter and the Na^+/Cl^- co-transporter, Cl^- is concentrated in the cytoplasm of the parietal cell. The Cl^- leaves the parietal cell at the apical membrane by facilitated transport.

Pepsins

The pepsins are a group of proteases secreted primarily by the chief cells of the gastric glands and often collectively referred to simply as *pepsin*. Pepsins are secreted as inactive proenzymes known as *pepsinogens*. Cleavage of acid-labile linkages converts pepsinogens to pepsins: the lower the pH, the more rapid the conversion. Pepsins also act proteolytically on pepsinogens to form more pepsins.

The pepsins have unusually low pH optima; they have their highest proteolytic activity at pH 3 and below. Pepsins may digest as much as 20% of the protein in a typical meal. When the duodenal contents are neutralized, pepsins are inactivated irreversibly.

Pepsinogens are contained in membrane-bound zymogen granules in the chief cells. The contents of the zymogen granules are released by exocytosis when the chief cells are stimulated to secrete.

into the cell, and the energy for the active transport of Cl^- comes from the downhill movement of HCO_3^-. The HCO_3^- that leaves the parietal cell is carried away in the blood, and it increases the pH of venous blood leaving the stomach. This phenomenon is called the *alkaline tide.*

A Na^+/Cl^- co-transport system in the basolateral

Intrinsic Factor

The function of *intrinsic factor,* a glycoprotein secreted by the parietal cells of the stomach that is required for the normal absorption of vitamin B_{12}, is discussed in Chapter 31. Intrinsic factor is released in response to the same stimuli that evoke the secretion of HCl by parietal cells.

Secretion of Mucus and Bicarbonate

Secretions that contain glycoprotein mucins are viscous and sticky and are collectively termed *mucus.* Mucus is secreted by mucous neck cells in the necks of gastric glands and by the surface epithelial cells of the stomach. Secretion of mucus is stimulated by sham feeding and by some of the same stimuli that enhance acid and pepsinogen secretion, especially by acetylcholine released from parasympathetic nerve endings near the gastric glands.

The surface epithelial cells also secrete watery fluid with Na^+ and Cl^- concentrations similar to plasma but with higher K^+ (four times) and HCO_3^- (two times) concentrations than in plasma. The high HCO_3^- con-

centration makes the mucus alkaline. Mucus is secreted by the resting mucosa and lines the stomach with a sticky, viscous, alkaline coat. When food is eaten, the rates of secretion of mucus and of HCO_3^- increase. Among the stimuli that enhance secretion are mechanical stimulation of the mucosa and stimulation of either sympathetic or parasympathetic nerves to the stomach.

The mucus forms a gel on the luminal surface of the mucosa. The gel protects the mucosa from mechanical damage from chunks of food. The alkaline fluid it entraps protects the mucosa against damage by HCl and pepsin. The mucus and alkaline secretions are part of the *gastric mucosal barrier* that prevents damage to the mucosa by gastric contents (Figure 30-13).

The mucus layer prevents the bicarbonate-rich secretions of the surface epithelial cells from rapidly mixing with the contents of the gastric lumen. Consequently the surfaces of gastric epithelial cells are bathed in their own bicarbonate-rich secretions. HCO_3^- buffers H^+ ions that diffuse from the lumen to the epithelial surface (Figure 30-13). In this way the surface of the epithelial cells can be maintained at

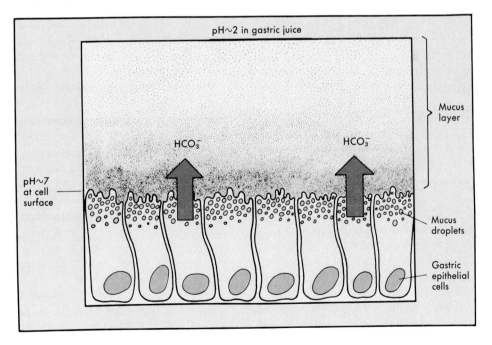

FIGURE 30-13   The protection provided to the mucosal surface of the stomach by bicarbonate-containing mucus layer known as the *gastric mucosal barrier.* Buffering by the bicarbonate-rich secretions of the surface epithelial cells and the restraint to convective mixing caused by the high viscosity of the mucus layer allow the pH at the cell surface to remain near 7, whereas the pH in the gastric juice in the lumen is 1 to 2.

nearly neutral pH, despite a luminal pH of about 2. The protection depends on both mucus and HCO_3^- secretion; neither mucus alone nor HCO_3^- alone can hold the pH at the epithelial cell surface near neutral. The unstirred layer provided by the mucus retards convective mixing of epithelial cell secretions with luminal contents and slows diffusion of H^+ to the surface and HCO_3^- into the lumen. When the unstirred layer is 1 mm thick, the diffusion times for H^+ and HCO_3^- are about 10 minutes, but this time delay would not prevent a rise in the concentration of H^+ at the epithelial surface without a continuous secretion of HCO_3^- to neutralize the H^+ as it arrives. It is still unclear how gastric juice is transported from the mouths of gastric glands to the lumen of the stomach without mixing with the mucus-bicarbonate layer.

Mucus is stored in large granules in the apical cytoplasm of mucous neck cells and surface epithelial cells. It is released by exocytosis, by dissolution of the apical membrane of the cell, or by exfoliation of an entire epithelial cell into the mucous coat.

The maximal rate of HCO_3^- secretion is about 10% of the maximal rate of HCl secretion. α-Adrenergic agonists diminish HCO_3^- secretion. This effect may play a role in the pathogenesis of *stress ulcers:* a chronically elevated level of circulating epinephrine may suppress HCO_3^- secretion sufficiently to decrease protection of the epithelial cell surface. Aspirin and other nonsteroidal antiinflammatory agents inhibit secretion of both mucus and HCO_3^-; prolonged use of these drugs may damage the mucosal surface.

Structure of Mucus. Mucins are the major components of mucus. The mucins produced by surface epithelial cells of the stomach are glycoproteins that are about 80% carbohydrate. Intact mucins have molecular weights of almost 2 million. They consist of four similar monomers with molecular weights of 500,000, linked together by disulfide cross-links (Figure 30-14). Each monomer is largely covered by carbohydrate side chains that protect it from proteolytic degradation. The portion of the monomer that participates in the disulfide cross-links is rich in cystine and free of carbohydrate; thus it is vulnerable to proteolysis. Pepsins cleave bonds in the central region of the tetramer and release fragments roughly as large as monomers. The tetrameric mucins form a gel that adds mechanical stability to the mucus-bicarbonate layer. The mucin monomers cannot form a gel. Proteolysis of mucins by pepsin thus dissolves the gel. Maintenance of the protective mucus layer requires that new tetrameric mucins be secreted to replace mucins that are cleaved by pepsins.

CONTROL OF GASTRIC ACID SECRETION

Control of HCl Secretion at the Level of the Parietal Cell

Acetylcholine, histamine, and gastrin each binds to a distinct class of receptors on the plasma membrane of the parietal cell and directly stimulates the parietal cell to secrete HCl (Figure 30-15). Acetylcholine is released near parietal cells by cholinergic nerve terminals. Gastrin, a hormone, is produced by G cells in the mucosa of the gastric antrum and the duodenum and reaches parietal cells via the bloodstream. Histamine is released from cells in the gastric mucosa and diffuses to the parietal cells.

Acetylcholine, histamine, and gastrin are important in the regulation of HCl secretion. Specific antagonists for acetylcholine (e.g., atropine) and histamine (e.g., cimetidine) not only block the effects of acetylcholine and histamine, respectively, but also inhibit acid secretion in response to any effective stimulus. Under many circumstances each of the primary agonists (acetylcholine, histamine, and gastrin) can potentiate the acid secretion elicited by one of the other two mediators.

Cellular Mechanisms of Parietal Cell Agonists Acetylcholine stimulates acid secretion by binding to muscarinic receptors on the basal membrane of the parietal cell (Figure 30-15). Binding of acetylcholine to its receptors opens Ca^{++} channels and allows Ca^{++} to enter the cell. The increased cytosolic level of free Ca^{++} augments the secretion of HCl.

Histamine binds to H_2 receptors on the parietal cell membrane and activates adenylate cyclase in the plasma membrane. This process increases the cytosolic level of cAMP, which leads to increased HCl secretion.

Histamine is a major physiologic mediator of HCl secretion. Cimetidine blocks a large portion of the acid secretion elicited by any known secretagogue. Mast cells are present in the gastric muscosa, and these cells synthesize and store histamine. On stimulation by acetylcholine or gastrin, the mast cells release histamine, which diffuses to nearby parietal cells to stimulate HCl secretion.

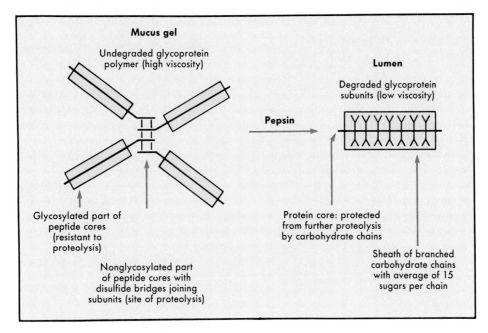

FIGURE 30-14   Schematic representation of the structure of gastric mucus glycoprotein before and after hydrolysis by pepsin. (Redrawn from Allen A: Br Med Bull 34:28, 1978.)

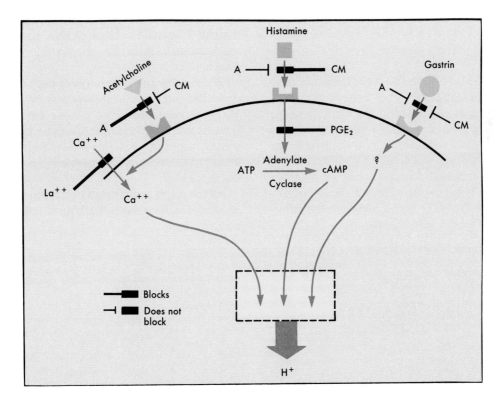

FIGURE 30-15   Mechanisms whereby secretagogues elicit acid secretion from the parietal cells. The action of blocking agents is also shown: atropine *(A)*, cimetidine *(CM)*, lanthanum ion (La^{+++}), and prostaglandin E_2 *(PGE$_2$)*. (Redrawn from Soll AH: Physiology of isolated canine parietal cells: receptors and effectors regulating function. In Johnson LR, editor: Physiology of the gastrointestinal tract, New York, 1981, Raven Press.)

Gastrin is not as potent a direct stimulant of parietal cells as acetylcholine and histamine. The direct actions of gastrin are not blocked by muscarinic antagonists or by H_2-receptor blockers, and the second messenger for gastrin action is not known. The physiologic response to elevated levels of gastrin in the blood is greatly attenuated by cimetidine. Thus a major component of the physiologic response to gastrin may result from gastrin-stimulated release of histamine.

In Vivo Control of Acid Secretion Rate When the stomach has been empty for several hours, HCl is secreted at a basal rate, which is approximately 10% of the maximal rate. The basal rate of HCl secretion varies diurnally, being highest in the evening and lowest in the morning. After a meal the rate of acid secretion by the stomach increases promptly. There are three phases of increased acid secretion in response to food: the *cephalic phase*, elicited before food reaches the stomach; the *gastric phase*, elicited by the presence of food in the stomach; and the *intestinal phase*, elicited by mechanisms originating in the duodenum and upper jejunum (Table 30-1).

Cephalic Phase

The cephalic phase of gastric secretion is normally elicited by the sight, smell, and taste of food. The cephalic phase may be studied in isolation by means of sham feeding. In sham feeding, food is chewed but not swallowed or is swallowed and then diverted to the outside of the body by an esophageal fistula.

The cephalic phase of gastric acid secretion is mediated entirely by impulses in the vagus nerves; cutting all branches of the vagus nerves to the stomach completely abolishes the cephalic phase. Cholinergic vagal fibers and cholinergic neurons of the intramural plexuses are the principal mediators of the cephalic phase. Acetylcholine released from these neurons directly stimulates parietal cells to secrete HCl. The acetylcholine also stimulates acid secretion indirectly by releasing gastrin from G cells in the antrum and duodenum and histamine from mast cells in the gastric mucosa.

The presence of a low pH in the antrum diminishes the amount of HCl secreted during the cephalic phase. In the absence of food in the stomach to buffer the acid secreted, the pH of the antral contents falls rapidly during the cephalic phase. The low pH limits the amount of acid secreted by a direct effect on parietal cells.

Gastric Phase

The gastric phase of gastric secretion is elicited by the presence of food in the stomach. The principal stimuli are distension of the stomach and the presence of amino acids and peptides resulting from the actions of pepsins. Most of the acid secreted in response to a meal is secreted during the gastric phase.

When either the body or the antrum of the stomach is distended, mechanoreceptors are stimulated. These mechanoreceptors initiate local and central reflexes that bring about HCl secretion. Both local and central responses are largely cholinergic. Afferent and efferent pathways of the central reflexes are in the vagus nerves.

All the responses elicited by gastric distension can be blocked effectively by bathing the mucosal surface

Table 30-1   Major Mechanisms for Stimulation of Gastric Acid Secretion

Phase	Stimulus	Pathway	Stimulus to Parietal Cell
Cephalic	Chewing, swallowing, etc.	Vagus nerve to:	
		1. Parietal cells	Acetylcholine
		2. G cells	Gastrin
Gastric	Gastric distension	Local and vagovagal reflexes to:	
		1. Parietal cells	Acetylcholine
		2. G cells	Gastrin
Intestinal	Protein digestion products in duodenum	1. Intestinal G cells	Gastrin
		2. Intestinal endocrine cells	Enterooxyntin

Modified from Johnson LR, editor: Gastrointestinal physiology, ed 3, St Louis, 1985, The CV Mosby Co. adapted from MI Grossman.

Table 30-2 Major Mechanisms for Inhibition of Gastric Acid Secretion

Region	Stimulus	Mediator	Inhibit Gastrin Release	Inhibit Acid Secretion by Partietal Cell
Antrum	Acid (pH < 3.0)	None, direct	+	
Duodenum	Acid	Secretin	+	+
		Bulbogastrone	+	+
		Nervous reflex		+
Duodenum and jejunum	Hyperosmotic solutions	Unidentified enterogastrone		+
	Fatty acids, monoglycerides	Gastric inhibitory peptide	+	+
		Cholecystokinin		+
		Unidentified enterogastrone		+

Modified from Johnson LR, editor: Gastrointestinal physiology, ed 3, St Louis, 1985, The CV Mosby Co; adapted from MI Grossman.

with an acid solution with a pH 2 or less. Once the buffering capacity of the gastric contents is saturated, gastric pH falls rapidly and further acid release is greatly inhibited. In this way the acidity of gastric contents regulates itself. In patients with duodenal ulcers, acid secretion is less inhibited by the presence of acid in the antrum (see Figure 30-10).

The presence of amino acids or peptides in the stomach elicits secretion of gastric acid by causing G cells to release gastrin. Intact proteins do not have this effect. The various amino acids differ greatly in their abilities to stimulate acid secretion; tryptophan and phenylalanine are particularly potent stimuli of gastrin secretion. Other frequently ingested substances that enhance gastric acid secretion include calcium ions, caffeine, and alcohol.

Intestinal Phase

The presence of chyme in the duodenum brings about neural and endocrine responses that first stimulate and later inhibit secretion of acid by the stomach. Early in gastric emptying, when the pH of gastric chyme is greater than 3, the stimulatory influences predominate. Later, when the buffer capacity of gastric chyme is exhausted and the pH of chyme emptied into the duodenum falls to less than 3, inhibitory influences prevail. Tables 30-1 and 30-2 summarize the major mechanisms that stimulate and inhibit gastric acid secretion.

Stimulation of Secretion Gastric secretion is enhanced by distension of the duodenum and by the presence of protein digestion products (peptides and amino acids) in the duodenum. The duodenum and proximal jejunum contain G cells that release gastrin when stimulated by peptides and amino acids. The gastrin is carried in the blood to the parietal cells and stimulates them to secrete acid. Amino acids in the blood can also stimulate release of gastrin.

Inhibition of Secretion Several different mechanisms that operate during the intestinal phase inhibit gastric secretion (Table 30-2). The stimuli for these mechanisms are the presence of acid, fat digestion products, and hypertonicity in the duodenum and proximal part of the jejunum. Acid solutions in the duodenum release the hormone *secretin* into the bloodstream. Secretin inhibits gastric acid secretion in two ways: inhibiting (1) gastrin release by G cells and (2) the response of parietal cells to secretagogues. Acid in the duodenum also inhibits gastric acid secretion via a local nervous reflex.

Acid in the duodenal bulb releases another hormone, *bulbogastrone,* which has not been chemically characterized. Bulbogastrone, as does secretin, inhibits acid secretion by the parietal cells.

Fatty acids with 10 or more carbons and monoglycerides, the major products of triglyceride digestion, in the duodenum and proximal part of the jejunum release two hormones: *gastric inhibitory peptide* and *cholecystokinin.* Gastric inhibitory peptide inhibits acid secretion by suppressing gastrin release and directly inhibiting secretion of acid from the parietal cells. Cholecystokinin also inhibits acid secretion by parietal cells.

Hyperosmotic solutions in the duodenum release another unidentified hormone that inhibits gastric acid secretion.

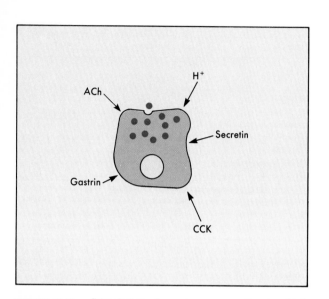

FIGURE 30-16   Stimulation of pepsinogen secretion by chief cells through various agonists: acetylcholine *(ACh)*, H^+, secretin, cholecystokinin *(CCK)*, and gastrin.

Pepsinogen Secretion

Most of the agents that stimulate parietal cells to secrete acid also elicit release of pepsinogens from chief cells (Figure 30-16). Thus the rates of release of acid and pepsinogens from the gastric glands are highly correlated. Acetylcholine is a potent stimulus for the chief cells to release pepsinogens. Gastrin also stimulates chief cells to secrete pepsinogens. Acid in contact with the gastric mucosa enhances the output of pepsinogens by a local reflex. Secretin and cholecystokinin, hormones released by the duodenal mucosa in response to acid and fat digestion products, respectively, stimulate chief cells to secrete pepsinogens.

PANCREATIC SECRETION

The human pancreas weighs less than 100 g, yet each day it secretes 1 L (10 times its mass) of pancreatic juice. The pancreas is unusual in having both endocrine and exocrine secretory functions. The exo-

crine secretions of the pancreas are important in digestion. Pancreatic juice is composed of an *aqueous component,* rich in bicarbonate, that helps to neutralize duodenal contents and an *enzyme component* that contains enzymes for digesting carbohydrates, proteins, and fats. Pancreatic exocrine secretion is controlled by both neural and hormonal signals, elicited primarily by the presence of acid and digestion products in the duodenum. *Secretin* plays a major role in eliciting secretion of the aqueous component, and *cholecystokinin* stimulates the secretion of pancreatic enzymes.

Structure and Innervation of the Pancreas

The structure of the exocrine pancreas resembles that of the salivary glands (see Figures 30-3 and 30-4). Microscopic, blind-ended tubules are surrounded by polygonal acinar cells whose primary function is to secrete the enzyme component of pancreatic juice. The acini are organized into lobules; the tiny ducts that drain the acini are called intercalated ducts. The intercalated ducts empty into somewhat larger intralobular ducts. The intralobular ducts of a particular lobule drain into a single extralobular duct that empties the lobule into still larger ducts. The larger ducts converge into a main collecting duct that drains the pancreas and enters the duodenum along with the common bile duct.

The endocrine cells of the pancreas reside in the *islets of Langerhans.* Although islet cells account for less than 2% of the volume of the pancreas, their hormones are essential in regulating metabolism. Insulin, glucagon, somatostatin, and pancreatic polypeptide are hormones released from cells of the islets of Langerhans (see Chapter 37).

The pancreas is innervated by preganglionic parasympathetic branches of the vagus nerve. Vagal fibers synapse with cholinergic neurons that are within the pancreas and that innervate both acinar and islet cells. Postganglionic sympathetic nerves from the celiac and superior mesenteric plexuses innervate pancreatic blood vessels. Secretion of pancreatic juice is stimulated by parasympathetic activity and inhibited by sympathetic activity.

Aqueous Component of Pancreatic Juice

The aqueous component of pancreatic juice is elaborated principally by the columnar epithelial cells that line the ducts. The Na^+ and K^+ concentrations of

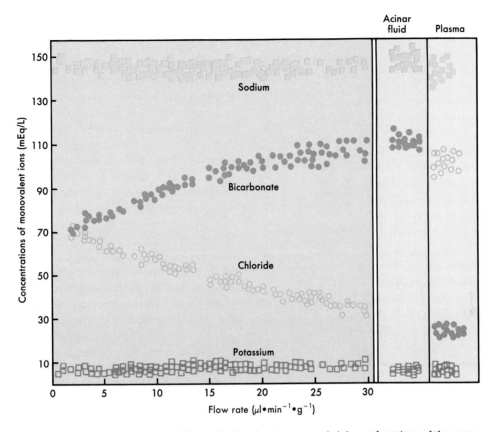

FIGURE 30-17   Concentrations of the major ions in rat pancreatic juice as functions of the secretory flow rate. The concentrations of the ions in acinar fluid and in plasma are shown for reference. Secretion was stimulated by intravenous injection of secretin. (Redrawn from Mangos JA and McSherry NA: Am J Physiol 221:496, 1971.)

pancreatic juice are similar to those in plasma. HCO_3^- and Cl^- are its major anions. The HCO_3^- concentration varies from approximately 70 mEq/L at low rates of secretion to more than 100 mEq/L at high secretory rates (Figure 30-17). HCO_3^- and Cl^- concentrations vary reciprocally. The aqueous component secreted by the duct cells is slightly hypertonic and has a high HCO_3^- concentration. As it flows down the ducts, water equilibrates across the epithelium to make the pancreatic juice isotonic, and some HCO_3^- exchanges for Cl^- (Figure 30-18).

Under resting conditions the aqueous component is produced primarily by the intercalated and other intralobular ducts. When secretion is stimulated by secretin, however, the additional flow comes mostly from the extralobular ducts (Figure 30-18). Secretin is the major physiologic stimulus for secretion of the aqueous component.

Enzyme Component of Pancreatic Juice

The secretions of the acinar cells comprise the enzyme component of pancreatic juice. The fluid that is secreted by the acinar cells resembles plasma in its tonicity and in the concentrations of various ions. The enzyme component contains enzymes important for the digestion of all the major classes of foodstuffs. In the complete absence of pancreatic enzymes, *malabsorption* of lipids, proteins, and carbohydrates occurs.

The proteases of pancreatic juice are secreted in inactive zymogen form. The major pancreatic proteases are *trypsin, chymotrypsin,* and *carboxypeptidase.* They are secreted as trypsinogen, chymotrypsinogen, and procarboxypeptidase, respectively. Trypsinogen is specifically activated by *enterokinase* (not a kinase, but a protease), which is secreted by

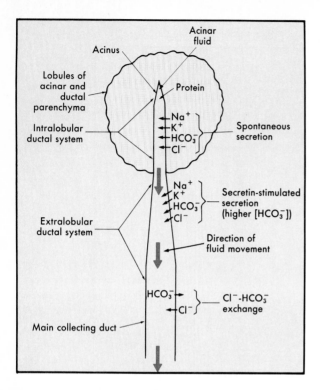

FIGURE 30-18   The locations of important transport processes involved in the elaboration of pancreatic juice. Acinar fluid is isotonic and resembles plasma in its concentrations of Na^+, K^+, Cl^-, and HCO_3^-. The secretion of acinar fluid and the proteins it contains is stimulated by cholecystokinin and acetylcholine. A spontaneous secretion that is produced by the intralobular ducts has higher concentrations of K^+ and HCO_3^- than plasma. The hormone secretin stimulates water and electrolyte secretion by the cells that line the extralobular ducts. The secretin-stimulated secretion is still richer in HCO_3^- than the spontaneous secretion. (Adapted from Swanson CH and Solomon AK: J Gen Physiol 62:407, 1973.)

the duodenal mucosa. Trypsin then activates trypsinogen, chymotrypsinogen, and procarboxypeptidase. *Trypsin inhibitor*, a protein present in pancreatic juice, prevents the premature activation of proteolytic enzymes in the pancreatic ducts.

Pancreatic juice contains an α-amylase that is secreted in active form. Pancreatic amylase cleaves starch molecules into oligosaccharides.

Pancreatic juice also contains a number of lipid-digesting enzymes, or lipases. Among the major pancreatic lipases are *triacylglycerol hydrolase, cholesterol ester hydrolase,* and *phospholipase A_2*.

Regulation of Pancreatic Exocrine Secretion

The secretory activities of duct and acinar cells of the pancreas are controlled by hormones and by substances released from nerve terminals. Stimulation of the vagal branches to the pancreas enhances the rate of secretion. Activation of sympathetic fibers inhibits pancreatic secretion, partly by decreasing blood flow to the pancreas. Secretin and cholecystokinin, hormones released from the duodenal mucosa in response to particular constituents of duodenal contents, stimulate secretion of the aqueous and enzyme components, respectively. Because the aqueous and enzyme components of pancreatic juice are separately controlled, the composition of the juice varies from less than 1% to as much as 10% protein. Substances other than secretin and cholecystokinin also modulate pancreatic exocrine function.

Cephalic Phase   Sham feeding induces the secretion of a low volume of pancreatic juice with a high protein content. *Gastrin* released from the mucosa of the gastric antrum in response to vagal impulses is a major mediator of pancreatic secretion during the cephalic phase. Gastrin is a member of the same class of peptides as cholecystokinin, but it is much less potent as a pancreatic secretagogue than cholecystokinin.

Gastric Phase   During the gastric phase of secretion, gastrin is released in response to gastric distension and the presence of amino acids and peptides in the antrum of the stomach. The gastrin released during the gastric phase enhances secretion by the pancreas. In addition, reflexes elicited by stretching either the fundus or the antrum of the stomach evoke secretion of small volumes of pancreatic juice with high enzyme content.

Intestinal Phase   In the intestinal phase of secretion, certain components of the chyme in the duodenum and upper jejunum evoke pancreatic secretion. Acid in the duodenum and upper jejunum elicits the secretion of a large volume of pancreatic juice rich in bicarbonate but poor in pancreatic enzymes. The hormone secretin is the major mediator of this response to acid. Secretin is released by certain cells in the mucosa of the duodenum and upper jejunum in response to acid in the lumen. Secretin is released when the pH of duodenal contents is 4.5 or less. Secretin directly stimulates the cells of the pancreatic ductular epithelium to secrete the bicarbonate-rich aqueous component of the pancreatic juice.

The presence of peptides and certain amino acids, especially tryptophan and phenylalanine, in the duodenum brings about the secretion of pancreatic juice rich in protein components. Fatty acids with chain lengths longer than eight carbon atoms and monoglycerides of these fatty acids also elicit secretion of protein-rich pancreatic juice. Cholecystokinin is the most important physiologic mediator of this response to the digestion products of proteins and lipids. Cholecystokinin is a hormone released by particular cells in the duodenum and upper jejunum in response to these digestion products. This hormone directly stimulates the acinar cells to release the contents of their zymogen granules.

Cholecystokinin has little effect on the ductular epithelium of the pancreas, but it potentiates the stimulatory effect of secretin on the ducts. Secretin is a weak agonist in acinar cells, but it potentiates the effect of cholecystokinin on acinar cells.

Cellular Mechanisms of the Mediators of Pancreatic Exocrine Secretion

Figure 30-19 shows six classes of receptors that mediate the responses of pancreatic acinar cells to secretagogues. Secretagogues that act via cAMP potentiate the effect of secretagogues using Ca^{++} as a second messenger, and vice versa. Secretagogues that act via a common second messenger do not potentiate one another's effects.

FUNCTIONS OF THE LIVER AND GALLBLADDER

Structure of the Liver

One view of the histology of the liver is shown in Figure 30-20. Each liver lobule is organized around a central vein. At the periphery of the lobule blood

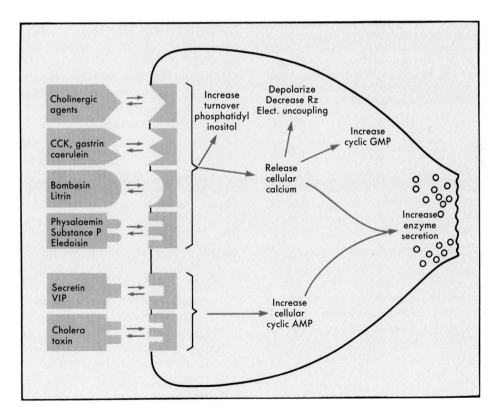

FIGURE 30-19   Representation of the cellular mechanisms of action of secretagogues on pancreatic acinar cells. Six classes of receptors for secretagogues are shown. Two receptor types are linked to adenylate cyclase and to increased intracellular cAMP. The other four receptor types are coupled to turnover of inositol phospholipids and to increased intracellular Ca^{++}. (Adapted from Jensen RT and Gardner JD: Adv Cyclic Nucl Res 17:375, 1984.)

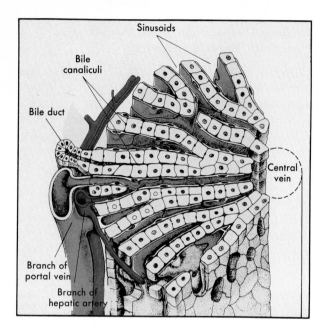

Sinusoids

Bile
canaliculi

Bile duct

Central
vein

Branch of
portal vein

Branch of
hepatic artery

FIGURE 30-20   Diagrammatic representation of a hepatic
lobule. A central vein is located in the center of the lobule
with plates of hepatic cells disposed radially. Branches of the
portal vein and hepatic artery are located on the periphery of
the lobule, and blood from both perfuses the sinusoids.
Peripherally located bile ducts drain the bile canaliculi that
run between the hepatocytes. (Adapted from Bloom W and
Fawcett DW: A textbook of histology, ed 10, Philadelphia,
1975, WB Saunders.)

enters the sinusoids from branches of the portal vein
and the hepatic artery. In the sinusoids blood flows
toward the center of the lobule between plates of
hepatic cells that are one or two hepatocytes thick.
Because of the large fenestrations between the endothelial cells that line the sinusoids, each hepatocyte is
in direct contact with sinusoidal blood. The intimate
contact of a large fraction of the hepatocyte surface
with blood contributes to the ability of the liver to
clear the blood effectively of certain classes of compounds. Biliary canaliculi lie between adjacent hepatocytes, and the canaliculi drain into bile ducts at the
periphery of the lobule.

Functions of the Liver

The liver performs many functions vital to the
health of the organism. The liver is essential in regulating metabolism, synthesizing many proteins, serving as a storage site for certain vitamins and iron,

degrading certain hormones, and inactivating and
excreting many drugs and toxins.

The liver regulates the metabolism of carbohydrates, lipids, and proteins. Liver and skeletal muscle
are the two major sites of glycogen storage in the
body. When the level of glucose in the blood is high,
glycogen is deposited in the liver. When blood glucose is low, liver glycogen is broken down to glucose
(glycogenolysis), and the glucose is then released into
the blood. In this way the liver helps to maintain a
relatively constant blood glucose level. The liver is
also the major site of *gluconeogenesis*, the conversion
of amino acids, lipids, or simple carbohydrate substances (e.g., lactate) into glucose. Carbohydrate
metabolism by the liver is regulated by several hormones (see Chapter 37).

The liver is also involved in lipid metabolism. As
described in Chapter 31, lipids absorbed by the intestine leave the intestine in chylomicrons in the lymph.
Lipoprotein lipase on the endothelial cell surface of
blood vessels hydrolyzes some of the triglyceride in
the chylomicrons, thereby allowing glycerol and fatty
acids to be taken up by adipocytes. This results in
formation of chylomicron remnants rich in cholesterol. Chylomicron remnants are taken up by hepatocytes and degraded. Hepatocytes synthesize and
secrete very-low-density lipoproteins. Very-low-density lipoproteins are then converted to the other types
of serum lipoproteins. These lipoproteins are the
major sources of cholesterol and triglycerides for most
other tissues of the body. Cholesterol present in bile
represents the only route of excretion of cholesterol.
Hepatocytes are thus a principal source of cholesterol
in the body and the major site of excretion of cholesterol. Thus, hepatocytes play an important role in regulation of serum cholesterol levels.

In certain physiologic and pathologic conditions
(e.g., diabetes mellitus), β-oxidation of fatty acids
provides the major source of energy for the body. In
the liver the oxidation of fatty acids produces acetoacetate, β-hydroxybutyrate, and acetone. These
three compounds are called *ketone bodies*. Ketone
bodies are released from hepatocytes and carried in
the circulation to other tissues, where they are
metabolized. The hepatic functions that regulate lipid
metabolism are also subject to endocrine control.

The liver is centrally involved in protein metabolism. When proteins are broken down (catabolized),
amino acids are deaminated to form ammonia (NH_3).
Ammonia cannot be further metabolized by most tis-

sues and becomes toxic at levels achievable by metabolism. Ammonia is dissipated by conversion to urea, mainly in the liver. The liver also synthesizes all the nonessential amino acids.

The liver synthesizes all the major plasma proteins, including the plasma lipoproteins, albumins, globulins, fibrinogens, and other proteins involved in blood clotting.

The liver stores certain substances important in metabolism. Next to hemoglobin in red blood cells, the liver is the most important storage site for iron. Certain vitamins, most notably A, D, and B_{12}, are stored in the liver. Hepatic storage protects the body from limited dietary deficiencies of these vitamins.

The liver is a major site for the degradation and excretion of hormones. Epinephrine and norepinephrine are inactivated by oxidation. Certain polypeptide hormones are degraded by liver cells. The liver also inactivates and excretes steroid hormones.

The liver transforms and excretes many drugs and toxins. These substances are frequently converted to inactive forms by reactions that occur in hepatocytes. The smooth endoplasmic reticulum of hepatocytes contains systems of enzymes and cofactors that are responsible for chemical transformations of many drugs. Certain other enzymes in the endoplasmic reticulum catalyze the conjugation of many compounds with glucuronic acid, glycine, or glutathione. The transformations that occur in the liver render many drugs more water soluble, and thus they are more readily excreted by the kidneys. Some drug metabolites are secreted into the bile.

Bile Secretion

The hepatic function most important to the digestive tract is the secretion of bile. Bile, elaborated by hepatocytes, contains bile acids, cholesterol, lecithin, and bile pigments. These constituents are all synthesized and secreted by hepatocytes into the bile canaliculi, along with an isotonic fluid that resembles plasma in its electrolyte concentrations. The bile canaliculi merge into ever larger ducts and finally into a single large bile duct. The epithelial cells that line the bile ducts secrete a watery fluid that is rich in bicarbonate and contributes to the volume of bile leaving the liver.

The secretory function of the liver resembles that of the exocrine pancreas. In both organs the major parenchymal cell type elaborates a primary secretion containing the substances responsible for the major digestive function of the organ. In both pancreas and liver the primary secretion is isotonic and contains Na^+, K^+, and Cl^- at concentrations near plasma levels, and the primary secretion is stimulated by cholecystokinin. In both pancreas and liver the epithelial cells that line the duct system modify the primary secretion. When stimulated by secretin, these epithelial cells contribute an aqueous secretion with a high bicarbonate concentration.

Between meals bile is diverted into the gallbladder. The gallbladder epithelium extracts salts and water from the stored bile, and the bile acids are thereby concentrated 5- to 20-fold. After an individual has eaten, the gallbladder contracts and empties its concentrated bile into the duodenum. The most potent stimulus for emptying of the gallbladder is cholecystokinin. From 250 to 1500 ml of bile enters the duodenum each day.

Bile acids emulsify lipids, thereby increasing the surface area available to lipolytic enzymes. Bile acids then form mixed micelles (see Chapter 31) with the products of lipid digestion. This process increases the transport of lipid digestion products to the brush border surface and thus enhances absorption of lipids by the epithelial cells. Bile acids are actively absorbed, primarily in the terminal part of the ileum. A small fraction of bile acids escapes absorption and is excreted. The returning bile acids are avidly taken up by the liver and are rapidly resecreted during the course of digestion. The entire bile acid pool is recirculated twice in response to a typical meal. The recirculation of the bile is known as the *enterohepatic circulation*. Approximately 20% of the bile acid pool is excreted in the feces each day and is replenished by hepatic synthesis of new bile acids. Figure 30-21 summarizes some major aspects of the enterohepatic circulation.

Fraction of Bile Secreted by Hepatocytes

Bile Acids Bile acids comprise about 50% of the dry weight of bile. Other important compounds secreted by the hepatocytes into the bile include lecithin, cholesterol, bile pigments, and proteins.

Bile acids have a steroid nucleus and are synthesized by the hepatocytes from cholesterol. The major bile acids synthesized by the liver are called *primary bile acids*. These are cholic acid (3-hydroxyl groups) and chenodeoxycholic acid (2-hydroxyl groups). The

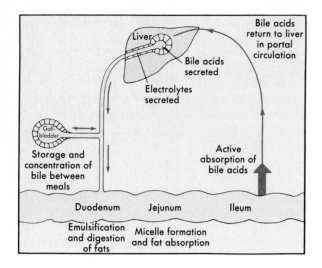

FIGURE 30-21   Overview of the enterohepatic circulation of bile.

presence of the carboxyl and hydroxyl groups makes the bile acids much more water soluble than the cholesterol from which they are synthesized.

Bacteria in the digestive tract dehydroxylate bile acids to form *secondary bile acids*. The major secondary bile acids are deoxycholic acid (from dehydroxylation of cholic acid) and lithocholic acid (from dehydroxylation of chenodeoxycholic acid). Bile contains both primary and secondary bile acids.

Bile acids normally are secreted conjugated with glycine or taurine. In conjugated bile acids the glycine or taurine is linked by a peptide bond between the carboxyl group of an unconjugated bile acid and the amino group of glycine or taurine. The pK's of the carboxyl groups of unconjugated bile acids are near neutral pH, but the pK's of conjugated bile acids are considerably lower. Thus, at the near neutral pH of the gastrointestinal tract, the conjugated bile acids are more completely ionized, and thus more water soluble, than the unconjugated bile acids. Conjugated bile acids therefore exist almost entirely as salts of various cations (mostly Na^+) and are often called *bile salts*.

The steroid nucleus of bile acids is roughly planar. In solution bile acids have their polar (hydrophilic) groups—the hydroxyl groups, the carboxyl moiety of glycine or taurine, and the peptide bond—all on one surface of the molecule. The other surface is quite hydrophobic. This makes the bile acid molecule *amphipathic,* that is, having both hydrophilic and

hydrophobic domains. Conjugated bile acids are more amphipathic than unconjugated ones. Because they are amphipathic, bile acids tend to form molecular aggregates, called micelles, by turning their hydrophobic faces inside and away from water and their hydrophilic surfaces toward the water. Whenever bile acids are present above a certain concentration, called the *critical micelle concentration,* bile acid micelles will form. Above this concentration any additional bile acid will go into the micelles exclusively and not into molecular solution. In bile, the bile acids are normally present at a concentration much greater than the critical micelle concentration.

Phospholipids in Bile   Hepatocytes also secrete phospholipids into the bile. Lecithins are the most prominent class of phospholipids in bile. Cholesterol is also secreted into the bile, and this is the major route for cholesterol excretion. Although lecithin and cholesterol are essentially insoluble in water, they dissolve in the bile acid micelles. The lecithin increases the amount of cholesterol that can be solubilized in the micelles. If more cholesterol is present in the bile than can be solubilized in the micelles, crystals of cholesterol will form in the bile. These crystals are important in formation of cholesterol gallstones (the most common gallstones) in the duct system of the liver or more often in the gallbladder.

Bile Pigments   When senescent red blood cells are degraded in reticuloendothelial cells, the porphyrin moiety of hemoglobin is converted to *bilirubin*. Bilirubin is released into the plasma, where it is bound to albumin. Hepatocytes efficiently remove bilirubin from blood in the sinusoids via a protein-mediated transport mechanism in the hepatocyte plasma membrane that faces the sinusoids. In the hepatocytes bilirubin is conjugated with one or two glucuronic acid molecules, and the bilirubin glucuronides are secreted into the bile. Unconjugated bilirubin is not secreted into the bile. Bilirubin is yellow and contributes to the yellow color of bile.

Secretion of Bile Duct Epithelium

The epithelial cells that line the bile ducts contribute an aqueous secretion that can account for about 50% of the total volume of the bile. The secretion of the bile duct epithelium is isotonic and contains Na^+ and K^+ at levels similar to plasma. However, the concentration of HCO_3^- is greater and the concentration of Cl^- is less than in plasma. The secretory activity of the bile duct epithelium is specifically stimulated by secretin.

Cellular Mechanism of Bile Formation

Secretion of Bile Acids Figure 30-22 illustrates the current understanding of the cellular mechanisms responsible for secretion of bile acids by hepatocytes into bile canaliculi. As is true for many epithelial cells, the plasma membrane of the hepatocyte is polarized such that the membrane facing the bile canaliculus is different from the basolateral membrane (the membrane facing the sinusoid plus the lateral cell membrane that faces adjacent hepatocytes). The basolateral membrane contains a bile acid transport protein that uses the energy of the Na^+ gradient to accumulate bile acids actively from the sinusoidal blood into the cytosol of the hepatocyte. The Na^+ gradient in turn, is created by the Na^+, Ka^+-ATPase that resides in the basolateral plasma membrane. In the hepatocyte most of the bile acids present are bound by specific cytosolic proteins.

Secretion of Electrolytes and Water Water and electrolytes enter the primary secretion in the bile

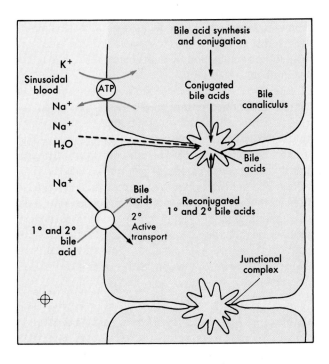

FIGURE 30-22 A representation of cellular mechanisms responsible for the secretion of bile acids by hepatocytes into the bile canaliculi.

canaliculi and are also secreted into the bile by the bile duct epithelium.

Bile Concentration and Storage in the Gallbladder

Between meals the tone of the sphincter of Oddi, which guards the entrance of the common bile duct into the duodenum, is high. Thus most bile flow is diverted into the gallbladder. The gallbladder is a small organ, having a capacity of 15 to 60 ml (average about 35 ml) in humans. Many times this volume of bile may be secreted by the liver between meals. The gallbladder concentrates the bile by absorbing Na^+, Cl^-, HCO_3^-, and water from the bile, such that the bile acids are concentrated from 5 to 20 times in the gallbladder. The active transport of Na^+ is the primary active process in the concentrating action of the gallbladder. Cl^- and HCO_3^- are absorbed to preserve electroneutrality.

Because of its high rate of water absorption, the gallbladder serves as a model for water and electrolyte transport by tight-junctioned epithelia. The *standing osmotic gradient mechanism* for fluid absorption was first proposed for the gallbladder. It was noted that during fluid reabsorption by the gallbladder, the lateral intercellular spaces between the epithelial cells were large and swollen. When fluid transport was blocked, the intercellular spaces almost disappeared. These observations strongly suggested that the intercellular spaces are a major route of fluid flow during absorption.

A current view is that Na^+ is actively pumped into the lateral intercellular spaces. The Na^+ pumps are believed to be especially dense near the mucosal (apical) end of the channel (Figure 30-23). Cl^- and HCO_3^- also are transported into the intercellular space, probably because of an electrical potential created by Na^+ transport. The high ion concentration near the apical end of the intercellular space causes the fluid there to be hypertonic. This produces an osmotic flow of water from the lumen via adjacent cells into the intercellular space. Water distends the intercellular channels because of increased hydrostatic pressure. As a result of water flow from adjacent cells, the fluid becomes less hypertonic as it flows down the intercellular channel. It is essentially isotonic when it reaches the serosal (basal) end of the channel. Ions and water flow across the basement membrane of the epithelium and are carried away by the capillaries.

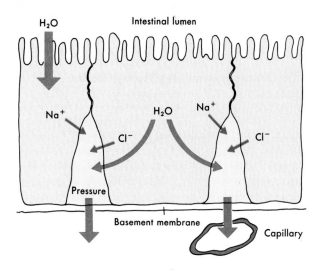

H₂O   Intestinal lumen

FIGURE 30-23   Water absorption from the gallbladder by the standing gradient osmotic mechanism. Na^+ is actively pumped into the lateral intercellular spaces; Cl^- follows. Water is drawn by osmosis to enter the intercellular spaces, elevating the hydrostatic pressure there. Water, Na^+, and Cl^- are filtered across the porous basement membrane and enter the capillaries.

Emptying of the Gallbladder

Emptying of the gallbladder contents into the duodenum begins several minutes after the start of a meal. Intermittent contractions of the gallbladder force bile through the partially relaxed sphincter of Oddi. During the *cephalic* and *gastric phases* of digestion, gallbladder contraction and relaxation of the sphincter are mediated by cholinergic fibers in branches of the vagus nerve. Stimulation of sympathetic nerves to the gallbladder and duodenum inhibits emptying of the gallbladder.

The highest rate of gallbladder emptying occurs during the *intestinal phase* of digestion; the strongest stimulus for the emptying is cholecystokinin. Cholecystokinin reaches the gallbladder via the circulation and causes strong contractions of the gallbladder and relaxation of the sphincter of Oddi. Substances that mimic the actions of cholecystokinin in promoting gallbladder emptying are *cholecystagogues,* such as gastrin. Gastrin has the same sequence of five amino acids at its *C.* terminal as does cholecystokinin; however, gastrin is not nearly as potent as cholecystokinin. Nevertheless, gastrin plays a role in eliciting gallbladder contractions during the cephalic and gastric phases.

Under normal circumstances the rate of gallbladder emptying is sufficient to keep the concentration of bile acids in the duodenum greater than the critical micelle concentration.

Intestinal Absorption of Bile Acids and Their Enterohepatic Circulation

The functions of bile acids in emulsifying dietary lipid and in forming mixed micelles with the products of lipid digestion are discussed in Chapter 31.

Normally, by the time chyme reaches the terminal part of the ileum, dietary fat is almost completely absorbed. Bile acids are then absorbed. The epithelial cells of the distal part of the ileum actively take up bile acids against a large concentration gradient. The active transport system has a higher affinity for conjugated than for unconjugated bile acids. Since bile acids are also lipid soluble, they can be taken up by simple diffusion as well. Bacteria in the terminal part of the ileum and colon deconjugate bile acids and also dehydroxylate them to produce secondary bile acids. Both deconjugation and dehydroxylation lessen the polarity of bile acids, enhancing their lipid solubility and their absorption by simple diffusion.

Typically about 0.5 g of bile acids escapes absorption and is excreted in the feces each day. This quantity is 15% to 35% of the total bile acid pool, and normally it is replenished by synthesis of new bile acids by the liver.

Bile acids, whether absorbed by active transport or simple diffusion, are transported away from the intestine in the portal blood, mostly bound to plasma proteins. In the liver, hepatocytes avidly extract the bile acids from the portal blood. In a single pass through the liver the portal blood is essentially cleared of bile acids. Bile acids in all forms, primary and secondary, both conjugated and deconjugated, are taken up by the hepatocytes. The hepatocytes reconjugate almost all the deconjugated bile acids and rehydroxylate some of the secondary bile acids. These bile acids are secreted into the bile along with newly synthesized bile acids.

Control of Bile Acid Synthesis and Secretion

The rate of return of bile acids to the liver affects the rate of synthesis and secretion of bile acids. Bile acids in the portal blood stimulate the uptake and resecretion of bile acids by the hepatocytes. This is called the *choleretic* effect of bile acids; substances that enhance bile acid secretion are known as *choleretics*. So powerful is the stimulus to resecrete the returning bile acids that the entire pool of bile acids (1.5 to 31.5 g) recirculates twice in response to a typical meal. In response to a meal with a very high fat content, the bile acid pool may recirculate five or more times.

Gallstones

The most common type of gallstone contains cholesterol as its major component. Cholesterol is essentially insoluble in water. When bile contains more cholesterol than can be solubilized in the bile acid-lecithin micelles, crystals of cholesterol form in the bile. Such bile is said to be *supersaturated* with cholesterol. The greater the concentration of bile acids and lecithin in bile, the greater is the amount of cholesterol that can be contained in the mixed micelles.

Bile pigment stones are the other major class of gallstones; their major constituent is the calcium salt of unconjugated bilirubin. Conjugated bilirubin is quite soluble and does not form insoluble calcium salts in bile. Bile may contain elevated levels of unconjugated bilirubin because hepatocytes are deficient in forming the glucuronide or because of excessive deconjugation by glucuronidase.

INTESTINAL SECRETIONS

The mucosa of the intestine, from the duodenum through the rectum, elaborates secretions that contain mucus, electrolytes, and water. The total volume of intestinal secretions is about 1500 ml/day. The mucus in the secretions protects the mucosa from mechanical damage. The nature of the secretions and the mechanisms that control secretion vary from one segment of the intestine to another.

Duodenal Secretions

The duodenal submucosa contains branching glands that elaborate a secretion rich in mucus. The glands have ducts that empty into Lieberkühn's crypts. The duodenal epithelial cells also contribute to duodenal secretions, but most of the secretions are produced by the glands. The duodenal secretion contains mucus and an aqueous component that does not differ greatly from plasma in its concentrations of the major ions.

Secretions of the Small Intestine

Goblet cells, which lie among the columnar epithelial cells of the small intestine, secrete mucus. During normal digestion an aqueous secretion is elaborated by the epithelial cells at a rate only slightly less than the rate of fluid absorption by the small intestine.

Secretions of the Colon

The secretions of the colon are smaller in volume but richer in mucus than small intestinal secretions. The mucus is produced by the numerous goblet cells of the colonic mucosa. The aqueous component of colonic secretions is rich in K^+ and HCO_3^-. The production of colonic secretions is stimulated by mechanical irritation of the mucosa and by activation of cholinergic pathways to the colon. Stimulation of sympathetic nerves to the colon decreases the rate of colonic secretion.

BIBLIOGRAPHY
Journal Articles

Allen A and Garner A: Mucus and bicarbonate secretion in the stomach and their possible role in mucosal protection, Gut 21:249, 1980.

Jensen RT and Gardner JD: The cellular basis of action of gastrointestinal peptides, Adv Cyclic Nucleotide Protein Phosphorylation Res 17:375, 1984.

Klaasen CD and Watkins III JB: Mechanisms of bile formation, hepatic uptake, and biliary excretion, Pharmacol Rev 36:1, 1984.

Putney Jr JW: Identification of cellular activation mechanisms associated with salivary secretion, Annu Rev Physiol 48:75, 1986.

Williams JA: Regulatory mechanisms in pancreas and salivary acini, Annu Rev Physiol 46:361, 1984.

Books and Monographs

Davenport HW: Physiology of the digestive tract, ed 5, Chicago, 1982, Year Book Medical Publishers, Inc.

Johnson RL, editor: Physiology of the gastrointestinal tract, ed 2, vols 1 and 2, New York, 1987, Raven Press.

CHAPTER

Digestion and Absorption

In most instances nutrients cannot be absorbed by the cells that line the gastrointestinal (GI) tract in the forms in which they are ingested. *Digestion* refers to the processes by which ingested molecules are cleaved into smaller ones by reactions catalyzed by enzymes in the lumen or on the luminal surface of the GI tract. As a result of digestion, ingested molecules are converted to forms that can be absorbed from the lumen of the GI tract. *Absorption* refers to the processes by which molecules are transported into the epithelial cells that line the GI tract and then enter the blood or lymph draining that region of the tract.

DIGESTION AND ABSORPTION OF CARBOHYDRATES

Carbohydrates in the Diet

No nutritional requirement exists for carbohydrate per se, but it is usually the principal source of calories. Plant starch, *amylopectin,* is the major source of carbohydrate in most human diets. Amylopectin is a high-molecular-weight ($>10^6$), branched molecule of glucose monomers. A smaller proportion of dietary starch is *amylose,* a smaller-molecular-weight ($>10^5$), linear, α-1,4-linked polymer of glucose. Cellulose is a β-1,4-linked glucose polymer. Intestinal enzymes cannot hydrolyze β-glycosidic linkages. Thus cellulose and other molecules with β-glycosidic linkages remain undigested and contribute to dietary "fiber." The amount of the animal starch, glycogen, typically ingested varies widely among cultures and among individuals within a given culture. Sucrose and lactose are the principal dietary disaccharides, and glu-

cose and fructose are the major monosaccharides. The capacity of a healthy digestive system to digest and absorb carbohydrates greatly exceeds the amount of carbohydrate normally presented to it.

Digestion of Carbohydrates

The structure of a branched starch molecule is depicted in the Figure 31-1. Starch is a polymer of glucose and consists of chains of glucose units linked by α-1,4 glycosidic bonds. The α-1,4 chains have branch points formed by α-1,6 glycosidic linkages, and the starch molecule is highly branched.

The digestion of starch begins in the mouth with the action of salivary amylase. This enzyme catalyzes the hydrolysis of the internal α-1,4 links of starch, but it cannot hydrolyze the α-1,6 branching links. As shown in Figure 31-1, the principal products of α-amylase digestion of starch are *maltose, maltotriose,*

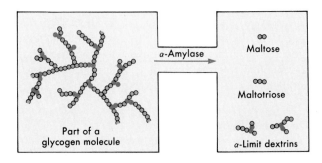

FIGURE 31-1 Structure of a branched starch molecule and the action of α-amylase. The circles represent glucose monomers. The red circles show glucose units linked by α-1,6 linkages at the branch points. The α-1,6 links and terminal α-1,4 bonds cannot be cleaved by α-amylase.

and branched oligosaccharides known as α-*limit dextrins*. Considerable digestion of starch by the salivary amylase may occur normally, but this enzyme is not required for the complete digestion and absorption of the starch ingested. After the salivary amylase is inactivated by gastric acid, no further processing of carbohydrate occurs in the stomach.

The pancreatic secretions contain a highly active α-amylase. The products of starch digestion by this enzyme are the same as for the salivary amylase, but the total activity of the pancreatic enzyme is considerably greater than the salivary amylase. The pancreatic amylase is most concentrated in the duodenum. Within 10 minutes after entering the duodenum, starch is entirely converted to the oligosaccharides shown in Figure 31-1. The further digestion of these oligosaccharides is accomplished by enzymes that reside in the brush border membrane of the epithelium of the duodenum and jejunum (Figure 31-2). The major brush border oligosaccharidases are *lactase*, which splits lactose into glucose and galactose; *sucrase*, which splits sucrose into fructose and glucose; α-*dextrinase* (also called *isomaltose*), which "debranches" the α-limit dextrins by cleaving the α-

1,6 linkages at the branch points; and *glucoamylase*, which breaks maltooligosaccharides down to glucose units. The activities of these four enzymes is highest in the brush border of the upper jejunum, and they gradually decline through the rest of the small intestine.

Absorption of Carbohydrates

The duodenum and upper jejunum have the highest capacity to absorb sugars. The capacities of the lower jejunum and ileum are progressively less. The only dietary monosaccharides that are well absorbed are *glucose*, *galactose*, and *fructose*. Glucose and galactose are actively taken up by the brush border epithelial cells through a Na^+-powered secondary active transport system. Glucose and galactose compete for entry; other sugars are less effective competitors.

Na^+ and glucose or galactose are transported into the cell by a common membrane protein, which has two Na^+-binding sites and one sugar-binding site. The presence of Na^+ in the lumen enhances the absorption of glucose and galactose, and vice versa.

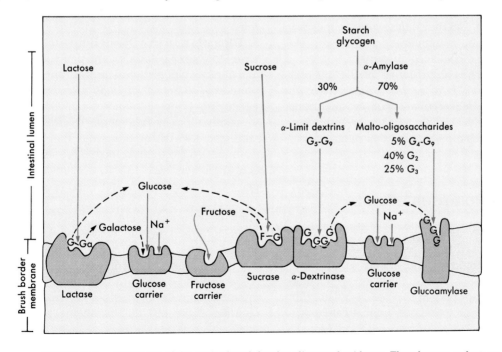

FIGURE 31-2   Functions of the major brush border oligosaccharidases. The glucose, galactose, and fructose molecules released by enzymatic hydrolysis are then transported into the epithelial cells by specific transport proteins in the brush border membrane. *G*, Glucose; *Ga*, galactose; *F*, fructose. (Redrawn from Gray GM: N Engl J Med 292:1225, 1975. Reprinted by permission of The New England Journal of Medicine.)

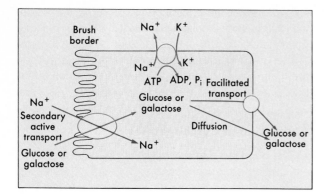

FIGURE 31-3 Major features of glucose and galactose absorption in the small intestine. Glucose and galactose enter the epithelial cell against a concentration gradient. The gradient of Na^+ provides the energy for sugar entry. Glucose and galactose leave the cell at the basolateral membrane by facilitated transport and by simple diffusion.

The energy released by Na^+ moving down its electrochemical potential gradient is harnessed to transport glucose or galactose into the cell against a concentration gradient of the sugar. Glucose and galactose leave the intestinal epithelial cell at the basal and lateral plasma membranes via facilitated transport and by simple diffusion, and they diffuse into the mucosal capillaries. Figure 31-3 summarizes some major features of glucose and galactose absorption.

Fructose does not compete well for the glucose-galactose system, and fructose transport is not known to be linked to Na^+ absorption. However, fructose is transported almost as rapidly as glucose and galactose and much more rapidly than other monosaccharides.

DIGESTION AND ABSORPTION OF PROTEINS

The amount of dietary protein varies greatly among cultures and among individuals within a culture. In poor societies it is difficult for an adult to obtain the amount of protein (0.5 to 0.7 g/day/kg of body weight) required to balance normal catabolism of proteins, and it is even more difficult for children to receive the relatively greater amounts of protein required to sustain normal growth. In wealthier societies a typical individual may ingest protein far in excess of the nutritional requirement. In addition to the protein ingested, the GI tract must deal with 10 to

30 g of protein per day contained in digestive secretions and a similar amount of protein in desquamated epithelial cells.

In normal humans essentially all ingested protein is digested and absorbed. Most of the protein in digestive secretions and desquamated cells is also digested and absorbed. The small amount of protein in the feces is derived principally from colonic bacteria, desquamated colonic cells, and proteins in mucous secretions of the colon. In humans ingested protein is almost completely absorbed by the time the meal has traversed the jejunum.

Digestion of Proteins

Digestion in the Stomach *Pepsinogen* is secreted by the chief cells of the stomach and is converted by hydrogen ions to the active enzyme *pepsin*. The extent to which pepsin hydrolyzes dietary protein is significant but highly variable. At most approximately 15% of dietary protein may be reduced to amino acids and small peptides by pepsin. The duodenum and small intestine have such a high capacity to process protein that the total absence of pepsin does not impair the digestion and absorption of dietary protein.

Digestion in the Duodenum and Small Intestine Proteases secreted by the pancreas play a major role in protein digestion. The most important of these proteases are *trypsin, chymotrypsin,* and *carboxypeptidase*. The pancreatic juice contains these enzymes in inactive, proenzyme forms. The enzyme *enterokinase,* secreted by the mucosa of the duodenum and jejunum, converts trypsinogen to trypsin. Trypsin acts autocatalytically to activate trypsinogen and also converts chymotrypsinogen and procarboxypeptidase to the active enzymes (Figure 31-4). The pancreatic proteases are present at high activity levels in the duodenum and rapidly convert dietary protein to small peptides. About 50% of the ingested protein is digested and absorbed in the duodenum.

The brush border of the duodenum and the small intestine contains a number of *peptidases*. These peptidases are integral membrane proteins whose active sites face the intestinal lumen. Figure 31-4 illustrates some major proteases in the small intestine.

The principal products of protein digestion by pancreatic proteases and brush border peptidases are small peptides and single amino acids. The small pep-

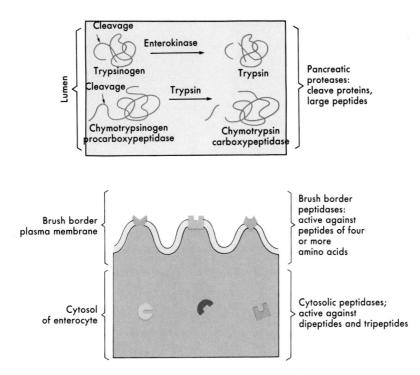

FIGURE 31-4   The major proteases and peptidases present in the lumen of the small intestine, on the brush border plasma membrane, and in the cytosol of the enterocytes of the small intestine.

tides (mainly dipeptides, tripeptides, and tetrapeptides) are about three or four times more concentrated than the single amino acids. As discussed next, small peptides and amino acids are transported across the brush border plasma membrane into intestinal epithelial cells. Small peptides are then hydrolyzed by peptidases in the cytosol of the epithelial cells; consequently, single amino acids and a few dipeptides appear in the portal blood. The cytosolic peptidases are particularly active against dipeptides and tripeptides, which are transported with high efficiency across the brush border membrane. The brush border peptidases, on the other hand, are mainly active against peptides of four or more amino acids.

Absorption of Protein Digestion Products

Intact Proteins and Large Peptides Intact proteins are not absorbed by humans to an extent that is nutritionally significant, but amounts sufficient to trigger an immunologic response can be absorbed. In ruminants and rodents, but not in humans, the neonatal intestine has a high capacity for the specific absorption of immune globulins present in colostrum.

This is vital in the development of normal immune competence in ruminants and rodents.

Absorption of Small Peptides Dipeptides and tripeptides are transported across the brush border membrane. The rate of transport of dipeptides or tripeptides usually exceeds the rate of transport of individual amino acids. For example, glycine is absorbed by the human jejunum less rapidly as the amino acid than it is from glycylglycine or from glycylglycylglycine. A single membrane transport system with broad specificity is probably responsible for absorption of small peptides. The transport system has high affinity for dipeptides and tripeptides, but very low affinities for peptides of four or more amino acid residues. Transport of dipeptides and tripeptides across the brush border plasma membrane is a secondary active transport process, powered by the electrochemical potential difference of Na^+ across the membrane. The total amount of each amino acid that enters intestinal epithelial cells in the form of dipeptides or tripeptides is considerably greater than the amount that enters as the single amino acid.

Absorption of Amino Acids With respect to amino acid transport, the brush border plasma membrane of small intestinal epithelial cells differs considerably from the basolateral plasma membrane. Normally amino acids are transported across the brush border plasma membrane into the enterocyte by way of certain specific amino acid transport systems. Transport of amino acids out of the epithelial cell across the basolateral membrane occurs by other transporters. Some of the transporters depend on the Na^+ gradient (as previously described for glucose and galactose absorption), whereas other transport systems are independent of Na^+. The Na^+-dependent brush border transport systems are found only in epithelial cells. Brush border membranes of the small intestine and proximal renal tubule are similar with respect to the nature of Na^+-dependent amino acid transport processes. The amino acid transporters in the brush border plasma membrane that do not depend on the Na^+ gradient are similar to amino acid transport systems found in nonepithelial cells.

Other transport systems are responsible for transporting amino acids across the basolateral plasma membrane out of the intestinal epithelial cell. As is true for the brush border membrane, some of the amino acid transporters present in the basolateral membrane depend on Na^+ and others do not. The basolateral membrane, however, is less highly differentiated than the brush border membrane, in that all of the amino acid transporters present in the basolateral membrane occur in certain nonepithelial cells as well. For most amino acids simple diffusion is a significant pathway across both brush border and basolateral membranes. The more hydrophobic the amino acid and the larger its concentration gradient across the membrane, the greater is the importance of diffusion.

INTESTINAL ABSORPTION OF WATER AND SALTS

Under normal circumstances humans absorb almost 99% of the water and ions presented to them in ingested food and in GI secretions. Thus net fluxes of water and ions are normally from the lumen of the gut to the blood. In most cases the net fluxes of water and ions are the differences between much larger unidirectional flows from lumen to blood and from blood to lumen.

Absorption of Water

Typically about 2 L of water are ingested each day, and approximately 7 L/day are contained in GI secretions. Only about 50 to 100 ml of water per day are lost in the feces. Thus the GI tract typically absorbs more than 8 L/day.

Very little net absorption occurs in the duodenum, but the chyme is brought to isotonicity here. The chyme that is delivered from the stomach is often hypertonic. The action of digestive enzymes creates still more osmotic activity. The duodenum is highly water permeable, and very large fluxes of water occur from lumen to blood and from blood to lumen. Usually

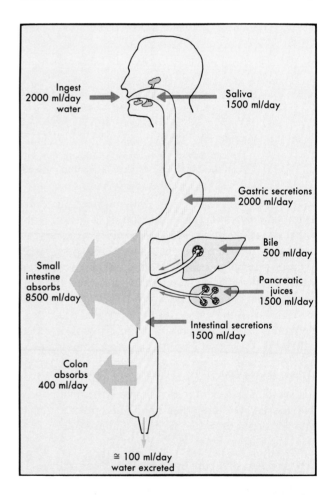

FIGURE 31-5 Overall fluid balance in the human gastrointestinal tract. Approximately 2 L of water is ingested each day, and 7 L of various secretions enters the GI tract. Of this total of 9 L, about 8.5 L is absorbed in the small intestine. Approximately 500 ml is passed on to the colon, which normally absorbs 80% to 90% of the water presented to it.

the net flux is from blood to lumen because of the hypertonicity of the chyme. Large net water absorption occurs in the small intestine: the jejunum is more active than the ileum in absorbing water. The net absorption that occurs in the colon is relatively small, approximately 400 ml/day. However, the colon can absorb water against a larger osmotic pressure difference than the rest of the GI tract. Figure 31-5 summarizes the handling of water by the GI tract.

Absorption of Na^+

Na^+ is absorbed along the entire length of the intestine. As occurs with water, net absorption is the result of large unidirectional fluxes from blood to lumen and from lumen to blood. The unidirectional fluxes are greater in the proximal gut than in the distal intestine. Na^+ crosses the brush border membrane down an electrochemical gradient and is actively extruded from the epithelial cells by the Na^+, K^+-ATPase in the basolateral plasma membrane. Normally the contents of the small bowel are isotonic to plasma. Luminal contents have about the same Na^+ concentration as plasma, so that Na^+ absorption normally occurs in the absence of a significant net concentration gradient. Na^+ absorption is active, however, and can occur against a small electrochemical potential difference for Na^+.

In the jejunum the net rate of absorption of Na^+ is highest. Here Na^+ absorption is enhanced by the presence in the lumen of glucose, galactose, and neutral amino acids. These substances and Na^+ cross the brush border membrane on the same transport pro-teins. Na^+ moves down its electrochemical potential gradient and provides the energy for moving the sugars (glucose and galactose) and neutral amino acids into the epithelial cells against a concentration gradient. Thus Na^+ enhances the absorption of sugars and amino acids, and vice versa.

In the ileum the net rate of Na^+ absorption is smaller. Na^+ absorption is only slightly stimulated by sugars and amino acids because the sugar and amino acid transport proteins are less concentrated in the ileum. The ileum can absorb Na^+ against a larger electrochemical potential than can the jejunum.

In the colon Na^+ is normally absorbed against a large electrochemical potential difference. Sodium concentrations in the luminal contents can be as low as 25 mM, compared with about 120 mM in the plasma.

Absorption of Cl^- and HCO_3^-

In the jejunum both chloride and bicarbonate are absorbed in large amounts. By the end of the jejunum most of the HCO_3^- of the hepatic and pancreatic secretions has been absorbed. In the ileum Cl^- is absorbed, but HCO_3^- is normally secreted. If the HCO_3^- concentration in the lumen of the ileum exceeds about 45 mM, the flux from lumen to blood exceeds that from blood to lumen, and net absorption occurs. In the colon the transport of these ions is similar to that in the ileum, since Cl^- is absorbed and HCO_3^- is usually secreted.

Table 31-1   Transport of Na^+, K^+, Cl^-, and HCO_3^- in the Large and Small Intestines

Segment of Intestine	Na^+	K^+	Cl^-	HCO_3^-
Jejunum	Actively absorbed; absorption enhanced by sugars, neutral amino acids	Passively absorbed when concentration rises resulting from absorption of water	Absorbed	Absorbed
Ileum	Actively absorbed	Passively absorbed	Absorbed, some in exchange for HCO_3^-	Secreted, partly in exchange for Cl^-
Colon	Actively absorbed	Net secretion occurs when K^+ concentration in lumen < 25 mM	Absorbed, some in exchange for HCO_3^-	Secreted, partly in exchange for Cl^-

Absorption of K⁺

As with the other ions, the net movement of potassium across the intestinal epithelium is the difference between large unidirectional fluxes from lumen to blood and from blood to lumen. In the jejunum and in the ileum the net flux is from lumen to blood. As the volume of intestinal contents is reduced because of the absorption of water, K^+ is concentrated, providing a driving force for the movement of K^+ across the intestinal mucosa and into the blood. Evidence for active transport of K^+ in the small intestine is lacking. In the colon K^+ may be secreted or absorbed. Net secretion occurs when the luminal concentration is less than about 25 mM; if greater than 25 mM, net absorption occurs. Under most circumstances net secretion of K^+ occurs in the colon; the secretory process may be active.

Because most absorption of K^+ results from its enhanced concentration in the lumen caused by the absorption of water, significant K^+ loss may occur in diarrhea. If diarrhea is prolonged, the K^+ level in the extracellular fluid compartment of the body falls. Maintaining normal K^+ levels is important, especially for the heart and other muscles. K^+ imbalance can have life-threatening consequences, such as cardiac arrhythmias.

Table 31-1 summarizes the transport of Na^+, K^+, Cl^-, and HCO_3^- in the small and large intestines.

Ion Transport by Intestinal Epithelial Cells

The ion transport processes that occur in the jejunum, ileum, and colon are shown in Figure 31-6.

Structural Considerations in Salt and Water Absorption by the Intestine

Tight Junctions The epithelial cells that line the intestine are connected to their neighbors by tight junctions near their luminal surfaces. The tight junctions leak the most in the duodenum, are somewhat tighter in the jejunum, are still tighter in the ileum, and are tightest in the colon.

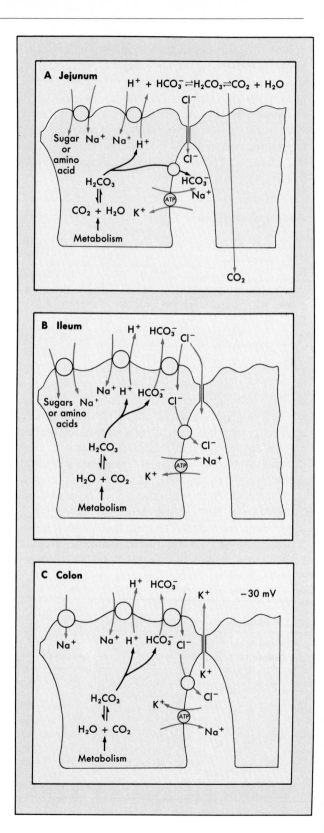

FIGURE 31-6   Summary of the major ion transport processes that occur in **A,** the jejunum; **B,** the ileum; and **C,** the colon. *ATP,* Adenosine triphosphate.

Transcellular Versus Paracellular Transport

Because the tight junctions leak, some fraction of the water and ions that traverses the intestinal epithelium passes between the epithelial cells rather than passing through them. Transmucosal movement through the tight junctions and the lateral intracellular spaces is called *paracellular* transport. Passage through the epithelial cells is termed *transcellular* transport.

Because the tight junctions in the duodenum leak considerably, significant proportions of the large unidirectional fluxes of water and ions that take place in the duodenum occur via the paracellular pathway. The proportions of water or a particular ion that pass through the transcellular and paracellular routes are determined by the relative permeabilities of the two pathways for the substance in question. Even in the ileum, where the junctions are much tighter than in the duodenum, the paracellular pathway contributes more to the total ionic conductance of the mucosa than does the transcellular pathway.

Mechanism of Water Absorption

The absorption of water depends on the absorption of ions, principally Na^+ and Cl^-. Water absorption in the small intestine normally occurs in the absence of an osmotic pressure difference between the luminal contents and the blood in the intestinal capillaries. Our current understanding is that water absorption occurs by a mechanism known as *standing gradient osmosis* (see Chapter 30 and Figure 30-23). The major features of the standing gradient osmotic mechanism are:

1. Active pumping of Na^+ into the lateral intercellular spaces by the Na^+, K^+-ATPase
2. Entry of Cl^- into the lateral intercellular spaces by flow from the lumen via the tight junctions or from the adjacent epithelial cells by facilitated transport
3. Presence of hypertonic NaCl near the luminal ends of the lateral intercellular spaces
4. Water flow by osmosis into the lateral intercellular spaces
5. Hydrostatic flow of water and ions down the lateral intercellular spaces and across the epithelial basement membrane

Because most water absorption occurs in the absence of a transmucosal osmotic pressure difference, the absorption of the end products of digestion, particularly sugars and amino acids, facilitates water absorption by allowing more water to be absorbed.

Control of Intestinal Electrolyte Absorption

Electrolyte transport in the intestine is regulated by certain hormones, neurotransmitters, and paracrine substances.

Autonomic Nervous System Stimulation of sympathetic nerves to the intestine or an elevated level of epinephrine increases the absorption of Na^+, Cl^-, and water. Stimulation of parasympathetic nerves to the gut decreases the net rate of ion and water absorption.

Adrenal Hormones *Aldosterone* strongly stimulates the secretion of K^+ and the absorption of Na^+ and water by the colon and to a much lesser extent by the ileum. Aldosterone acts by increasing the number of Na^+ channels in the luminal membrane of the colonic epithelial cells (Figure 31-6, *C*) and the number of active Na^+, K^+-ATPase molecules in the basolateral membrane. Aldosterone has similar effects on the epithelial cells of the distal tubule of the kidney (see Chapter 33). The enhanced absorption of NaCl and water induced by aldosterone in the colon and kidney is an important mechanism in the body's compensatory response to dehydration. *Glucocorticoids* also increase the content of Na^+, K^+-ATPase in the basolateral membrane and thereby enhance Na^+ and water absorption and K^+ secretion in the colon.

Secretion of Electrolytes and Water The normal net absorption of Na^+, Cl^-, and water is the result of large unidirectional fluxes from lumen to blood and from blood to lumen. Mature intestinal epithelial cells near the tips of the villi are active in net absorption, whereas the more immature epithelial cells in Lieberkühn's crypts function as net secretors (Figure 31-7). The secretory activities of the crypt cells are also subject to physiologic and pharmacologic regulation. Secretion by the crypt cells is a normal physiologic function. In certain diarrheal diseases, such as cholera, the electrogenic Cl^- efflux across the brush border membrane is specifically stimulated. This results in secretion of Cl^-, Na^+, and water into the intestinal lumen by the cells in Lieberkühn's crypts.

Absorption of Ca^{++}

Calcium ions are actively absorbed by all segments of the intestine. The duodenum and jejunum are

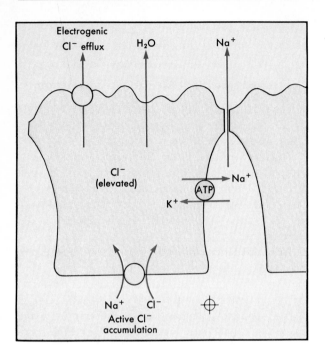

FIGURE 31-7   The ion transport processes involved in secretion of Cl^-, Na^+, and water by the epithelial cells in Lieberkühn's crypts in the small intestine.

especially active and can concentrate Ca^{++} against a greater than tenfold concentration gradient. The rate of absorption of Ca^{++} is much greater than that of any other divalent ion but still 50 times slower than Na^+ absorption.

The ability of the intestine to absorb Ca^{++} is regulated. Animals receiving a calcium-deficient diet increase their ability to absorb Ca^{++}. Animals receiving high-calcium diets are less able to absorb Ca^{++}. Intestinal absorption of Ca^{++} is stimulated by vitamin D and slightly stimulated by parathyroid hormone.

Brush Border Membrane A current view of the cellular mechanism of Ca^{++} absorption by the epithelial cells of the small intestine is shown in Figure 31-8. Ca^{++} moves down its electrochemical potential gradient across the brush border membrane into the cytosol. An integral protein of the brush border plasma membrane, called the *intestinal membrane calcium-binding protein (IMCal)*, may function as a membrane transporter for Ca^{++}.

Epithelial Cell Cytosol The cytosol of the intestinal epithelial cells contains *calcium-binding protein (CaBP, also known as calbindin)*. In mammals the CaBP has a molecular weight of about 10,000 and binds two calcium ions with high affinity. The level of CaBP in the epithelial cells correlates well with the capacity to absorb Ca^{++}. CaBP appears to be an essential component of Ca^{++} absorption. It has been proposed that CaBP allows large amounts of Ca^{++} to traverse the cytosol, while avoiding concentrations of free Ca^{++} high enough to form insoluble salts with intracellular anions.

Basolateral Membrane The basolateral plasma membrane contains two transport proteins capable of ejecting Ca^{++} from the cell against its electrochemical potential gradient. A Ca^{++}-ATPase in the basolateral membrane is a primary active transport protein that splits adenosine triphosphate (ATP) and uses the energy to transport Ca^{++}. The Na^+/Ca^{++} exchanger present in the basolateral membrane uses the energy of the Na^+ gradient to extrude Ca^{++} by secondary active transport. The Na^+/Ca^{++} exchanger is more effective at high levels of free intracellular Ca^{++}, whereas at low levels of free intracellular Ca^{++}, the Ca^{++}-ATPase is the major mechanism for Ca^{++} extrusion.

Actions of Vitamin D Vitamin D is essential for normal levels of calcium absorption by the intestine. In rickets, a disease caused by vitamin D deficiency, the rate of absorption of Ca^{++} is very low.

The actions of vitamin D are discussed in Chapter 38. Vitamin D may stimulate each phase of absorption of Ca^{++} by the epithelium of the small intestine: passage across the brush border membrane, traversal of the cytosol, and active extrusion across the basolateral membrane. Vitamin D_3 induces the synthesis of the cytosolic CaBP. In addition, vitamin D increases the level of the basolateral Ca^{++}-ATPase that actively pumps calcium out of the enterocyte.

Absorption of Iron

A typical adult in Western societies ingests approximately 15 to 20 mg of iron daily. Of this amount, only 0.5 to 1 mg is absorbed by normal adult men and 1 to 1.5 mg is absorbed by premenopausal adult women. Iron depletion, (e.g., caused by hemorrhage) increases iron absorption. Growing children and pregnant women absorb greater amounts of iron.

Iron absorption is limited because iron tends to form insoluble salts with hydroxide, phosphate, bicarbonate, and other anions present in intestinal secretions. Iron also forms insoluble complexes with other

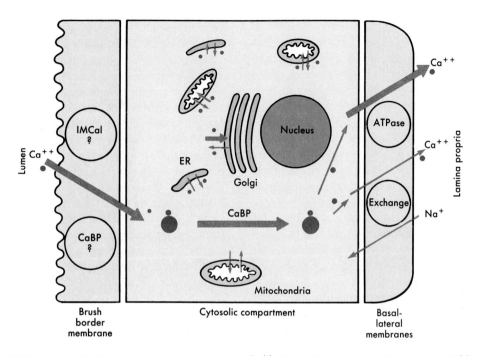

FIGURE 31-8   Cellular mechanisms involved in Ca^{++} absorption in the small intestine. Ca^{++} crosses the brush border membrane by processes that may involve the intestinal membrane calcium-binding protein *(IMCal)* or calcium-binding protein *(CaBP)*. In the cytosol of the entero-cyte, Ca^{++} is bound to CaBP. Ca^{++} is extruded across the basolateral membrane by a Ca^{++}-ATPase and by a Na^{+}/Ca^{++} exchange mechanism. Small dots represent Ca^{++} ions. Large dots represent cytosolic CaBP. *ER*, Endoplasmic reticulum. (Redrawn from Wasserman RH and Full-mer CS: Annu Rev Physiol 45:375, 1983. Reproduced, with permission, from the Annual Review of Physiology, vol 45, © 1983 by Annual Reviews Inc.)

substances typically present in food, such as phytate, tannins, and the fiber of cereal grains. These iron complexes are more soluble at low pH. Therefore HCl secreted by the stomach enhances iron absorption, whereas iron absorption is usually low in individuals deficient in HCl secretion. Ascorbate effectively pro-motes iron absorption by forming a soluble complex with iron and by reducing Fe^{+++} to Fe^{++}. Fe^{++} has much less tendency to form insoluble complexes than Fe^{+++}, and thus Fe^{++} is better absorbed.

Heme iron is absorbed relatively well; about 20% of the heme ingested is absorbed. Proteolytic enzymes release heme groups from proteins in the intestinal lumen. Heme is probably taken up by facilitated transport by the epithelial cells that line the upper small intestine. In the epithelial cell iron is split from the heme by reactions involving xanthine oxidase. No intact heme is transported into the portal blood.

Cellular Mechanism of Iron Absorption   A current view of iron absorption is depicted in Figure 31-9. The epithelial cells of the duodenum and jeju-

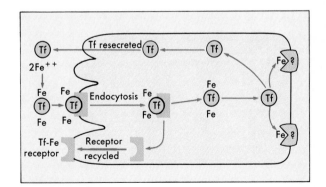

FIGURE 31-9   Current view of the mechanism of iron absorp-tion by the epithelial cells of the small intestine. *Tf* denotes a form of transferrin that is secreted into the lumen of the intestine, where it binds Fe^{++}.

num release an iron-binding protein into the lumen. the protein is called *transferrin* and is very similar, but not identical, to the transferrin that is the principal iron-binding protein of plasma. In the lumen of the duodenum and jejunum, transferrin binds iron; two iron ions can be bound by each transferrin molecule. Receptors on the brush border surface of the duodenum and jejunum bind the transferrin-iron complex, and the complex is taken up into the epithelial cell by receptor-mediated endocytosis. In the cytosol of the intestinal epithelial cell, transferrin acts as a soluble iron carrier. Much of the transferrin, after it releases its bound iron, is resecreted into the lumen. Iron ultimately appears in plasma bound to plasma transferrin. Transport of iron across the basolateral membrane requires metabolic energy but is otherwise uncharacterized.

Regulation of Iron Absorption Iron absorption is regulated in accordance with the body's need for iron. In chronic iron deficiency or after hemorrhage, the capacity of the duodenum and jejunum to absorb iron is increased. The intestine also protects the body from the consequences of absorbing too much iron.

An important mechanism for preventing excess absorption of iron is the almost irreversible binding of iron to ferritin in the intestinal epithelial cell. Iron bound to ferritin is not available for transport into the plasma (Figure 31-10), but it is lost into the intestinal lumen and excreted in the feces when the intestinal epithelial cell is sloughed off. The amount of ferritin present in the intestinal cells may determine how much iron can be trapped in this nonabsorbable pool. The synthesis of ferritin is stimulated at the translational level by iron, and this protects against absorption of excessive amounts of iron.

The capacity of the duodenum and jejunum to absorb iron increases after a hemorrhage, with a time lag of 3 to 4 days. The intestinal epithelial cells require this time to migrate from their sites of formation in Lieberkühn's crypts to the tips of the villi, where they are most involved in absorptive activities. The iron-absorbing capacity of the epithelial cells appears to be programmed when the cells are in Lieberkühn's crypts. The brush border membranes of the duodenum and jejunum of an iron-deficient animal have an increased number of receptors for the complex of iron with transferrin and thus absorb the iron-transferrin complex from the lumen more rapidly.

Absorption of Other Ions

Magnesium Magnesium is absorbed along the entire length of the small intestine, with about half the normal dietary intake being absorbed and the rest excreted.

Phosphate Phosphate is absorbed all along the small intestine. Some phosphate may be absorbed by active transport.

Copper Copper is absorbed in the jejunum, with approximately 50% of the ingested load being absorbed. Copper is secreted in the bile bound to certain bile acids; this copper is lost in the feces.

ABSORPTION OF WATER-SOLUBLE VITAMINS

Most water-soluble vitamins can be absorbed by simple diffusion if they are taken in sufficiently high doses. Nevertheless, specific transport mechanisms play important roles in the normal absorption of most water-soluble vitamins. Table 31-2 summarizes current knowledge of these transport mechanisms.

Absorption of Vitamin B₁₂

A specific active transport process has been implicated in the absorption of vitamin B_{12}. In the absence of vitamin B_{12} the maturation of red blood cells is retarded, and *pernicious anemia* ensues. Because of its medical importance, much attention has focused

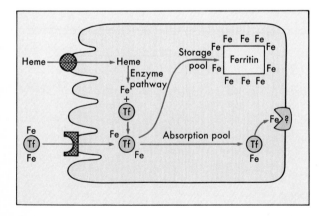

FIGURE 31-10   In the enterocytes of the small intestine, iron that becomes bound to ferritin is unavailable for transport across the basolateral membrane. Iron bound to ferritin is lost into the lumen when the enterocyte is sloughed off.

Table 31-2 Intestinal Absorption of Vitamins

Vitamin	Species	Site of Absorption	Transport Mechanism	Maximal Absorptive Capacity in Humans (Per Day)	Dietary Requirement in Humans (Per Day)
Ascorbic acid (C)	Humans, guinea pig	Ileum	Active	> 5000 mg	< 50 mg
Biotin	Hamster	Upper small intestine	Active	?	?
Choline	Guinea pig, hamster	Small intestine	Facilitated	?	?
Folic acid					
Pteroylglutamate	Rat	Jejunum	Facilitated	> 1000 μg/dose	100-200 μg
5-Methyltetrahydrofolate	Rat	Jejunum	Diffusion		
Nicotinic acid	Rat	Jejunum	Facilitated	?	10-20 mg
Pantothenic acid		Small intestine	?	?	(?) 10 mg
Pyridoxine (B_6)	Rat, hamster	Small intestine	Diffusion	> 50 mg/dose	1-2 mg
Riboflavin (B_2)	Humans, rat	Jejunum	Facilitated	10-12 mg/dose	1-2 mg
Thiamin (B_1)	Rat	Jejunum	Active	8-14 mg	$\approx$ 1 mg
Vitamin B_{12}	Humans, rat, hamster	Distal ileum	Active	6-9 μg	3-7 μg

Data from Matthews DM: In Smyth DH, editor: Intestinal absorption, vol 4B: Biomembranes, London, 1974, Plenum Press; and Rose RC: Annu Rev Physiol 42:157, 1980.

on the absorption of vitamin B_{12}. The dietary requirement for B_{12} is fairly close to maximal absorptive capacity for the vitamin (Table 31-2). Enteric bacteria synthesize vitamin B_{12} and other B vitamins, but the colonic epithelium lacks specific mechanisms for their absorption.

Storage in Liver The liver contains a large store of vitamin B_{12} (2 to 5 mg). Vitamin B_{12} is normally present in the bile (0.5 to 5 μg daily), but approximately 70% of this is normally reabsorbed. Because only about 0.1% of the store is lost daily, even if absorption totally ceases, the store will last for 3 to 6 years.

Gastric Phase Most of the vitamin B_{12} present in food is bound to proteins. The low pH in the stomach and the digestion of proteins by pepsin release free vitamins B_{12}, which is then rapidly bound to a class of glycoproteins known as *R proteins*. R proteins are present in saliva and in gastric juice and bind vitamin B_{12} tightly over a wide pH range.

Intrinsic factor (IF) is a vitamin B_{12}–binding protein secreted by the gastric parietal cells. IF binds vitamin B_{12} with less affinity than the R proteins; thus in the stomach most of the vitamin B_{12} present in food is bound to R proteins.

Intestinal Phase Pancreatic proteases degrade the complex between R proteins and vitamin B_{12}, which causes vitamin B_{12} to be released. The free vitamin B_{12} is taken up by IF, which is highly resistant to digestion by pancreatic proteases.

Specific Absorption Sequence Figure 31-11 summarizes a current view of the mechanism of vitamin B_{12} absorption. The normal absorption of vitamin B_{12} depends on the presence of IF. Two IF molecules bind two vitamin B_{12} molecules. The brush border plasma membranes of the epithelial cells of the ileum contain a receptor protein that recognizes and binds the IF-B_{12} complex. Free IF does not compete for binding, and the receptor does not recognize free vitamin B_{12}. Binding to the receptor is required for B_{12} uptake. Following its absorption, vitamin B_{12} appears in the portal blood bound to a protein called *transcobalamin II*.

Absorption in the Absence of Intrinsic Factor In the complete absence of IF, approximately 1% to 2% of the vitamin B_{12} ingested will be absorbed. If large doses of B_{12} are taken (about 1 mg/day), enough can be absorbed to treat pernicious anemia.

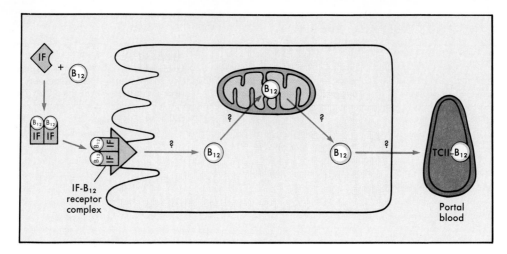

FIGURE 31-11   Mechanism of absorption of vitamin B_{12} absorption by epithelial cells in the ileum. *IF*, Intrinsic factor.

DIGESTION AND ABSORPTION OF LIPIDS

The primary lipids of a normal diet are *triglycerides.* The diet contains smaller amounts of *sterols, sterol esters,* and *phospholipids.* Because lipids are only slightly soluble in water, they pose special problems to the GI tract at every stage of their processing. In the stomach lipids tend to separate out into an oily phase. In the duodenum and small intestine lipids are emulsified with the aid of bile acids. The large surface area of the emulsion droplets allows access of the water-soluble lipolytic enzymes to their substrates. The digestion products of lipids form small molecular aggregates, known as *micelles,* with the bile acids. The micelles are small enough to diffuse among the microvilli and allow absorption of the lipids from molecular solution along the entire surface of the intestinal brush border.

Lipids in the Stomach

Because fats tend to separate out into an oily phase, they usually are emptied from the stomach later than the other gastric contents. Despite the presence of gastric lipase, little digestion of lipids occurs in the stomach. Any tendency to form emulsions with phospholipids or other natural emulsifying agents is inhibited by the high acidity. Fat in the duodenum strongly inhibits gastric emptying. This ensures that the fat is not emptied from the stomach more rapidly

than it can be accommodated by the duodenal mechanisms that provide for emulsification and digestion.

Digestion of Lipids and Micelle Formation

The lipolytic enzymes of the pancreatic juice are water-soluble molecules and thus have access to the lipids only at the surfaces of the fat droplets. The surface available for digestion is increased many thousand times by emulsification of the lipids (Figure 31-12). Bile acids themselves are rather poor emulsifying agents. However, with the aid of lecithin, which is present in high concentration in the bile, the bile acids emulsify dietary fats. The emulsion droplets are approximately 1 μm in diameter and have a large surface area on which the digestive enzymes can work.

Pancreatic Secretions   Pancreatic juice contains the major lipolytic enzymes responsible for digestion of lipids. The most important digestive enzymes and their actions are:

1. *Glycerol ester hydrolase,* also called *pancreatic lipase,* cleaves the 1 and 1' fatty acids preferentially off a triglyceride to produce two free fatty acids and one 2-monoglyceride. *Colipase,* a small protein present in pancreatic juice, is essential for the function of glycerol ester hydrolase. Colipase is required for glycerol ester hydrolase to bind to the surface of the emulsion droplets in the presence of bile acids.

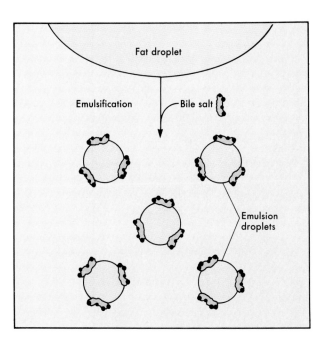

FIGURE 31-12   Emulsification of fats by bile salts and lecithin greatly increases the surface area available to the fat-digesting enzymes. (From Vander AJ, et al: Human physiology, ed 4, New York, 1985, McGraw-Hill.)

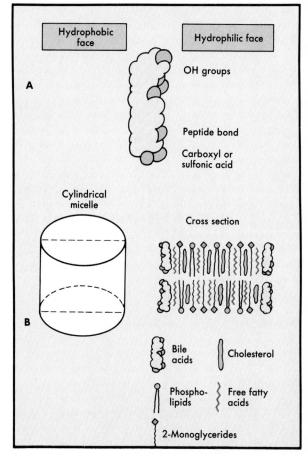

FIGURE 31-13   Structure of bile acids and micelles. **A,** A bile acid molecule is *amphipathic* because it has a hydrophobic face and a hydrophilic face. **B,** Model of the structure of a bile acid–lipid mixed micelle.

2. *Cholesterol esterase* cleaves the ester bond in a cholesterol ester to give one fatty acid and free cholesterol.

3. *Phospholipase A₂* cleaves the ester bond at the 2 position of a glycerophosphatide to yield, in the case of lecithin, one fatty acid and one lysolecithin.

Formation of Micelles Bile acids form micelles with the products of fat digestion, especially 2-monoglycerides. The micelles are multimolecular aggregates (about 5 nm in diameter) containing approximately 20 to 30 molecules (Figure 31-13). Bile acids are flat molecules that have a polar face and nonpolar face. Much of the surface of the micelles is covered with bile acids, with the nonpolar face toward the lipid interior of the micelle and the polar face toward the outside. Extremely hydrophobic molecules, such as long-chain fatty acids, cholesterol, and certain fat-soluble vitamins, tend to partition into the interior of the micelle. Phospholipids and monoglycerides tend

to have their more polar ends facing the outside aqueous phase. Micelles contain almost no intact triglyceride.

Bile acids must be present at a certain minimal concentration, called the *critical micelle concentration,* before micelles will form. Conjugated bile acids have a much lower critical micelle concentration than the unconjugated forms. In the normal state bile acids are always present in the duodenum at greater than the critical micelle concentration.

Absorption of Lipid Digestions Products

Transport into the Intestinal Epithelial Cell
The micelles are important in the absorption of the products of lipid digestion and in the absorption of

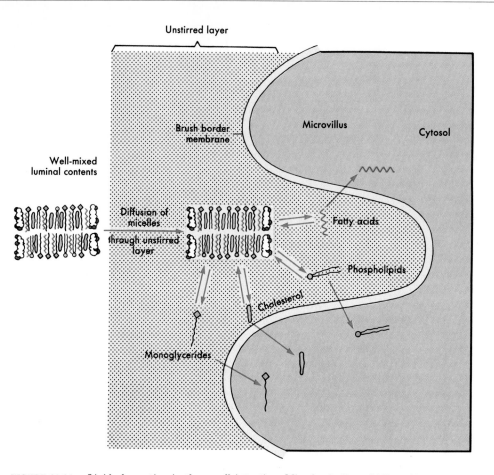

FIGURE 31-14   Lipid absorption in the small intestine. Mixed micelles of bile acids and lipid digestion products diffuse through the unstirred layer. As lipid digestion products are absorbed from free solution, more lipids partition out of the micelles.

most other fat-soluble molecules (e.g., fat-soluble vitamins). The micelles diffuse among the microvilli that form the brush border. This diffusion allows the huge surface area of the brush border membrane to participate in lipid absorption (Figure 31-14). The presence of micelles tends to keep the aqueous solution along the brush border saturated with fatty acids, 2-monoglycerides, cholesterol, and other micellar contents.

Because of their high lipid solubility, the fatty acids, 2-monoglycerides, cholesterol, and lysolecithin can diffuse rapidly across the brush border membrane. The main limitation to the rate of lipid uptake by the epithelial cells of the upper small intestine is the diffusion of the mixed micelles through an *unstirred layer* on the luminal surface of the brush border plasma membrane (Figure 31-14). Partly because of the convoluted surface of the intestinal mucosa, the

fluid in immediate contact with the epithelial cell surface is not readily mixed with the bulk of the luminal contents. Thus this fluid forms an effective unstirred layer. The average thickness of the unstirred layer ranges from 200 to 500 μm. Nutrients present in the well-mixed contents of the intestinal lumen must diffuse through the unstirred layer to reach the plasma membrane of the brush border.

The duodenum and jejunum are most active in fat absorption, and most of the ingested fat is absorbed by the time chyme reaches midjejunum. The fat present in normal stools is not ingested fat (which is completely absorbed), but fat from colonic bacteria and desquamated intestinal epithelial cells.

Fat in the Intestinal Epithelial Cell The products of lipid digestion are taken up by the smooth endoplasmic reticulum. A cytoplasmic fatty acid–binding protein transports fatty acids to the smooth

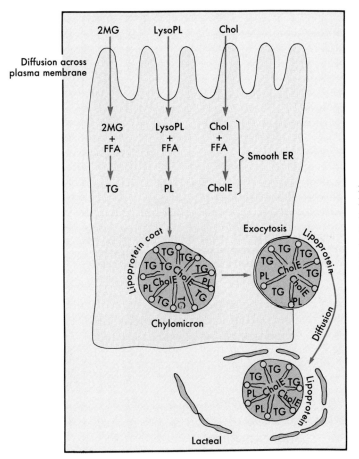

FIGURE 31-15   Resynthesis of lipids in the epithelial cells of the small intestine, formation of chylomicrons, and subsequent transport of chylomicrons into the lymphatic vessels. *FFA*, Free fatty acid; *2MG*, 2-monoglyceride; *TG*, triglyceride; *PL*, phospholipid; *lysoPL*, lysophospholipid; *Chol*, cholesterol; *CholE*, cholesterol ester; *ER*, endoplasmic reticulum.

endoplasmic reticulum. In the smooth endoplasmic reticulum, which is engorged with lipid after a meal, considerable chemical reprocessing occurs (Figure 31-15). The 2-monoglycerides are reesterified with fatty acids at the 1 and 1′ carbons to reform triglycerides. Lysophospholipids are reconverted to phospholipids. Cholesterol is reesterified to a considerable extent.

Chylomicron Formation and Transport The reprocessed lipids accumulate in the vesicles of the smooth endoplasmic reticulum. Phospholipids cover the external surfaces of these lipid droplets. The hydrophobic acyl chains of the phospholipids lie in the fatty interior of the droplets, and their polar head groups face toward the aqueous exterior. The lipid droplets, approximately 10 nm in diameter at this point, are known as *chylomicrons*. About 10% of their surface is covered by β-lipoprotein, which is

synthesized in the intestinal epithelial cells.

Chylomicrons are ejected from the epithelial cell by exocytosis (Figure 31-15). The chylomicrons leave the cells at the level of the nuclei and enter the lateral intercellular spaces. Chylomicrons are too large to pass through the basement membrane that invests the mucosal capillaries. However, they do enter the lacteals, which have sufficiently large fenestrations for the chylomicrons to pass through. The chylomicrons leave the intestine with the lymph, primarily via the thoracic lymphatic duct, and flow into the venous circulation.

Absorption of Bile Acids

The absorption of dietary lipids is typically complete by the time chyme reaches midjejunum. By contrast, bile acids are absorbed largely in the terminal

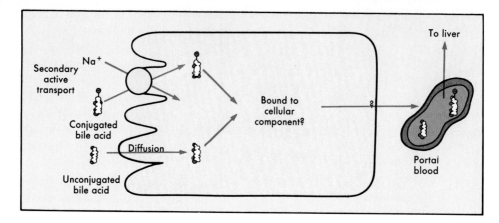

FIGURE 31-16   Absorption of bile acids by epithelial cells of the terminal ileum. Bile acids are absorbed both by simple diffusion and by Na$^+$-powered secondary active transport. Conjugated bile acids are absorbed mainly by active transport; unconjugated bile acids are absorbed chiefly by diffusion.

part of the ileum. As for other fat-soluble substances, the unstirred layer is an important barrier to bile acid absorption. Bile acids cross the brush border plasma membrane by two routes: an active transport process and simple diffusion (Figure 31-16). The active process is secondary active transport, powered by the Na$^+$ gradient across the brush border membrane. In this process one Na$^+$ ion is co-transported across the brush border plasma membrane with one bile acid molecule. Conjugated bile acids are the principal substrates for active absorption; deconjugated bile acids have poor affinity for the transporter. Deconjugation and dehydroxylation make bile acids less polar, and thus they are better absorbed by simple diffusion.

Absorbed bile acids are carried away from the intestine in the portal blood. Hepatocytes avidly extract bile acids, essentially clearing them from the blood in a single pass through the liver. In the hepatocytes most deconjugated bile acids are reconjugated, and some secondary bile acids are rehydroxylated. The reprocessed bile acids, together with newly synthesized bile acids, are secreted into bile.

Absorption of Fat-Soluble Vitamins

Because of their solubility in nonpolar solvents, the fat-soluble vitamins (A, D, E, and K) partition into the mixed micelles formed by the bile acids and lipid digestion products. As with other lipids, the fat-soluble vitamins enter the intestinal epithelial cell by diffusing across the plasma membrane of the brush border. The presence of bile acids and lipid digestion products enhances the absorption of fat-soluble vitamins. In the intestinal epithelial cell the fat-soluble vitamins enter the chylomicrons and leave the intestine in the lymph. In the absence of bile acids a significant fraction of the ingested load of a fat-soluble vitamin may be absorbed and leave the intestine in the portal blood.

BIBLIOGRAPHY
Journal Articles

Seetharam B and Alpers DH: Absorption and transport of cobalamin (vitamin B$_{12}$), Annu Rev Nutr 2:343, 1982.

Shiau YF: Mechanisms of intestinal fat absorption, Am J Physiol 240:G1, 1981.

Stevens BR et al: Intestinal transport of amino acids and sugars: advances using membrane vesicles, Annu Rev Physiol 46:417, 1984.

Wasserman RH and Fullmer CS: Calcium transport proteins, calcium absorption, and vitamin D, Annu Rev Physiol 45:375, 1983.

Wilson FA: Intestinal transport of bile acids, Am J Physiol 241:G83, 1981.

Young S and Bomford A: Transferrin and cellular iron exchange, Clin Sci 67:273, 1984.

Books and Monographs

Davenport HW: Physiology of the digestive tract, ed 5, Chicago, 1982, Year Book Medical Publishers, Inc.

Johnson LR, editor: Gastrointestinal physiology, ed 3, St Louis, 1985, The CV Mosby Co.

Johnson LR, editor: Physiology of the gastrointestinal tract, ed 2, vols 1 and 2, 1987, New York, Raven Press.

RENAL SYSTEM

BRUCE A. STANTON
BRUCE M. KOEPPEN

PART VII

Elements of Renal Function

FUNCTIONAL ANATOMY

The kidneys are regulatory organs that help preserve the volume and composition of the body fluid compartments. To accomplish this, the kidneys produce an ultrafiltrate of plasma and, through reabsorptive and secretory processes, selectively excrete and conserve water and solutes. To understand how the kidneys accomplish their important tasks, the relationships between the structure and function of its components must first be considered.

Gross Features

The kidneys are paired organs that lie behind the peritoneum on the posterior wall of the abdomen on each side of the vertebral column. In the adult human the weight of each kidney ranges from 115 to 170 g. Each kidney is approximately 11 cm in length, 6 cm in width, and 3 cm in thickness.

The gross anatomic features of the mammalian kidney are illustrated in Figure 32-1. The medial side of each kidney contains an indentation through which pass the renal artery and vein, nerves, and pelvis. On the cut surface of a bisected kidney, two regions are evident: a pale outer region called the *cortex*, and a darker inner region called the *medulla*. The cortex and medulla are composed of *nephrons*, (the functional units of the kidney), blood vessels, lymphatics, and nerves. The medulla in the human kidney is divided into 8 to 18 conical masses, the *renal pyramids*. The base of each pyramid originates at the corticomedullary border, and the apex terminates in the papilla, which lies within the *pelvic space*. The *pelvis* represents the upper expanded region of the *ureter*, which conducts urine from the pelvic space to the urinary bladder. In the human kidney the pelvis divides into two or three open-ended pouches, the *major calyces*, which extend outward from the dilated end of the pelvis. Each major calyx divides into *minor calyces*, which collect the urine expressed from each papilla. The walls of the calyces, pelvis, and the ureters contain smooth muscle that contracts to propel the urine toward the bladder.

The blood flow to the two kidneys is equal to 25% (1.25 L·min^{-1}) of the cardiac output in resting subjects. However, the kidneys constitute less than 0.5% of the total body weight. The *renal artery* enters the kidney alongside the ureter, and it branches to form progressively the *interlobar artery*, the *arcurate artery*, the *interlobular artery*, and the *afferent arteriole*, which leads into the *glomerular capillaries* (Figure 32-2, *B*). The glomerular capillaries coalesce to form the *efferent arteriole*, which leads into a second capillary network, the *peritubular capillaries*, which supply blood to the nephron. The vessels of the venous system run parallel to the arterial vessels and form progressively the *interlobular vein*, the *arcuate vein*, the *interlobar vein*, and the *renal vein*, which courses beside the ureter.

The Nephron

The functional unit of the kidney is the nephron. Each human kidney contains approximately 1.2 million nephrons. The nephron consists of a *renal corpuscle (glomerulus)*, a *proximal tubule*, a *Henle's loop*, a

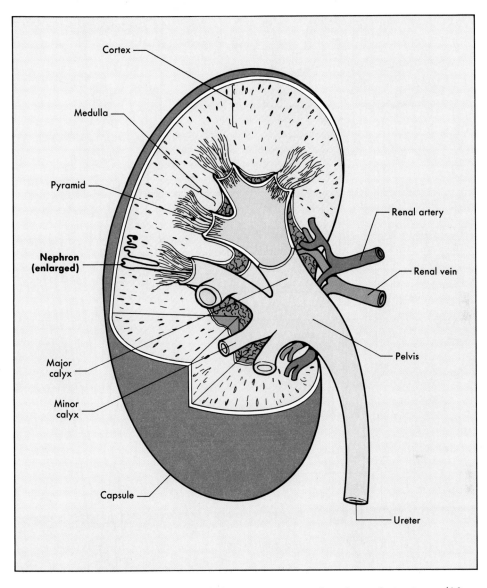

FIGURE 32-1   Structure of the human kidney, cut open to show internal structures. (After Marsh, DJ: Renal physiology, New York, 1983, Raven Press.)

distal tubule, and a *collecting duct* (Figure 32-2, *A*). The renal glomerulus consists of glomerular capillaries and *Bowman's capsule.*

Each nephron segment is composed of cells that are uniquely suited to perform specific transport functions (Figure 32-3). Proximal tubule cells have an extensively amplified apical membrane (the urine side of the cell) called the *brush border.* The basolateral membrane (the blood side of the cell) is highly invaginated. These invaginations contain a high density of mitochondria. Together with the large surface

area of the apical and basolateral membranes, the mitochondria enable the cells to reabsorb approximately two thirds of the sodium and water filtered by the glomeruli. In contrast, the thin ascending limb and thin descending limb of Henle's loop have poorly developed apical and basolateral surfaces and few mitochondria. Active sodium absorption by these segments is negligible. On the other hand, the thick ascending limb and the distal tubule cells have abundant mitochondria and extensive infoldings of the basolateral membrane. Accordingly, these tubular

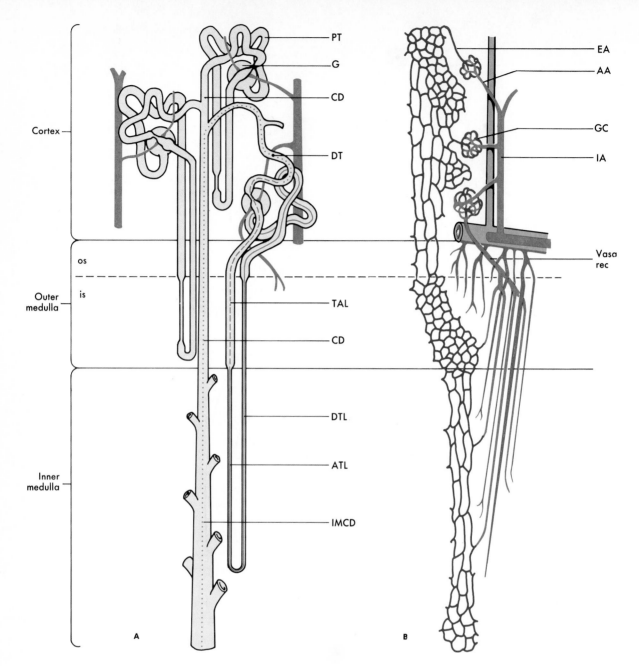

FIGURE 32-2 A, Organization of the nephrons of the human kidney. Three types of nephrons are shown: superficial *(left)*, superficial with short Henle's loops *(middle)*, and juxtamedullary *(right)*. The outer medulla is divided into an outer stripe *(os)* and inner stripe *(is)*. The renal corpuscles (glomeruli, *G*) and the proximal tubules *(PT)* are shown in brown. The descending thin limb *(DTL)* and the ascending thin limb *(ATL)* of Henle's loop are shown in pale brown. The thick ascending limbs of Henle's loop *(TAL)* are shown in yellow (macula densa indicated by a bulge in the tubule where it touches the glomerulus). The distal tubules *(DT)* are shown in orange and the collecting ducts *(CD)* in green; *IMCD,* inner medullary collecting duct. **B,** Organization of the vascular system of the human kidney. Segments of the arcuate artery (shown in red) and the arcuate vein (shown in blue) are shown at the boundary between the cortex and outer medulla. The interlobular artery *(IA)* arises from the arcuate artery, ascends toward the renal capsule, and branches to form afferent arterioles *(AA)*. The afferent arterioles branch to form glomerular capillary *(GC)* networks, which coalesce to form the efferent arterioles *(EA)*. The efferent arterioles of the outer cortical nephrons form capillary networks that suffuse the cortical labyrinth. The efferent arterioles of the juxtaglomerular nephrons divide into descending vasa recta, which form capillary networks that supply blood to the inner stripe *(is)* and the inner medulla.

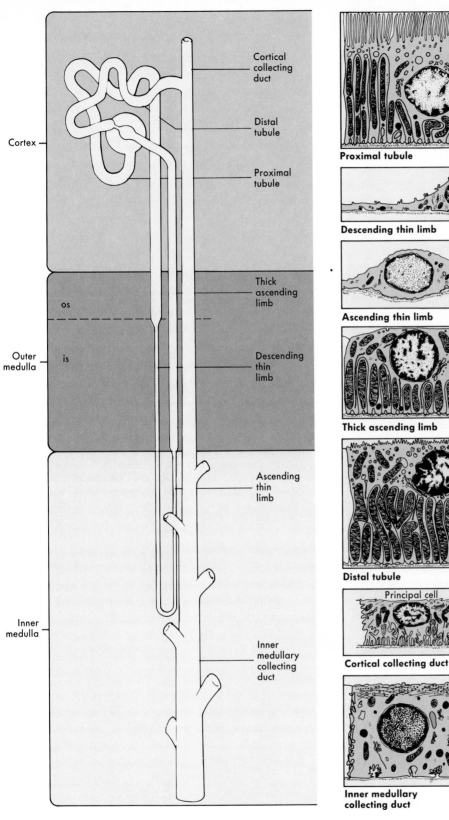

FIGURE 32-3 Diagram of a nephron, including the ultrastructure of each cell type in successive tubular segments.

segments together can reabsorb 25% to 30% of the filtered sodium. The collecting duct is composed of two cell types: principal cells and intercalated cells. *Principal cells* have a moderately invaginated basolateral membrane, contain few mitochondria, and reabsorb 2% to 5% of the filtered sodium. *Intercalated cells* have a high density of mitochondria and are thought to secrete protons.

Nephrons may be subdivided into *superficial* and *juxtamedullary* types (Figure 32-2, *A*). The glomerulus of each superficial nephron lies in the more central regions of the cortex. Its Henle's loop is short, and its efferent arteriole branches into peritubular capillaries that surround the tubular segments of its own and adjacent nephrons. This capillary network conveys oxygen and important nutrients to the tubular segments, delivers substances to the tubules for secretion, and serves as a pathway for the return of reabsorbed water and solutes to the circulatory system. A few species, including humans, also possess very short superficial nephrons whose Henle's loops never enter the medulla.

The glomerulus of each juxtamedullary nephron arises from the region of the cortex adjacent to the medulla (Figure 32-2, *A*). In comparison with the superficial nephrons, the juxtamedullary nephrons differ anatomically in three important ways: (1) the renal glomerulus is larger, (2) the Henle's loop is longer and extends deeper into the medulla, and (3) the efferent arteriole forms not only a network of peritubular capillaries, but also a series of vascular loops called the *vasa recta*. The vasa recta descend into the medulla where they form capillary networks that surround the collecting ducts and ascending limbs of Henle's loop (Figure 32-2, *B*). The blood returns to the cortex in ascending vasa recta. Although less than 0.7% of the renal blood flow enters the vasa recta, they provide important nutrients to the medulla and, as discussed in Chapter 33, are instrumental in concentration and dilution of the urine.

Glomerulus

The first step in urine formation begins with the ultrafiltration of plasma across the glomerular capillaries. The term *ultrafiltration* refers to the passive movement of protein-free fluid from the glomerular capillaries into Bowman's space. To appreciate the process of ultrafiltration, it is useful to describe the anatomy of the glomerulus (Figure 32-4). The glomer-

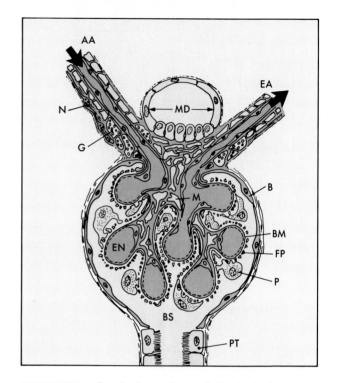

FIGURE 32-4   Renal glomerulus and the juxtaglomerular apparatus. *AA*, Afferent arteriole; *EA*, efferent arteriole; *N*, sympathetic nerve terminals; *G*, granular cells of afferent and efferent arterioles; *MD*, macula densa; *B*, Bowman's capsule; *M*, mesangial cell; *BM*, basement membrane; *FP*, foot processes; *P*, podocyte cell body (visceral cell layer); *EN*, fenestrated endothelium; *PT*, proximal tubule cells; *BS*, Bowman's space.

ulus consists of a network of capillaries supplied by the afferent arteriole and drained by the efferent arteriole. During development the glomerular capillaries press into the blind end of the proximal tubule. The capillaries are covered by epithelial cells called *podocytes*, which form the *visceral layer* of Bowman's capsule. The visceral cells are reflected at the vascular pole to form the *parietal layer* of Bowman's capsule. The space between the visceral layer and the parietal layer is called *Bowman's space*, which becomes the lumen of the proximal tubule at the urinary pole of the glomerulus.

The wall of the glomerular capillaries consists of three layers: (1) endothelium, (2) basement membrane, and (3) podocytes, with foot processes. These three layers form the *filtration barrier*. The fenestrated *endothelium* is freely permeable to water; to small solutes such as sodium, urea, and glucose; and even to small protein molecules. Because the fenestrations are relatively large (70 nm wide), the endothelium acts as a filtration barrier only to cells. The *basement membrane* consists of three layers (lamina rara interna, lamina densa, and lamina rara externa) and is the **main filtration barrier to plasma proteins** larger than 7 to 10 nm. The podocytes, which are phagocytic, have long fingerlike processes that completely encircle the outer surface of the capillaries. The processes interdigitate to cover the basement membrane and are separated by gaps called *filtration slits.* Each filtration slit is bridged by a thin diaphragm, which contains pores with dimensions of 4×14 nm. Therefore the filtration slits filter some macromolecules that pass through the endothelium and basement membrane. Because all three structures of the glomerular wall contain negatively charged glycoproteins, the glomerular capillary wall filters substances not only on the basis of size, but also on the basis of charge. For molecules with an effective molecular radius between 2 and 4 nm, cationic molecules are filtered more readily than anionic molecules.

Another important component of the glomerulus is the *mesangium,* which consists of *mesangial cells* and the *mesangial matrix* (Figure 32-4). Mesangial cells provide structural support for the glomerular capillaries, secrete the extracellular matrix, exhibit phagocytic activity, and secrete prostaglandins. Because mesangial cells exhibit contractile activity, they may influence glomerular filtration rate by regulating blood flow through glomerular capillaries. Mesangial cells located outside the glomerulus (between the afferent and efferent arterioles) are called *extraglomerular mesangial cells,* or *lacis cells.*

Anatomy of the Juxtaglomerular Apparatus

The structures that comprise the juxtaglomerular apparatus include (1) the *macula densa* of the thick ascending limb, (2) the extraglomerular mesangial cells, and (3) the renin-producing *granular cells* of the afferent and efferent arterioles (Figure 32-4). The macula densa cells represent a morphologically distinct region of the thick ascending limb that passes through the angle of the afferent and efferent arterioles. The macula densa cells contact the extraglomerular mesangial cells and the granular cells of the afferent and efferent arterioles. The granular cells of the afferent and efferent arterioles are modified smooth muscle cells that manufacture, store, and release *renin.* Renin is involved in the formation of *angiotensin II* and ultimately in the secretion of aldosterone. The juxtaglomerular apparatus is one component of an important feedback mechanism involved in the autoregulation of renal blood flow and the glomerular filtration rate. Details of this feedback mechanism are discussed below.

Innervation

Renal nerves help regulate renal blood flow, glomerular filtration rate, and salt and water reabsorption by the nephron. The nerve supply to the kidney consists of sympathetic nerve fibers that originate mainly from the *celiac plexus. Adrenergic fibers* innervating the renal parenchyma release norepinephrine and dopamine. The adrenergic fibers lie adjacent to the smooth muscle cells of the major branches of the renal artery, including the arcuate arteries, interlobular arteries, and the afferent and efferent arterioles. Moreover, the renin-producing granular cells of the afferent and efferent arterioles are innervated; renin secretion is elicited by increased sympathetic activity. Nerve fibers also innervate the proximal tubule, Henle's loop, distal tubule, and collecting duct. Activation of these nerves enhances sodium absorption by these tubular segments.

Anatomy and Physiology of the Lower Urinary Tract

Once urine enters the renal pelvis, it flows through the ureters and enters the bladder, where urine is stored (Figure 32-5). The ureters are muscular tubes 30 cm long; they enter the bladder on its posterior aspect near the base and the *internal urethral meatus.* The triangular region of the posterior bladder wall, lying above the entrance to the posterior urethra (called the *internal meatus*) and below the point where the ureters enter the bladder, is called the *trigone.* The bladder is composed of two parts: the *fundus,* or body that stores urine; and the *neck,* which is funnel shaped and connects with the urethra. The bladder neck, which is 2 to 3 cm long, is also called the *posterior urethra.* In females the posterior urethra is the end of the urinary tract and the point of exit of urine from the body. In males urine flows through the posterior urethra into the *anterior urethra,* which extends through the penis. Urine leaves the urethra

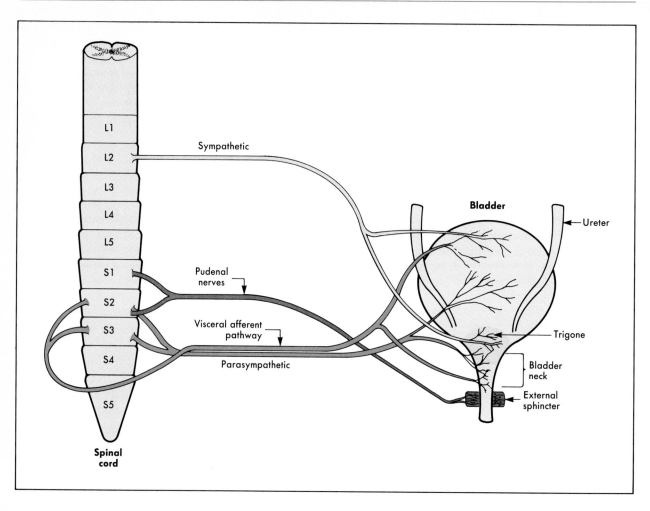

FIGURE 32-5   The urinary bladder and its innervation.

through the *external meatus.*

The renal pelvis, ureter, bladder, and posterior urethra are lined with a transitional epithelium that is surrounded by several layers of smooth muscle. The anterior urethra is lined with pseudostratified columnar epithelium that is surrounded by several layers of smooth muscle. In the ureters the muscle layers have a spiral arrangement, except where they enter the bladder. The wall of the ureters within the bladder is composed of three layers: from the lumen outward, the layers are longitudinal, circular, and longitudinal. Bladder smooth muscle is known as the *detrusor muscle.* These smooth muscle cells form a basketlike mesh of interconnecting fibers that are not arranged in layers. The fibers converge at the neck of the bladder to form a more longitudinal pattern. The muscle cells in this area form the *internal sphincter,* which is not under conscious control. Its inherent tone prevents emptying of the bladder until appropriate stimuli initiate urination. The urethra passes through the *urogenital diaphragm,* which contains a layer of skeletal muscle called the *external sphincter.* This muscle is under voluntary control and can be used to prevent or even voluntarily interrupt urination, especially in males. In females the external sphincter is poorly developed; thus, it is less important in voluntary bladder control.

The smooth muscle cells in the lower urinary tract are electrically coupled, exhibit action potentials, contract when stretched, and respond to neurotransmitters. These smooth muscle cells are similar to cardiac cells in that they generate spontaneous action potentials; regions of specialized cells within the muscle wall of the renal pelvis exhibit pacemaker activity.

The walls of the ureters, bladder, and urethra are

highly folded and thereby very distensible. In the bladder and urethra these folds are called *rugae*. As the bladder fills with urine, the rugae flatten out and the volume of the bladder increases with very little change in intravesical pressure. The volume of this structure can increase from a minimal volume of 10 ml following urination to 400 ml with a pressure change of only 5 cm H_2O. This illustrates the highly compliant nature of the bladder.

Innervation of the Lower Urinary Tract Innervation of the bladder and urethra is important in controlling urination. The smooth muscle of the bladder fundus and neck receives sympathetic innervation from the hypogastric nerves. β-*Adrenergic receptors,* located primarily in the fundus of the bladder, cause relaxation; α-*adrenergic receptors,* located primarily in the bladder neck and the urethra, cause contraction. Stimulation of these receptors facilitates storage of urine by causing relaxation of the fundus and facilitating closure of the urethra. Parasympathetic (muscarinic) fibers innervate the body of the bladder and cause a sustained bladder contraction. The *pudendal nerve* innervates the skeletal muscle fibers of the external sphincter in the urogenital diaphragm and causes contraction.

Passage of Urine from the Renal Pelvis to the Bladder Distention of the renal pelvis as it fills with urine promotes the inherent pacemaker activity in this structure. The pacemaker activity initiates a peristaltic contraction beginning in the pelvis that spreads along the length of the ureter, forcing urine from the renal pelvis toward the bladder. Transmission of the peristaltic wave is caused by action potentials, generated by the pacemaker, passing along the smooth muscle syncytium of the ureter. Sympathetic innervation of the ureter decreases the frequency of the peristaltic waves, and parasympathetic innervation increases the frequency. The ureters are also innervated with sensory nerve fibers (pelvic nerves). When the ureter is blocked, as with a renal stone, reflex constriction of the ureter around the stone will cause obstruction, proximal distention of the ureter and pelvis, and visceral pain *(renal colic)*.

Micturition Micturition is the process of emptying the urinary bladder. Two processes are involved: (1) progressive filling of the bladder until the pressure rises to a critical value and (2) a neuronal reflex called the *micturition reflex,* which empties the bladder. The micturition reflex is an **automatic spinal cord reflex;** however, it can be inhibited or facilitated by centers in the brainstem and the cerebral cortex.

Filling of the bladder stretches the bladder wall and causes it to contract. Contractions are the result of a reflex initiated by stretch receptors in the bladder. Sensory signals from the bladder enter the spinal cord via pelvic nerves and return directly to the bladder through parasympathetic fibers in the same nerves. Stimulation of parasympathetic fibers causes intense stimulation of the detrusor muscle. Because the smooth muscle in the bladder is a syncytium, stimulation of the detrusor also causes the muscle cells in the neck of the bladder to contract. The muscle cells of the bladder neck, in the region of the internal meatus, are oriented longitudinally and radially. When they contract, they open the internal meatus and allow urine to flow through the internal sphincter. A voluntary relaxation of the external sphincter, via cortical inhibition of the pudendal nerve, permits the flow of urine through the external meatus. **Voluntary relaxation of the external sphincter is required and may be the event that initiates micturition.**

ASSESSMENT OF RENAL FUNCTION

The amount of a substance appearing in the urine reflects the coordinated action of the nephron's various portions and represents three general processes: *(A)* glomerular filtration, *(B)* reabsorption of the substance from the lumen of the nephron back into the blood, and *(C)* secretion of the substance from the blood into the tubular fluid. These three processes are illustrated in Figure 32-6, in which the entire nephron population of both kidneys is represented by a single nephron.

This section develops the concept of renal clearance to quantitate these processes. To illustrate the principles of renal clearance, the renal handling of three solutes is considered: (1) inulin, whose excretion occurs solely by glomerular filtration; (2) glucose, which is filtered and subsequently reabsorbed; and (3) *p*-aminohippuric acid, which is both filtered and secreted.

Concept of Renal Clearance

The concept of renal clearance is based on the principle of mass balance. Figure 32-7 illustrates the various factors required to describe the mass balance relationships for the kidney. The renal artery is the single input source to the kidney, whereas the renal

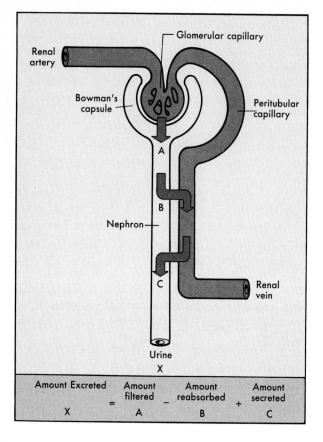

Amount Excreted		Amount filtered		Amount reabsorbed		Amount secreted
X	=	A	−	B	+	C

FIGURE 32-6   Schematic representation of the entire nephron population of both kidneys depicting the three general processes that determine and modify the composition of the urine: glomerular filtration *(A)*, tubular reabsorption *(B)*, and tubular secretion *(C)*.

vein and the ureter constitute the two output routes. Maintaining mass balance, the following relationship can be derived:

$$P_x^a \cdot RPF^a = (P_x^v \cdot RPF^v) + (U_x \cdot V) \qquad (1)$$

where P_x^a and P_x^v are the concentrations of substance x in the renal artery plasma and renal vein plasma, respectively; RPF_x^a and RPF_x^v are the renal plasma flow rates in the artery and vein, respectively; U_x is the concentration of x in the urine; and V is the urine flow rate per minute. Using this relationship, it is possible to describe the handling of any substance by the kidney.

The principle of renal clearance (C_x) emphasizes the excretory function of the kidney; it considers only the rate ($U_x \cdot V$) at which the substance is excreted into the urine and the concentration (P_x) of this substance in the plasma.

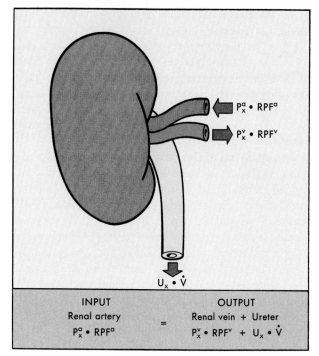

INPUT		OUTPUT
Renal artery		Renal vein + Ureter
$P_x^a \cdot RPF^a$	=	$P_x^v \cdot RPF^v + U_x \cdot \dot{V}$

FIGURE 32-7   Mass balance relationships for the kidney. See text for definition of symbols.

$$C_x = \frac{U_x \cdot \dot{V}}{P_x} \qquad (2)$$

Although the clearance equation is derived from the principle of mass balance, several differences between equation 1 and 2 deserve emphasis. Because clearance involves the ability of the kidney to remove substances from the body, the clearance does not explicitly consider the output route represented by the renal vein. P_x, which is obtained by sampling blood from a peripheral vein, is assumed to be equal to P_x^a. Finally, although clearance has the dimensions of volume/time, it does not represent a physically definable volume of fluid. Instead, **clearance is a volume of plasma from which all the substance has been removed and excreted into the urine.** This last point is best illustrated by considering the following example.

If a substance appears in the urine at a rate of 100 mg·min^{-1} ($U_x \cdot \dot{V}$), and if this substance is found in the plasma at a concentration of 1 mg·ml^{-1} (P_x), then according to equation 2, 100 ml (1 dl) of plasma have been cleared of that substance each minute ($C_x = 100$ ml·min^{-1}). It should be apparent that the same result (i.e., excretion of 100 mg·min^{-1} of x) could have been accomplished by the kidney removing only 0.5 mg of x from 200 ml of plasma. However, by convention clearance is defined as volume of plasma from which **all** of x is removed. Thus clearance defines the smallest volume of plasma from which x can be cleared.

Measurement of Glomerular Filtration: Clearance of Inulin

Inulin is a polyfructose molecule (molecular weight, 5000). It is not produced endogenously by the body and therefore must be administered intravenously. Inulin is freely filtered at the glomerulus and is not transported or metabolized by the cells of the nephron. Because of these unique properties, and as illustrated in Figure 32-8, the amount of inulin excreted in the urine per minute equals the amount of inulin filtered at the glomerulus each minute:

$$\text{Amount filtered} = \text{Amount excreted}$$
$$GFR \cdot P_{in} = U_{in} \cdot V \tag{3}$$

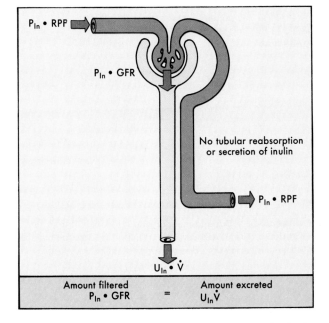

$$\frac{\text{Amount filtered}}{P_{in} \cdot GFR} = \frac{\text{Amount excreted}}{U_{in} \dot{V}}$$

FIGURE 32-8   Renal handling of inulin. Inulin is freely filtered at the glomerulus and is neither reabsorbed nor secreted by the nephron. P_{in}, Plasma inulin concentration; *RPF*, renal plasma flow; *GFR*, glomerular filtration rate; U_{in}, urinary concentration of inulin; $\dot{V}$, urine flow rate. See text for details.

where GFR is the *glomerular filtration rate*, P_{in} and U_{in} are the plasma and urine concentrations of inulin, and V is the urine flow rate. If equation 3 is solved for the GFR:

$$GFR = \frac{U_{in} \cdot \dot{V}}{P_{in}} \tag{4}$$

This equation has the same form as that for clearance. Thus the clearance of inulin provides a means for determining the GFR.

Note that not all the inulin coming to the kidney through the renal artery is filtered at the glomerulus. The portion of plasma that is filtered is termed the *filtration fraction* and is determined as:

$$\text{Filtration fraction} = GFR/RPF \tag{5}$$

where, again, RPF is renal plasma flow. Under normal conditions the filtration fraction averages 0.15 to 0.20.

Measurement of Tubular Reabsorption: Clearance of Glucose

Glucose excretion by the kidney is determined by the amount filtered at the glomerulus minus the amount subsequently reabsorbed from the tubular fluid by the nephron. The amount of glucose filtered at the glomerulus each minute *(filtered load)* is determined by the GFR and the plasma glucose concentration (P_G):

$$\text{Filtered load} = GFR \cdot P_G \tag{6}$$

The transport mechanism for glucose reabsorption has a maximal rate. This rate is termed the *tubular transport maximum* (T_m) and is the maximal amount of glucose that can be reabsorbed per minute. The T_m normally has a value of 375 mg·min^{-1}. This means that when the filtered load of glucose is less than the T_m, all the glucose will be reabsorbed from the tubular fluid and none will appear in the urine. When the filtered load of glucose exceeds the T_m, however, 375 mg·min^{-1} will be reabsorbed and the remainder excreted in the urine. Figure 32-9 illustrates these two cases.

Figure 32-10 depicts the relationships between the plasma glucose concentration and its filtered load, excretion rate, and tubular reabsorption rate. Two features of these relationships deserve special comment. First, the plasma glucose concentration at which glucose first appears in the urine is called the

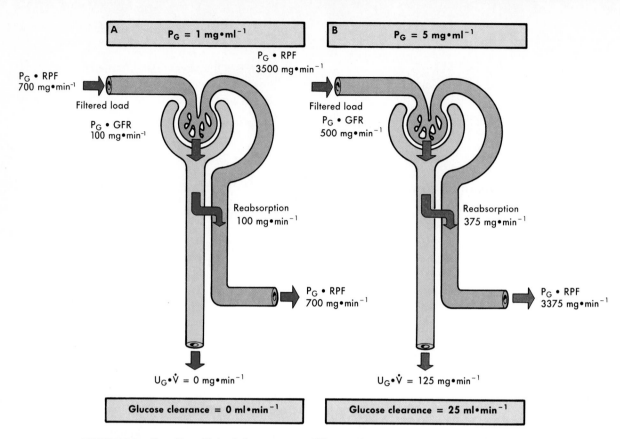

FIGURE 32-9 Renal handling of glucose at two different plasma concentrations. **A,** -The filtered load is less than the tubular transport maximum (T_m) for glucose. **B,** The filtered load exceeds the T_m for glucose. For both cases the renal plasma flow *(RPF)* is 700 ml·min^{-1}, the glomerular filtration rate *(GFR)* is 100 ml·min^{-1}, and the glucose T_m is 375 mg·min^{-1}. P_G, Plasma glucose concentration; U_G, urine glucose concentration; $\dot{V}$, urine flow rate.

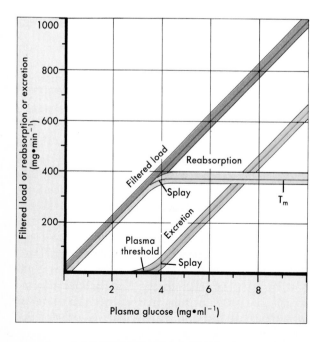

FIGURE 32-10 Dependence of the glucose filtered load, excretion rate, and reabsorptive transport rate on the plasma glucose concentration. See text for details.

plasma threshold. Second, the excretion and reabsorption curves each display the phenomenon of *splay*. Splay represents the nonlinear portions of these curves and reflects heterogeneity in the glucose reabsorptive rates of individual nephrons. In other words, the T_m is not identical for all nephrons but represents an average value for them.

Measurement of Tubular Secretion: Clearance of *p*-Aminohippuric Acid

p-Aminohippuric acid (PAH) is an organic acid that is excreted into the urine by the processes of glomerular filtration and tubular secretion. As with inulin, PAH is not produced in the body and therefore must be infused. As with glucose, the secretory

mechanism for PAH has a transport maximum. It is not the filtered load of PAH, however, that determines whether or not the secretory mechanism is saturated, but rather the amount of PAH delivered to the peritubular capillaries. The normal PAH T_m is 80 mg·min^{-1}. Thus, when plasma PAH concentration reaches approximately 0.12 mg·ml^{-1}, the PAH T_m is exceeded, and significant amounts of PAH will begin to appear in the renal vein (Figure 32-11). Figure 32-12 depicts PAH handling by the kidney and shows the relationships between the plasma PAH concentration and its filtered load, excretion rate, and secretion rate. Similar to what was observed with the renal handling of glucose, splay appears in the excretion and secretion curves for PAH.

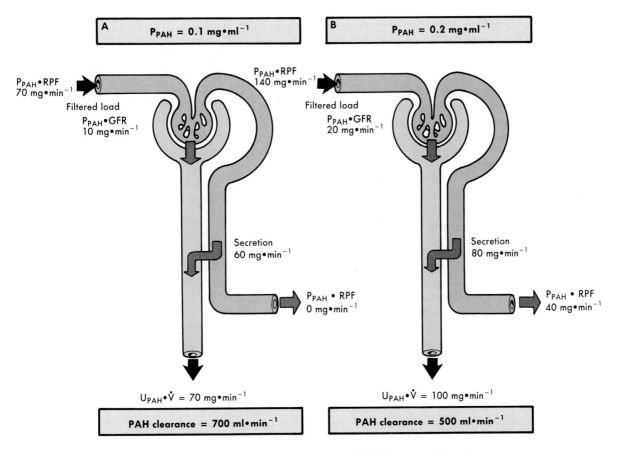

FIGURE 32-11   Renal handling of *p*-aminohippuric acid *(PAH)* at two different plasma concentrations *(P$_{PAH}$)*. **A,** The P$_{PAH}$ is less than the value that would be expected to saturate the PAH secretory mechanism. **B,** The elevated P$_{PAH}$ results in the delivery of PAH to the secretory mechanism exceeding the transport maximum (T$_m$). For both cases the renal plasma flow *(RPF)* is 700 ml·min^{-1}, the glomerular filtration rate *(GFR)* is 100 ml·min^{-1}, and the T$_m$ for the PAH secretory mechanism is 80 mg·min^{-1}. *U$_{PAH}$,* Urine PAH concentration; *V̇,* urine flow rate.

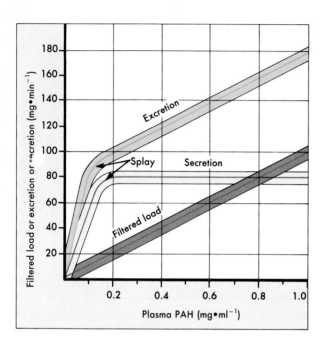

FIGURE 32-12   Dependence of PAH filtered load, excretion rate, and secretory transport rate on the plasma PAH concentration.

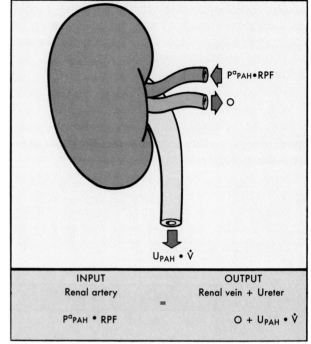

FIGURE 32-13   The use of PAH clearance to measure the renal plasma flow *(RPF)*. P^a_{PAH}, Plasma PAH concentration; U_{PAH}, urine PAH concentration; $\dot{V}$, urine flow rate.

Measurement of Renal Plasma Flow

When the plasma PAH concentration is so low that the T_m of the secretory mechanism is not exceeded, its clearance provides a measure of the renal plasma flow (see Figure 32-11, *A*). The reason for this is illustrated in Figure 32-13.

The amount of PAH arriving at the kidneys per minute is simply the product of the plasma PAH concentration (P_{PAH}) and the renal plasma flow (RPF). Since all of the PAH is excreted in the urine, and none is returned to the systemic circulation in the renal vein, the following relationship is true:

$$RPF \cdot P_{PAH} = U_{PAH} \cdot \dot{V} \qquad (7)$$

where U_{PAH} is urine PAH concentration and $\dot{V}$ is urine flow rate. Rearranging and solving for RPF, the following equation is obtained:

$$RPF = \frac{U_{PAH} \cdot \dot{V}}{P_{PAH}} \qquad (8)$$

This is the same as the general clearance equation (equation 2). Thus, at low plasma PAH concentra-

tions, the PAH clearance is equal to the renal plasma flow. At high plasma PAH concentrations, however, the PAH secretory mechanism will be saturated, and a significant amount of PAH will appear in the renal venous blood. Under this condition the PAH clearance **will not** equal the renal plasma flow.

The relations described here are idealized. Even at plasma concentrations that do not exceed the T_m for PAH, some PAH still appears in the renal vein plasma. The reason for this is related to the tubular and vascular anatomy of the kidney. The PAH secretory mechanism is located in the proximal tubule. Consequently, if all the PAH is to be secreted into the tubular fluid, all the blood coming to the kidney must flow through the peritubular capillaries that surround the proximal tubules; approximately 90% to 95% of all blood does. Because 5% to 10% of the blood does not perfuse these capillaries, the PAH in this blood cannot be secreted into the tubular fluid. Thus the value determined by the clearance of PAH actually underestimates the true RPF by 5% to 10%. For this reason the clearance of PAH is said to measure the *effective renal plasma flow (ERPF).*

Once the ERPF is determined, and the hematocrit (Hct) of the blood is known, the *renal blood flow (RBF)* can be calculated as:

$$RBF = \frac{RPF}{1 - Hct} \qquad (9)$$

Using Clearance to Estimate Transport Mechanisms

As depicted in Figure 32-6, the excretion rate for any substance can be determined as:

Excretion = Filtered load − Reabsorption + Secretion **(10)**

Most substances are filtered and either secreted or reabsorbed. The notable exceptions are K^+, urea, and uric acid, all of which undergo filtration, as well as reabsorption and secretion at different points along the nephron. If it is known that a substance is freely filtered at the glomerulus, a comparison of its renal clearance to that of inulin will indicate how the substance is handled by the kidney. Thus:

1. If its clearance is less than the inulin clearance, the substance (e.g., glucose) is reabsorbed by the nephron.
2. If its clearance is greater than the inulin clearance, the substance (e.g., PAH) is secreted.
3. If its clearance equals the inulin clearance, the substance is only filtered.

For those substances both secreted and reabsorbed by the nephron (e.g., K^+, urea, and uric acid), their clearance will reflect the dominant transport mechanism.

One final point about the use of clearance to assess renal function relates to the dependency of clearance on the plasma concentration of the solute. This is illustrated in Figure 32-14 for the clearances of inulin, PAH, and glucose. The figure shows that the clearance of inulin is constant as the plasma inulin concentration is varied. This reflects that the glomerular filtration rate is constant. In contrast, the clearances of glucose and PAH vary with their plasma concentrations. At low plasma concentrations the clearance of PAH exceeds that of inulin, whereas the clearance of glucose is less than the inulin clearance. As the plasma concentrations for these substances are increased, their clearances asymptotically approach that of inulin. The reason for this convergence of the PAH and glucose clearances is that at high plasma concentrations the reabsorptive and secretory mech-

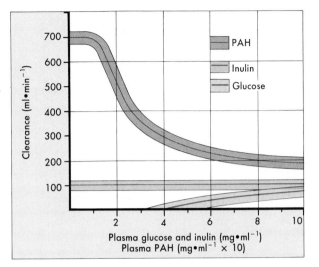

FIGURE 32-14   The dependency of clearance of PAH, inulin, and glucose on the plasma concentration.

anisms are saturated (T_m is exceeded), and the filtered load becomes a much larger fraction of the total amount of the substance excreted in the urine.

GLOMERULAR FILTRATION AND RENAL BLOOD FLOW

The first step in the formation of urine is the production of an ultrafiltrate of the plasma at the glomerulus. The ultrafiltrate is devoid of cellular elements and is virtually protein free. The concentrations of salts and of organic molecules, such as glucose and amino acids, are similar in the plasma and ultrafiltrate. Ultrafiltration is driven by Starling forces (see below and Chapter 21) across the glomerular capillaries, and changes in these forces and in renal blood flow (RBF) alter the glomerular filtration rate (GFR). GFR and RBF are normally held within very narrow ranges by the phenomenon of autoregulation (see below and Chapter 22). This section reviews the composition of the glomerular filtrate, the dynamics of its formation, and the relationship between RBF and GFR. In addition, the factors that contribute to the autoregulation of GFR and RBF are discussed.

Glomerular Filtration

The unique structure of the glomerular filtration barrier (capillary endothelium, basement membrane,

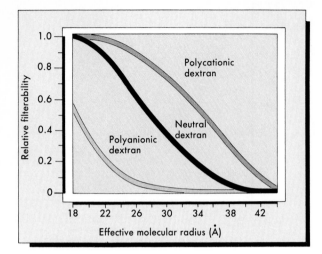

FIGURE 32-15   Influence of size and electrical charge of dextran on its filterability. A value of 1 indicates that dextran is freely filtered, whereas a value of 0 indicates that it is not filtered.

and filtration slits of the podocytes) determines the composition of the filtrate. The glomerular filtration barrier filters molecules on the basis of **size and electrical charge.** Figure 32-15 illustrates how these factors affect the filtration of dextrans by the glomerulus. Dextrans are a family of exogenous polysaccharides produced in various molecular weights, as well as in an electrically neutral form or with negative charges (polyanionic) or positive charges (polycationic). At constant charge, as the size (i.e., effective molecular radius) increases, filtration decreases. For any given molecular radius, cationic molecules are more readily filtered than are anionic molecules. The restriction of anionic molecules is explained by the presence of negatively charged glycoproteins on the surface of all components of the glomerular filtration barrier. These charged glycoproteins repel similarly charged molecules. Because most plasma proteins are negatively charged, the negative charge on the filtration barrier restricts the filtration of proteins.

Figures 32-15 also indicates that for the normal adult kidney, molecules (e.g., water, electrolytes, amino acids, and glucose) with a molecular radius less than 18 angstroms (Å) that are not protein bound are freely filtered. Molecules (e.g., proteins larger than albumin, such as IgG, IgA) larger than 44 Å are not filtered, and molecules (e.g., small proteins) between 18 and 44 Å are filtered to various degrees. Albumin, an anionic protein that has an effective molecular

radius of 35 Å, is poorly filtered by the glomerulus. Approximately 7 g of albumin are filtered each day. However, because albumin is avidly reabsorbed by the proximal tubule, almost none appears in the urine.

The forces responsible for the glomerular filtration of plasma are the same as those involved in fluid exchange across all capillary beds. Ultrafiltration occurs because of *Starling forces* that drive fluid from the lumen of glomerular capillaries, across the filtration barrier, and into Bowman's space. As shown in Figure 32-16, Starling forces across glomerular capillaries are similar to the forces that promote filtration across other capillary beds and include hydrostatic and oncotic pressures (see Chapter 21). Both the hydrostatic pressure in the glomerular capillary (P_{GC})

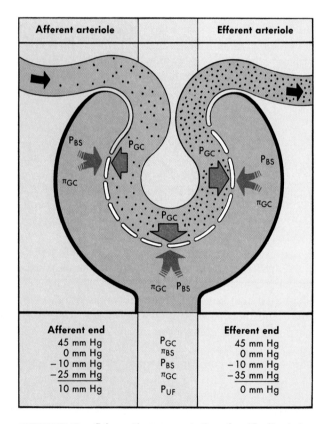

Afferent end		Efferent end
45 mm Hg	P_{GC}	45 mm Hg
0 mm Hg	π_{BS}	0 mm Hg
−10 mm Hg	P_{BS}	−10 mm Hg
−25 mm Hg	π_{GC}	−35 mm Hg
10 mm Hg	P_{UF}	0 mm Hg

FIGURE 32-16   Schematic representation of an idealized glomerular capillary and the Starling forces across the filtration barrier. P_{UF}, Net ultrafiltration pressure; P_{GC}, glomerular capillary hydrostatic pressure; P_{BS}, Bowman's space hydrostatic pressure; π_{GC}, glomerular capillary oncotic pressure; π_{BS}, Bowman's space oncotic pressure; arrows indicate direction of forces.

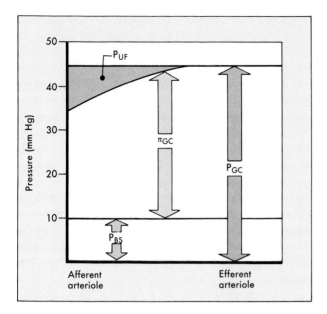

FIGURE 32-17   Relationship between the hydrostatic pressure in the glomerular capillary (P_{GC}) and Bowman's space (P_{BS}) and the oncotic pressure in the glomerular capillary (π_{GC}) along the length of an idealized glomerular capillary. P_{UF}, Net ultrafiltration pressure.

and the oncotic pressure in Bowman's space (π_{BS}) are oriented so as to promote the movement of fluid from the glomerular capillary into Bowman's space. Because the glomerular filtrate is essentially protein free, π_{BS} is near zero. Therefore P_{GC} is the only force that favors filtration, and it is opposed by the hydrostatic pressure in Bowman's space (P_{BS}) and the oncotic pressure in the glomerular capillary (π_{GC}).

As illustrated in Figure 32-17, a net ultrafiltration pressure (P_{UF}) of 10 mm Hg exists at the afferent end of the glomerulus, whereas at the efferent end the P_{UF} is zero. Thus, at the efferent end of the glomerulus, filtration equilibrium has been achieved and net ultrafiltration stops. Two additional points concerning Starling forces are illustrated in Figure 32-17. First, the hydrostatic pressure within the capillary is virtually constant along its length. The only force that changes during the process of ultrafiltration is π_{GC}. The increase in π_{GC} results from the filtration of water. Because water is filtered and protein is retained in the glomerular capillary, the protein concentration in the capillary rises and π_{GC} increases.

The GFR is proportional to the P_{UF} that exists across the capillaries times the *ultrafiltration coefficient*, K_f:

$$GFR = K_f \left[(P_{GC} - P_{BS}) - (\pi_{GC} - \pi_{BS}) \right] \quad \textbf{(11)}$$

The K_f is related to the intrinsic permeability of the capillary and to the surface area available for filtration. Although the net filtration pressure is similar in glomerular capillaries and in other capillary beds, the rate of filtration is considerably greater in glomerular capillaries because K_f is approximately 100 times higher.

Plasma flow is an important determinant of GFR. Figure 32-18 shows that the P_{UF}, and thereby GFR, depends on the plasma flow through the glomerular capillaries. If the plasma flow increases, more filtrate will be formed before the capillary oncotic pressure rises enough to stop filtration (i.e., $[P_{GC} - P_{BS} - \pi_{GC}] = 0$). Conversely, if the plasma flow decreases, a smaller rate of filtration will be required to reduce the driving force for filtration to zero.

The GFR can be altered by changing K_f or by changing any of the Starling forces. Physiologically, however, GFR is usually affected in two primary ways:

1. An increase in P_{GC} enhances GFR, and a decrease in P_{GC} depresses GFR. Changes in arterial pressure are the most frequent cause of variations in P_{GC}.
2. Variations in RBF. As afferent arteriolar plasma flow increases, GFR rises. As plasma flow in the capillaries increases, π_{GC} rises more slowly. Accordingly, the P_{UF} increases. A fall in plasma flow decreases GFR.

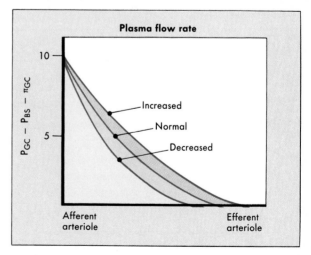

FIGURE 32-18   Relationship between plasma flow rate along an idealized glomerular capillary and the net ultrafiltration pressure $(P_{GC} - P_{BS} - \pi_{GC})$. See Figure 32-17 legend for abbreviations.

Pathologic conditions and drugs may also affect GFR, primarily by changing π_{GC}, P_{BS} and K_f. Thus GFR may change by three additional mechanisms:

1. Changes in π_{GC}. An inverse relationship exists between π_{GC} and GFR. Alterations in π_{GC} result from changes in protein metabolism outside the kidney.

2. Changes in K_f. Increased K_f enhances GFR, whereas decreased K_f reduces GFR. Some kidney diseases reduce K_f, and some vasoactive drugs increase K_f.

3. Changes in P_{BS}. Increased P_{BS} reduces GFR, whereas decreased P_{BS} facilitates GFR. Acute obstruction of the urinary tract (e.g., a kidney stone occluding the ureter) increases P_{BS}.

Renal Blood Flow and Autoregulation of Glomerular Filtration Rate.

The kidney, as with most other organs, regulates its blood flow by adjusting the vascular resistance in response to changes in arterial pressure (see Chapter 22). As illustrated in Figures 32-19, this adjustment in resistance is so precise that the blood flow remains constant as arterial blood pressure changes between 90 and 180 mm Hg. GFR is also regulated over the same range of arterial pressures. The phenomena whereby RBF and GFR are maintained constant is called *autoregulation*. As the term is meant to indicate, autoregulation is achieved by changes exclusively within the kidney. Because both GFR and RBF are regulated over the same range of pressures, and because RBF is an important determinant of GFR, it is not surprising that the same mechanisms regulate both flows.

Two mechanisms are responsible for autoregulation of RBF and GFR: one that responds to changes in arterial pressure and another that responds to changes in flow. The pressure-sensitive mechanism, the *myogenic mechanism,* (see Chapter 22), is related to an intrinsic property of all vascular smooth muscle: the tendency to contract when stretched. Accordingly, when arterial pressure rises and the renal afferent arteriole is stretched, the smooth muscle contracts. Because the increase in the resistance (ΔR) of the arteriole offsets the increase in pressure (ΔP), blood flow (Q) and therefore GFR remain constant (i.e., Q = $\Delta P / \Delta R$).

The second mechanism responsible for autoregulation of GFR and RBF, the flow-dependent mechanism, is known as *tubuloglomerular feedback*. This mechanism involves a feedback loop in which the flow of tubular fluid (or some other factor such as the solute composition of tubular fluid) is sensed by the macula densa of the juxtaglomerular apparatus (JGA). The macula densa in turn governs the filtration rate of the glomerulus to which it is apposed. When the flow of tubular fluid is increased at the macula densa, the filtration rate of that nephron is reduced. This occurs because the resistance of the efferent arteriole decreases, which reduces the hydrostatic pressure in the glomerular capillaries. The opposite prevails when tubular flow rate past the macula densa is decreased; efferent arteriole resistance increases, P_{GC} increases, and filtration rate rises. The major unknowns about tubuloglomerular feedback concern the variable that is sensed at the JGA and the effector substance that alters the resistance of the efferent arteriole. It has been suggested that macula densa cells may sense the osmolality of the tubular fluid or the concentrations of calcium, chloride, or sodium in tubular fluid. The effector mechanism may involve the renin-angiotensin system (see Chapter 33) or other vasoactive substances, such as prostaglandins, catecholamines, or kinins.

Because animals engage in many activities that change arterial blood pressure, independent of alterations in fluid and solute balance, it is highly desirable to have mechanisms that uncouple RBF and GFR from blood pressure. If RBF and GFR were to rise or fall suddenly in proportion to changes in blood pres-

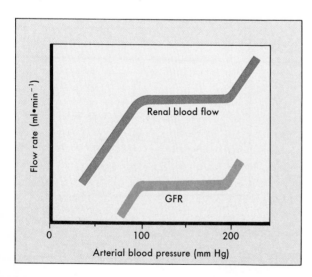

FIGURE 32-19   Relationships between arterial blood pressure and renal blood flow/glomerular filtration rate *(GFR)*.

Table 32-1   Flitration, Excretion, and Reabsorption of Water, Electrolytes, and Solutes in an Adult

Substance	Measure	Filtered*	Excreted	Reabsorbed	Percent of Filtered Load Reabsorbed
Water	L/day	180	1.5	178.5	99.2
Na^+	mEq/day	25,200	150	25,050	99.4
K^+	mEq/day	720	100	620	86.1
Ca^{++}	mEq/day	540	10	530	98.1
HCO_3^-	mEq/day	4320	2	4218	99.9+
Cl^-	mEq/day	18,000	150	17,850	99.2
Glucose	mmol/day	800	0.5	799.5	99.9+
Urea	g/day	56	28	28	50.0

*The filtered amount of any substance is calculated by multiplying the concentration of that substance in the ultrafiltrate by the glomerular filtration rate; for example, the filtered load of Na^+ is calculated as: $[Na^+]$ ultrafiltrate (140 mEq·L^{-1}) × Glomerular filtration rate (180 L/day) = 25,200 mEq/day.

sure, the urinary excretion of fluid and solutes would not match their intake. Accordingly, fluid and solute balance would change. Autoregulation of GFR and RBF provides an effective means for uncoupling renal function from arterial pressure and ensures that both fluid and solute excretion remain constant.

TUBULAR FUNCTION

Formation of urine involves three basic processes: ultrafiltration of plasma by the glomeruli, reabsorption of water and solutes from the ultrafiltrate, and secretion of selected solutes into the tubular fluid. Although 180 L of protein-free fluid are filtered by the human glomeruli each day, only 1% to 2% of the water, less than 1% of the filtered sodium, and variable amounts of the other solutes are excreted in the urine (Table 32-1). By the processes of reabsorption and secretion, the renal tubules modulate the volume and composition of the urine (Table 32-2). Consequently the tubules control precisely the volume, osmolality, composition, and pH of the intracellular and extracellular fluid compartments.

Because of the importance of tubular reabsorption and secretion, the first part of this section defines some basic transport mechanisms along the nephron. Then water and solute transport along the renal tubules and some of the factors and hormones that regulate transport are discussed. Details on specific electrolytes and solutes are provided in Chapter 34.

Table 32-2   Composition of the Urine of an Adult*

Substance	Concentration
Na^+	50-130 mEq·L^{-1}
K^+	20- 70 mEq·L^{-1}
NH_4^+	30- 50 mEq·L^{-1}
Ca^{++}	5- 12 mEq·L^{-1}
Mg^{++}	2- 18 mEq·L^{-1}
Cl^-	50-130 mEq·L^{-1}
$H_2PO_4^-$	20- 40 mEq·L^{-1}†
$SO_4^=$	30- 45 mEq·L^{-1}
Urea	200-400 mM
Creatinine	6- 20 mM
pH	5.0-7.0
Osmolality	500-800 mOsm/Kg H_2O

†When urine pH is 6 or lower, almost all phosphate exists in the monovalent form.
*Numbers represent average values. Water excretion ranges from 0.5 to 1.5 L/day.

Basic Transport Mechanisms

Solutes may be transported across cell membranes by passive or active mechanisms (see Chapter 1). The movement is *passive* if it develops spontaneously and does not require metabolic energy. Passive transport of solutes or water always occurs from an area of higher concentration to an area of lower concentration. In addition to concentration gradients, the passive movement of ions (but not uncharged solutes, such as glucose and urea) is affected by the electrical potential differences across cell membranes and across the renal tubules. Cations (sodium, potassium, etc.) tend to move to the negative side of the membrane, whereas anions (chloride, bicarbonate, etc.) tend to move to the positive side of the membrane.

Transport is *active* if it depends on energy derived from metabolic processes. Active transport of solutes usually occurs from an area of low concentration to an area of high concentration. In the kidney the *primary active transport mechanism* is the Na^+,K^+-*ATPase* (or sodium pump), which is located in the basolateral membrane of all cells. This mechanism actively pumps Na^+ out of the cell and pumps K^+ into the cell. In mammals solute movement occurs by both passive and active mechanisms, whereas all water movement is passive.

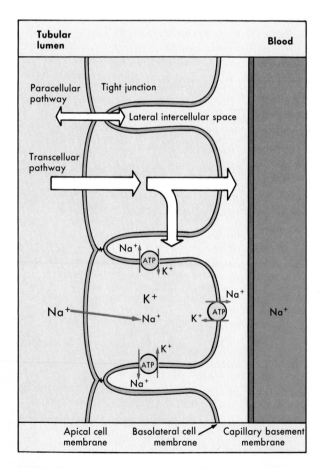

FIGURE 32-20   Schematic representation of transport pathways in an idealized proximal tubule. *ATP,* Adenosine triphosphate.

Renal tubular epithelia, as with other epithelia (e.g., the intestine), can transport solutes and water from one side of the tubule to the other. Renal cells are held together by tight junctions and are separated by lateral intercellular spaces (Figure 32-20). The tight junctions separate the apical membranes from the basolateral membranes. *Reabsorption* is the net transport of a substance from the tubular lumen into the blood, whereas *secretion* is the net transport in the opposite direction. A substance may be reabsorbed or secreted by the renal tubules across cells, the *transcellular pathway,* or between cells, the *paracellular pathway* (Figure 32-20).

Sodium reabsorption by the transcellular pathway in the renal tubule depends on the operation of the Na^+,K^+-ATPase (Figure 32-20). The Na^+,K^+-ATPase moves sodium out of the cell into the blood. Potassium enters the cell by the action of the Na^+,K^+-ATPase pump. Thus the operation of the Na^+,K^+-ATPase lowers intracellular Na^+ concentration and increases intracellular K^+ concentration. Because intracellular Na^+ is low (12 mEq·L^{-1}) and the Na^+ concentration in tubular fluid is high (140 mEq·L^{-1}), Na^+ moves across the apical cell membrane down an electrochemical gradient from the tubular lumen into the cell. The Na^+,K^+-ATPase senses the addition of Na^+ to the cell and is stimulated to increase its rate of Na^+ extrusion into the blood and thereby returns intracellular Na^+ to normal levels. Thus Na^+ reabsorption by the renal tubule is a two-step process: (1) movement across the apical membrane into the cell down an electrochemical gradient established by the Na^+,K^+-ATPase and (2) movement across the basolateral membrane against an electrochemical gradient via the Na^+,K^+-ATPase.

Proximal Tubule

The proximal tubule reabsorbs approximately 67% of the filtered water, sodium, chloride, potassium, and other solutes and thereby critically influences the volume and composition of the urine (Figure 32-21). In addition, virtually all the glucose and amino acids are reabsorbed. The key element in proximal tubule reabsorption is the Na^+,K^+-ATPase in the basolateral membrane. The reabsorption of every substance, including water, is linked to the operation of this enzyme.

The filtered water and solutes are reabsorbed by the proximal tubule in two phases: (1) reabsorption of

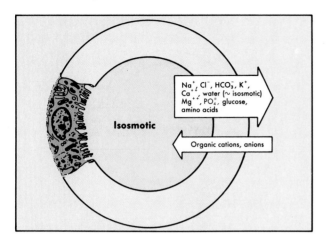

FIGURE 32-21 Schematic representation of a cell in the proximal tubule and the primary transport characteristics. Tubular fluid is isosmotic.

Na^+ with glucose, amino acids, and bicarbonate (HCO_3^-) in the first half of the proximal tubule, and (2) reabsorption of Na^+ with chloride in the second half of the proximal tubule.

During the first phase of proximal tubule reabsorption Na^+ entry into the cell across the apical membrane is mediated by specific transport proteins and not by simple diffusion. An important property of these transport proteins is that they couple the movement of Na^+ with the movement of other solutes. Each of these transport proteins uses the potential energy released by the downhill movement of Na^+ to power the uphill movement of the other solute. Examples include the *co-transport proteins* that couple the uptake of Na^+ with that of either glucose, amino acids, phosphate, chloride, or lactate (Figure 32-22). Sodium entry is also coupled with hydrogen ion (H^+) extrusion from the cell by a Na^+-H^+ *antiporter*. Proton (H^+) secretion, via the Na^+-H^+ antiporter, results in bicarbonate reabsorption (Figure 32-22; see Chapter 34). The Na^+ that enters the cell across the apical membrane leaves the cell across the basolateral membrane via the Na^+,K^+-ATPase. The other solutes that enter the cell with Na^+ exit across the basolateral membrane down their electrochemical gradients.

The reabsorption of Na^+ and the other solutes just described increases the osmolality of the lateral intercellular space. Because the lateral intercellular space is slightly hyperosmotic (~3 mOsm/kg H_2O) with respect to tubular fluid, and because the proximal tubule is highly permeable to water, water will flow by osmosis across both the tight junctions and the prox-

imal tubular cells into this hyperosmotic compartment (Figure 32-22). Accumulation of fluid within the lateral intercellular space increases the hydrostatic pressure in this compartment and thereby drives fluid into the capillaries. Thus **water reabsorption follows solute transport.** The reabsorbed fluid is essentially *isosmotic* to plasma. An important consequence of osmotic water flow across the proximal tubule is that some solutes, especially potassium, calcium, and magnesium, are entrained in the reabsorbed fluid and are thereby reabsorbed by the process of *solvent drag* (Figure 32-22).

Because the reabsorption of virtually all solutes and water are coupled to Na^+ reabsorption, changes in

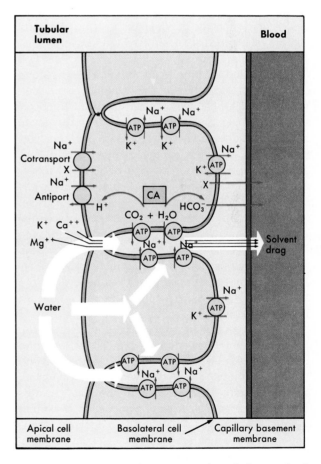

FIGURE 32-22 Schematic representation of the proximal tubule. For the Na^+-X co-transport protein, X represents either glucose, amino acids, phosphate, chloride, or lactate. CO_2 and H_2O combine inside the cells to form H^+ and HCO_3^- in a reaction facilitated by the enzyme carbonic anhydrase *(CA)*. *ATP,* Adenosine triphosphate.

Na$^+$ reabsorption will influence the reabsorption of water and other solutes by the proximal tubule. Thus, in the first phase of proximal reabsorption, reabsorption of Na$^+$ is coupled to that of bicarbonate, glucose, amino acids, phosphate, and lactate.

The second phase of proximal tubular reabsorption involves the reabsorption of Na$^+$ with Cl$^-$ in the second half of the proximal tubule. This occurs because in the first half of the proximal tubule, Na$^+$ is reabsorbed with bicarbonate as the primary accompanying anion, leaving behind a solution that becomes enriched in Cl$^-$. The rise in Cl$^-$ concentration in the tubular fluid creates a gradient that favors the diffusion of Cl$^-$ from the tubular lumen across the tight junctions and into the lateral intercellular space. Movement of the negatively charged chloride ions attracts the positively charged sodium ions. Thus, in the second half of the proximal tubule, some Na$^+$ and Cl$^-$ are reabsorbed across the tight junctions by passive diffusion.

Sodium and chloride reabsorption by the second half of the proximal tubule also occurs by a transcellular route. The pathway for Na$^+$ and Cl$^-$ transport across the apical membrane is unknown. Evidence suggests that they may be coupled via a co-transport protein that uses the energy in the Na$^+$ gradient across the apical membrane to drive the uphill movement of Cl$^-$ into the cell. The Na$^+$ that enters the cell is pumped into the blood by the Na$^+$,K$^+$-ATPase. Although the movement of Cl$^-$ across the basolateral membrane is down its electrochemical gradient, the mechanism is currently unknown. Thus, during the second phase of proximal tubule reabsorption, Na$^+$ and Cl$^-$ are reabsorbed by two routes: across the tight junction by passive mechanisms and via the cell requiring the Na$^+$,K$^+$-ATPase.

In addition to reabsorbing solutes and water, the proximal tubule also secretes organic cations and anions (see Figure 32-21). These substances are end products of metabolism in the liver, and they circulate in the plasma. The proximal tubule also secretes numerous exogenous organic compounds, including p-aminohippurate and penicillin. Because these organic compounds are bound to plasma proteins, they are not readily filtered. Therefore excretion by filtration alone eliminates from the body only a small portion of these potentially toxic substances. High excretion rates of these compounds are obtained by secretion from the peritubular capillaries into the tubular fluid. Because the kidney removes virtually all

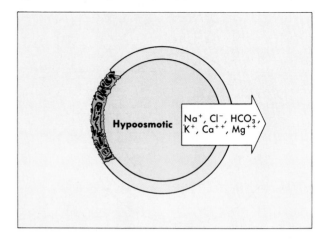

FIGURE 32-23   Tubular profile of the thick ascending limb of Henle's loop illustrating the cellular ultrastructure and the primary transport characteristics. Tubular fluid is hypoosmotic.

organic ions and drugs from the plasma that enters the kidney, these secretory mechanisms evidently are very powerful. The concentration of these compounds in the renal vein is close to zero.

Henle's Loop

Henle's loop reabsorbs approximately 20% of the filtered sodium, chloride, and potassium (Figure 32-23). Calcium, bicarbonate (HCO$_3^-$), and magnesium are also reabsorbed in Henle's loop. This reabsorption occurs almost exclusively in the thick ascending limb. By comparison, the descending thin limb and the ascending thin limb do not reabsorb significant amounts of solutes.

Henle's loop reabsorbs approximately 17% of the filtered water. This reabsorption, however, occurs exclusively in the descending thin limb. **The ascending limb is impermeable to water.**

The key element in solute reabsorption by the thick ascending limb is the Na$^+$,K$^+$ATPase in the basolateral membrane (Figure 32-24). As with reabsorption in the proximal tubule, the reabsorption of every substance by the thick ascending limb is linked to the sodium pump.

The operation of the Na$^+$,K$^+$-ATPase in the thick ascending limb maintains a low cell Na$^+$ concentration, which provides a favorable chemical gradient for the movement of Na$^+$ from the tubular fluid into the cell. The movement of Na$^+$ across the apical membrane is mediated by a specific transport protein that

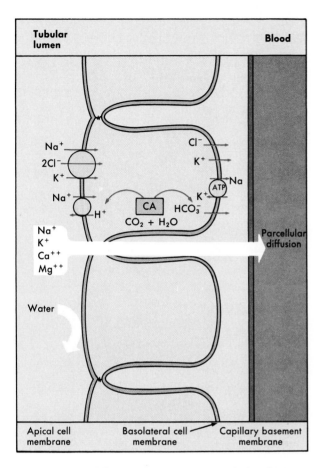

Tubular lumen

Blood

Na^+
$2Cl^-$
K^+

Na^+

Na^+
K^+
Ca^{++}
Mg^{++}

Water

Cl^-
K^+

Na

K^+

CA HCO_3^-
$CO_2 + H_2O$

$-H^+$

Parcellular diffusion

Apical cell membrane   Basolateral cell membrane   Capillary basement membrane

FIGURE 32-24   Schematic representation of the thick ascending limb of Henle's loop and the key transport pathways. *CA,* Carbonic anhydrase; *ATP,* adenosine triphosphate.

the thick ascending limb occurs by two routes: the transcellular and paracellular pathways.

Because the thick ascending limb is very impermeable to water, reabsorption of Na^+, Cl^-, and other solutes reduces the osmolality of tubular fluid to less than 150 mOsm/kg H_2O.

Distal Tubule and Collecting Duct

The distal tubule and the collecting duct are important in determining the amount of Na^+, K^+, H^+, and water excreted by the kidney. These segments reabsorb approximately 12% of the filtered Na^+ and Cl^-, secrete K^+ and protons, and reabsorb some 15% of the water filtered by the glomerulus.

The initial segment of the distal tubule reabsorbs Na^+, Cl^-, and Ca^{++} and is impermeable to water. Thus this segment continues the active dilution of the tubular fluid begun in the thick ascending limb (Figure 32-25).

The last segment of the distal tubule and the collecting duct are composed of two cell types, principal cells and intercalated cells. As illustrated in Figure 32-26, principal cells reabsorb Na^+ and water and secrete K^+. Intercalated cells secrete protons, reabsorb HCO_3^-, and thus are important in regulating acid-base balance (see Chapter 34). Both Na^+ reabsorption and K^+ secretion by principal cells depend on the activity of the Na^+, K^+-ATPase in the basolateral membrane (Figure 32-27). This enzyme maintains a

couples Na^+ with Cl^- and K^+ uptake into the cell. This co-transport protein uses the potential energy released by the downhill movement of Na^+ to power the uphill movement of Cl^- and K^+ into the cell. A Na^+-H^+ antiporter protein in the apical cell membrane also mediates Na^+ reabsorption, as well as HCO_3^- reabsorption in the thick ascending limb (Figure 32-24; see Chapter 34). Na^+ leaves the cell across the basolateral membrane via the Na^+, K^+-ATPase. K^+, Cl^-, and HCO_3^- leave the cell across the basolateral membrane via separate pathways.

The voltage across the thick ascending limb is positive on the tubule-lumen side because of the unique permeability characteristics of this tubular segment. This voltage drives the reabsorption of several cations, including Na^+, K^+, Ca^{++}, and Mg^{++}, across the paracellular pathway. Thus salt reabsorption across

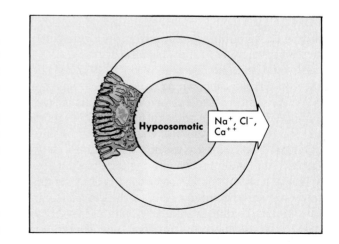

Hypoosomotic Na^+, Cl^-, Ca^{++}

FIGURE 32-25   Tubular profile of the initial segment of the distal tubule illustrating the cellular ultrastructure and the primary transport characteristics. Tubular fluid is hypoosmotic.

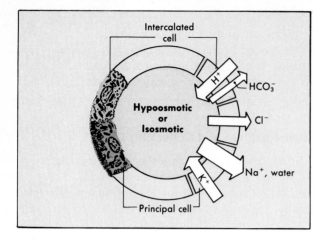

FIGURE 32-26   Schematic representation of the last segment of the distal tubule and the collecting duct illustrating the cellular ultrastructure and the primary transport characteristics. Tubular fluid is hypoosmotic or isosmotic.

low cell Na^+ concentration, which provides a favorable chemical gradient for the movement of Na^+ from the tubular fluid into the cell. Because Na^+ enters the cell across the apical membrane by diffusion, the negative potential inside the cell facilitates Na^+ entry. Sodium leaves the cell across the basolateral membrane and enters the blood via the Na^+,K^+-ATPase. Although the collecting duct reabsorbs significant amounts of Cl^-, the mechanism and route of transport are unknown.

Potassium is secreted by principal cells from blood into the tubular fluid in two steps (Figure 32-27). K^+ uptake across the basolateral membrane is mediated by the Na^+,K^+-ATPase. Because the K^+ concentration inside the cells is high (140 $mEq \cdot L^{-1}$) and the K^+ concentration in tubular fluid is low ($\sim$10 $mEq \cdot L^{-1}$), this ion will diffuse down its concentration gradient across the apical cell membrane into the tubular fluid. Although the negative potential inside the cells tends to retain K^+ within the cell, the combined electrochemical gradient across the apical membrane favors K^+ secretion from the cell into the tubular fluid.

Water reabsorption by the last segment of the distal tubule and the entire collecting duct is facilitated by antidiuretic hormone (ADH). When ADH levels are high, water reabsorption is increased. In contrast, when ADH levels are low, water reabsorption declines (see Chapter 33).

Intercalated cells secrete protons and reabsorb HCO_3^-. Proton secretion across the apical membrane is against an electrochemical gradient and is medi-

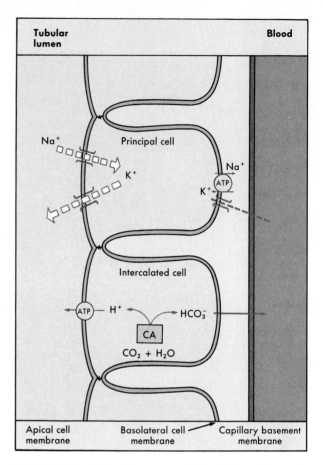

FIGURE 32-27   Schematic representation of the cell types in the last segment of the distal tubule and the collecting duct and the key transport pathways in each cell type. Although some potassium taken up by principal cells via the Na^+-K^+-ATPase diffuses back into the blood, the majority leaves the cell across the apical membrane due to a more favorable electrochemical gradient across this membrane. *CA,* Carbonic anhydrase; *ATP,* adenosine triphosphate.

ated by a H^+-ATPase transport mechanism (Figure 32-27). The generation of H^+ inside the cell is facilitated by the enzyme carbonic anhydrase, and it results in the production of HCO_3^-. For each H^+ secreted into the tubular fluid, one HCO_3^- leaves the cell across the basolateral membrane. Additional details of H^+ secretion by the collecting duct are considered in Chapter 34.

Regulation of Tubular Solute and Water Reabsorption

Many factors and hormones influence renal excretion of solutes and water by altering tubular reabsorption and secretion. This section reviews the primary

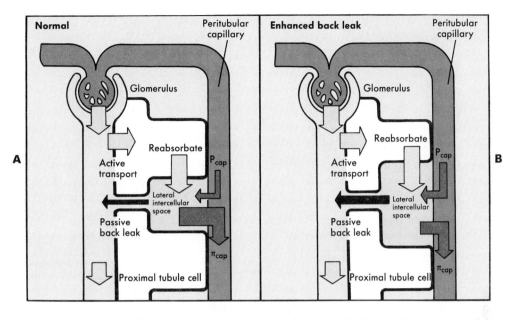

FIGURE 32-28   Effects of Starling forces on the passive back leak of solutes and water across the proximal tubule. **A,** Normal; **B,** enhanced back leak. An increase in the hydrostatic pressure of the peritubular capillary (P_{cap}) and a fall in the oncotic pressure of the peritubular capillary (π_{cap}) both enhance passive back leak of solutes and water across the tight junction. Arrows indicate direction of forces.

factors and hormones that regulate these functions.

Starling Forces   The reabsorption of solutes and water by the proximal tubule and Henle's loop are regulated by Starling forces. These forces do not affect transport by the distal tubule and collecting duct.

As illustrated in Figure 32-28, Starling forces across the wall of the peritubular capillaries in the proximal tubule influence the reabsorption of sodium and water. Once the reabsorbed solutes and water are deposited in the lateral intercellular space, they rapidly mix with the contents of the interstitium. The Starling forces that favor movement from the interstitium into the peritubular capillaries are the capillary oncotic pressure (π_{cap}) and the hydrostatic pressure in the lateral intercellular space. The opposing Starling forces are the interstitial oncotic pressure and the capillary hydrostatic pressure (P_{cap}). Normally the Starling forces favor movement of solute and water from the interstitium into the capillary.

Starling forces in the peritubular capillary are readily altered. Dilation of the efferent arteriole increases P_{cap}, whereas constriction of the efferent arteriole decreases P_{cap}. An increase in peritubular capillary pressure inhibits reabsorption, whereas a decrease in peritubular capillary pressure stimulates reabsorption.

The π_{cap} is determined by the formation of the ultrafiltrate across the glomerular capillaries. As more ultrafiltrate is formed (i.e., as glomerular filtration rate, GFR, increases) from a constant amount of plasma (constant renal plasma flow, RPF), the plasma proteins become more concentrated, thus raising the glomerular and peritubular capillary oncotic pressure. Thus the **peritubular oncotic pressure is directly related to the filtration fraction (FF=GFR/RPF).** A rise in the FF increases the π_{cap} and thereby enhances the peritubular capillary uptake of sodium and water. A fall in the FF has the opposite effect. When peritubular capillary uptake decreases, some of the fluid and solute that accumulates in the lateral intercellular space returns to the tubular lumen by passive back leak across the tight junctions (Figure 32-28). Thus the net reabsorption of solute and water by the proximal tubule is reduced.

The importance of Starling forces in regulating solute and water reabsorption by the proximal tubule is underscored by the phenomenon of *glomerulotubular balance (G-T balance).* Spontaneous changes in GFR greatly alter the filtered load of Na$^+$. Unless such changes are rapidly accompanied by adjustments in Na$^+$ reabsorption, urine Na$^+$ excretion would fluctuate widely and disturb whole body Na$^+$ balance.

Table 32-3   Effect of Hormones on Sodium, Chloride, and Water Transport by Nephron Segments

Segment	Hormone	Effects
Proximal tubule	Parathyroid hormone	Decreases NaCl, and water reabsorption
	Angiotensin	Increases NaCl, and water reabsorption
Thick ascending limb	Aldosterone, calcitonin, glucagon, vasopressin, parathyroid hormone	Increases NaCl reabsorption
Distal tubule and collecting duct	Calcitonin	Increases NaCl reabsorption
	Vasopressin	Increases permeability to water and NaCl reabsorption
	Aldosterone	Increases NaCl reabsorption
	Prostaglandins	Decreases NaCl reabsorption
	Bradykinin	Decreases NaCl reabsorption

However, spontaneous changes in GFR do not alter Na^+ balance because of G-T balance, which means that in the steady state, a constant fraction of the filtered Na^+ and water are reabsorbed in the proximal tubule, despite variations in GFR. The net result of G-T balance is to reduce the impact of GFR changes on the amount of Na^+ and water delivered from the proximal tubule and Henle's loop into the collecting duct. Because the collecting duct reabsorbs the excess Na^+ and water delivered to this segment when GFR rises, urinary excretion of these substances is unchanged.

Two mechanisms are responsible for G-T balance. One is related to the π_{cap} and P_{cap}; the other to the filtered load of glucose and amino acids. As an example of the first mechanism, an increase in GFR (at constant RPF) raises the protein concentration above normal in the glomerular capillary plasma. This protein-rich plasma leaves the glomerular capillaries, flows through the efferent arteriole, and enters the peritubular capillaries. The increased π_{cap} augments the movement of Na^+ and fluid from the lateral intercellular space into the peritubular capillaries. This mechanism thereby regulates proximal Na^+ and water reabsorption and matches solute and water reabsorption to the GFR.

The second mechanism responsible for G-T balance is initiated by an increase in the filtered load of glucose and amino acids. As discussed earlier, the reabsorption of Na^+ is coupled to that of glucose and amino acids. The rate of Na^+ reabsorption therefore depends partly on the filtered load of glucose and ami-

no acids. As GFR and the filtered load of glucose and amino acids increase, Na^+ and water reabsorption also rise.

In addition to G-T balance, another physiologic mechanism maintains a constant Na^+ delivery into the collecting ducts. As described earlier, an increase in GFR, and thus in the amount of Na^+ filtered by the glomerulus, activates the tubuloglomerular feedback mechanism, which returns GFR and the filtration of Na^+ to normal values. Thus spontaneous changes in GFR only temporarily increase the amount of Na^+ filtered. Until GFR returns to normal values, the mechanisms that underlie G-T balance maintain urinary Na^+ excretion constant.

Hormones   Table 32-3 summarizes the effects of several hormones on NaCl, and water reabsorption by the renal tubules. The regulation by these hormones of potassium, calcium, phosphate, and magnesium reabsorption are described in Chapter 34.

Sympathetic Nervous System   The sympathetic nervous system also regulates NaCl, and water reabsorption by the proximal tubule and Henle's loop. Activation of sympathetic nerves (e.g., after volume depletion) stimulates reabsorption, whereas inhibition of these nerves has the opposite effect.

BIBLIOGRAPHY
Journal Articles

Kriz W and Bankir L: A standard nomenclature for structures of the kidney, Am J Physiol 254:F1, 1988.

Navar LE: Renal autoregulation: perspectives from whole-kidney and single-nephron studies, Am J Physiol 234:F357, 1978.

Wein AJ: Physiology of micturition, Clin Geriatr Med 2(4):689, 1986.

Wright FS and Brigs JP: Feedback control of glomerular blood flow, pressure, and filtration rate, Physiol Rev 59:958, 1979.

Books and Monographs

Berry CA: Transport functions of the renal tubules. In Brenner BM et al, editors: Renal physiology in health and disease, Philadelphia, 1987, WB Saunders Co.

Burg MB: Renal handling of sodium, chloride, water, amino acids, and glucose. In Brenner BM and Rector FC Jr, editors: Philadelphia, 1986, WB Saunders Co.

Dworkin LD and Brenner BM: Biophysical basis of glomerular filtration. In Seldin DW and Giebisch G, editors: The kidney: physiology and pathophysiology, vol 1, New York, 1985, Raven Press.

Frohnert PP: Glomerular filtration. In Knox F, editor: Textbook of renal pathophysiology, New York, 1978, Harper & Row, Publishers, Inc.

Kassier JP and Harrington JT: Laboratory evaluation of renal function. In Schrier RW and Gottschalk CW, editors: Diseases of the kidney, ed 4, Boston, 1988, Little, Brown & Co.

Koushanpour E and Kriz W: Renal physiology—principles structure and function, ed 2, Berlin, 1986, Springer-Verlag.

Kriz W and Kaissling B: Structural organization of the kidney. In Seldin DW and Giebisch G, editors: The kidney: physiology and pathophysiology, vol 1, New York, 1985, Raven Press.

Marsh DJ: Renal physiology, New York, 1983, Raven Press.

Schuster VL and Seldin DW: Renal clearance. In Seldin DW and Giebisch G, editors: The kidney: physiology and pathophysiology, vol 1, New York, 1985, Raven Press.

Tisher CC and Madsen KM: Anatomy of the kidney. In Brenner BM and Rector FC Jr, editors: The Kidney, ed 3, vol 1, Philadelphia, 1986, WB Saunders Co.

CHAPTER

Control of Body Fluid Volume and Osmolality

The kidney maintains the osmolality and volume of the body fluids within a very narrow range by regulating its excretion of water and NaCl. This chapter discusses regulation of renal water excretion (urine concentration and dilution) and renal NaCl excretion. As an introduction, the normal volumes and compositions of the various body fluid compartments are reviewed.

THE BODY FLUID COMPARTMENTS

Measurement of Body Fluid Volumes

The volume of a body fluid compartment can be measured by adding a marker to the compartment and measuring its concentration once equilibrium has been reached. For example, if 5 g of glucose are added to an unknown volume of fluid in a beaker, and if the glucose concentration after complete equilibration is 10 $g \cdot L^{-1}$, the volume of water in the beaker is:

$$\text{Volume} = \frac{\text{Amount of solute}}{\text{Concentration}} = \frac{5 \text{ g}}{10 \text{ g} \cdot L^{-1}} = 0.5 \text{ L} \quad \textbf{(1)}$$

To qualify as a good marker for a particular fluid compartment, a substance must be confined to and distributed evenly throughout that compartment. Various markers have been used to measure the volume of the extracellular fluid and the total body water. However, no markers are available for measuring the volume of the intracellular fluid compartment. The volume of this compartment is calculated from the difference between the volume of the total body water and the volume of the extracellular fluid.

Volumes of Body Fluid Compartments

Water accounts for approximately 60% of the body weight. The water content of different individuals varies with the amount of adipose tissue; the greater the amount of adipose tissue, the less is the fraction of body weight attributable to water.

The *total body water (TBW)* is contained within two major compartments, which are divided by the cell membranes. The larger of these compartments is the *intracellular fluid (ICF)*, which represents approximately two thirds of the TBW. The remaining one third is the *extracellular fluid (ECF)*.

The extracellular water is subdivided into several compartments. The largest of these is the *interstitial fluid (ISF)*, which represents the fluid surrounding the cells in the various tissues of the body. The ISF comprises approximately three quarters of the ECF volume. Included within this compartment is the water contained within bone and dense connective tissue. The remaining one quarter of the ECF represents the *plasma*. Figure 33-1 summarizes the relationships among these various fluid compartments and provides estimates of their volumes.

Composition of Body Fluid Compartments

The concentrations of the major cations and anions in the ECF are illustrated in Figure 33-2, *A*. Sodium (Na^+) and its anions chloride (Cl^-) and bicarbonate (HCO_3^-) are the major ions of the ECF. The compositions of the two major divisions of the ECF (ISF and plasma) are very similar. Because these two compartments are separated only by the capillary endothelium, and because this barrier is freely permeable to

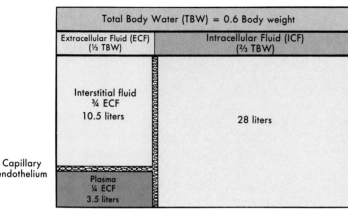

FIGURE 33-1 Volumes of the major body fluid compartments calculated for a 70 kg individual.

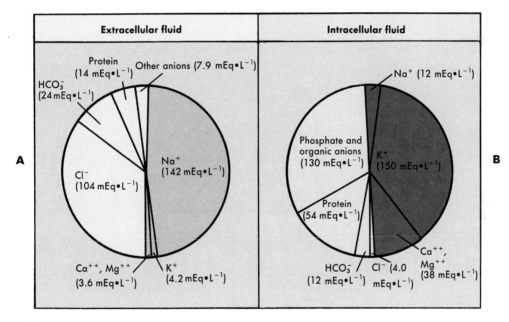

FIGURE 33-2 Concentrations of the major cations and anions in **A,** the extracellular (plasma) and **B,** the intracellular (skeletal muscle) fluids. The concentrations for Ca^{++} and Mg^{++} are the sum of these two ions. Concentrations represent the total of free and complexed ions.

small ions, the major difference between the ISF and plasma is that plasma contains significantly more protein. The presence of protein in the plasma can also affect the ionic composition of the ISF and plasma by the Gibbs-Donnan effect (see Chapter 2). The Gibbs-Donnan effect is normally quite small, and for most purposes the ionic composition of the plasma and ISF can be considered as identical.

Na^+ and its attendant anions (Cl^- and HCO_3^-) are the major determinants of the osmolality of the ECF. A rough estimate of the ECF osmolality can be obtained by simply doubling the sodium concentration [Na^+]. For example, if the plasma [Na^+] is 145 $mEq \cdot L^{-1}$, the osmolality of the plasma and ECF can be estimated as:

$$\text{Plasma osmolality} = 2 \,(\text{plasma } [Na^+]) \qquad \textbf{(2)}$$
$$= 290 \text{ mOsm/kg } H_2O$$

The composition of the ICF is more difficult to measure and can vary considerably from one tissue to another. Figure 33-2, *B,* provides information on the intracellular composition of skeletal muscle cells. In contrast to the ECF, the [Na$^+$] of the ICF is extremely low, whereas potassium (K$^+$) is the predominant intracellular cation. As explained in Chapter 2, this distribution of Na$^+$ and K$^+$ is maintained by the action of the Na$^+$,K$^+$-ATPase. The anion composition of the ICF also differs greatly from that of the ECF. The major intracellular anions are phosphates and organic anions, whereas the Cl$^-$ concentration is relatively low.

Fluid Exchange Between Compartments

Water moves readily between the various body fluid compartments. Two forces determine this movement: hydrostatic pressure and osmotic pressure. Hydrostatic pressure and the osmotic pressure generated by proteins (oncotic pressure) are important determinants of fluid movement across the capillary endothelium (see Chapter 21), whereas osmotic pressure differences between the ICF and ECF are responsible for water movement across the cell membranes. Because the plasma membranes of cells are highly permeable to water, a change in the osmolality of either the ICF or the ECF will result in a rapid movement of water between these compartments. Thus, except for transient changes, the ICF and ECF compartments are in osmotic equilibrium.

Although water can freely cross cell membranes, the movement of ions varies, depending on the type of cell and on the presence of specific pump and leak pathways. Consequently, fluid shifts between the ICF and ECF compartments occur primarily by the movement of water and not ions. This can best be illustrated by considering the consequences of adding either water or NaCl to the ECF (Figure 33-3).

When water is added to the ECF, the osmolality of this compartment is reduced. For the example in Figure 33-3, 2 L of water are added to the extracellular

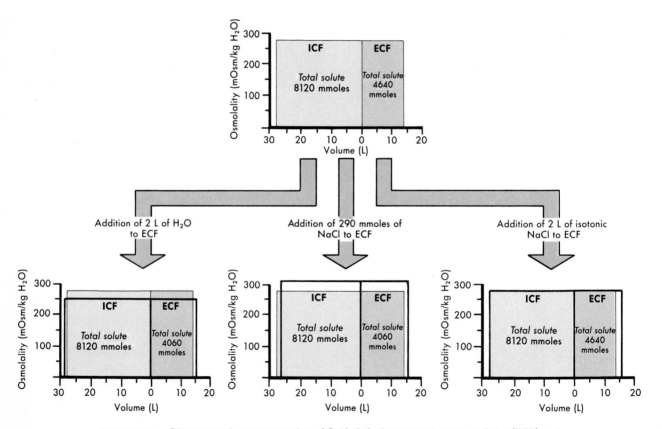

FIGURE 33-3   Diagrammatic representation of fluid shifts between the extracellular (ECF) and intracellular (ICF) fluid compartments following the addition of water, NaCl, and isotonic NaCl to the extracellular fluid.

space. If no fluid shifts were to occur, the osmolality of the ECF compartment would decrease from 290 to 254 mOsm/kg H_2O. However, the cell membranes are freely permeable to water. The osmotic gradient resulting from the addition of water to the ECF compartment will cause water to move into the cells. Figure 33-3 illustrates the volumes and osmolalities of the ICF and ECF compartments once osmotic equilibration has been reestablished. Both fluid compartments are increased in volume, and the osmolality of both compartments is reduced.

Figure 33-3 also illustrates the consequences of adding NaCl to the ECF compartment. Because of the presence of Na^+,K^+-ATPase in the cells, the added NaCl is effectively restricted to the extracellular space. If 290 mmoles of NaCl are added, this will add 580 mOsm of osmotically active particles. Consequently the ECF osmolality will increase from 290 to 311 mOsm/kg H_2O. This increase in ECF osmolality will cause fluid to move out of the cells. When osmotic equilibrium is again established, the volume of the ICF will be reduced, whereas that of the ECF will be increased. The osmolality of both compartments will be increased.

If both water and NaCl are added to the ECF compartment as a solution with the same osmolality as the body fluids (290 mOsm/kg H_2O), no osmotic gradients exist and no fluid shifts will occur. As a result, all the administered fluid will remain in the ECF compartment and increase its volume, but not its osmolality.

CONTROL OF BODY FLUID OSMOLALITY: URINE CONCENTRATION AND DILUTION

The kidney is the major route for the elimination of water from the body. Moreover, the renal excretion of water is regulated to maintain the osmolality of the body fluids constant. When water intake is low, or when water is lost from the body by other routes (e.g., perspiration, diarrhea), the kidney conserves water by producing a small volume of urine that is hyperosmotic with respect to plasma. When water intake is high, a large volume of hypoosmotic urine is produced. In a normal individual the urine osmolality can vary from approximately 50 to 1200 mOsm/kg H_2O, and the urine volume can vary from 0.5 to 20 L/day.

This section discusses the mechanism by which the kidney excretes either a hypoosmotic (dilute) or hyperosmotic (concentrated) urine. In addition, the secretion and action of the hormone vasopressin is reviewed.

Vasopressin

Vasopressin, or antidiuretic hormone (ADH), acts on the kidney to regulate the osmolality and volume of the urine. When plasma ADH levels are low, a large volume of urine is excreted *(diuresis)* and the urine is dilute. When plasma levels of ADH are elevated, a small volume of urine is excreted *(antidiuresis)* and the urine is concentrated.

ADH is a small peptide, nine amino acids in length. It is synthesized in cells of the hypothalamus (Figure 33-4) and is stored and released from axon terminals located in the neurohypophysis (posterior lobe of the pituitary gland).

The secretion of ADH from the posterior lobe of the pituitary is mainly regulated by the osmolality of the plasma, but both blood volume and pressure also have an influence. Cells *(osmoreceptors)* within the hypothalamus sense changes in the osmolality of the plasma. These cells are distinct from those that produce ADH. They appear to behave as osmometers and sense changes in plasma osmolality by either shrinking or swelling. The osmoreceptors are very sensitive, and a change in plasma osmolality of only 1% changes ADH secretion significantly. When plasma osmolality is elevated, ADH secretion increases. Conversely, when plasma osmolality is reduced, secretion is inhibited. Because ADH is rapidly degraded in the plasma, circulating levels can be reduced to zero within minutes after secretion is inhibited. As a result the ADH system can respond rapidly to fluctuations in plasma osmolality. Figure 33-5, *A* illustrates the effect of changes in plasma osmolality on plasma ADH levels.

A decrease in blood volume or pressure also stimulates ADH secretion. The receptors responsible for this response are located in both the low-pressure (left atrium) and the high-pressure (aortic arch and carotid sinus) sides of the circulatory system. Signals from these receptors are relayed to the ADH secretory cells via afferent fibers in the vagus and glossopharyngeal nerves. The sensitivity of this baroreceptor system is less than that of the osmoreceptors, and a 5% to 10% change in volume is required to alter ADH secretion.

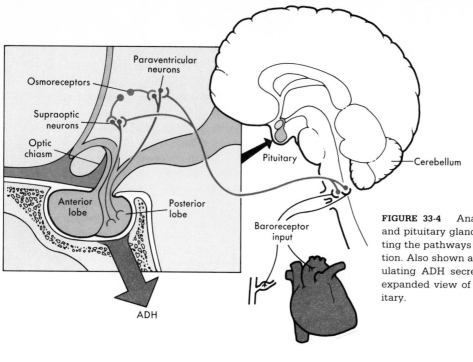

FIGURE 33-4 Anatomy of the hypothalamus and pituitary gland (midsagittal section) depicting the pathways for vasopressin (ADH) secretion. Also shown are pathways involved in regulating ADH secretion. Closed box illustrates expanded view of the hypothalamos and pituitary.

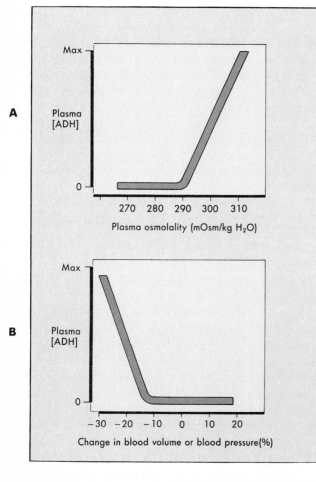

FIGURE 33-5 **A,** Effect of changes in plasma osmolality on plasma ADH levels. **B,** Effect of changes in the blood volume or blood pressure on plasma ADH levels.

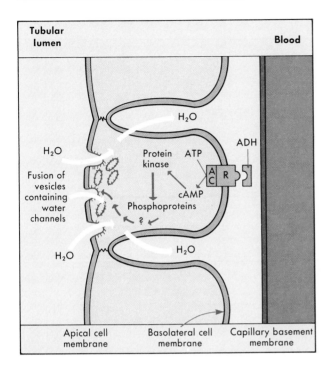

FIGURE 33-6   Action of ADH to increase the water permeability of the apical membrane of cells of the collecting duct. For water to be absorbed from the tubular lumen, an osmotic gradient must exist (luminal osmolality < blood osmolality). *ATP,* Adenosine triphosphate; *cAMP,* cyclic adenosine monophosphate; *AC,* adenylate cyclase.

Figure 33-5, *B* illustrates the effect of changes in blood volume or pressure on plasma levels of ADH.

In addition to affecting the secretion of ADH, changes in the osmolality and volume or pressure of the blood control the thirst center of the brain. When plasma osmolality is increased, or when the blood volume or pressure is decreased, the person experiences the desire to drink. If the individual has free access to water, the intake of water is increased; together with decreased renal excretion of water, the plasma osmolality and blood volume or pressure are restored to their normal values. A decrease in the plasma osmolality or an increase in the blood volume or pressure suppresses the urge to drink.

The primary action of ADH is to increase the permeability of the collecting duct to water. A simplified scheme for this action is shown in Figure 33-6. ADH binds to a receptor on the basolateral membrane of the cell. Binding to this receptor, which is coupled to *adenylate cyclase,* increases the intracellular levels of *cyclic adenosine monophosphate (cAMP).* The rise in

intracellular cAMP activates one or more *protein kinases,* which in turn increase the permeability of the cell's apical membrane to water. This increase in the membrane's permeability to water appears to result from the insertion of water channels into the membrane. In the absence of ADH, these same water channels are removed from the apical membrane, and the apical membrane once again becomes impermeable to water.

ADH also acts on the medullary portion of the thick ascending limb of Henle's loop and on the collecting duct to increase the active reabsorption of NaCl. Furthermore, ADH acts on the inner medullary portion of the collecting duct to increase its permeability to urea. The mechanisms by which these actions augment the ability of the kidney to produce a hyperosmotic urine are considered next.

Countercurrent Multiplication by Henle's Loop

The production of urine that is either hypoosmotic or hyperosmotic with respect to plasma requires that solute be separated from water at some point along the nephron. Henle's loop is the main nephron site where this separation of solute and water occurs, through a process termed *countercurrent multiplication* (Figure 33-7).

Henle's loop consists of two parallel limbs, with tubular fluid flowing in opposite directions (countercurrent flow). The ascending limb is impermeable to water and reabsorbs solute (NaCl) from the luminal fluid. Thus the fluid within the lumen of the ascending limb is diluted. The solute reabsorbed by the ascending limb accumulates in the surrounding interstitial fluid, raising its osmolality. Because the descending limb is highly permeable to water, the elevated interstitial osmolality pulls water from the descending limb, concentrating the fluid within its lumen. The net effect of this process is the establishment of a 200 mOsm/kg H_2O osmotic gradient between the ascending and descending limbs. This osmotic gradient (separation of solute and water) is termed the *single effect* of the countercurrent multiplication process. As illustrated in Figure 33-7, countercurrent multiplication in the steady state will produce tubular fluid in the ascending limb that is hypoosmotic (100 mOsm/kg H_2O) and tubular fluid at the bend of the loop that is hyperosmotic (1200 mOsm/kg H_2O) with respect to plasma.

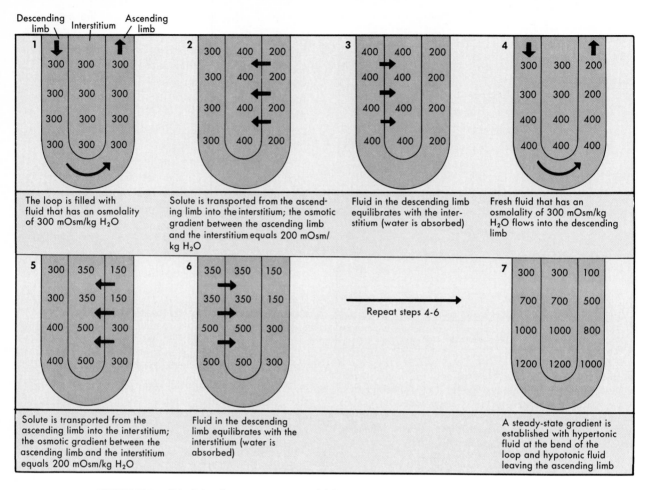

FIGURE 33-7   Principle of countercurrent multiplication by Henle's loop. See text for details.

Although countercurrent multiplication is central to the kidney's ability to concentrate and dilute the urine, a complete understanding of this process requires knowledge of the following:

1. The transport properties of the various portions of Henle's loop, distal tubule, and collecting duct
2. The role of the medullary interstitium osmolality, particularly the role of urea
3. The action of the vasa recta as countercurrent exchangers
4. The effects of ADH

Transport Properties of Nephron Segments

As discussed in Chapter 32, the proximal tubule reabsorbs solute and water by a process that is essentially isomotic. Moreover, this applies regardless of whether a dilute or concentrated urine is produced. Thus no separation of solute from water takes place in the proximal tubule. The ability to separate solute from water begins with the thin ascending limb of Henle's loop and continues throughout the remainder of the nephron.

Table 33-1 summarizes the transport and passive permeability properties of the nephron segments involved in the process of concentrating and diluting the urine. As already noted, a dilute urine is produced when plasma ADH levels are low; inspection of Table 33-1 indicates how this occurs. In the absence of ADH, all tubule segments, except for the thin descending limb of Henle's loop, are impermeable to water. Reabsorption of NaCl by these segments will therefore dilute the tubule fluid. Because of its large transport capacity, the thick ascending limb of Henle's loop is the major site where dilution of the

Table 33-1 Transport and Permeability Properties of Nephron Segments Involved in Urine Concentration and Dilution

| Tubule Segment | Active Transport | Passive Permeability* | | | Effect of ADH |
		NaCl	Urea	H₂O	
Henle's loop					
Thin descending limb	0	+	+	+++	None
Thin ascending limb	0	+++	+	0	None
Thick ascending limb	+++	+	+	0	Increased active NaCl reabsorption
Distal convoluted tubule	+	+	0	0	None
Collecting Duct					
Cortex	+	+	0	0	Increased water permeablility and NaCl reabsorption
Outer medulla	+	+	0	0	Increased water permeability and NaCl reabsorption
Inner medulla	+	+	++	0	Increased permeability to water and urea

*Permeability is proportional to the number of + indicated: +, low permeability; +++, high permeability; 0 = impermeable.

tubule fluid occurs. Indeed, the thick ascending limb of Henle's loop is often referred to as the *diluting segment.*

Medullary Interstitial Fluid Measurements of the composition of the medullary interstitial fluid have shown that its principal components are NaCl and urea and that the distribution of these solutes is not uniform throughout the medulla. At the junction of the medulla and the cortex, the interstitial fluid has an osmolality of approximately 300 mOsm/kg H₂O, with virtually all osmoles attributable to NaCl. The concentrations of both NaCl and urea increase progressively with increasing depth into the medulla, and at the papilla the osmolality of the interstitial fluid is approximately 1200 mOsm/kg H₂O. Of this value, 600 mOsm/kg H₂O are attributed to NaCl and 600 mOsm/kg H₂O to urea.

This medullary gradient for NaCl and urea is established by two processes:

1. As NaCl is reabsorbed from the tubule fluid by the ascending limb of Henle's loop, it will accumulate in the medullary interstitium.
2. Urea accumulation is more complex.

As indicated in Table 33-1, the permeability of the nephron segments to urea is relatively low, except for the inner medullary collecting duct. Given these permeability properties, urea, which enters the tubule fluid by glomerular filtration, is essentially trapped within the tubule lumen and becomes concentrated

as water is reabsorbed along the nephron (except for the thick segment of Henle's loop) until it reaches the terminal portion of the collecting duct. At this point it diffuses along its concentration gradient into the medullary interstitium, where it accumulates. The changes in tubular urea concentration are described at the end of this section on urine concentration and dilution.

As already described, the hyperosmotic medullary interstitium is important for the process of countercurrent multiplication and specifically for the reabsorption of water from the descending limb of Henle's loop. The hyperosmotic medullary interstitium is also essential for concentrating the tubular fluid within the collecting duct. In the presence of a hyperosmotic interstitium, and in the presence of ADH (see following discussion), water can be reabsorbed from the collecting duct, and thus a concentrated urine is excreted. Because water reabsorption is a passive process driven by an osmotic gradient, the maximal concentration that the urine can attain is equal to that of the medullary interstitium at the papilla (approximately 1200 mOsm/kg H₂O). Because a hyperosmotic medullary interstitium is essential for urine concentration, any condition that reduces this gradient will impair the ability of the kidney to concentrate the urine maximally. An important example of such a concentrating defect is produced by a protein-deficient diet. When protein intake is inadequate, urea production in the body is decreased. Therefore the urea content, and thus the osmolality of the medullary interstitium, are reduced. This in turn will impair water reabsorption from the collecting duct and thereby reduce the concentrating ability of the kidney.

Countercurrent Exchange by the Vasa Recta
The vasa recta are highly permeable to solute and water. As with Henle's loops, the vasa recta form a parallel set of hairpin loops within the medulla (see Chapter 32). The vasa recta function not only to bring nutrients and oxygen to the tubules within the medulla, but more importantly as *countercurrent exchangers* to maintain medullary hyperosmolality. This process of countercurrent exchange is illustrated in Figure 33-8.

Blood coming into the medulla via the descending vasa recta has an osmolality equal to the systemic blood (300 mOsm/kg H_2O). As the blood flows deeper into the medulla, it equilibrates with the surrounding interstitial fluid by losing water and gaining solute. If the vasa recta were to exit the medulla at this equilibrium stage, they would rapidly dissipate the gradient by removing hyperosmotic fluid from the medul-

lary interstitium. Because of the countercurrent flow arrangement, however, the blood reequilibrates as it ascends back through the medulla. As the blood exits the medulla in the ascending vasa recta, its osmolality is only slightly higher than that of the descending vasa recta. Therefore such an arrangement effectively traps the hyperosmotic interstitial fluid deep within the medulla. It should be emphasized that the ability of the vasa recta to maintain the medullary interstitial gradient is flow dependent. Despite the countercurrent arrangement, a substantial increase in blood flow through the vasa recta will ultimately dissipate the medullary gradient. This often occurs when water intake is elevated for prolonged periods.

Normally the portions of the nephron within the medulla continuously add solute and water to the medullary interstitium. The vasa recta serve to carry this excess solute and fluid out of the medulla.

Role of Antidiuretic Hormone (ADH) In the absence of ADH, the separation of solute and water by the nephron leads to the excretion of hypoosmotic urine. The only change that must occur to allow the excretion of hyperosmotic urine is that ADH be present and effective.

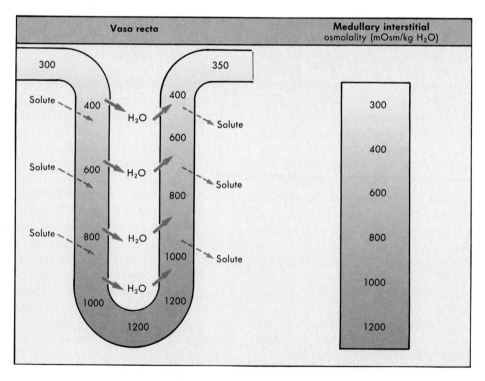

FIGURE 33-8   Mechanism of countercurrent exchange by the vasa recta. See text for details.

ADH has the following actions in the kidney:

1. ADH increases the permeability of the collecting duct to water.
2. ADH increases the reabsorption of NaCl by the collecting duct.
3. ADH stimulates the active reabsorption of NaCl by the medullary portion of the thick ascending limb of Henle's loop.
4. ADH increases the permeability of the inner medullary collecting duct to urea.

Together these actions allow the excretion of a concentrated urine.

As indicated in Table 33-1, the thin and thick ascending limbs of Henle's loop and the distal tubule are impermeable to water, both in the presence and in the absence of ADH. Thus, regardless of the plasma levels of ADH and therefore also regardless of the urine osmolality, fluid within these nephron segments will be diluted by the removal of solute. Fluid exiting the ascending limb of Henle's loop has an osmolality of approximately 100 mOsm/kg H_2O. When this dilute fluid reaches the collecting duct and ADH levels are high, water will diffuse out of the tubule lumen into the surrounding interstitial tissue because of the osmotic gradient. With maximal levels of ADH the tubular fluid will come to osmotic equilibrium with the interstitial fluid as the tubular fluid flows down the collecting duct. Thus in the cortex the tubular fluid will be concentrated to a value of approximately 300 mOsm/kg H_2O, and at the papilla it will have an osmolality of about 1200 mOsm/kg H_2O.

ADH also maintains the gradient between interstitium and collecting duct at a time when water reabsorption from the collecting duct could dissipate the gradient. For example, ADH stimulates the active reabsorption of NaCl by the medullary portion of the think ascending limb of Henle's loop, and thus it adds extra NaCl to the medullary interstitium. Also, ADH increases the permeability of the inner medullary portion of the collecting duct to urea. This allows more urea to accumulate in this area of the medullary interstitium which also maintains the medullary gradient.

Integrated View of the Urine Concentrating Process

Figure 33-9 summarizes the essential features of the urine concentrating process. It should be emphasized that in the absence of ADH, the same process would allow the production of a dilute urine. The following numbers 1 to 6 refer to those in Figure 33-9.

1. Fluid entering the thin descending limb of Henle's loop from the proximal tubule is isosmotic with respect to plasma. This reflects the essentially isosmotic nature of solute and water reabsorption in the proximal tubule (see Chapter 32). Thus the tubule fluid has an osmolality of approximately 300 mOsm/kg H_2O, with the majority of these osmoles ascribable to NaCl.

2. The thin descending limb is highly permeable to water and much less so to NaCl and urea. Consequently, as the fluid descends deeper into the hyperosmotic medulla, water is reabsorbed. By this process, fluid at the bend of the loop will have an osmolality equal to that of the surrounding interstitial fluid (1200 mOsm/kg H_2O). Again, most of the osmoles are attributable to NaCl, although the other components (urea, etc.) of the tubular fluid will be concentrated as well.

3. The thin ascending limb is impermeable to water but highly permeable to NaCl. As the NaCl-rich tubular fluid moves up the ascending limb, NaCl will be passively reabsorbed (luminal [NaCl] > interstitial [NaCl]). This passive reabsorption of NaCl, without concomitant water reabsorption, will begin to dilute the tubular fluid.

4. The thick ascending limb of Henle's loop is also impermeable to water. This portion of the nephron actively reabsorbs NaCl and further dilutes the tubular fluid. Dilution occurs to such a degree that fluid leaving the thick ascending limb is hypoosmotic with respect to plasma (approximately 100 mOsm/kg H_2O). In the presence of ADH the active reabsorption of NaCl is stimulated.

5. Fluid reaching the collecting duct is hypoosmotic to the surrounding interstitial fluid. Therefore, in the presence of ADH, water diffuses out of the tubule lumen; this begins the process of urine concentration. The maximal osmolality that the fluid in the cortical collecting duct can attain is 300 mOsm/kg H_2O, which is the osmolality of the surrounding interstitial fluid. Although this fluid has the same osmolality as it did when it first entered Henle's loop, its composition has changed dramatically. Because of the large quantities of NaCl reabsorbed up to this point of the nephron, only a small fraction of the total urine osmolality is attributable to NaCl. Instead, the high urine osmolality reflects the presence of many other substances

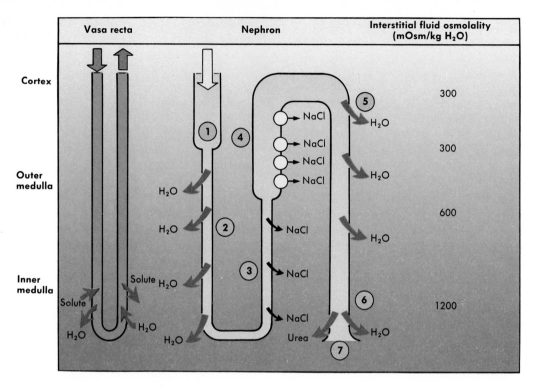

FIGURE 33-9   Mechanism for the excretion of concentrated urine. Plasma ADH levels are maximal. See text for details.

that have escaped reabsorption or have been secreted into the tubular fluid. Of these other solutes, urea is the most abundant.

6. As fluid continues through the medulla, water is reabsorbed. This increases the osmolality of the tubular fluid, which at this point is caused mainly by urea. In the inner medulla the collecting duct is normally permeable to urea; moreover, ADH increases this permeability. Because the urea concentration of the tubular fluid has been increased, urea will now diffuse out of tubule and into the medullary interstitium. The final urine will have an osmolality of 1200 mOsm/kg H_2O and contain high concentrations of urea and other nonreabsorbed solutes.

CONTROL OF EXTRACELLULAR FLUID VOLUME

The major solutes of the ECF are the salts of sodium. Of these, the most abundant is NaCl. Because the ADH and thirst systems are so sensitive that they maintain the osmolality of the ECF within a very narrow range, changes in the total amount of NaCl present in this compartment will lead to parallel changes in ECF volume. For example, addition of NaCl (without water) to the ECF will increase the osmolality of this compartment. This increase in osmolality will in turn stimulate ADH secretion and thirst. The ADH-induced decrease in urinary water excretion, together with the increased water intake, will restore the ECF osmolality to normal. However, the volume of this compartment will be increased in proportion to the amount of added NaCl. Conversely, a decrease in the ECF NaCl content will reduce the volume of this compartment.

The maintenance of a constant ECF volume depends on the body's ability to regulate the amount of NaCl in this compartment. This is accomplished by the kidney, which regulates its excretion of NaCl to match precisely the amount ingested in the diet. If an imbalance exists between the amount of NaCl ingested and the amount excreted, and if the ADH and thirst centers are intact, the volume of the ECF will be altered. To defend against such changes in ECF volume, the body relies on a system to monitor the volume of this compartment and then to signal the kidney so that appropriate adjustments in NaCl excretion can be made.

This section reviews the physiology of the volume receptors and considers the various signals that act on the kidney to regulate NaCl excretion. Finally the response of the nephron to these signals is discussed.

Volume Receptors

Volume receptors have been identified in both the low-pressure and high-pressure sides of the vascular tree. These receptors act as simple stretch receptors and respond to both volume and pressure. The low-pressure receptors are found in the pulmonary vasculature and cardiac atria (see Chapters 22 and 26). High-pressure baroreceptors have been identified in the aortic arch, carotid sinus, and the afferent arterioles of the kidney.

In response to changes in the ECF volume, these receptors signal the kidneys to make appropriate adjustments in NaCl excretion (also water) that tend to return volume to normal. Accordingly, when the ECF volume is expanded, renal NaCl and water excretion are increased. Conversely, when the ECF volume is contracted, renal NaCl and water excretion are reduced. The signals involved in coupling the volume receptors to the kidney are both neural and hormonal.

Renal Sympathetic Nerves

As described in Chapter 32, sympathetic fibers innervate the afferent and efferent arterioles, as well as the cells of the renal tubule. These fibers can be activated by a decrease in the ECF volume, which results in the following responses:

1. Constriction of the afferent arteriole that causes a decrease in the glomerular filtration rate.
2. Renin secretion by cells of the afferent and efferent arterioles is increased.
3. NaCl reabsorption by the proximal tubule and Henle's loop is increased.

These responses are mediated via activation of α-adrenergic receptors. The net result is to reduce the excretion of NaCl. The mechanisms responsible for this decrease in renal NaCl excretion are considered later.

Other neural pathways are also affected by a change in ECF. One of these pathways involves the control of ADH secretion. As already described, a 5% to 10% reduction in ECF volume stimulates ADH secretion. ADH acting on the collecting duct increases water reabsorption. This, together with the reduced excretion of NaCl, restores the ECF volume to normal.

When the volume of the ECF is increased, sympathetic nerve activity is reduced. This generally reverses the effects just described, with the result that NaCl and water excretion are enhanced and ECF volume is restored to normal.

Renin-Angiotensin-Aldosterone

Cells located in the afferent and efferent arterioles are the site of synthesis and release of renin (Figure 33-10). Three factors have been identified as important regulators of renin secretion:

1. As already noted, the afferent arteriole behaves as a baroreceptor. When renal perfusion pressure is reduced, renin secretion is stimulated. Conversely, an increase in perfusion pressure inhibits renin release.
2. The increased activity of sympathetic nerve fibers that innervate afferent and efferent arterioles stimulates renin release.
3. The macula densa, which helps regulate the glomerular filtration rate (see Chapter 32), probably also regulates renin secretion. For example, if NaCl delivery to the macula densa is decreased, renin secretion increases. Conversely, increased NaCl delivery to the macula densa inhibits renin release.

Renin alone does not have a physiologic function; it functions solely as a proteolytic enzyme. Its substrate is a circulating protein, *angiotensinogen,* which is produced by the liver. Angiotensinogen is cleaved by renin to yield a 10–amino acid peptide, *angiotensin I.* Angiotensin I also has no known physiologic function and is further cleaved to an 8–amino acid peptide, *angiotensin II,* by a converting enzyme found in high concentration in the lung. Angiotensin II has several important physiologic functions, including (1) stimulation of aldosterone secretion by the adrenal cortex, (2) arteriolar vasoconstriction, (3) stimulation of ADH secretion and thirst, and (4) enhancement of NaCl reabsorption by the proximal tubule.

Aldosterone is a steroid hormone with many

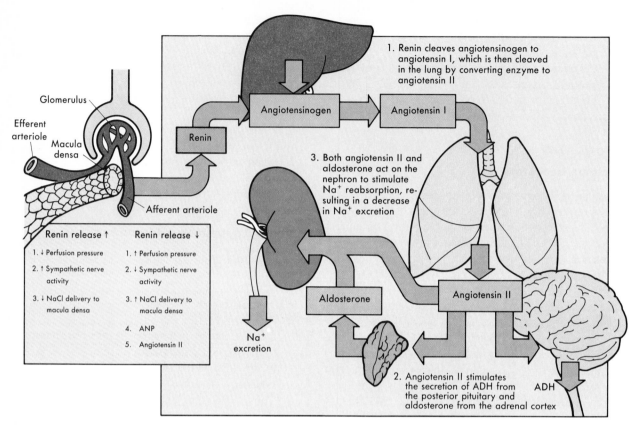

FIGURE 33-10   The renin-angiotensin-aldosterone system, and the factors that control the secretion of renin.

important actions in the kidney (see Chapters 32 and 34). With regard to the regulation of ECF volume, aldosterone acts to reduce NaCl excretion by stimulating its reabsorption by the collecting duct and Henle's loop.

The actions of aldosterone on the sodium-transporting cells of the collecting duct (principal cells) are summarized in Figure 33-11. Aldosterone *(A)* enters the cell and binds to a cytoplasmic receptor *(R)*. The hormone-receptor complex *(AR)* interacts with specific binding sites on the deoxyribonucleic acid *(DNA)* to effect messenger ribonucleic acid *(mRNA)* transcription and ultimately new protein synthesis. Some of these aldosterone-induced proteins may represent new apical membrane sodium channels, enzymes needed for the synthesis of adenosine triphosphate *(ATP)*, and the Na^+, K^+-ATPase. By these actions

sodium entry into the cell across the apical membrane is enhanced, as is sodium extrusion from the cell across the basolateral membrane; chloride reabsorption is increased as well, but by different mechanisms. Thus reabsorption of NaCl from the tubular fluid into the blood is enhanced. The precise cellular mechanisms by which aldosterone stimulates NaCl reabsorption by Henle's loop (thick ascending limb) have not yet been elucidated.

Atrial Natriuretic Peptide

The myocytes of the cardiac atria produce a peptide hormone that promotes water and NaCl excretion by the kidney. This hormone, which is 28 amino acids in length, is called atrial natriuretic peptide (ANP). ANP is released in response to atrial stretch, such as

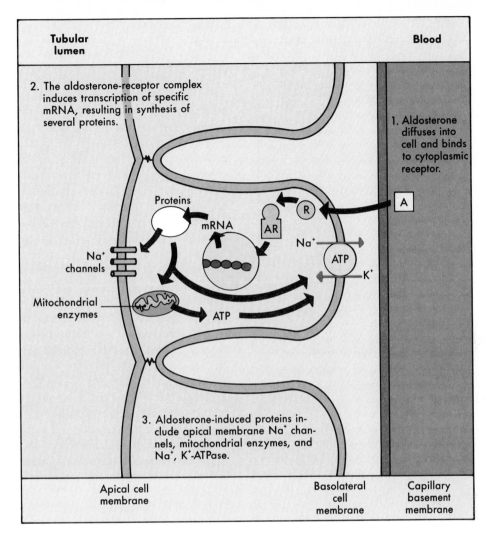

FIGURE 33-11 Cellular actions of aldosterone on principal cells of the collecting duct. See text for abbreviations and details.

occurs when the ECF volume is increased, and elicits various effects throughout the body. In general, ANP antagonizes the effects of the renin-angiotensin-aldosterone system. Actions of ANP include the following:

1. ANP increases renal NaCl and water excretion by renal and extrarenal actions.
2. ANP dilates the afferent and efferent arterioles and thereby increases the glomerular filtration rate and thus the filtered load of NaCl.
3. ANP directly inhibits NaCl reabsorption by the collecting duct and indirectly inhibits NaCl reabsorption by the collecting duct by its actions on the renin-angiotensin-aldosterone system.
4. ANP reduces plasma aldosterone levels by inhibiting the release of renin and directly inhibiting the secretion of aldosterone from the adrenal cortex.

5. ANP promotes the excretion of water, because it inhibits the secretion of ADH.

Control of Na$^+$ Excretion with Normal Extracellular Fluid Volume

The maintenance of a normal ECF volume *(euvolemia)* requires a precise balance between the amount of NaCl ingested and the amount lost from the body. Because the kidney is the major route for NaCl excretion, it adjusts the amount of NaCl excreted in the urine to the amount ingested in the diet during euvolemia. To understand how renal NaCl excretion is regulated, the general features of renal Na$^+$ handling must be understood. Figure 33-12 summarizes the contribution of each nephron segment to the handling of the filtered load of Na$^+$ (see Chapter

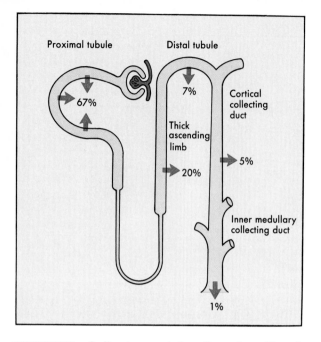

FIGURE 33-12   Sodium transport along the nephron. The values represent the percentage of the filtered load (i.e., GFR × P_{Na^+}) reabsorbed by each segment.

32). The following discussion considers only the renal handling of Na^+. Although not specifically addressed, Cl^- reabsorption occurs in parallel.

In a normal adult the filtered load of Na^+ is approximately 25,000 mEq/day. With a typical diet 1% or less of this filtered load is excreted in the urine ($\simeq$ 150 mEq/day). Because of this large filtered load of Na^+, it is important to recognize that small changes in Na^+ reabsorption by the nephron have a large effect on the volume of the ECF. For example, an increase in the excretion of Na^+ from 1% to 3% of the filtered load would represent an additional Na^+ loss of approximately 500 mEq/day. Because the ECF Na^+ concentration is 140 mEq·L^{-1}, such a Na^+ loss would decrease the ECF by more than 3 L.

During euvolemia the collecting duct is the main nephron segment where Na^+ reabsorption is adjusted to maintain excretion at a level appropriate for dietary intake. This does not mean, however, that the other portions of the nephron are not important in this process. Because the reabsorptive capacity of the collecting duct is limited, these other portions of the nephron must reabsorb the bulk of the filtered load of Na^+. As indicated in Figure 33-12, Na^+ reabsorption by the proximal tubule, Henle's loop, and distal tubule

results in the delivery of only 6% of the filtered load of Na^+ to the collecting duct. Importantly, this delivery is maintained relatively constant despite small variations in Na^+ intake. Two mechanisms operate to ensure this constant delivery. First, the glomerular filtration rate, and thus the filtered load of Na^+, is maintained relatively constant by the process of autoregulation (see Chapter 32). Second, even in the presence of small changes in the filtered load of Na^+, reabsorption by the proximal tubule is adjusted in parallel. The mechanism responsible for this phenomenon is termed *glomerulotubular balance* (see Chapter 32).

With a constant delivery of Na^+, small adjustments in collecting duct reabsorption are sufficient to balance excretion with intake. Aldosterone is the primary regulator of collecting duct Na^+ reabsorption under this condition (see earlier discussion). Other hormones also influence collecting duct Na^+ reabsorption. For example, ANP inhibits Na^+ reabsorption. Na^+ reabsorption is also inhibited by prostaglandins and bradykinin, but their roles in regulating Na^+ excretion have not been resolved. Finally, ADH not only stimulates water reabsorption, but also stimulates Na^+ reabsorption by the collecting duct.

As long as variations in the dietary intake of NaCl are small, the mechanisms just described can regulate renal Na^+ excretion appropriately and thereby maintain the volume of the ECF. Large changes in NaCl intake, however, cannot be handled effectively by these mechanisms. Consequently the volume of the ECF will be altered. When this occurs, additional factors act on the kidney to adjust Na^+ reabsorption to reestablish the euvolemic state.

Control of Na^+ Excretion With Increased Extracellular Fluid Volume

When the ECF volume is increased, the various signals just described act on the kidney to increase Na^+ excretion. The integrated response to an increase in the ECF volume is illustrated in Figure 33-13.

The important difference between this condition and that described earlier for the euvolemic state is that the renal response is not limited to the collecting duct; rather, it involves the entire nephron. Three general responses to an increase in ECF volume occur (Figure 33-13):

1. *The glomerular filtration rate increases.* This is

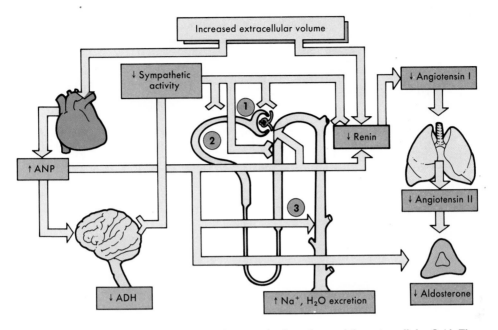

FIGURE 33-13   Integrated response to an increase in the volume of the extracellular fluid. The numbers along the nephron indicate sites where sodium handling is regulated. See text for details.

achieved mainly by decreased activity of the sympathetic nerve fibers innervating the afferent and efferent arterioles, which leads to vasodilation. ANP may also contribute to this response by dilating afferent and efferent arterioles. Because the glomerular filtration rate is increased, the filtered load of Na^+ increases in parallel.

2. *Reabsorption of Na^+ decreases in the proximal tubule and Henle's loop.* Various mechanisms cause this inhibition of reabsorption, but controversy exists regarding the precise role of each. Because activation of the sympathetic fibers that innervate these portions of the nephron stimulates Na^+ reabsorption, the decreased sympathetic activity engendered by the expansion of the ECF may contribute to the observed decrease in Na^+ reabsorption. Also, dilation of the afferent and efferent arterioles increases the hydrostatic pressure in the peritubular capillaries surrounding the proximal tubule. This increase in pressure inhibits the uptake of NaCl and water by the capillary and thereby inhibits their reabsorption (see Chapter 32). Note that proximal tubule reabsorption is reduced even though the glomerular filtration rate is increased. This shows that glomerulotubular balance only occurs during euvolemia, not when the ECF volume is altered. Finally, reduced levels of aldosterone may reduce NaCl reabsorption by Henle's loop.

3. *Na^+ reabsorption decreases in the collecting duct.* Both the increase in filtered load and the decrease in proximal tubule reabsorption result in the delivery of a large amount of Na^+ to the collecting duct. Thus the delivery of Na^+ to the collecting duct is no longer a constant fraction of the filtered load, as is the case during euvolemia. Instead, Na^+ delivery varies in parallel with the degree of ECF volume expansion. This large load of Na^+ overwhelms the reabsorptive capacity of the collecting duct. The reabsorptive capacity is further reduced by the action of ANP and by the diminution in the circulating levels of aldosterone. Thus, by altering Na^+ reabsorption along the entire length of the nephron, the capacity to excrete Na^+ is increased greatly above that seen during euvolemia.

One final component to be considered in the response to ECF expansion is the excretion of water. As Na^+ excretion is increased, plasma osmolality begins to fall. This inhibits the secretion of ADH and leads to an increase in water excretion. An additional mechanism by which ADH secretion is suppressed is through the action of ANP (see earlier section). Be-

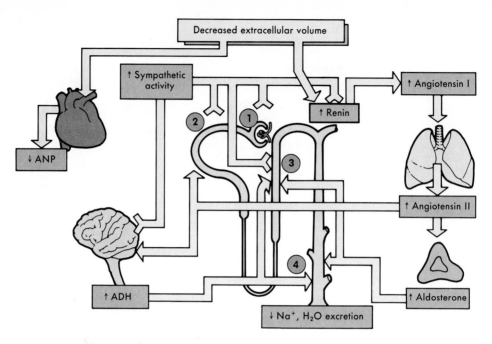

FIGURE 33-14   Integrated response to a decrease in the volume of the extracellular fluid. Symbols are the same as in Figure 33-13.

cause of the sensitivity of the ADH system, the excretion of Na$^+$ occurs in parallel to that of water. Thus the ECF volume will be restored to normal, and its osmolality will be unchanged.

Control of Na$^+$ Excretion With Decreased Extracellular Fluid Volume

When the volume of the ECF is decreased, Na$^+$ and water excretion by the kidney are reduced. The mechanisms involved are essentially the opposite of those just described and are summarized in Figure 33-14. Again, the entire nephron contributes to this response. The three general responses to a decrease in the volume of the ECF are:

1. *The glomerular filtration rate, and therefore the filtered load of Na$^+$, is reduced.* This results from afferent and efferent arteriolar constriction mediated by the sympathetic nerves.

2. *Na$^+$ reabsorption by the proximal tubule and Henle's loop is stimulated* (note that glomerulotubular balance does not occur). This stimulation is the result of increased sympathetic nerve activity and the direct action of angiotensin II. Na$^+$ reabsorption by Henle's loop is also stimulated, both by increased

sympathetic nerve activity and by elevated levels of aldosterone and ADH.

3. *Na$^+$ reabsorption by the collecting duct is enhanced, mainly by the action of aldosterone.* Other hormones may also contribute to this response. In particular, the circulating levels of ANP are reduced, and ADH levels are elevated. ADH not only stimulates Na$^+$ reabsorption but also increases the reabsorption of water by the collecting duct.

The decreased filtered load of Na$^+$ together with enhanced reabsorption by the proximal tubule and Henle's loop result in a severe decrease in Na$^+$ delivery to the collecting duct. Because Na$^+$ reabsorption by the collecting duct is increased, the urine can be rendered virtually sodium free. This retention of Na$^+$, and the concurrent conservation of water, will restore the ECF volume.

BIBLIOGRAPHY
Journal Articles

Abramow M et al: Cellular events in vasopressin action, Kidney Int 32:S56, 1987.

Jamison RJ: The renal concentrating mechanism, Kidney Int 32:S43, 1987.

Marsden PA and Skorecki KL: Afferent limb of volume homeostasis, Cont Iss Nephrol—Body Fluid Homeostasis 16:1, 1987.

Needleman P and Greenwald JE: Atriopeptin: a cardiac hormone intimately involved in fluid, electrolyte, and blood-pressure homeostasis, N Engl J Med 314:828, 1986.

Raymond KH and Stein JH: Efferent limb of volume homeostasis, Cont Iss Nephrol—Body Fluid Homeostasis 16:33, 1987.

Robertson GL: Physiology of ADH secretion, Kidney Int 32:S20, 1987.

Skorecki K and Brenner B: Body fluid homeostasis in man, Am J Med 70:77, 1981.

Books and Monographs

Fanestil DD: Compartmentation of body water. In Maxwell MH et al, editors: Clinical disorders of fluid and electrolyte metabolism, ed 4, New York, 1987, McGraw-Hill Book Co.

Hogg RJ and Kokko JP: Urine concentrating and diluting mechanisms in mammalian kidneys. In Brenner BM and Rector FC Jr, editors: The Kidney, ed 3, Philadelphia, 1986, W B Saunders Co.

Oh MS and Carroll HJ: Regulation of extra- and intracellular fluid composition and content. In Arieff AI and DeFronzo RA, editors: Fluid electrolyte and acid-base disorders, New York, 1985, Churchill Livingstone.

Rose BD: Clinical physiology of acid-base and electrolyte disorders, ed 2, New York, 1984, McGraw-Hill Book Co.

Seifter JL et al: Control of extracellular fluid volume and pathophysiology of edema formation. In Brenner BM and Rector FC Jr, editors: The kidney, ed 3, Philadelphia, 1986, WB Saunders Co.

Valtin H: Physiological effects of vasopressin on the kidney. In Gash D M and Boer GJ, editors: Vasopressin, New York, 1987, Plenum Publishing Corp.

Regulation of Potassium, Calcium, Magnesium, Phosphate, and Acid-Base Balance

POTASSIUM

Potassium (K^+) is one of the most abundant cations in the body, and the maintenance of K^+ homeostasis is critical for life. This section focuses on the distribution of K^+ between the intracellular fluid (ICF) and extracellular fluid (ECF) compartments *(internal potassium balance)* and the balance between dietary K^+ intake and urinary K^+ excretion *(external potassium balance)*. In addition, the regulation and control of these two elements of K^+ homeostasis are reviewed.

Overview of K^+ Homeostasis

The distribution of K^+ in the body is shown in Figure 34-1. Total body K^+ has been estimated at 50 mEq/kg of body weight, or 3500 mEq for a 70 kg individual. Ninety-eight percent of the K^+ in the body lies within cells, where its average concentration is 150 $mEq \cdot L^{-1}$. A high intracellular concentration of K^+ is required for many cell functions, including cell growth and division and volume regulation. Only 2% of total body K^+ is located in the ECF, where its normal concentration is 4 $mEq \cdot L^{-1}$. The large concentration difference of K^+ across cell membranes (146 $mEq \cdot L^{-1}$) is maintained by the operation of the Na^+,K^+-ATPase. This gradient is important in maintaining the potential difference across cell membranes. Thus K^+ is critical for the excitability of nerve and muscle cells, as well as for the contractility of cardiac, skeletal, and smooth muscle cells (see Chapters 2 and 16).

When the K^+ concentration in the ECF exceeds 5.5 $mEq \cdot L^{-1}$, an individual is *hyperkalemic*. Hyperkalemia diminishes the resting membrane potential of cardiac cells and increases cell excitability. Severe increases in plasma K^+ can lead to cardiac arrest and death. When the K^+ concentration of the ECF is less than 3.5 $mEq \cdot L^{-1}$, an individual is *hypokalemic*. A decline in extracellular K^+ hyperpolarizes the resting cell membrane of nerve and muscle cells and reduces excitability. Severe hypokalemia can lead to paralysis, cardiac arrhythmias, and death. Therefore maintenance of a high intracellular and a low extracellular K^+ concentration, as well as a high K^+ concentration gradient across cell membranes, is essential for a number of cellular functions.

Figure 34-1 illustrates the importance of the kidneys in maintaining K^+ homeostasis. Each day the kidneys excrete approximately 92% of the ingested K^+. K^+ excretion is essentially equivalent to dietary K^+ intake, even when intake increases by as much as 20-fold. This equivalence between urinary excretion and dietary intake underscores the critical role played by the kidneys in maintaining K^+ homeostasis. Although some K^+ is excreted from the body in the stool and sweat ($\sim$ 8 mEq/day), the amount is essentially constant and therefore relatively much less important than K^+ excretion by the kidneys.

Internal K^+ Balance

After a meal the K^+ absorbed by the gastrointestinal tract rapidly enters the ECF. If the K^+ ingested during a normal meal ($\sim$50 mEq) were to remain in

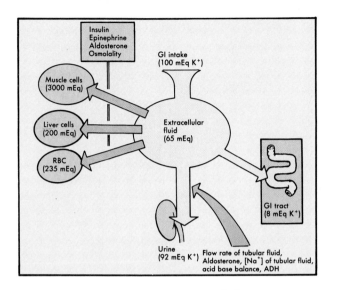

FIGURE 34-1   Overview of potassium (K^+) homeostasis. Internal K^+ balance, the distribution of K^+ between the intracellular fluid and the extracellular fluid, is regulated by insulin, epinephrine, and aldosterone. These hormones are secreted when the K^+ concentration of the extracellular fluid rises, and they stimulate K^+ uptake by muscle, liver, and red blood cells (RBCs). The total amount of K^+ in the body is determined by the amount of K^+ ingested in the diet and the amount excreted in the urine and stool. An individual is in balance when dietary intake and urinary output (and output by the gastrointestinal tract) are equal. The excretion of K^+ by the kidneys is regulated by several hormones and factors shown in the figure. Excretion by the GI tract is relatively constant. *ADH,* Antidiuretic hormone.

the ECF compartment, plasma K^+ concentration would increase by a potentially lethal 3.6 mEq·L^{-1} (50 mEq added to 14 L of ECF). This rise in plasma K^+ is prevented by the rapid uptake of K^+ into cells. Because the excretion of K^+ by the kidneys after a meal is relatively slow (hours), the buffering of K^+ by cells is essential to prevent life-threatening hyperkalemia.

Several hormones enhance the uptake of K^+ into skeletal muscle, liver, bone, and red blood cells and thereby prevent hyperkalemia. As illustrated in Figure 34-1, these hormones include *insulin, aldosterone,* and *epinephrine.* Hyperkalemia stimulates insulin secretion from the pancreas, aldosterone release from the adrenal cortex, and epinephrine secretion from the adrenal medulla. All three hormones enhance K^+ movement into cells and thereby decrease the K^+ concentration of the ECF.

The pH of the ECF also affects the K^+ concentra-

tion of the ECF. During an acidosis of the ECF produced by inorganic acids (e.g., HCl, H_2SO_4), but not by organic acids (e.g., lactic acid, acetic acid), the high extracellular concentration of hydrogen ions (H^+) promotes movement of H^+ into cells and the reciprocal movement of K^+ out of cells. The converse occurs during alkalosis. The mechanism for these shifts is not fully understood. It has been proposed that the movement of H^+ occurs as the cells buffer changes in the pH of the ECF. As H^+ moves across the cell membranes, K^+ moves in the opposite direction, and thus no gain or loss of cations occurs across the cell membranes. It should be noted, however, that changes in pH do not restore plasma K^+ levels to normal values. Instead, alterations in pH are often responsible for producing hyperkalemia or hypokalemia. Accordingly, acid-base balance should not be considered a mechanism that regulates K^+ levels.

Finally the osmolality of the plasma also influences the distribution of K^+ across cell membranes. An increase in the osmolality of the ECF enhances K^+ uptake by cells and thus lowers the extracellular K^+ concentration. Hypoosmolality has the opposite action. It is important to recognize that, as with pH, changes in osmolality do not regulate the distribution of K^+ across cell membranes. Frequently, alterations in osmolality lead to either hyperkalemia or hypokalemia.

External K^+ Balance: K^+ Excretion by the Kidneys

Because K^+ is not bound to plasma proteins, it is freely filtered by the glomeruli. The normal plasma K^+ is 4 mEq·L^{-1}, and urinary K^+ excretion is 15% of the amount filtered. Accordingly, K^+ must be reabsorbed along the nephron under normal conditions. When dietary K^+ uptake is augmented, however, K^+ excretion can exceed the amount filtered, indicating K^+ may also be secreted.

The direction and magnitude of K^+ transport along the nephron is illustrated in Figure 34-2. The proximal tubule reabsorbs 67% of the filtered K^+ under most conditions. Approximately 20% of the filtered K^+ is reabsorbed by Henle's loop and, as with the proximal tubule, reabsorption is a constant fraction of the amount filtered. In contrast to these segments, the distal tubule and the collecting duct have the dual capacity to reabsorb and secrete K^+. Furthermore,

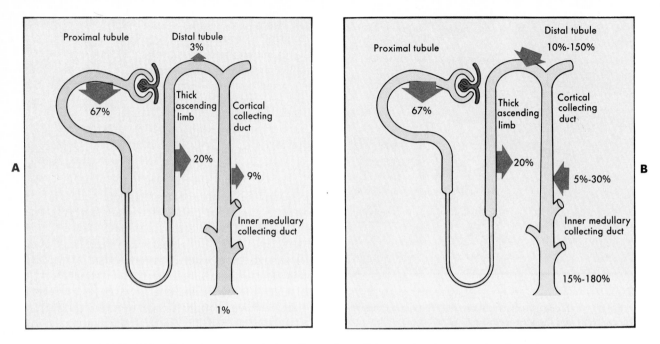

FIGURE 34-2   Potassium transport along the nephron. Urinary potassium excretion depends on the rate and direction of potassium transport by the distal tubule and collecting duct. Percentages refer to the amount of the filtered potassium reabsorbed or secreted by each nephron segment. **A,** Dietary potassium depletion. **B,** Normal and increased dietary potassium intake.

the rate of K^+ reabsorption or secretion depends on a variety of hormones and factors. For example, when K^+ intake is normal, K^+ is secreted by the distal tubule and the collecting duct. A rise in K^+ intake increases K^+ secretion, such that the amount of K^+ appearing in the urine may approach 180% of the amount filtered (Figure 34-2, B). In contrast, a low-potassium diet activates K^+ reabsorption along the distal tubule and collecting duct, such that urinary excretion falls to 1% of the K^+ filtered by the glomeruli (Figure 34-2, A).

Because the magnitude and direction of K^+ transport by the distal tubule and collecting duct are variable, the overall rate of urinary K^+ excretion is determined by these tubular segments. The remainder of this section focuses on the mechanisms of K^+ transport by the distal tubule and the collecting duct and on the factors and hormones that regulate potassium transport by these segments.

K^+ Transport by the Distal Tubule and Collecting Duct

As illustrated in Figure 34-3, K^+ secretion by principal cells in the distal tubule and collecting duct is a two-step process. K^+ uptake across the basolateral membrane is mediated by Na^+,K^+-ATPase. The operation of this pump creates a high intracellular K^+ concentration, which provides the chemical driving force for K^+ exit across the apical membrane. K^+ leaves the cell across the apical membrane and enters the tubular fluid for two reasons. First, the electrochemical gradient of K^+ across the apical membrane favors the downhill movement into the tubular fluid. Second, the permeability of the apical membrane to K^+ is greater than that of the basolateral membrane. Therefore K^+ preferentially diffuses across the apical membrane into the tubular fluid. The four principal determinants of K^+ secretion by the distal tubule and the collecting duct are:

1. The activity of the Na^+,K^+-ATPase
2. The K^+ concentration inside the cell
3. The electrochemical driving force for K^+ across the apical membrane
4. The permeability of the apical membrane to K^+ (Figure 34-3)

Every change in K^+ secretion can be explained in terms of one or more of these principal determinants.

The cellular pathways and mechanisms of K^+

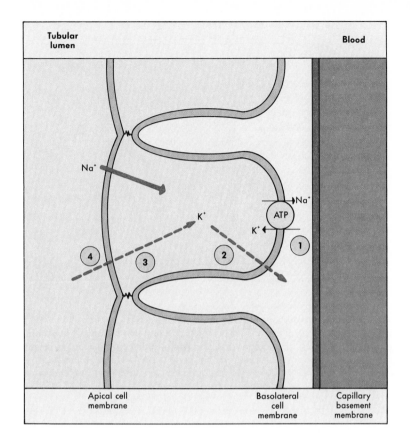

FIGURE 34-3 Cellular mechanism of potassium secretion by the principal cells in the distal tubule and collecting duct. The numbers indicate the sites where potassium secretion is regulated: *(1)* Na^+,K^+-ATPase, *(2)* cell K^+ concentration, *(3)* electrochemical gradient across the apical membrane, and *(4)* the permeability of the apical membrane to K^+. Because the electrochemical gradient of K^+ across the apical membrane is larger than the electrochemical gradient of K^+ across the basolateral membrane, K^+ preferentially leaves the cell and enters the tubular fluid.

reabsorption in the distal tubule and collecting duct are not completely understood. Some evidence suggests that K^+ may be reabsorbed by intercalated cells.

Regulation of K^+ Excretion

Figure 34-4 illustrates the primary factors and hormones that regulate K^+ secretion by the distal tubule and collecting duct. **The flow rate of tubular fluid is one of the most important determinants of K^+ secretion.** A rise in flow stimulates secretion, whereas a fall in flow reduces secretion. Alterations in flow influence K^+ secretion by changing the electro-

chemical gradient of K^+ across the apical membrane. As K^+ is secreted into the tubular fluid, the concentration of K^+ in the fluid increases. This will decrease the favorable electrochemical gradient for K^+ across the apical membrane and retard secretion. An increase in flow will minimize the rise in tubular fluid K^+ concentration as the secreted K^+ is washed downstream. As a result K^+ secretion is stimulated at high flow rates. Because diuretic drugs increase urine flow rate, they also enhance urinary K^+ excretion.

A second important factor that regulates K^+ secretion is the sodium concentration $[Na^+]$ of the tubular fluid. A rise in $[Na^+]$ stimulates secretion, whereas a fall has the opposite affect. Na^+ stim-

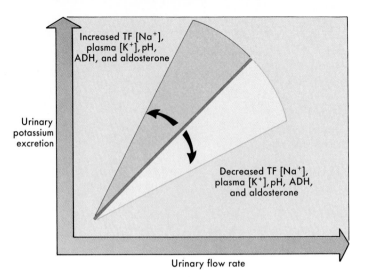

Increased TF [Na$^+$], plasma [K$^+$], pH, ADH, and aldosterone

Urinary potassium excretion

Decreased TF [Na$^+$], plasma [K$^+$], pH, ADH, and aldosterone

Urinary flow rate

FIGURE 34-4   Relationship between urinary flow rate, which is proportional to the flow rate of fluid through the distal tubule and the collecting duct, and urinary potassium excretion. Increased tubular fluid *(TF)* sodium concentration *[Na$^+$]*; plasma potassium concentration *[K$^+$]*; and plasma pH, aldosterone, and antidiuretic hormone *(ADH)* levels all increase K$^+$ excretion at any flow rate. The red line indicates the normal situation, the blue line indicates increased secretion, and the yellow line indicates decreased secretion.

ulates K$^+$ secretion by increasing the activity of the Na$^+$,K$^+$-ATPase. An increase in [Na$^+$] enhances Na$^+$ movement across the apical membrane of principal cells and increases intracellular Na$^+$. This stimulates the Na$^+$,K$^+$-ATPase and accelerates K$^+$ uptake across the basolateral membrane. The rise in intracellular K$^+$ also enhances K$^+$ secretion.

Plasma potassium also regulates K$^+$ secretion by the distal tubule and collecting duct. Hyperkalemia stimulates secretion, and hypokalemia inhibits secretion. Several mechanisms are involved. First, plasma K$^+$ concentration directly regulates the Na$^+$,K$^+$-ATPase. Hyperkalemia stimulates the operation of the Na$^+$,K$^+$-ATPase, stimulates K$^+$ uptake across the basolateral membrane, and increases intracellular K$^+$ concentration. Second, hyperkalemia also increases the permeability of the apical membrane to K$^+$. Third, hyperkalemia stimulates aldosterone secretion by the adrenal cortex; as discussed below, aldosterone independently stimulates K$^+$ secretion.

Another important factor that regulates K$^+$ secretion is the acid-base balance of the ECF. Acute alterations (over hours) in the pH of the plasma influence K$^+$ secretion. Alkalosis increases secretion, and acidosis decreases secretion. A fall in pH reduces

intracellular K$^+$ concentration, possibly by inhibiting the operation of the Na$^+$,K$^+$-ATPase, and it reduces the permeability of the apical membrane to K$^+$. A rise in pH has the opposite effects. Prolonged acidosis (over days) may actually stimulate K$^+$ secretion by increasing the flow of tubular fluid through the distal tubule and collecting duct. Thus acidosis may decrease or stimulate K$^+$ secretion, depending on the duration of the disturbance.

Both aldosterone and antidiuretic hormone (ADH) stimulate K$^+$ secretion by the distal tubule and collecting duct. Aldosterone enhances secretion by stimulating the Na$^+$,K$^+$-ATPase, thereby increasing cell K$^+$. Aldosterone also increases the electrochemical gradient of K$^+$ across the apical membrane and increases the permeability of the apical membrane to K$^+$. ADH promotes secretion by enhancing the electrochemical gradient for K$^+$ across the apical membrane.

MULTIVALENT IONS

Calcium (Ca^{++}), magnesium (Mg^{++}), and inorganic phosphate (PO$_4^{\equiv}$) are multivalent ions present in the body fluids. These ions subserve many complex functions. In a normal adult the kidneys excrete each day as much calcium, magnesium, and phosphate as the intestinal tract absorbs. If dietary intake of any mineral declines substantially, both intestinal absorption and renal tubular reabsorption increase to a rate such that endogenous stores are maintained constant. During growth and pregnancy, intestinal absorption exceeds urinary excretion, and thus these ions accumulate in newly formed tissue and bone. In contrast, bone disease or a decline in lean body mass increases urinary mineral loss without a change in intestinal absorption.

This section focuses on calcium, magnesium, and phosphate handling by the kidney, with a special emphasis on the hormones and factors that regulate urinary excretion. Information on intestinal absorption is presented in Chapter 31, and an overview on the regulation of the plasma levels of these minerals is provided in Chapter 38.

Calcium

Calcium ions play a major role in many cell functions, including cell division and growth, blood coagulation, hormone-response coupling, and electrical stimulus-response coupling (e.g., muscle contraction and neurotransmitter release). The total concentration of Ca^{++} in plasma is about 5 mEq·L^{-1}. Approximately 40% is bound to plasma proteins, primarily albumin, and this fraction is not filtered by the glomeruli. The filterable fraction (60%) consists of an ionized and a nonionized moiety. The nonionized moiety is complexed to several anions, including bicarbonate, citrate, phosphate, and sulfate.

Renal Ca^{++} Excretion Normally 99% of the filtered Ca^{++} is reabsorbed by the renal tubules. This is equivalent to the amount of Na$^+$ reabsorbed by the nephron. As illustrated in Figure 34-5, the proximal tubule reabsorbs 60% of the filtered Ca^{++}. Another 20% is reabsorbed in Henle's loop (mainly the thick ascending limb), 10% to 15% is reabsorbed by the distal tubule, and 5% is reabsorbed by the collecting duct. About 1% (10 mEq/day) is excreted in the urine. This moiety is equal to the amount absorbed daily by the gastrointestinal tract.

Ca^{++} reabsorption by the proximal tubule occurs by two pathways: transcellular and paracellular (Figure 34-6). Ca^{++} reabsorption across the cellular pathway accounts for one third of proximal reabsorption

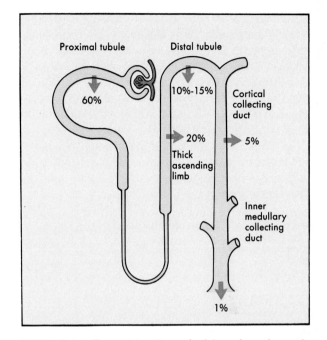

FIGURE 34-5   Transport pattern of calcium along the nephron. Calcium is reabsorbed by all nephron segments. Percentages refer to the amount of the filtered calcium reabsorbed by each nephron segment.

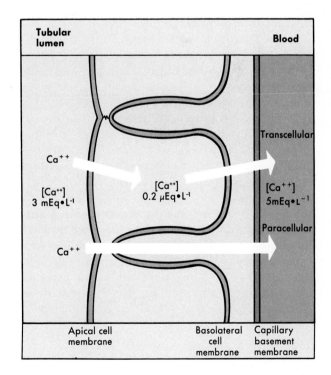

FIGURE 34-6   Cellular schema of calcium reabsorption by the proximal tubule. Calcium is reabsorbed by transcellular and paracellular routes.

and occurs in two steps. Ca^{++} diffuses across the apical membrane into the cell down its electrochemical gradient. This gradient is exceptionally steep because the cell Ca^{++} concentration is only 0.2 $\mu Eq \cdot L^{-1}$, about 10,000-fold less than in the tubular fluid (3 $mEq \cdot L^{-1}$). The cell interior is electrically negative with respect to the luminal side of the apical membrane, and this also favors Ca^{++} entry into the cell. Ca^{++} is extruded across the basolateral membrane against its electrochemical gradient, and therefore extrusion is an active process. The mechanism for the active extrusion of Ca^{++} has not been defined, but could occur by either a Ca^{++}-ATPase or a Na^+-Ca^{++} antiporter. Ca^{++} is also reabsorbed across the paracellular pathway by passive diffusion and by solvent drag.

Ca^{++} reabsorption by Henle's loop is restricted to the thick ascending limb. Ca^{++} is reabsorbed via cellular and paracellular routes by mechanisms similar to those described for the proximal tubule, except that Ca^{++} is not reabsorbed by solvent drag in this segment (recall that the thick ascending limb is impermeable to water). In both the proximal tubule and the thick ascending limb, Ca^{++} and Na^+ reabsorption parallel each other. Therefore, **anything that changes sodium reabsorption will also alter calcium reabsorption** by these tubular segments.

In the distal tubule and collecting duct, where the voltage in the tubular fluid is electrically negative with respect to the interstitial space, Ca^{++} reabsorption is entirely active because Ca^{++} is reabsorbed against its electrochemical gradient. These tubular segments determine the final amount of Ca^{++} that appears in the urine. Most of the factors and hormones that regulate urinary Ca^{++} excretion affect the distal tubule and collecting duct.

Regulation of Ca^{++} Excretion Figure 34-7 illustrates the sites of control and the primary factors that regulate Ca^{++} reabsorption along the nephron. **Parathyroid hormone (PTH) exerts the most powerful control on renal calcium excretion.** This hormone stimulates overall Ca^{++} reabsorption. Although PTH inhibits Ca^{++} reabsorption by the proximal tubule it dramatically stimulates Ca^{++} reabsorption in the distal tubule and thick ascending limb of Henle's loop. As a result urinary Ca^{++} excretion falls.

Changes in the ECF volume alter Ca^{++} excretion mainly by affecting Na^+ and fluid reabsorption in the proximal tubule. Contraction of the ECF volume

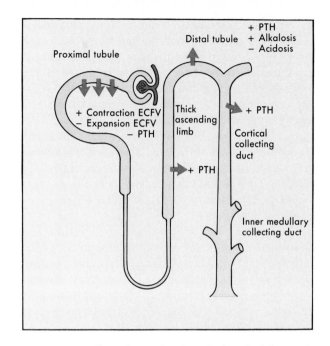

FIGURE 34-7   Sites of control and regulation of calcium reabsorption along the nephron. The + indicates a positive effect on reabsorption; the − indicates a negative effect. *PTH*, Parathyroid hormone; *ECFV*, extracellular fluid volume.

increases Na^+ and water reabsorption by the proximal tubule and thereby enhances Ca^{++} reabsorption. Accordingly, urinary Ca^{++} excretion declines. Expansion of the ECF volume has the opposite effect.

Phosphate infusion increases PTH levels and thereby decreases Ca^{++} excretion, whereas phosphate depletion has the opposite effect. Finally, acidosis increases Ca^{++} excretion, whereas alkalosis decreases excretion. The regulation of Ca^{++} reabsorption by pH occurs in the distal tubule by an unknown mechanism.

Magnesium

Magnesium is the second most common intracellular electrolyte. This divalent cation is a major cofactor in almost all cellular functions. Mg^{++} has many biochemical roles in cells, including activation of enzymes and regulation of protein synthesis. The plasma concentration of Mg^{++} is about 1.7 $mEq \cdot L^{-1}$. Approximately 20% is protein bound and therefore unavailable for ultrafiltration by the glomeruli. The

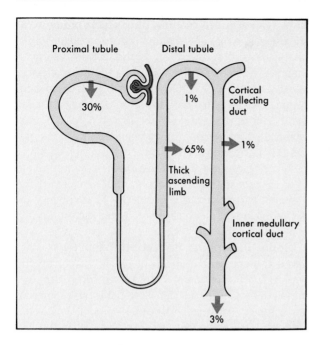

FIGURE 34-8   Transport pattern of magnesium along the nephron. Magnesium is reabsorbed mainly by the proximal tubule and the thick ascending limb of Henle's loop. Percentages refer to the amount of the filtered magnesium reabsorbed by each nephron segment.

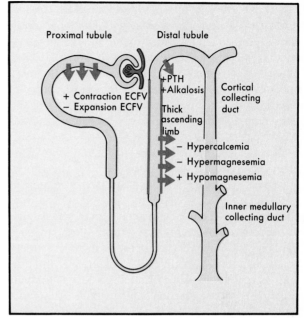

FIGURE 34-9   Sites of control and regulation of magnesium reabsorption along the nephron. The + indicates a positive effect on reabsorption; the − indicates a negative effect. *PTH*, Parathyroid hormone; *ECFV*, extracellular fluid volume.

Mg^{++} that is filtered consists of an ionized moiety and a nonionized component that is complexed to bicarbonate, citrate, phosphate, and sulfate. Accordingly, the Mg^{++} concentration in the ultrafiltrate is 20% less than the Mg^{++} concentration in the plasma. The kidneys are important in Mg^{++} homeostasis and excrete 3% of the filtered Mg^{++}. Urinary Mg^{++} excretion is equal to the amount of Mg^{++} absorbed by the gastrointestinal tract (8 mEq/day).

Renal Mg^{++} Excretion   The pattern of Mg^{++} transport along the nephron is shown in Figure 34-8. Approximately 30% of the filtered Mg^{++} is reabsorbed by the proximal tubule. The thick ascending limb of Henle's loop is the major site of Mg^{++} transport, reabsorbing 65% of the filtered Mg^{++}. Reabsorption is passive and is driven by the lumen-positive transepithelial voltage across the thick ascending limb. Alterations in urinary Mg^{++} excretion usually arise from changes in Mg^{++} reabsorption by the thick ascending limb. Very little Mg^{++} is reabsorbed by the distal tubule and collecting duct.

Regulation of Mg^{++} Excretion   Figure 34-9

illustrates the sites of control and the principal factors that regulate Mg^{++} transport along the nephron. Most of the regulation occurs at the thick ascending limb. Mg^{++} reabsorption is increased by elevated PTH levels, reduced ECF volume, and alkalosis. In contrast, Mg^{++} reabsorption is decreased by a fall in PTH levels, a rise in ECF volume, and acidosis.

Two additional factors regulate Mg^{++} transport: Mg^{++} balance and plasma Ca^{++} levels. Mg^{++} depletion, which leads to hypomagnesemia, stimulates Mg^{++} reabsorption by the thick ascending limb and decreases urinary Mg^{++} excretion. Mg^{++} loading has the opposite effect. Increased plasma Ca^{++} inhibits Mg^{++} reabsorption. Hypocalcemia has the opposite affect. The cellular mechanisms involved in the regulation of Mg^{++} transport are poorly understood.

Phosphate

Phosphate is an important component of many organic molecules, including deoxyribonucleic acid

(DNA), ribonucleic acid (RNA), adenosine triphosphate (ATP), and intermediates of metabolic pathways. The plasma concentration of phosphate is about 2 mEq·L^{-1}. Approximately 20% of the phosphate in the plasma is protein bound and therefore unavailable for ultrafiltration by the glomeruli. Accordingly, the phosphate concentration in the ultrafiltrate is 20% less than the phosphate concentration in plasma.

The kidneys are important in maintaining phosphate homeostasis; they excrete 20% of the filtered phosphate. Urinary phosphate excretion is equivalent to the amount absorbed by the gastrointestinal tract (72 mEq/day). The tubular mechanisms for phosphate reabsorption share many properties with the glucose transport system. These properties include a transport maximum (T_m) and Na$^+$ dependence. In addition, as with the glucose titration curve (see Chapter 32), the phosphate titration curve exhibits splay and a threshold (Figure 34-10).

However, the titration curves for glucose and phosphate are significantly different in several respects. First, the T_m for phosphate is only slightly above the normal filtered load (Figure 34-10). Accordingly, **the kidneys regulate plasma phosphate concentration.** For example, a small increase in plasma phosphate concentration increases the filtered load such that the T_m is exceeded, and thus the urinary phosphate excretion rises (Figure 34-10). This in turn causes plasma phosphate to fall. In contrast, because the T_m for glucose is considerably above the normal filtered load of glucose, small changes in plasma glucose do not change glucose excretion or plasma glucose concentration. Thus the kidneys do not regulate plasma glucose levels. A second unique aspect of the phosphate titration curve is that the T_m for phosphate is variable and it is regulated by dietary phosphate. In contrast, the T_m for glucose is relatively stable.

Renal Phosphate Excretion Figure 34-11 illustrates the transport pattern of phosphate along the nephron. The proximal tubule reabsorbs approximately 67% of the amount filtered by the glomeruli, and Henle's loop reabsorbs about 10% of the filtered amount. Little phosphate transport occurs in the distal tubule and the collecting duct. Therefore about 20% of the filtered load is excreted.

Phosphate reabsorption by the proximal tubule is reabsorbed mainly, if not exclusively, by a transcellu-

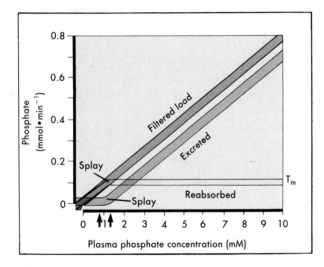

FIGURE 34-10   Titration curve for phosphate. The data in this figure were obtained by infusing phosphate into subjects and measuring the amount excreted in the urine ($U_{PO_4} \times \dot{V}$) and the amount filtered (glomerular filtration rate $\times$ [PO$_4$] in the ultrafiltrate). The amount reabsorbed was calculated as the difference between the amount filtered minus the amount excreted. The arrows on the x axis indicate the normal range of plasma phosphate concentrations. T_M, Transport maximum.

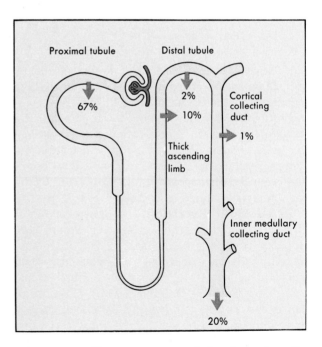

FIGURE 34-11   Transport pattern of phosphate along the nephron. Phosphate is reabsorbed mainly by the proximal tubule and Henle's loop. Percentages refer to the amount of the filtered phosphate reabsorbed by each nephron segment.

lar route. Phosphate uptake across the apical membrane occurs by a sodium-phosphate co-transport mechanism, and its exit across the basolateral membrane occurs by a passive mechanism. The cellular mechanisms of phosphate reabsorption by Henle's loop, the distal tubule, and the collecting duct are poorly understood.

Regulation of Phosphate Excretion Figure 34-12 illustrates the sites of control and the major factors that regulate phosphate transport along the nephron. Urinary phosphate excretion is mainly determined by phosphate transport along the proximal tubule. **Parathyroid hormone is the most important hormone that controls phosphate reabsorption.** PTH increases cyclic adenosine monophosphate (cAMP) production and inhibits phosphate reabsorption by the proximal tubule.

Dietary phosphate intake regulates phosphate excretion by mechanisms that are unrelated to changes in PTH levels. Phosphate loading reduces reabsorption by the proximal tubule, whereas phosphate depletion stimulates reabsorption. Changes in the ECF volume also affect phosphate reabsorption by the proximal tubule. Volume expansion reduces

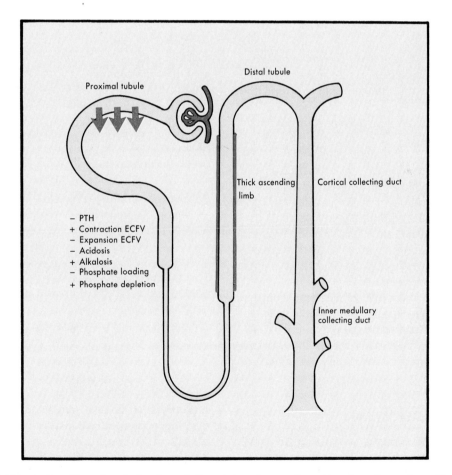

FIGURE 34-12 Sites of control and regulation of phosphate reabsorption along the nephron. The + indicates a positive effect on reabsorption; the − indicates a negative effect. All hormones and factors control urinary phosphate excretion by regulating reabsorption by the proximal tubule. *PTH,* Parathyroid hormone; *ECFV,* extracellular fluid volume.

phosphate reabsorption by a mechanism independent of PTH. Finally, acidosis inhibits proximal phosphate reabsorption, and alkalosis stimulates it. The pH of the tubular fluid directly regulates the sodium-phosphate co-transport protein in the apical membrane; an acidic pH reduces the activity of the co-transporter.

ACID-BASE BALANCE

The concentration of H^+ in the body fluids is low compared to that of other ions. For example, Na^+ is present at a concentration approximately 1 million times greater than that of H^+ ($[Na^+] = 140$ mEq·L^{-1}; $[H^+] = 40$ nEq·L^{-1}). Because of the low H^+ concentration of the body fluids, it is commonly expressed as the negative logarithm, or pH. The normal pH of the body fluids is 7.40.

Many of the metabolic functions of the body are exquisitely sensitive to pH. Thus the pH of the body fluids must be maintained within a very narrow range. This is accomplished by buffers within the body fluids, as well as the coordinated action of the lungs and kidneys. With a normal diet the body produces approximately 15 to 20 moles of acid per day. Virtually all this acid is derived from CO_2 (CO_2 + H_2O = H_2CO_3) and is therefore called *volatile acid*. It is the responsibility of the lungs, by excreting CO_2, to handle this potential acid load. In addition, the metabolism of certain foodstuffs produces acids that cannot be excreted by the lungs. These acids are termed *nonvolatile acids* and are produced at a rate of approximately 50 to 100 mEq/day. The kidneys must excrete these nonvolatile acids.

This section discusses the general mechanisms involved in the maintenance of acid-base balance. Special emphasis is placed on the role of the kidney, particularly the mechanisms involved in the renal excretion of acid. The role of the lungs is described in Chapters 27 and 28.

The CO_2/HCO_3^- Buffer System

Quantitatively, bicarbonate (HCO_3^-) is the most important buffer in the extracellular fluid (normal plasma $[HCO_3^-] = 24$ mEq·L^{-1}). The CO_2/HCO_3^- buffer system differs from the other buffer systems of the body because it is regulated by both the lungs and the kidneys. This is best illustrated by considering the following reaction:

$$CO_2 + H_2O \overset{CA}{\rightleftharpoons} H_2CO_3 \rightleftharpoons H^+ + HCO_3^- \quad (1)$$

The first reaction (hydration/dehydration of CO_2) is the rate-limiting step. This reaction, which is normally slow, is greatly accelerated in the presence of the enzyme carbonic anhydrase (CA). The ionization of H_2CO_3 is virtually instantaneous.

The dissociation constant for the previous reaction can be written as:

$$K' = \frac{[H^+][HCO_3^-]}{[CO_2][H_2CO_3]} \quad (2)$$

Because the reaction includes not only the hydration and dehydration of CO_2 but also the ionization of H_2CO_3, K' is not a true dissociation constant. For plasma at 37° C, this apparent dissociation constant equals $10^{-6.1}$ ($pK' = 6.1$).

The terms in the denominator of equation 2 represent the total amount of CO_2 dissolved in solution. Most of this CO_2 is in the gas form, with only 0.3% being H_2CO_3. Because the amount of CO_2 in solution depends on its partial pressure (P_{CO_2}) and its solubility (α), equation 2 can be rewritten as:

$$K' = \frac{[H^+][HCO_3^-]}{\alpha P_{CO_2}} \quad (3)$$

For plasma at 37° C, α equals 0.03.

A more useful form of this equation is obtained by solving for $[H^+]$:

$$[H^+] = \frac{K'\alpha P_{CO_2}}{[HCO_3^-]} \quad (4)$$

Taking the negative logarithm of both sides of this equation yields:

$$-\log[H^+] = \frac{-\log[K'] - \log \alpha P_{CO_2}}{-\log[HCO_3^-]} \quad (5)$$

$$pH = pK' + \log \frac{[HCO_3^-]}{\alpha P_{CO_2}} \quad (6)$$

Equation 6 is the *Henderson-Hasselbalch equation*. Inspection of this equation clearly shows that the pH of the ECF varies when either the $[HCO_3^-]$ or the P_{CO_2} is altered. Disturbances of acid-base balance that result from a change in the ECF $[HCO_3^-]$ are termed *metabolic*, whereas those resulting from a change in the P_{CO_2} are termed *respiratory*. As discussed shortly, the kidneys mainly control the $[HCO_3^-]$, whereas the lungs control the PCO_2.

Production of Nonvolatile Acid

As already mentioned, a tremendous amount of potential acid is produced each day in the form of CO_2. This is generated from the metabolism of carbohydrates and fats. In addition, the metabolism of certain other foodstuffs produces nonvolatile acids. The bulk of these nonvolatile acids is produced from the metabolism of certain amino acids. The sulfur-containing amino acids cysteine and methionine yield sulphuric acid when metabolized, whereas hydrochloric acid results from the metabolism of lysine, arginine, and histidine. In addition to the nonvolatile acid produced by these amino acids, the ingestion of phosphate $(H_2PO_4^-)$ increases the dietary acid load.

A portion of this nonvolatile acid load is offset by the production of HCO_3^- through the metabolism of the amino acids aspartate and glutamate, as well as by metablism of certain organic anions (e.g., citrate). On balance, acid production exceeds HCO_3^- production, with the net effect being the addition of approximately 1 mEq/kg body weight of nonvolatile acid to the body each day.

These nonvolatile acids do not circulate throughout the body but are immediately buffered.

$$H_2SO_4 + 2NaHCO_3 \rightleftharpoons Na_2SO_4 + 2CO_2 + 2H_2O \quad \textbf{(7)}$$

$$HCl + NaHCO_3 \rightleftharpoons NaCl + CO_2 + H_2O \quad \textbf{(8)}$$

Thus this titration process yields the sodium salts of the strong acid anions, and removes HCO_3^- from the ECF. The kidney must excrete these sodium salts and replenish the HCO_3^- lost by titration.

Renal Acid Excretion

To maintain acid-base balance, the kidneys must excrete an amount of acid equal to the nonvolatile acid production. In addition, they must prevent the loss of HCO_3^- in the urine. This latter task is quantitatively more important, since the filtered load of HCO_3^- is approximately 4500 mEq/day compared to the 50 to 100 mEq/day of nonvolatile acid.

Both the reabsorption of the filtered load of HCO_3^- and the excretion of acid are accomplished through the process of H^+ secretion by the nephrons. Thus in a single day the nephrons secrete approximately 4600

mEq of H^+. Most of these H^+ are not excreted in the urine but serve to reclaim the filtered load of HCO_3^-. Only 50 to 100 mEq/day of H^+ are excreted. As a result of this acid excretion, the urine is normally acidic.

Theoretically the kidney could excrete the nonvolatile acids and replenish the HCO_3^- lost during their titration by reversing the reactions shown in equations 7 and 8. Because the pKs of these acids are so low, however, this process would require a urine pH of 1.0. Because the minimal urine pH attainable by the kidney is only 4.0 to 4.5, the kidney must instead excrete the sodium salts of these acids (e.g., Na_2SO_4, NaCl), whereas the H^+ is excreted with other buffers. The two major urinary buffers are ammonia (NH_3) and phosphate $(HPO_4^=)$. The urinary phosphate and some other buffer species (e.g., creatinine) are collectively termed *titratable acid.*

The overall process of acid excretion by the kidney can be quantitated as follows:

$$\text{Net acid excretion rate} = [(U_{NH_4^+} \cdot \dot{V}) + \quad \textbf{(9)}$$
$$(U_{TA} \cdot \dot{V})] - (U_{HCO_3^-} \cdot \dot{V})$$

where $U_{NH_4^+} \cdot \dot{V}$ and $U_{TA} \cdot \dot{V}$ are the rates of H^+ excretion (mEq/day) as NH_4^+ and titratable acid (TA), and $U_{HCO_3^-} \cdot \dot{V}$ is the amount of HCO_3^- loss in the urine (equivalent to adding H^+ to the body). To maintain acid-base balance, the net acid excretion must equal the nonvolatile acid production.

Bicarbonate Reabsorption Along the Nephron

Glomerular filtration delivers 4500 mEq/day of HCO_3^- to the proximal tubule. Approximately 85% of this HCO_3^- is reabsorbed by this segment. The general process by which this reabsorption of HCO_3^- occurs is illustrated in Figure 34-13 (see Chapter 32).

The apical membrane of the proximal tubule cell contains a Na^+-H^+ antiporter. Driven by the lumen-to-cell Na^+ gradient, this antiporter secretes H^+ into the tubule lumen. Recent evidence indicates that a small portion of H^+ secretion is also mediated by a H^+-ATPase. Within the cell H^+ and HCO_3^- are produced in a reaction catalyzed by carbonic anhydrase (equation 1). The H^+ is secreted into the tubule lumen, whereas the HCO_3^- exits the cell across the basolateral membrane. Although the electrochemical gradient for HCO_3^- would allow passive exit from the

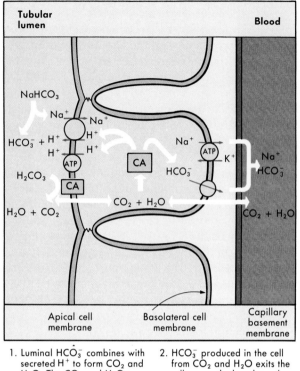

1. Luminal HCO_3^- combines with secreted H^+ to form CO_2 and H_2O. The CO_2 and H_2O are reabsorbed.

2. HCO_3^- produced in the cell from CO_2 and H_2O exits the cell across the basolateral membrane and is returned to the blood.

FIGURE 34-13 Cellular mechanism of bicarbonate (HCO_3^-) reabsorption by the proximal tubule. See text for details. *CA*, Carbonic anhydrase.

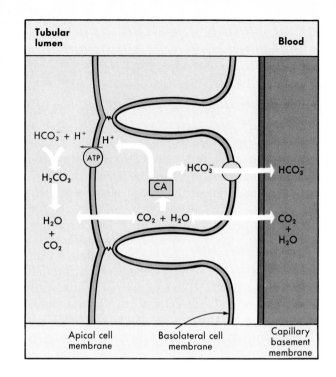

FIGURE 34-14 Cellular mechanism of HCO_3^- reabsorption by the intercalated cell of the collecting duct. *CA*, Carbonic anhydrase.

cell across the basolateral membrane, HCO_3^- movement is coupled to that of other ions (e.g., Na^+, Cl^-). Within the tubule lumen the secreted H^+ combines with the filtered HCO_3^- to form H_2CO_3. This is rapidly converted to CO_2 and H_2O by the carbonic anhydrase present in the apical membrane of the cell. Because the tubule is highly permeable to both CO_2 and H_2O, they are rapidly reabsorbed. The net effect of this process is that for each HCO_3^- removed from the tubule lumen, one HCO_3^- appears in the peritubular blood.

An additional 10% to 15% of the filtered load of HCO_3^- is reabsorbed by Henle's loop. Most of this is reabsorbed by the thick ascending limb, which, as with the proximal tubule, has a Na^+-H^+ antiporter in its apical membrane.

The distal tubule and collecting duct reabsorb the small amount of HCO_3^- that escapes reabsorption by the proximal tubule and Henle's loop (~5% of the filtered load). The mechanism whereby this occurs does not depend on Na^+ and occurs solely by a H^+-ATPase.

Another difference between the collecting duct and the more proximal nephron segments is that not all cells of the collecting duct are involved in H^+ secretion. Only the intercalated cells secrete H^+. The mechanism by which this occurs is illustrated in Figure 34-14 (see Chapter 32). Within the intercalated cell, H^+ and HCO_3^- are produced by the hydration of CO_2. This reaction is catalyzed by carbonic anhydrase. The H^+ is secreted into the tubule lumen by the H^+-ATPase, and the HCO_3^- exits the cell across the basolateral membrane. The tubule cells of the collecting duct are highly impermeable to H^+; thus the pH of the luminal fluid can be rendered quite acidic. Indeed, the most acidic fluid along the nephron (pH = 4.0 to 4.5) is produced at this site. In comparison, the permeability of the proximal tubule to H^+ and HCO_3^- is much higher, and the pH of the tubule fluid falls to only 6.5.

Regulation of Bicarbonate Reabsorption

HCO_3^- reabsorption is regulated by several factors, which act at both the proximal tubule and the collecting duct. (Table 34-1).

Because of the phenomenon of glomerulotubular

Table 34-1 Factors Regulating H$^+$ Secretion by the Nephron

Factor	Nephron Site of Action
INCREASING H$^+$ SECRETION	
Increase in filtered load of HCO$_3^-$	Proximal tubule
Decrease in extracellular fluid volume	Proximal tubule
Decrease in extracellular [HCO$_3^-$]	Proximal tubule and collecting duct
Increase in blood PCO_2	Proximal tubule and collecting duct
Aldosterone	Collecting duct
DECREASING H$^+$ SECRETION	
Decrease in filtered load of HCO$_3^-$	Proximal tubule
Increase in extracellular fluid volume	Proximal tubule
Increase in extracellular [HCO$_3^-$]	Proximal tubule and collecting duct
Decrease in blood PCO_2	Proximal tubule and collecting duct

balance (see Chapter 32), any change in the filtered load of HCO$_3^-$ is matched by an appropriate change in HCO$_3^-$ reabsorption by the proximal tubule. Recall that the bulk of proximal tubule HCO$_3^-$ reabsorption occurs by the Na$^+$- H$^+$ antiporter. Consequently, factors that affect Na$^+$ reabsorption will alter HCO$_3^-$ reabsorption secondarily. Thus expansion of the ECF volume inhibits HCO$_3^-$ reabsorption, and the opposite occurs when the ECF volume is decreased. Changes in systemic acid-base balance also affect HCO$_3^-$ reabsorption in the proximal tubule. Systemic acidosis, whether produced by a decrease in the plasma bicarbonate concentration [HCO$_3^-$] (metabolic) or by an increase in the plasma PCO_2 (respiratory), stimulates H$^+$ secretion by the proximal tubule cells. This stimulation is the result of acidification of the ICF of the tubule cell. When the ICF is acidic, a more favorable cell-to-lumen H$^+$ gradient exists, and H$^+$ secretion is stimulated. Conversely, metabolic and respiratory alkalosis inhibit H$^+$ secretion in the proximal tubule.

The reabsorption of HCO$_3^-$ by the thick ascending limb and collecting duct is also modulated by systemic acid-base balance; acidosis stimulates and alkalo-

sis inhibits this process. An additional factor that regulates HCO$_3^-$ reabsorption in the collecting duct is aldosterone. When aldosterone levels are elevated, H$^+$ secretion by the intercalated cell is stimulated. Conversely, when aldosterone levels are reduced, H$^+$ secretion is decreased.

Formation of New Bicarbonate

As discussed previously, the reabsorption of the filtered load of HCO$_3^-$ is important for the maintenance of acid-base balance. HCO$_3^-$ loss in the urine would decrease the plasma [HCO$_3^-$], and thus it would be equivalent to the addition of acid to the body. However, HCO$_3^-$ reabsorption alone does not replenish the HCO$_3^-$ that was lost during the titration of the nonvolatile acids. For acid-base balance to be maintained, the kidney must replace this lost HCO$_3^-$. The production of new HCO$_3^-$ is critically dependent on the availability of urinary buffers. Figure 34-15 illustrates how the titration of urinary buffers results in the formation of new bicarbonate.

As mentioned, the two major urinary buffers are ammonia (NH$_3$) and phosphate (HPO$_4^=$). Phosphate is derived solely from the diet. The amount excreted as titratable acid therefore depends on the filtered load

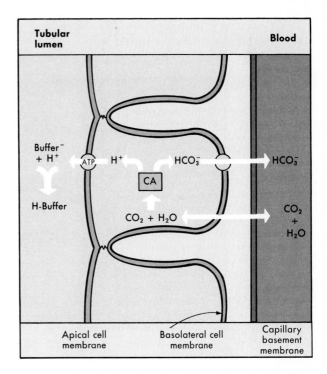

FIGURE 34-15   Production of new bicarbonate by the titration of urinary buffers. *CA,* Carbonic anhydrase.

minus the amount reabsorbed by the nephron. Ammonia is produced by the kidney, and its synthesis and subsequent excretion can be regulated in response to the acid-base requirements of the body. Because of this, NH_3 is the more important urinary buffer.

Figute 34-16 illustrates the general features of NH_3 production and secretion by the tubule cells. Ammonia is produced from glutamine. Each molecule of glutamine produces two molecules of NH_4^+ and a divalent anion. Ultimately it is the complete metabolism of this divalent anion that provides the new bicarbonate.

$$\text{Glutamine} \rightleftharpoons 2NH_4^+ + \text{Anion}^= \rightleftharpoons 2HCO_3^- + 2NH_4^+ \quad \textbf{(10)}$$

The mechanisms for the secretion of NH_4^+ into the tubule lumen vary with the nephron segment. NH_4^+ can be secreted in exchange for Na^+ (proximal tubule) or by the processes of *nonionic diffusion* and *diffusion trapping* (collecting duct). These latter processes show that the renal tubular cell membrane is highly permeable to NH_3, whereas it it relatively impermeable to NH_4^+ (Figure 34-16).

As already indicated, an important feature of the ammonia buffer system is that it can be regulated. With systemic acidosis, the enzymes responsible for the metabolism of glutamine are stimulated. This stimulation involves the synthesis of more enzyme, a step that requires several days. Because of this adaptive response, the kidney can increase the excretion of H^+ and produce more new bicarbonate to defend against the acidosis.

Renal Response to Acid-Base Disorders

The pH of the body fluids is maintained within a very narrow range (pH = 7.40 ± 0.02). *Acidosis* is

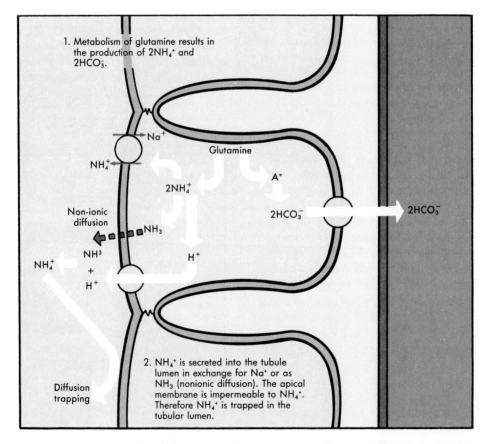

FIGURE 34-16   Production of ammonium (NH_4^+) and new bicarbonate by a renal tubule cell. NH_4^+ can be secreted into the tubule lumen by a Na^+-NH_4^+ antiporter (proximal tubule) and by nonionic diffusion with diffusion trapping of NH_3 (collecting duct).

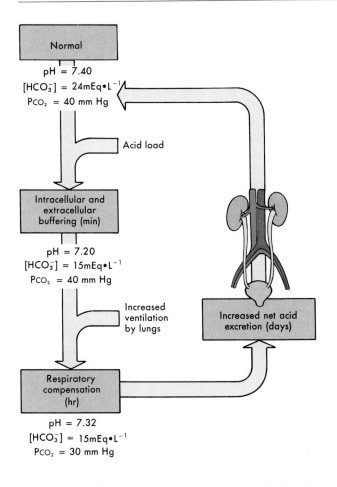

Normal

pH = 7.40
$[HCO_3^-]$ = 24mEq•L^{-1}
P_{CO_2} = 40 mm Hg

Acid load

Intracellular and extracellular buffering (min)

pH = 7.20
$[HCO_3^-]$ = 15mEq•L^{-1}
P_{CO_2} = 40 mm Hg

Increased ventilation by lungs

Increased net acid excretion (days)

Respiratory compensation (hr)

pH = 7.32
$[HCO_3^-]$ = 15mEq•L^{-1}
P_{CO_2} = 30 mm Hg

FIGURE 34-17 Mechanisms of defense against a fixed acid load.

said to exist when the blood pH falls below 7.40, whereas *alkalosis* exists when the blood pH is greater than 7.40. When the acid-base disorder results from a primary change in the $[HCO_3^-]$, it is termed a *metabolic* disorder. When the primary disturbance is an alteration in the blood P_{CO_2}, it is termed a *respiratory* disorder.

The body has three general mechanisms to defend against acid-base disturbances. These are illustrated in Figure 34-17, in which the defense against an acid load (metabolic acidosis) is depicted.

The first line of defense is extracellular and intracellular buffering. The response of the extracellular buffers (mainly HCO_3^-) is almost instantaneous. The movement of H^+ into the cells (intracellular buffering) is somewhat slower, but it is still complete within a few minutes. Although these processes prevent a dramatic fall in the blood pH, they do not contribute to the elimination of the acid load from the body.

The lungs represent the second line of defense. As discussed in Chapter 28, a fall in the blood pH stimulates ventilation. As a result of the increased pulmonary gas exchange, the blood P_{CO_2} will fall. According to the Henderson-Hasselbalch equation (equation 6), this fall in the P_{CO_2} will ameliorate the decrease in blood pH (Figure 34-17). The respiratory response may require several hours to complete, and as with the buffer systems, it prevents large changes in blood pH but does not eliminate the acid load.

The third and final line of defense is the kidney. In

Table 34-2   Mechanisms of Defense Against Acid-Base Disturbances

Type of Disturbance	Primary Alteration	Defense Mechanisms
Metabolic acidosis	Decrease in plasma [HCO_3^-]	Intracellular and extracellular buffers Hyperventilation to decrease P_{CO_2}
Metabolic alkalosis	Increase in plasma [HCO_3^-]	Intracellular buffers Hypoventilation to increase P_{CO_2}
Respiratory acidosis	Increase in blood P_{CO_2}	Intracellular buffers Increased renal acid excretion
Respiratory alkalosis	Decrease in blood P_{CO_2}	Intracellular buffers Decreased renal acid excretion

response to acidosis, the entire filtered load of HCO_3^- is reabsorbed, and the production and excretion of NH_4^+ are stimulated. As a result, net acid excretion is increased, and new bicarbonate is returned to the body to correct the acidosis. The renal response requires several days to correct the acidosis.

The primary acid-base disorders and the appropriate defense mechanisms are listed in Table 34-2. As can be seen, the lungs compensate for metabolic disorders and the kidneys compensate for respiratory disorders. It should be emphasized that these compensatory mechanisms do not correct the underlying disorder, but simply reduce the magnitude of the change in blood pH. Complete recovery from the acid-base disorder requires correction of the underlying cause.

BIBLIOGRAPHY
Books and Monographs

Alpern RJ et al: Renal acidification mechanisms. In Brenner BM and Rector FC Jr, editors: The kidney, ed 3, Philadelphia, 1986, WB Saunders Co.

Giebisch G et al: Renal transport and control of potassium excretion. In Brenner BM and Rector CF Jr. editors: The kidney, ed 3, Philadelphia, 1986, WB Saunders Co.

Knox FG and Haramati A: Renal regulation of phosphate excretion. In Seldin DW and Giebisch G, editors: The kidney: physiology and pathophysiology, vol 2, New York, 1985, Raven Press.

Koeppen B et al: Mechanism and regulation of renal tubular acidification. In Seldin DW and Giebisch G, editors: The kidney: physiology and pathophysiology, vol 2, New York, 1985, Raven Press.

Rose DB: Clinical physiology of acid-base and electrolyte disorders, ed 2, New York, 1984, McGraw-Hill Book Co.

Stanton BS and Giebisch G: Renal potassium transport. In Windhager E, editor: Handbook of physiology: renal physiology, ed 2, New York, 1990, Oxford University Press.

Sutton RAL and Dirks JH: Calcium and magnesium: renal handling and disorders of metabolism. In Brenner BM and Rector FC Jr, editors: The kidney ed 3, Philadelphia, 1986, WB Saunders Co.

Wright FA and Giebisch G: Regulation of potassium secretion. In Seldin DW and Giebisch G, editors: The kidney: physiology and pathophysiology, vol 2, New York, 1985, Raven Press.

Valtin H and Gennari FJ: Acid-base disorders: basic concepts and clinical management, Boston, 1987, Little, Brown & Co.

ENDOCRINE SYSTEM

SAUL M. GENUTH

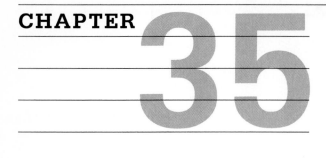

PART VIII

General Principles of Endocrine Physiology

The endocrine system is a key component in the adaptation of the human organism to changes in the internal and external environment. This system acts to maintain a stable internal milieu when confronted by changes in inflow or outflow of substrates, minerals, water, environmental molecules, heat, and so on. Specific endocrine cells usually grouped in glands sense the disturbance and respond by secreting chemical substances called *hormones* into the blood-stream. These special molecules are carried via the circulation to various tissues, where they signal and act on their target cells. As a result the target cells respond in a manner that usually opposes the direction of change that evoked the secretion of hormone, thereby restoring the organism toward its original stable state. In addition to this fundamental role in maintaining *homeostasis*, the endocrine system also helps to initiate, mediate, and regulate the processes

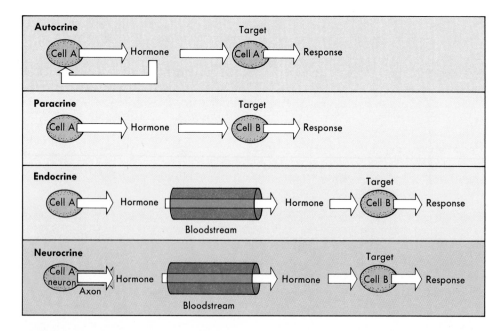

FIGURE 35-1 Schematic representation of mechanisms for cell-to-cell signaling via hormone molecules. In *autocrine* function the hormone signal acts back on the cell of origin or adjacent identical cells. In *paracrine* function the hormone signal is carried to an adjacent target cell over short distances via the interstitial fluid. In *endocrine* function the signal is carried to a distant target via the bloodstream. In *neurocrine* function the hormone signal originates in a neuron and, after axonal transport to the bloodstream, is carried to a distant target cell.

of growth, development, maturation, reproduction, and senescence.

A hormone was originally defined as a substance that was elaborated by one type of cell, and that carried a signal *through the bloodstream* to distant target cells. However, this sophisticated method of signaling probably evolved from a more primitive one (Figure 35-1). Hormone molecules secreted by endocrine cells can also reach and act on target cells within the same locale simply by diffusing through the interstitial fluid separating them; this is called *paracrine* function. Also, evidence suggests that hormone molecules can even act back on their cells of origin to modulate their own secretion or other intracellular processes; this is called *autocrine* function.

The endocrine system may act independently of or may be integrated with the nervous system, which is the other major component in the organism's adaptability to internal or external change (Figure 35-2). These two signaling systems have several characteristics in common:

1. Both neurons and endocrine cells are capable of secreting.
2. Both endocrine cells and neurons generate electrical potentials and can be depolarized.
3. In certain instances the same molecule serves as both a neurotransmitter and a hormone.
4. The mechanism of action of both hormones and neurotransmitters requires interaction with specific receptors in target cells.

Although the endocrine system responds more often to chemical stimuli and the nervous system more often to physical or mechanical stimuli, considerable overlap exists. For example, changes in the quantity of light and changes in plasma sustrate concentrations may evoke responses by both systems. Interaction between the two systems takes several forms:

1. Some stimuli to hormone release are first sensed by the nervous system, which in turn signals the appropriate endocrine cell to respond.
2. Some neurons extend their axons in bundles or tracts that terminate adjacent to capillaries. Stimulation causes release of their neurotransmitters into the bloodstream. This hybrid form of signal transmission is called *neurocrine* function (Figure 35-1), and the signaling molecules are called *neurohormones*.
3. Some stimuli evoke integrated endocrine and nervous system responses that augment each other in restoring homeostasis.

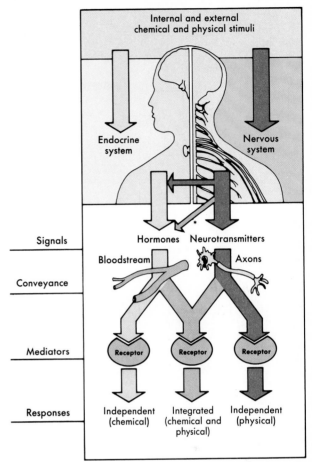

FIGURE 35-2   Overview of the relationship between the endocrine system and the nervous system. Similar stimuli may elicit activity of both systems. Hormones secreted by endocrine cells and conveyed via the bloodstream are analogous to neurotransmitters released by neurons after being conveyed by their axons. Responses are mediated by receptors in each system and may consist of either chemical or physical changes.

The principle of chemical homeostasis and the fundamental relationship of the endocrine system to the nervous system are well illustrated by the response of the organism to an abrupt lowering of the plasma concentration of glucose (*hypoglycemia*) (Figure 35-3). Because a supply of glucose is absolutely required to sustain brain function, hypoglycemia cannot be tolerated for long. Endocrine cells in the pancreas respond to hypoglycemia by secreting a hormone called *glucagon* that stimulates the release of stored glucose from the liver. Other endocrine cells in the pancreas respond in the opposite way to hypoglycemia by

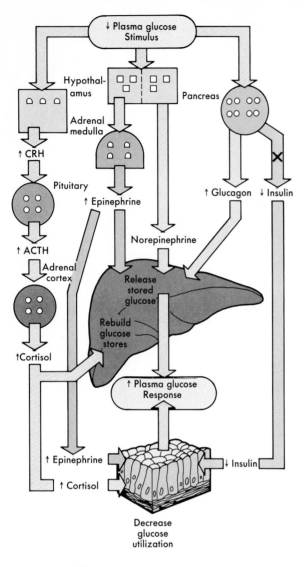

FIGURE 35-3   The integrated endocrine and neural response to hypoglycemia. The anterior pituitary gland, the adrenal cortex, the adrenal medulla, and the pancreatic islets participate in the main endocrine components of the response. The hypothalamus and the sympathetic nervous system participate in the neural components of the response. See text for details on how each component acts to increase the plasma glucose concentration. *CRH,* Corticotropin-releasing hormone; *ACTH,* adrenocorticotropic hormone.

diminishing the secretion of the hormone *insulin,* thereby reducing utilization of glucose by tissues other than the brain.

Certain neurons in the hypothalamus sense hypoglycemia and augment the release of stored glucose directly by transmitting sympathetic nervous system impulses to liver cells and indirectly by transmitting

sympathetic nervous system impulses to the adrenal medulla. This neuroendocrine gland secretes the hormone *epinephrine,* which acts on the liver to release stored glucose and on other tissues to reduce glucose utilization. Finally, other neurons in the hypothalamus also sense hypoglycemia and, via combined neurocrine and endocrine pathways, stimulate the adrenal cortex to secrete the hormone *cortisol.* This hormone augments synthesis of glucose in the liver to maintain the supply in case initial stores become depleted. Cortisol also inhibits the insulin-stimulated utilization of glucose by tissues other than the brain. Together these endocrine and neural responses to hypoglycemia raise plasma glucose levels back to normal within 60 to 90 minutes.

PATTERNS OF HORMONE SYNTHESIS, STORAGE, AND SECRETION

Hormones are synthesized, stored, and secreted in a variety of ways. *Peptide* and *protein hormones* are synthesized by a general process that characterizes the synthesis of all secreted proteins (Figure 35-4). The gene or deoxyribonucleic acid (DNA) molecule that directs hormone synthesis transcribes a messenger ribonucleic acid (mRNA) molecule. The latter traverses the nuclear membrane to the cytoplasm, where it translates its message on ribosomes by directing the assembly of the correct sequence of amino acids into a primary product. This is larger than the hormone itself and is called a *preprohormone.* At the N terminal a signal peptide directs transfer of the preprohormone from the ribosome into the endoplasmic reticulum. During this process the signal peptide is degraded, leaving a *prohormone.* This molecule contains the hormone as well as other peptide sequences.

The prohormone is transferred to the Golgi apparatus, where it undergoes further processing. Depending on the particular hormone, this may include cleavage, the addition of carbohydrate units, or the combination of separate subunits derived from different genes. In the Golgi apparatus the hormone and its peptide coproducts are packaged together within a *secretory granule.*

On stimulation of the endocrine cell the contents of secretory granules are released by the process of *exocytosis* into the extracellular fluid and then into adjacent bathing capillaries (Figure 35-5). By contraction

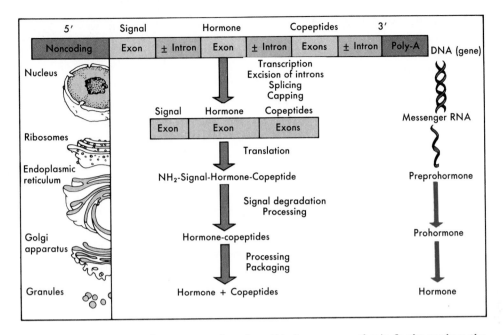

FIGURE 35-4   A schematic representation of peptide hormone synthesis. In the nucleus the primary gene transcript undergoes excision of introns (noncoding regions), splicing of the exons (coding regions), and capping. The resultant mature messenger RNA enters the cytoplasm, where it directs the synthesis of a precursor peptide sequence (preprohormone) on ribosomes. In this process the N-terminal signal is removed, and the resultant prohormone is transferred into the endoplasmic reticulum. The prohormone undergoes further processing and is then packaged into secretory granules in the Golgi apparatus. After final cleavage of the prohormone within the granules, the hormone and copeptides are ready for secretion by exocytosis.

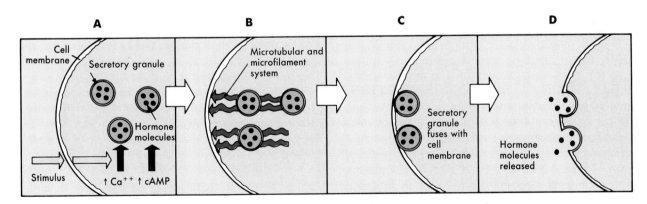

FIGURE 35-5   Secretion of peptide hormones via exocytosis. **A,** Secretion is initiated by application of a stimulus that raises cytosolic Ca^{++} and also usually raises intracellular cyclic adenosine monophosphate *(cAMP)* levels. **B,** The secretory granules are lined up and translocated to the plasma membrane via activation of a microtubular and microfilament system. **C,** The membrane of the secretory granule fuses with that of the cell. **D,** The common membrane is lysed, releasing the hormone into the interstitial space.

of microfilaments and guidance by microtubules, the secretory granules move to the plasma membrane of the cell and fuse with it. The mechanisms of granule release require an increase in intracytoplasmic calcium (Ca^{++}) concentration; the Ca^{++} is derived from both extracellular fluid and from intracellular stores within the endoplasmic reticulum and other organelles. Exocytosis is usually accompanied by increases in cyclic adenosine monophosphate (cAMP) concentration.

Catecholamine hormones (epinephrine, norepinephrine, dopamine) are synthesized from the amino acid tyrosine through a series of enzymatic reactions. However, they are are stored in secretory granules and secreted from the cell in a process similar to that for peptide hormones.

Thyroid hormones (thyroxine, triidothyronine) are synthesized from tyrosine and iodide in a series of reactions that occur with the amino acid already incorporated via peptide linkage into a large protein molecule. The hormones are then sequestered within the protein molecule in a storage space (follicle) shared by a group of surrounding endocrine cells. Secretion of thyroid hormone requires retrieval from the follicle and enzymatic release from its protein storage form.

Steroid hormones (cortisol, aldosterone, androgens, estrogens, progestins) are synthesized from cholesterol by a series of enzymatic reactions. However, they are not stored to any appreciable degree within the gland of origin. Thus, to increase the secretion of a steroid hormone, the entire biosynthetic sequence from cholesterol must be activated. In effect, the storage form of all steroid hormones is the intracellular depot of cholesterol.

REGULATION OF HORMONE SECRETION

The secretion of hormones is related to their roles in maintaining homeostasis. Therefore the dominant mechanism of regulation is that of *negative feedback* (Figure 35-6). If hormone A acts to raise the plasma concentration of substrate B, a decrease in substrate B will stimulate hormone A secretion, whereas an increase in substrate B will suppress hormone A secretion. In essence, physiologic conditions that require the action of a hormone also stimulate its release; conditions or products resulting from prior hormone action suppress hormone release. This

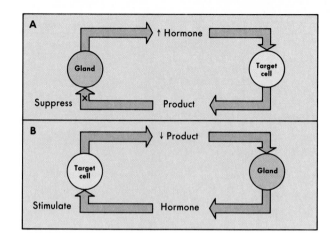

FIGURE 35-6   Negative feedback principle. **A,** If an increase in hormone secretion stimulates a greater output of product from the target cell, the product feeds back on the gland to suppress further hormone secretion. In this way hormone excess is limited or prevented. **B,** If a decrease in output of product from the target cell stimulates hormone secretion, the hormone in turn will stimulate output of product by the target cell. In this way product deficiency is limited or corrected.

homeostatic partnership may exist between a hormone and one or more substrates, minerals, other hormones, or even physical factors such as fluid volume.

Occasionally, *positive feedback* is observed. In such circumstances a product of hormone action stimulates further hormone secretion initially. When the product eventually reaches appropriate concentrations, it may then exert negative feedback on hormone secretion. This mechanism of regulation is seen when a biologic process begins at a very low level, yet must reach rather high levels in the course of normal physiologic function.

Feedback regulation, whether negative or positive, may be exerted at all levels of endocrine cell function, that is, at transcription of the hormone gene, at translation of the gene message, and at release of stored hormone.

Engrafted on homeostatic feedback are patterns of hormone release dictated by diurnal (daily) or ultradian (within a day) rhythms, stages of sleep, seasonal variation, and stages of development (fetal, neonatal, pubertal, senescent). In addition, pain, emotion, fright, injury, physical or mental stress, and sexual arousal may evoke or shut off the release of hormones via complex neural pathways.

HORMONE TURNOVER

After secretion into the blood, different hormones enter pools of varying mass size, volume of distribution, and rate of turnover. Catecholamine, peptide, and protein hormones generally circulate unbound to other plasma constituents. In contrast, thyroid and steroid hormones circulate largely bound to specific globulins as well as to albumin. The extent of protein binding greatly influences the rates at which hormones exit from plasma into interstitial fluid, where they interact with target cells. The plasma half-life of a hormone (the time required for a 50% decrease in concentration) is positively correlated with the percentage of protein binding. Larger protein hormones with carbohydrate components have longer half-lives than do smaller protein and peptide hormones. Exiting from plasma does not have to be entirely irreversible; hormone molecules may return to plasma from other compartments via lymphatic channels, sometimes after dissociation from target cells.

Irreversible removal of hormone from the body results from target cell uptake, metabolic degradation, and urinary or biliary excretion. The sum of all removal processes is expressed in the concept of *metabolic clearance rate* (MCR). This is defined as the volume of plasma cleared of hormone per unit of time. In a steady state this equals the mass of hormone removed per unit of time divided by its plasma concentration:

$$MCR = \frac{mg \cdot min^{-1} \text{ (removed)}}{mg \cdot ml^{-1} \text{ (plasma)}} = \frac{\frac{ml \text{ plasma (cleared)}}{min}}{} \qquad (1)$$

MCR is an expression of the efficiency with which a hormone is removed from plasma, irrespective of the mechanism (just as renal clearance rate is an expression of the efficiency with which a substance is specifically removed from plasma by urinary excretion). MCR is most conveniently determined by infusing a hormone at a constant rate until a new steady-state plasma concentration is reached. At this point the rate of removal from plasma must equal the rate of infusion. Therefore,

$$MCR = \frac{\text{Hormone infused (milligrams/minute)}}{\text{Hormone concentration (milligrams/milliliter)}}$$

$$= \frac{\text{Milliliters}}{\text{Minute}} \qquad (2)$$

The kidney and liver are the major sites of the metabolic degradation of hormones. Renal clearance of a hormone is extremely low if it is bound to specific plasma globulins (e.g., thyroid hormones). Although peptide and smaller protein hormones are filtered to some degree by the renal glomeruli, they usually undergo tubular reabsorption and subsequent degradation within the kidney so that only a minute amount appears in the urine.

Metabolic degradation of hormones occurs by enzymatic processes that include proteolysis, oxidation, reduction, hydroxylation, decarboxylation, and methylation. In addition, hormones or their metabolites may be conjugated to glucuronic acid and sulfate and the conjugates subsequently excreted in the bile or urine. Some hormonal degradation also occurs during interactions with specific target cells.

Quantitation of Hormone Secretion

Under research conditions, it is possible to measure the total amount of hormone secreted into the blood stream per unit time, using isotopic techniques. For clinical purposes, plasma or urine measurements must usually suffice. However, they are valid indices of hormone production if certain conditions exist. In a steady state, the amount of hormone entering the plasma is equal to the amount of hormone exiting the plasma.

$$\text{Secretion rate} = \text{Disposal rate} \qquad (1)$$

$$\text{Disposal Rate } (mg \cdot ml^{-1}) =$$
$$\text{Metabolic Clearance Rate (MCR) } (ml \cdot min^{-1}) \times$$
$$\text{Plasma Concentration (P) } (mg \cdot ml^{-1}) \qquad (2)$$

$$S = MCR \times P \qquad (3)$$

If the MCR is within normal limits and *can be taken as a constant*, then S is proportional to P.

This is the theoretical basis for employing plasma hormone measurement alone as an index of secretory rate. However, the secretion of many hormones is characterized by diurnal variation and episodic spurts. In such instances, multiple plasma samples may be necessary for a valid estimate.

Similarly, hormone secretion can sometimes be assessed by measuring its *urinary excretion* in accurately timed collections. This offers the advantage of, in effect, averaging out plasma fluctuations over the collection interval. The urinary excretion is a valid index of secretion rate when kidney function and kidney handling of the hormone are normal.

HORMONE ACTION

Three major sequential steps are involved in eliciting target cell responses (Figure 35-7):

1. The hormone must be recognized.
2. An intracellular signal must then be generated.
3. One or more intracellular processes must be increased or decreased (e.g., enzyme reactions, ion movements, cytoskeletal rearrangements).

Hormone Recognition: Receptors

Recognition takes place by binding of the hormone to a specific receptor located within the plasma membrane, the cytoplasm, the nucleus, or possibly other intracellular organelles of the target cell. The receptor has a complementary binding site with high affinity for the hormone. The two molecules associate in reversible fashion to form a hormone-receptor complex. The presence of this receptor confers *specificity* to the interaction of a hormone with its target cell. Only cells that have the hormone's receptor can respond to the hormone; only hormones for which the cell possesses receptors can affect the cell.

Receptors are large protein molecules that may also contain carbohydrate units. The genes for a number of receptor molecules have been cloned and the structure of both the gene and the primary gene product determined. Those receptors incorporated into plasma membranes may resemble immunoglobulins in structure, and significant analogies exist between hormone-receptor and antigen-antibody reactions.

The reaction between hormone and receptor not only imparts specificity but is also the initial determinant of the rate of hormone action. This reaction can

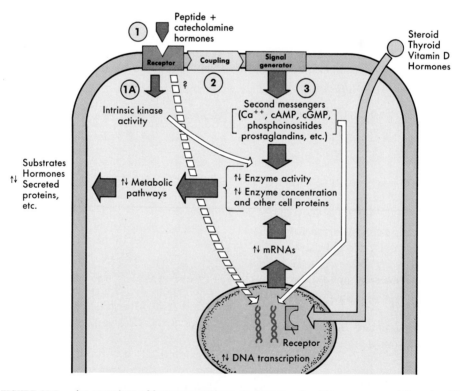

FIGURE 35-7   An overview of hormone actions on target cells. Hormones may interact with either plasma membrane or intracellular receptors. Hormones may generate second messengers within the plasma membrane, the cytoplasm, or the nucleus. Metabolic pathways may be regulated by altering the activities or the concentrations of enzymes.

be expressed in classical chemical terms.

$$[H] + [R] \rightleftharpoons [HR] \qquad (4)$$

$$K = \frac{[HR]}{[H][R]} \qquad (5)$$

$$[HR] = K\,[H][R] \qquad (6)$$

Where [H] equals the free hormone concentration; [R] equals the free or unoccupied receptor concentration; [HR] equals the hormone receptor complex or bound hormone concentration; and K equals the affinity (association) constant.

The amount of receptor occupied by hormone, [HR], is the critical component that governs the magnitude of hormone action at the receptor step. As can be seen from equation 6, [HR] is increased when the receptor has a high affinity for the hormone, when the cell is exposed to higher hormone concentrations, and when the receptor number is high.

In some cases [HR] appears to be the rate limiting step in the whole sequence of hormone action; therefore the maximum biologic response to these hormones is directly proportional to the number of receptors. In other cases, [HR] is not usually rate limiting and maximum biologic responses can be seen when only a small proportion of the available receptors is occupied by hormone. Peptide hormones appear to have such "spare receptors."

Receptor molecules are continually synthesized, translocated to sites of association with hormone molecules, and degraded. These processes can be influenced by their respective hormone partners. Some hormones decrease or "down regulate" the number of their own receptors, thereby limiting their action on the cell. Other hormones recruit their own receptors, thereby amplifying hormone action on the cell.

Signal Generation

Signal generation is the next step in hormone action. When hormone receptor association occurs within the plasma membrane of the cell, the resultant complex is coupled to other plasma membrane components. These act to generate within the cytoplasm a variety of signal molecules, or "second messengers," which then influence metabolic processes within the cell. In this situation the essential information for triggering the cell's response actually resides in the receptor molecule; this information is transmitted to the cytoplasm when the hormone occupies and changes the conformation of the membrane receptor. The hormone is essentially an extracellular signal.

In contrast, when hormone receptor association occurs within the cytoplasm or nucleus, the hormone-receptor complex ultimately interacts with specific DNA molecules and alters gene expression. Here the second messengers are transcribed RNA molecules that direct the synthesis of protein molecules. In this situation essential information for triggering the cell's response resides in the hormone molecule itself as well as in the receptor molecule. The hormone is an intracellular signal.

Membrane-Generated Second Messengers
The major membrane-generated second messengers include the following:

1. *Cyclic adenosine monophosphate* (cAMP). Several peptide hormone receptors are coupled to the plasma membrane enzyme *adenylate cyclase*. This enzyme complex consists of a *stimulating* regulatory subunit, a *catalytic* subunit, and an *inhibiting* regulatory subunit (Figure 35-8). The formation of a hormone-receptor complex causes the stimulating subunit to activate the catalytic subunit. This increases the rate of formation of cAMP from magnesium–adenosine triphosphate (Mg-ATP). Other hormone receptor complexes interact with the inhibitory subunit, which deactivates the catalytic subunit. This decreases the rate of formation of cAMP from ATP. Cytoplasmic cAMP, the second messenger, activates the enzyme protein kinase A in the cytoplasm. Protein kinase A then phosphorylates various enzyme kinases, ultimately either increasing or decreasing the activity of numerous enzymes in various metabolic pathways. The final result of changing cAMP levels is a cascade of reactions that increases, decreases, or alters the direction of substrate flux within the cell and then often within the whole body.

2. *Calcium-calmodulin*. Plasma membranes contain channels that allow graded influx of extracellular Ca^{++} into the cytoplasm. Hormone occupancy of plasma membrane receptors can somehow open these calcium channels and can also mobilize intracellular Ca^{++} bound to endoplasmic reticulum (Figure 35-9). As free cytoplasmic Ca^{++} concentration rises, it is associated with its binding protein, *calmodulin*. The calcium-calmodulin complex is a powerful regulator of many enzymes' activity and thus is a second messenger that can transduce the extracellular hormone signal into a change in substrate flow.

FIGURE 35-8   For legend see opposite page.

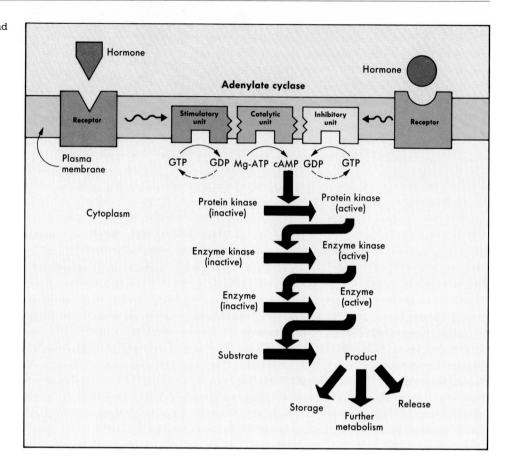

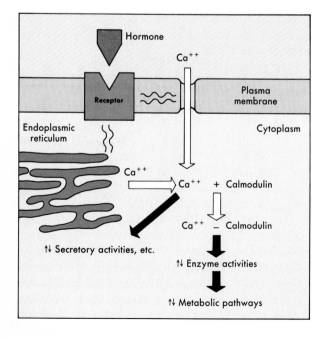

FIGURE 35-9   Mechanism of hormone action via calcium as a second messenger. Binding of hormone to receptor leads to opening of calcium channels in the cell membrane. Influx of Ca^{++} is followed by its binding to calmodulin, which produces a complex that can activate or deactivate various enzymes and metabolic pathways. The intracellular pool of Ca^{++} can also be increased by release of Ca^{++} from the endoplasmic reticulum.

FIGURE 35-8   Mechanism of hormone action via cyclic adenosine monophosphate *(cAMP)* as a second messenger. The hormone-receptor complex interacts with a regulatory unit (either stimulatory or inhibitory) of adenylate cyclase. Guanosine triphosphate *(GTP)* is bound, and the catalytic unit then catalyzes the synthesis of cAMP from adenosine triphosphate *(ATP)*. The cAMP activates protein kinase A, which then phosphorylates various enzyme kinases. These kinases activate enzymes, and ultimately this cascade of effects leads to increases in the intracellular levels of various substances. If the target enzymes of the hormone are inactivated by phosphorylation, inhibition of metabolic pathways rather than stimulation will result. *GDP,* Guanosine diphosphate.

3. *Phospholipid products.* Ten percent of the lipid portion of the plasma membrane consists of *phosphatidylinositols.* Formation of a hormone-receptor complex activates a membrane-bound phospholipase that splits phosphatidylinositol 4,5-biphosphate into two products: a *diacylglycerol,* often containing arachidonic acid as a major fatty acid component, and *inositol triphosphate* (Figure 35-10). These two second messengers, together with Ca^{++}, activate the enzyme *protein kinase C.* This kinase in turn phosphorylates other enzymes in the cytoplasm and thereby alters flow in the metabolic pathways that they catalyze.

4. *Receptor kinases.* In some instances the plasma membrane receptor itself acts as a second messenger. After binding to hormone, some receptors undergo autophosphorylation of specific tyrosine residues. The altered receptor molecule now possesses kinase activity toward other proteins and may itself initiate a cascade of enzyme phosphorylations that regulate metabolic processes.

5. None of the membrane generated second messengers is unique to any particular hormone and a single hormone may operate through multiple messengers. Furthermore, one hormone-receptor unit can generate many molecules of a second messenger; hence these systems *amplify* the original hormonal signal.

Nuclear Second Messengers   Hormones that freely enter the cell (steroids, vitamin D, thyroid hormones) combine with receptors that are large protein molecules located in the cytoplasm and nucleus (Figure 35-11). This initial hormone-receptor complex requires an activation or transformation process in the nucleus. After transformation, the complex can associate with specific DNA molecules.

One specific domain of the receptor molecule binds the hormone. Another domain in the middle portion of the receptor binds to DNA. For all the hor-

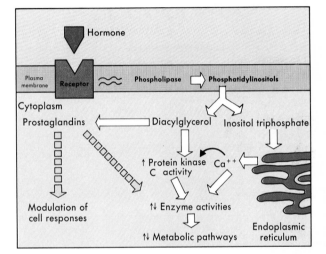

FIGURE 35-10   Mechanism of hormone action via membrane phospholipids. Binding of hormone to receptor causes membrane phospholipase to split phosphatidylinositol 4,5-biphosphate into diacylglycerol and inositol triphosphate components. Diacylglycerol activates protein kinase C by enhancing its binding to Ca^{++}. Ca^{++} is mobilized from intracellular stores by inositol triphosphate. Protein kinase C then activates or deactivates target enzymes via phosphorylation. As a subsidiary effect, the diacylglycerol yields arachidonic acid for synthesis of prostaglandins, which also modulate intracellular processes.

mones just listed, this second domain exhibits considerable amino acid homologies among the receptors, and it is similar to oncogene products. The DNA site with which the hormone-receptor complex interacts is termed the *regulatory site;* it is usually upstream from the basal promotor site at the 5' end of the gene. After the hormone-receptor complex has been bound, transcription of the primary gene message by RNA polymerase is either induced or repressed. Thus, by raising or lowering the levels of specific RNA molecules, the hormone increases or

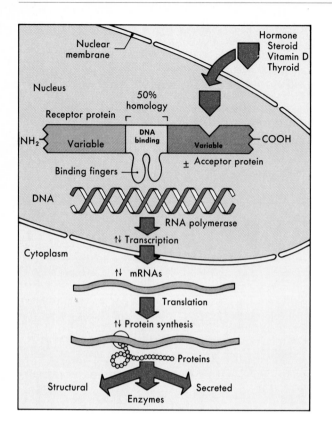

FIGURE 35-11   Mechanism of action of vitamin D, steroid, and thyroid hormones. The hormone combines with a nuclear protein receptor. The carboxy-terminal portion of the receptor varies for each hormone. The midportion of the receptor molecule has considerable similarity among hormones. This midportion contains DNA-binding fingers. Binding of the hormone-receptor complex to target DNA molecules either stimulates or suppresses gene transcription. The result is increased or decreased synthesis of cell proteins.

decreases the concentration of specific cell proteins. When the latter are enzymes, the rates of specific metabolic reactions are likewise increased or decreased by the hormone.

The onset of hormone actions mediated by nuclear second messengers is generally slower than that mediated by cytoplasmic second messengers. There is also a graded response to increasing hormone concentrations, rather than amplification, since each hormone-receptor unit engages a single DNA molecule. The magnitude of action may be influenced by the concentrations of RNA polymerase, of the enzymes of protein synthesis or processing, of transfer RNA, of amino acids, or of the number and activity of ribosomes. Gene expression can also be regulated by hormones whose receptors are mainly located in the plasma membrane. To explain this, either membrane-generated second messengers, internalized hormone-receptor complexes, or internalized free hormone molecules must also reach and modulate the regulatory sites of target DNA molecules.

Outcome of Hormone Action

In quantitative terms the final outcome of the interaction of a hormone with its target cells depends on several factors. These include hormone concentration, receptor number, duration of exposure to hormone, intervals between consecutive exposures, intracellular conditions such as concentrations of rate-limiting enzymes, cofactors or substrates, and the concurrent effects of antagonistic or synergistic hormones. Hormonal effects are not "all-or-none" phenomena.

The dose-response curve for the action of a hormone is generally complex and often exhibits a sigmoidal shape (Figure 35-12). An intrinsic basal level of cell activity may be observed independent of added hormone and long after any previous exposure. A certain minimal *threshold* concentration of hormone then is required to elicit a measurable response. The effect that is obtained at saturating doses of hormone is defined as the *maximal responsiveness* of the target cells. The concentration of hormone required to

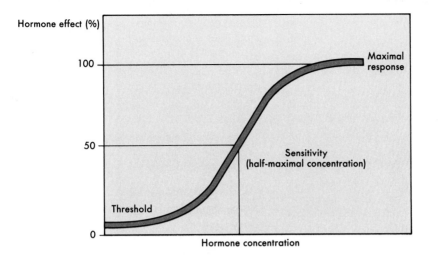

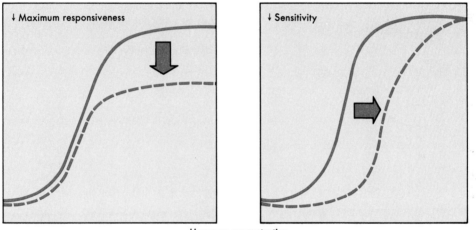

FIGURE 35-12   The general shape of a hormone dose-response curve. Alterations in this curve can take the form of a change in maximal responsiveness *(left lower panel)* or a change in sensitivity *(right lower panel).*

elicit a half-maximal response is an index of the *sensitivity* of the target cells.

Alterations in the dose-response curve in vivo can take two general forms (Figure 35-12):

1. A decrease in maximal responsiveness may be caused (a) by a decrease in the number of functional target cells, in the total number of receptors per cell, or in the concentration of an enzyme activated by the hormone; or (b) by an increase in the concentration of a noncompetitive inhibitor.

2. A decrease in hormone sensitivity may be caused (a) by a decrease in hormone receptors, (b) by an increase in the rate of hormone degradation, or (c) by an increase in the concentration of antagonistic hormones.

BIBLIOGRAPHY
Journal Articles

Alford FP et al: Temporal patterns of circulating hormones as assessed by continuous blood sampling, J Clin Endocrinol Metab 36:108, 1973.

Catt K and Dufau M: Introduction: the clinical significance of peptide hormone receptors, J Clin Endocrinol Metab 12:xi, 1983.

Chambon P et al: Promoter elements of genes coding for proteins and modulation of transcription by estrogens and progesterone, Recent Prog Horm Res 40:1, 1984.

Gordon P et al: Internalization of polypeptide hormones: mechanism, intracellular localization and significance, Diabetologia 18:263, 1980.

Lacy PE: Beta cell secretion—from the standpoint of a pathobiologist, Diabetes 19:895, 1970.

Lefkowitz R et al: Mechanisms of membrane-receptor regulation: biochemical, physiological, and clinical insights derived from studies of the adrenergic receptors, N Engl J Med 310:1570, 1984.

Spelsberg TC et al: Role of specific chromosomal proteins and DNA sequences in the nuclear binding sites for steroid receptors, Recent Prog Horm Res 39:463, 1983.

Tait JF: The use of isotopic steroids for the measurement of production rates in vivo, J Clin Endocrinol 23:1285, 1963.

Walters MR: Steroid hormone receptors and the nucleus, Endocr Rev 6:512, 1985.

Zor U: Role of cytoskeletal organization in the regulation of adenylate cyclase–cyclic adenosine monophosphate by hormones, Endocr Rev 4:1, 1984.

Books and Monographs

Exton JH and Blackmore PF: Calcium-mediated hormonal responses. In DeGroot LJ, editor: Endocrinology, ed 2, Philadelphia, 1989, WB Saunders Co.

Habener JF: Genetic control of hormone formation. In Wilson JD and Foster DF, editors: Textbook of endocrinology, ed 7, Philadelphia, 1985, WB Saunders Co.

Roth J and Grunfeld C: Mechanism of action of peptide hormones and catecholamines. In Wilson JD and Foster DF, editors: Textbook of endocrinology, ed 7, Philadelphia, 1985, WB Saunders Co.

CHAPTER 36

Whole Body Metabolism

Metabolism may be broadly defined as the sum of all the chemical (and physical) processes involved (1) in producing energy from exogenous and endogenous sources, (2) in synthesizing and degrading structural and functional tissue components, and (3) in disposing of resultant waste products. Regulating the rate and direction of many basic aspects of metabolism is one of the major functions of the endocrine system. Therefore a firm grasp of the fundamentals of metabolism is essential if one is to understand the important influence of hormones on body functions.

ENERGY METABOLISM

Balance

The laws of thermodynamics require that energy balance be constantly maintained in living organisms. However, energy may be obtained in various forms, may be stored in other forms, and may be expended in many different ways. Therefore numerous interconversions of chemical, mechanical, and thermal energy are possible within the basic rule that

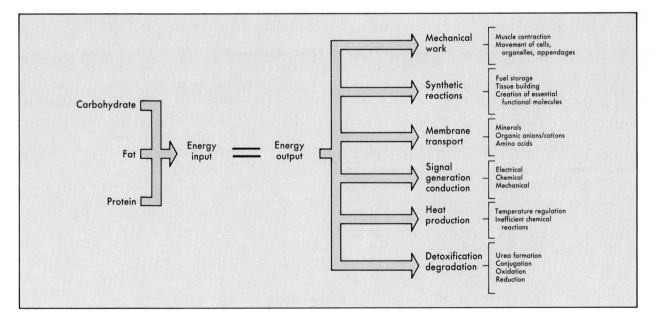

FIGURE 36-1 Overview of energy balance. In a steady state, input as caloric equivalents of food equals output as caloric equivalents of various forms of mechanical and chemical work and heat.

in the steady state, energy input must always equal energy output. Figure 36-1 illustrates this overall flow of energy through the human organism.

Energy Input

Energy input consists of foodstuffs, which are classified into three major chemical categories: carbohydrate, fat, and protein. The complete combustion of each chemical type yields characteristic amounts of energy, expressed as joules or kilocalories per gram (1 kcal = 4184 joules). Combustion also requires characteristic amounts of oxygen, depending on the proportions of carbon, hydrogen, and oxygen in the substance. However, for each class of foodstuff, the energy yield per liter of oxygen used is quite similar because the ratio of carbon to hydrogen atoms is also similar in each class. Within the body the carbon skeletons of carbohydrate and protein can be converted to fat, and their potential energy can be stored more efficiently in that manner. The carbon skeletons of protein can be converted to carbohydrate when that energy source is specifically needed. However, there is no significant conversion of carbon atoms from fat to carbohydrate.

Energy Output

Energy output can be divided into several distinct and measurable components.

1. At rest, energy is expended in a myriad of synthetic and degradative chemical reactions; in generating and maintaining gradients of ions and other molecules across cell and organelle membranes; in the creation and conduction of signals, particularly in the nervous system; in the mechanical work of respiration and circulation of the blood; and in obligate heat loss to the environment. This absolute minimal energy expenditure is called the *basal* or *resting metabolic rate* (BMR or RMR). In the adult human BMR amounts to an average daily expenditure of 20 to 25 kcal (84 to 105 kjoules)/kg body weight (or 1.0 to 1.2 kcal/minute) and requires the utilization of approximately 200 to 250 ml oxygen/minute.

The BMR is linearly related to lean body mass and body surface area. It declines in the elderly, partly because lean body mass declines with age. BMR is increased by raising environmental temperature. During sleep BMR falls 10% to 15%. Studies in identical twins and families suggest that some of the variation in BMR is genetically determined.

2. Ingestion of food causes a small obligate increase in energy expenditure referred to as *diet-induced thermogenesis*. This is explained by the increased rate of reactions involved in the disposition of the ingested calories, such as storage of glucose in the large molecule, glycogen.

3. *Facultative thermogenesis*, or nonshivering thermogenesis, comes into play in such circumstances as exposure to cold, when energy may be expended specifically to produce heat.

4. Energy is also expended by sedentary individuals in spontaneous physical activity such as "fidgeting," at least some of which is unconscious and seemingly purposeless.

5. The additional energy expended in occupational labor and purposeful exercise varies greatly among individuals as well as from day to day and from season to season. This component generates the greatest need for variation in daily caloric intake and underscores the importance of energy stores to buffer temporary discrepancies between energy output and intake.

Of a total average daily expenditure of 2300 kcal (9700 kjoules) in a sedentary human, basal metabolism accounts for 75%, dietary thermogenesis for 7%, and spontaneous physical activity for 18%. Up to an additional 3000 kcal may be used in daily physical work. During short periods of occupational or recreational exercise, energy expenditure can increase more than tenfold over basal levels.

ENERGY GENERATION

The basic chemical currency of energy in all living cells are the two high-energy phosphate bonds contained in adenosine triphosphate (ATP). To a much lesser extent, other purine and pyrimidine nucleotides (guanosine triphosphate, cytosine triphosphate, uridine triphosphate, inosine triphosphate) also serve as energy sources after the energy from ATP is transferred to them. In muscle, creatine phosphate is a high-energy molecule of particular importance.

The two terminal P-O bonds of ATP each contain about 12 kcal of potential energy per mole under physiologic conditions. These bonds are in constant flux. They are generated by oxidative reactions and are consumed as the energy is either (1) transferred into other high-energy bonds involved in synthetic reactions (e.g., amino acid + ATP → amino acyl

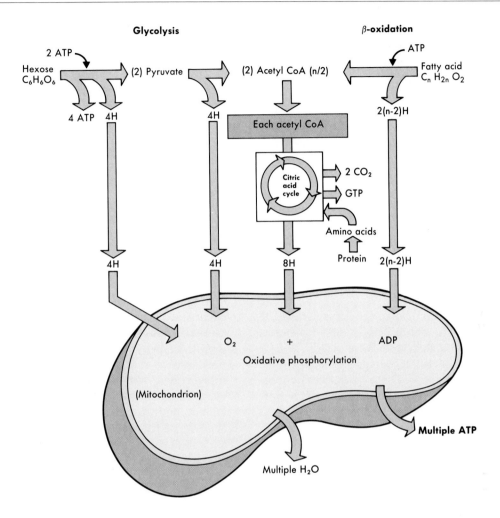

FIGURE 36-2   An overview of energy production. Glycolysis and each turn of the citric acid cycle supply only 2 ATP equivalents apiece. ATP is generated mainly when hydrogens removed from carbohydrate, fat, or protein substrates are oxidized in the mitochondria. *ATP,* Adenosine triphosphate; *ADP,* adenosine diphosphate; *GTP,* guanosine triphosphate.

AMP), (2) expended in creating lower-energy phosphorylated metabolic intermediates (e.g., glucose + ATP → glucose-6-phosphate), or (3) converted to mechanical work (e.g., propulsion of spermatozoa). Because the production and transfer of energy are not 100% efficient, about 18 kcal of substrate are required to generate each terminal P-O bond of ATP. In a normal day when 2300 kcal are turned over, about 128 moles, or 63 kg, of ATP (a mass approximating body weight) are generated and expended. An overview of energy production with generation of ATP from the major substrates is shown in Figure 36-2.

The combustion of *carbohydrates,* chiefly glucose with lesser amounts of fructose and galactose, includes two major phases:

1. At the end of an anaerobic phase known as *glycolysis* (Embden-Meyerhof pathway), each glucose molecule has yielded two molecules of pyruvate but only 8% of its energy content. Glycolysis can serve as a sole source of energy only briefly because (a) the supply of glucose is limited, and (b) the accumulated pyruvate must be syphoned off by reduction to lactate, a metabolite that is ultimately noxious.

2. During an aerobic phase the two pyruvate molecules are decomposed to CO_2 via the citric acid cycle (Krebs cycle), and the remaining energy is liberated. In this pathway acetyl coenzyme A (acetyl CoA), initially formed by oxidative decarboxylation of pyruvate, is condensed

with *oxaloacetate* to form citrate. Through a cyclic series of reactions the carbons of acetyl CoA appear as CO_2, and oxaloacetate is regenerated.

The combustion of *fatty acids,* the major energy component of fats, proceeds through a repetitive biochemical sequence known as β-*oxidation.* This process releases two carbons at a time as acetyl CoA until the entire fatty acid molecule is broken down. The resultant acetyl CoA is disposed of via the citric acid cycle, as already described. A variable portion of fatty acid oxidation in the liver stops at the last four carbons and yields acetoacetic and β-hydroxybutyric acids. These water-soluble ketoacids are released by the liver to be oxidized in other tissues as additional energy substrates.

The combustion of *protein* first requires hydrolysis to its component amino acids. Each of these undergoes degradation by individual pathways, which ultimately lead to intermediate compounds of the citric acid cycle and then to acetyl CoA and CO_2.

The combustion of all foodstuffs yields large numbers of hydrogen atoms. These hydrogens are oxidized to H_2O in the mitochondrion in linkage with phosphorylation of ADP to ATP (Figure 36-2). In this process, 3 high energy P-O bonds are formed for each atom of oxygen used. This yields an overall efficiency of 60% to 65% for the recovery of usable chemical energy.

Respiratory Quotient

In the process of oxidizing substrates to meet basal energy needs, the proportion of carbon dioxide produced ($\dot{V}CO_2$) to oxygen used ($\dot{V}O_2$) varies according to the fuel mix. The ratio of $\dot{V}CO_2$ to $\dot{V}O_2$ is known as the *respiratory quotient* (RQ). As indicated by the following equations, RQ equals 1.0 for oxidation of carbohydrate (e.g., glucose), whereas RQ equals 0.70 for oxidation of fat (e.g., palmitic acid).

For carbohydrates:

$$C_6H_{12}O_6 + 6\ O_2 \rightarrow 6\ CO_2 + 6\ H_2O$$
Glucose

$$RQ = \frac{6\ CO_2}{6\ O_2} = 1.0 \tag{1}$$

For fats:

$$C_{15}H_{31}COOH + 23\ O_2 \rightarrow 16\ CO_2 + 16\ H_2O$$
Palmitic acid

$$RQ = \frac{16\ CO_2}{23\ O_2} = 0.70 \tag{2}$$

The RQ for protein reflects that of the individual RQs of the amino acids and averages 0.80. Ordinarily, protein is a minor energy source. The small contribution of protein oxidation to the overall RQ can be corrected for by measuring the urinary excretion of the nitrogen that results from the metabolism of amino acids.

ENERGY STORAGE AND TRANSFERS

The intake of energy in the form of food is periodic. Its time course does not match either the constant rate of energy expenditure in the basal state or that expended during intermittent muscle work. Therefore the organism must have mechanisms for storing ingested energy for future use. The greatest part of these energy reserves (75%) is in the form of fat as triglycerides, stored in adipose tissue. In normal-weight humans fat constitutes 10% to 30% of body weight, but it can reach 80% in very obese individuals. Fat is a particularly efficient storage fuel because of its high caloric density (i.e., 9 kcal/g) and because it engenders little additional weight as intracellular water. Fat stores can supply energy needs for up to 2 months in totally fasted individuals of normal weight. Triglycerides are formed by esterification of *free fatty acids,* largely derived from the diet, with α-glycerol phosphate. However, free fatty acids can also be synthesized from acetyl CoA derived from oxidation of glucose; thus carbohydrate can be converted to fat in liver and adipose tissue and its energy stored in that more efficient form (Figure 36-3).

Protein (4 kcal/g) constitutes almost 25% of the potential energy reserves, and the component amino acids can contribute to the glucose supply. However, virtually all proteins serve some vital structural and functional role. Therefore their use as a major source of energy is deleterious and only arises as a last resort before death from fasting.

Carbohydrate (4 kcal/g) in the form of a glucose polymer, glycogen, forms less than 1% of total energy reserves. However, this portion is critical for support of central nervous system metabolism and for short bursts of intense muscle work. Approximately one fourth of the glycogen stores (75 to 100 g) is in the

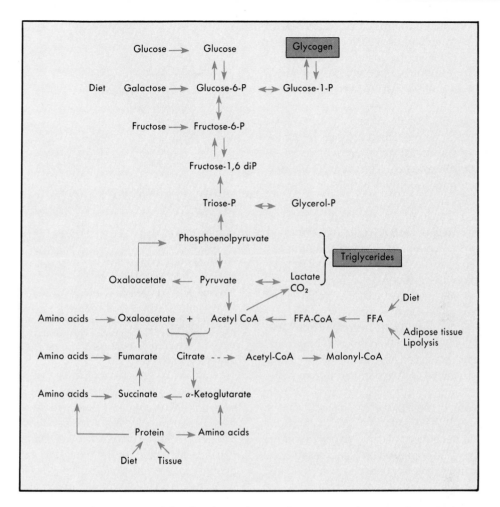

FIGURE 36-3   An overview of the chemical pathways of energy transfer and storage. Carbohydrates funnel through glucose-6-phosphate to be stored as glycogen. Alternatively, they can undergo glycolysis to pyruvate and be used for synthesis of fatty acids. The latter, regardless of source, are esterified with glycerol-phosphate and stored as triglycerides. Amino acids from endogenous or exogenous sources are converted to glucose via oxaloacetate or pyruvate.

liver, and about three fourths (300 to 400 g) is in the muscle mass. Liver glycogen can be made available to other tissues by the process of *glycogenolysis* and glucose release. Muscle glycogen can only be used by muscle, since this tissue lacks the enzyme glucose-6-phosphatase, which is required for release of glucose into the bloodstream.

Glycogen can be formed from all three major dietary sugars. In addition, in the liver (and to a much lesser extent in the kidney) glucose itself can also be synthesized de novo from the three carbon precursors pyruvate, lactate, and glycerol and from parts of the carbon skeleton of all 20 amino acids in protein except leucine. This process, known as *gluconeogenesis,* converts two pyruvate molecules to glucose, but it is not a simple reversal of all the reactions of glycolysis (Figure 36-3). The chemical free energy change is too large to permit efficient backward flow of the glycolytic reactions at three steps: (1) pyruvate to phospho*enol*pyruvate, (2) fructose-1,6-diphosphate to fructose-6-phosphate, and (3) glucose-6-phosphate to glucose. Substitution of simple phosphatase reactions reverses the last two steps. However, the first step requires energy input in the form of ATP and guanosine triphosphate (GTP). It is also important to realize that **net glucose synthesis cannot occur**

from acetyl CoA, even though carbon atoms from acetyl CoA can become part of oxaloacetate and then part of glucose molecules via the citric acid cycle. Thus fat can only contribute to carbohydrate stores by way of the 3 carbon glycerol moiety of triglycerides, a minor source.

The processes of energy storage and transfer themselves expend energy. This partly accounts for the stimulation of oxygen utilization after a meal (i.e., diet-induced thermogenesis). The cost of storing dietary fatty acids as triglycerides in adipose tissue is only 3% of the original calories, and the cost of storing glucose as glycogen is only 7% of the original calories. In contrast, conversion of carbohydrate to fat uses up 23% of the original calories, and a similar amount is expended in storing dietary amino acids as protein or in converting them to glycogen.

Because glucose and fatty acids are alternative and in effect competing energy substrates, some relationship between their utilization and their synthesis and storage within cells could be expected. For example, when dietary glucose is plentiful, glycolysis is augmented, more acetyl CoA is generated from pyruvate, and more citrate is formed (Figure 36-3). Citrate is a potent activator of the first step in the synthesis of fatty acids (acetyl CoA → malonyl CoA). In addition, glycolysis will produce more glycerol phosphate from triose phosphates. The combination of increased fatty acid synthesis and glycerol phosphate availability results in accentuated synthesis of triglycerides and reduced oxidation of fat. Thus increased carbohydrate utilization shifts fat metabolism from oxidation to storage. Conversely, under circumstances where fatty acid supply is augmented, β-oxidation increases. Several of its products then retard glycolysis and enhance gluconeogenesis. Thus glucose and glycogen synthesis are increased, and increased use of fat as fuel shifts carbohydrate metabolism from oxidation to storage. Many of these intrinsic chemical checks and balances are also reinforced by hormonal signals.

In addition to these intracellular relationships, transfer of energy between organs is another important metabolic process (Figure 36-4). The stored energy contained within adipose tissue triglycerides is transported as free fatty acids to the liver. There part of the **energy** (not the carbon atoms) is effectively transferred to glucose molecules, because as fatty acids are oxidized, gluconeogenesis is stimulated concurrently, as previously described. The newly

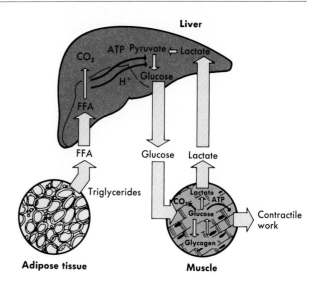

FIGURE 36-4   Interorgan energy transfers. Energy contained in free fatty acids *(FFA)* can be transferred to energy contained in glucose in the liver. Lactate released from muscle glycogen by glycolysis can carry energy back to the liver, where the lactate is built back into new glucose molecules (and glycogen).

synthesized glucose molecules in turn can then be transported to muscle tissue, where their energy is released during glycolysis and applied to muscle contraction. Furthermore, if the lactate produced exceeds the ability of the muscle to oxidize it rapidly enough in the citric acid cycle, the lactate can be returned to the liver, where it may again be built back up into glucose molecules. From this viewpoint the liver is a flexible and versatile organ that can transmute and transfer energy from fuel depots to working tissues.

CARBOHYDRATE METABOLISM

Dietary carbohydrates give rise to various sugars (hexoses), the most important of which are glucose, fructose, and galactose. In addition to serving as energy sources, sugars are also components of glycoproteins, glycopeptides, and glycolipids that have structural and functional roles. These roles include basement membrane collagen, mucopolysaccharides, nerve cell myelin, hormones, and hormone receptors.

Glucose is the central molecule in carbohydrate metabolism. Other sugars are metabolized through the glucose pathways. Postabsorptive plasma glucose concentration averages 80 mg·dl^{-1} (4.5 mM·L^{-1}), with a range of 60 to 115 mg·dl^{-1}. When the plasma glucose level falls below 60 mg·dl^{-1}, brain uptake of the sugar and brain utilization of oxygen decrease in parallel. Central nervous system function becomes progressively impaired, and death may ensue. The major products of glycolysis, lactate and pyruvate, circulate at average concentrations of 0.7 and 0.07 mM, respectively. This 10:1 ratio of lactate to pyruvate prevails even when the glycolytic rate changes. However, when tissues are deprived of oxygen, the equilibrium between the two shifts toward lactate, the reduced molecule; plasma concentration ratios as high as 30:1 may then be observed. Very high concentrations of lactate produce metabolic acidosis.

In the basal state glucose turnover is about 2 mg·kg^{-1}·min^{-1} (11 mmole·kg^{-1}·min^{-1}), which is equivalent to about 9 g·hour^{-1} or 225 g·day^{-1} in adults. Approximately 55% of glucose utilization results from terminal oxidation, of which the brain accounts for the greatest part. Another 20% is caused by glycolysis; the resulting lactate then returns to the liver for resynthesis into glucose (Cori cycle). Reuptake by the liver and other splanchnic tissues accounts for the remaining 20% of glucose utilization. Most of glucose use (about 70%) in the basal state is independent of insulin, a hormone with otherwise important regulatory effects on glucose metabolism.

The circulating pool of glucose is only slightly larger than the liver output in 1 hour. This pool is only sufficient to maintain brain oxidation for 3 hours, even if all other glucose utilization ceased. This emphasizes the crucial importance of continuous hepatic production of glucose in the fasting state. About 80% of this production results from glycogenolysis and 20% from gluconeogenesis. Hepatic uptake and utilization of circulating lactate accounts for more than half the glucose supplied by gluconeogenesis. The remainder is largely accounted for by amino acids, especially alanine. The supply of lactate comes from glycolysis in muscle, red blood cells, white blood cells, and a few other tissues. The amino acid precursors come from proteolysis of muscle.

When an individual ingests glucose after overnight fasting, approximately 70% of the load is assimilated by peripheral tissues, mainly muscle, and about 30% by splanchnic tissues, mainly liver. Only 20% to 30% of a glucose load is oxidized during the 3 to 5 hours required for its absorption from the gastrointestinal tract. The remainder is stored as glycogen, partly in muscle and partly in liver. Glucose initially stored as muscle glycogen can later be transferred to the liver by undergoing glycolysis to lactate, which is released into the circulation; the lactate is then taken up by the liver, rebuilt into glucose, and stored as glycogen in that organ. During the period of peak absorption of exogenous glucose, hepatic output of the sugar is largely unnecessary and is greatly reduced from basal levels. These metabolic adaptations are largely facilitated by coordinated secretion of the pancreatic islet hormones, insulin and glucagon.

PROTEIN METABOLISM

The average adult body contains 10 kg of protein, of which about 6 kg is metabolically active. Approximately 50 g of amino acids are released daily by proteolysis from muscle, the main endogenous repository. Therefore daily dietary intake of 50 g of protein is ordinarily sufficient. The minimal daily requirement is only 0.5 g·kg^{-1} because some of the endogenously generated amino acids are rapidly reused for protein synthesis. When accretion of lean body mass is taking place (for example, in growing children, pregnant women, persons recovering from prior weight loss), daily protein requirements increase to 1.5 to 2.0 g·kg^{-1}.

All proteins are composed of the same 20 amino acids. Half of these are called *essential amino acids* because their carbon skeletons, the corresponding α-ketoacids, cannot be synthesized by humans. Once present, however, they can be converted to the essential amino acids by transamination. The other half, the *nonessential amino acids,* can be synthesized endogenously because the appropriate carbon skeletons can be built from glucose metabolites in the citric acid cycle. The essential amino acids must be supplied in the diet, with individual minimal requirements ranging from 0.5 to 1.5 g·day^{-1}. All 20 amino acids are required for normal protein synthesis; therefore a deficiency of even one essential amino acid disrupts this process.

Protein sources vary greatly in their biologic effectiveness, depending in part on the ratio of essential to nonessential amino acids. Milk and egg proteins are of the highest quality in this regard. During infancy and childhood about 40% of the protein intake should consist of essential amino acids in order to support growth. In adults this requirement falls to 20%. In addition to their incorporation into proteins, many of the amino acids, including some essential ones, are precursors for important molecules, such as purines, pyrimidines, polyamines, phospholipids, creatine, carnitine, methyl donors, thyroid and catecholamine hormones, and neurotransmitters.

All 20 amino acids are completely oxidized to CO_2 and H_2O after removal of the amino group. Each traverses a specific degradative pathway. (Refer to standard biochemistry textbooks for details.) However, all these pathways converge into three general metabolic processes: gluconeogenesis, ketogenesis, and ureagenesis. Except for leucine, all the amino acids can contribute carbon atoms for the synthesis of glucose. Five ketogenetic amino acids give rise either to acetoacetate or its CoA precursors. In the degradation of all amino acids, ammonia is released. Ammonia, incorporated mainly into glutamine and alanine molecules, is then transported to the liver. In the liver ammonia is "detoxified" by incorporation into urea, a metabolically inert molecule. The synthesis of urea via the Krebs-Henseleit cycle is depicted in Figure 36-5. The urea resulting from protein degradation is excreted by the kidney (see Chapter 32).

In the healthy adult under steady-state conditions, the total daily nitrogen excreted in the urine as urea plus ammonia, along with minor losses of nitrogen in the feces (0.4 g·day^{-1}) and skin (0.3 g·day^{-1}), is equal to the nitrogen released during metabolism of exoge-

nous and endogenous protein. Such an individual is said to be in *nitrogen balance*. When there is no dietary protein intake, the sum of urea plus ammonia nitrogen in the urine reflects almost quantitatively the rate of endogenous protein degradation. When protein breakdown is greatly accelerated by tissue trauma or disease, urinary urea plus ammonia nitrogen may exceed protein nitrogen intake. In these two cases the individual is said to be in *negative nitrogen balance*. In a growing child or in a previously malnourished individual undergoing protein repletion with gain in body mass, urinary urea plus ammonia nitrogen excretion is less than the intake of protein nitrogen. This individual is said to be in *positive nitrogen balance*.

Healthy adults who receive isocaloric diets containing adequate protein and who are in nitrogen balance synthesize and degrade body protein at a rate of 3 to 4 g·kg^{-1}·day^{-1}. Approximately 5% of this total is accounted for by hepatic synthesis of albumin, a protein that turns over rapidly. When the diet is severely deficient in energy, total protein, or one of the essential amino acids, the rate of total body protein synthesis diminishes. In compensation, protein degradation also diminishes, but not to the same extent as synthesis, so that net loss of body protein results.

FAT METABOLISM

Fat represents almost half the total daily substrate for oxidation (about 100 g, or 900 kcal). As noted previously, fat is the major and most advantageous form of stored fuel. The usual daily intake in the United States is also approximately 100 g, or 40% of total calories. The major component of both dietary and storage fat is triglycerides. These largely consist of long-chain saturated and monounsaturated fatty acids (chiefly palmitic, stearic, and oleic acids) esterified to glycerol. Because these fatty acids can also be synthesized in the liver and adipose tissue, in an overall sense, no strict dietary requirement exists for fat. However, about 3% to 5% of fatty acids are polyunsaturated and cannot be synthesized in the body. These are termed *essential dietary fatty acids* (chiefly linoleic and linolenic) because they are required as precursors for certain membrane phospholipid and glycolipid substances, as well as for important intracellular mediators known as *prostaglandins*. Another component of fat is the steroid molecule *cholesterol*,

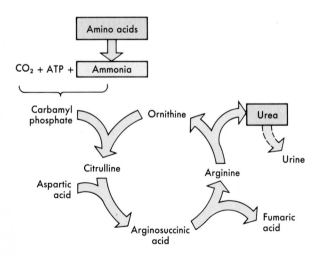

FIGURE 36-5   The Krebs-Henseleit urea cycle for disposal of amino acid ammonia.

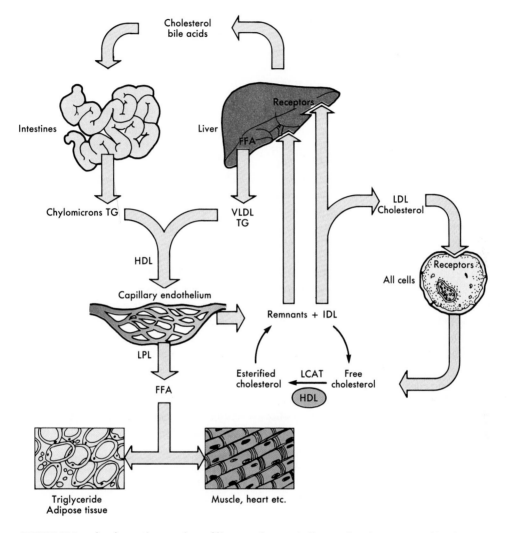

FIGURE 36-6   A schematic overview of lipoprotein metabolism and major aspects of lipid turnover in humans. Exogenous triglycerides (*TG;* chylomicrons absorbed from the intestine) and endogenous triglycerides (very-low-density lipoproteins, *VLDL,* produced in the liver) both give rise to free fatty acids *(FFA)* for storage in adipose tissue and oxidation in muscle. High-density lipoprotein particles, *(HDL)* facilitate and the enzyme lipoprotein lipase *(LPL)* directly catalyzes liberation of the free fatty acids from triglycerides. The resultant particles, called remnants, from chylomicrons and intermediate-density lipoproteins *(IDL)* from VLDL, undergo further change in the circulation, which also is facilitated by HDL. The ratio of esterified cholesterol to free cholesterol is increased in the remnant and IDL particles by the enzyme lecithin-cholesterol acyltransferase *(LCAT).* The remnant particles are then taken up by the liver for further metabolism. The IDL particles are partly taken up by the liver and partly converted to cholesterol-rich low-density lipoprotein *(LDL)* particles. The latter are then taken up by virtually all cells after interaction with specific LDL receptors. Cholesterol, either synthesized in the liver or extracted from remnant and IDL particles, is also excreted into the intestine, partly as bile acids.

which serves a variety of specific functions in membranes and is the precursor for bile acids and steroid hormones. Cholesterol is both ingested (300 to 600 mg·day^{-1}) and synthesized by most cells.

Transport of the nonpolar lipids in plasma requires that they be incorporated into a variety of complex *lipoprotein* particles, which turn over at different rates and also interact with each other. The protein portion of each particle is derived from several apoproteins synthesized in the liver and intestine. These apoproteins serve catalytic functions and interact with specific cell receptors. Figure 36-6 summarizes the metabolic pathways and interactions of the lipoprotein particles.

Chylomicrons, formed from dietary fat (see Chapter 31), are the lowest-density lipoproteins. Following their absorption from the gastrointestinal tract, they disappear from plasma rapidly. Their major component is triglycerides, which are partly hydrolyzed by the key enzyme *lipoprotein lipase* on capillary endothelial surfaces. This enzyme is activated by apoprotein C-II and transferred to the chylomicron by high-density lipoprotein (HDL) particles (see following discussion). The resultant free fatty acids are taken up both by adipose cells for resynthesis into triglycerides and storage and by other cells for oxidation. The residual lipoprotein particles, now relatively higher in cholesterol content and known as chylomicron *remnants,* are taken up by the liver for further degradation.

In contrast, very-low-density lipoprotein (VLDL) particles are formed by endogenous synthesis in the liver, and to a lesser extent in the intestine, in the postabsorptive state. They are denser, contain somewhat more cholesterol than chylomicrons, and have a longer plasma half-life. Their rate of formation varies from as little as 15 to as much as 90 g·day^{-1}. The initial metabolism of VLDL is the same as that of chylomicrons. The product of lipoprotein lipase action consists of particles called intermediate-density lipoprotein (IDL). About half the IDLs return to the liver and are taken up, as occurs with chylomicron remnants. The other half of the IDLs are further enriched with cholesterol to form low-density lipoprotein (LDL) particles. Circulating LDL is responsible for transferring cholesterol into other cells. Uptake of LDL, IDL, and remnant particles is via interaction between their apoproteins and specific cell receptors followed by endocytosis.

The uptake of LDL cholesterol by cells has important regulatory actions on intracellular cholesterol metabolism. Cholesterol uptake from plasma downregulates the LDL receptor, thereby reducing further entrance of the sterol. Uptake of cholesterol also suppresses its own de novo synthesis.

HDLs are synthesized in the liver and intestine and have a long half-life. These particles facilitate the major steps in chylomicron, VLDL, IDL, and LDL movement just described. HDLs exchange key apoproteins with the other lipoprotein particles. They also accept free-cholesterol molecules, esterify them via the enzyme *lecithin-cholesterol acyltransferase* (LCAT), and transfer them back to other particles. The net effect of HDL is to accelerate clearance of triglycerides from plasma and regulate the ratios of free to esterified cholesterol.

Free fatty acids circulate in an average concentration of 400 μmol·L^{-1} bound to albumin molecules. They undergo a very rapid turnover of approximately 8 g·hour^{-1}. Half of this represents oxidation and half reesterification to triglycerides. The plasma concentration of total cholesterol (average 185 mg·dl^{-1}), and especially of LDL cholesterol (average 120 mg·dl^{-1}), is a very important risk factor for *atherosclerosis* and death from cardiovascular events. On the other hand, HDL cholesterol exerts a protective effect against cardiovascular disease. Certain genetic abnormalities in apoprotein or LDL receptor synthesis can lead to elevated LDL cholesterol levels, with premature development of *coronary artery disease*.

METABOLIC ADAPTATIONS

Fasting

In the fasting state the individual totally depends on endogenous substrates for energy. Mobilization of glucose provides essential fuel for the central nervous system; release of free fatty acids provides for the oxidative needs of the other tissues. An increase in protein degradation to amino acids is also a fundamental feature of this response. The fasting individual is said to be in a state of *catabolism* because carbohydrate, fat, and protein stores are all decreasing.

The liver supplies glucose to the circulation initially by augmenting glycogenolysis. After 12 to 15 hours of fasting, however, hepatic glycogen stores are almost depleted, and a rapid enhancement of gluconeogenesis fills the void. To supply glucose precursors, 75 to 100 g of muscle protein are broken down daily during the first few days. This is reflected in a rising excretion of nitrogen in the urine. Gluconeogenesis is also supported by the provision of 15 to 20 g of glycerol daily, which is released during the accelerated lipolysis of triglycerides in adipose tissue. Glucose oxidation in muscle and liver is spared as increasing quantities of free fatty acids become available. A portion of fatty acid oxidation in liver yields the ketoacids β-*hydroxybutyrate* and *acetoacetate*. These can also be oxidized by muscle cells, further sparing utilization of glucose. The net shift away from glucose and toward fatty acid oxidation lowers the respiratory quotient. These adaptations are also reflected in changing plasma concentrations of substrates. The concentrations of glucose and the major gluconeogenic amino acid alanine decrease, whereas the concentrations of free fatty acids, glycerol, and branch-chain amino acids such as leucine increase. High levels of the strong ketoacids produce a tendency to metabolic acidosis and a slight reduction in blood pH.

As fasting is prolonged beyond a few days, other important adaptations occur. Total energy expenditure, reflected in the BMR, decreases 10% to 20%, limiting the drain on energy stores. The central nervous system no longer depends entirely on glucose as an energy source, and two thirds of its needs are eventually met by the ketoacids. As less glucose is needed for oxidation, gluconeogenesis diminishes and protein breakdown declines to 25 to 30 $g \cdot day^{-1}$. In long-term fasting body weight diminishes by an average of 300 $g \cdot day^{-1}$, of which two-thirds is accounted for by fat and one-third by lean tissue. Of the latter, 25% constitutes protein and 75% intracellular water and electrolytes. In long-term fasting fatty acids provide 90% of the total energy expenditure. As long as sufficient fluid is ingested, an individual of normal weight can survive up to 60 days. About that time fat stores are almost exhausted, protein degradation suddenly accelerates, and death follows.

Exercise

The metabolic response to exercise resembles the response to fasting, in that the mobilization and generation of fuels for oxidation are dominant factors. The type and amounts of substrate vary with the intensity and duration of the exercise (Figure 36-7). For very intense, short-term exercise (e.g., a 10- to 15-second sprint), stored creatine phosphate and ATP provide the energy at a rate of approximately 50 $kcal \cdot min^{-1}$. When these stores are depleted, additional intensive exercise for up to 2 minutes can be sustained by breakdown of muscle glycogen to glucose-6-phosphate, with glycolysis yielding the necessary energy

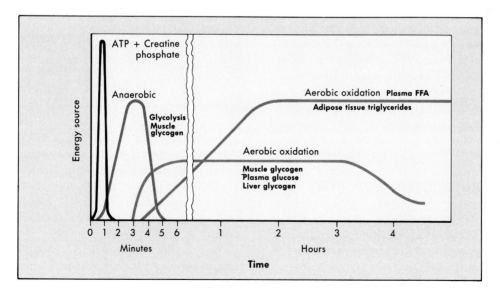

FIGURE 36-7 Energy sources during exercise. Note the sequential use of stored high-energy phosphate bonds, glycogen, circulating glucose, and circulating free fatty acids *(FFA)*. The latter dominate in sustained exercise.

(at a rate of 30 kcal·min^{-1}). This *anaerobic* phase is not limited by depletion of muscle glycogen at this point, but rather by the accumulation of lactic acid in the exercising muscles and the circulation.

After several minutes of exhaustive anaerobic exercise, an oxygen debt of 10 to 12 L can be built up. This must be repaid before the exercise can be repeated. From 6 to 8 L are required either to rebuild the accumulated lactic acid back into glucose in the liver or to oxidize it to CO_2. About 2 L are required to replenish normal muscle ATP and creatine phosphate content. A further 2 L will replenish the oxygen present in the lungs and body fluids and oxygen bound to myoglobin and hemoglobin.

For less intense but longer periods of exercise, *aerobic* oxidation of substrates is required to produce the necessary energy (at about 12 kcal·min^{-1}). Substrates from the circulation are added to muscle glycogen. After a few minutes glucose uptake from the plasma increases dramatically, up to thirty-fold in some muscle groups. To offset this drain, hepatic glucose production increases up to five-fold. Initially this is largely from glycogenolysis. With exercise of long duration, gluconeogenesis becomes increasingly important as liver glycogen stores become depleted. However, endurance can be improved by high-carbohydrate feedings for several days before prolonged exercise (e.g., a marathon run), since this increases both liver and muscle glycogen stores. To support gluconeogenesis, amino acids are increasingly released by muscle proteolysis. Eventually, fatty acids, liberated from adipose tissue triglycerides, form the predominant substrate, supplying two thirds of the energy needs during sustained exercise. Except for increases in circulating pyruvate and lactate resulting from enhanced glycolysis, the pattern of change in plasma substrates is similar to that of fasting, only telescoped in time.

REGULATION OF ENERGY STORES

As stated, the preponderance of stored energy consists of fat. What determines the proper quantity of this energy reserve, and what regulates it? Does an ideal relationship exist between fat mass and either total body weight or lean body mass? Clear-cut answers to these questions are not yet available.

A genetic influence on fat mass is suggested by (1) the greater similarity of adipose stores in identical twins than in fraternal twins and (2) the tendency for body mass of adopted children to resemble that of their biological parents more than that of their adoptive parents. Environmental influences, specifically the quality and quantity of the food available, are suggested by the excessive weight gain of certain laboratory animals in response to presentation of high-fat or "junk food" diets, as well as by the much greater prevalence of obesity in affluent westernized societies than in other populations. Also, the human species has more energy-storage adipose cells per unit body mass than any other species except whales, which may contribute to this propensity for obesity.

Some data suggest the existence of a particular set point for energy stores in each individual. Once adult weight is reached, it tends to be constant, at least until middle age, at which point most humans incur at least a modest weight gain to a new, higher constant level with a higher proportion of body fat. Normal laboratory animals subjected to overfeeding or underfeeding experiments will return to their original weight and degree of fatness when again allowed free access to food. They will do this not only by adjusting food intake, but also by adjusting energy expenditure in the appropriate direction.

Decreases in energy stores caused by excessive expenditure are compensated by increased caloric intake. Control of appetite appears to reside in the hypothalamus. In rodents evidence exists for both a *hunger center* in the lateral hypothalamus and a *satiety center* in the ventromedial hypothalamus. Exactly what signals arising from the periphery cause compensatory hypothalamic responses is not yet known.

Increases in energy stores resulting from excessive food intake can be compensated for by increased energy expenditure. This may occur via thermogenic processes, such as "futile cycles" that are wasteful of ATP (e.g., glucose → glucose-6-phosphate → glucose), ion pumping via the membrane enzyme Na^+,K^+-ATPase, or uncoupling of ATP formation from mitochondrial oxidation. Various hormones as well as the sympathetic nervous system can regulate energy expenditure.

Pathologic accumulation of energy stores as fat (i.e., obesity) is a major health problem in many countries. A body weight 20% above average increases the risk of disorders such as diabetes and hypertension, and a body weight 50% above average greatly increases the risk of death. The cause of human obesity is not known. Some obese individuals behave as if they have an elevated set point for energy stores; they defend this set point tenaciously by decreasing energy expenditure when caloric intake is reduced by dieting. Other obese individuals behave as though appetite and caloric intake are uncoupled from any perception of energy stores; these individuals may regulate appetite abnormally as opposed to having an elevated set point. In obese humans the profile of hormones involved in regulation of fat metabolism generally favors deposition rather than mobilization of fat. Adipose tissue of obese humans also contains elevated levels of lipoprotein lipase, the key enzyme that transfers circulating triglycerides into cells. Although abnormalities of this sort are intriguing, none has been proved to be the primary cause of human obesity.

BIBLIOGRAPHY
Journal Articles

Bogardus C et al: Familial dependence of the resting metabolic rate, N Engl J Med 315:96, 1986.

Cahill GF: Starvation in man, N Engl J Med 282:668, 1970.

Felig P et al: Amino acid metabolism during prolonged starvation, J Clin Invest 48:584, 1969.

Ferrannini E et al: The disposal of an oral glucose load in health subjects: a quantitative study, Diabetes 34:580, 1985.

Foster D: From glycogen to ketones—and back, Banting Lecture 1984, Diabetes 33:1188, 1984.

Morley JE and Levine AS: Nutrition: the changing scene—the central control of appetite, Lancet 1:398, 1983.

Ravussin E et al: Determinants of 24-hour energy expenditure in man: methods and results using a respiratory chamber, J Clin Invest 78:1568, 1986.

Reeds P and James W: Nutrition: the changing scene—protein turnover, Lancet 1:571, 1983.

Schaefer E et al: Pathogenesis and management of lipoprotein disorders, N Engl J Med 312:1300, 1985.

Sims EAH and Danforth E Jr: Expenditure and storage of energy in man, J Clin Invest 79:1019, 1987.

Monograph

Flatt JP: The biochemistry of energy expenditure. In Bray G, editor: Recent advances in obesity research: II, Proceedings of the 2nd International Congress on Obesity, Los Angeles, 1978, Newman Publishing.

Hormones of the Pancreatic Islets

The major pancreatic islet hormones, *insulin* and *glucagon,* are rapid and powerful regulators of metabolism. Together they coordinate the disposition of nutrient input from meals as well as the flow of endogenous substrates by actions on the liver, adipose tissue, and muscle mass. Their cells of origin are intimately interspersed in anatomic islets that comprise 1% to 2% of the mass of the pancreas and are scattered throughout that organ. These islets are composed of 60% β-*cells,* the source of insulin, and 25% α-*cells,* the source of glucagon. The remaining islet cells secrete the peptides *somatostatin (δ-cells)* and *pancreatic polypeptide* (F-cells).

The strategic location of the islets (Figure 37-1) reflects their functional role. Insulin and glucagon are secreted in response to nutrient inflow and gastrointestinal secretagogues, as are the enzymes of the acinar pancreas (see Chapter 30). The islet hormones may have paracrine effects on each other through tight junctions and gap junctions between the endocrine cells or on nearby acinar cells. Most importantly the location of the islets dictates secretion of their hormones into the pancreatic veins and then into the portal vein. This arrangement permits the liver, the central organ in nutrient traffic, to be exposed to higher concentrations of insulin and glucagon than peripheral tissues. It also permits the liver to modulate the availability of insulin and glucagon to peripheral tissues by extracting variable amounts of these hormones during first passage through that organ.

Insulin and glucagon are often secreted and act reciprocally. When one is needed, the other usually is not. Therefore the *ratio* of insulin to glucagon concentrations may be more critical than the absolute concentrations of each hormone. The consequences of isolated insulin deficiency—the common disease *diabetes mellitus*—are so devastating that insulin has tended to dominate our physiologic thinking. In contrast, isolated glucagon deficiency is virtually unknown in medicine; moreover, it can be compensated for by other mechanisms.

INSULIN

Synthesis and secretion

Insulin is a peptide hormone with a molecular weight of 6000 and is composed of two chains linked by disulfide bridges. The B chain contains the core of biologic activity, whereas the A chain contains most of the species-specific sites that generate immune responses. Human, beef, pork, and fish insulin have essentially equivalent biologic activity on a molar basis. All induce formation of antibodies when injected subcutaneously for treatment of diabetes mellitus, although human insulin is by far the least immunogenic.

The synthesis of insulin by the β-cell follows the general pattern for peptide hormones described in Chapter 35. The gene on chromosome 11 directs the synthesis of a preprohormone from which the signal peptide is cleaved to yield the single-chain proinsulin. Establishment of the disulfide linkages is followed by excision of a connecting peptide known as C-peptide. Insulin and C-peptide are packaged together in the secretory granules by the Golgi apparatus. These granules also contain zinc, which acts to join six insu-

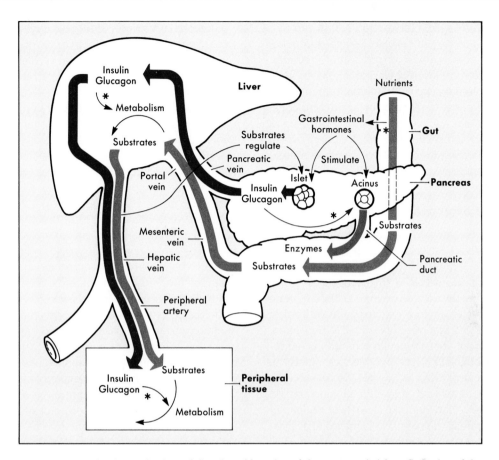

FIGURE 37-1   A schematic view of the pivotal location of the pancreatic islets. Secretion of the islet hormones insulin and glucagon is coordinated with secretion of exocrine pancreatic enzymes. Both are stimulated by entry of nutrients into the gastrointestinal tract and by gastrointestinal hormones. Islet hormones are secreted into the portal vein and thereby reach the liver with the substrate products of nutrient digestion. Within the liver they affect the metabolism of the ingested substrates. Islet hormones that pass through the liver with substrates affect disposition of these substrates by peripheral tissues. In turn these substrates feed back on the pancreatic islets to modulate the secretion of insulin and glucagon.

lin molecules into hexamers. Crystalline zinc insulin is the basic pharmaceutic preparation for treatment of diabetes mellitus.

Insulin is secreted by exocytosis (see Figure 35-5). Insulin-containing granules are arrayed in parallel with microtubules in the β-cell cytoplasm. The microtubules are associated with a web of microfilaments containing myosin and actin near the plasma membrane. On application of a stimulus, contraction of microfilaments draws the granules to the plasma membrane, where they fuse with it, rupture, and release equimolar amounts of insulin and C-peptide.

Many agents can stimulate insulin secretion. Indi-

vidual β-cell receptors for each of these stimuli have not yet been identified. The biochemical process of stimulation requires an increase in cytosolic calcium. In conjunction with insulin release, cyclic adenosine monophosphate (cAMP) levels usually increase, and activation of the phosphatidylinositol–protein kinase C system (see Chapter 35) also may be involved.

The most important stimulator of insulin secretion is *glucose*. Much evidence indicates that glucose must be phosphorylated and metabolized further before causing insulin release. The phosphorylating enzyme glucokinase may thus participate in regulating β-cell function. The coupling of nutrient metabolism to exocytosis may be via increasing β-cell adenosine triphosphate (ATP) concentrations. In the process of insulin release the cell also becomes depolarized, which suggests a role for potassium channels. Sufficient concentrations of potassium and calcium in the interstitial fluid are essential to maintain maximal β-cell responsiveness to glucose. In addition to causing insulin release, glucose stimulates synthesis of the hormone by increasing the transcription rate of the insulin gene and the translation rate of its mature messenger ribonucleic acid (mRNA).

Regulation of Secretion

In the broadest sense, insulin secretion is governed by a feedback relationship with the exogenous nutrient supply (Figure 37-2). When this supply is abundant, insulin is secreted in response; the hormone then stimulates utilization of these same incoming nutrients while simultaneously inhibiting mobilization of endogenous substrates. When nutrient supply is low or absent, insulin secretion is dampened and mobilization of endogenous fuels is enhanced.

The central regulating molecule is glucose. At plasma levels less than 50 mg·dl^{-1}, little or no insulin is secreted, whereas the response is maximal at plasma levels greater than 250 mg·dl^{-1}. Brief exposure of the β-cell to glucose induces a rapid but transient release of insulin. With continuous glucose exposure, this initial response fades, later to be replaced by a slower second phase, which probably reflects glucose stimulation of insulin synthesis.

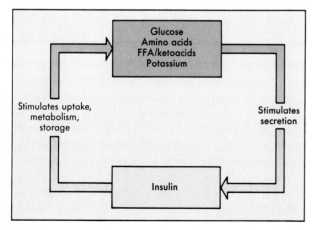

FIGURE 37-2 Feedback relationship between insulin and nutrients. Those nutrients that stimulate insulin secretion are the same nutrients whose disposal is facilitated by insulin. *FFA*, Free fatty acids.

Glucose entry from the gastrointestinal tract follows digestion of the carbohydrate components of a meal. Under these circumstances insulin release is greater than can be accounted for by the degree to which plasma glucose levels rise. This results from release of gastrointestinal peptide hormones such as gastric inhibitory peptide (GIP), gastrin, secretin, cholecystokinin, and glucagon-like peptides from intestinal cells, all of which can stimulate insulin secretion. In addition, digestion of the protein in a meal yields amino acids, some of which synergize with glucose in stimulating the β-cells. Lipids and their products contribute little directly to the β-cell response to a meal. When digestion and absorption of the nutrients are completed, plasma glucose and amino acid levels return to baseline, and insulin secretion subsides to a rate that is maintained steadily during the usual overnight fasting period.

If fasting is extended for days, insulin secretion declines below the basal rate and then resets at a lower level. In this state secretion is maintained by lower but still stimulatory plasma glucose levels, with contributions from greatly elevated ketoacid and free fatty acid levels. Insulin secretion is also modulated by cholinergic and β-adrenergic stimulatory and α-adrenergic inhibitory influences. The usual physiologic fluctuations in peripheral plasma insulin levels and the average equivalent rates of insulin delivery into the peripheral circulation are summarized in Figure 37-3.

Insulin concentration in the portal vein ranges from two- to tenfold higher than in the peripheral circulation. The liver extracts about half the insulin reaching it, but this varies with the nutritional state. Thus actual β-cell secretory rates are better estimated by measurement of plasma C-peptide, since this cosecreted molecule is not removed by the liver. Such estimates yield values for insulin secretion of 1.0 to 2.5 mg·dl^{-1}. These estimates are similar to amounts of exogenous insulin required to normalize plasma glucose levels in individuals who do not produce any endogenous insulin. C-peptide and the small quantity of proinsulin secreted by β-cells have no known biologic actions.

Insulin has a short plasma half-life, mainly because of specific degradation in the kidney and liver. However, some insulin is also degraded in conjunction with its actions on other target cells after receptor binding and internalization of the hormone. Very little intact insulin is excreted in the urine.

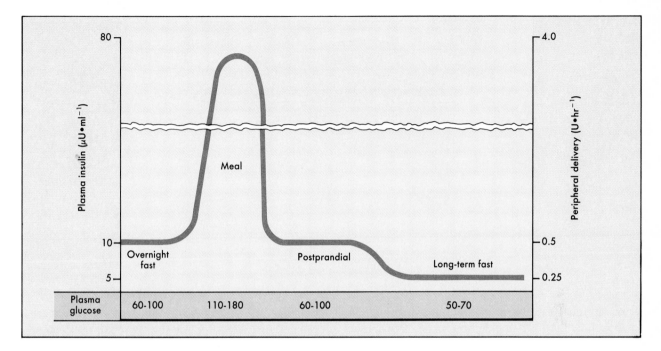

FIGURE 37-3   The pattern of plasma insulin levels and the corresponding insulin delivery rates to the peripheral circulation in various physiologic states ($10 \, \mu U \cdot ml^{-1} = 7 \times 10^{-11} M$). The usual prevailing plasma glucose levels are also indicated.

Actions of Insulin

Fuel Turnover The overall thrust of insulin action is to facilitate storage of substrates and inhibit their release (Figure 37-4). As a result, secreted or administered insulin **decreases** the plasma concentrations of glucose, of *free fatty acids* and *ketoacids,* and of predominantly the *essential branch-chain amino acids* (leucine, isoleucine, valine). The major sites of insulin action are liver, muscle, and adipose tissue. In each target tissue, carbohydrate, lipid, and protein metabolism are regulated coordinately.

Under maximal insulin stimulation, the usual rate of glucose utilization by peripheral tissues is increased five- to sixfold. Simultaneously the output of glucose by the liver drops to considerably below half. Most of the extra glucose uptake occurs in muscle, with a very small fraction in adipose tissue. Approximately 75% of this glucose is converted to glycogen, and only 20% to 30% undergoes glycolysis and terminal oxidation to CO_2. The absolute rate of glucose oxidation, however, is increased three fold by insulin.

The basal rate of free fatty acid inflow to plasma—equivalent to the rate of release from adipose tissue—is decreased two-thirds by insulin. Simultaneously, inflow of glycerol, the other product of triglyceride hydrolysis, is decreased sharply. As a result of this reduction in free fatty acid availability, insulin reduces the basal rate of lipid oxidation more than 90%.

The action of insulin on protein turnover has been assessed indirectly by determining the hormone's effect on the flux of the essential amino acid leucine. Maximal doses decrease the rate of inflow of leucine into plasma to almost half. Because the only source of leucine in the basal state is endogenous protein, this action of insulin must be caused by inhibition of proteolysis. In addition, insulin decreases the rate of oxidation of leucine. The result of these insulin effects is a net gain in body protein.

Carbohydrate Metabolism In muscle and adipose tissue, insulin stimulates the transport of glucose from the plasma, across the cell membrane, and into the cytoplasm, where it is rapidly phosphory-

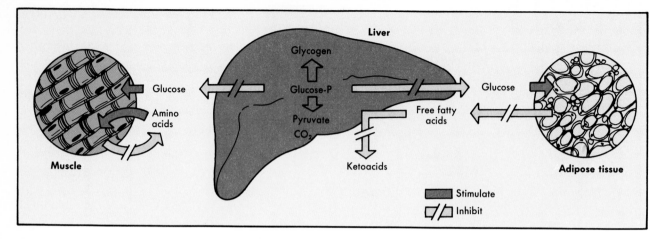

FIGURE 37-4   Effect of insulin on the overall flow of fuels. Tissue uptake of glucose, free fatty acids, and amino acids results in a decrease in their plasma levels.

lated. In muscle and liver, insulin largely stimulates glycogen formation from glucose-6-phosphate and, to a much lesser extent, glycolysis and oxidation. In adipose tissue, the most important effect of insulin is to stimulate production of α-glycerol phosphate from the triose phosphate intermediates of glycolysis. The α-glycerol phosphate is used to esterify free fatty acids, thus storing them as triglycerides.

Insulin stimulates conversion of glucose to glycogen and inhibits the reverse reaction, breakdown of glycogen to glucose, in several ways. The equilibrium between intracellular glucose and glucose-6-phosphate is shifted toward the latter because the phosphorylating enzyme glucokinase is increased by insulin. In liver only, the dephosphorylating enzyme glucose-6-phosphatase is decreased by insulin. Insulin increases the enzyme (glycogen synthase) that polymerizes phosphorylated glucose to glycogen, whereas insulin decreases the enzyme (phosphorylase) that degrades glycogen. Insulin also shifts the balance between glycolysis and gluconeogenesis toward the former and away from the latter. Glycolysis is accelerated because insulin increases the key enzymes phosphofructokinase, pyruvate kinase, and pyruvate dehydrogenase; gluconeogenesis is retarded because insulin decreases the enzymes fructose-1,6-diphosphatase, phospho*enol*pyruvate carboxykinase, and pyruvate carboxylase.

In liver, glucose availability reinforces those effects of insulin that lead toward glycogen storage or glycolysis and away from glucose release. Conversely, as plasma glucose levels decline, these effects of insulin become attenuated by intrahepatic autoregulatory phenomena as well as by the secretion of hormones (glucagon, epinephrine, cortisol, growth hormone) with actions antagonistic to insulin.

Fat Metabolism In adipose tissue, insulin facilitates transfer of circulating fat into the adipose cell (see Chapter 36) by increasing the enzyme lipoprotein lipase. More free fatty acid is thereby liberated from circulating triglyceride and is rapidly taken up into the adipose cell, where it is reesterified with α-glycerol phosphate. Thus dietary fat not needed for immediate energy generation is stored. Of equal or greater importance, insulin profoundly inhibits the reverse reaction (i.e., lipolysis of stored triglyceride) by decreasing the necessary enzyme, hormone-sensitive adipose tissue lipase. In this manner free fatty acid release and delivery to other tissues is greatly suppressed.

In liver, insulin favors shunting of incoming free fatty acids away from β-oxidation and toward esterification, again by increasing production of α-glycerol phosphate. Because β-oxidation is diminished, less β-hydroxybuturate and acetoacetate are produced. Thus insulin is powerfully antiketogenic. Insulin also stimulates de novo synthesis of free fatty acids from pyruvate-derived acetyl coenzyme A (acetyl CoA) by increasing the enzymes acetyl CoA carboxylase and fatty acid synthase. In addition, insulin stimulates synthesis of cholesterol from acetyl CoA by increasing the key enzyme hydroxymethylglutaryl CoA reductase. The net effect of insulin is therefore to increase the fat content of the liver and in some circumstances the release of very-low-density lipoprotein from that organ.

Protein Metabolism In muscle, insulin stimulates the transport of certain amino acids from the plasma, across the cell membrane, and into the cytoplasm in a manner that is analogous to, but independent of, the transport of glucose. The overall synthesis of proteins from amino acids is also increased by stimulation of transcription and translation. These anabolic effects are reinforced by anticatabolic effects, that is, inhibition of the enzymes of proteolysis and inhibition of amino acid release from the cell. Other examples of anabolic effects are observed in the liver and exocrine pancreas, where the synthesis of albumin and amylase, respectively, is increased by insulin. Moreover, in cartilage and osseous tissue, insulin and structurally related peptides called *somatomedins* enhance the general synthesis of proteins as well as DNA, RNA, and other macromolecules. Thus insulin is an important contributor to growth, to tissue regeneration, and to bone remodeling.

Other Effects Both glycogen and protein synthesis require concurrent cellular uptake of potassium, phosphate, and magnesium. The translocation of all three of these electrolytes from the extracellular to the intracellular space is stimulated by insulin. Therefore, when insulin is secreted or administered, the plasma concentrations of potassium, phosphate, and magnesium all decrease. Insulin also stimulates the reabsorption of potassium, phosphate, and sodium by the renal tubules. Preventing urinary loss of potassium and phosphate contributes to anabolism, whereas conservation of sodium may be related to the need for additional extracellular fluid formation to accompany the expansion of lean body mass.

Insulin has incompletely defined actions that are relevant to total body energy turnover. Diet-induced thermogenesis, particularly following carbohydrate ingestion, is enhanced by insulin, probably through stimulation of glycogen formation. Insulin also increases energy expenditure by stimulating Na^+,K^+-ATPase. Although most parts of the brain are unresponsive to insulin, considerable evidence points to an action of the hormone on the hypothalamus. Possibly by facilitating glucose uptake across specialized capillaries in the hypothalamus, insulin may act directly to increase satiety signals and dampen hunger when plasma glucose is elevated. However, when plasma glucose levels drop below those needed for normal brain metabolism (i.e., less than 50 mg·dl^{-1}),

hunger is stimulated, even if the hypoglycemia was induced by insulin administration.

Molecular Mechanisms The initial step in all actions of insulin is binding of the hormone to its plasma membrane receptor (Figure 37-5). This receptor is a glycoprotein composed of two symmetric units connected by disulfide bonds. Each of the units is made up of an α-subunit that extends exterior to the cell membrane and a β-subunit that extends inward through the cell membrane and terminates in an intracytoplasmic tail. Insulin receptors cluster in coated pits on the cell surface and exhibit mobility within the cell membrane. After binding with the hormone, they are internalized. Some receptor molecules are degraded, whereas others return to an intracytoplasmic pool from which they can again be recruited into the plasma membrane. Insulin down-regulates the number of its own receptors by increasing the rate of their degradation.

After the outer α-subunit of the receptor binds insulin, a single tyrosine residue on the innter β-subunit undergoes autophosphorylation with ATP. The receptor thus acquires kinase activity and becomes capable of phosphorylating other proteins in the cell membrane or cytoplasm. This may initiate a cascade that rapidly activates some target enzymes by phosphorylating them (e.g., acetyl CoA carboxylase) and others by dephosphorylating them (e.g., pyruvate dehydrogenase). In addition, insulin binding may initiate cleavage of its receptor or nearby membrane components and generate another second messenger that activates the enzyme glycogen synthase. Insulin also decreases cyclic AMP levels in some target cells (e.g., adipose tissue) by inhibiting protein kinase A and by stimulating the enzyme that degrades cyclic AMP, *phosphodiesterase*. Decreased cyclic AMP levels mediate the inhibitory effect of insulin on hormone-sensitive adipose tissue lipase activity.

The most exclusive effect of insulin is to stimulate a specific glucose carrier system within the plasma membrane of muscle and adipose tissue cells. This carrier facilitates diffusion (not active transport) of extracellular glucose into the cytosol, down an already existing, large concentration gradient. A family of such carrier proteins has been isolated and their genes cloned. Insulin appears to increase the synthesis of a glucose carrier protein as well as its transfer from cytoplasmic depots to strategic locations within the plasma membrane. The crucial importance of this insulin action is that transport of glucose into muscle

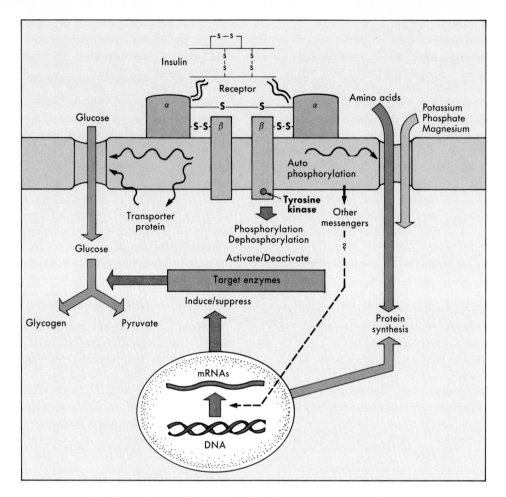

FIGURE 37-5   Insulin action on cells. Binding of insulin to its receptor causes auotphosphorylation of the receptor, which then itself acts as a tyrosine kinase capable of phosphorylating other cell proteins and enzymes. Numerous target enzymes are ultimately activated or inactivated, and the result is to shift the metabolism of glucose toward glycogen and pyruate. A glucose transporter protein is also recruited to the plasma membrane, where it facilitates glucose entry into the cell. By the way of less well-defined messengers, insulin facilitates transport of amino acids, certain cations, and phosphate into the cell and also modulates gene transcription.

and adipose tissue cells is normally the rate-limiting step in its utilization. By comparison, phosphorylation of glucose is so rapid that intracellular concentrations of free glucose are negligible. (Liver cells are an exception because they freely permit entry of glucose. Therefore hepatic intracellular glucose concentrations are essentially equilibrated with extracellular concentrations.) After entry of glucose the rate and direction of its intracellular metabolism is affected by rapid *activation* or *deactivation* of numerous enzymes, as already described. In addition, however, the *synthesis* of specific regulatory enzymes is also

either induced (e.g., glucokinase) or suppressed (e.g., phospho*enol*pyruvate carboxykinase) by altering the rate of transcription of the respective genes. The link between the insulin receptor complex in the plasma membrane and the nuclear actions of the hormone is unknown.

Correlation of Insulin Action and Secretion

The major actions of insulin relate well to variations in its plasma concentration. The low insulin concentrations that prevail in the overnight fasting state are partly able to restrain and thereby regulate endogenous release of free fatty acids and amino acids. Somewhat higher insulin concentrations elicited by incoming nutrients are required to shut off glucose production by the liver. The peak insulin concentrations elicited by a meal greatly stimulate glucose and amino acid uptake by peripheral tissues, especially muscle. This uptake process ensures that these subtrates may be stored for future use. The key metabolic role of insulin means that its absence causes a dramatic distortion of homeostasis. Plasma levels of glucose, free fatty acids, and ketoacids rise to extreme heights. This causes plasma pH and bicarbonate to fall. Extreme loss of adipose mass and lean body mass occurs. Without insulin replacement, death inevitably ensues.

GLUCAGON

Synthesis and Secretion

Although glucagon was discovered shortly after insulin, its physiologic importance has only recently been demonstrated convincingly. It is now clear that glucagon is an important regulator of intrahepatic glucose and free fatty acid metabolism.

Glucagon is a single straight-chain peptide with a molecular weight of 3500. The N-terminal residues 1 to 6 are essential for biologic activity. The glucagon gene, located on human chromosome 2, directs the synthesis of a preproglucagon in the α-cells of the pancreatic islets. This preproglucagon is processed to a prohormone that subsequently yields glucagon and other peptides of still unknown function. In certain cells of the intestinal tract alternative processing of preproglucagon yields glucagon-like peptides. In contrast to insulin, glucagon synthesis is inhibited by high glucose levels and stimulated by low glucose levels.

The secretion of glucagon is related in feedback fashion to the principal function of the hormone—stimulating glucose output by the liver and sustaining plasma glucose levels (Figure 37-6). Thus hypoglycemia promptly evokes a two- to fourfold increase in plasma glucagon from basal levels of about 100 $pg \cdot ml^{-1}$ (3×10^{-11} M), whereas hyperglycemia suppresses glucagon secretion by more than 50%. These

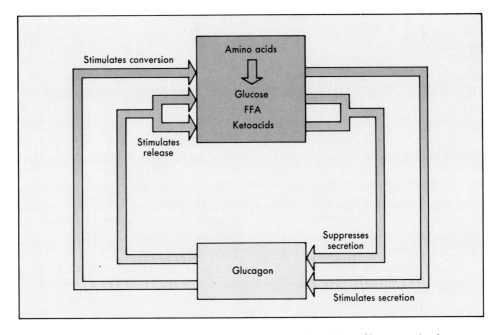

FIGURE 37-6   Feedback relationship between glucagon and nutrients. Glucagon stimulates production and release of glucose, free fatty acids *(FFA)*, and ketoacids, which in turn suppress glucagon secretion. Amino acids stimulate glucagon secretion, and glucagon in turn stimulates amino acids' conversion to glucose.

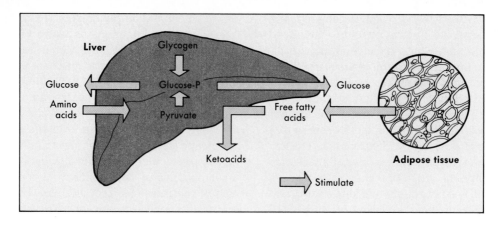

FIGURE 37-7   Effect of glucagon on the overall flow of fuels. Tissue release of glucose, free fatty acids, and ketoacids raises their plasma levels, whereas liver uptake of amino acids lowers their plasma levels.

effects of glucose are independently reinforced by insulin, possibly through a paracrine action within the islet. Thus insulin directly inhibits glucagon secretion; conversely, when insulin is absent, the stimulatory effect of low glucose levels on glucagon secretion is exaggerated. The other major energy substrate, free fatty acids, also suppresses glucagon release, whereas a sharp decline in plasma free fatty acid levels is stimulatory. A protein meal and amino acids, the substrates for glucose production, stimulate glucagon secretion, but this response is dampened by concurrent glucose or insulin action. As a result, the usual mixed meal produces only small and variable increases in plasma glucagon levels, in contrast to the large and consistent increases in plasma insulin levels.

Prolonged fasting and sustained exercise—circumstances that require glucose mobilization—increase glucagon secretion. Under stressful conditions such as major infection or surgery, glucagon secretion is often greatly augmented. This probably occurs through sympathetic nervous system stimulation of the α-cells via α-adrenergic receptors.

Glucagon is extracted by the liver on the first pass, as is insulin, and it has a short half-life in peripheral plasma. As with insulin, glucagon is degraded in the kidney and liver, and very little is secreted in the urine.

Actions of Glucagon

In almost all respects the actions of glucagon are opposite to those of insulin. Glucagon promotes mobilization rather than storage of fuels, especially glucose (Figure 37-7). Both hormones act at numerous similar control points for glucose metabolism in the liver. Indeed, glucagon may be the primary hormone that regulates hepatic glucose production (and ketogenesis), with insulin's main role being that of a glucagon antagonist.

Glucagon exerts an immediate and profound glycogenolytic effect through activation of glycogen phosphorylase. Simultaneously, resynthesis of phosphorylated glucose molecules back to glycogen is prevented by inhibition of glycogen synthase. Glucagon also stimulates gluconeogenesis by at least several mechanisms. The hepatic extraction of amino acid precursors is increased. The activities of key gluconeogenic enzymes—pyruvate carboxylase, phospho-*enol*pyruvate carboxykinase, and fructose-1,6-diphosphatase—are increased, whereas activities of key glycolytic enzymes, phosphofructokinase and pyruvate kinase, are decreased. The enzyme pair phosphofructokinase/fructose-1,6-diphosphatase determines the flow between fructose-6-phosphate and fructose-1,6-diphosphate. Thus this enzyme pair determines the relative rates of gluconeogenesis and glycolysis. The activities of these two enzymes in turn are reciprocally related by the hepatic level of another metabolite, fructose-2,6-biphosphate. This metabolite is decreased by glucagon, an action that favors flow from fructose-1,6-diphosphate to fructose-6-phosphate, thereby stimulating glucoenogenesis. (Insulin has the opposite effect, probably by inhibiting the action of glucagon.)

The importance of glucagon is shown by a decline

of 75% in hepatic glucose output if glucagon secretion is inhibited. Increasing glucagon concentrations from 1.5- to 5.0-fold powerfully stimulates glycogenolysis and rapidly raises the plasma glucose level. This occurs even in the presence of modestly elevated insulin levels. However, this hyperglycemic action of glucagon is transient. After the initial increase, hepatic glucose production wanes during continuous glucagon administration, probably because of glucose autoregulation within the liver and stimulation of insulin release. However, if glucagon is given in a more physiologic fluctuating pattern, each increment in the hormone increases hepatic glucose output.

Another intrahepatic action of glucagon is to direct incoming free fatty acids away from triglyceride synthesis and toward β-oxidation. Thus glucagon is a ketogenic hormone. The mechanism involves the intermediate malonyl CoA, which is an inhibitor of free fatty acid transfer into the mitochondria. Glucagon suppresses the synthesis of malonyl CoA by inhibiting the enzyme acetyl-CoA carboxylase. The lower levels of malonyl CoA then allow a higher rate of influx of free fatty acids into the mitochondria for conversion to ketoacids.

Glucagon actions on adipose tissue or muscle are insignificant unless insulin is virtually absent. Peripheral glucose utilization is largely unaffected by glucagon. However, glucagon can activate hormone-sensitive adipose tissue lipase and thereby increase lipolysis, the delivery of free fatty acids to the liver, and ketogenesis. Another action of glucagon, opposite to that of insulin, is to inhibit renal tubular sodium resorption and thus to cause natriuresis.

The molecular mechanism of glucagon action begins with binding to a plasma membrane receptor in the liver. The glucagon receptor complex causes a rapid increase in intracellular cAMP (see Chapter 35). This is followed by a specific enzymatic cascade (Figure 35-8). Protein kinase A activity increases, converting inactive phosphorylase kinase to active phosphorylase kinase. The latter then converts inactive phosphorylase to active phosphorylase, and glycogenolysis results. Several other enzymes whose functional state depends on addition or removal of phosphate are likewise regulated by glucagon.

INSULIN/GLUCAGON RATIO

It should now be apparent that substrate fluxes are very sensitive to the relative availability of insulin and glucagon. The usual molar ratio of insulin to glucagon in plasma is about 2.0. Under circumstances that require mobilization and increased utilization of endogenous substrates, the insulin/glucagon ratio drops to 0.5 or less. This is seen in fasting, in prolonged exercise, and in the neonatal period when the infant is abruptly cut off from maternal fuel supplies but is not yet able to assimilate exogenous fuel efficiently. The ratio drops because of both decreased insulin secretion and increased glucagon secretion. Conversely, under circumstances in which substrate storage is advantageous, as after a pure carbohydrate load or a mixed meal, this ratio rises to 10 or more, mainly because of increased insulin secretion. An interesting example of only a small and insignificant change in the insulin/glucagon ratio occurs after ingesting a pure protein meal. In this situation insulin secretion increases, facilitating muscle uptake of amino acids and their synthesis into proteins. At the same time glucagon secretion also increases. This prevents the immediate decrease in hepatic glucose output and hypoglycemia that would ensue if the extra insulin action were unopposed.

ISLET SOMATOSTATIN

Somatostatin is a neuropeptide hormone originally discovered in the hypothalamus, where it serves as an inhibitor of growth hormone secretion from the anterior pituitary gland (see Chapter 39). In the δ-cells of the islets a large preprohormone is processed to two somatostatin peptides, one of 28 amino acids and one of 14 amino acids. Their secretion is stimulated by glucose, amino acids, free fatty acids, glucagon, and several gastrointestinal hormones such as cholecystokinin and vasoactive intestinal peptide. Somatostatin secretion is probably inhibited by insulin. After a mixed meal, plasma somatostatin concentration increases 50% to 100%.

The overall thrust of somatostatin action is to decrease the rate of digestion and absorption of nutrients from the gastrointestinal tract and their subsequent utilization. Thus this neuropeptide inhibits gastric, duodenal, and gallbladder motility; it reduces the secretion of hydrochloric acid, pepsin, gastrin, secretin, intestinal juice, and pancreatic enzymes. Somatostatin also inhibits the absorption of glucose and triglycerides across the intestinal mucosal membrane. Finally, it powerfully inhibits insulin and glucagon secretion.

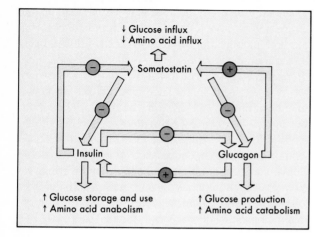

FIGURE 37-8   The interrelationships between somatostatin, insulin, and glucagon effects on each other's secretions and their effects on glucose and amino acid metabolism. (Modified from Unger RH et al. Reproduced with permission from the *Annual Review of Physiology,* volume 40. Copyright © 1978 by Annual Reviews, Inc.)

Somatostatin of islet origin probably participates in a feedback arrangement whereby entrance of food into the gut stimulates the release of the hormone so as to prevent rapid nutrient overload. The anatomic relationships between α-, β-, δ-cells and the existence of tight junctions and gap junction between them suggest that all three islet hormones—somatostatin, insulin, and glucagon—may be influencing each other's secretion by paracrine effects (Figure 37-8). This may improve coordination between the bulk movement, digestion, and absorption of nutrients, with the insulin and glucagon responses necessary for proper disposition of nutrients in the liver and other tissues.

BIBLIOGRAPHY
Journal Articles

Cheng JS and Kalant N: Effects of insulin and growth hormone on the flux rates of plasma glucose and plasma free fatty acids in man, J Clin Endocrinol 31:647, 1970.

Czech MP: The nature and regulation of the insulin receptor: structure and function, Annu Rev Physiol 47:357, 1985.

Genuth SM: Plasma insulin and glucose profiles in normal, obsese, and diabetic persons, Ann Intern Med 79:812m 1973.

Granner DK and Andreone TL: Insulin modulation of gene expression, Diabetes Metab Rev 1:139, 1985.

Katz L et al: Splanchnic and peripheral disposal of oral glucose in man, Diabetes 32:675, 1983.

Liljenquist JE et al: Evidence for an important role of glucagon in the regulation of hepatic glucose production in normal man, J Clin Invest 59:369, 1977.

Nair S et al: Effect of intravenous insulin treatment on in vivo whole body leucine kinetics and oxygen consumption in insulin deprived Type I diabetic patients, Metabolism 36:491, 1987.

Nurjhan N et al: Insulin dose-response characteristics for suppression of glycerol release and conversion to glucose in humans, Diabetes 35:1326, 1986.

Thiebaud D et al: The effect of graded doses of insulin on total glucose uptake, glucose oxidation, and glucose storage in man, Diabetes 31:957, 1982.

Unger RH et al: Insulin, glucagon, and somatostatin secretion in the regulation of metabolism, Annu Rev Physiol 40:307, 1978.

Unger RH et al: Glucagon and the A cell. I. Physiology and pathophysiology, N Engl J Med 304:1518, 1981.

Weigle DS: Pulsatile secretion of fuel regulatory hormones, Diabetes 36:764, 1987.

Books and Monographs

Matchinsky FM and Bedoya FJ: Metabolism of pancreatic islets and regulation of insulin and glucagon secretion. In DeGroot LJ, editor: Endocrinology, Philadelphia, 1989, WB Saunders Co.

Steiner DF et al: Chemistry and biosynthesis of pancreatic protein hormones. In DeGroot LJ editor: Endocrinology, Philadelphia, 1989, WB Saunders Co.

Yalow RS and Bauman WA: Plasma insulin in health and disease. In Ellenberg M and Rifkin H, editors: Diabetes mellitus, New Hyde Park, 1983, Medical Examination Publishing Co, Inc.

CHAPTER 38

Endocrine Regulation of the Metabolism of Calcium and Related Minerals

Calcium, phosphate, and magnesium homeostasis are essential for health and life. A complex system acts to maintain normal body contents and extracellular fluid levels of these minerals in the face of environmental (e.g., diet) and internal (e.g., pregnancy) changes. The key elements in the system are *vitamin D* and *parathyroid hormone* (PTH), with subsidiary participation by *calcitonin* (CT) and other hormones. The intestinal tract, the kidneys, the skeleton, the skin, and the liver are all involved in the homeostatic regulation of calcium, phosphate, and magnesium metabolism.

CALCIUM, PHOSPHATE, AND MAGNESIUM TURNOVER

Calcium

The calcium ion (Ca^{++}) is of fundamental importance to all biologic systems. Calcium, usually complexed to *calmodulin* (see Chapter 1), participates in numerous enzymatic reactions of metabolic importance. It is a vital component in the mechanisms of hormone secretion and hormone action. Calcium is intimately involved in neurotransmission, muscle contraction, mitosis and cell division, fertilization, and blood clotting. Ca^{++} is the major cation in the crystalline structure of bone and teeth. For these reasons, it is vital that cells be bathed with fluid in which the calcium concentration is kept within narrow limits.

Calcium metabolism may be viewed as having two parts: an intracellular microcomponent and an extracellular macrocomponent. Each is regulated differently and somewhat independently. The crucial intracellular calcium functions are carried out at an average basal cytosolic free calcium concentration of 10^{-7} molar (range 5×10^{-8} to 3×10^{-7} molar). In contrast, the free calcium concentration in extracellular fluid is approximately 10^{-3} molar, or 10,000-fold higher. This large extracellular/intracellular gradient of calcium is maintained by a low permeability of the plasma cell membrane to Ca^{++} and by the regulated activities of a Ca^{++}-ATPase pump and a Ca^{++}-Na^{+} exchange system (see Chapter 1). Within the cell a much larger store of Ca^{++} is bound to various proteins and membranes, to the endoplasmic reticulum, and within the mitochondria. The total Ca^{++} content is equivalent to an intracellular concentration of 10^{-2} molar.

The cytosolic free calcium concentration can be altered as needed, both by regulating influx from outside the cell and by mobilizing intracellular stores. When calcium is required to function as a second messenger, the free calcium concentration can rise from 10^{-7} molar to as high as 10^{-5} molar. In absolute terms, however, this represents the movement of only small amounts of extra calcium into the cytoplasmic fluid. Such changes are transient (seconds to minutes). The excess cytosolic calcium is either rapidly extruded from the cell or returned to the intracellular reservoirs. The influx and efflux of calcium are so finely balanced that this ubiquitous ion can serve as an internal cell signal with a large dynamic range of gain and sensitivity.

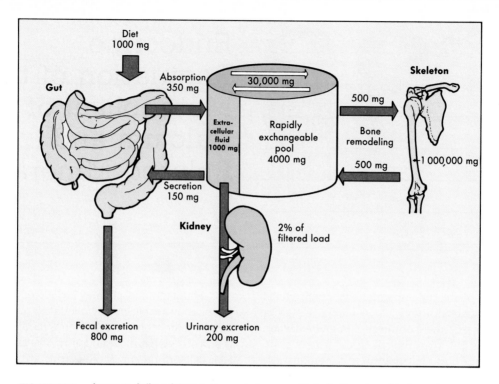

FIGURE 38-1   Average daily calcium turnover in humans. Note both external balance between intake and excretion and internal balance between entry into and exit from bone.

The concentration of calcium in the extracellular fluid and plasma normally fluctuates little. This clearly helps to maintain intracellular calcium at a proper level. When plasma and extracellular fluid calcium is either greater or less than the normal range, intracellular function can be widely and severely affected. Abnormalities in neurotransmission and in the growth and renewal of the skeleton are prominent examples.

The normal range of total calcium concentration in plasma is 8.6 to 10.6 mg·dl^{-1}, or 2.15 to 2.65 × 10^{-3} M. Because Ca^{++} is a divalent ion, this is equivalent to 4.3 to 5.3 mEq·L^{-1}. Individual day-to-day variation is less than 10%. Approximately 50% of total plasma calcium is in the ionized form, Ca^{++}, which is biologically active; 40% is bound to proteins, mainly albumin; 10% is complexed in nonionic but ultrafilterable forms, such as calcium bicarbonate. The equilibrium between ionized and protein-bound calcium depends on blood pH. Alkalosis increases the protein-bound and decreases the ionized Ca^{++} concentration, whereas acidosis has the opposite effect. The total plasma calcium concentration rises or falls with plasma albumin levels, but this has no biologic conse-

quence as long as the ionized Ca^{++} concentration remains in the normal range.

Figure 38-1 details the normal turnover of calcium in the body. Daily dietary calcium intake may range from 200 to 2000 mg. The percentage of dietary calcium absorbed from the gut is inversely related to the intake in a curvilinear manner. Thus, in the face of dietary calcium deprivation, one important mechanism for maintaining normal plasma calcium concentration and normal body calcium stores is an adaptive increase in fractional absorption. Conversely, in the face of dietary calcium excess, overload is prevented by an adaptive decrease in absorption. At a daily intake of 1000 mg, about 35% is absorbed. The same amount of calcium, 350 mg, is excreted. In a steady state approximately 150 mg is secreted into intestinal juices and excreted in the stools, along with the unabsorbed fraction from the diet. The remaining 200 mg is excreted in the urine. Although the kidney filters about 10,000 mg of non–protein-bound Ca^{++} per day, approximately 98% is reabsorbed by the renal tubules. Therefore alteration in renal tubular Ca^{++} transport provides another sensitive means for maintaining calcium balance.

The extracellular pool of calcium is only 1000 mg. The largest store of calcium, about 1.2 kg, is in the skeleton. Of this, 4000 mg is available for rapid exchange with the extracellular pool and for buffering of plasma calcium. Bone is a dynamic tissue that undergoes daily turnover. In this process approximately 500 mg of calcium is extracted from the extracellular pool as new bone is formed, and a like amount is returned to this pool as old bone is broken down.

Phosphate

The phosphate ion (PO_4^-) is also of critical importance to all biologic systems and is the major intracellular anion. Phosphate is a component of all intermediates in glucose metabolism. It is part of the structure of all high-energy transfer compounds, such as adenosine triphosphate (ATP); of cofactors such as nicotinic acid dinucleotide; and of lipids, such as phosphatidylcholine. Phosphate functions as a covalent modifier of numerous enzymes. Phosphate also forms part of the crystalline structure of bone.

The normal concentration of phosphate in the plasma is 2.4 to 4.5 mg·dl^{-1}, or 0.81 to 1.45 × 10^{-3} M. Because the valence of phosphate changes with pH, it is less useful to express normal concentrations in mEq·L^{-1}. The turnover of phosphate is shown in Figure 38-2. In contrast to calcium, the percentage of phosphate absorbed from the diet is relatively constant, and thus the net absorption of phosphate from the gut is linearly related to intake. Therefore urinary excretion provides the major mechanism for regulating phosphate balance. The daily filtered load is approximately 6000 mg, but renal tubular reabsorption can vary from 70% to 100%, giving the needed flexibility to compensate for fluctuation in dietary intake. Large soft tissue stores of phosphate, as in muscle, are a source for rapid regulation of the plasma concentration. Approximately 250 mg of phosphate enters and leaves the extracellular fluid daily in the course of bone turnover. Severe depletion of phosphate can result in serious cardiac and skeletal muscle dysfunction and in abnormal bone growth.

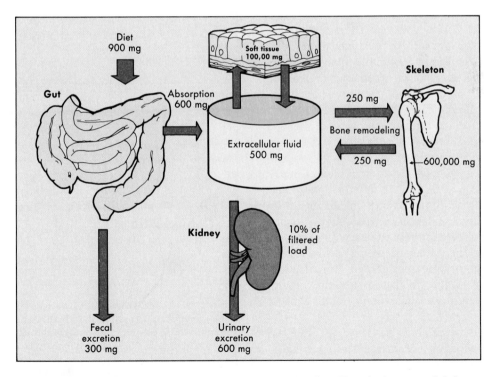

FIGURE 38-2   Average daily phosphate turnover in humans. Note both external balance between intake and exretion and internal balance between entry into and exit from bone.

Magnesium

The divalent cation magnesium (Mg^{++}) is related in some metabolic respects to calcium and phosphate. Mg^{++}, a major intracellular cation, is essential to neuromuscular transmission and is a cofactor in numerous reactions, most notably those involving energy transfers via ATP and those concerned with protein synthesis. The normal range of magnesium in plasma is 1.8 to 2.4 $mg \cdot dl^{-1}$, or 0.75 to 1.00×10^{-3} M, which equals 1.5 to 2.0 $mEq \cdot L^{-1}$. One third of plasma magnesium is bound to protein. The average daily intake is about 300 mg, of which 40% is absorbed and (in a steady state) excreted in the urine. The body content of magnesium is about 25 g, 50% of which is present in the skeleton. Severe depletion of magnesium can produce abnormal neuromuscular transmission and serious irregularity of the heart rhythm.

BONE TURNOVER

As already indicated, bone is a major and dynamic reservoir for calcium and phosphate. It is therefore essential to understand those aspects of bone structure and function that are pertinent to endocrine regulation.

Bone is broadly divided into two types. *Cortical*, or compact, bone represents 80% of the total and is typified by the thick shafts of the appendicular skeleton (arms and legs). *Trabecular*, or spongy, bone constitutes 20%; it makes up most of the axial skeleton (vertebrae, skull, ribs) and bridges the center of the long bones. The fivefold greater surface area of trabecular bone gives it disproportionate significance in regulation of calcium metabolism, despite its lesser mass.

Bone formation occurs on the outer surface of cortical bone, whereas bone resorption occurs on its inner surface. Both formation and resorption also take place in specialized nutrient canals within cortical bone and on the surfaces of trabecular bone. Throughout life, the processes of bone formation and resorption are tightly regulated. During growth phases, formation exceeds resorption and the skeletal mass increases.

Linear growth occurs between the heads and the shafts of long bones in specialized areas known as *epiphyseal growth plates*. These close off at the end of puberty when adult height is reached. Width increases by adding bone to the outer surfaces. Total

bone mass reaches a peak between ages 20 and 30 years. Thereafter, equal rates of formation and resorption prevail until ages 40 to 50 years, at which time resorption begins to exceed formation and the total bone mass slowly decreases. The process of bone turnover in the adult, known as *remodeling*, may involve up to 15% of the total bone mass per year. Endocrine disturbances that disrupt the coupling of formation and resorption are particularly harmful when they are superimposed on either the growth phase or the senescent phase of life. Women are especially affected because they have a smaller bone mass and a more rapid rate of senescent loss than men.

Bone Formation

Three major cell types exist in bone: osteoblasts, osteocytes, and osteoclasts (Figure 38-3). The first two arise from primitive mesenchymal cells, called *osteoprogenitor cells*, within the investing connective tissue. Various bone proteins attract osteoprogenitor cells, direct their differentiation into osteoblasts, and stimulate their further growth. Osteoclasts arise from

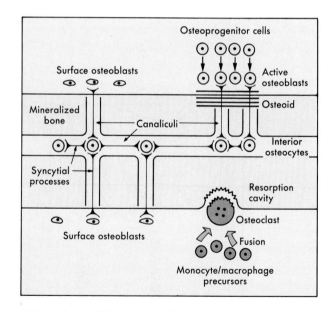

FIGURE 38-3 The relationships between bone cells and bone remodeling. Note that the canaliculi provide channels through which calcium and phosphate can be transferred from the interior to the exterior of bone. These minerals may be used for new bone formation or may be transported into the circulation. (Redrawn from Avioli LV et al: Bone metabolism and disease. In Bundy PK and Rosenberg LE: Metabolic control and disease, Philadelphia, 1980, WB Saunders Co.)

the same precursors as do circulating monocytes and tissue macrophages. Together these three cell types form the *osteon,* or bone modeling unit.

Bone formation is carried out by active *osteoblasts,* which synthesize and secrete type 1 collagen. The collagen fibrils line up in regular arrays. This process creates an organic matrix known as *osteoid,* within which calcium and phosphate are subsequently deposited in amorphous masses. The slow addition of hydroxide and bicarbonate ions to the mineral phase produces mature *hydroxyapatite* crystals. As the completely mineralized bone accumulates and surrounds the osteoblast, that cell loses its synthetic activity and becomes an interior *osteocyte* (Figure 38-3). Osteoblastic activity therefore is observed only on the surfaces of bone, along which resting cells wait to be activated.

The mineralization process requires normal plasma concentrations of calcium and phosphate. The enzyme *alkaline phosphatase* and other proteins from the osteoblast also participate. *Osteocalcin,* which forms 1% to 2% of all bone protein, has a strong affinity for calcium and for uncrystallized hydroxyapatite. Osteocalcin may therefore actually function as a local feedback inhibitor of mineralization. Alkaline phosphatase and osteocalcin circulate in plasma, and their concentrations correlate well with histologic evidence of osteoblastic activity.

Within each osteon, minute fluid-containing channels, called *canaliculi,* traverse the mineralized bone; through these channels the interior osteocytes remain connected with surface osteocytes and osteoblasts via syncytial cell processes (Figure 38-3). This arrangement provides an enormous surface area for the transfer of calcium from the interior to the exterior of the osteons and from there to the extracellular fluid. This transfer process, carried out by the osteocytes, is known as *osteocytic osteolysis.* It probably does not decrease mature bone mass, but simply removes calcium from the most recently formed crystals.

Bone Resorption

The process of bone resorption does not merely extract calcium; it destroys the entire organic matrix as well, thereby diminishing the mature bone mass. The cell responsible is the *osteoclast,* which is a giant multinucleated cell formed by fusion of several precursors (Figure 38-3). The osteoclast contains large numbers of mitochondria and lysosomes. It attaches

to the surface of the osteon and creates at this point a ruffled border by infolding of its plasma membrane. In this zone the process of dissolution is carried out by collagenase and other enzymes. During this process the osteoclast literally tunnels its way into the mineralized bone. Calcium, phosphate, magnesium, and the constituent amino acids—including hydroxyproline and hydroxylysine, which are unique to collagen—are released into the extracellular fluid.

As already emphasized, resorption and formation of bone are closely coordinated locally. The osteoclast may be activated by a product of nearby resting osteoblasts. Furthermore, the resorption cavity created by the osteoclast is the usual site of subsequent osteoblastic activity, which fills in the recently formed cavity with new bone. Thus bone resorption may actually be the trigger for bone formation.

The recruitment of osteoblasts and osteoclasts from precursors and the activity of each cell type are regulated by various local factors, including lymphokines such as interleukin-1, prostaglandins, and an array of hormones. As a general principle, whether the primary effect of a hormone is on the formation or the resorption of bone, the phenomenon of coupling will secondarily alter the other process in the same direction. Therefore the net effect of a hormone excess or deficiency will depend on the degree to which the coupling phenomenon defends the total bone mass.

VITAMIN D AND ITS METABOLISM

Vitamin D, through its active metabolites, is a major regulator of calcium and phosphate metabolism. Its actions help to sustain normal plasma concentrations of calcium and phosphate by increasing their inflow from the intestinal tract and bone reservoirs. Vitamin D is a hormone in the sense that it is synthesized in the body, although not by an endocrine gland; after further processing, it is transported via the circulation to act on target cells. It is a vitamin in the sense that when it cannot be synthesized in sufficient quantities, it must be ingested in minimal amounts for health to be maintained. Deficiency of vitamin D causes failure of bone mineralization and results in the classic disease of *rickets* in children and softening of the bones *(osteomalacia)* in adults.

The sterol structure of the synthesized form of vitamin D (D_3) (Figure 38-4) differs slightly from the form

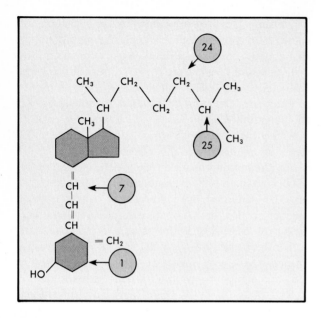

FIGURE 38-4   The structure of vitamin D. Positions 1, 24 and 25 are important sites of hydroxylation that affect biological activity.

usually ingested (D_2). The latter can be prepared by irradiating plant or milk ergosterol. Vitamins D_3 and D_2 are essentially prohormones that undergo identical processing that converts them to molecules with identical qualitative and quantitative actions. Henceforth, the term *vitamin D* will be employed to indicate both.

The minimum daily requirement of vitamin D is approximately 2.5 μg (100 units). Endogenously, it is synthesized in the skin by ultraviolet irradiation of the precursor *7-dehydrocholesterol*. Exogenously, vitamin D is available in fish, liver, and milk, and it is absorbed from the gut just as fats are absorbed. Vitamin D deficiency states can therefore result from lack of sunlight, lack of dietary intake, diseases that impair its absorption from the gut, or inadequate hepatic or renal metabolism to active forms. Vitamin D is stored in adipose tissue in amounts normally sufficient for several months, requirements.

Once vitamin D enters the circulation from the skin or the gut, it is concentrated in the liver. There it is hydroxylated to 25-OH-D. This molecule is transported to the kidney where it undergoes alternative fates (Figure 38-5). Hydroxylation in the the 1-position produces the metabolite, 1,25-$(OH)_2$-D, which unquestionably expresses most if not all of the biologic activity of vitamin D. Alternatively, 25-OH-D may be

hydroxylated in the 24-position. 24, 25-$(OH)_2$-D is only 1/20th as potent as 1,25-$(OH)_2$-D and mainly serves to dispose of excess vitamin D.

Feedback control of vitamin D activation occurs through regulation of the renal 1-hydroxylase and 24-hydroxylase activities (Figure 38-5). 25-OH-D is preferentially directed toward the active metabolite, 1,25-$(OH)_2$-D, whenever calcium, phosphate, or vitamin D itself is lacking. Calcium deprivation leads to compensatory secretion of parathyroid hormone. This hormone then stimulates 1-hydroxylation. A lowering of plasma phosphate and renal phosphate content also augments 1-hydroxylase activity. In addition, 1,25-$(OH)_2$-D is a feedback inhibitor of its own synthesis; hence in vitamin D deficiency, 1-hydroxylase activity is enhanced. By contrast, 24-hydroxylase activity is stimulated by normal to elevated calcium or phosphate concentrations, and by 1,25-$(OH)_2$-D. The net result of this regulation is that the supply of active 1,25-$(OH)_2$-D is increased (and that of inactive 24,25-$(OH)_2$-D is decreased) whenever homeostasis requires increasing calcium and phosphate absorption from dietary sources and mobilizing these minerals from skeletal stores.

Vitamin D, 25-OH-D and 1,25-$(OH)_2$-D circulate bound to a protein carrier. 1,25-$(OH)_2$-D has by far the lowest concentration and the shortest half-life of the three. However, regulation is powerful enough to maintain the appropriate concentration of 1,25-$(OH)_2$-D even when the concentrations of its precursors are very reduced.

Actions of Vitamin D

The active form of vitamin D (1,25-$(OH)_2$-D) acts through the general mechanism outlined for steroid hormones (see Chapter 35). After binding to a cytosolic receptor, the hormone receptor complex enters the nucleus, where it stimulates transcription of messenger ribonucleic acid (mRNA) for at least one identified product. This is a calcium-binding protein found in cells of the intestinal mucosa, bone, kidney, and parathyroid glands. It has significant homology with calmodulin and a high affinity for calcium. This calcium-binding protein is not essential for vitamin D stimulation of calcium transport because the latter precedes the appearance of the protein in intestinal cells. However, the calcium-binding protein may protect the cells from high cytoplasmic concentrations of calcium during enhanced transport.

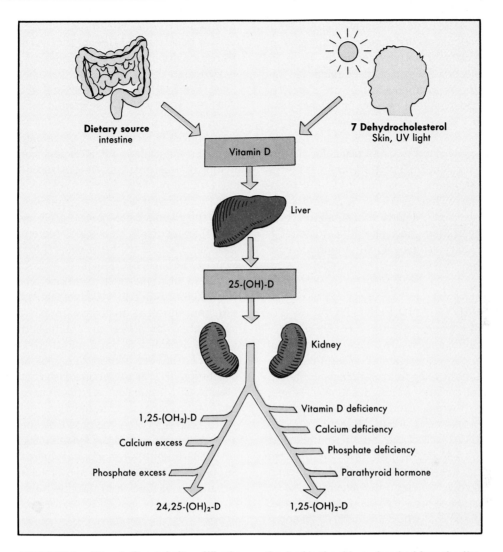

FIGURE 38-5   Vitamin D metabolism. Whether synthesized in the skin or absorbed from the diet, vitamin D undergoes 25 hydroxylation in the liver. In the kidney, it is further hydroxylated in the 1 position when more biological activity is required or in the 24 position when less biologic activity is required.

The major action of 1,25-(OH)$_2$-D is to stimulate absorption of calcium from the intestinal lumen against a concentration gradient (see Chapter 31). 1,25-(OH)$_2$-D localizes in the nuclei of intestinal villus and crypt cells, where it acts on the brush border. It probably stimulates production of a plasma membrane protein in the calcium channel because inhibition of protein synthesis prevents hormone action on calcium transport. 1,25-(OH)$_2$-D is responsible for the adaptation previously described whereby intestinal absorption of calcium increases in response to decreases in its dietary intake. 1,25-(OH)$_2$-D also augments active absorption of phosphate and magnesium across the intestinal cell membrane.

1,25-(OH)$_2$-D also stimulates bone resorption. Osteoclasts have no receptors for 1,25-(OH)$_2$-D, but osteoblasts do; therefore the resorptive action of the osteoclast may be driven by an osteoblast—derived factor. This effect of 1,25-(OH)$_2$-D is physiologically important in sensitizing the bone to the resorptive effects of PTH.

The normal mineralization of newly formed osteoid

along a regular front is critically dependent on vitamin d. In its absence, excess osteoid accumulates from continued osteoblastic activity, and the bone so formed is weakened. The major mechanism for this Vitamin D action is augmentation of the supply of calcium and phosphate. However, osteoblasts can also respond to 1,25-$(OH)_2$-D by decreasing collagen synthesis.

Skeletal muscle is another target tissue for vitamin D. It increases calcium transport and uptake by the sarcoplasmic reticulum, as well as uptake of phosphate. Deficiency of vitamin D leads to muscle weakness, electrophysiologic evidence of abnormal contraction and relaxation, and altered cytoarchitecture. Vitamin D may also play a role in immunoregulation.

PARATHYROID GLAND FUNCTION

The parathyroid glands are major regulators of plasma calcium concentration and calcium flux. The parathyroid glands develop from branchial pouches at 5 to 14 weeks of gestation. They descend to lie just posterior to the thyroid gland. The total weight of adult parathyroid tissue is about 130 mg.

The predominant cell of the parathyroid gland is known as the *chief cell*. These cells are present throughout life and are the source of PTH. A second cell type of unknown function, the *oxyphil cell,* first appears at puberty and increases in number with age. Active chief cells have a large convoluted Golgi apparatus with vacuoles and vesicles, and a granular endoplasmic reticulum. During hormone secretion numerous granules may be seen undergoing exocytosis.

The paramount effect of PTH is to sustain or increase the plasma calcium level. This is accomplished by stimulating entry of calcium into plasma from bone, tubular urine, and the intestinal tract. An important second effect is to decrease or prevent an undue rise in the plasma phosphate level by stimulating excretion of phosphate into the urine.

Synthesis and Secretion of Parathyroid Hormone

PTH is a single-chain protein (9600 molecular weight) that contains 84 amino acids. The biologic activity of the hormone resides in the N-terminal portion of the molecule within amino acids 1 to 34. The function of the larger carboxyl portion is not known.

The gene for PTH directs the synthesis of prepro-PTH. As the peptide chain grows to its complete length on the ribosomes, 25 amino acids are enzymatically cleaved from the N-terminal end, leaving pro-PTH. Pro-PTH is then transported to the Golgi apparatus, where another six amino acids are cleaved. The resulting PTH is packaged for storage in secretory granules. Degradation of PTH also occurs within the gland; therefore not all synthesized molecules reach the circulation.

The dominant regulator of parathyroid gland activity is the plasma calcium level. PTH and Ca^{++} form a negative feedback pair, and secretion of PTH is inversely related to the plasma calcium concentration (Figure 38-6). Maximal secretory rates are achieved when plasma ionized calcium concentration falls below 3.5 mg·dl^{-1}. Conversely, as ionized calcium concentration increases to 5.5 mg·dl^{-1}, PTH secretion is progressively diminished. However, it reaches a persistent basal rate that is not suppressible by further elevation of the ambient calcium concentration. PTH secretion increases within minutes if plasma Ca^{++} is selectively decreased by chelation and total plasma calcium remains unchanged.

Release of PTH is mediated by the activation of adenylate cyclase and the resultant rise in intracellular cyclic adenosine monophosphate (cAMP) levels. Suppression of PTH secretion by calcium represents a remarkable exception to the rule that calcium influx

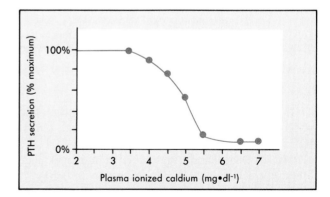

FIGURE 38-6   The inverse relationship between parathyroid hormone secretion and plasma ionized calcium concentration in humans. (Redrawn from Brent GA, et al: Relationship between the concentration and rate of change of calcium and serum intact parathyroid hormone levels in normal humans, J Clin Endocrinol Metab 67:944, 1988. Copyright 1988 by The Endocrine Society.)

into endocrine cells stimulates hormone secretion. The mechanism underlying this paradoxic effect in the parathyroid glands is still uncertain. Preliminary evidence for an external Ca^{++} sensor in the plasma membrane of parathyroid cells has been presented.

Calcium also modulates PTH turnover within the gland. Prolonged exposure to a high calcium concentration lowers the rate of PTH synthesis and stimulates the intraglandular degradation of PTH. In addition, calcium regulates the size and number of parathyroid cells. The net effect of calcium excess is to decrease both the glandular stores and the release rates of PTH. Conversely, a decrease in calcium increases PTH stores, secretory rates, and ultimately gland size as well.

Hypomagnesemia stimulates PTH secretion in a manner analogous to that of hypocalcemia, although it is much less potent. Therefore it is of little importance in the normal range of plasma magnesium con-centrations. However, chronic magnesium depletion ultimately inhibits PTH synthesis, perhaps by diminishing energy production from ATP-related reactions.

Phosphate exerts no direct effect on PTH secretion in vitro. However, by complexing calcium and decreasing Ca^{++} concentration, a rise in plasma phosphate concentration indirectly causes a transient increase in PTH secretion. 1,25-$(OH)_2$-D directly feeds back on the parathyroid gland to decrease the level of prepro-PTH mRNA and eventually reduce PTH secretion.

Actions of Parathyroid Hormone

Intracellular Effects The overall effect of PTH is to increase plasma calcium levels and decrease plasma phosphate levels by acting on three major target organs: kidney, bone, and indirectly the gastrointes-

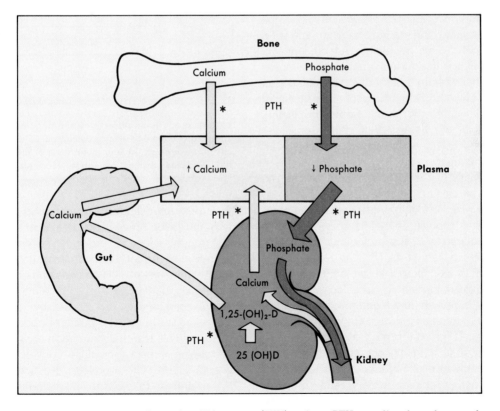

FIGURE 38-7   Overview of parathyroid hormone *(PTH)* actions. PTH acts directly on bone and kidney to increase calcium influx into plasma. By stimulating 1,25-$(OH)_2$-D synthesis, PTH indirectly also increases calcium absorption from the gut. Thus plasma calcium level increases. In contrast, PTH inhibits renal tubular resorption of phosphate, thereby increasing urinary phosphate excretion. This effect quantitatively offsets entry of phosphate from bone and gut. Therefore plasma phosphate level decreases.

tinal tract. Actions on all three targets ultimately increase calcium influx into the plasma and raise its concentration (Figure 38-7). In contrast, PTH action on the kidney, which increases phosphate exit from plasma, overwhelms the actions on bone and gut, which increase phosphate entry into the plasma; therefore plasma phosphate concentration falls (Figure 38-7).

PTH action is initiated by binding to a plasma membrane receptor. In all target cells, activation of adenylate cyclase and an increase in cAMP follow. The second messenger then triggers a protein kinase cascade (see Chapter 35), which ultimately leads to phosphorylation of proteins necessary for expression of PTH action.

Independent of cAMP, PTH also stimulates the uptake of calcium into the cytoplasm of its target cells. Whether this calcium is itself another intracellular second messenger or whether it only acts to modulate the adenylate cyclase system remains moot. However, the initial uptake of calcium is reflected in a slight transient *hypocalcemia,* which immediately follows PTH administration and precedes the classic *hypercalcemia.* The presence of $1,25\text{-}(OH)_2\text{-}D$ is required for maximal responsiveness to PTH. Whether vitamin D has a direct cellular action or whether it only augments the pool of extracellular calcium from which the rapid early cytoplasmic uptake occurs is not certain. A sufficient intracellular concentration of magnesium is also necessary for PTH to act maximally.

Renal Effects In the kidney PTH increases the reabsorption of calcium in the distal tubule and decreases the reabsorption of phosphate in the proximal and distal tubules (see Chapter 34). These effects are mediated by PTH stimulation of cAMP production at the capillary surface of the renal tubular cell. The cAMP is transported to the luminal surface, where it activates protein kinases located in the membranes of the brush border and involved in calcium and phosphate reabsorption. During this process cAMP is released into the tubular lumen. Therefore the earliest observable renal effect of PTH in vivo is a dramatic increase in cAMP excretion in the urine.

The relationship between urinary calcium excretion and plasma calcium concentration is shifted to the right by PTH (Figure 38-8). At any given plasma calcium concentration, PTH diminishes the amount of calcium lost in the urine and thus counters hypocalcemia. Conversely, suppression of PTH secretion

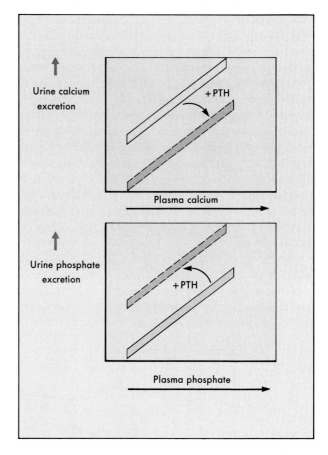

FIGURE 38-8   Renal effects of parathyroid hormone *(PTH).* At any given level of plasma calcium, PTH **decreases** urinary calcium excretion. At any given level of plasma phosphate, PTH **increases** urinary phosphate excretion.

by an excess calcium load increases calcium excretion and helps prevent hypercalcemia.

The net effect of prolonged alterations in PTH secretion on urinary calcium excretion is dominated by the influence of PTH on bone and gut. Excess PTH eventually elevates the plasma calcium level as well as the load of calcium filtered by the glomerulus. Therefore the absolute amount of calcium excreted in the urine will eventually increase, despite stimulation of tubular reabsorption. The converse sequence of events occurs with a prolonged deficiency of PTH.

In contrast to calcium, the relationship between urinary phosphate excretion and plasma phosphate level is shifted to the left by PTH (Figure 38-8). This phosphaturic effect of PTH allows disposition of the extra phosphate that is released when the hormone stimulates bone resorption (see next section). Otherwise, PTH would simultaneously elevate plasma calcium and phosphate levels and thereby create the danger of precipitating calcium-phosphate complexes in tissue.

PTH also inhibits the reabsorption of sodium and bicarbonate in the proximal tubule in parallel with phosphate (see Chapter 34). This action may prevent metabolic alkalosis, which could result from the release of bicarbonate during the dissolution of hydroxyapatite crystals in bone (see following discussion).

PTH directly stimulates the synthesis of 1,25-$(OH)_2$-D from 25-$(OH)_2$-D. The decrease in plasma and renal phosphate content caused by PTH further augments this direct action on 1-hydroxylase activity. The increase in 1,25-$(OH)_2$-D stimulates calcium absorption from the gut and raises plasma calcium levels (see Figure 38-4). This important indirect effect of PTH on the intestinal tract again serves the major function of the hormone.

Skeletal Effects The major action of PTH on bone is to accelerate removal of calcium. The initial effect of PTH is to stimulate osteocytic osteolysis, causing a transfer of calcium from the bone canalicular fluid into the osteocyte and then out the opposite side into extracellular fluid. Replenishment of calcium in the canalicular fluid probably then occurs from the surface of partially mineralized bone.

A second, more slowly developing effect of PTH is to stimulate the osteoclasts to resorb completely mineralized bone. In this process both calcium and phosphate are released for transfer into the extracellular fluid; in addition, the organic bone matrix is hydrolyzed by PTH activation of collagenase and of lysosomal enzymes. PTH initially increases the active resorptive ruffled border of the osteoclasts. This is followed by PTH stimulation of osteoclast size, number of nuclei, fusion, and proliferation. PTH also induces increases in the enzymes acid phosphatase and carbonic anhydrase and the accumulation of lactic and citric acids. The resultant lowering of bone pH contributes to the resorptive process. Because of collagen degradation, PTH increases the release of hydroxyproline into plasma and subsequently into the urine.

The dramatic effects of PTH on osteoclasts in vitro are not evident in the absence of osteoblasts. Furthermore, no PTH receptors exist on osteoclasts. Therefore an initial integral action of PTH on osteoblasts may be required to evoke a local factor that second-arily stimulates the osteoclasts. PTH receptors are present on osteoblasts, and when exposed to the hormone, these cells immediately show marked changes in shape due to aggregation of actin-myosin fibrils. Later PTH inhibits the synthesis of collagen by the osteoblasts, probably at the level of transcription. Stimulation of osteoclastic bone resorption and inhibition of osteoblastic bone formation are achieved by the elevated concentrations of hormone that result from stimulation of the parathyroid glands through hypocalcemia. Thus these actions of PTH are part of its general mission to restore plasma calcium level rapidly to normal.

In some circumstances PTH can increase bone formation. This occurs when plasma calcium level is normal or even increased, but hypersecretion of PTH has existed for a long time. Histologic examination may show an increase in new bone adjacent to the areas of increased resorptive activity. The plasma level of alkaline phosphatase, an osteoblastic enzyme whose activity parallels bone formation, is often increased by PTH. Thus the net effect of sustained exposure to PTH may be either a decrease or an increase in total skeletal mass, depending on concomitant factors that affect bone remodeling, such as the availability of calcium, phosphate, or vitamin D.

CALCITONIN

Synthesis and Secretion

Another peptide hormone, calcitonin (CT), decreases plasma calcium levels by antagonizing the actions of PTH on bone. CT is secreted by a small population of neuroendocrine cells known as *C cells,* or *parafollicular cells,* in the thyroid gland. These relatively large cells contain small secretory granules enclosed in membranes. C-cell neoplasms and other tumors of neural crest origin often secrete great amounts of CT. Although uncertainty exists about the significance of its role in normal human physiology, CT is an important regulator of plasma calcium levels in lower animals that live in an aquatic environment high in calcium.

Calcitonin, a straight-chain peptide of 32 amino acids, has a molecular weight of 3400. The biologically active core of the molecule probably resides in its central region. Fish calcitonins are active in humans. Synthesis proceeds from a preprohormone through a prohormone to CT, which is packaged in granules along with N-terminal and C-terminal copeptides. The latter also has a calcium-lowering action and is present in the plasma of humans.

Calcitonin is also present in nervous tissue, where it may function as a neuromodulator. The gene for CT illustrates well the relationship between the endocrine and nervous systems. In some cells the primary transcript of the gene is processed to a messenger RNA that directs synthesis of calcitonin. However, in other cells the same primary transcript is processed to a different RNA that directs the synthesis of an alternative peptide product. This molecule, known as *calcitonin gene-related peptide* (CGRP), circulates in human plasma and probably arises from perivascular nerve axons. It is a potent vasodilator and a cardiac inotropic agent.

The major stimulus to CT secretion is a rise in plasma calcium concentration. However, the degree of response seen in various species is related to their need to prevent hypercalcemia. The first vertebrates to develop CT did so after migrating from fresh water of low calcium concentration into the sea, with a high calcium concentration. When vertebrates moved to land, the emphasis in calcium economy shifted toward defense against a low calcium concentration. PTH was then developed, and the importance of CT in plasma calcium regulation probably declined.

CT circulates in humans at concentrations of 10 to 100 pg·ml^{-1} (10^{-11} M). It increases considerably when plasma calcium level is raised as little as 1 mg·dl^{-1}. The stimulating effect of calcium on CT secretion involves an increase in cAMP. Ingestion of food also increases CT secretion, a response mediated by gastrin and other gastrointestinal hormones.

Actions of Calcitonin

The major effect of CT is to decrease plasma calcium levels. Binding of CT to its plasma membrane receptors stimulates adenylate cyclase and elevates cAMP. This second messenger initiates at least a portion of CT action in all target cells, but the subsequent intracellular events are obscure. The magnitude of the fall in plasma calcium concentration is directly proportional to the baseline rate of bone turnover. Thus young growing animals are most affected by CT, whereas adults with more stable skeletons respond minimally to the hormone. The hypocalcemic action is caused by inhibition both of osteocytic

osteolysis and osteoclastic bone resorption, particularly when these are stimulated by PTH. Continued exposure to CT eventually decreases the number of osteoclasts and alters their morphology. As bone formation is also stimulated, denser bone with fewer resorption cavities eventually results.

CT is clearly a physiologic antagonist to PTH with respect to calcium. However, with respect to phosphate, it has the same net effect as PTH; that is, CT decreases plasma phosphate concentration and increases urinary phosphate excretion slightly.

The importance of CT in humans is controversial. CT deficiency does not lead to hypercalcemia and CT hypersecretion does not produce hypocalcemia. It may be that abnormal CT secretion is easily compensated for by adjustment in PTH and Vitamin D levels. A role for CT in fetal bone development and in the declining bone mass of aging has been proposed.

INTEGRATED REGULATION OF CALCIUM AND PHOSPHATE

An integrated system maintains normal concentrations of calcium and phosphate. Calcium deprivation (Figure 38-9) stimulates PTH secretion. The PTH increases urinary phosphate excretion and thereby decreases plasma and renal cortical phosphate content. Excess PTH secretion together with the decreased phosphate concentration stimulate the production of 1,25-(OH)$_2$-D. The sterol hormone raises the plasma calcium concentration back toward normal by increasing the absorption of calcium from the gut. PTH also increases bone resorption and calcium reabsorption from the renal tubular urine. Together then, PTH and 1,25-(OH)$_2$-D respond to calcium deprivation by increasing the flux of calcium into plasma. Simultaneously the extra phosphate that enters with the calcium is eliminated in the urine by PTH action.

Phosphate deprivation (Figure 38-10) directly stimulates 1,25-(OH)$_2$-D production. The latter increases the flux of phosphate into plasma by stimulating bone resorption and phosphate absorption from the gut. The extra calcium that enters simultaneously will raise the plasma calcium level. This suppresses PTH secretion, and the absence of PTH causes urinary phosphate excretion to diminish, thus aiding in the restoration of plasma phosphate levels back to normal. At the same time suppression of PTH diminishes

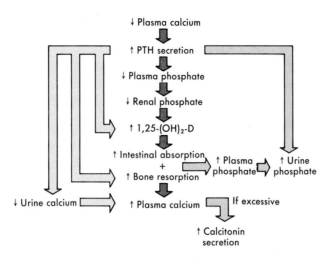

FIGURE 38-9   The compensatory response to a decrease in plasma calcium concentration.

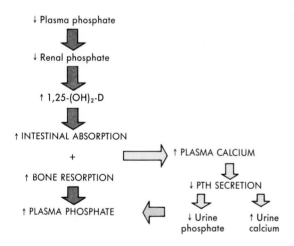

FIGURE 38-10   The compensatory response to a decrease in plasma phosphate concentration.

high to very low levels, the rate of urinary calcium excretion can rise 25-fold and that of phosphate excretion can fall to almost 0. The gastrointestinal responses of this integrated system are slower and narrower in range. Bone responses to regulation by PTH and 1,25-$(OH)_2$-D are rapid when they are produced by osteocytic osteolysis and relatively slow when caused by osteoclastic resorption. However, the capacity for calcium and phosphate uptake and release by the skeleton is enormous.

The compensatory responses of the kidney and the gut defend the total body and bone stores of calcium and phosphate against erosion. In contrast, the skeletal mechanisms that defend the plasma calcium and phosphate levels have the important disadvantage that they eventually sacrifice the chemical and structural integrity of the bone mass.

renal tubular calcium reabsorption and increases urinary calcium excretion. Thus the extra calcium that was mobilized is eliminated more easily.

Therefore this combined arrangement of dual hormone regulation and dual hormone action by PTH and vitamin D permits selective defense of either plasma calcium level or plasma phosphate level, without creating a circulatory excess of the other. The same principles apply in reverse when excess loads of calcium or phosphate are imposed on the body.

The renal responses to PTH provide the most rapid (within minutes) defense against sudden changes in calcium or phosphate levels. As PTH ranges from very

BIBLIOGRAPHY
Journal Articles

Avioli L and Haddad J: The vitamin D family revisited, N Engl J Med 311:47, 1984.

Burger E et al: In vitro formation of osteoclasts from long-term cultures of bone marrow mononuclear phagocytes, J Exp Med 156:1604, 1982.

Canalis E: The hormonal and local regulation of bone formation, Endocr Rev 4:62, 1983.

Cheung WY: Calmodulin plays a pivotal role in cellular regulation, Science 207:19, 1979.

DeLuca H and Schnoes H: Vitamin D: recent advances, Annu Rev Biochem 52:411, 1983.

Mawer EB: Clinical implications of measurements of circulating vitamin D metabolites, Clin Endocrinol Metab 9:63, 1980.

Nijweide PJ et al: Cells of bone: proliferation, differentiation and hormonal regulation, Physiol Rev 66:885, 1986.

Norman A et al: The vitamin D endocrine system: steroid metabolism, hormone receptors, and biological response (calcium binding proteins), Endocr Rev 3:331, 1982.

Parfitt AM: The actions of parathyroid hormone on bone: relation to bone remodeling and turnover, calcium homeostasis, and metabolic bone disease. I. Mechanisms of calcium transfer between blood and bone and their cellular basis: morphologic and kinetic approaches to bone turnover, Metabolism 25:809, 1976.

Parfitt AM: The actions of parathyroid hormone on bone: relation to bone remodeling and turnover, calcium homeostasis, and metabolic bone disease. II. PTH and bone cells: bone turnover and plasma calcium regulation, Metabolism 25:909, 1987.

Parthemore JG et al: Calcitonin secretion in normal human subjects, J Clin Endocrinol Metab, 47:184, 1978.

Raisz LG: Direct effects of vitamin D and its metabolites on skeletal tissue, Clin Endocrinol Metab 9:27, 1980.

Books and Monographs

Avioli LV et al: Bone metabolism and disease. In Bondy PK and Rosenberg LE, editors: Metabolic control and disease, Philadelphia, 1980, WB Saunders Co.

Neer RM: Calcium and inorganic phosphate homeostasis. In Degroot LJ, editor: Endocrinology, New York, 1989, Grune & Stratton, Inc.

Rosenblatt M et al: Parathyroid hormone: physiology, chemistry, biosynthesis, secretion, metabolism, and mode of action. In Degroot, LJ, editor: Endocrinology, New York, 1989, Grune & Stratton, Inc.

CHAPTER 39

The Hypothalamus and Pituitary Gland

The pituitary gland, once called "the master gland," retains a preeminent position in endocrinology even though it is now known to be under neural control from the hypothalamus and under feedback control by products of its target glands. The pituitary gland and hypothalamus, with vascular connections, form a complex functional unit that epitomizes the close, subtle interrelationship between the endocrine system and the nervous system. This unit regulates water metabolism, milk secretion, body growth, reproduction, lactation, and the growth and secretory activities of the thyroid, adrenal, and reproductive glands.

The neurons of the hypothalamus synthesize and secrete neurohormones (Figure 39-1). Two of these neurohormones are stored in secretory vesicles in terminal swellings of the axons within the *posterior pituitary gland,* also known as the *neurohypophysis.* From there they are released into the bloodstream to act on distant target cells (neurocrine function). Other hypothalamic neurohormones are transported down axons that end in a neurovascular region known as the *median eminence* just below the hypothalamus (Figure 39-1). From storage vesicles, these neurohormones are released into the bloodstream and stimulate or inhibit proximal endocrine target cells in the *anterior pituitary gland,* also known as the *adenohypophysis* (again, neurocrine function). The endocrine cells in the adenohypophysis synthesize, store, and secrete a variety of peptide and protein hormones that are released into the bloodstream to act on distant peripheral target cells (endocrine function). In addition, the hormones of these closely intertwined endocrine cells may act on neighboring target cells within the adenohypophysis (paracrine function).

The pituitary gland sits beneath the hypothalamus in a socket of bone (sella turcica) within the skull. The gland represents a fusion of two tissues. The posterior portion, or neurohypophysis, develops as a downward outpouching of neuroectoderm from brain tissue in the floor of the third ventricle. This differentiates into the neurons of the hypothalamus. The lower part of the downward-growing neural stalk forms the bulk of the posterior pituitary. The upper part of the neural stalk expands to form the median eminence. Both the posterior pituitary and the median eminence consist largely of the terminals of various hypothalamic neurons. Both tissues are highly vascularized, and their capillaries contain fenestrations (intercellular windows) that allow influx and efflux of protein molecules.

The posterior pituitary is supplied by the inferior hypophyseal artery, whose capillary plexus invests the terminal swellings of axons from the *supraoptic* and *paraventricular* areas of the hypothalamus. These terminals are the immediate source of the peptide neurohormones *antidiuretic hormone* (ADH) and *oxytocin* (OCT); ADH is also known as *arginine vasopressin* (AVP). These neurohormones are released into this capillary plexus, which carries them into the systemic circulation via draining veins (Figure 39-1).

The anterior pituitary, or adenohypophysis, develops from an upward outpouching of ectoderm from the floor of the oral cavity. After pinching off, the pouch becomes separated from the mouth by the sphenoid bone of the skull. At the junction of the

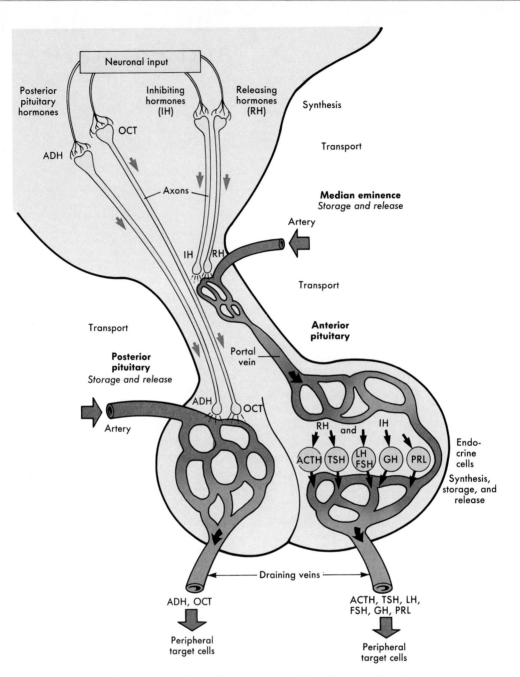

FIGURE 39-1   A schematic overview of the anatomic and functional relationships between the hypothalamus and the pituitary gland. Note that the posterior pituitary gland is an extension of neural tissue that stores neurohormones and has its own arterial blood supply. In contrast the anterior pituitary gland is endocrine tissue with a blood supply derived from veins that first drain neural tissue in the median eminence. By this arrangement the endocrine cells are exposed to high concentrations of neurohormones originating in the hypothalamus and stored in the median eminence. The hormones exiting the posterior and anterior pituitary gland reach and act on peripheral target cells. *ADH*, Antidiuretic hormone; *OCT*, oxytocin; *ACTH*, adrenocorticotropic hormone; *TSH*, thyroid-stimulating hormone; *LH*, leutinizing hormone; *FSH*, follicle-stimulating hormone; *GH*, growth hormone; *PRL*, prolactin.

anterior and posterior lobes of the pituitary gland is an intermediate zone, miniscule in humans but well developed in animals, from which another peptide hormone, *melanocyte-stimulating hormone* (MSH), is produced.

The median eminence is supplied mainly by the superior hypophyseal artery (and to a lesser extent the inferior hypophyseal artery). Its capillary plexus invests terminal swellings of axons from a variety of hypothalamic neurons. These neurons are the source of releasing hormones and inhibiting hormones that regulate anterior pituitary function. The capillary plexus of the median eminence forms a set of portal veins that descend into the anterior pituitary (Figure 39-1). These veins then give rise to a second fenestrated capillary plexus, which serves a dual role. Hypothalamic releasing and inhibiting hormones carried down from the median eminence exit the second plexus and regulate the secretion of the endocrine cells in the anterior pituitary. The hormone products of these cells then enter the same capillary plexus and are delivered via the circulation to distant target cells.

The anterior pituitary therefore has essentially no direct arterial blood supply. Furthermore, its endocrine cells lie outside the blood-brain barrier. Reversal of flow upward in the portal veins may permit high concentrations of hormones from the anterior pituitary to reach the median eminence and hypothalamus, where they could feed back on neurons without impedance from the blood-brain barrier.

HYPOTHALAMIC FUNCTION

A comprehensive discussion of the hypothalamus is given in Chapter 9. From an endocrine standpoint, however, the hypothalamus may be viewed as a central relay station for collecting and integrating signals from diverse sources and funneling them to the pituitary (Figure 39-2). The hypothalamus receives input from the thalamus, the reticular activating substance, the limbic system (amygdala, olfactory bulb, hippocampus, and habenula), the eyes, and remotely from the neocortex. Through this input pituitary function can be influenced by sleep or wakefulness, pain, emotion, fright, rage, smell, light, and possibly even thought. It can be coordinated with such other behavior as mating responses. Other axonal connections with nearby hypothalamic neurons allow the output of pituitary hormones to respond to changes in autonomic nervous system activity and to the needs of temperature regulation, water balance, and energy requirements.

From a teleologic standpoint the proximity of these various functional areas of the hypothalamus to each other is logical. As one example, hormones of the thyroid gland increase energy expenditure, metabolic rate, and thermogenesis. The neurons that ultimately control thyroid gland output are thus close to neurons that regulate temperature and energy intake via appetite control. As the hypothalamic-pituitary unit is systematically studied, other examples become apparent.

Separation of the hypothalamus into individual nuclei or discrete anatomic centers of endocrine function is imprecise, with two exceptions. The supraoptic nucleus is a collection of large neurons that secrete mainly ADH, and the paraventricular nucleus is a similar collection of neurons that secrete mainly OCT. These two neuronal pools overlap only slightly. In contrast, the small neurons that secrete the hypothalamic releasing and inhibiting hormones are more loosely aggregated in various areas, and they overlap more. As a general rule, only one cell type secretes each individual neurohormone, although in one instance the same hypothalamic neuron has been reported to contain two. In some hypothalamic neurons, monoamine neurotransmitters are also produced.

Each hypothalamic anterior pituitary releasing or inhibiting hormone can be assigned a primary target for which it has been named, such as, thyrotropin-releasing hormone (a hormone that releases another hormone which stimulates the thyroid gland) or somatostatin (*soma*, referring to body growth; *statin*, referring to halting of function). However, some of these neuropeptides act on more than one anterior pituitary cell.

In addition to those neurons whose axons end in the posterior pituitary and median eminence, other neurons have axons that project to different parts of the brain. In these instances the same hypothalamic peptides serve as neurotransmitters. (Furthermore, these neuropeptides have also been found in sites outside the brain, including the spinal cord, the sympathetic ganglia, sensory neurons, pancreatic islets, and neuroendocrine cells of the gastrointestinal tract.)

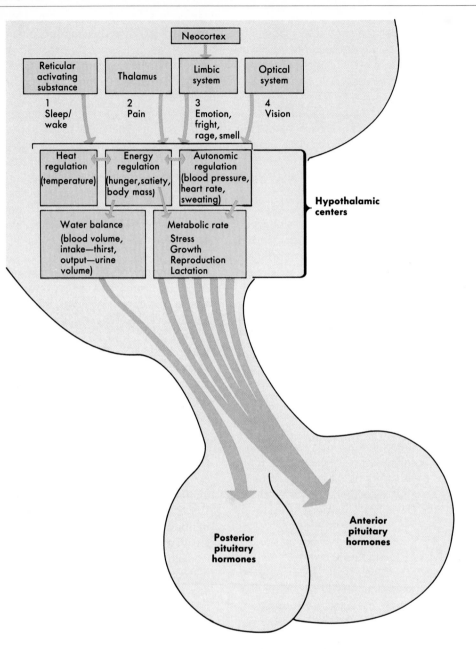

FIGURE 39-2   Interrelationships among various hypothalamic regulatory centers, their inputs from various parts of the brain, and their outputs to the pituitary gland.

Hypothalamic neurohormones are synthesized from preprohormones. Copeptide products of processing, predicted from the gene structures, are now being identified and their functional roles investigated. Hypothalamic neurohormones are typically secreted in pulses generated by an intrinsic neural oscillator (Figure 39-3). Optimal effects on target cells result from this pulsatile pattern of signaling. These hormones react with plasma membrane receptors in anterior pituitary cells, and calcium, phosphatidylinositol products, and cyclic adenosine monophosphate (cAMP) are generated as second messengers. The releasing hormones all stimulate exocytosis of granules containing tropic hormone. In addition, they

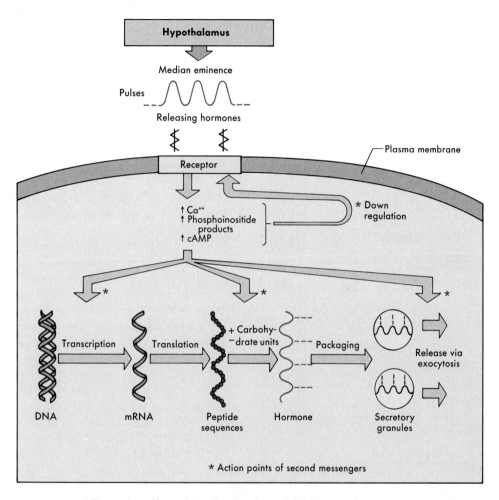

FIGURE 39-3   The action of hypothalamic releasing or inhibiting hormones on anterior pituitary cells. Characteristically the neurohormones are released in pulses, bind to plasma membrane receptors, and act through calcium ions *(Ca++)* and other second messengers. They regulate gene expression, posttranslational processes, and secretion of anterior pituitary tropic hormones. cAMP, Cyclic adenosine monophate; *DNA,* deoxyribonucleic acid; *mRNA,* messenger ribonucleic acid.

stimulate synthesis of the tropic hormones at the level of gene transcription and often enhance their biologic activity by modulating posttranslational modification. Hypothalamic releasing hormones can also regulate their own receptors.

Afferent impulses to hypothalamic neurons are transmitted via norepinephrine, serotonin, acetylcholine, and γ-aminobutyric acid (GABA). From some such hypothalamic neurons, dopamine and β-endorphin transmit signals to other neurons via intrahypothalamic tracts and to the median eminence via efferent tracts; such signals modulate the discharge of releasing and inhibiting hormones. In addition, neu-

rotransmitters from the hypothalamus may themselves reach the portal vein blood and directly influence the output of anterior pituitary hormones.

The pituitary hypothalamic axis is under feedback control from its peripheral targets (Figure 39-4). Virtually all the tropic hormones from the adenohypophysis cause changes in the concentrations of (1) hormones secreted by the thyroid, adrenal, and reproductive glands; (2) peripheral peptide products; or (3) substrates such as glucose or free fatty acids. These in turn regulate the output of both the hypothalamus and the anterior pituitary. This is known as *long-loop feedback* and is usually negative, although it can

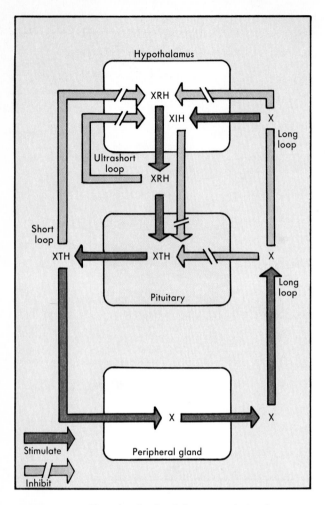

Hypothalamus

XRH

XIH

X

Long loop

Ultrashort loop

XRH

Short loop

XTH

XTH

X

Long loop

Pituitary

X

X

Stimulate

Peripheral gland

Inhibit

FIGURE 39-4   Negative feedback loops regulating hormone secretion in a typical hypothalamus-pituitary-peripheral gland axis. *X,* Peripheral gland hormone; *XTH,* pituitary tropic hormone; *XRH,* hypothalamic releasing hormone; *XIH,* hypothalamic inhibiting hormone.

transiently be positive. Negative feedback can also be exerted by the pituitary hormones themselves on the synthesis or discharge of the related hypothalamic releasing or inhibiting hormones. This is known as *short-loop feedback.* Because these hormones do not ordinarily cross the blood-brain barrier, short-loop feedback may occur either via fenestrated cells of the capillaries that bathe hypothalamic neurons or via retrograde flow through pituitary portal veins. Finally a hypothalamic releasing hormone may even inhibit its own synthesis and discharge or stimulate that of a paired hypothalamic inhibiting hormone. This is called *ultrashort-loop feedback.* It may occur by axo-

nal connections between the pertinent neurons or via the neurohormones gaining access to the cerebrospinal fluid.

POSTERIOR PITUITARY FUNCTION

ADH and OCT, two small peptides with molecular weights of approximately 1000 and with homologous structure, are secreted by the posterior pituitary gland. The primary role of the more important ADH is to conserve water and regulate the tonicity of body fluids (see Chapter 33). A secondary role is to help maintain vascular volume. The primary role of OCT is to eject milk from the lactating mammary gland; a secondary role is to stimulate contraction of the uterus. Although their functions are very different, both hormones are synthesized, stored, and secreted in similar fashion.

The genes that direct synthesis of the preprohormones for ADH and OCT are very similar and are probably mutated from a common ancestor gene. In addition to the two neuropeptides, the gene products include distinctive proteins, known as *neurophysins,* with molecular weights of 10,000. Neurophysin-1 for OCT and neurophysin-2 for ADH are very similar in amino acid sequence. After processing, ADH and OCT are packaged with their respective neurophysins in neurosecretory granules. The neurophysins may serve as carrier proteins during transport of the neuropeptides down the axons to the posterior pituitary gland.

Release of ADH or OCT occurs when an electrical discharge is transmitted from the cell body in the hypothalamus down its axon, where it depolarizes the neurosecretory vesicle in the posterior pituitary. An influx of calcium into the vesicles releases hormone by exocytosis. During this process each hormone dissociates from its neurophysin, and the two molecules enter the circulation separately.

Arginine Vasopressin

Regulation of Secretion The secretion of ADH illustrates the homeostatic principle that the release of a hormone is stimulated by conditions that require its action (Figure 39-5). Water deprivation raises plasma osmolality, which evokes release of ADH. In turn ADH causes retention of free water by the kidney and an increase in urine osmolality, with the result that

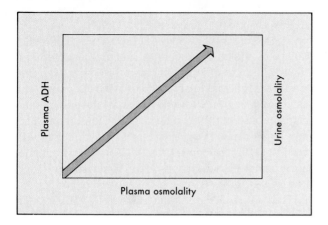

FIGURE 39-5   Positive correlation between the stimulus of plasma osmolality, the response of ADH secretion, and the hormone's effect on urine osmolality.

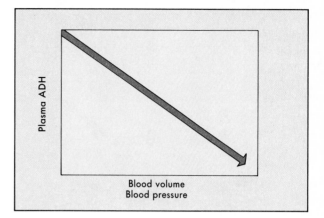

FIGURE 39-6   Negative correlation between the stimulus of blood volume or blood pressure and the response of ADH secretion.

plasma osmolality declines to normal (see Chapter 33). Conversely, ingestion of a water load decreases plasma osmolality. This suppresses ADH release, which increases water excretion and raises plasma osmolality to normal. Thus water and ADH form a negative feedback loop.

The direct physiologic stimulus to ADH release is an increase in the osmolality of fluids that bathe osmoreceptor neurons in the hypothalamus. This creates a gradient for water movement out of the neurons, and the consequent rise in intracellular osmolality triggers release of ADH. Any administered solute such as sodium that does not readily penetrate cell membranes creates the same osmotic disequilibrium and stimulates ADH secretion. By contrast, solutes such as urea that freely enter cells do not stimulate ADH release.

The hypothalamic osmoreceptors respond to changes in plasma osmolality of only 1% to 2%. The osmolar threshold for ADH release is approximately 280 mOsm/kg of body weight. Plasma ADH then increases about 1 $pg \cdot ml^{-1}$ for each 3 $mOsm \cdot kg^{-1}$ increase in plasma osmolality. Generation of sufficient ADH to produce maximal retention of water and maximal urinary osmolality occurs when plasma osmolality reaches 294 $mOsm \cdot kg^{-1}$. In contrast, the osmolar threshold for stimulation of thirst is 290 to 295 $mOsm \cdot kg^{-1}$. This fact emphasizes the preeminence of ADH secretion over thirst in defending normal body water content and tonicity.

ADH release is also stimulated by hypovolemia. This is a much less sensitive response, since a decrease of 5% to 10% in blood volume, cardiac output, or blood pressure is required (Figure 39-6). Hemorrhage, quiet standing, and positive pressure breathing, all of which reduce cardiac output and central blood volume, increase ADH secretion. Conversely, increasing central blood volume by administration of blood or isotonic saline solution suppresses ADH release. Hypovolemia is perceived by several pressure (rather than volume) sensors (see Chapters 18 and 22). These include carotid and aortic baroreceptors, stretch receptors in the walls of the left atrium and pulmonary veins, and possibly the juxtaglomerular apparatus of the kidney. Normally the pressure receptors tonically inhibit ADH release. A reduction in circulating volume, and therefore of pressure on the baroreceptors, reduces the flow of inhibitory impulses (via the brainstem) to the hypothalamus. This increases ADH secretion. Hypovolemia also stimulates the generation of renin and angiotensin directly within the brain. The angiotensin augments the release of ADH and also stimulates thirst. Plasma ADH concentration rises much more in response to hypovolemia than in response to hyperosmolality. This correlates with the lesser sensitivity of the vascular system than of the kidney to hormone action.

The two major stimuli of ADH release interact (Figure 39-7). Increases or decreases in volume reinforce the osmolar responses by raising or lowering, respectively, the threshold for osmotic release of ADH. Thus hypovolemia sensitizes the system to hyperosmolarity. When hypovolemia is severe, baroregulation overrides osmotic regulation. Consequently, ADH secre-

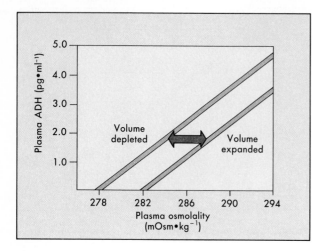

FIGURE 39-7 Regulation of ADH secretion by the interaction between plasma osmolality and blood volume. A reduction in blood volume sensitizes the hypothalamus–posterior pituitary gland so that ADH is secreted at a lower threshold of plasma osmolality. (Modified from Robertson GL et al: J Clin Edocrinol Metab 42:613, 1976. The Endocrine Society.)

tion is stimulated even though plasma osmolality may be below the threshold of 280 mOsm·kg^{-1}.

Pain, emotional stress, nausea and vomiting, heat, and a variety of drugs also stimulate ADH release. Ethanol, on the other hand, is a frequently encountered inhibitor; 30 to 90 ml of whiskey is sufficient to suppress secretion completely and cause a diuresis. Cortisol and thyroid hormone may restrain ADH release; when they are deficient, ADH may be secreted even though plasma osmolality is low.

ADH circulates at basal concentrations of about 1 pg·ml^{-1} (10^{-12} M). Although the plasma half-life is only 6 to 10 minutes, the half-life of biologic action may be longer. During water deprivation, ADH secretion increases three- to fivefold, and synthesis of the hormone is augmented. Transient increases in plasma ADH of fiftyfold can occur with hemorrhage, severe pain, or nausea. Plasma levels of neurophysin-2 also rise and fall in parallel with ADH.

Actions of Arginine Vasopressin The major action of ADH is on the renal tubular mechanism for concentrating the urine, that is, for reabsorbing osmotically unencumbered water from the glomerular filtrate (see Chapter 33). ADH stimulates the two phases in the countercurrent concentrating mechanism. First, the hormone modestly increases the transport of sodium out of the thick ascending portion of Henle's loop into the medullary interstitium, thus

helping to create the osmotic gradient for water. Second and more importantly, ADH increases the permeability of the collecting duct membranes to water, thus facilitating back diffusion of water into the medulla. The maximal effect of ADH increases the osmolality of urine to a value four-fold higher than plasma, or about 1200 mOsm·kg^{-1}. As noted in Figure 39-5, urine osmolality correlates directly with plasma ADH concentration. Without the hormone, urine osmolality falls to less than 100 mOsm·kg^{-1}, and free water clearance reaches 10 to 15 ml·min^{-1}.

The intracellular mechanism of ADH action requires binding to a plasma membrane receptor, activation of adenylate cyclase, generation of cAMP, and subsequent phosphorylation of proteins mediated by protein kinase A. In the cells of the ascending limb of Henle's loop, this results in an increase in the number of sodium, chloride, and potassium transport units in the apical membrane. In the collecting ducts, activation of microtubules and microfilaments leads to aggregation of granular subunits and their insertion into the cell membrane. Presumably these components then alter permeability to water.

Several factors can blunt the action of ADH on tubular cells. These include solute diuresis, chronic water loading (which reduces medullary hyperosmolality), potassium deficiency, calcium excess, cortisol excess, and lithium.

In addition to its major role in water metabolism, ADH may subserve other functions. It contributes in a minor way to increasing vascular tone in response to hemorrhage. When administered systemically in large doses, it elevates the blood pressure and constricts the coronary and splanchnic beds. This action requires binding to a different receptor in vascular cells and is mediated by the phosphatidylinositol–protein kinase C second messenger system. ADH functions as a hypothalamic releasing factor via axons projecting to the median eminence. ADH also serves neurotransmitter functions elsewhere within the brain, for example, facilitating long-term memory.

Oxytocin

Regulation of Secretion OCT, known biologically as the *milk letdown factor,* is secreted within seconds in response to suckling. Sensory receptors in the nipple generate afferent impulses, which reach the hypothalamic paraventricular and supraoptic nuclei via various relays. A final cholinergic synapse causes discharge of OCT and neurophysin-1 from the posterior pituitary in a manner similar to ADH. Continued suckling further stimulates OCT synthesis and transport to the posterior pituitary. In humans there is neither crossover secretion of ADH with suckling nor OCT secretion with an increase in plasma osmolality. OCT secretion can also be stimulated by sexual activity and inhibited by emotional distress.

Actions of Oxytocin OCT causes the myoepithelial cells of the alveoli in the breast to contract. This forces the milk from the alveoli into the ducts, from where it is extracted by the infant. OCT acts via plasma membrane receptors and cAMP generation in target cells. Binding to the receptor is increased by estrogen. Although basal plasma levels of OCT are similar in men and women, no role for the circulating hormone in men is known.

OCT also stimulates contraction of the uterus. Lower doses cause rhythmic contractions, whereas higher doses cause sustained tetanic contraction. There is little evidence that OCT is essential for normal labor in humans, but the sustained contractions it produces may be important in reducing blood loss from the uterus after delivery of the conceptus. OCT is often used therapeutically to induce labor or to stop excessive postpartum bleeding. OCT and its receptor are also present in the human ovary and testis, where the locally produced hormone may play a role in reproduction.

ANTERIOR PITUITARY FUNCTION

The anterior pituitary gland, or adenohypophysis, makes up most of the 500 mg of pituitary tissue. It contains at least five types of endocrine cells, each being the source of a different hormone with a distinct function. These cell types, their relative proportions in the pituitary, and their major secretory products are shown in Figure 39-8. Although the five types of cells aggregate to some extent on a functional basis, they do not form unique enclaves but are also interspersed among each other. They vary somewhat in size and in characteristics of their secretory granules, but they can only be identified with certainty by immunohistochemical staining of the hormones within. Other cells, called null cells, are also present. These have all the cytoplasmic organelles needed for protein hormone synthesis and contain a few secretory granules. However, their products, if any, have not yet been identified.

Each anterior pituitary cell is regulated by one or more hypothalamic neurohormones that reach them by portal veins, as described previously. Three of the cell types produce hormones that regulate the function of the thyroid gland (thyroid-stimulating hormone, TSH), the adrenal glands (adrenocorticotropic hormone, ACTH), and the gonads (leutinizing and follicle-stimulating hormones, (LH, FSH), respectively. For purposes of better integration, the synthesis,

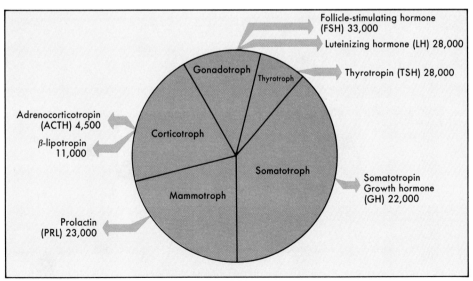

FIGURE 39-8 The relative proportions of cell types in the anterior pituitary gland, their major hormonal products, and the molecular weights of the latter.

secretion, and actions of these tropic hormones are presented in conjunction with their major peripheral target glands (see Chapters 40, 41, and 43). In this section only the function of those cells that secrete growth hormone, which acts on numerous peripheral tissues, and prolactin, which acts on the mammary glands, is presented.

Growth Hormone (Somatotropin)

Growth hormone (GH) stimulates postnatal somatic growth and development. Without GH, children and young animals grow at a much reduced rate, and sexual maturation is delayed. Lean body mass is diminished, whereas fat mass is increased (Figure 39-9). If GH deficiency develops in normal adults, the biologic consequences appear minimal and are difficult to detect. In contrast, an excess of GH produces spectacular effects in both children and adults. In children, gigantism results; in adults, accretion of bone and soft tissue causes dramatic changes in appearance and serious metabolic derangements, including diabetes.

Synthesis and Secretion Somatotrophs are the most numerous cells of the pituitary (see Figure 39-8). Their product, GH, is a single, large, polypeptide chain with 191 amino acids and two disulfide bridges. The exact site necessary for biologic activity is still unknown. The human genome contains multiple genes coding for a family of closely related GH molecules. Only one of these genes is expressed as normal GH. The normal messenger ribonucleic acid (mRNA) directs synthesis of a prehormone. Following removal of a signal peptide, the complete hormone is stored in granules. GH synthesis is increased by thyroid hormone, cortisol, and the specific hypothalamic releas-

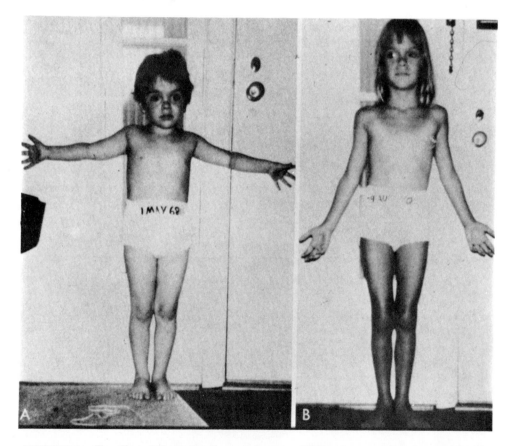

FIGURE 39-9   The effect of 15 months of growth hormone (GH) replacement on a 6-year-old child with GH deficiency. Note that GH increases linear growth and decreases adiposity. (Reprinted with permission from Foster D and Wilson J, editors: Williams textbook of endocrinology, Philadelphia, 1985, WB Saunders Co.)

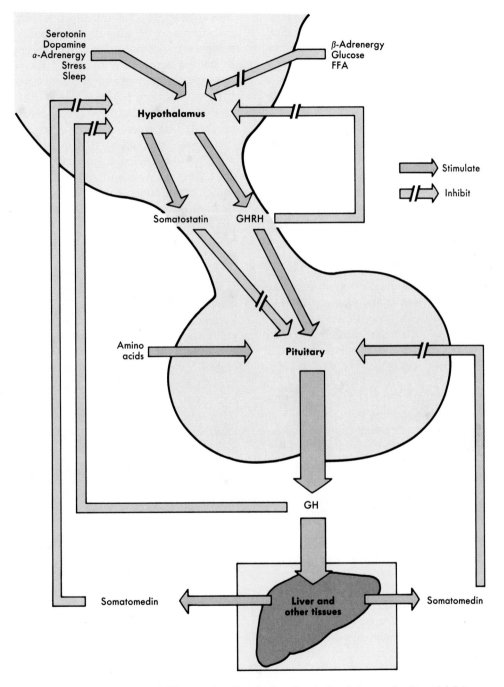

FIGURE 39-10   Regulation of GH secretion. Note both a direct stimulatory and a direct inhibitory influence from the hypothalamus. Negative feedback by the peripheral product is exerted at the hypothalamic and the pituitary level. *GHRH,* Growth hormone–releasing hormone; *FFA,* free fatty acids.

ing hormone, *growth hormone–releasing hormone* (GHRH).

GH secretion by exocytosis is stimulated by GHRH, a hypothalamic peptide with 44 amino acids. GHRH interacts with its plasma membrane receptor, following which calcium, phosphatidylinositol products, and cAMP are generated as second messengers. Only the role of calcium has proved to be essential.

Somatostatin, a hypothalamic peptide with 14 or 28 amino acids, is a powerful inhibitor of GH release. Somatostatin blocks GHRH stimulation noncompetitively. The inhibitor acts through its own plasma membrane receptor, in part by decreasing both calcium entry into the cells and cAMP levels. GH is secreted in pulses that are mainly caused by intermittent release of GHRH into portal pituitary vein blood. Somatostatin diminishes the response to GHRH.

The secretion of GH is influenced by many factors (Figure 39-10). However, the final common pathway for most stimulators of GH is an increase of GHRH, a decrease in somatostatin, or both. Conversely, suppressors of GH either decrease GHRH, increase somatostatin, or both. However, some agents can alter GH secretion by direct effects on the somatotroph.

GH release is regulated metabolically by the energy substrates glucose and free fatty acids and by amino acids. A sharp drop in either glucose or free fatty acid levels stimulates a two- to tenfold increase in plasma GH, whereas elevation of glucose or free fatty acid levels reduces plasma GH by at least 50%. Protein ingestion or intravenous amino acid infusion stimulates GH release. Arginine is especially effective. Both short-term fasting and prolonged protein-calorie deprivation increase GH secretion. In contrast, obesity reduces GH responses to all stimuli, including GHRH.

Central nervous system regulation takes several forms. A nocturnal surge in GH occurs 1 to 2 hours After the onset of deep sleep. Conversely, light sleep, associated with rapid eye movements (REM sleep), inhibits GH release. Various stresses, including trauma, surgery, anesthesia, fever, or even simple venipuncture, elevate plasma GH. Exercise is also a potent stimulant. These conditions influence hypothalamic GHRH and somatostatin neurons through a variety of monoamine neurotransmitters (Figure 39-10).

Age, gender, and other hormonal influences also alter GH secretion. Children secrete somewhat more GH than adults, especially during puberty. In aged individuals GH secretion declines. Females are more responsive to GH stimuli than males, except during pregnancy. GH secretion is reduced by a deficiency of thyroid hormone or by an excess of cortisol. The average normal daily secretion is 0.5 mg, and basal plasma levels are 1-5 $ng \cdot ml^{-1}$ ($10^{-10}M$).

Feedback regulation of GH occurs at all levels (Figure 39-10). Long-loop negative feedback is exerted by a peripheral product of GH action, known as somatomedin (see following discussion). This peptide inhibits GHRH release and its action on the pituitary somatotroph, and it also stimulates somatostatin release. Short-loop negative feedback is exerted by GH itself, by stimulating somatostatin release. Ultrashort-loop negative feedback is exerted by GHRH, possibly via synapses with somatostatin neurons.

Actions of Growth Hormone GH interacts with several distinct plasma membrane receptors in target cells throughout the body. Thus far, none of the known membrane-generated second messengers appears to mediate its intracellular actions. However, the growth-promoting effect of GH requires the generation of an entirely different family of peptides known as *somatomedins*. These peptides, with a molecular weight of 7000, resemble proinsulin in structure. They were originally discovered in plasma and were termed *insulin growth factors* (IGFs).

Two principal IGFs, their receptors, and the respective genes are well characterized. IGF-1 has 50% and IGF-2 has 70% amino acid homology with the A and B chains of insulin. Somatomedins or IGFs are probably produced by many tissues in response to GH. However, circulating somatomedins originate mainly in the liver, and the lag between administration of GH and the subsequent increase in plasma IGF-1 and IGF-2 is about 12 hours. Both peptides circulate bound to a specific large carrier protein. This accounts for their relatively stable concentration and much longer half-lives than that of GH itself. Both somatomedins, but especially IGF-1, are greatly reduced in the plasma of GH-deficient subjects.

Although somatomedins may function as circulating hormones in classic endocrine fashion, they probably also function as locally produced hormones in paracrine and even autocrine fashion. GH probably induces differentiation of precursor cells in target tissues, such as cartilage, into mature cells (chondrocytes), which then express the IFG-1 gene under further GH stimulation. IGF-1 acts through plasma membrane receptors with structural similarity to the insulin receptor (see Chapter 37). The IGF-2 receptor is dissimilar to those of IGF-1 and insulin.

Somatomedins mediate the typical GH responses of cartilage, bone, muscle, adipose tissue, fibroblasts, and tumor cells in vitro. Individuals who lack the ability to produce somatomedins show retarded growth despite high GH levels. Although fetal GH is not required for intrauterine growth, somatomedins generated in the placenta or other fetal tissues probably participate in regulating prenatal growth.

During adolescence, plasma IGF-1 levels rise because of increases in GHRH release and GH secretion. The progression of pubertal growth correlates with the increase in plasma IGF-1. In states of fasting and protein-calorie malnutrition, somatomedin levels in plasma are diminished, which correlates with the negative nitrogen balance in these conditions.

Because GH levels are elevated in these catabolic states, factors other than GH must also regulate somatomedin production. In turn, the high GH levels most likely result from negative feedback caused by low somatomedin levels (Figure 39-10). Somatomedin production is also diminished by cortisol and estrogens, hormones that antagonize GH action.

GH, via somatomedins, is a hormone with overall anabolic action. When it is administered to GH-deficient individuals, it decreases plasma amino acid levels and urea production, since the amino acids are shunted toward protein synthesis and away from oxidative degradation (see Chapter 36). Total body nitrogen balance becomes positive, along with the related balances of the intracellular minerals potassium and phosphate.

The multiplicity of GH targets and effects is indicated in Figure 39-11. The most striking and specific effect is the acceleration of linear growth (see Figure

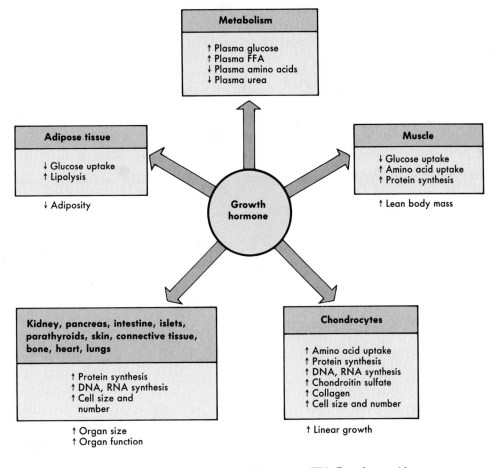

FIGURE 39-11   An overview of GH actions. *FFA*, Free fatty acids.

39-9) that results from GH action on the epiphysial cartilage growth centers of long bones. All aspects of the metabolism of chondrocytes, the cartilage-forming cells, are stimulated. This includes the synthesis of collagen and of proteoglycan chondroitin, which together form the resilient extracellular matrix of cartilage. In addition, GH stimulates the synthesis of proteins, RNA, and DNA in these cells, as well as their proliferation. In support of the augmented protein synthesis, GH also stimulates cellular uptake of amino acids.

Many tissues share in the anabolic response to GH. The width of bones increases, as well as their length. Visceral organs (liver, kidney, pancreas, intestines), endocrine glands (adrenals, parathyroids, pancreatic islets), skeletal muscle, heart, skin, and connective tissue all undergo enlargement. This is reflected in enhanced function of these organs. For example, glomerular filtration, cardiac output, and hepatic metabolic capacity are all increased by GH.

GH affects carbohydrate and lipid metabolism. It stimulates expression of the insulin gene; without GH, insulin secretion declines. A prolonged **excess** of GH increases insulin secretion. However, this occurs also in compensation because GH induces resistance to the action of insulin; GH inhibits glu-

cose uptake by muscle and adipose cells and raises plasma glucose levels. In addition, GH enhances lipolysis and antagonizes insulin-stimulated lipogenesis. These actions increase plasma free fatty acid levels and decrease adipose tissue. If insulin secretion is deficient because of pancreatic β-cell damage, GH can even cause ketosis. Thus, on balance, GH is a *diabetogenic* hormone.

Correlation of Growth Hormone and Insulin Actions The secretion and actions of GH and insulin are metabolically coordinated in the following three ways (Figure 39-12).

1. When protein and energy intake are ample, amino acids can be used for protein synthesis and growth. Both GH and insulin secretion are stimulated by protein ingestion, and together they augment the production of somatomedins. The latter in turn stimulate accretion of lean body mass. At the same time the insulin-antagonistic effect of GH helps to prevent hypoglycemia, which might otherwise result from the increased insulin in the absence of ingested carbohydrate.

2. When carbohydrate is ingested alone, insulin secretion is increased but GH secretion is suppressed. In this situation accelerated generation of somatomedins is not advantageous in the absence of

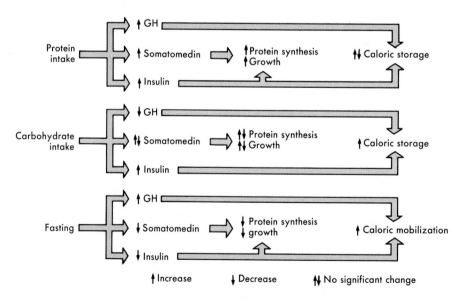

FIGURE 39-12   Complementary regulation of GH and insulin secretion. Both hormones are increased by protein intake, and both likewise stimulate anabolic processes. Insulin and GH secretion are regulated in opposite directions under circumstances where caloric storage (facilitated by insulin) or caloric mobilization (facilitated by GH) are required.

amino acids. Insulin antagonism also is not necessary; on the contrary, unrestrained expression of insulin action permits efficient storage of the excess carbohydrate calories.

3. With fasting, insulin secretion falls, GH secretion rises, but somatomedins still decline. This combination seems appropriate in a situation where protein synthesis cannot be sustained and protein catabolism is essential. However, the increase in GH may still be beneficial because it enhances lipolysis and decreases peripheral tissue glucose utilization. This helps to mobilize free fatty acids for oxidative purposes and to provide glucose for central nervous system needs.

Prolactin

Prolactin (PRL) is a protein hormone principally concerned with stimulating breast development and milk production in women. In addition, it may play some role in reproductive function; in humans this is manifest when excess prolactin is present. The PRL-producing cells, called mammotrophs, are the second most prevalent in the pituitary gland (see Figure 39-8). They increase in number during pregnancy and lactation.

Synthesis and Secretion PRL is a single-chain protein with 198 amino acids and three disulfide bridges. It is structurally similar to GH, and the two genes are thought to have arisen from a common ancestor. Synthesis of PRL proceeds in the manner described for GH via a prehormone. A small number of pituitary cells, called mammosomatotrophs, actually secrete both hormones. Transcription of the PRL gene is regulated by the same factors that regulate secretion of the hormone (see following discussion).

The most important influence on prolactin secretion is the combination of pregnancy, estrogens, and nursing (Figure 39-13). Consistent with its essential role in lactation, PRL secretion increases steadily during pregnancy to a twentyfold plasma elevation. This is probably mediated by the large increase in estrogen levels, which stimulates hyperplasia of mammotrophs and transcription of the PRL gene. Although estrogen does not directly stimulate the release of prolactin, it does enhance responsiveness to other stimuli. If a new mother fails to nurse her child, the plasma PRL level declines to the range that prevails in nonpregnant women by 6 weeks after delivery. Nursing, however, maintains elevated PRL levels. This

effect may be mimicked by mechanical stimulation of the breast.

PRL secretion, as with that of GH, rises at night and in conjunction with major stresses. The functional significance of increased PRL in these situations is unclear.

Unique among the anterior pituitary hormones, secretion of PRL is predominantly under **inhibition** by hypothalamic factors (Figure 39-13). Disruption of the connections to the hypothalamus leads to a great increase in PRL secretion, whereas the secretion of all other anterior pituitary hormones decreases. *Dopamine* released from the median eminence into the portal veins is the major hypothalamic inhibitory factor. This catecholamine neurohormone dramatically suppresses the release and synthesis of PRL. An additional prolactin-inhibiting factor may be a copeptide synthesized with the hypothalamic peptide, LH-releasing hormone. Short-loop negative feedback also operates as PRL inhibits its own secretion by stimulating the synthesis and release of hypothalamic dopamine (Figure 39-13).

The hypothalamus is also the source of PRL-releasing factors. *Thyrotropin-releasing hormone* (TRH) strongly stimulates PRL synthesis and release by acting through its receptors in mammotrophs. However, TRH is not the mediator of the PRL response to nursing. Other hypothalamic peptides with PRL-releasing activity include *vasoactive intestinal peptide* (VIP).

Actions of Prolactin PRL participates in stimulating the original differentiation of breast tissue and its further expansion during pregnancy. It is the principal hormone responsible for lactogenesis (milk production). PRL, together with estrogen, progesterone, cortisol, and GH, stimulate proliferation and branching of the breast ducts. During pregnancy, PRL, estrogen, and progesterone cause development of glandular tissue (alveoli), within which milk production will occur. After parturition, milk synthesis and secretion require PRL, along with cortisol and insulin.

The action of PRL begins by combination with a plasma membrane receptor. None of the known cytoplasmic second messengers appears to mediate the hormone's action. PRL rapidly induces transcription of RNAs for the milk proteins casein and lactalbumin. Concurrently, PRL induces enzymes necessary for the synthesis of lactose, the major sugar in milk.

A second area of PRL action may be on the reproductive axis. An excess of PRL blocks the synthesis and release of LH-releasing hormone, which inhibits

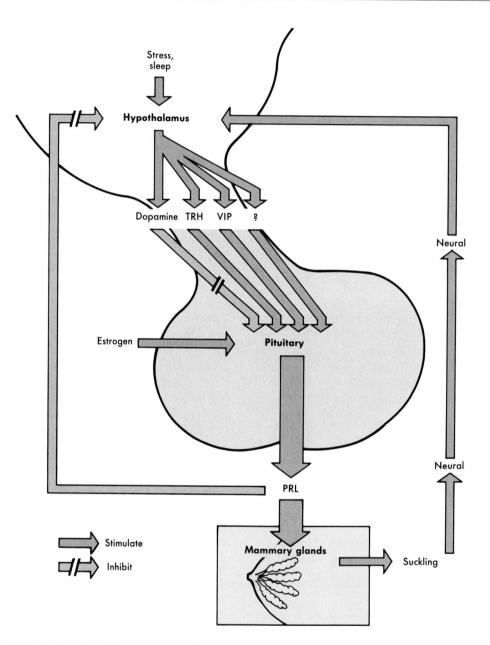

FIGURE 39-13   Regulation of prolactin *(PRL)* secretion. The predominant hypothalmic influence is normally inhibitory via dopamine. Pregnancy and lactation are the major physiologic stimulators. *TRH,* Thyrotropin-releasing hormone; *VIP,* vasoactive intestinal peptide.

gonadotropin secretion. This prevents ovulation and spermatogenesis. High PRL concentrations inhibit synthesis of gonadal steroid hormones in both women and men; however, low PRL concentrations help to sustain progesterone secretion by the ovary. Certain behavioral effects of PRL have been described, including inhibition of libido in humans and stimulation of parental protective behavior toward the newborn in animals. At present the importance of normal PRL secretion to human reproduction is uncertain.

BIBLIOGRAPHY
Journal Articles

Argente J et al: Relationship of plasma growth hormone–releasing hormone levels to pubertal changes, J Clin Endocrinol Metab 63:680, 1986.

Frohman L: Growth hormone–releasing factor: a neuroendocrine perspective, J Lab Clin Med 103:819, 1984.

Grossman A: Neuroendocrinology of opioid peptides, Br Med Bull 39:82, 1983.

Hall K and Sara V: Somatomedin levels in childhood, adolescence and adult life, Clin Endocrinol Metab 13:91, 1984.

Pelletier G et al: Identification of human anterior pituitary cells by immunoelectron microscopy, J Clin Endocrinol Metab 46:534, 1978.

Sklar A and Schrier R: Central nervous system mediators of vasopressin release, Physiol Rev 63:1243, 1983.

Snyder SH: Brain peptides as neurotransmitters, Science 209:976, 1980.

Weitzman RE et al: The effect of nursing on neurohypophyseal hormone and prolactin secretion in human subjects, J Clin Endocrinol Metab 41:836, 1980.

Books and Monographs

Daughaday WH: The anterior pituitary. In Foster D and Wilson J, editors: Williams textbook of endocrinology, Philadelphia, 1985, WB Saunders Co.

Frohman LA et al: The physiological and pharmacological control of anterior pituitary hormone secretion. In Dunn, A and Nemeroff C, editors: Behavioral neuroendocrinology, New York, 1983, Spectrum Publications, Inc.

Guillemin R: Neuroendocrine interrelations. In Body P and Rosenberg LE, editors: Metabolic control and disease, Philadelphia, 1980, WB Saunders Co.

Kato Y et al: Regulation of prolactin secretion. In Imura H: The pituitary gland, New York, 1985, Raven Press.

Reichlin S: Neuroendocrinology. In Foster D and Wilson J, editors: Williams textbook of endocrinology, Philadelphia, 1985, WB Saunders Co.

Riskind PN and Martin JB: Functional anatomy of the hypothalamic–anterior pituitary complex. In Degroot LJ, editor: Endocrinology, Philadelphia, 1989, WB Saunders Co.

Seo H: Growth hormone and prolactin: chemistry, gene organization, biosynthesis, and regulation of gene expression. In Imura H: The pituitary gland, New York, 1985, Raven Press.

Verbalis J and Robinson A: Neurophysin and vasopressin: newer concepts of secretion and regulation. In Imura H: The pituitary gland, New York, 1985, Raven Press.

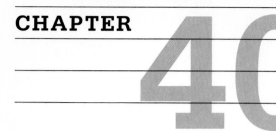

CHAPTER 40

The Thyroid Gland

The thyroid gland was the first endocrine gland to be recognized as such. The correlation between absence or enlargement of the thyroid gland with altered biology at distant body sites provided the clue that the gland was producing a substance that affected distant targets. Extracts of the thyroid gland were subsequently shown to correct the abnormal state resulting from its absence.

The thyroid gland produces two hormones, thyroxine (T_4) and triiodothyronine (T_3), at a rather steady pace. These hormones increase the rate of basal oxygen utilization and metabolism and the consequent rate of heat production, so as to adjust them to alterations in energy need, caloric supply, and thermal environment. Thyroid hormones also increase the delivery of substrates and of oxygen needed to sustain the appropriate metabolic rate. Finally, their actions are critical for normal growth and maturation of the fetus and the child.

ANATOMY

The thyroid gland develops from endoderm of the pharyngeal gut. It descends to the anterior neck, where half of the gland lies on each side of the trachea. By 12 weeks of human gestation, the gland is capable of synthesizing and secreting thyroid hormones under the stimulus of the fetal hypothalamus and pituitary gland. This entire axis is required for subsequent normal intrauterine development of the central nervous system and skeleton because neither maternal thyroid hormone nor its pituitary-stimulating hormone can cross the placenta.

The thyroid gland in adults weighs approximately 20 g. The histologic structure is shown schematically in Figure 40-1. The endocrine cells are surrounded by a basement membrane, and they form single-layered circular *follicles*. The lumens of the follicles contain thyroid hormones stored in the form of a *colloid* material. When stimulated, the endocrine cells enlarge and assume a columnar shape, with their nuclei at the base. The colloid material in the lumen appears scalloped because it is undergoing proteolysis. Also scattered within the thyroid gland are the parafollicular cells, or C cells, which secrete calcitonin (see Chapter 38).

SYNTHESIS AND SECRETION OF THYROID HORMONES

Thyroid hormones are unique in that they incorporate an inorganic element, *iodine,* into an organic structure made up of two molecules of the amino acid tyrosine. The secretory products of the thyroid gland are known as *iodothyronines*. The major product is 3,5,3′,5′-tetraiodothyronine, known as *thyroxine* and referred to as T_4. This molecule functions largely as a circulating prohormone. Secreted in much less quantity is 3,5,3′-triiodothyronine, known simply as *triiodothyronine* and referred to as T_3. This molecule, which provides virtually all thyroid hormone activity in target cells, is actually produced mostly in peripheral tissues from the prohormone T_4. A trivial secretory product with no identified hormonal action is 3,3′,5′-triiodothyronine. This is known as *reverse T_3,* or rT_3, because it differs from T_3 only in the location of

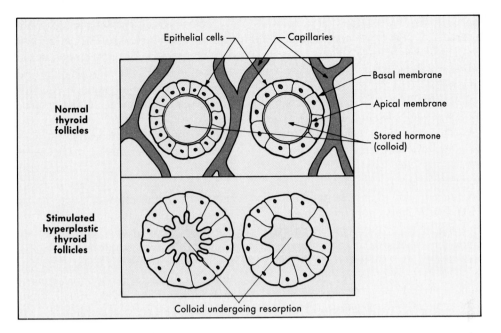

FIGURE 40-1   Schematic representation of the basic thyroid unit. A normal follicle consists of a central core of colloid material surrounded by a single layer of cuboidal cells. When stimulated by thyrotropin, the cells elongate and the central core becomes scalloped because of resorption of the colloid.

one of the three iodine atoms. This inactive molecule is an alternate product of the prohormone T_4, produced when less thyroid hormone action is needed. The structures of T_4, T_3, and rT_3 are shown in Figure 40-2.

Three major steps are involved in the synthesis of thyroid hormones: (1) uptake and concentration of iodide within the gland, (2) oxidation and incorporation of the iodide into the phenol ring of tyrosine, and (3) coupling of two iodinated tyrosine molecules to form either T_4 or T_3. Of particular note is that neither iodination nor coupling occurs with tyrosine free in solution. Rather, the tyrosine molecules must first be incorporated by standard peptide linkages into a protein known as *thyroglobulin*. Thyroglobulin is the substance actually iodinated on specific constituent tyrosines, and the latter are brought into proximity for coupling by the three-dimensional structure of the protein. The thyroid hormones formed remain in peptide linkage within thyroglobulin, and their release into the circulation requires proteolytic cleavage.

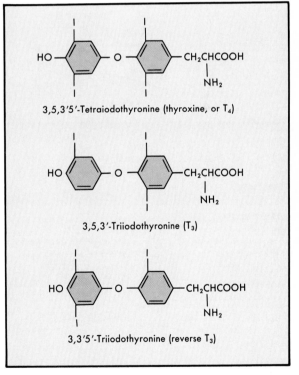

FIGURE 40-2   The structures of thyroxine (T_4), triiodothyronine (T_3), and reverse T_3 (rT_3). Note T_3 and rT_3 differ only in the position from which an iodine atom was removed from T_4.

Step One: Iodination

Iodide is an essential dietary element because of its thyroid role. The minimal daily requirement for hormone synthesis is 75 μg. In the United States the average daily intake is 300 to 400 μg, and almost the same amount is excreted in the urine. From an extracellular iodide pool, which averages 400 μg in size, about 80 μg (or 20%) is taken up daily by the gland. With iodide deficiency the pool size shrinks but the gland can increase the daily percentage uptake to 80% to 90%, thereby still acquiring sufficient iodide for hormone synthesis. Under steady-state conditions, 75 μg of iodide is released from the gland daily, most in the form of T_4. The content of iodide within the thyroid gland is 100 times greater than the amount needed daily for hormone production. Because all this is stored in the form of iodotyrosines and iodothyronines, the human is protected for approximately 2 months from the effects of iodide deficiency.

Iodide is actively transported into the thyroid gland against chemical and electrical gradients. The normal ratio of free iodide concentration in the gland to plasma iodide is 30:1. The trapping mechanism for iodide requires energy generation via oxidative phosphorylation, but its chemical nature remains obscure. Some evidence links it to a Na+,K+-ATPase. Various anions, such as thiocyanate (CNS$^-$), perchlorate (HClO$_4^-$), and pertechnetate (TcO$_4^-$), act as competitive inhibitors of iodide transport.

Small increases in dietary iodide intake lead to increases in the rate of thyroid hormone synthesis. However, as the dosage of iodide exceeds 2000 μg, the intraglandular concentration of free iodide or of some iodinated product reaches a point that inhibits the iodide trap and the biosynthetic mechanism, and thus the hormone production declines back to normal. Although the iodide trap is stimulated by iodine deficiency, a severe lack ultimately depletes the avail-

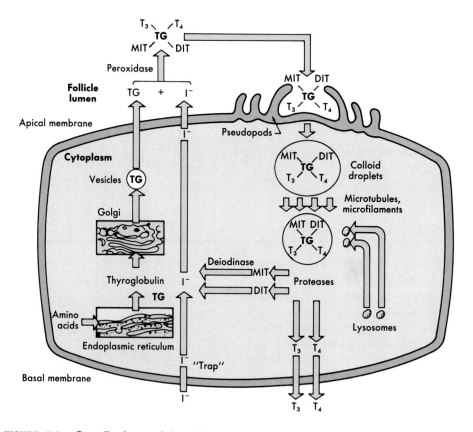

FIGURE 40-3 Overall schema of thyroid hormone synthesis and release. T_4 and T_3 synthesis occurs within the protein molecule thyroglobulin *(TG)* at the border of the cytoplasm and the follicle lumen. Retrieval of stored hormone requires endocytosis of the colloid followed by intracytoplasmic proteolysis by lysosomes. Iodide in the precursor molecules monoiodotyrosine *(MIT)* and diiodotyrosine *(DIT)* is recovered by the action of the enzyme deiodinase.

able pool, and hormone synthesis becomes inadequate.

Step 2: Iodination of Tyrosine within Thyroglobulin

Thyroglobulin is a large glycoprotein synthesized as two separate peptide units. These combine and are then glycosylated in transit to the Golgi apparatus. The completed protein, incorporated in small vesicles, moves to the apical membrane and then into the adjacent lumen of the follicle (Figure 40-3).

Just inside the follicle lumen iodide is incorporated into thyroglobulin. An enzyme complex known as *thyroid peroxidase* is bound to the apical membrane. This enzyme catalyzes simultaneously the oxidation of iodide and its substitution for a hydrogen in the benzene ring of tyrosine. The immediate oxidant of iodide is hydrogen peroxide (H_2O_2), which is probably generated via the reduction of O_2 by reduced nicotinamide adenosine dinucleotide phosphate (NADPH) and flavoproteins. Either monoiodotyrosine (MIT) or diiodotyrosine (DIT) results from iodination.

Step 3: Coupling

The coupling step is also carried out by peroxidase by juxtaposing one DIT molecule either with another DIT molecule within thyroglobulin to form T_4 or with an MIT molecule to form T_3.

The usual ratio of T_4 to T_3 in the gland is $10:1$. When iodide availability is restricted or when the thyroid gland is hyperstimulated, the formation of T_3 is favored, thus providing relatively more active hormone.

Retrieval

Once thyroglobulin has been iodinated, it is stored within the follicle as colloid. Release of the peptide-linked T_4 and T_3 into the bloodstream requires retrieval of the thyroglobulin. The latter is transferred from the lumen of the follicle into the endocrine cell by endocytosis (Figure 40-3). The cell membrane forms pseudopods that engulf a pocket of colloid. This is pinched off by the cell membrane and becomes a colloid droplet within the cytoplasm. The droplet moves in a basal direction, probably as a result of microtubule and microfilament function. At the same time, lysosomes move from the base toward the apex of the cell and fuse with the colloid droplets. Lysosomal pro-

teases then release free T_4 and T_3, which leave the cell through the basal membrane and enter the adjacent capillary blood (Figure 40-3).

The MIT and DIT molecules, which are also released from thyroglobulin, are rapidly deiodinated within the cell by the enzyme *deiodinase* (Figure 40-2). Because these compounds are metabolically useless and, if secreted, would be lost in the urine, their deiodination conserves iodide for recycling into T_4 and T_3 synthesis. Normally, only minor amounts of intact thyroglobulin leave the cell.

Any step in the sequence from iodide trapping to thyroglobulin proteolysis may be defective in congenital biosynthetic disorders, and these defects result in thyroid hormone deficiency. A group of drugs known as *thiouracils* block the enzyme thyroid peroxidase and are very useful in treating states of thyroid hyperfunction. A large excess of iodide itself, its competitive anion perchlorate, or lithium are also effective inhibitors of T_4 synthesis.

REGULATION OF THYROID GLAND ACTIVITY (HYPOTHALAMIC-PITUITARY-THYROID AXIS)

The thyroid gland is the effector component of a classic hypothalamic–anterior pituitary–peripheral gland axis (Figure 40-4). (See Chapter 39). The major stimulator of thyroid hormone secretion is *thyrotropin*, or thyroid-stimulating hormone (TSH), which is secreted by the anterior pituitary gland. The direct stimulator of TSH secretion is *thyrotropin-releasing hormone* (TRH) from the hypothalamus. The thyroid hormones T_4 and T_3 in turn inhibit TSH release from the pituitary and possibly TRH secretion from the hypothalamus by negative feedback.

Thyrotropin-Releasing Hormone

TRH is a tripeptide, pyroglutamine-histadine-proline-amide. Its synthesis in the hypothalamus is directed by a gene that codes for a large precursor molecule containing five repeating sequences of glutamine-histadine-proline-glycine. After translation of this primary gene product, the glutamic acid undergoes cyclization, and the terminal glycine is replaced with an amino group. TRH is stored in the median eminence and reaches its target cells via the pituitary

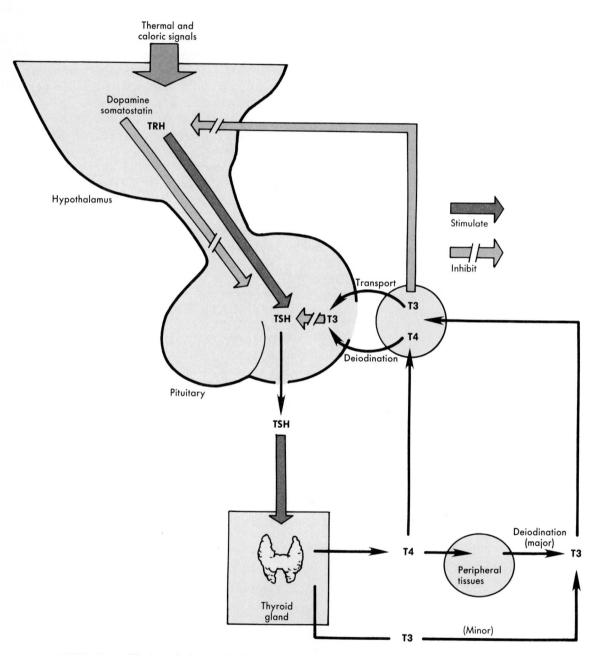

FIGURE 40-4   The hypothalamic–pituitary gland–thyroid gland axis. Thyrotropin-releasing hormone *(TRH)* stimulates thyrotropin *(TSH)* release from the pituitary gland. TSH stimulates T_4 and to a minor degree T_3 secretion by the thyroid gland. T_3 arising from T_4 in peripheral tissues or within the pituitary gland itself blocks the effect of TRH and suppresses TSH release by negative feedback. Dopamine and somatostatin also tonically inhibit TSH release.

portal vein. There TRH interacts with specific plasma membrane receptors on the thyrotroph cell. This triggers an influx of calcium and increases in phosphatidylinositol products, which act as second messengers. TSH is then released by exocytosis. Prolonged stimulation with TRH also increases TSH synthesis and its bioactivity through posttranslational modification. TRH eventually down-regulates its own receptors; thus the releasing hormone loses effectiveness.

Thyrotropin

TSH is a glycoprotein hormone of 28,000 molecular weight. It is composed of two peptide subunits, each of which is coded for by separate genes on two different chromosomes. The α-subunit is "nonspecific" because it is also part of three unrelated hormones with reproductive function (leutinizing and follicle-stimulating hormones from the pituitary and chorionic gonadotropin from the placenta). In contrast, the β-subunit of TSH is completely different and contains the specific biologically active sites of the hormone.

Nonetheless, by noncovalent forces, the β-subunit must be combined with the α-subunit for TSH to stimulate thyroid cells.

TSH circulates in concentrations of the order of 10^{-11} M. For technical reasons, these are usually reported in units of biologic activity, the normal range being approximately 0.5 to 6.0 $\mu U \cdot ml^{-1}$. The α-subunit also circulates.

TSH acts on the follicular cells of the thyroid gland to produce many effects, which are summarized in Figure 40-5. The process of iodide trapping and of each step in T_4 and T_3 synthesis, as well as the endocytosis of colloid and the proteolytic release of T_4 and T_3 from the gland, are all rapidly stimulated by TSH. Sustained exposure to TSH leads to hyperplasia of the follicular cells (see Figure 40-1), accompanied by increases in endoplasmic reticulum, ribosomes, the size and complexity of the Golgi apparatus, and DNA synthesis. In the absence of TSH the gland atrophies, although it still maintains a low basal level of thyroid hormone secretion. The trophic effects of TSH on the thyroid gland may be mediated by local generation of

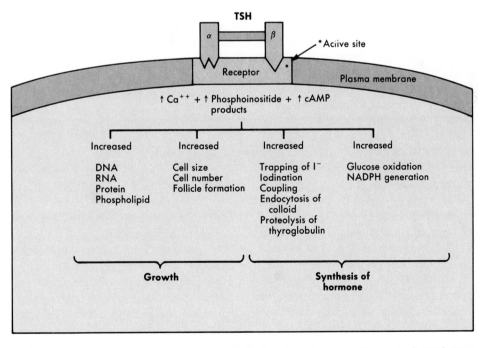

FIGURE 40-5   TSH actions on the thyroid cell. Cyclic adenosine monophosphate *(cAMP)* along with calcium ions *(Ca⁺⁺)* and phosphoinositol products act as second messengers generated by TSH binding to its receptor. All steps in thyroid hormone production, as well as many aspects of thyroid cell metabolism and growth, are stimulated by TSH.

insulin growth factors 1 and 2 or epidermal growth factor.

The initial step in TSH action is binding to a plasma membrane receptor. This transmembrane molecule consists of two distinct TSH-binding components. One, functionally linked to adenylate cyclase, increases cyclic adenosine monophosphate (cAMP), which then mediates stimulation of iodide uptake by the cell, perhaps by inducing synthesis of the carrier protein. The other binding component, acting through the phosphatidylinositol system, along with cAMP, stimulates the subsequent steps in thyroid hormone secretion. Within minutes of thyroid cell exposure to TSH, colloid droplets from the follicle are transferred into the cell. Shortly thereafter, iodide uptake and peroxidase activity increase. Concurrently, TSH also stimulates glucose oxidation, which may be the means for generating the NADPH needed for the peroxidase reaction. After several hours, TSH increases nucleic acid, protein, and phospholipid synthesis. These actions underlie the growth-promoting effects of TSH.

Feedback

Thyroid hormone output is under sensitive feedback control, and therefore plasma T_4, T_3, and TSH levels show minimal variations. Changes in thyroid hormone levels of only 10% to 30% are enough to change TSH levels in the opposite direction. Negative feedback is exerted predominantly at the pituitary level (Figure 40-4). This is well demonstrated by the results of repeated stimulation of the axis with TRH injections. The initially brisk increase in TSH is progressively dampened as T_4 levels rise in response to the TSH. These slightly elevated T_4 levels then feedback to inhibit the responsiveness of the pituitary thyrotroph to later TRH injections. The reverse sequence can be seen with maneuvers that primarily lower plasma T_4 and T_3 levels; in this case plasma TSH response to TRH stimulation is enhanced. Individuals with thyroid disease that results in chronic deficiency of thyroid hormone usually have enlarged pituitary glands, with increased numbers of thyrotroph cells and elevated TSH content as well as high plasma TSH levels. Conversely, a pathologic excess of thyroid hormone causes atrophy of the thyrotroph cells and very low plasma TSH levels.

The effector molecule of negative feedback is T_3.

This can either enter the thyrotroph cell from the plasma or can be generated within the pituitary by deiodination of T_4 taken up from the plasma (see Figure 40-4). T_3 suppresses not only TSH release but also its synthesis by inhibiting expression of the TSH gene. T_3 further blunts TSH release by decreasing the number of TRH receptors.

TSH secretion is also tonically inhibited by dopamine and somatostatin from the hypothalamus. Cortisol and growth hormone reduce TSH secretion as well, the latter probably by stimulating somatostatin release (see Chapter 39).

The net results of thyroid gland regulation are a slightly pulsatile plasma TSH level and a steady plasma T_4 and T_3 level. This befits hormones, whose actions on metabolism are gradual and wax and wane slowly. Physiologic conditions that alter TSH levels, and therefore T_4 and T_3 levels, are in accordance with the action of thyroid hormones on energy utilization and thermogenesis. During total fasting, TSH responsiveness to TRH stimulation and possibly TRH release itself are diminished; T_3 levels also fall. This coincides with an advantageous decrease in resting metabolic rate (see Chapter 36). In contrast, ingestion of excess calories, especially carbohydrate, tends to increase T_3 availability. In animals, exposure to cold increases TSH and thyroid hormone secretion. In humans, this is only seen the first few hours after birth, when the change in temperature from the maternal to external environment is accompanied by a sharp rise in plasma TSH and T_4.

METABOLISM OF THYROID HORMONES

T_4, the dominant secreted and circulating form, serves as a prohormone for T_3, in addition to providing some intrinsic intracellular action of its own. Average daily secretion of T_4 is 90 μg. The **circulating storage** function of plasma T_4 is reflected in its large pool size and long half-life. In contrast, the major portion of T_3 (and virtually all rT_3) comes from deiodination of circulating T_4 rather than from thyroid gland secretion. T_3, the active metabolite, has a much smaller pool size and shorter half-life. Average plasma concentrations are: T_4, 8 μg·dl^{-1}; T_3, 0.12 μg·dl^{-1}; rT_3, 0.04 μg·dl^{-1}.

Protein Binding

T_4 and T_3 circulate almost entirely bound to proteins. The major binding protein is *thyroxine-binding globulin* (TBG), a glycoprotein that is synthesized in the liver. Each TBG molecule binds one molecule of T_4. About 75% of T_4 and T_3 is bound to TBG. The remainder is bound to albumin and other proteins. Ordinarily, only changes in TBG concentrations significantly alter total plasma T_4 and T_3 levels.

Two biologic functions can be ascribed to TBG. First, by creating a circulating reservoir of T_4, it buffers against acute changes in thyroid gland function. Even the sudden addition to the plasma of an entire day's thyroid gland output would cause only a 10% increase in the total T_4 concentration. After removal of the gland, it would take nearly 1 week for the plasma T_4 concentration to fall 50%. Second, by binding T_4 and T_3, TBG prevents their glomerular filtration and urinary excretion.

Only 0.03% of total T_4 and 0.3% of total T_3 are in the free state. However, these are the critical **biologically active** fractions. They not only exert thyroid hormone effects on target tissues, but are also responsible for pituitary feedback. Thus the chemical equilibrium between T_4 and TBG governs the distribution of the hormone between the free $[T_4]$ and bound $[T_4 \cdot TBG]$ fractions:

$$T_4 + TBG \rightleftharpoons T_4 \cdot TBG \qquad (1)$$

$$K_{eq} = \frac{[T_4 \cdot TBG]}{[T_4][TBG]} \qquad (2)$$

$$\frac{Free\ T_4}{Bound\ T_4} = \frac{[T_4]}{[T_4 \cdot TBG]} = \frac{1}{K_{eq}[TBG]} \qquad (3)$$

where K_{eq} is the equilibrium constant.

A temporary decrease in free T_4, caused by a decrease in thyroid gland secretion, can be rapidly corrected by disassociation of bound T_4 (equation 1). Likewise, a temporary increase in free T_4 can be rapidly compensated for by association of the excess with TBG, because only 30% of the binding sites on TBG are normally occupied. Sustained decreases or increases in T_4 supply resulting from thyroid disease, however, must eventually lead to sustained decreases or increases in total T_4 and in the bound and free fractions.

A primary change in TBG concentration itself disturbs the ratio of free to bound T_4 (equation 3). In this situation the normal thyroid gland must increase or decrease its rate of hormone secretion appropriately until the new equilibrium state restores the absolute free T_4 level to normal. For example, TBG concentration can decrease because of reduced hepatic synthesis. The absolute free T_4 concentration would then temporarily increase (equation 3) and suppress pituitary TSH secretion by negative feedback. Thyroid gland output would then decrease until the new lower steady level of bound T_4 yielded a normal free T_4 level.

Conversely, TBG concentration is increased by pregnancy. Here the absolute free T_4 concentration decreases temporarily, and this stimulates secretion of TSH. Consequently, T_4 output by the thyroid gland increases until the new higher level of bound T_4 is sufficient to restore the free T_4 level to normal. Identical considerations govern the levels of free and bound T_3.

Metabolic Pathways

The liver, kidney, and skeletal muscle are the major sites of degradation of thyroid hormones. The rate of disposal of T_4 is proportional to the free T_4 concentration in plasma.

T_4 is largely a prohormone, only 25% as active as T_3. Therefore the initial step of converting it either to the active metabolite T_3 (by outer ring deiodination) or the inactive metabolite rT_3 (by inner ring deiodination) is an important means of adjusting thyroid hormone action on tissues. Normally the split between T_3 and rT_3 is equal. When it is physiologically desirable to have less thyroid hormone action, as in fasting, less T_3 and more rT_3 is generated.

ACTIONS OF THYROID HORMONE

Intracellular Mechanism

T_4 and T_3 enter target cells, possibly by facilitated transport, where most of the T_4 undergoes deiodination to T_3 (Figure 40-6). Both are transferred to the nucleus, where **T_3 binds to a nuclear receptor with much greater affinity than does T_4.** The T_3 receptor complex interacts with DNA to stimulate or inhibit transcription of messenger RNAs, as described in Chapter 35. The latter then direct increased or decreased synthesis of many specific proteins in different tissues. This sequence of events is illustrated by the effect of T_3 on the pituitary gland synthesis of growth hormone. T_3 stimulates parallel increases in the nuclear content of the primary RNA transcript of the growth hormone gene and the cytoplasmic content of its RNA. This is followed by an increase in synthesis of growth hormone itself.

The responsiveness of tissues to T_3 correlates well with their nuclear receptor capacity and with the degree of receptor occupancy. In the normal state about half the available receptor sites are occupied by T_3. In humans, whole body indices of T_3 action correlate with the calculated percentage occupancy of nuclear receptors. T_3 also may down-regulate its own receptor by inhibiting its synthesis. Because T_3 acts

largely through gene transcription, a 12 to 48-hour delay occurs before its effects become evident in vivo. Several weeks of hormone replacement are required before all the consequences of a deficiency state are corrected.

Thyroid hormone stimulates oxygen consumption (Figure 40-6). A single intracellular mechanism for this effect has still not been established. T_3 stimulates the activity of Na^+, K^+ ATPase, an enzyme responsible for membrane cation transport (see Chapter 1). Because large amounts of adenosine triphosphate (ATP) are consumed and much adenosine diphosphate (ADP) is correspondingly generated by Na^+, K^+ ATPase, the extra ADP could be the "messenger" by

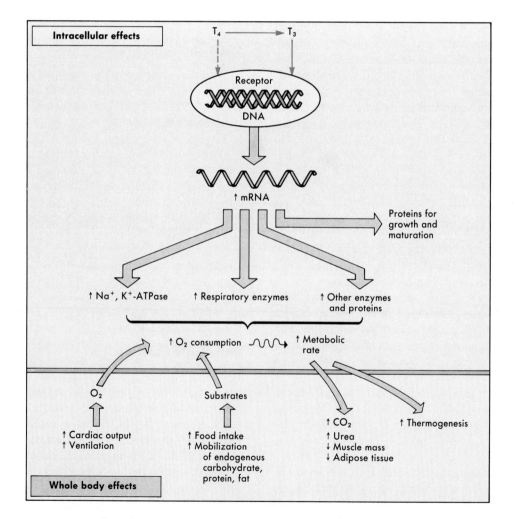

FIGURE 40-6   Overall schema of thyroid hormone effects. The upper portion represents intracellular actions; the lower portion, whole body effects.

which thyroid hormone stimulates mitochondrial O_2 utilization. Quantitatively, however, this does not account for all the hormone's effect. Another possibility is that thyroid hormone simultaneously stimulates fatty acid synthesis and oxidation—in effect, a futile energy cycle that requires energy and generates heat. Other related actions of thyroid hormone are to increase the activities of certain respiratory enzymes as well as enzymes involved with glucose oxidation and gluconeogenesis. In tissues such as the brain, in which O_2 consumption is not stimulated by T_3, the hormone increases the synthesis of specific structural or functional proteins.

Whole Body Actions

The most obvious effect of thyroid hormone is to increase the whole body rate of O_2 consumption and heat production (Figure 40-6). O_2 use at rest in humans is approximately 225 to 250 ml·min^{-1}. It falls to about 150 ml·min^{-1} in the absence of thyroid hormone and can increase to 400 ml·min^{-1} with thyroid hormone excess. Thus, the basal metabolic rate ranges from -40% to $+80\%$ of normal at these extremes of thyroid function. Of necessity, thermogenesis increases or decreases concomitant with O_2 use. In turn, increases and decreases in body temperature parallel fluctuations in thyroid hormone effect. These changes, however, are moderated by compensatory increases or decreases in heat loss through appropriate changes in blood flow, sweating, and ventilation mediated by thyroid hormone.

Thyroid hormone could not augment O_2 utilization for long without augmenting O_2 supply to the tissues (Figure 40-6). Thus thyroid hormone increases the resting rate of ventilation sufficiently to maintain a normal arterial oxygen pressure (P_{O_2}) with increased O_2 utilization, and a normal carbon dioxide pressure (P_{CO_2}) with increased carbon dioxide production. In addition, the O_2-carrying capacity of the blood is enhanced by a small increase in red cell mass, probably through stimulation of erythropoeitin production.

An important action of thyroid hormone is to increase cardiac output, ensuring sufficient O_2 delivery. The resting heart rate and stroke volume are both increased, and the speed and force of myocardial contractions are enhanced (see Chapter 18). These effects are partly indirect, via adrenergic stimulation. However, thyroid hormone directly increases myocardial Ca^{++} uptake, adenylate cyclase activity, and the active form of myosin-stimulated ATPase. Systolic blood pressure rises and diastolic blood pressure falls, reflecting the combined effects of the increased stroke volume with a substantial reduction in peripheral vascular resistance. The latter results from blood vessel dilation produced by the increased tissue metabolism.

Stimulation of O_2 utilization also depends on the provision of substrates for oxidation. Thyroid hormone potentiates the stimulatory effects of other hormones on glucose absorption from the gastrointestinal tract, on gluconeogenesis, on lipolysis, on ketogenesis, and on proteolysis of the labile protein pool. The overall metabolic effect of thyroid hormone has therefore aptly been described as accelerating the response to starvation.

Thyroid hormone also stimulates the biosynthesis of cholesterol, its oxidation, its conversion to bile acids, and its biliary secretion. The net effect is to decrease the body pool and plasma level of cholesterol. The rate of metabolic disposal of steroid hormones, B vitamins, and many administered drugs is increased. Therefore, to maintain effective plasma levels of these substances in the presence of increased thyroid hormone, their endogenous production or their exogenous administration must be increased.

Thyroid Hormone and Sympathetic Nervous System Activity

The activity of the sympathetic nervous system is diminished by thyroid hormone, as evidenced by decreased plasma levels and urinary excretion of the specific neurotransmitter, norepinephrine. On the other hand, the sensitivity of tissues to certain effects of the catecholamine hormones (see Chapter 42) is enhanced by thyroid hormones. These include the thermogenic, lipolytic, glycogenolytic, gluconeogenic, and insulin secretory effects of epinephrine and norepinephrine. With regard to cardiovascular responses to catecholamines, a modest reinforcing effect of thyroid hormone may exist. The mechanism of enhanced sensitivity to catecholamines appears to lie in the ability of thyroid hormone to increase the number of β-adrenergic receptors and to couple them to adenylate cyclase. Thus levels of cAMP, the β-adrenergic second messenger, are increased by thyroid hormone.

Effects on Growth and Development

A major effect of thyroid hormone is on growth and maturation. For example, in amphibians, thyroid hormone levels are very low until just before the major stage of metamorphosis. At this point the levels increase sharply, paralleling the rapid and spectacular change from the larval to the adult form. Addition of thyroid hormone to the fluid bathing tadpoles accelerates limb growth, tail resorption, shortening of the gastrointestinal tract, and induction of hepatic urea synthesis. These effects are accompanied by thyroid hormone–induced increases in protein and nucleic acid synthesis in the limb buds, increases in proteolytic and hydrolytic enzyme activities in the tail, and increases in the hepatic content of carbamyl phosphate synthase, the rate-limiting enzyme in the urea cycle.

In humans, thyroid hormone stimulates linear growth, development, and maturation of bone. A direct effect of T_3 on the activity of chondrocytes in the growth plate of bone may initiate this process. T_3 also accelerates growth by stimulating secretion of growth hormone. Interestingly, thyroid hormone is not required for linear growth until after birth, even though it is already essential for maturation of the growth centers in the bones of the fetus. The regular progression of tooth development and eruption is dependent on thyroid hormone, as is the normal cycle of renewal of the epidermis and hair follicles. Because thyroid hormone stimulates degradative processes in the structural and integumentary tissues, it can also cause resorption of bone and accelerated shedding of skin and hair. The synthesis of mucopolysaccharides that form the intercellular ground substance is inhibited by thyroid hormone.

Normal skeletal muscle function also requires thyroid hormone. This may be related to the regulation of energy production and storage in this tissue. Concentrations of creatine phosphate are reduced by an excess of thyroid hormone; the inability of muscle to take up and phosphorylate creatine leads to an increase in its urinary excretion.

Thyroid hormone has critical effects on the development of the central nervous system. If thyroid hormone is deficient in utero, growth of the cerebral and cerebellar cortex, proliferation of axons and branching of dendrites, and myelinization are all impaired. Irreversible brain damage results when the deficiency of thyroid hormone is not recognized and treated immediately after birth. These anatomic defects are paralleled by several biochemical abnormalities. Without thyroid hormone, cell size, RNA and protein content, protein synthesis, the enzymes necessary for DNA synthesis, protein and lipid content of myelin, neurotransmitter receptors, and neurotransmitter synthesis are decreased in various areas of the brain. In children and adults, thyroid hormone enhances the speed and amplitude of reflexes, wakefulness, alertness, responsiveness to various stimuli, awareness of hunger, memory, and learning capacity. Normal emotional tone also depends on appropriate thyroid hormone levels.

Thyroid hormone contributes to the regulation of reproductive function in both genders. The normal process of sperm production; the ovarian cycle of follicular development, maturation, and ovulation; and the maintenance of a healthy pregnant state are all disrupted by significant deviations of thyroid hormone levels from normal. In part these may be caused by alterations in the metabolism of steroid hormones.

THYROID DYSFUNCTION

Excess thyroid hormone presents a striking clinical picture. The increase in metabolic rate causes the highly characteristic combination of weight loss and increased intake of food. The increased heat generated causes discomfort in warm environments, fever if the condition is severe, excessive sweating, a greater intake of water, and increased ventilation. The increase in β-adrenergic responsivity is manifest by a rapid heart rate, tremor, nervousness, and an anxious stare. Weakness is caused by loss of muscle mass. The thyroid gland is usually enlarged.

The opposite picture characterizes a state of thyroid deficiency. The lowered metabolic rate leads to weight gain without an appreciable increase in food intake. The decreased thermogenesis lowers body temperature and causes discomfort with cold, decreased sweating, and dry skin. Decreased β-adrenergic responsivity is manifest by a slowed heart rate; slowed movement, speech, and thought; and sleepiness. An excess of ground substance mucopolysaccharides causes a concomitant accumulation of fluid. This produces puffy features and swelling of many areas in the body. The notable effects of thyroid deficiency in infancy or childhood include growth retardation; immaturity of bone; delayed developmental milestones such as sitting, standing, and walking; and in severe cases, irreversible mental retardation. Such individuals are known as *cretins*.

BIBLIOGRAPHY

Journal Articles

Bantle JP et al: Common clinical indices of thyroid hormone action: relationships to serum free 3,5,3'-triiodothyronine concentration and estimated nuclear occupancy, J Clin Endocrinol Metab 50:286, 1980.

Chin WW: Hormonal regulation of thyrotropin and gonadotropin gene expression, Clin Res 36:484, 1988.

Danforth E Jr: The role of thyroid hormone and insulin in the regulation of energy metabolism, Am J Clin Nutr 38:1006, 1983.

Everett AW et al: Change in synthesis rates of α- and β-myosin heavy chains in rabbit heart after treatment with thyroid hormone, J Biol Chem 285:2421, 1983.

Izumo S et al: All members of the MHC multigene family respond to thyroid hormone in a highly tissue-specific manner, Science 231:597, 1986.

Klein I and Levey GS: New perspective on thyroid hormone, catecholamines, and the heart, Am J Med 76:167, 1984.

Larsen PR et al: Relationships between circulating and intracellular thyroid hormones, physiological and clinical implications, Endocr Rev 2:87, 1981.

Martial JA et al: Regulation of growth hormone gene expression: synergistic effects of thyroid and glucocorticoid hormones, Proc Natl Acad Sci USA 74:4293, 1977.

Oppenheimer JH et al: Advances in our understanding of thyroid hormone action at the cellular level, Endocr Rev 8:288, 1987.

Schimmel M et al: Thyroidal and peripheral production of thyroid hormones, Ann Intern Med 87:760, 1977.

Sestoft L: Metabolic aspects of the calorigenic effect of thyroid hormone in mammals, Clin Endocrinol 13:489, 1980.

Taylor T and Weintraub B: Thyrotropin (TSH)-releasing hormone regulation of TSH subunit biosynthesis and glycosylation in normal and hypothyroid rat pituitaries, Endocrinology 116:1968, 1985.

Books and Monographs

Galton VA: Thyroid hormone action in amphibian metamorphosis. In Oppenheimer JH and Samuels HH: Molecular basis of thyroid hormone action, New York, 1983, Academic Press, Inc.

Greer MA et al: Thyroid secretion. In Handbook of physiology, section 7: Endocrinology, vol III: Thyroid, Baltimore, 1974, American Physiological Society.

Reichlin S: Neuroendocrine control of thyrotropin secretion. In Ingebar SH and Braverman LE, editors: Werner's the thyroid, Philadelphia, 1986, JB Lippincott Co.

Schwartz HL: Effect of thyroid hormone on growth and development. In Oppenheimer JH and Samuels HH: Molecular basis of thyroid hormone action, New York, 1983, Academic Press, Inc.

Taurog A: Hormone synthesis: thyroid iodine metabolism. In Ingbar SH and Braverman LE, editors, Wener's the thyroid, Philadelphia, 1986, JB Lippincott Co.

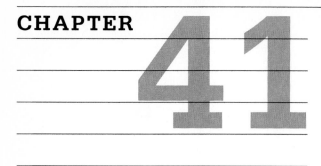

The Adrenal Cortex

The adrenal glands are multifunctional endocrine organs that secrete a variety of hormones. Experimental and clinical observations in the mid-1800s demonstrated that the adrenal glands were essential to life. The adrenal hormones subserve a wide variety of physiologic functions, including blood glucose regulation, protein turnover, sodium and potassium metabolism, survival in the face of stress, and modulation of tissue response to injury or infection.

The adrenal glands are located just above each kidney, and their total weight is 6 to 10 g. They are really a combination of two separate functional entities

(Figure 41-1). The outer zone, or *cortex,* comprises 80% to 90% of the weight. It is derived from mesodermal tissue and is the source of corticosteroid hormones. The inner zone, or *medulla,* comprises the other 10% to 20%. It is derived from neuroectodermal cells of the sympathetic ganglia and is the source of catecholamine hormones. The adrenal glands have one of the highest rates of blood flow per gram of tissue. Arterial blood enters the outer cortex and breaks up into capillaries; the blood then drains down veins into the medulla. This exposes the inner cells of the cortex and the cells of the medulla to high con-

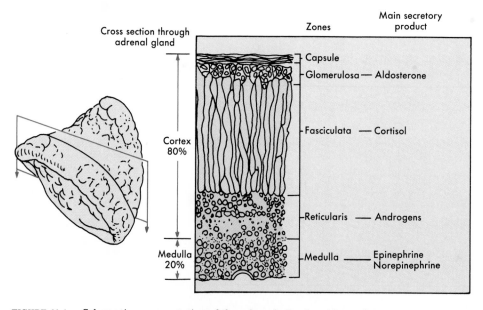

FIGURE 41-1   Schematic representation of the adrenal gland and its main secretory products.

centrations of steroid hormones from the outer cortex.

The outermost *zona glomerulosa* of the adrenal cortex is only a few cells thick (Figure 41-1). The middle *zona fasciculata* is the widest and consists of long cords of columnar cells. The inner most *zona reticularis* contains networks of interconnecting cells. Steroid-secreting cells are usually rich in lipid droplets and contain numerous large mitochondria with vesicles in their membranes.

The major hormones of the cortex are (1) the *glucocorticoid, cortisol,* which is critical to life because of its effect on carbohydrate and protein metabolism and its role in adaptation to stress; (2) the *mineralocorticoid, aldosterone,* which is vital to maintaining normal extracellular fluid volume and potassium levels; and (3) *sex steroid precursors,* which contribute to maintaining secondary sexual characteristics. Medical interest in cortisol was greatly heightened by the discovery of its potent antiinflammatory effects. In very high doses various glucocorticoids are used to treat numerous diseases.

SYNTHESIS OF CORTICOSTEROID HORMONES

The precursor for all adrenocortical hormones is *cholesterol,* which is taken up from the plasma via a specific low-density lipoprotein receptor in the plasma membrane. After transfer into the cell, the cholesterol is largely esterified and stored in cytoplasmic vacuoles. Under basal conditions cholesterol just taken up from plasma is immediately used for hormone synthesis. However, when hormone production is stimulated, stored cholesterol becomes increasingly important. Cholesterol can also be synthesized within the cell, but this is a minor source.

Most of the reactions from cholesterol to corticosteroid hormones are catalyzed by *cytochrome P-450 enzymes.* The genes that direct their synthesis have considerable similarity, even though they may be located on different chromosomes. A single P-450 enzyme may catalyze more than one reaction, depending on its location in the cortex and on substrate availability. These enzymes catalyze hydroxylations of the steroid nucleus, employing molecular oxygen, NADPH, a flavoprotein and an iron-containing protein called adrenoxin in the reactions.

Glucocorticoids

The synthesis of cortisol, the major glucocorticoid in humans (Figure 41-2), occurs largely in the zona fasciculata. The initial reaction that converts cholesterol to pregnenolone is catalyzed by the side-chain cleavage complex P-450$_{SCC}$ (also known as *20,22-desmolase*). This is the rate-limiting step carried out in the mitochondria to which cholesterol has been transferred. The product is then converted to progesterone, following which hydroxyls are successively added at the 17 and 21 positions. These steps take place within the endoplasmic reticulum. The resultant 11-deoxycortisol is transferred to the mitochrondria and hydroxylated in the 11 position, the critical step in creating a glucocorticoid molecule.

Neither the final product, cortisol, nor its precursors are stored in the adrenocortical cell. Thus an acute need for increased cortisol secretion requires rapid activation of the initial controlling, rate-limiting reaction.

Mineralocorticoids

The synthesis of aldosterone, the major mineralocorticoid (Figure 41-2), is carried out exclusively in the zona glomerulosa. The sequence from cholesterol to corticosterone is identical to that in the zona fasciculata. In the subsequent key step, the C_{18} methyl group of corticosterone is oxidized (by the same mitochondrial enzyme that catalyzes 11-hydroxylation) to yield aldosterone. Deoxycorticosterone and its 18-hydroxy derivative are other steroids that have mineralcorticoid activity and that are synthesized in small quantities in the zona fasciculata.

Androgens and Estrogens

The synthesis of the sex steroids occurs largely in the zona reticularis. The potent androgen *testosterone* and the potent estrogen *estradiol* are normally secreted only in trace amounts by the adrenal cortex. However, substantial amounts of precursor steroids with weak androgenic activity are secreted and converted to testosterone and estradiol by peripheral tissues. These precursors, androstenedione and dehydroepiandrosterone (DHEA), are synthesized from 17-OH-progesterone or 17-OH-pregnenolone, respectively, as shown in Figure 41-2.

In women the adrenal precursors supply 50% of the

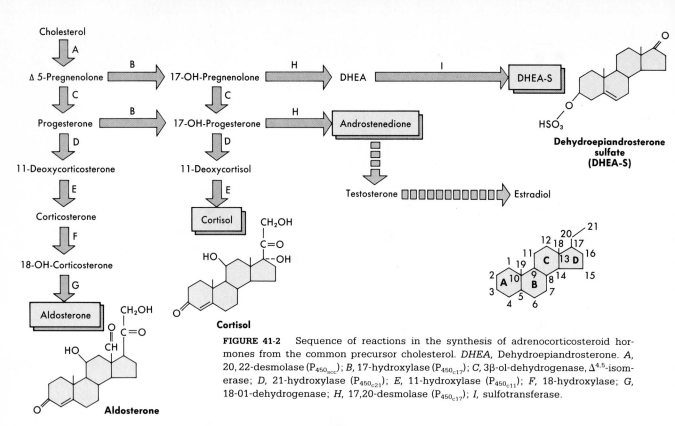

FIGURE 41-2   Sequence of reactions in the synthesis of adrenocorticosteroid hormones from the common precursor cholesterol. *DHEA*, Dehydroepiandrosterone. *A*, 20,22-desmolase ($P_{450_{scc}}$); *B*, 17-hydroxylase ($P_{450_{c17}}$); *C*, 3β-ol-dehydrogenase, $\Delta^{4,5}$-isomerase; *D*, 21-hydroxylase ($P_{450_{c21}}$); *E*, 11-hydroxylase ($P_{450_{c11}}$); *F*, 18-hydroxylase; *G*, 18-01-dehydrogenase; *H*, 17,20-desmolase ($P_{450_{c17}}$); *I*, sulfotransferase.

androgenic hormone requirements. In men they are unimportant because the testes produce testosterone. After menopause the estrogens that arise directly or indirectly from the adrenal cortex become the only source for this biologic activity in women.

CORTICOSTEROID HORMONE METABOLISM

Basal morning plasma cortisol concentrations are 5 to 20 μg·dl^{-1}. The hormone circulates 90% bound to a specific corticosteroid-binding globulin called *transcortin*. The concentration of transcortin and likewise total plasma cortisol are increased during pregnancy and estrogen administration. However, because bound cortisol is biologically inactive, the physiologic effects of an increase in transcortin are determined by principles similar to those discussed with regard to thyroxine binding (see Chapter 40). The plasma half-life of cortisol is about 70 minutes. Free (unbound) cortisol is filtered by the kidney, but less than 1% of daily cortisol secretion is excreted in the urine. Nevertheless, when kidney function is normal, urinary cortisol excretion is a valid index of secretion.

Cortisol is in equilibrium with its 11-keto analogue, *cortisone*, which has no intrinsic biologic activity. However, the interconversion is catalyzed by an enzyme present in many tissues, rendering exoge-

nous cortisone an effective source of cortisol activity. Almost all of the cortisol and cortisone is metabolized in the liver; the metabolites are conjugated and excreted in the urine as glucuronides. The measurement of these urinary metabolites, known generally as *17-hydroxycorticoids*, also provides a reasonable index of cortisol secretion as long as hepatic and renal functions are intact.

Aldosterone circulates bound to a specific *aldosterone-binding globulin,* to transcortin, and to albumin. The plasma half-life of aldosterone is only 20 minutes. Aldosterone and its liver-generated metabolites are excreted in the urine as glucuronides.

Adrenal androgen precursors are also metabolized in the liver and excreted in the urine in a fraction known as *17-ketosteroids*. However, these products are not specific for the adrenal gland because they also arise from gonadal androgens. Measurement of plasma or urinary dehydroepiandrosterone sulfate (DHEA-S) gives the best index of activity of the adrenal zona reticularis.

REGULATION OF CORTISOL SECRETION

The pattern of cortisol secretion is very complex (Figure 41-3). The immediate stimulator of cortisol secretion is *adrenocorticotropin* (ACTH) from the anterior pituitary gland. The most important immedi-

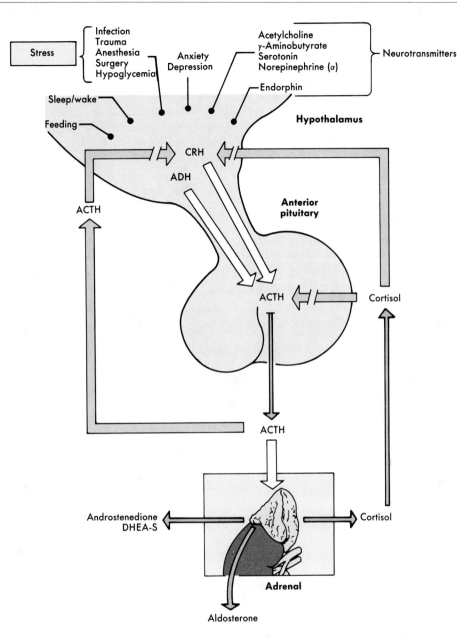

FIGURE 41-3 The regulation of cortisol secretion by the hypothalamic-pituitary-adrenal axis. A variety of inputs to the hypothalamus stimulate corticotropin—releasing hormone *(CRH)* secretion, and in turn adrenocorticotropin *(ACTH)* and cortisol secretion. Cortisol exerts negative feedback at both the hypothalamic and the pituitary levels. *ADH,* Antidiuretic hormone; *DHEA-S,* dehydroepiandrosterone sulfate.

ate stimulator of ACTH secretion is the neuropeptide *corticotropin-releasing hormone* (CRH) from the hypothalamus. Thus a hypothalamic–anterior pituitary–adrenal cortex axis exists, and the three hormones just listed form a classic negative feedback loop (Figure 41-3). Cortisol (or any synthetic glucocorticoid analogue, e.g., *dexamethasone, prednisone*):

1. Feeds back within minutes on the pituitary gland to inhibit the release of ACTH by blocking the stimulatory action of CRH on the corticotroph cells
2. Feeds back more slowly (within hours) to inhibit the synthesis of ACTH by blocking transcription of its gene

3. Feeds back on the hypothalamus to block release and probably synthesis of CRH

The operation of this long-loop feedback is shown in Figure 41-3. Short-loop feedback also exists as ACTH inhibits CRH release. In addition to CRH, the neuropeptide arginine vasopressin (AVP) (see Chapter 39) augments ACTH secretion and therefore cortisol secretion in certain situations. Cortisol also feeds back to restrain AVP release.

Synthetic glucocorticoids and, to a lesser extent, ACTH are used to treat various nonendocrine diseases. When given for more than a week in very high doses, they profoundly suppress the hypothalamic CRH neurons, the pituitary corticotroph cells, and consequently the cells of the zona fasciculata. After withdrawal from such therapy, full recovery of the atrophied hypothalamic-pituitary-adrenal axis can take up to 1 year.

Corticotropin-Releasing Hormone

CRH is a 41–amino acid peptide synthesized from a prepro-CRH that has considerable homology to prepro-AVP. CRH enters the portal veins and travels to the pituitary conticotroph cells. After binding to a plasma membrane receptor, CRH stimulates the release of ACTH from its secretory granules via calcium and cyclic adenosine monophosphate (cAMP) as second messengers, and further stimulates ACTH synthesis. In addition to this endocrine role, and possibly functionally related to it, CRH exhibits diverse other actions in the central nervous system. These include stimulating sympathetic nervous system activity, decreasing elevated body temperatures, suppressing reproductive function and sexual activity, suppressing growth hormone release, and altering behavior. Peripheral plasma CRH levels are very low, but they mirror negative feedback since they are slightly increased by cortisol deficiency and decreased by glucocorticoid administration.

Adrenocorticotropin

ACTH is a 39 amino acid peptide that increases the synthesis and immediate release of cortisol, adrenal androgens, their precursors, and aldosterone. However, only cortisol feeds back negatively on the hypothalamus and the pituitary gland. ACTH is synthesized via a large precursor called propiomelanocortin, which gives rise to a number of cosecreted products, including β-endorphin. After binding to its adrenal plasma membrane receptor, ACTH stimulates the generation of cAMP, which is the major second messenger for its actions (Figure 41-4). Calcium and phosphatidylinositol products may play adjunctive roles. The ultimate effects of ACTH presumably reflect a cascade of enzyme phosphorylations catalyzed by protein kinases A and C. ACTH acutely stimulates cholesterol uptake by the cell, cholesterol ester hydrolysis, cholesterol transfer to the mitochondria, the rate-limiting $P-450_{SCC}$ desmolase reaction, and the critical 11-hydroxylation step in cortisol synthesis. ACTH also alters the shape of the adrenocortical cell by affecting its cytoskeleton and bringing the cholesterol vacuoles into contact with the mitochondria.

Patterns of Cortisol Secretion

Cortisol is secreted in pulses and in a diurnal pattern (Figure 41-5). The pulses of cortisol are induced by preceding pulses of ACTH, which in turn are caused by the pulsatile release of CRH. These 7 to 13 episodes of cortisol secretion each day form a circadian pattern. The peak plasma ACTH and cortisol levels are achieved about 2 hours before awakening, at 4 to 6 AM; the nadir of plasma ACTH and cortisol is reached just before falling asleep. The morning peak of cortisol constitutes 50% of the daily total secretion. The clock time of this peak can be altered by systematically shifting the sleep-wake cycle. This phenomenon is of occupational significance, such as in transoceanic airline flights. The circadian rhythm is generated within the hypothalamus, probably by stimulation from other brain centers and not by negative feedback. Nevertheless, a small dose of exogenous glucocorticoid suppresses, and prior cortisol deficiency accentuates, the early-morning ACTH peak. Loss of consciousness and constant exposure to either dark or light also blunts this circadian rhythm.

Cortisol is a hormone that is required for survival of the stressed organism. Accordingly, a variety of major stresses greatly stimulate the triad of sequential CRH, ACTH, and cortisol release (see Figure 41-3). In addition, severe pain and prolonged exercise also cause release of cortisol, whereas the state of analgesia induced by endorphins blocks the cortisol response. Stress can override the diurnal pattern of cortisol secretion and the suppressive effects of negative feedback. Several neurotransmitters mediate the

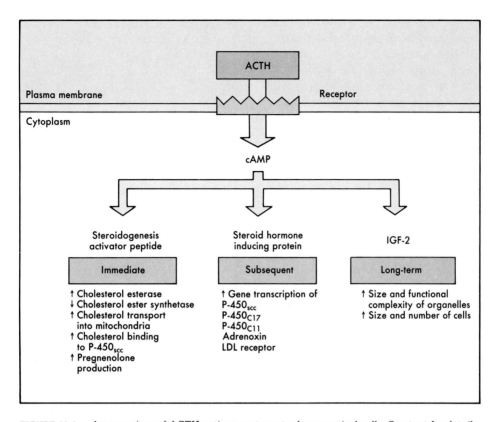

FIGURE 41-4   An overview of ACTH actions on target adrenocortical cells. See text for details. *cAMP,* Cyclic adenosine monophosphate; IGF-2, insulin growth factor 2; *LDL,* low-density lipoprotein.

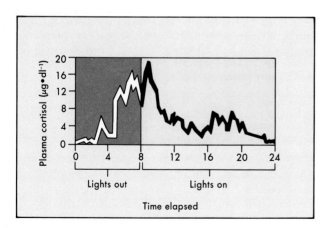

FIGURE 41-5   Pulsatile and diurnal nature of cortisol secretion. (Redrawn from Weitzman ED et al: Twenty-four hour pattern of the episodic secretion of cortisol in normal subjects, J Clin Endocrinol Metab 33:14, 1971, © by The Endocrine Society)

stressful inputs that stimulate CRH (plus AVP) release (see Figure 41-3). Prolonged stress causes hyperplasia of the adrenal cortex because of continuous ACTH stimulation.

Activation of the process of cell-mediated immunity also increases ACTH and cortisol release. The lymphokines interleukin-1 and interleukin-2 both stimulate ACTH secretion. Because stresses such as infection and tissue trauma are accompanied by cell-mediated immune responses, and because cortisol is an important modulator of those responses (see following discussion), a significant feedback relationship between the immune and endocrine systems has now been revealed.

ACTIONS OF CORTISOL (GLUCOCORTICOIDS)

Cortisol is a hormone essential for life. Human beings cannot survive complete removal of the adrenal glands for long without glucocorticoid replacement. The exact reasons for this life-preserving action are not certain, despite knowledge of many important effects of cortisol. Most clearly, the hormone is required to sustain glucose production from protein and to support vascular responsiveness. In addition, the hormone affects fat metabolism, central nervous system function, skeletal turnover, hematopoiesis, muscle function, renal function, and immune responses. The term *permissive* has been used to describe cortisol's action, implying that the hormone may not directly **initiate,** so much as **allow,** critical processes to occur. The following facts may better define this concept:

1. Cortisol can amplify the effect of another hormone on a process that cortisol itself does not affect. For example, cortisol does not stimulate glycogenolysis. However, if cortisol is present, glycogenolysis stimulated by glucagon is enhanced.
2. Cortisol can facilitate induction of an enzyme by its substrate. For example, tyrosine only induces the enzyme tyrosine transaminase if cortisol is present.

Intracellular Mechanisms

Most effects of cortisol are mediated via transcriptional mechanisms. Cortisol enters target cells by facilitated diffusion and binds to its receptor in the cytoplasm and/or nucleus. The cortisol-receptor complex must undergo an activation process before it can bind to a specific DNA molecule. Hormonal action is directly proportional to the degree of DNA binding, and the final response is an increase or decrease in gene transcription of specific messenger ribonucleic acids (mRNAs). Although other steroids can bind to the cortisol receptor and other steroid receptors may bind to a similar site on the same DNA molecule, the specific combination of cortisol, its receptor, and the responsive DNA molecule are required to elicit the cortisol action.

Effects on Metabolism

The most important overall effect of cortisol is to stimulate the conversion of protein to glucose and the storage of glucose as glycogen; thus the term *glucocorticoid* (Figure 41-6). All phases of this process are augmented, including mobilization

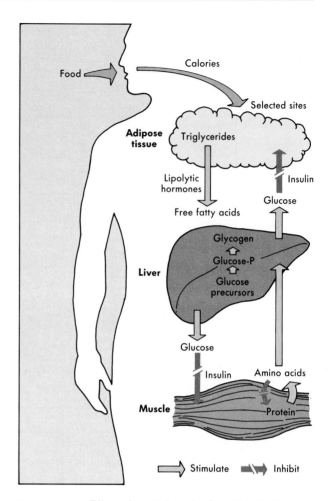

FIGURE 41-6   Effect of cortisol on the flow of fuels. Cortisol stimulates mobilization of amino acids and their conversion to glucose. The glucose is preferentially but not exclusively stored as glycogen. Insulin-mediated glucose uptake by peripheral tissues is inhibited. Cortisol facilitates storage of fat in selected adipose tissue sites but also facilitates release of free fatty acids.

of protein from muscle stores, entrance of the released amino acids into the hepatic gluconeogenetic pathway, conversion of pyruvate to glycogen, and disposition of the ammonia released from metabolism of the precursor amino acids. Cortisol increases the activity of enzymes involved in each of these steps. In some of these instances, the hormone "permits" substrate induction of its enzyme; in others, cortisol directly increases transcription of the target enzyme gene.

If the glucocorticoid effect is continued beyond the point where it is physiologically beneficial, the continuous drain on body protein produces serious dele-terious effects. Most notably, muscle, bone, connec-tive tissue, and skin lose mass. This is further exac-erbated by the inhibitory effects of cortisol on the syn-thesis of constitutive proteins such as collagen.

The glucocorticoid effect is essential for mainte-nance of plasma glucose levels and for survival during prolonged fasting. Without cortisol, death may occur from hypoglycemia once glycogen stores are gone. However, only a minor increase in corticol secretion occurs with fasting, and thus ordinary levels of the hormone can effectively mobilize amino acids and promote gluconeogenesis. On the other hand, plasma cortisol levels do increase sharply in response to acute hypoglycemia. In this situation cortisol amplifies the glycogenolytic actions of glucagon and epinephrine and synergizes with them in rebuilding liver glycogen stores.

Consonant with its role in preventing hypoglyce-mia, cortisol is a strong antagonist to insulin (Figure 41-6). Cortisol inhibits insulin-stimulated glucose uptake by muscle and adipose tissue and blocks the suppressive effect of insulin on hepatic glucose out-put. The interaction between cortisol and insulin is complex. Both hormones favor hepatic glycogen stor-age by increasing glycogen synthase activity (see Figure 36-3). However, they have opposite effects on expression of the genes for the gluconeogenetic enzyme, phospho*enol*pyruvate carboxykinase, and the glucose-releasing enzyme, glucose-6-phospha-tase. Thus cortisol favors glucose output by the liver, whereas insulin inhibits it. The net result of prolonged exposure to an excess of cortisol is a rise in plasma glucose concentration and a compensatory increase in plasma insulin levels.

Cortisol also plays a complex role in fat metabolism

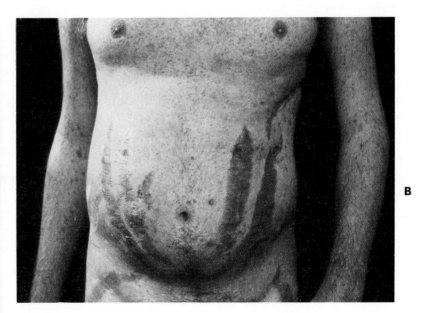

FIGURE 41-7   An individual suffering from Cushing's syndrome, an excess of cortisol. **A,** Note the selective accumulation of abdominal fat and the loss of musculature in the extremities. **B,** The extreme thinness of the skin reveals blood flowing through under-lying capillaries.

(Figure 41-6). The presence of cortisol "permits" maximal stimulation of fat mobilization by growth hormone, epinephrine, and other lipolytic factors during fasting. In cortisol-deficient individuals total fat stores may be relatively increased and ketogenesis may be impaired. However, the hormone also greatly increases appetite and stimulates lipogenesis in certain adipose tissue depots. Therefore an excess of cortisol also results in accumulation of fat, but the obesity has a peculiar distribution, favoring the face and trunk but sparing the extremities (Figure 41-7).

Thus **cortisol is a catabolic, antianabolic, and diabetogenic hormone.** In stress situations cortisol accentuates hyperglycemia produced by other hormones while greatly accelerating loss of body protein. These actions are amplified if insulin secretion is simultaneously deficient, as in diabetes mellitus.

Effects on Tissues and Organs

Muscle Basal levels of cortisol are required for maintenance of normal contractility and maximal work output of skeletal and cardiac muscle. The inotropic effects are probably exerted at the myoneural junctions. In contrast, excess cortisol produces muscle atrophy and weakness through protein wastage (Figure 41-7).

Bone The major effect of cortisol is to decrease bone formation. Less prominently, cortisol increases bone resorption. The net outcome of cortisol excess can be a profound reduction in bone mass and, in children, a reduction in linear growth as well. Several actions contribute to this outcome. Cortisol decreases the synthesis of 1,25-$(OH)_2$ vitamin D and blocks its action; therefore calcium absorption from the gastrointestinal tract is defective. At the same time, urinary calcium excretion is increased. Thus less calcium is available for mineralization of bone. Cortisol also inhibits the differentiation of mesenchymal precursors into osteoblasts and the synthesis of collagen by these cells.

Connective Tissue Cortisol causes thinning of the skin and the walls of capillaries with consequent fragility and easy rupture (Figure 41-7). These effects are related to cortisol's inhibition of the synthesis of collagen and glycosaminoglycans for ground substance.

Vascular System Cortisol is required for the maintenance of normal blood pressure. The hormone permits enhanced responsiveness of arterioles to the constrictive action of adrenergic stimulation and also optimizes myocardial performance. Cortisol helps to maintain blood volume by decreasing the permeability of the vascular endothelium.

Kidney Cortisol increases the rate of glomerular filtration. The hormone is also essential for the rapid excretion of a water load. In the absence of cortisol the secretion of ADH is increased and its action on the renal tubule enhanced. This diminishes free water clearance and prevents maximal dilution of the urine.

Central Nervous System A certain type of receptor for cortisol is present throughout the brain and concentrated in the hippocampus, reticular-activating substance, and autonomic nuclei of the brainstem. Cortisol somehow modulates perceptual and emotional functioning. A deficiency of cortisol accentuates auditory, olfactory, and gustatory acuity, suggesting that the hormone may normally have a damping effect. The diurnal increase in CRH pulses and cortisol just before awakening are essential for normal arousal and initiation of daytime activity. An excess of cortisol interferes with normal sleep and can either elevate the mood strikingly or depress it.

Fetus Cortisol has important permissive effects that facilitate in utero maturation of the gastrointestinal tract and lungs, the central nervous system, retina, and skin. Cortisol facilitates timely preparation of the fetal lung to permit satisfactory breathing immediately after birth. The rate of development of the pulmonary alveoli, flattening of the lining cells, and thinning of the lung septa are increased. Most importantly, the synthesis of surfactant, a phospholipid vital for maintaining alveolar surface tension, is increased. Cortisol also facilitates maturation of the enzyme capacity of the intestinal mucosa from a fetal to an adult pattern. This permits the newborn to digest the disaccharides present in milk.

Inflammatory and Immune Responses Cortisol has an important influence on the set of reactions evoked by tissue trauma, chemical irritants, foreign proteins, and infection. The overriding effect is to inhibit every step in the response to tissue injury (see box). The consequences are to impede the ability of tissues either to eliminate immediately noxious substances and invaders or to wall them off from the rest of the body. The mechanisms by which cortisol suppresses these responses vary, as shown by the following examples.

1. Cortisol induces a phosphoprotein called *lipocortin* that inhibits the enzyme *phospholipase A_2*. This enzyme generates arachidonic acid. Because the latter serves as the precursor for synthesis of prostaglandins and related compounds, the production of these mediators of inflammation is reduced.
2. Cortisol decreases the production of interleukin-1 by repressing expression of the lymphokine gene. In this way cortisol blocks the entire cascade of cell-mediated immunity as well as the generation of fever.
3. Cortisol stabilizes lysosomes, thereby reducing release of enzymes capable of degrading foreign substances.
4. Cortisol blocks recruitment of leukocytes by inhibiting their ability to bind chemotactic peptides.

At this point a paradox is evident. On the one hand, augmentation of cortisol secretion is essential to the survival of severely stressed, traumatized, or infected individuals through its metabolic actions. On the other hand, many of the tissue defense mechanisms evoked by such conditions are inhibited by elevated cortisol levels. To explain this, it has been suggested that permissive lower levels of cortisol may be required for the initial responses to stress. Subsequently, even higher levels of cortisol may be secreted to limit cellular and tissue reactions so that they do not themselves seriously damage the individual.

In this regard the therapeutic use of glucocorticoids may require administration of high doses for some time. Therefore, they represent a two-edged sword. Glucocorticoids are dramatically beneficial when the reactions to tissue injury are so severe as to be functionally disabling or life threatening, or when the rejection of transplanted tissues must be prevented. However, if glucocorticoids are administered for long periods, they may increase the susceptibility to infections or allow their dissemination, and they may prevent normal wound healing after injury. These adverse effects, along with induction of diabetes, osteoporosis, and psychiatric disorders, require physicians to prescribe glucocorticoids cautiously and only when no safer form of treatment exists. This injunction does not apply to the use of replacement doses of cortisol in individuals who lack adrenocortical function.

REGULATION OF ALDOSTERONE SECRETION

Aldosterone, the major product of the zona glomerulosa, has two principal functions: (1) to sustain extracellular fluid volume by conserving body sodium and (2) to prevent overload of potassium by accelerating its excretion. (See also Chapters 33 and 34.) Thus aldosterone is largely secreted in response to a reduction in circulating fluid volume and increases in plasma potassium (Figure 41-8).

When sodium is depleted, the fall in extracellular fluid and plasma volume causes a decrease in renal arterial blood flow and pressure. The juxtaglomerular cells of the kidney respond by secreting the enzyme *renin* into the peripheral circulation (see Chapter 33). Renin acts on its substrate, *angiotensinogen,* to form

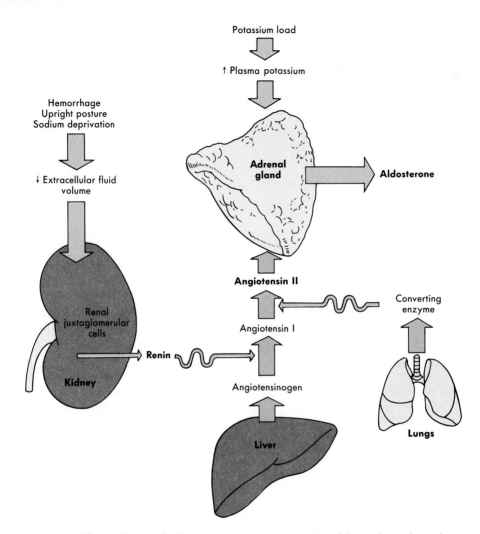

FIGURE 41-8   The regulation of aldosterone secretion. Activation of the renin-angiotensin system in response to hypovolemia is the predominant stimulus to aldosterone production. Elevation of plasma potassium is the other major stimulus.

angiotensin I. The latter is then further cleaved by a converting enzyme to the potent vasoconstrictor *angiotensin II* and to *angiotensin III.* These bind to specific receptors in the zona glomerulosa and stimulate the key enzymatic steps in the synthesis and release of aldosterone (Figure 41-8). Calcium and the phosphatidylinositol messenger system are mediators.

Basal plasma aldosterone ranges from 5 to 15 ng·dl^{-1}. When hypovolemia is produced rapidly by hemorrhage or by acute diuresis, or slowly by chronic sodium deprivation, aldosterone concentration increases up to fourfold. Conversely, when excess sodium is ingested and extracellular fluid volume expands, renin release, angiotensin generation, and

aldosterone secretion are all suppressed. **Thus the juxtaglomerular cells and the zona glomerulosa form a physiologic feedback system to defend extracellular fluid volume.** Sodium loss induces hypersecretion of renin and then aldosterone. When the additional aldosterone has caused sufficient sodium retention to restore extracellular fluid volume to normal, renin and aldosterone hypersecretion ceases.

Other factors that influence renin release secondarily affect aldosterone secretion. β-Adrenergic stimulation of the kidney in response to hypovolemia increases the output of renin and aldosterone. Certain prostaglandins produced within the kidney increase renin release; therefore frequently used inhibitors of prostaglandin synthesis depress aldosterone secretion. The atrial naturiuretic peptide hormones, synthesized and released by atrial myocytes in response to changes in vascular volume, decrease aldosterone secretion. They do so directly by acting on the zona glomerulosa through specific receptors and indirectly by reducing renin release.

Aldosterone also participates in a vital feedback relationship with potassium (Figure 41-8). Aldosterone facilitates the clearance of potassium from the extracellular fluid, and concordantly potassium is an important stimulator of aldosterone secretion. In humans raising the plasma potassium concentration only $0.5 \, mEq \cdot L^{-1}$ immediately increases plasma aldosterone levels threefold, and major increases in dietary potassium increase daily aldosterone secretion sixfold. Conversely, potassium depletion lowers aldosterone secretion. Potassium has a direct effect on the zona glomerulosa to depolarize the cell membrane. This action allows influx of calcium and activation of aldosterone biosynthesis.

Aldosterone secretion is also stimulated by ACTH, but this effect of ACTH wanes after several days. The diminished response is probably ascribable to appropriate compensatory decreases in renin levels and to increases in atrial natriuretic hormone release, since sodium is retained and extracellular fluid volume rises. The physiologic role of ACTH in regulating aldosterone output appears to be limited to a tonic one; when ACTH is deficient, the response to the primary stimulus of sodium depletion is diminished.

The major factors that stimulate aldosterone secretion interact with one another. A low sodium intake potentiates aldosterone responsiveness to angiotensin, potassium, and ACTH. The increased sensitivity to angiotensin is caused partly by increased binding of the hormone to its adrenal receptor. Conversely, if adrenal cell potassium content is depleted, the responses to angiotensin and ACTH are diminished.

ACTIONS OF ALDOSTERONE (MINERALOCORTICOIDS)

The kidney is the major site of mineralocorticoid activity. Aldosterone binds to a receptor identical to one type of cortisol receptor in renal tubular cells. mRNAs and proteins of still undetermined nature are induced and apparently mediate the hormone's actions. A lag of hours is required between exposure toaldosterone and onset of effect. Aldosterone stimulates active reabsorption of sodium from the distal tubular urine; the sodium is transported through the tubular cell and back into the capillary blood (Figure 1-9) (see Chapter 33). Thus net urinary sodium excretion is diminished, and the vital extracellular cation is conserved. Because water is passively reabsorbed with the sodium, plasma sodium concentration increases only slightly, and extracellular fluid volume expands isotonically. Although only 3% of total reabsorption is regulated by aldosterone, deficiency of the hormone produces a critical negative sodium balance.

Aldosterone acts at various loci in distal renal tubular and collecting duct cells: (1) at the apical (luminal) surface to increase the number of membrane channels through which sodium enters the cell along an electrochemical gradient; (2) at the basal (capillary) surface of the cell, to activate Na^+, K^+-ATPase, which pumps the sodium into the interstitial fluid, and (3) in the mitochondria, stimulating Kreb's cycle reactions that help generate the energy needed for extrusion of sodium.

Aldosterone also stimulates the active secretion of potassium out of the tubular cell and into the urine concurrently with sodium reabsorption (Figure 41-9). This does not constitute a simple stoichiometric *exchange* of potassium for sodium. Nonetheless, the reabsorption of sodium creates in the tubular lumen an electronegative condition that facilitates the transfer of potassium into the tubular urine. Therefore the extent to which aldosterone affects potassium secretion is greatly dependent on the delivery of sodium to the distal tubule. Aldosterone cannot significantly increase potassium excretion in a sodium-depleted subject; conversely, a high-sodium intake exaggerates the urinary potassium loss caused by aldosterone. Potassium flux, unlike sodium flux, does not entrain the movement of water. Therefore potassium retention because of aldosterone deficiency can result in a dangerous rise in plasma potassium, and an excess of aldosterone can cause a serious decrease in plasma potassium (Figure 12). Aldosterone also enhances tubular secretion of hydrogen ion in conjunction with sodium reabsorption. Therefore aldosterone causes a mild systemic metabolic alkalosis.

Continued administration of aldosterone produces only a limited retention of sodium, which then ceases. This escape is caused by expansion of the extracellu-

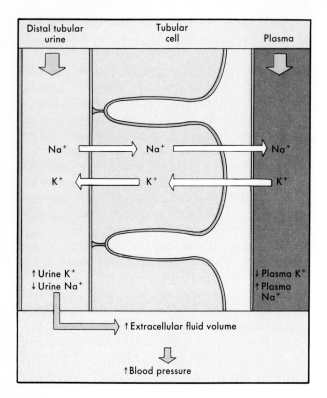

FIGURE 41-9 The action of aldosterone on the renal tubule. Sodium reabsorption from tubular urine is stimulated. Simultaneously, potassium secretion into the tubular urine is increased.

lar fluid and is mediated in part by atrial natriuretic hormones. In contrast, the potassium loss induced by aldosterone continues because sodium delivery to the distal tubule is maintained.

Aldosterone significantly affects sodium and potassium exchange across muscle cells. The net result is to increase potassium content of the intracellular space, another effect that helps prevent hyperkalemia. Aldosterone also modestly stimulates sodium reabsorption from the gastrointestinal tract and enhances potassium excretion in the feces.

An excess of aldosterone increases blood pressure because of sodium retention, expansion of extracellular fluid volume, and a slight increase in cardiac output. Agents such as *spironolactone* that compete with aldosterone for binding to its receptor are useful in treating fluid overload and hypertension.

ADRENOCORTICAL DYSFUNCTION

Destruction of the adrenal cortex is ultimately incompatible with life. A lack of cortisol leads to loss of appetite, weight loss, malaise, fatigue, muscle weakness, fever, poor tolerance of medical or surgical stress, and hypoglycemia. In females the loss of adrenal androgen precursors may contribute to anemia and a reduction in sexual hair. Negative feedback leads to hypersecretion of ACTH and all proopiomelanocortin products; their melanocyte-stimulating activity produces hyperpigmentation of the skin. Deficiency of aldosterone causes natriuresis, dehydration, hypotension, hyperkalemia, hyponatremia, and metabolic acidosis. Death occurs unless cortisol and sodium in isotonic fluids are provided.

Hypersecretion of cortisol produces the striking clinical picture shown in Figure 41-7. It also induces osteoporosis, muscle weakness, hyperglycemia, central nervous system disturbances, and impaired response to infections. Aldosterone excess leads to hypertension, hypokalemia, and metabolic alkalosis. Muscle weakness and neuromuscular irritability can result.

When cortisol deficiency results from congenital defects in its biosynthesis, overproduction of androgens typically results from accumulation of 17-hydroxylated precursors and negative feedback–induced ACTH hypersecretion. This may result in intrauterine masculinization of the external genitalia of a female fetus or in virilization of a female adult.

BIBLIOGRAPHY
Journal Articles

Fauci AS et al: Glucocorticosteroid therapy: mechanisms of action and clinical considerations, Ann Intern Med 84:304, 1976.

Gustafsson J et al: Biochemistry, molecular biology and physiology of the glucocorticoid receptor, Endocr Rev 8:185, 1987.

Jackson R et al: Synthetic ovine corticotropin-releasing hormone: simultaneous release of propiolipomelanocortin peptides in man, J Clin Endocrinol Metab. 58:740, 1984.

Miller WL: Molecular biology of steroid hormone synthesis, Endocr Rev 9:295, 1988.

Munck A et al: Physiological functions of glucocorticoids in stress and their relation to pharmacological actions, Endocr Rev 5:25, 1984.

Quinn SJ and Williams GH: Regulation of aldosterone secretion, Annu Rev Physiol 50:409, 1988.

Rizza RA et al: Cortisol-induced insulin resistance in man: impaired suppression of glucose production and stimulation of glucose utilization due to a postreceptor defect of insulin action, J Clin Endocrinol Metab 54:131, 1982.

Simpson ER and Waterman MR: Regulation of the synthesis of steroidogenic enzymes in adrenal cortical cells by ACTH, Annu Rev Physiol 50:427, 1988.

Suda T et al: Immunoreactive corticotropin-releasing factor in human plasma, J Clin Invest 76:2026, 1985.

Taylor AL and Fishman LM: Corticotropin-releasing hormone, N Engl J Med 319:213, 1988.

Thompson EB et al: Unlinked control of multiple glucocorticoid-induced processes in HTC cells, Mol Cell Endocrinol 15:135, 1979.

Books and Monographs

Baxter JD et al: Glucocorticoid hormone action, New York, 1979, Springer-Verlag New York, Inc.

Crabbe J: Mechanism of action of aldosterone. In Degroot LJ, editor, Endocrinology, Philadelphia, 1989, WB Saunders Co.

Keith LD and Kendall JW: Regulation of ACTH secretion. In Imura H: The pituitary gland, New York, 1985, Raven Press.

Meikle AW: Secretion and metabolism of the cortiosteroids and adrenal function and testing. In Degroot LJ, editor: Endocrinology, Philadelphia, 1989, WB Saunders Co.

Numa S and Imura H: ACTH and related peptides: gene structure and biosynthesis. In Imura H: The pituitary gland, New York, 1985, Raven Press.

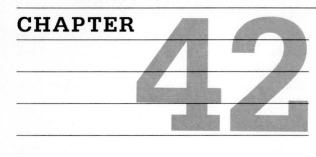
The Adrenal Medulla

The adrenal medulla is the source of the circulating catecholamine hormone *epinephrine*. It also secretes lesser amounts of *norepinephrine,* nominally a neurotransmitter, which can function as a hormone. The catecholamine hormones are important mediators of rapid fuel mobilization; they increase both glucose and free fatty acids, especially during acute stress. They also stimulate the cardiovascular system and cause contraction or relaxation of smooth muscles in the respiratory, gastrointestinal, and genitourinary tracts.

The adrenal medulla is essentially a specialized sympathetic ganglion. However, the neuronal cells of the medulla do not have axons; instead, they discharge their products directly into the bloodstream, and thus they function in true endocrine fashion. The medulla is usually activated in association with the sympathetic nervous system and acts in concert with it in the "fight-or-flight" reaction. Many of the neurotransmitter actions of norepinephrine are duplicated and amplified by the hormone actions of epinephrine, which reaches similar targets via the circulation. However, epinephrine has other effects of its own, some of which modulate those of norepinephrine.

The adrenal medulla is formed in parallel with the peripheral sympathetic nervous system. In the fetus, at about 7 weeks of gestation, neuroectodermal cells invade the adrenal cortex, where they develop into the medulla. By birth the medulla is completely functional. The development of this tissue and induction of hormone synthesis is stimulated by *nerve growth factor,* a somatomedin-like peptide (see Chapter 39).

The adult adrenal medulla weighs about 1 g and is composed of *chromaffin cells.* These are organized in cords in intimate relationship with venules that drain the adrenal cortex and with nerve endings from cholinergic preganglionic fibers of the sympathetic nervous system. Within the chromaffin cells are numerous granules similar to those found in postganglionic sympathetic nerve terminals. These contain catecholamines, adenosine triphosphate (ATP), enkephalins, β-endorphin, and other proopiomelanocortin peptides. Approximately 85% of the chromaffin granules store epinephrine, and 15% store norepinephrine.

SYNTHESIS, STORAGE, AND SECRETION OF MEDULLARY HORMONES

The catecholamines are synthesized by a series of reactions shown in Figure 42-1. The intermediates move back and forth in sequence between the cytoplasm and the storage granules. The first, rate-limiting step occurs in the cytoplasm. Here the conversion of tyrosine to dihydroxyphenylalanine (DOPA) requires molecular oxygen, a tetrahydropteridine, and NADPH. The subsequent decarboxylation of DOPA to dopamine in the cytoplasm employs pyridoxal phosphate as a cofactor. The dopamine must then be taken up into the chromaffin granule before it can be acted on by the next enzyme in the sequence, dopamine β-hydroxylase, which is present exclusively within the granule. It catalyzes the formation of norepinephrine from dopamine, molecular oxygen, and a hydrogen donor. In a few granules the sequence ends here and the norepinephrine remains stored.

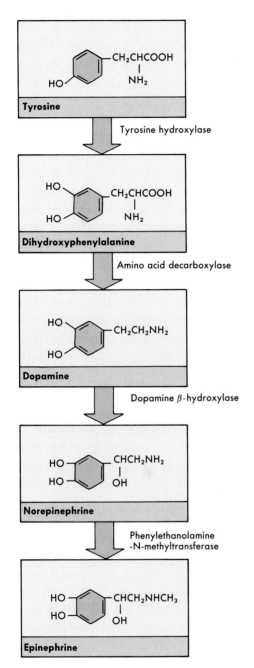

FIGURE 42-1   Pathway of catecholamine hormone synthesis in the adrenal medulla.

medullary hormone. Granule uptake of catecholamines and their storage at high concentrations requires ATP. One mole of the nucleotide complexes with 4 moles of catecholamine hormone and a specific protein known as *chromogranin*.

The synthesis of epinephrine and norepinephrine is regulated by several factors. Acute sympathetic stimulation of the medulla activates the initial rate-limiting step; chronic stimulation induces increased concentrations of the first two enzymes, thereby maintaining catecholamine output in the face of continuous demand. Cortisol specifically induces the last enzyme in the sequence, N-methyltransferase, thereby selectively stimulating epinephrine synthesis. The perfusion of the adrenal medulla with blood from the cortex, which contains a high concentration of cortisol, facilitates this induction.

The effector pathway for release of adrenal medullary hormones consists of cholinergic preganglionic fibers in the splanchnic nerves. On nerve stimulation, acetylcholine released from the nerve terminals depolarizes the chromaffin cell membrane by increasing its permeability to sodium. In turn this induces an influx of calcium ions, which probably causes microfilament contraction and draws the granules to the cell membrane. Exocytosis follows, with secretion of epinephrine, norepinephrine, ATP, dopamine β-hydroxylase, opioid peptides, and chromogranin.

METABOLISM OF CATECHOLAMINE HORMONES

Essentially all the circulating epinephrine is derived from adrenal medullary secretion. Basal plasma epinephrine levels are 25 to 50 $pg \cdot ml^{-1}$. In contrast, almost all the circulating norepinephrine is derived from sympathetic nerve terminals and from the brain, having escaped immediate local reuptake from synaptic clefts. Basal norepinephrine levels are 100 to 350 $pg \cdot ml^{-1}$. Both catecholamines have plasma half-lives of about 2 minutes, which allows rapid turn-off of their dramatic effects. Only 2% to 3% of catecholamines are secreted unchanged in the urine; the daily total is about 50 μg, of which 20% is epinephrine and 80% is norepinephrine.

Most of the epinephrine is metabolized before secretion within the chromaffin cell itself because synthesis exceeds the capacity for storage. Circulating epinephrine and norepinephrine are metabolized

In most granules norepinephrine diffuses back into the cytoplasm, where it is N-methylated with S-adenosylmethionine as the methyl donor. The resultant epinephrine is then taken back up into the chromaffin granule, where it is stored as the predominant adrenal by O-methylation and oxidative deamination, predominantly in the liver and kidney. The major end products, vanillylmandelic acid (VMA) and metanephrines, are excreted in the urine. They serve as indices of sympathetic nervous system activity or of pathological hypersecretion of epinephrine.

REGULATION OF ADRENAL MEDULLARY SECRETION

As previously noted, secretion from the adrenal medulla is part of the fight-or-flight reaction (Figure 41-2). Thus perception or even anticipation of danger,

trauma, pain, hypovolemia, hypotension, anoxia, hypothermia, hypoglycemia, and intense exercise causes rapid secretion of epinephrine and norepinephrine. These stimuli are sensed at various levels in the sympathetic nervous system, and responses are initiated in the hypothalamus and brainstem (see Chapter 9). Often epinephrine secretion follows activation of the sympathetic nervous system by more intense stimuli. However, in response to mild hypoglycemia, moderate hypoxia, and fasting, epinephrine secretion increases, whereas sympathetic nervous system activity remains constant or decreases.

Mild hypoglycemia causes a five- to tenfold

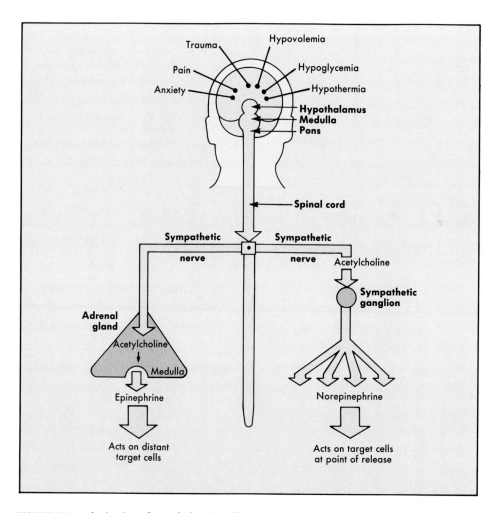

FIGURE 42-2   Activation of catecholamine effects via the sympathetic nervous system and the adrenal medulla. Note the adrenal medulla is homologous to a sympathetic ganglion, but the adrenal medulla releases its catecholamine into the bloodstream rather than into a synaptic cleft.

increase in plasma epinephrine concentration but little change in norepinephrine concentration. The resultant epinephrine concentration can stimulate a compensatory increase in the plasma glucose level; the norepinephrine level cannot. The reduction in central venous pressure produced by assuming the upright position increases both plasma epinephrine and norepinephrine concentrations twofold. However, only the epinephrine concentration is high enough to increase heart rate and blood pressure. Thus epinephrine functions as a true hormone in both situations, whereas norepinephrine does not. Norepinephrine, however, contributes as a neurotransmitter to the compensatory responses to hypovolemia and to more severe hypoglycemia, because the higher concentration necessary for its action is generated locally at the effector site (Figure 42-2). In states of major metabolic decompensation, the circulating concentrations of norepinephrine can also be high enough to evoke responses.

ACTIONS OF CATECHOLAMINE HORMONES

Intracellular Mechanisms

Epinephrine and norepinephrine exert their many effects via several plasma membrane receptors, designated β_1, β_2, α_1, and α_2. The β_1-, β_2-, and α_2-receptors are structurally similar glycoproteins. Each winds in and out of the plasma membrane so that more than one surface is presented extracellularly for hormone binding and intracellularly for signal generation. The two β-receptors are coupled to the stimulating unit of membrane adenylate cyclase, and hormone binding increases cyclic adenosine monophosphate (cAMP) levels. In contrast, the α_2-receptor is coupled to the inhibiting unit of adenylate cyclase, and hormone binding decreases cAMP levels. Catecholamine hormones therefore either trigger (β_1 or β_2) or repress (α_2) a cascade of protein kinase A−catalyzed protein phosphorylations. The α_1-receptor is structurally different and is coupled to calcium and phosphatidylinositol products as second messengers.

Continuous exposure to catecholamines eventually down-regulates the number of receptors and induces partial refractoriness to hormone action. A distinctly different phenomenon—acute desensitization to suc-

cessive doses of catecholamine hormones—is caused by protein kinase A− or protein kinase C−mediated phosphorylation of the receptor molecules themselves. This desensitization process constitutes a form of rapid intracellular negative feedback that almost immediately limits hormone actions.

Whole Body Effects

The metabolic effects of catecholamines are characterized by fuel mobilization (Figure 42-3). Glycogenolysis in the liver is stimulated via cAMP-mediated activation of phosphorylase, and glucose output increases. Glycogenolysis in muscle is similarly stimulated. This increases muscle glucose supply, but the lactate released by glycolysis then can serve as a substrate for hepatic gluconeogensis. Through α_1-receptors in the liver, gluconeogenesis is also directly stimulated by catecholamines. Plasma glucose levels also rise because catecholamines inhibit insulin secretion as well as glucose uptake by muscle tissue. The availability of free fatty acids is increased by activating the enzyme, adipose tissue lipase. The enhanced lipolysis in turn leads to increased free fatty acid oxidation in the liver and ketogenesis.

Catecholamines increase the basal metabolic rate by stimulating facultative, or nonshivering, thermogenesis (see Chapter 36). It is an important part of the response to cold exposure. In neonates of many species brown adipose tissue is an important site where catecholamines increase heat production. Here the hormones uncouple ATP synthesis from oxygen utilization in the mitochondria. Catecholamines also increase diet-induced thermogenesis, thereby helping to regulate overall energy balance and stores.

Sympathetic nervous system activity decreases during fasting and increases after feeding. In this way norepinephrine adapts total energy utilization to energy availability. In contrast, epinephrine secretion increases slightly during prolonged fasting and also 4 to 5 hours after a meal, when plasma glucose is declining. This response serves the different purpose of sustaining glucose production for use by the central nervous system.

The cardiovascular and visceral effects of epinephrine are consonant with its metabolic actions (Table 42-1). For example, during exercise epinephrine increases cardiac output due to an increase in heart rate and contractile force (see Chapter 46). At the same time, muscle arterioles dilate, whereas renal,

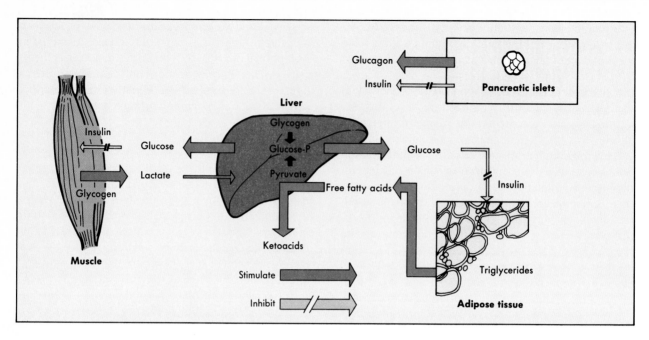

FIGURE 42-3   Metabolic effects of epinephrine. The hormone stimulates glucose production and inhibits glucose uptake. Lipolysis and ketogenesis are stimulated. The net result is an increase in plasma glucose, free fatty acids, and ketoacids.

Table 42-1   Actions of Catecholamine Hormones

	β	α
Metabolic	↑ Glycogenolysis ↑ Glucose utilization ↑ Lipolysis and ketosis (β_1) ↑ Calorigenesis (β_1) ↑ Insulin secretion (β_2) ↑ Glucagon secretion (β_2) ↑ Muscle K^+ uptake (β_2)	↑ Gluconeogenesis (α_1) ↓ Insulin secretion (α_2)
Cardiovascular	↑ Cardiac contractility (β_1) ↑ Heart rate (β_1) ↑ Conduction velocity (β_1) ↑ Arteriolar dilation (β_2) (muscle) ↓ Blood pressure	↑ Arteriolar vasoconstriction (α_1) (splanchnic, renal, cutaneous, genital) ↑ Blood pressure
Visceral	↑ Muscle relaxation (β_2) Gastrointestinal Urinary Bronchial	↑ Sphincter contraction (α_1) Gastrointestinal Urinary
Other		Sweating (adrenergic) Dilation of pupils Platelet aggregation (α_2)

↑, Increased; ↓, decreased.

splanchnic, and cutaneous arterioles constrict. Systolic blood pressure increases, and diastolic blood pressure decreases slightly or remains unchanged. The net effect is to shunt blood to exercising muscles and away from other tissues, while maintaining essential coronary and cerebral blood flow. This guarantees delivery of oxygen and substrate for energy production to the critical tissues in situations of danger.

During exposure to cold, constriction of cutaneous vessels helps to conserve heat, which reinforces epinephrine's thermogenic action. The relaxation of bronchioles to improve alveolar gas exchange and the dilation of pupils to permit better distant vision assist the threatened individual. The inhibition of temporarily unneeded gastrointestinal and genitourinary motor activity is also beneficial.

The catecholamines have significant actions on mineral metabolism. They increase sodium reabsorption by the kidney by stimulating renal tubular sodium transport and by stimulating renin and therefore aldosterone secretion. They also stimulate influx of potassium into muscle cells via β_2-receptors and help to prevent hyperkalemia.

PATHOLOGIC HYPERSECRETION OF CATECHOLAMINES

Hypersecretion of epinephrine and norepinephrine results in a distinctive, dangerous syndrome. Bursts of catecholamine release can cause sudden tachycardia, extreme anxiety with a sense of impending

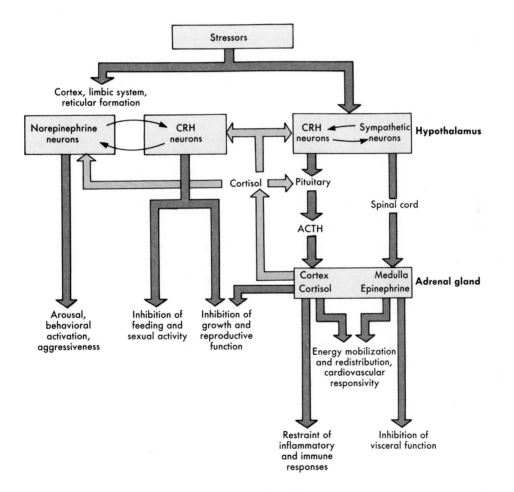

FIGURE 42-4   Integrated responses to stress mediated by the sympathetic nervous system and the hypothalamic-pituitary-adrenocortical axis. The responses are mutually reinforcing, both at the central and peripheral levels. Negative feedback by cortisol also can limit an overresponse that might be harmful to the individual. *Red arrows,* stimulation; *blue arrows,* inhibition; *CRH,* corticotropin-releasing hormone; *ACTH,* adrenocorticotropic hormone.

death, cold perspiration, skin pallor resulting from vasoconstriction, blurred vision, headache, and chest pain. Blood pressure may rise greatly and cause stroke or left-sided heart failure. If epinephrine is secreted, the heart rate increases; if norepinephrine is secreted, the heart rate decreases because of the baroreceptor reflex in response to the hypertension. In addition to such episodes, chronic catecholamine excess may produce weight loss, as a result of an increased metabolic rate, and hyperglycemia.

INTEGRATION OF THE RESPONSE TO STRESS

The adrenal medulla and adrenal cortex are both major participants in the adaptation to stress. Their intimate anatomic juxtaposition mirrors a more fundamental functional relationship between the adrenergic nervous system and the corticotropin-releasing hormone (CRH)–adrenocorticotropic hormone (ACTH)–cortisol axis. Although our knowledge of stress and the human body's adaptation to it is still incomplete, recent advances justify presenting an integrated overview (Figure 42-4).

Stress can be perceived by many areas of the brain, from the cortex down to the brainstem. Major stresses almost simultaneously activate CRH neurons and adrenergic neurons in the hypothalamus. The activation is mutually reinforcing because norepinephrine input increases CRH release and CRH increases adrenergic discharge (Figure 42-4). CRH release ultimately elevates plasma cortisol levels; adrenergic stimulation elevates plasma catecholamine levels. Together these hormones increase glucose production; epinephrine does so rapidly by activating glycogenolysis, and cortisol more slowly by providing amino acid substrate for gluconeogenesis. Together they shift glucose utilization toward the central nervous system and away from peripheral tissues. Epinephrine also rapidly augments free fatty acid supply to the heart and to muscles, and cortisol facilitates the lipolytic response. Both hormones, catecholamines directly and cortisol permissively, raise blood pressure and cardiac output and improve delivery of substrates to tissues that are critical to the immediate defense of the organism. If the stress involves tissue trauma or invasion, high cortisol levels eventually act to restrain the initial inflammatory and immune responses so that they do not lead to irreparable damage.

The same signaling molecules, the neurotransmitter norepinephrine and the neuropeptide CRH, can produce other adaptive responses to stress. A general state of arousal and vigilance, an activation of defensively useful behavior, and appropriate aggressiveness result from adrenergic stimuli to the pertinent brain centers. At the same time CRH input to other hypothalamic neurons inhibits growth hormone and gonadotropin release, presumably because growth and reproduction are not useful functions during stress. This is reinforced by the excess of cortisol, which also suppresses growth and ovulation. In addition, CRH inhibits sexual activity and feeding, again inappropriate activities when the organism perceives itself to be in immediate serious danger. Thus the adaptation to stress represents a prime example of the integration of the nervous system and the endocrine system.

BIBLIOGRAPHY
Journal Articles

Clutter W et al: Epinephrine plasma metabolic clearance rates and physiologic thresholds for metabolic and hemodynamic actions in man, J Clin Invest 66:94, 1980.

Cryer PE: Physiology and pathophysiology of the human sympathoadrenal neuroendocrine system, N Engl J Med 303:436, 1980.

Landsberg L and Young JB: The role of the sympathetic nervous system and catecholamines in the regulation of energy metabolism, Am J Clin Nutr 36:1018, 1983.

Lefkowitz RJ and Caron MG: Adrenergic receptors: molecular mechanisms of clinically relevant recognition, Clin Res 33:395, 1985.

Santiago JV et al: Epinephrine, norepinephrine, glucagon, and growth hormone release in association with physiological decrements in the plasma glucose concentration in normal and diabetic man, J Clin Endodrinol Metab 51:877, 1980.

Silverberg A et al: Norepinephrine: hormone and neurotransmitter in man, Am J Physiol 234:E252, 1978.

Wortsman J et al: Adrenomedullary response to maximal stress in humans, Am J Med 77:779, 1984.

CHAPTER 43

Overview of Reproductive Function

The endocrine glands, as discussed in previous chapters, are essential to maintenance of the life and well being of the individual. In contrast, the endocrine function of the gonads is primarily concerned with maintaining the life and well-being of the species. Human reproduction requires highly complex patterns of gonadal function. These ensure the development and maintenance of mature gametes—ova and spermatozoa—from primordial germ cells, their subsequent successful union (fertilization), and finally the growth and development of the conceptus within the body of the mother. Many obvious differences exist between male and female gonadal function, but important conceptual similarities and operational homologies are also present. The gender differences are better appreciated if the common aspects of gonadal function are first understood.

The gonad, whether ovary or testis, consists of two distinct anatomic and functional parts. One part encloses the developing germ cell line with specialized membrane and cytoplasmic barriers that prevent its indiscriminate exposure to the general constituents of plasma and interstitial fluid. In the ovary the germ cell enclosure is the *follicle;* in the testis it is the *spermatogenic (seminiferous) tubule.* The other part is composed of surrounding endocrine cells that secrete sex steroid hormones, protein hormones, and other products necessary for germ cell development. The most important sex steroids are *estradiol* and *progesterone* in the female and *testosterone* in the male. Protein hormones produced by the gonads include inhibin, activin, follistatin, antimüllerian hormone, oocyte meiosis inhibitor, as well as various derivatives of proopiomelanocortin (see Chapter 41).

The gonadal hormones have various purposes.

Acting locally in paracrine and autocrine fashion, they stimulate the development of the respective germ cells into ova and spermatozoa. Acting peripherally in endocrine fashion, these hormones:

1. Stimulate the development and function of the secondary sexual organs essential for support and delivery of the ova and spermatozoa to the site of fertilization
2. Regulate the secretion of hypothalamic-pituitary hormones essential to gonadal function
3. Modify somatic shape and certain physiologic functions within each gender
4. Support the conceptus in the early phase of pregnancy in the female

There are two principal types of endocrine cells in the gonads. Those immediately adjacent to the germ cells are called *granulosa cells* in the ovary and *Sertoli cells* in the testis. Those more distant from the germ cells and separated from them by a basement membrane are called *theca,* or *interstitial, cells* in the ovary and *Leydig cells* in the testis. The homologous granulosa and Sertoli cells mainly secrete estrogens, whereas the homologous theca and Leydig cells secrete androgens. Progesterone is secreted in large amounts only in females by transformed granulosa and theca cells, known as *luteal cells.* The protein products come mostly from granulosa and Sertoli cells

SYNTHESIS OF SEX STEROID HORMONES

Biosynthesis of gonadal steroid hormones follows a common pathway in both genders (Figure 43-1). The enzymes, their organelle localization, and cofactor

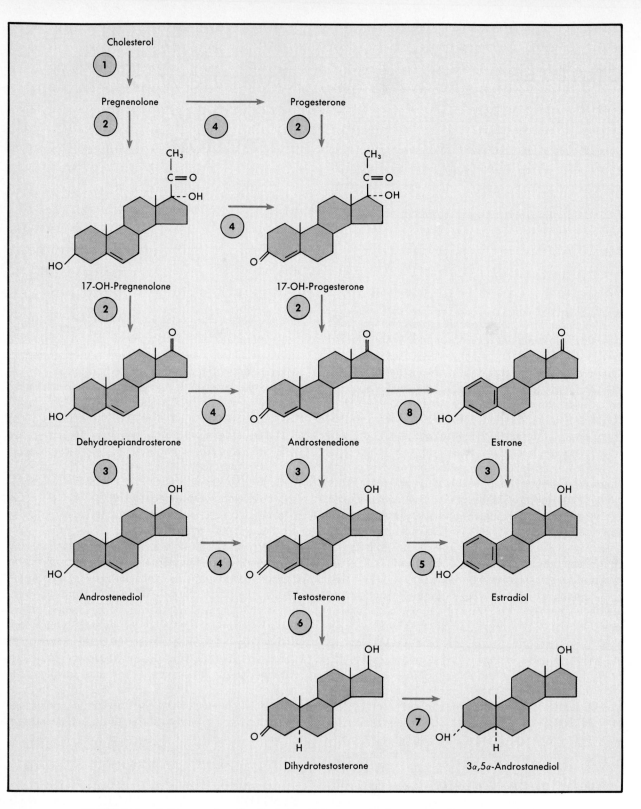

FIGURE 43-1 Pathways of synthesis of steroid hormone in the gonads. Testosterone is the major product of the testis. Estradiol and progesterone are the major products of the ovary. The enzymes are: *1*, 20,22 desmolase (P-450$_{scc}$); *2*, 17-hydroxylase/17,20 desmolase; *3*, 17 β-OH-steroid dehydrogenase; *4*, 3 β-ol-dehydrogenase/Δ 4,5-isomerase; *5*, aromatase; *6*, 5 α-reductase; *7*, 3 α-reductase.

requirements are those described for the adrenal cortex (see Chapter 41). Furthermore, the gonadal enzymes are identical to the adrenal enzymes, and the same genes direct their synthesis. Cholesterol, either synthesized in situ from acetylcoenzyme A (acetyl CoA) or taken up from the plasma low-density lipoproteins (LDLs), is the starting compound. P-450$_{SCC}$ (20,22-desmolase) catalyzes side-chain cleavage of cholesterol and is the rate-limiting step for synthesis of progesterone; the androgens testosterone, *dihydrotestosterone* (DHT), and *androstanediol;* and the estrogens estradiol and *estrone.* Within the testes a small quantity of testosterone undergoes 5α-reduction to DHT, a potent androgen. However, a much larger and more important conversion of testosterone to DHT occurs in target tissues, catalyzed by the enzyme 5α-reductase. Estrogens are synthesized from androgens by the P-450 aromatase enzyme complex. This sequentially catalyzes the hydroxylation and oxidation of the 19-methyl group, the creation of a 1-2 double bond, the decarboxylation of position 19, and the formation of the characteristic benzene ring of estrogens.

REGULATION OF GONADAL STEROID HORMONE SECRETION

A hypothalamic–anterior pituitary–gonadal axis, analogous to that involved in thyroid and adrenal function, is the basis for gonadal regulation (Figure 43-2). The components are a hypothalamic releasing hormone and two pituitary gonadotropins designated *luteinizing hormone* (LH) and *follicle-stimulating hormone* (FSH). A single pituitary cell type, the gonadotroph, generally produces both LH and FSH, although occasionally the gonadotroph contains only one or the other.

Gonadotropin-Releasing Hormone

The single hypothalamic releasing hormone is known as both *gonadotropin-releasing hormone* (GnRH) and *luteinizing hormone–releasing hormone* (LHRH) (Figure 43-2). It stimulates both LH and FSH secretion, but LH more so. The existence of a separate FSH-releasing hormone is suspected but unproved. GnRH (or LHRH), a decapeptide synthesized from a much larger preprohormone, is produced in two clusters of hypothalamic neurons in the arcu-

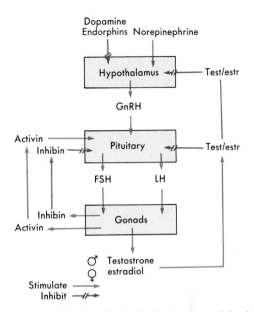

FIGURE 43-2   The hypothalamic-pituitary-gonadal axis. The hypothalamic peptide gonadotropin-releasing hormone (*GnRH*) stimulates release of two gonadotropins, luteinizing hormone, (*LH*) and follicle-stimulating hormone (*FSH*), from the pituitary gland. The gonadotropins in turn stimulate gonadal secretion of primarily testosterone in males and primarily estradiol in females. These feed back negatively at both the pituitary and hypothalamic levels to inhibit LH and FSH secretion. In addition, FSH stimulates release from the gonad of inhibin, which feeds back negatively to block FSH release preferentially. By contrast, estradiol (in women) and the gonadal protein activin have positive feedback effects on pituitary secretion.

ate and preoptic areas. From here the hormone is transported axonally for storage in the median eminence. Input from other areas of the brain allows reproduction to be influenced by light-dark cycles, by olfactory stimuli, via airborne molecules known as *pheromones,* and by stress. Dopaminergic and endorphinergic tracts within the hypothalamus and the median eminence transmit important inhibitory influences on GnRH release (Figure 43-2). In adults GnRH is released into the pituitary portal veins in a pulsatile pattern. Men have 8 to 10 pulses per day, whereas in women the frequency and periodicity of pulses varies with the menstrual cycle. In children pulsatility is greatly reduced or absent.

GnRH binds to its plasma membrane receptor and causes it to microaggregate. This leads to an influx of extracellular calcium into the gonadotroph cells. Complexed to calmodulin, the calcium acts as the major second messenger, and phosphatidylinositol products play a subsidiary role. GnRH stimulates the simultaneous release of LH and FSH from their secretory granules. The ratio of FSH to LH increases when the frequency of GnRH pulses declines. GnRH also stimulates the transcription of genes that direct the synthesis of the two gonadotropins and their subsequent processing via glycosylation. Thus a GnRH infusion typically produces a biphasic LH response. However, prolonged stimulation by GnRH causes down-regulation of its receptor, desensitization of the gonadotroph, and profound inhibition of gonadotropin secretion.

Gonadotropins

LH and FSH are glycoproteins that, as with thyroid-stimulating hormone (TSH), consist of two subunits. The α-subunit of all three hormones is identical, whereas their respective β-subunits are different and are determined by unique genes. The α- and β-subunit in each gonadotropin is required for binding to its gonadal receptor, and proper carbohydrate components are necessary for full biologic activity.

LH primarily stimulates the theca cells of the female and the Leydig cells of the male to synthesize and secrete androgens and, to a lesser extent, estrogens. In addition, LH stimulates granulosa cells, once LH receptors have been expressed by these cells during the female cycle. cAMP is the major second messenger for LH actions. Continuous stimulation by LH down-regulates its receptor and reduces responsivity to the hormone.

Analogous to adrenocorticotropic hormone (ACTH), LH stimulates cholesterol availability to the mitochondria and its conversion to pregnenolone. Subsequently the levels of steroidogenic enzymes and adrenoxin are increased by stimulating transcription of their genes. Most importantly, LH increases the level of 17-hydroxylase/17,20-desmolase, the essential step in testosterone synthesis.

FSH stimulates granulosa and Sertoli cells to secrete estrogens. Acting via its plasma membrane receptor and cyclic adenosine monophosphate (cAMP) as a second messenger, FSH increases transcription of the gene for aromatase, the enzyme specific to estradiol synthesis. Another important effect of FSH is to increase the number of LH receptors in target cells, thereby amplifying their sensitivity to LH. FSH also stimulates the secretion of inhibin and other proteins.

Feedback Regulation

Regulation of sex steroid hormone secretion and other aspects of gonadal function is complex, and those aspects distinctive to each gender are described in later sections. However, certain common principles do exist (Figure 43-2). Testosterone in men and estradiol in women inhibit secretion of LH and FSH. This basic negative feedback loop acts at the pituitary level by blocking the actions of GnRH on gonadotropin release and synthesis, and it acts at the hypothalamic level by decreasing GnRH. Both the frequency and the amplitude of plasma LH and FSH pulses are thereby diminished. In women a specific *positive feedback* effect of estradiol on LH secretion is included in the basic framework. (see Chapter 45). This effect depends on the dose and time of exposure to the sex steroid.

Another negative feedback loop relates *inhibin* from granulosa and Sertoli cells to FSH secretion. Inhibin reduces GnRH release and also specifically blocks the stimulatory effect of GnRH on FSH secretion. In contrast, from the same gonadal cells, *activin* exerts a positive feedback effect to stimulate FSH secretion. Thus the output of LH and FSH from the pituitary gland can be exquisitely and differentially regulated by both positive and negative feedback by the hypothalamus and the gonads. At various times the dominant or critical influence may be from either site. In this sense the gonad can be viewed as a much more self-regulatory gland than either the adrenal cortex or the thyroid. This is most apparent in women, as discussed later.

AGE-RELATED CHANGES IN REPRODUCTION

The hypothalamic-pituitary-gonadal axis is unique in that it undergoes extreme changes throughout the human life span. Although the patterns of females and males differ, again certain common aspects bear emphasis (Figure 43-3).

Intrauterine and Childhood Pattern

In humans GnRH is present in the hypothalamus, and FSH and LH are present in the pituitary gland by 10 to 12 weeks of gestation. A broad peak of gonadotropin concentrations occurs in fetal plasma at midgestation. After the concentrations drop to low levels before birth, they transiently increase (more prolong-

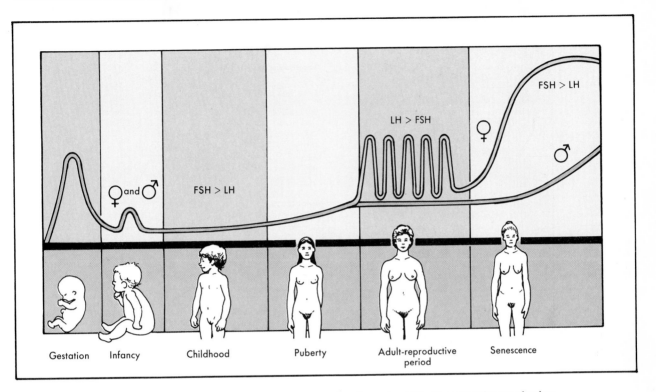

FIGURE 43-3   The pattern of gonadotropin secretion throughout life. Note transient peaks during gestation and early infancy and low levels thereafter in childhood. Women subsequently develop monthly cyclic bursts, with luteinizing hormone (*LH*) exceeding follicle-stimulating hormone (*FSH*); men do not. Both genders show increased gonadotropin production after age 50 years, with FSH exceeding LH. (Reprinted, by permission of The New England Journal of Medicine, from Boyar RM et al: 287:582, 1972.)

ed in females) again at about 2 months of age. For the rest of childhood both gonadotropins are secreted at very low levels. These changes are mirrored by fluctuations of plasma testosterone in males and estradiol in females.

Puberty

The transition from a nonreproductive to a reproductive state during puberty requires maturation of the entire hypothalamic-pituitary-gonadal axis. Before the child reaches age 10 years, plasma LH and FSH levels are low despite very low concentrations of gonadal hormones. Therefore, either the negative feedback system is inoperative or the hypothalamus and pituitary gland are exquisitely sensitive to testosterone, estradiol, and inhibin. One factor in puberty may thus be the gradual maturing of hypothalamic neurons that leads to an increased synthesis and release of GnRH. The time and rate of onset of this maturational process may well be genetically prepro-

grammed because familial patterns are apparent. Other central nervous system components (e.g., the pineal gland) influence this process. As puberty approaches, a pulsatile pattern of LH and FSH secretion appears. The ratio of plasma LH to FSH rises as the pulse frequency increases. Furthermore, during early and middle puberty, but at no other time of life, a nocturnal peak in LH secretion is observed. This then disappears as adult status is reached (Figure 43-4). The gonad itself is not necessary for these changes in GnRH and gonadotropins to occur.

During early puberty the responsiveness of the pituitary gland to GnRH changes so that LH exceeds FSH output. This may result from increased synthesis and storage of LH in response to pulsatile GnRH secretion because the latter allows better maintenance of GnRH receptors. Although the gonadal target cells respond to LH in childhood, their responsiveness is augmented during puberty. Thus this period can be viewed as a cascade of increasing maturation from the hypothalamic to the pituitary to the gonadal

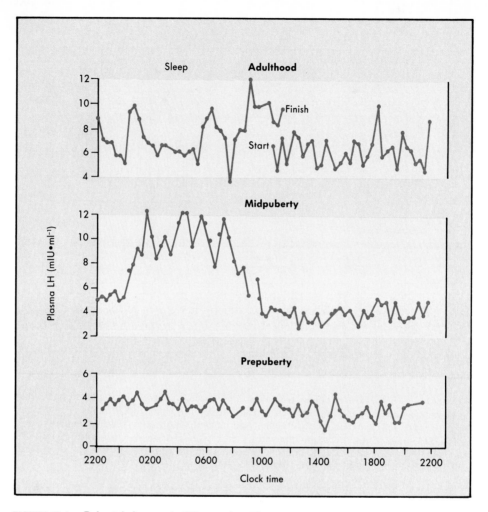

FIGURE 43-4   Pubertal changes in LH secretion. The pattern of secretion becomes much more pulsatile. In addition, a nocturnal peak in LH appears early in puberty and then disappears when puberty is completed. Males and females both show these changes. (Redrawn from Boyar RM et al: N Engl J Med 287:582, 1972.)

level. Once the adult pattern of gonadotropin secretion is established, the basal plasma concentrations of LH and FSH (approximately 10^{-11} molar) are similar in men and women. An important distinguishing feature between the genders is the additional establishment of a dramatic monthly gonadotropin cycle in females only (see Chapter 45), with the LH bursts greatly exceeding the FSH bursts (see Figure 43-3).

Climacteric

In both genders a loss of gonadal responsiveness to gonadotropin stimulation occurs after the fifth decade of life. In males this is gradual, and some reproductive capacity usually persists into the eighth decade. In females reproductive capacity is lost completely, and menopause occurs. In both genders, however, negative feedback leads to elevated plasma gonadotropins levels. The FSH level rises more than the LH level, and the increase is more distinct in females (see Figure 43-3).

SEXUAL DIFFERENTIATION

The most fundamental and obvious difference between the genders lies in the anatomy and consequent physiology of their reproductive tracts. During

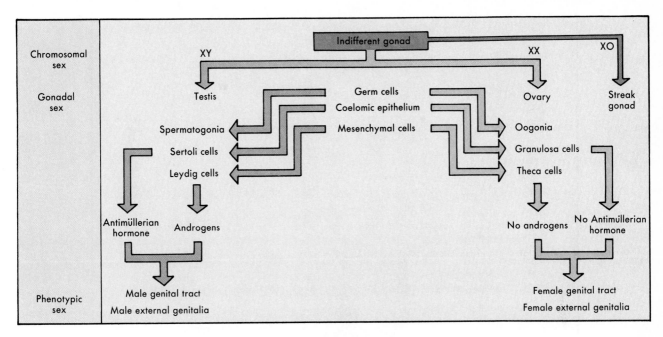

FIGURE 43-5   An overview of the development of the cells of the ovary and testis from the primitive indifferent gonad. The hormonal products from the testis and the absence of these products from the ovary determine the gender differences in the internal genital tracts and the external genitalia.

the first 5 weeks of gestation, however, the gonads of males and females are indistinguishable, and their genital tracts are unformed. From this stage of the "indifferent gonad" to that of the completed normal individual of either gender lies the process of sexual differentiation (Figures 43-5 and 43-6). The final maleness or femaleness is best characterized in terms of differences in genetic sex, in gonadal sex, and in genital (phenotypic) sex.

Genetic Sex

Maleness is determined positively by the presence of the Y chromosome. The normal male has a chromosome complement of 44 autosomes and two sex chromosomes, XY. Without the Y chromosome, neither testicular development nor masculinization of the genital tracts and external genitalia can occur (with rare exceptions). In the presence of a Y chromosome, extra X chromosomes do not alter the fundamental maleness, even though the gonads are dysfunctional. The organization of the indifferent gonad into the characteristic spermatogenic tubules of the male is directed by a gene product from the short arm of the Y chromosome (Figures 43-5 and 43-6, *B*). Considerable evidence supports the involvement of a substance

known as the *H-Y antigen*, or one closely linked to it. This glycoprotein is found on the surface of all male cells and its presence correlates almost perfectly with the presence of a testis in all species. Although synthesis of the H-Y antigen appears to be normally under control of the Y chromosome, regulatory influences may also occur from the X chromosome or from autosomes. Even though it is essential, by itself the Y chromosome is not sufficient for maleness. Genetic material located on the X chromosome sensitizes the genital ducts and the external genitalia to the masculinizing effects of androgens. Autosomal genes may also participate in directing the organization of the primitive gonad into a functioning reproductive gland.

In contrast, femaleness is partly determined positively by the presence of an X chromosome but partly and importantly negatively by the absence of a Y chromosome. The normal female chromosome complement is 44 autosomes and two sex chromosomes, XX. Both X chromosomes are active in germ cells and are essential for the genesis of a normal ovary (Figures 43-5 and 43-6, *A*). However, female differentiation of the genital ducts and external genitalia requires that only a single X chromosome be active in directing

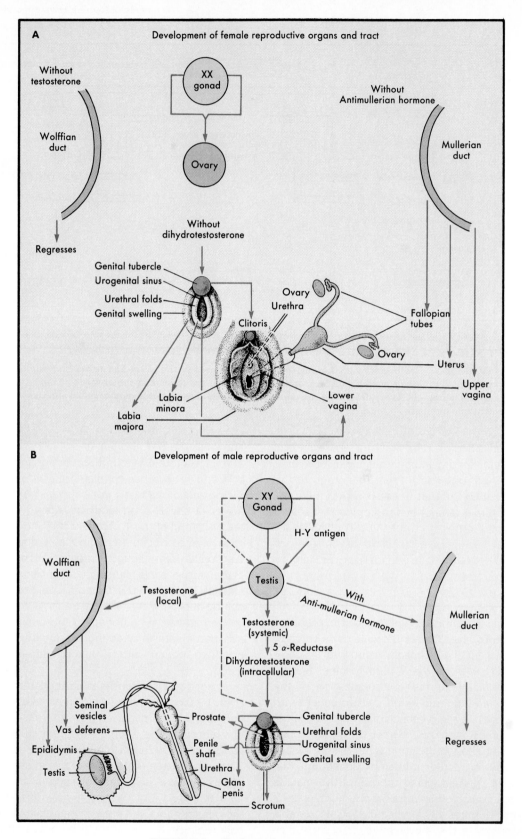

FIGURE 43-6 For legend see opposite page.

FIGURE 43-6   A, Development of the female reproductive organs. Note that this development does not require hormonal products from the ovary. Therefore, in the absence of gonads, the female pattern results. **B,** Development of the male reproductive organs. Note that the complete male pattern requires secretion of testosterone, reduction of testosterone to dihydrotestosterone, and the secretion of antimüllerian hormone.

transcription within their constituent cells. This is so because the second X chromosome of a normal XX female is genetically inactive in all tissues outside the gonad. Thus, if an abnormality in meiosis or in early mitosis produces an individual with only a single sex chromosome (an X), that individual will undergo normal female genital development even though her gonad is abnormal and without function (Figure 43-5).

Gonadal Sex

The indifferent gonad consists of a primordial mesonephric ridge with several components: celomic epithelium, the precursor of granulosa and Sertoli cells; mesenchymal stromal cells, the precursors of theca and Leydig cells; and germ cells that have migrated there from primitive ectoderm (Figure 43-5). This assembly is organized as an outer *cortex* and an inner *medulla*.

In a normal male fetus the spermatogenic tubules begin to form at 6 weeks, followed by differentiation of the Sertoli cells at 7 weeks and the Leydig cells at 8 to 9 weeks. At that point the testis is recognizable; the germ cells have become enclosed within the medulla, whereas the cortex, including its Leydig cells, has regressed. Neither androgens nor any other known hormonal influences are required for this differentiation of the indifferent gonad into a testis.

In a normal female fetus differentiation of the indifferent gonad into an ovary does not start until 9 weeks of age. At this time both X chromosomes within the germ cells become activated. The germ cells begin to undergo *mitosis*, giving rise to daughter cells called *oogonia*, which continue to proliferate. Shortly thereafter, *meiosis* is initiated in some oogonia, and each becomes surrounded by differentiating granulosa cells and precursor theca cells to form a follicle. The germ cells, now known as primary oocytes, remain in the first stage, or prophase, of meiosis until activated many years later. In contrast to the male, the cortex (which contains the follicles) predominates in the developed ovary, whereas the medulla regresses. The primitive ovary begins to synthesize estrogenic hor-

mones concurrent with these developments, and these hormones may contribute to later ovarian differentiation (in contrast to the male situation).

Genital (Phenotypic) Sex

Up to this point in fetal development, sexual differentiation is largely independent of known hormonal products. However, **differentiation of the genital ducts and the external genitalia requires specific hormonal signals from the gonad to produce the masculine format. Without such input, the feminine format will result.**

During the sexually indifferent stage, from 3 to 7 weeks, two different genital ducts develop on each side. In the male, at about 9 to 10 weeks, the *wolffian*, or *mesonephric*, *ducts* on each side begin to grow and give rise to the epididymis, the vas deferens, the seminal vesicles, and the ejaculatory duct by 12 weeks (Figure 43-6, *B*). This constitutes the system for delivering sperm from the male to the female. The growth and differentiation of the wolffian ducts in the male is induced by testosterone. In fact, the testosterone produced by each testis acts unilaterally on its own wolffian duct. Testosterone is not converted to its active metabolite, dihydrotestosterone, before acting on the wolffian duct cells, as it must be in other genital tissues. In the normal female the wolffian ducts regress at 10 to 11 weeks because the ovary does not secrete testosterone.

The two *müllerian ducts* arise parallel to the wolffian ducts on each side. In the male the müllerian ducts begin to regress at 7 to 8 weeks, about the same time that the Sertoli cells of the testis appear. These cells produce a glycoprotein, antimüllerian hormone (AMH), which causes atrophy of the müllerian ducts. AMH also initiates descent of the testes into the inguinal area. Although the homologous granulosa cells of the ovary produce AMH, they do not do so until **after** the müllerian ducts have already developed to the point where AMH can no longer cause their regression. Therefore in the female these ducts grow and differentiate into fallopian tubes at their upper ends and join at their lower ends to form a single uterus, cervix, and upper vagina (Figure 43-6, *A*). **This process does not require any known ovarian hormone**.

The external genitalia of both genders begin to differentiate at 9 to 10 weeks. They are derived from the same primitive structures: the genital tubercle, the genital swelling, the urethral or genital folds, and the urogenital sinus (Figure 43-6). In the male testosterone must be secreted into the fetal circulation and subsequently converted to dihydrotestosterone within these tissues in order for masculine differentiation of the external genitalia to occur. With androgenic stimulation the genital tubercle grows into the glans penis, the genital swellings fold and fuse into the scrotum, the urethral folds enlarge and enclose the penile urethra and corpora spongiosa, and the urogenital sinus gives rise to the prostate gland (Figure 43-6, *B*). In the normal female or in the individual with only one sex chromosome (an X) the external genitalia develop without significant hormonal influence into the clitoris, labia majora, labia minora, and lower vagina (Figure 43-6, *A*). The critical importance of androgens themselves to the male phenotype of external genitalia is emphasized by the normal female urogenital tract having adequate levels of androgen receptors but lacking the hormonal stimulus. However, if the female is pathologically exposed to an excess of testosterone or other androgens early in gestation, a masculine phenotype of external genitalia can result. If exposure to androgen occurs after the female phenotype has already been established, it cannot reverse the pattern to that of the male, but it can cause the clitoris to enlarge. Conversely, in males biosynthetic defects in androgen synthesis or lack of androgen receptors can result in a feminine phenotype of external genitalia. Such individuals are known as *pseudohermaphrodites*.

The initial androgen production necessary for male sexual differentiation does not seem to depend on fetal pituitary gonadotropins. An LH-like hormone, chorionic gonadotropin from the placenta, stimulates early testosterone production by the Leydig cells of the testis. On the other hand, the continued growth of the male genitalia in the last 6 months does require fetal pituitary LH to support the necessary testicular androgen production. Similarly, the later moulding of the female genitalia in utero may be stimulated by ovarian estrogen production, which also depends on pituitary gonadotropins.

Other aspects of phenotypic sexual differentiation are not evident until long after birth. These include differences between the constant pattern of gonadotropin secretion in the male versus the cyclic pattern in the female, the degree of breast development, and psychologic identification with one gender. It is not certain what factors imprint or regulate these traits in humans. Evidence from rodents suggest that circulating testosterone induces the fetal hypothalamus to program the ultimate noncycling pattern of gonadotropin secretion of the postpubertal male. (To do so, testosterone may paradoxically require conversion to estradiol within the target neurons.) Without androgens, the ultimate cyclic pattern of the female results. This would constitute another instance in which the female pattern was the "neutral pattern," whereas the male pattern required an action derived from the Y chromosome.

Mammary gland development in the rodent embryo also is clearly regulated by androgen. In its absence a normal female breast develops; in its presence the elaborated ductal system is suppressed. In the human, however, male/female differences in breast tissue are not apparent until puberty. At that time the hormonal milieu in the female induces growth and differentiation of breast tissue, whereas that in the male suppresses it.

A large body of evidence suggests that psychologic gender identification is mostly independent of hormonal regulation or even of the phenotype of the genitalia; instead, it appears to depend on rearing cues. However, exceptions to this have been noted in certain cases of XY male pseudohermaphrodites raised as girls. In such individuals significant growth of the penis at puberty can reverse the psychosocial gender from female to male.

BIBLIOGRAPHY
Journal Articles

George FW and Wilson JD: Hormonal control of sexual development, Vitam Horm 43:145, 1986.

Josso N: AntiMüllerian hormone: new perspectives for a sexist molecule, Endocr Rev 7:421, 1986.

Matsumoto A and Bremner W: Modulation of pulsatile gonadotropin secretion by testosterone in man, J Clin Endocrinol Metab 58:609, 1984.

Naftolin F and Butz E: Sexual dimorphism, Science 211:1263, 1981.

Ohno S: The role of H-Y antigen in primary sex determination, JAMA 239:217, 1978.

Scott R and Burger H: An inverse relationship exists between seminal plasma inhibin and serum follicle-stimulating hormone in man, J Clin Endocrinol Metab 52:796, 1981.

Veldhuis J et al: Endogenous opiates modulate the pulsatile secretion of biologically active luteinizing hormone in man, J Clin Invest 72:2031, 1983.

Wilson JD: Sexual differentiation, Annu Rev Physiol 40:279, 1978.

Ying SY: Inhibins, activins, and follistatins: gonadal proteins modulating the secretion of follicle-stimulating hormone, Endocr Rev 9:267, 1988.

Books and Monographs

Chin W: Organization and expression of glycoprotein hormone genes. In Imura H: The pituitary gland, New York, 1985, Raven Press.

Frohman LA et al: The physiological and pharmacological control of anterior pituitary hormone secretion. In Dunn A and Nemeroff C, editors: Behavioral neuroendocrinology, New York, 1983, Spectrum Publications, Inc.

Knobil E and Neill JD: The physiology of reproduction, New York, 1988, Raven Press.

Savoy-Moore RT et al: Differential control of FSH and LH secretion. In Greep RO, editor: Reproductive physiology, Baltimore, 1980, University Park Press.

CHAPTER 44

Male Reproduction

ANATOMY

The testes are situated in the scrotum, where they are maintained below the body core temperature. Each adult testis weighs about 40 g and has a long diameter of 4.5 cm. Eighty percent of the testis is made up of the spermatogenic tubules; the remaining 20% is composed of connective tissue containing the Leydig cells. The spermatogenic tubules, a coiled mass of loops, empty into a ductal system that eventually drains into the *epididymis,* a maturation and storage site for spermatozoa. From there, spermatozoa are carried via the *vas deferens* and *ejaculatory duct* into the penis to be emitted during copulation.

The structure of the spermatogenic tubule is shown schematically in Figure 44-1. Each tubule is bounded by a basement membrane separating it from the Leydig cells and adjacent capillaries. Beneath this membrane are Sertoli cells and immature germ cells, the *spermatogonia.* As the spermatogonia divide and develop around the circumference of the tubule, columns of maturing germ cells are formed below them. These columns reach from the basement membrane to the lumen and culminate in the spermatozoa. The columns lie between the cytoplasms of two adjoining Sertoli cells, each of which extends from the basement membrane to the lumen (Figure 44-1). Special processes of the Sertoli cell cytoplasms fuse into *tight junctions* that create two compartments of intercellular space between the basement membrane and the lumen. The spermatogonia lie within the basal compartment, whereas their descendents resulting from subsequent stages in spermatozoan development lie in the adluminal compartment. This compartmentalization accomplishes two important functions. The basement membrane and the Sertoli cell cytoplasm together form a blood-testis barrier. This barrier can exclude harmful circulating substances from the intercellular fluid bathing the maturing germ cells and from the tubular fluid. Conversely, products from the later stages of spermatogenesis are prevented from affecting earlier stages or from diffusing back into the bloodstream and producing antibodies. This arrangement also provides high **local** concentrations of testosterone from the Leydig cells and estradiol and protein products from the Sertoli cells. Such high concentrations are essential for spermatogenesis.

THE BIOLOGY OF SPERMATOGENESIS

Sperm production continues throughout the male's reproductive life. At peak, 100 to 200 million sperm can be produced daily. To generate this large number, the spermatogonia must also renew themselves by cell division. This differs from the situation in the female, who at birth has a fixed number of germ cells, which continually decrease throughout her life.

The descendents of the spermatogonia undergo an extraordinary metamorphosis to spermatozoa as they move from the basement membrane to the tubule. By two mitotic divisions a spermagonium first gives rise

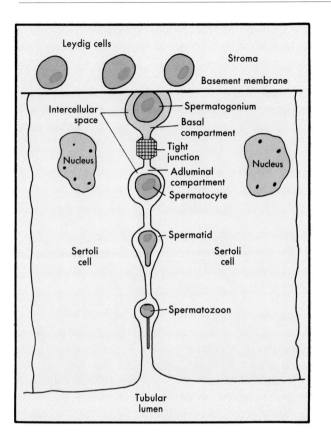

FIGURE 44-1 Schematic representation of the relationship between the germ cells and the Sertoli cells within the spermatogenic tubule and the Leydig cells outside the tubule. The developing germ cells are completely surrounded by the cytoplasm of the Sertoli cells. In addition, a tight junction between the two adjacent Sertoli cells separates the early spermatogonium from its descendant spermatocytes and spermatids. The basement membrane constitutes a blood-testis barrier that does, however, permit testosterone from the Leydig cells to reach the spermatogenic tubule. (Redrawn from Fawcett DW: Ultrastructure and function of the Sertoli cell. In Hamilton DW and Greep RO, editors: Handbook of physiology, section 7, vol 5, Bethesda, Md, 1975, The American Physiological Society.)

to three active cells and a single resting cell; the latter will serve as the ancestor of a later generation of spermatozoa. The active cells divide further to yield type B spermatogonia, which then give rise to a number of primary spermatocytes (Figure 44-2). These enter the prophase of meiosis, the first reduction division, in which they remain for about 20 days.

The complex process of chromosomal reduplication, synapsis, crossover, division, and separation completes meiosis. The daughter cells, secondary

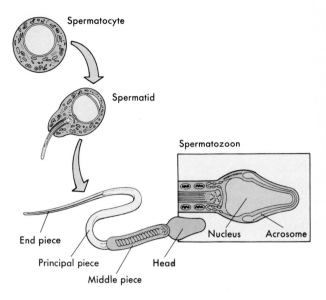

FIGURE 44-2 Schematic representation of the morphologic alterations in the development of the spermatozoon from the spermatocyte.

spermatocytes, immediately divide again. The products, called spermatids (Figure 44-2), now contain 22 autosomes and either an X or a Y chromosome. The spermatids lie near the lumen of the tubule, attached to the abutting Sertoli cells and connected to each other through intercellular bridges. They undergo nuclear condensation, shrinkage of cytoplasm, formation of an *acrosome,* development of a tail, and then emerge as flagellated spermatozoa (Figure 44-2). In the end 64 spermatozoa arise from each spermatogonium. The spermatozoa are then extruded into the lumen and lose most of their cytoplasm, which is phagocytized and degraded by the Sertoli cells. The latter also generate a fluid drive that sweeps the spermatozoa down the tubular system into the epididymis.

The spermatozoa are now linear structures with several components (Figure 44-2). The *head* contains the nucleus and an acrosomal cap in which are concentrated hydrolytic and proteolytic enzymes that will facilitate penetration of the ovum. The *middle piece,* or body, contains mitochondria, which generate the motile energy of the spermatozoa. The *chief,* or *principal, piece* contains stored adenosine triphosphate (ATP) and pairs of contractile microtubules down its entire length. An ATPase transfers the stored energy to the microtubules, which then impart flagellar motion to the spermatozoa. Both Ca^{++} and cyclic adenosine monophosphate (cAMP) are involved in regulating motility of the spermatozoa.

Approximately 70 days are required for this entire sequence of development. However, individual resting spermatogonia do not begin the process of spermatogenesis randomly. Groups of adjacent spermatogonia initiate a cycle of development about every 16 days, thus constituting one "generation." At about the same time that the primary spermatocytes of one cycle enter prophase, a second cycle of spermatogonia is activated. A third cycle begins about the time spermatids from the first cycle appear. When these spermatids have completed their transformation into spermatozoa, a fourth cycle of spermatogonia has begun. Around the circumference of any individual spermatogenic tubule, several cycles may be in process simultaneously. This gives rise to several specific cellular constellations existing side by side.

The individual descendants of any one type B spermatogonium lying within the adluminal compartment of the tubule may not be totally separated. Continuity of cytoplasm and possibly cell-to-cell intercommunication may exist. Because of this and because of the regular topographic association of particular stages of spermatogenesis in neighboring cycles, certain products of germ cells in one stage of spermatogenesis may initiate or regulate events in other stages.

After the spermatozoa are ejected, they reach the epididymis, which they traverse in 2 to 4 weeks. During this time they lose their remaining cytoplasm and acquire motility. This process may depend partly on specific secretions of the epididymal cells and partly on preprogramming within the spermatozoa. After reaching the vas deferens, they may be stored and remain viable for several months.

DELIVERY OF SPERMATOZOA

The process of ejaculation delivers spermatozoa from the vas deferens into the female genital tract. This requires erection of the penis, a process that results from filling of its venous sinuses. This is accomplished through simultaneous dilation of arterioles and constriction of veins and is under parasympathetic control. Ejaculation is then effected by sympathetic activation. Just before ejaculation, successive fluids are added to the contents of the vas deferens. The initial secretions are from the prostate gland, the alkalinity of which helps neutralize the acid pH of the female genital secretions. The terminal portion of the ejaculate is composed of secretions from the seminal vesicles. These contain fructose, an important oxidative substrate for the spermatozoa, and prostaglandins. The latter may stimulate contractions within the female tract that help propel the spermatozoa toward the ovum. Seminal fluid also contains luteinizing and follicle-stimulating hormones (LH, FSH), prolactin, testosterone, estradiol, inhibin, and endorphins. Their exact source and role in fertilization remain to be determined.

A typical seminal emission contains 200 to 400 million spermatozoa in a volume of 3 to 4 ml. Once within the vagina, the spermatozoa move inward at a rate up to 44 mm·min^{-1}. Their life span in the female genital tract is approximately 2 days. Transport of sperm to the ovum requires mechanical assistance by smooth muscle contractions of the female reproductive organs.

In vivo, human sperm cannot fertilize an ovum until they have been in contact with the female reproductive tract for several hours; the process is termed *capacitation*. In vitro, however, fertilization can occur after the ejaculated sperm have been washed free of seminal fluid. This observation suggests that washing has removed inhibitory substances. Although the process of capacitation is poorly understood, it increases motility and enhances the ability of sperm to penetrate the ovum. This involves a reaction in which the acrosomal membrane and the outer sperm membrane fuse to create pores through which the enclosed enzymes can escape.

MALE PUBERTY

Beginning at an average age of 10 to 11 years and ending at about 15 to 17 years, males develop adult levels of androgenic hormones and full reproductive function. Activation of the testes results in adult size and function of the other organs of reproduction, complete secondary sexual characteristics, and adult musculature. Boys undergo a linear growth spurt, and the epiphyseal growth centers close when adult height is attained. A composite picture of the sequence is shown in Figure 44.3.

Enlargement of the testes is the first physical sign of puberty. This principally represents an increase in the volume of the spermatogenic tubules and is preceded by small increases of plasma FSH. Leydig cells

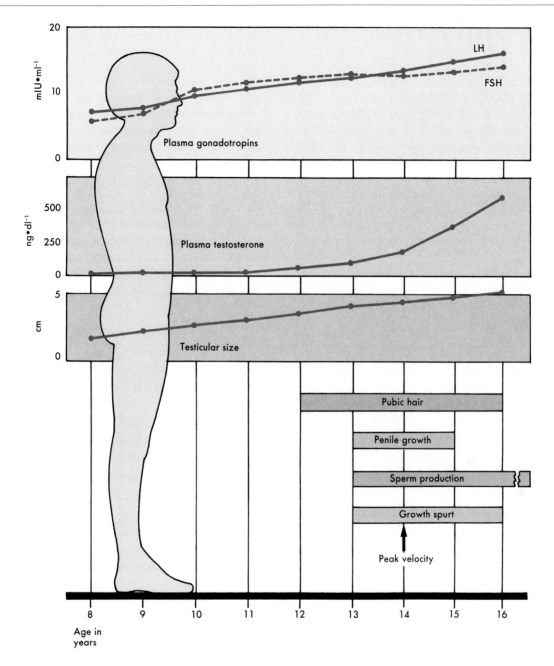

FIGURE 44-3   Average chronologic sequence of hormonal and biologic events in normal male puberty. (Redrawn from Marshall WA and Tanner JM: Arch Dis Child 45:13, 1970; and Winter JSD et al: Pediatr Res 6:126, 1972.)

appear and testosterone secretion rises as the plasma LH level then increases. Plasma testosterone level climbs rapidly over a 2-year period, during which time pubic hair appears, the penis enlarges, and peak velocity in linear growth is achieved (Figure 44-3). When the boy is about age 13, sperm production

begins. Growth ceases 1 to 2 years after adult testosterone levels are reached. In about one third of boys a transient stimulation of breast growth occurs, probably reflecting increased production of estradiol. As testosterone levels become dominant, the breast tissue regresses.

REGULATION OF SPERMATOGENESIS

For various reasons hormonal regulation of spermatogenesis is less completely understood than that of oogenesis. Several hormones and hormonal products are likely involved, and the possibilities of paracrine and autocrine actions among Leydig, Sertoli, and germ cells are numerous. Definitive in vivo studies of the testis are difficult to design, and tissue for in vitro study is not usually available. Finally, comparative studies of reproductive function in other subprimate species show sufficient variation to make extrapolation to the human uncertain. Thus the following discussion is more descriptive than mechanistic.

FSH, LH, testosterone, and possibly estradiol are coordinated in the regulation of spermatogenesis. In normal men experimental production of simultaneous FSH and LH deficiency by suppression of the pituitary gland with high doses of testosterone almost completely inhibits sperm production (Figure 44-4). Selective replacement of either FSH or LH then reinitiates sperm production, but not to normal levels. Other studies show that a suitable period of exposure to FSH is essential to spermatogenesis, but after that it can sometimes be maintained adequately by LH alone.

During early development of the fetal testis, testosterone may stimulate the transformation of primordial germ cells into resting spermatogonia. From then until puberty the spermatogonia normally remain dormant, presumably because gonadotropin secretion and testosterone levels in the testis are low. Activation of the spermatogonia then starts shortly after FSH secretion begins to undergo its pubertal increase. The completion of the long prophase of the

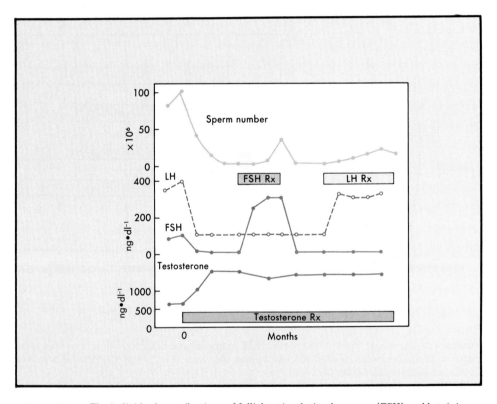

FIGURE 44-4   The individual contributions of follicle-stimulating hormone (*FSH*) and luteinizi. hormone (*LH*) to sperm production in normal men. When endogenous FSH and LH secretion are suppressed by administration of exogenous testosterone (negative feedback), sperm production declines to very low levels. Selective restoration of either FSH or LH by exogenous administration raises sperm production, but neither gonadotropin alone returns sperm counts to normal. (Redrawn from Matsumoto AM et al: J Clin Invest 72:1005, 1983; J Clin Endocrinol Metab 59:882, 1984, © by The Endocrine Society.)

primary spermatocytes seems to depend on the high **intratesticular** concentration of testosterone, normally 100-fold higher than that in plasma. This is produced by the specific action of LH on the Leydig cells. (In men who lack LH, the substitution of testosterone in amounts sufficient to raise plasma levels to normal is unable to promote spermatogenesis.) The critical action of LH on the Leydig cells may also be facilitated by prolactin. To what extent testosterone regulates spermatogenesis directly or via conversion to dihydrotestosterone (DHT) or estradiol is also uncertain.

FSH acts on the Sertoli cells, whose role is vital and complex. After these cells secrete antimüllerian hormone in early fetal life, their subsequent function until puberty is not known. After puberty the Sertoli cells do not undergo cell division. In association with the cycle of spermatogenesis, however, they show changes in the activity and shape of the nucleus; in the size, shape, and branching of the cytoplasmic processes; in concentrations of lipid and glycogen; in mitochondrial function; and in enzyme content. These changes somehow relate to the process of germ cell development. For example, because Sertoli cell cytoplasm acts as a conduit through which the germ cells migrate during developmental passage (see Figure 44-1), their tight junctions must open regularly to permit maturing primary spermatocytes to pass and then must close again behind them.

Sertoli cells are stimulated by FSH to synthesize estradiol from testosterone, which is provided by the Leydig cells. Conceivably, both sex steroids then have access to the developing spermatocytes. FSH also stimulates synthesis of *androgen-binding protein,* a unique Sertoli cell product, which complexes with high affinity to testosterone, DHT, and estradiol. This protein may serve to concentrate these sex steroids in the Sertoli cells and thereby create a storage form for controlled release during appropriate stages of spermatogenesis. Since androgen-binding protein is also secreted into the tubular fluid, it may prevent reabsorption of the sex steroids from the epididymis and thus ensure their availability to the spermatozoa during the latter's maturational transit. *Transferrin* (an iron-binding globulin), plasminogen activator (important in fibrinolysis), and peptides resembling gonadotropin-releasing hormone (GnRH) are also produced by Sertoli cells, and these presumably also have roles in spermatogenesis.

Numerous local feedback loops operate within and between the Sertoli and Leydig cells. FSH stimulates inhibin and estradiol production, but inhibin then blocks the critical aromatase reaction in estradiol synthesis. Testosterone from the Leydig cells stimulates inhibin secretion by the Sertoli cells, whereas activin from the Sertoli cells blocks testosterone synthesis in the Leydig cells. The timing of such actions must somehow be coordinated to produce an optimal balance of the hormones that foster spermatogenesis.

SECRETION AND METABOLISM OF ANDROGENS

Testosterone, the major androgenic hormone, is synthesized as described in Chapter 43. In adults plasma testosterone levels show small pulses throughout the day that correspond to those of LH. A small diurnal trend also exists, with lower levels in the evening. Testosterone is in part only a circulating prohormone. Much of androgen action is supplied by the reduction of testosterone to DHT in peripheral tissues. The two estrogens, estradiol and estrone, are also produced by males in significant amounts, mostly derived from aromatization of circulating testosterone and androstenedione, respectively, in such peripheral sites as adipose tissue and liver. In certain sites estradiol may even be the actual mediator of the apparent testosterone action.

Testosterone and DHT circulate mostly bound to a sex steroid–binding globulin (SSBG), which is similar but not identical to the androgen-binding protein of Sertoli cell origin. The remainder is bound to albumin. Only the free and loosely bound albumin fractions of the androgens are biologically active. Thus SSBG-bound fractions serve as circulating androgen reservoirs, similar to those of thyroid hormone and cortisol. SSBG concentration is itself decreased by androgens and increased by estrogens. Thus androgens increase their own biologic availability by increasing the percentage of free hormone that circulates. Most testosterone is metabolized to products oxidized at the 17-position and excreted in the urine; these products constitute 30% of the 17-ketosteroid fraction (see Chapter 41). The rest arises from adrenal sources.

Plasma testosterone levels vary throughout life. As shown in Figure 44-5, the plasma testosterone level rises to adult values in the fetus at the time the external genitalia are undergoing differentiation. By birth,

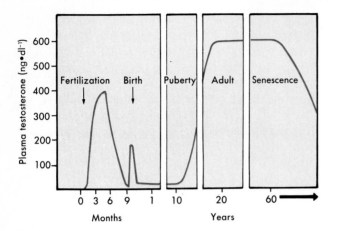

FIGURE 44-5 Plasma testosterone profile during the life span of a normal male. (Redrawn from Griffin JE et al. In Bondy PK and Rosenberg LE: Metabolic control and disease, Philadelphia, 1980, WB Saunders Co; and Winter JSD et al: J Clin Endocrinol Metab 42:679, 1976. Copyright 1976. Reproduced by permission.)

however, the levels have declined greatly. Very soon thereafter, plasma testosterone levels rise temporarily, producing a new transient peak of unknown physiologic significance. After 2 months of age, plasma testosterone (and LH) values fall to low levels throughout childhood, and Leydig cells cannot even be identified in the testis. At about age 11 years, Leydig cells reappear, and plasma testosterone concentration begins a steep rise and reaches a plateau of approximately 600 ng·dl^{-1} at about age 17 (Figure 44-5). This plateau is sustained for some 50 years; this time course corresponds with the reproductive role of the hormone. During the seventh and eighth decades of life, the plasma testosterone level gradually declines, this time because the Leydig cells lose their responsiveness to LH stimulation. Because of negative feedback, plasma LH levels rise slowly. Although decreasing testosterone levels may be associated with a decline in libido and sperm production, spermatogenesis still occurs in most octogenarians.

ACTIONS OF ANDROGENS

The effects of androgens on tissues outside the testis can be divided into two major categories: those pertaining to reproductive function and secondary sexual characteristics and those pertaining to stimulation of somatic growth and maturation.

Testosterone diffuses into target cells, where it is usually reduced to DHT. That testosterone itself can initiate androgen effects is demonstrated by the absence of 5α-reductase activity and DHT production in certain responsive cells. A single receptor binds androgens with a greater affinity for DHT than testosterone. The crucial requirement for the androgen receptor is shown by individuals who lack the gene for receptor synthesis failing to undergo masculinization of the genital ducts or external genitalia.

The androgen receptor complex interacts with DNA molecules (see Figure 35-12), probably assisted by nuclear proteins. This results in stimulation of RNA polymerase, induction of messenger RNAs, and their translation into proteins. Virtually all actions of androgens are blocked by inhibitors of either RNA or protein synthesis. Androgens stimulate remarkable growth of the male accessory organs of reproduction, characterized by hypertrophy and hyperplasia of the epithelial cells, stromal components, and blood vessels.

The major androgen effects, classified according to the probable effector molecule, are shown in Figure 44-6. DHT is specifically required in the fetus for the differentiation of the penis, scrotum, penile urethra, and prostate. DHT is required again during puberty for growth of the scrotum and prostate and stimulation of other prostatic secretions. DHT stimulates the hair follicles to produce the typical masculine beard growth, diamond-shaped pubic hair, and recession of the temporal hairline. Growth of the sebaceous glands and their production of sebum also results from DHT action. Testosterone stimulates fetal differentiation of the epididymis, vas deferens, and seminal vesicles (Figure 44-6). During puberty, testosterone and DHT cause enlargement of the penis and seminal vesicles and stimulate the latter to secrete. Although spermatozoa can be produced by adults who secrete testosterone but who lack DHT because of 5α-reductase deficiency, both androgens may normally participate in spermatogenesis and epididymal sperm maturation.

Testosterone first stimulates the pubertal growth spurt and then causes cessation of linear growth by closure of the epiphyseal growth centers. Potentiation of growth hormone secretion by testosterone, however, may be mediated through prior conversion to estradiol. Testosterone causes enlargement of the muscle mass in boys during puberty by increasing the size of muscle fibers. In subsequent adult life,

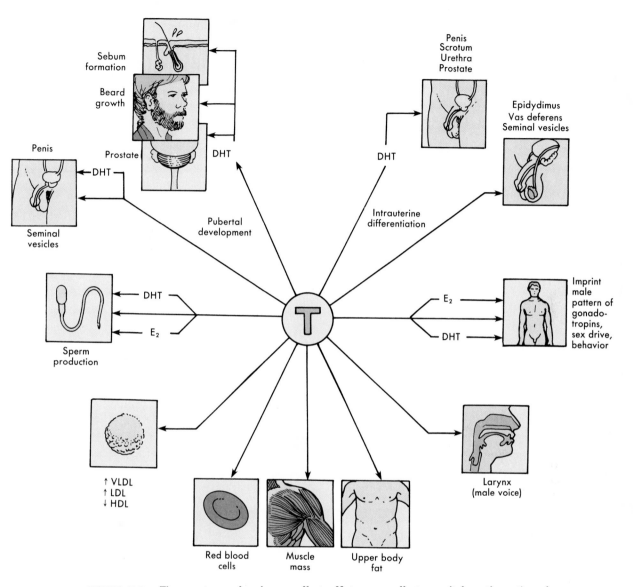

FIGURE 44-6   The spectrum of androgen effects. Note some effects result from the action of testosterone (*T*) itself, whereas others are mediated by dihydrotestosterone (*DHT*) and possibly estradiol (*E$_2$*) after they are produced from testosterone. *VLDL, LDL,* and *HDL* are very-low-density, low-density, and high-density lipoproteins.

administration of testosterone to both genders causes nitrogen retention, which reflects protein anabolism. Finally, testosterone enlarges the larynx and thickens the vocal cords, thereby deepening the voice.

Androgens also increase red blood cell mass by stimulating erythropoietin synthesis (see Chapter 14) and by a direct effect on maturation of erythroid precursors. Androgens regulate the synthesis of many hepatic proteins, decreasing all hormone-binding globulins. Importantly, plasma levels of very-low-density lipoproteins are increased, whereas plasma levels of high-density lipoproteins are decreased by androgen action. This may be partly responsible for the much higher risk of coronary artery disease in men. On the other hand, androgens create a greater bone mass in men than women, a protective effect against osteoporosis.

BIBLIOGRAPHY
Journal Articles

Forest MG et al: Kinetics of human chorionic gonadotropin–induced steroidogenic response of the human testis. II. Plasma 17α-hydroxyprogesterone, δ^4-androstenedione, estrone and 17β-estradiol: evidence for the action of human chorionic gonadotropin on intermediate enzymes implicated in steroid biosynthesis, J Clin Endocrinol Metab 49:284, 1979.

Harman SM et al: Reproductive hormones in aging men. I. Measurement of sex steroids, basal luteinizing hormone, and Leydig cell response to human chorionic gonadotropin, J Clin Endocrinol Metab 51:35, 1980.

Lipsett MB: Physiology and pathology of the Leydig cell, N Engl J Med 303:682, 1980.

Mooradian AD et al: Biological actions of androgens, Endocr Rev 8:1, 1987.

Overstreet JW and Blazak WF: The biology of human male reproduction: an overview, Am J Ind Med 4:5, 1983.

Pardridge WM et al: Androgens and sexual behavior, Ann Intern Med 96:488, 1982.

Books and Monographs

Bardin CW: Pituitary-testicular axis. In Yen SSC and Jaffe FB: Reproductive endocrinology, Philadelphia, 1986, WB Saunders Co.

Fawcett DW: Ultrastructure and function of the Sertoli cell. In Hamilton DW and Greep RO, editors: Handbook of physiology, section 7, vol 5, Bethesda, Md, 1975, The American Physiological Society.

Griffin JE and Wilson DJ: Disorders of the testes and male reproductive tract. In Wilson DJ and Foster DW, editors: Textbook of endocrinology, Philadelphia, 1985, WB Saunders Co.

Griffin JE et al: The testis. In Bondy PK and Rosenberg LE: Metabolic control and disease, Philadelphia, 1980, WB Saunders Co.

Steinberger E and Steinberger A: Hormonal control of spermatogenesis. In Degroot LJ et al, editors: Endocrinology, vol 3, New York, 1989, Grune & Stratton, Inc.

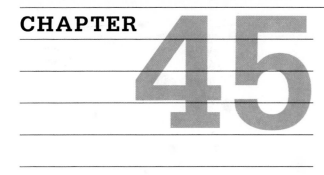

Female Reproduction

The ovaries, along with the fallopian tubes and uterus, are situated in the pelvis. Each ovary weighs approximately 15 g and consists of three zones. The dominant zone is the *cortex*, which is lined by germinal epithelium and contains all the oocytes. Each oocyte is enclosed in a *follicle* formed by surrounding endocrine cells. Follicles in various stages of development and regression are present throughout the cortex (Figure 45-1). The surrounding stroma is composed of connective tissue elements and *interstitial cells*. The other two zones of the ovary, the *medulla* and the *hilum*, contain scattered steroid-producing cells whose function is unknown.

The *granulosa and theca cells* of the ovary produce steroid hormones that function locally to modulate the development of the ovum and its extrusion from the follicle. These hormones are also secreted into the blood and act on the fallopian tubes, uterus, vagina, breasts, hypothalamus, pituitary gland, adipose tissue, liver, kidney, and bones. Many of these distant effects are closely related to the sequence of reproduction.

To facilitate understanding of the complex process of female reproduction, the biology of ova formation and the cyclic changes in hormone secretion characteristic of the female are described separately. These biologic factors and cyclic changes are then combined in a discussion of the hormonal regulation of oogenesis.

BIOLOGY OF OOGENESIS

Oogonia arise from primordial germ cells that migrate to the genital ridge at 5 to 6 weeks of gesta-tion. There, in the developing ovary, they undergo mitosis until 20 to 24 weeks, when the total number of oogonia has reached a maximum of 7 million. Beginning at 8 to 9 weeks and continuing until 6 months after birth, some oogonia start into the prophase of meiosis, becoming *primary oocytes*. The latter grow from 10 to 25 μm in diameter, when meiosis begins, to 50 to 120 μm at maturity. The process of meiosis remains suspended in prophase at least until sexual maturation of the individual. In some primary oocytes prophase can be maintained by an inhibitory hormonal milieu until menopause. Therefore an individual primary oocyte may have a life span of up to 50 years.

From the start of oogenesis a process of oocyte attrition occurs so that by birth only 2 million primary oocytes exist, and by the onset of puberty only 400,000 remain. This constitutes the entire supply of potential ova for the woman's reproductive life because no new oogonia can be formed. With continuing attrition, few if any oocytes are left when menopause begins, and reproductive capacity ends. This contrasts sharply with the male, in whom the supply of spermatogonia is continually being renewed.

DEVELOPMENT OF THE OVARIAN FOLLICLE

First Stage

The follicle develops in distinct stages. The first stage parallels the prophase of the oocyte and occurs very slowly. It begins in utero and ends at any time

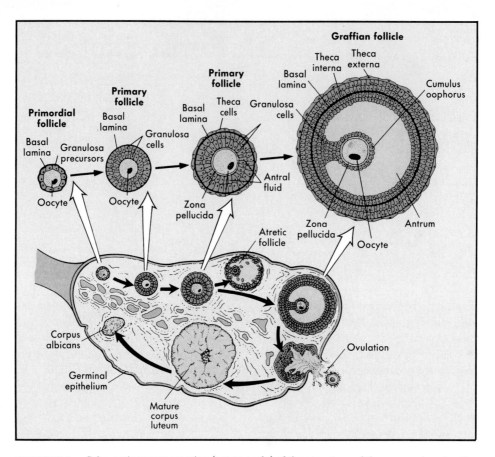

FIGURE 45-1   Schematic representation (not to scale) of the structure of the ovary, showing the various stages in the development of the follicle and its successor structure, the corpus luteum. The follicle grows from a primordial size of 25 μm to an ovulatory size of 10 to 20 mm. The oocyte is shielded from indiscriminate exposure to interstitial fluid contents by the basal lamina and the cytoplasm of the surrounding granulosa cells. The hormones and other constituents of the antral fluid are important regulators of follicular development. (Redrawn from Ham AW and Leeson TS: Histology, ed 4, Philadelphia, 1968, JB Lippincott Co.)

during reproductive life. As an oocyte enters meiosis, it induces a single layer of spindle-shaped cells from the stroma to surround it completely. These cells are *precursors of the granulosa cells*. Their cytoplasmic processes attach to the plasma membrane of the oocyte. Simultaneously a membrane called the *basal lamina* forms outside the granulosa cells. This delimits the *primordial follicle*, which has a diameter of 25 μm, from the surrounding stroma (Figure 45-1).

At 5 to 6 months of gestation the spindle-shaped granulosa cells in some of the follicles become cuboidal and begin to divide, creating several layers around the oocyte. This complex is now called a *primary follicle*. The granulosa cells secrete mucopolysaccharides, which form a protective halo, the *zona pellucida*, around the oocyte (Figure 45-1). However, the

cytoplasmic processes of the granulosa cells continue to penetrate the zona pellucida and provide nutrients and hormonal signals to the maturing primary oocytes. The cytoplasm of the granulosa cells also forms a filter through which plasma substances must pass before reaching the germ cell (compare to Sertoli cells and spermatocytes in Figure 44-1).

The primary follicle grows to a diameter of about 150 μm. Concurrently a new layer of cells from the stroma is recruited outside the basal lamina as theca cell precursors, and the granulosa cells begin to extrude small collections of fluid between themselves. This completes the first, or preantral, stage and is the maximal degree of follicular development ordinarily found in the ovary before establishment of menstrual cycles.

Second Stage

The second stage of follicular development begins only after the onset of menstrual cycling and requires 10 to 28 days for completion. During each cycle 6 to 12 primary follicles are recruited for further development. The small collections of follicular fluid coalesce into a single area called the *antrum* (Figure 45-1). The fluid of the *antral follicle* contains a complex of substances, some secreted by the granulosa and theca cells and some transferred from the plasma through the granulosa cell cytoplasm. Included are mucopolysaccharides, plasma proteins, enzymes of steroid synthesis, steroid hormones, follicle-stimulating hormone (FSH) and luteinizing hormone (LH), inhibin, oxytocin and arginine vasopressin, proopiomelanocortin derivatives, and other granulosa cell products. The steroid hormones reach the antrum by secretion from granulosa cells and by diffusion from theca cells. Nonsteroidal substances capable of inhibiting oocyte meiosis are also secreted into the antral fluid.

As the granulosa cells proliferate, they displace the oocyte into an eccentric position on a stalk, where it is surrounded by a distinctive layer, which is two to three cells thick and is called the *cumulus oophorus* (Figure 45-1). The layer of theca cells just outside the basal lamina also proliferates and is transformed into cuboidal steroid-secreting cells, the *theca interna*. Additional layers of spindle cells from the stroma form a vascularized layer, called the *theca externa*, around the theca interna. At the end of this stage the entire complex, called a preovulatory, or graafian, follicle (Figure 45-1), has reached an average diameter of 5 mm.

Third stage

The final stage of follicular development is completed within 2 to 3 days of the midpoint of the cycle. A single graafian follicle is "selected" by day 5 to 7 of the cycle and dominates the other second-stage follicles. This follicle now undergoes rapid expansion. The granulosa cells greatly increase the production of antral fluid. The colloid osmotic pressure of this fluid increases because of depolymerization of the mucopolysaccharides. The granulosa cells spread apart, the cumulus oophorus loosens, and the vascularity of the theca layers increases greatly. The total size of the dominant follicle reaches 10 to 20 mm. At a critical point the basal lamina adjacent to the surface of the ovary undergoes proteolysis. The follicle gently ruptures, releasing the oocyte with its adherent cumulus

oophorus into the peritoneal cavity. At this point the initial meiotic division of the oocyte is completed. The resultant secondary oocyte is drawn into the closely approximated fallopian tube, and a first polar body is discarded. In the fallopian tube sperm penetration causes completion of the second meiotic division, resulting in the haploid (23-chromosome) ovum and the second polar body. The remaining unsuccessful follicles from that cycle undergo atresia within the ovary (Figure 45-1).

CORPUS LUTEUM FORMATION

The residual elements of the ruptured dominant follicle next form a new endocrine unit, the corpus luteum (Figure 45-1). This provides the necessary steroid hormone balance that optimizes conditions for implantation of a fertilized ovum and for subsequent maintenance of the zygote until the placenta can assume this function. The corpus luteum is made up mainly of granulosa cells. These hypertrophy and form rows while their mitochondria develop dense matrices, their endoplasmic reticulum decreases, and numerous lipid droplets form within their cytoplasm. This process, called *luteinization*, begins just before ovulation and is greatly accelerated by the exit of the oocyte from the follicle. The rest of the corpus luteum consists of theca cells, arranged in folds along its outer surface. They luteinize less dramatically. Importantly the basal lamina between the theca and granulosa cells disappears, allowing ingrowth of blood vessels that supply the granulosa cells directly.

The corpus luteum regresses after a 14-day life span if conception does not follow ovulation. In this process of regression, known as *luteolysis*, the granulosa and theca cells undergo necrosis, and the structure is invaded by leukocytes, macrophages, and fibroblasts. Gradually the corpus luteum degenerates to an avascular scar, known as the *corpus albicans* (Figure 45-1).

ATRESIA OF FOLLICLES

During an average woman's reproductive life span, only 400 to 500 oocytes (one per month) will undergo the complete sequence that culminates in ovulation. The remaining millions of oocytes disappear in a process called *atresia*, which begins almost with the appearance of the initial primordial follicles. In first-stage follicles the oocyte simply becomes necrotic and the granulosa cells degenerate. This accounts for almost all oocytes. In second-stage follicles atresia may be more complex (Figure 45-1). In some the granulosa cells farthest from the oocyte first undergo necrotic changes. Loss of their function may precipitate a resumption of meiosis in the oocyte to the point of extrusion of the first polar body. However, the granulosa cells in the cumulus oophorus also eventually die, the unsupported oocyte degenerates, and everything inside the basal lamina collapses into a scar.

HORMONAL PATTERNS DURING THE MENSTRUAL CYCLE

The menstrual cycle is divided physiologically into three phases (Figure 45-2) that correspond with the dominant events in the development of each monthly ovum, as just described. The *follicular phase* begins with the onset of menstrual bleeding and is of variable length. The succeeding *ovulatory phase* lasts only 1 to 3 days. The final *luteal phase* lasts 13 to 14 days and ends with the onset of menstrual bleeding. A normal menstrual cycle may range from 21 to 35 days, depending on the length of the follicular phase.

A series of cyclic changes in ovarian steroid hormone production, pituitary LH and FSH secretion, and probably hypothalamic gonadotropin-releasing hormone (GnRH) pulses characterizes normal reproductive function. Both negative and positive feedback loops are involved in creating this complex pattern. The profile of cyclic changes of the hormones in plasma are described first before relating them to follicular development.

The critical regulators of the ovarian cycle are FSH and LH. Just before the start of the follicular phase, plasma FSH and LH concentrations are at their lowest levels (Figure 45-2), and the LH/FSH ratio is slightly greater than 1. The FSH level begins to rise gradually 1 day before menses begins, and it continues to do so through the first half of the follicular phase. The level of LH rises later. Then, during the second half of the follicular phase, the FSH level falls slightly, whereas the LH level continues to rise so that the LH/FSH ratio reaches about 2. Stimulated by the FSH, plasma estradiol concentration also increases gradually during the critical first 6 to 8 days. Later in the follicular phase the plasma estradiol level increases much more sharply, reaching a peak just before the ovulatory

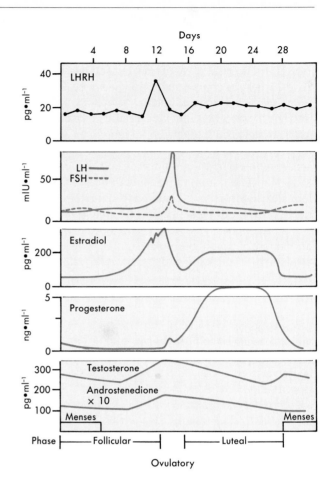

FIGURE 45-2   Profile of plasma hormone levels throughout the menstrual cycle. Note ovulatory surges of luteinizing hormone *(LH)* and follicle-stimulating hormone *(FSH)*, preceded by increases in estradiol and gonadotropin–releasing hormone (GnRH). The broad peaks of progesterone and estradiol in the luteal phase result from secretion by the corpus luteum.

phase (Figure 45-2). This estradiol is secreted by the granulosa cells of the dominant follicle. The higher estradiol level, along with increased ovarian secretion of inhibin, feed back to decrease the plasma FSH concentration during the second half of the follicular phase.

The next ovulatory phase is uniquely characterized by a very large but transient spike in the plasma LH level with a lesser spike in the FSH level (Figure 45-2). This surge in gonadotropin is preceded first by the "saw-tooth" estradiol peak of the late follicular phase and then by an increase in GnRH pulses (Figure 45-2). At the same time the plasma progesterone level rises slightly. Together these changes suggest that both the ovary and the hypothalamus contribute to the ovulatory surge of LH and FSH.

After ovulation the LH and FSH levels decline during the luteal phase to reach their nadirs toward its end (Figure 45-2). This decrease is caused by greater negative feedback from steroid hormones produced by the corpus luteum. The most distinctive feature of the luteal phase is a tenfold increase in plasma progesterone concentration that results from secretion by the corpus luteum. Estradiol, also originating from the corpus luteum, increases again. If pregnancy does not occur and the corpus luteum degenerates, progesterone and estradiol levels decrease dramatically to their lowest levels at the end of the luteal phase, FSH secretion increases, and menstrual bleeding starts.

HORMONAL REGULATION OF OOGENESIS

Primary Follicle Formation

The initial growth of the primordial follicle appears to be a local phenomenon in which factors from the oocyte stimulate granulosa cell development. In turn the granulosa cells initiate formation of the theca and then stop maturation of the oocyte once it reaches 80 μm in diameter. The transient surge of FSH and LH in midgestation and even the low levels of gonadotropins secreted during childhood are necessary for an adequate rate of follicular growth throughout the rest of life. Nonetheless the first stage from primordial to primary follicle continues to occur until menopause, independent of the presence or the state of reproductive cycling.

Follicular Development

The second stage of follicular development is directly stimulated by FSH acting on granulosa cells during the first half of the follicular phase (Figure 45-3). From a small group of FSH-responsive follicles, the single dominant follicle, normally in only one ovary, reaches its peak size between day 10 and 14 of the cycle and undergoes ovulation.

FSH initially stimulates granulosa cell growth and aromatase activity so that estradiol synthesis from androgens is enhanced (Figure 45-3). The increasing *local estradiol* then induces increases in its own receptors as well as FSH receptors. This sensitizes the granulosa cells to both hormones, and the result is even more follicular growth and a further boost in estradiol production. Thus, once started, second-stage follicular development becomes a self-propelling mechanism that combines endocrine, autocrine, and paracrine effects and that requires fine coordination between the pituitary gland and the ovary. The outcome is an **exponential** rate of follicular growth.

Two subsequent actions contribute to continuing follicular development (Figure 45-3):

1. FSH, along with estradiol, induces LH receptors on the granulosa and theca cells.
2. The slowly rising *plasma estradiol* level conditions the GnRH-gonadotropin axis so as to decrease FSH secretion but still permit a slight increase in LH secretion.

Pituitary stores of LH are also built up, thereby creating a supply for the coming ovulatory phase surge of LH. The estradiol effect partly occurs on the gonadotroph cells and is partly caused by interaction with dopaminergic and endorphinergic neurons that inhibit GnRH release.

The important role of LH in the second half of the follicular phase is to stimulate the theca cells to produce increasing amounts of androstenedione and testosterone. These androgens diffuse across the basal lamina into the granulosa cells, where they serve as vital precursors to estradiol. In addition, LH increasingly stimulates granulosa cell production of progesterone, which can diffuse back into the theca cells to serve as additional substrate for androgen synthesis. Thus, between theca and granulosa cells, a **two-way traffic of steroids** greatly increases the final efficiency of the follicle to produce estradiol, the critical element in its own development.

Both ovaries receive similar inputs of FSH and LH. Therefore the emergence of one dominant follicle may result from its possessing more FSH receptors and greater aromatase activity at the outset. Such characteristics would permit this particular follicle to exceed the others in estradiol production. Conversely, atresia of all the other second-stage follicles in a cycle appears to be related to their relatively low ratio of estradiol to androgen concentrations in the antral fluid. This is probably caused by declining FSH availability, because estradiol and inhibin secreted by the dominant follicle progressively reduce FSH secretion. The result is less aromatase activity and diminished estradiol production from androgens in the atretic follicles.

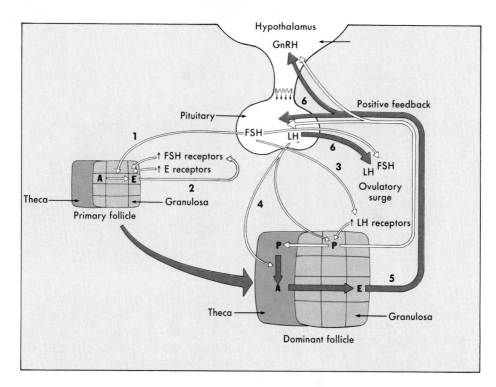

FIGURE 45-3   Hormonal regulation of follicular development. *1,* FSH stimulates granulosa cell growth and estradiol *(E)* synthesis in certain primary follicles. *2,* The local estradiol increases its own receptors and FSH receptors, amplifying both hormones' effects. Thus a self-propelling mechanism is set into motion. *3,* FSH later increases LH receptors, augmenting granulosa and theca cell responsiveness to LH. *4,* LH stimulates theca cell growth and androgen *(A)* production. Androgen is then converted to estradiol in the granulosa cells. LH also stimulates progesterone *(P)* production in the granulosa cells. *5,* As a result of two-way steroid traffic, the dominant follicle emerges as a very efficient secretor of estradiol. *6,* Rising estradiol, with late potentiation by progesterone, feeds back positively on the pituitary gland and hypothalamus to evoke the preovulatory surge of LH and FSH.

Ovulation

The ovulatory surge of LH and FSH is triggered by a **positive feedback** effect of estradiol (Figure 45-3). A critical plasma estradiol level of at least 200 pg·ml^{-1} sustained for at least 2 preceding days is required. The small preovulatory increase in plasma progesterone synergizes with estradiol to amplify and prolong the gonadotropin surge. This positive feedback effect is at both pituitary and hypothalamic levels. The pituitary gland, appropriately primed by the preceding pattern of ovarian steroid exposure, now responds to repetitive GnRH pulses in exaggerated fashion. Furthermore, the secreted LH molecules are more biologically active. In addition, estradiol and progesterone augment the flow of GnRH pulses from the hypothalamus to the pituitary gland.

Thus the hypothalamic-pituitary unit is condi-tioned by ovarian steroids to provide a sudden increase mainly in gonadotropin (mainly LH) stimulation of the dominant follicle. This triggers ovulation 12 hours later by a multicomponent mechanism:

1. LH neutralizes the action of oocyte maturation inhibitor, allowing completion of meiosis.
2. Stimulation of progesterone synthesis by LH enhances proteolytic enzyme activity, which loosens the wall and increases distensibility of the follicle.
3. Local synthesis of prostaglandins, some of which are required for follicular rupture, greatly increases.
4. FSH stimulates production of proteolytic enzymes, which catalyze final breakdown of the follicular wall.

Immediately after the gonadotropin surge, LH and

FSH receptors are temporarily down-regulated. This desensitizes the follicular cells to the gonadotropins. The resultant rapid fall in estradiol production also contributes to loss of integrity in the follicle.

Corpus Luteum Function

The organization of the remaining follicle into the corpus luteum and the growth and secretory pattern of the corpus luteum are under hormonal control. The ovulatory LH surge stimulates the luteinization of the granulosa cells. LH subsequently maintains a very high rate of progesterone production by the corpus luteum and a lower rate of estradiol production. Exposure to proper amounts of FSH in the preceding follicular phase ensures the reappearance of sufficient receptors for LH action. The vascular ingrowth into the corpus luteum is also important to deliver the LH and cholesterol necessary to sustain progesterone secretion. If the declining LH levels of the late luteal phase are not replaced by the equivalent placental hormone, human chorionic gonadotropin, the corpus luteum regresses and its secretion of progesterone and estradiol ceases completely by 14 days.

The broad luteal peaks of progesterone and estradiol concentrations, reinforced by inhibin, exert negative feedback on pituitary FSH and LH secretion. The consequent low level of gonadotropins withdraws support from any postovulatory follicles that had entered second-stage development and leads to their atresia. Without conception, the corpus luteum begins to regress after the eighth postovulatory day. *Luteolysis* is mediated by local prostaglandins. By the twelfth postovulatory day, corpus luteum secretion has fallen low enough to release the pituitary gland from feedback inhibition and allow the FSH rise of the next cycle to begin.

Origin of the Menstrual Cycle

Substantial evidence supports the thesis that, in humans, the monthly cycle of the LH/FSH surge and consequent ovulation is an **inherent ovarian rhythm** rather than the result of a primary central nervous system generator. No cycle of LH/FSH release is observed in the absence of functional ovaries. The gonadotropin surge does not occur until the dominant follicle has reached the preovulatory stage of development, irrespective of the number of days required for this to occur. Estradiol itself, administered in a proper fashion, can induce an LH surge.

Finally, if the pituitary gland is severed from the hypothalamus and GnRH pulses of appropriate frequency and amplitude are provided externally to the pituitary gland in a **fixed** pattern, a preovulatory surge of LH and ovulation occur without altering the profile of the GnRH input.

However, the ovarian signals to induce ovulation can by overridden by other influences. Loss of cyclic gonadotropin secretion can occur with caloric deprivation, habitual strenuous exercise, stress, and emotional disturbance. Such inhibitory influences on the hypothalamus may be mediated by endorphins, dopamine, or corticotropin-releasing hormone (CRH), and in some instances by changing the levels of cortisol, androgens, or thyroid hormone.

HORMONAL REGULATION OF REPRODUCTIVE TRACT FUNCTION

The cyclic changes in ovarian estradiol and progesterone secretion affect other reproductive tissues that facilitate conception.

Fallopian Tubes

Fertilization normally occurs in the fallopian tubes. They emerge from the uterus, and each tube ends in fingerlike projections called *fimbriae,* which lie close to the adjacent ovary. The tubes consist of a muscular layer in an epithelial lining that contains secretory and ciliated cells. During the follicular phase of the cycle, estradiol increases the number of cilia and their rate of beating and the number of active secretory cells. At ovulation the fimbriae undulate so as to draw the shed ovum into the tube, and tubal contractions move the ovum toward incoming sperm. During the luteal phase, progesterone maximizes this ciliary beat, facilitating movement of any fertilized ovum toward the uterus. Estradiol and progesterone also regulate tubal secretion of fluids, ions, and substrates supportive to the ovum, to sperm, or to a zygote.

Uterus

The function of the uterus is to house and nurture the developing conceptus and ultimately to evacuate the mature fetus. This muscular organ encloses a cavity that is lined with a mucous membrane, called the *endometrium.* At the start of each menstrual cycle,

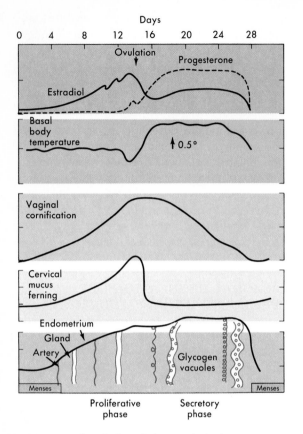

Days

FIGURE 45-4   Correlation of changes in body temperature, vaginal cytology, and endometrial structure and function with the profiles of plasma estradiol and progesterone concentrations. (Redrawn by permission from Odell WD: The reproductive system in women. In DeGroot LJ et al, editors: Endocrinology, vol 3, New York, 1989, Grune & Stratton, Inc.)

the endometrium is thin and its glands are sparse and straight with a narrow lumen (Figure 45-4); it exhibits few mitoses and is incapable of receiving a conceptus. After menstruation has ceased, the rise in plasma estradiol concentration during the follicular phase increases endometrial thickness three- to fivefold. Mitoses appear in the glands and stroma, the glands become tortuous, and the spiral arteries that supply the endometrium elongate. This is the characteristic appearance of the *proliferative phase* of the endometrium. Accompanying these specific tissue changes are increases in glucose, lipid, and amino acid metabolism, and in RNA/DNA and protein synthesis. Estradiol also changes the mucus elaborated by the cervix (the opening to the uterus) from a scant, very viscous material to a copious, more watery but more elastic substance. This mucus can be stretched into a long, fine thread, and it produces a characteristic fernlike

pattern when dried. Such cervical mucus creates channels that facilitate entrance of the sperm into the uterine cavity.

Shortly after ovulation, the rise in plasma progesterone concentration greatly alters the endometrium and produces the characteristic appearance of the *secretory phase* (Figure 45-4). The rapid growth and mitotic activity of the endometrium is inhibited. The glands becomes much more tortuous and accumulate glycogen. As the luteal phase of the cycle progresses, the glycogen vacuoles move from the base toward the lumen, and the glands greatly increase their secretion. The stroma of the endometrium becomes edematous; the spiral arteries elongate further and coil. These changes enable the endometrium to accept, implant, and nourish a conceptus. At the same time progesterone decreases the quantity of cervical mucus and returns it to its thick, nonelastic, and nonferning state.

If conception does not occur, the abrupt loss of progesterone and estradiol causes spasmic contractions of the spiral arteries, mediated by increased prostaglandins. The resultant loss of blood supply produces tissue death, and the superficial endometrial cells are shed along with clotted blood. This comprises the menstrual flow.

Vagina

The vaginal canal is lined with a stratified squamous epithelium that is highly sensitive to estradiol. In its absence, only a layer of basal cells is present. In the early follicular phase the epithelium is thin, and the cells have vesicular nuclei. As the ovulatory phase approaches, more layers of epithelium are added, and the maturing vaginal cells accumulate glycogen. They become large and cornified, and their nuclei strink or disappear. The percentage of such cells is a quantitative index of estrogenic activity (Figure 45-4). In the luteal phase progesterone reduces the percentage of cornified cells. Vaginal secretions are also increased by estradiol, and they also enhance the prospect for fertilization.

Sexual Functioning

Several processes combine to accomplish acceptance and inward transmission of sperm. In some women the desire for sexual activity is increased just before ovulation by the midcycle rise in plasma testosterone concentration (see Figure 45-2). During sexual intercourse, vascular erectile tissue beneath the clitoris is activated by parasympathetic impulses. This causes the vagina to be tightened around the penis. Simultaneously the glands beneath the labia and in the vaginal entrance secrete copious amounts of mucus. The secretions lubricate the vagina and help it produce a massaging effect on the penis. These glands are maintained by estradiol action.

Orgasm results from spinal cord reflexes that are similar to those involved in male ejaculation. Orgasm consists of involuntary contractions of the skeletal muscle of the perineum; of the musculature of the vagina, uterus, and fallopian tubes; and of the rectal sphincter. After orgasm the cervix remains widely patent for up to 30 minutes. This permits rapid entrance of a first wave of sperm into the uterus. However, these sperm are incapable of fertilizing an ovum.

Many spermatozoa are trapped and within a few hours are destroyed in the vagina. The remainder reach the cervix, where they dwell in storage sites formed by the estrogen-stimulated convoluted mucosa and its mucus. From this reservoir, spermatozoa migrate into the uterine cavity and fallopian tubes over 24 to 48 hours. Of these, as few as 50 to 100 spermatozoa eventually reach an ovum, but they are sufficient for fertilization.

Breasts

The mammary glands consist of lobular ducts lined by an epithelium capable of secreting milk. These ducts empty into larger conduits that converge at the nipple. The glandular structures are embedded in supporting adipose and connective tissue. Before puberty, the breasts grow only in proportion to the rest of the body. The development of adult breasts depends on estradiol, but progesterone, growth hormone, cortisol, and prolactin have synergistic effects. After puberty, estradiol stimulates growth of the lobular ducts in the area around the nipple. Estradiol also selectively increases the adipose tissue, giving the breast its distinctive female shape. Progesterone stimulates outpouching of the lobular ducts to form numerous alveoli capable of milk secretion.

Ovarian Steroid Effects on Other Tissue

During puberty, estradiol is to the female what testosterone is to the male. Estradiol causes almost all the somatic changes that result in the female adult appearance. In addition to stimulating growth of the internal reproductive organs and breasts, estrogens cause pubertal enlargement of the labia majora and labia minora. Linear growth is accelerated by estradiol; however, because the epiphyseal growth centers are more sensitive to estradiol than to testosterone, they close sooner. For this reason the average height of women is less than that of men. The hips enlarge and the pelvic inlet widens, facilitating future pregnancy. The predominance of estradiol over testosterone as a gonadal hormone is responsible for the total body adipose mass of women being twice as large as that of men, whereas muscle and bone mass are only two-thirds that of men.

The adult skeleton, the kidney, and the liver are also target tissues of estrogens. Estradiol inhibits bone resorption; loss of this important action can contribute to a declining bone mass and an increased fracture rate. Estradiol stimulates reabsorption of sodium from the renal tubules, and this may contribute to cyclic fluid retention. Estradiol increases hepatic synthesis of binding proteins for thyroid and steroid hormones, of the renin substrate angiotensinogen, and of very-low-density lipoproteins.

Progesterone produces the 0.5° C rise in body temperature that occurs shortly after ovulation (Figure 45-4). Central nervous system actions of progesterone include an increase in appetite, a decrease in wakefulness, and a heightened sensitivity of the respiratory center to carbon dioxide.

MECHANISMS OF ACTION OF OVARIAN STEROIDS

Estrogens and progesterone enter cells freely and bind to cytoplasmic nuclear receptors, whose structures have been established and whose genes have been cloned. The sex steroid–receptor complex undergoes an activation step that enhances its binding to specific DNA molecules. The estrogen receptor can also be phosphorylated by a protein kinase dependent on cyclic adenosine monophosphate (cAMP), thereby increasing its binding activity. By stimulating transcription of the respective genes, estradiol increases the synthesis of ovulbumin and ovomucoid. Progesterone increases the synthesis of uteroglobulin and avidin. These and other induced proteins have reproductive functions.

Because spare receptors generally are not present, the responsiveness of various tissues to ovarian steroids is proportional to receptor concentration. Estradiol and progesterone can fortify or inhibit each other's actions through receptor recruitment. Estradiol increases its own receptor and that of progesterone in the uterus during the latter part of the proliferative phase. Conversely, progesterone decreases estradiol receptors and therefore estrogen action of the endometrium during the secretory phase.

METABOLISM OF OVARIAN STEROIDS

Estradiol and estrone bind to sex steroid–binding globulin, but their affinities are much lower than that of testosterone. They also circulate largely bound to albumin, and this fraction, along with free estradiol, is biologically active. In women who menstruate, most of the circulating estrogen is estradiol from the ovaries. In postmenopausal women the dominant estrogen is estrone, which is produced in peripheral tissues from adrenal and thecal cell androgen precursors.

Estrogens are excreted in the urine as sulfate and glucuronate conjugates. An additional pathway of estradiol metabolism involves hydroxylation of the 2-position, resulting in the so-called catechol estrogens. These compounds resemble the neurotransmitters norepinephrine and dopamine. Because catechol estrogens are formed in the hypothalamus, they might mediate estradiol effects on GnRH release.

Progesterone can bind to cortisol-binding globulin but circulates largely bound to albumin. It is reduced to pregnanediol and excreted in the urine. During the follicular phase of the cycle, half the circulating progesterone is secreted by the ovary and half by the adrenal glands. During the luteal phase almost all the progesterone originates from the corpus luteum.

FEMALE PUBERTY

The general process of initiation of puberty is described in Chapter 43. Reproductive function begins after gonadotropin secretion increases from the low levels of childhood (Figure 45-5). Females exhibit an earlier rise in FSH than in LH more clearly than do males (compare Figure 44-3 with Figure 45-5). Budding of the breasts, the first physical sign of puberty, coincides with the first detectable increase in plasma estradiol concentration. The onset of men-

ses occurs approximately two years later, at 11 to 15 years of age, after LH levels have risen more sharply. This appears to depend on achieving either a critical body weight or a critical ratio of adipose mass to lean body mass.

The positive feedback effect of estradiol on gonadotropin secretion is the last step in the maturation of the hypothalamic-pituitary ovarian axis; thus ovulation usually does not occur in the first few menstrual cycles. These are irregular in length because the bleeding is induced by withdrawal of estrogen secretion from graafian follicles undergoing atresia.

The growth spurt and the peak velocity of growth are characteristically attained earlier in girls than in boys. Height increase usually stops to 1 to 2 years after the onset of menses. The development of pubic hair precedes menses and correlates best with rising levels of adrenal dehydroepiandrosterone sulfate (DHEA-S). The onset of all stages of female puberty vary widely; the timing is influenced by race, individual heredity, adiposity, and climate.

MENOPAUSE

The reproductive capacity of women wanes in the fifth decade of life, and menses terminate at an average age of 50. For several years before, the frequency of ovulation decreases. The menses occur at variable intervals, and the decreased menstrual flow is caused by irregular peaks of estradiol secretion and inadequate secretion of progesterone in the luteal phase. With the disappearance of almost all follicles, ovarian secretion of estrogens eventually ceases.

As menopause approaches, follicular sensitivity to gonadotropin stimulation diminishes, and plasma FSH and LH levels gradually increase in compensation. Once menopause occurs, loss of negative feedback from estradiol and inhibin increase plasma gonadotropins to levels 4 to 10 times those that are characteristic of the follicular phase, and FSH exceeds LH (see Figure 43-3). Although the cycle of gonadotropin secretion is lost, pulsatility persists.

The menopausal decrease in estrogen causes thinning of the vaginal epithelium and loss of its secretions. A decrease in breast mass, an accelerated loss of bone, and thinning of the skin also occur. Vascular flushing, emotional lability, and a sharp increase in the incidence of coronary artery disease are related to estrogen deficiency as well.

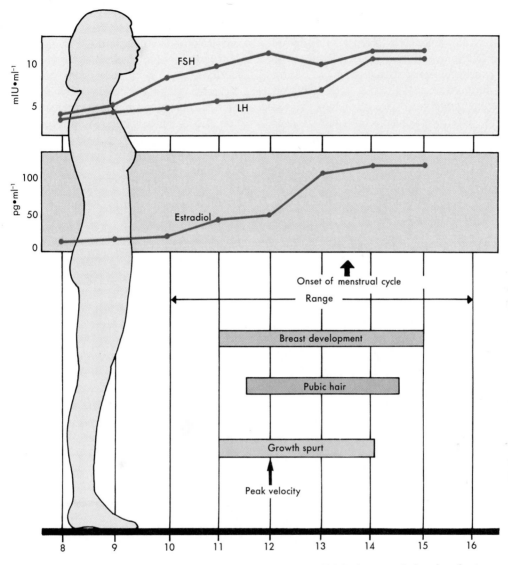

FIGURE 45-5   Average chronologic sequence of hormonal and biologic events in female puberty. (Redrawn from Lee PA et al: Puberty in girls, J Clin Endocrinol Metab 43:775, 1976, © by The Endocrine Society; and Marshall WA and Tanner JM: Arch Dis Child 45:13, 1970.)

PREGNANCY

Fertilization

After the ovum enters the widened proximal end of the fallopian tube (ampulla), it is transported down to the junction with the isthmus. There it must encounter sperm within 12 to 24 hours for fertilization to occur. The sperm in turn must reach the ovum within 48 hours after entering the vagina. Contact between the sperm and ovum is facilitated by a mixing of motion of the fallopian tube. Access of the sperm to the ovum begins with dispersal of the granulosa cells of the cumulus oophorus (see Figure 45-1). Dispersal is achieved through the action of hyaluronidase and a corona-dispersing enyme, both of which are contained in the acrosomal cap of the sperm (see Figure 44-2). The underlying zona pellucida of the ovum (see Figure 45-1) contains species-specific receptors for sperm. The single fertilizing sperm penetrates this barrier by releasing *acrosin,* a proteolytic enzyme. Penetration then releases materials contained in granules within the ovum, which block entrance of other sperm. This prevents *polyploidy,* which is the production of an individual with more than two sets of homologous chromosomes. The polar body resulting from the second reduction division is then ejected from the ovum, leaving a female pronucleus with 23 chromosomes. After fusion of their respective membranes, the DNA of the sperm head is engulfed by the ovum and forms the male pronucleus with 23 chromosomes. The two pronuclei then generate a spindle on which the chromosomes are arranged, and a zygote with 46 chromosomes is created.

Implantation

The zygote develops into a blastocyst, which traverses the fallopian tube in about 3 days. Within another 2 or 3 days, implantation in the uterus begins. The requisite dissolution of the zona pellucida is initiated by alternate contraction and expansion of the blastocyst, as well as by the action of lytic substances in the uterine secretions. These and other maternal factors necessary for implantation depend on adequate progesterone levels. From the initial solid mass of cells, a layer of *trophoblasts* separates. Microvilli of these cells interdigitate with those of endometrial cells, and junctional complexes form between the cell membranes. Once firmly attached, trophoblast cells burrow between and beneath endometrial cells, lyse the intercellular matrix with a variety of enzymes, and phagocytize and digest dead endometrial cells.

The depth of penetration by the trophoblasts is limited by changes in the endometrium. Late in the luteal phase, uterine stromal cells enlarge and accumulate glycogen and lipid. Now called *decidual* cells, they disappear unless pregnancy supervenes and the corpus luteum is maintained. In this situation, however, continuing progesterone and estrogen stimulation rapidly change the entire stroma into a sheet of decidual cells. This *decidua* functions initially as a source of nutrients for the embryo, until vascular connections between the fetus and the mother have been established. Thereafter the decidua may provide a mechanical and an immunologic barrier to further invasion of the uterine wall. The decidua also functions as an endocrine organ, secreting prolactin, relaxin and prostaglandins.

Functions of the Placenta

Pregnancy is marked by the development of a unique organ with a limited life span, the *placenta.* This organ serves as the fetal gut and nutrient supply, as the fetal lung in exchanging oxygen and carbon dioxide, and as the fetal kidney in regulating fluid volumes and disposing of waste metabolites. In addition, the placenta is an extraordinarily versatile endocrine gland, capable of synthesizing and secreting numerous protein and steroid hormones that affect maternal and fetal metabolism. These hormones can be found in fetal plasma and amniotic fluid and exhibit characteristic concentration profiles in maternal plasma (Figure 45-6).

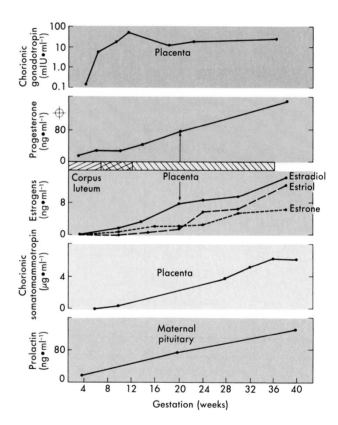

FIGURE 45-6   Profile of maternal plasma hormone changes in human pregnancy. Note logarithmic scale for human chorionic gonadotropin (HCG). Between 6 and 12 weeks, the source of estrogens and progesterone shifts from the corpus luteum to the placenta. (Redrawn from Goldstein DP et al: Am J Obstet Gynecol 102:110, 1968; Rigg LA et al: Am J Obstet Gynecol 129:454, 1977; Selenkow HA et al: Measurements and pathophysiologic significance of human placental lactogen. In Pecile A and Frinzi C: The foetoplacental unit, Amsterdam, 1969, Excepta Medica; and Tulchinski D et al: Am J Obstet Gynecol 112:1095, 1972.)

Human Chorionic Gonadotropin This hormone (HCG) is the first key hormone of pregnancy. Secreted by the placental *synctiotrophoblast* cells, HCG can be detected in maternal plasma and urine within 9 days of conception. HCG is a glycoprotein with two sub-units. The α-subunit is identical to that of thyroid-stimulating hormone (TSH), FSH, and LH. The β-sub-unit is 80% homologous with that of LH, and the two hormones have indistinguishable biologic actions. The secretion of HCG may be stimulated by GnRH produced in adjacent placental cells. Plasma HCG concentration increases at an exponential rate, reaches a peak at 9 to 12 weeks of gestation, and then declines to a stable plateau for the remainder of pregnancy (Figure 45-6).

HCG maintains the function of the corpus luteum that would otherwise degenerate in the absence of pregnancy. It stimulates the corpus luteum to secrete progesterone and estradiol by mechanisms identical to those of LH. Later, when the placenta itself synthe-sizes these steroids in adequate amounts, HCG secretion declines and the corpus luteum regresses. HCG also has other actions. It stimulates essential DHEA-S production by the fetal zone of the adrenal gland (see following discussion). In males HCG stim-ulates the early secretion of testosterone by the Ley-dig cells, which is critical to masculine genital tract differentiation.

Progesterone Progesterone is essential for suc-cessful implantation and initial sustenance of the fetus. It stimulates the endometrial glands to secrete nutrients on which the early zygote depends. There-after, progesterone maintains the decidual lining of the uterus, where it induces prolactin synthesis. The latter may inhibit maternal immune responses to fetal antigens and thereby prevent fetal rejection. Proges-terone transferred to the fetus is the substrate for syn-thesis of cortisol and aldosterone by the fetal adrenal cortex. The latter cannot itself synthesize progester-one because it lacks 3 β-OL dehydrogenase $\Delta^{4,5}$ isomerase activity.

Progesterone quiets uterine muscle activity and prevents premature expulsion of the fetus. Also, it stimulates mammary gland development and greatly enhances the eventual capacity to secrete milk. Final-ly, progesterone increases maternal ventilation needed for removal of the increased load of carbon dioxide created by pregnancy.

The placenta begins to synthesize progesterone at about 6 weeks, and by 12 weeks it is producing

enough to replace the corpus luteum source. During the transition period the otherwise progressive rise in plasma progesterone concentration temporarily reaches a plateau (Figure 45-6). Cholesterol extracted from maternal plasma serves as the major precursor for placental progesterone. The synthetic pathway is identical to that of the adrenal gland and ovary. By term, progesterone production reaches a level that is tenfold greater than peak production by the corpus luteum.

Estrogens Progressive increases in estradiol, estrone, and *estriol* occur throughout pregnancy. Estrogens stimulate continuous growth of the uterine muscles necessary for labor. They also cause relax-ation and softening of the pelvic ligaments and junc-tion of the pelvic bones; this allows better accommo-dation of the expanding uterus. They also augment growth of the ductal system of the breast to prepare for lactation.

Estrogens are initially produced by the corpus lute-um. The placenta subsequently assumes this role, but because it lacks 17-hydroxylase/17, 20-desmolase activity, it requires an androgen precursor from the maternal and fetal compartments to complete estro-gen synthesis. This exemplifies coordinated mater-nal-placental-fetal function. Thus the placenta ex-tracts DHEA-S derived from the maternal and fetal adrenal glands, removes the sulfate, and aromatizes the androgen to estradiol and estrone (Figure 45-7). In the case of estriol, the fetal liver must also 16-hydrox-ylate DHEA-S before the placenta acts on the precur-sor.

Human Chorionic Somatomammotropin A protein hormone, unique to pregnancy, is human chorionic somatomammotropin (HCS), also called *human placental lactogen*. Its structure is similar to growth hormone and prolactin. Synthesized by pla-cental trophoblasts within 4 weeks, maternal plasma HCS concentration rises steadily throughout preg-nancy. The peak HCS production rate of 1 to 2 g/day far exceeds that of any other human protein hor-mone.

HCS stimulates lipolysis and, as with growth hor-mone, is an insulin antagonist. Thus HCS raises maternal free fatty acid and glucose levels. As dis-cussed later, a major function of HCS is that of direct-ing maternal metabolism to maintain a continuous flow of substrates, especially glucose, to the fetus.

Other Placental Hormones The placenta produces several hypothalamic or pituitary-like pep-

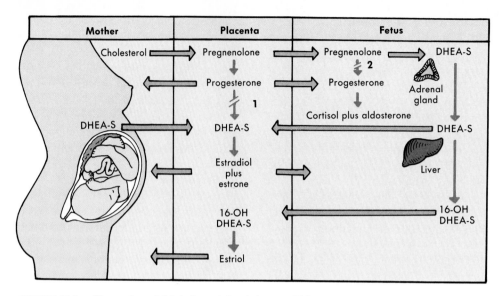

FIGURE 45-7   The maternal-fetal-placental unit in steroid hormone synthesis. Progesterone is synthesized in the placenta from maternal cholesterol. In turn this progesterone acts on the mother and also serves as the precursor to fetal cortisol and aldosterone synthesis. Estradiol and estrone are synthesized in the placenta from maternal and fetal DHEA-S and estriol from fetal 16-OH-DHEAS. *DHEA-S*, Dehydroepiandrosterone sulfate; *16-OH-DHEA-S*, 16α-hydroxydehydroepiandrosterone sulfate; *1*, 17-hydroxylase/17,20-desmolase; *2*, 3β-OL-dehydrogenase-$\Delta^{4,5}$ isomerase.

tides. These include GnRH, TRH, CRH, and somatostatin, as well as adrenocorticotropic hormone (ACTH), β-endorphin, β-lipotropin, TSH, and growth hormone. The placental releasing/inhibiting hormones may function in paracrine fashion to modulate secretion of placental pituitary-like hormones. Placental ACTH and TSH probably help to augment maternal adrenal and thyroid gland activity. The placenta also synthesizes 1,25-$(OH)_2$-vitamin D, which helps regulate calcium homeostasis and skeletal formation in the fetus.

Hormones of Maternal Origin

Prolactin Prolactin secretion from the maternal pituitary gland increases greatly during pregnancy (see Figure 45-6), stimulated by the high estrogen levels. Prolactin is essential for expression of the effects of estrogens and progesterone on the breasts, and it specifically stimulates the lactogenic apparatus (see Chapter 39). During pregnancy, however, lactation itself is inhibited by the great excess of estrogen and progesterone. After delivery of the fetus, true milk synthesis is initiated by the precipitous drop in steroid hormone levels, and it is maintained in a nursing

mother by prolactin. Insulin and cortisol also facilitate milk synthesis. Although basal prolactin concentrations gradually decline by 8 weeks after delivery, they are repetitively elevated during each period of suckling. This helps to sustain milk secretion.

Prolactin also helps suppress reproductive function in the nursing mother. During the first 7 to 10 days after delivery, plasma FSH and LH levels remain low. FSH levels then rise, but LH levels do not; this pattern simulates the situation in early puberty. In the nursing mother this persists because of inhibitory effects of prolactin on GnRH secretion. A decrease in circulating prolactin that follows either cessation of nursing or administration of a dopaminergic agonist (see Figure 39-12) triggers LH release and initiates menstrual cycling.

Relaxin Relaxin is a peptide hormone with structural similarity to proinsulin. It is secreted by the corpus luteum and the decida under HCG stimulation. Maternal plasma relaxin levels rise early in pregnancy, peak in the first trimester, and then decline somewhat. This hormone relaxes the mother's pelvic outlet, inhibits uterine muscle contractions, and softens the cervix. Thus relaxin acts to maintain uterine quiescence and prevent early abortion but facilitates easier passage of the fetus into the birth canal once labor has begun.

Other Hormonal Changes

The pregnant state induces important other changes in maternal endocrine function. Insulin secretion increases after the third month in response to glucose challenge or meals. It peaks during the last trimester and acts to compensate for the insulin resistance caused by HCS.

Aldosterone secretion increases throughout pregnancy because of estrogen-induced augmentation of renin and angiotensinogen levels. This induces a positive sodium balance that is needed to support a high maternal plasma volume and build the extracellular fluid of the fetus.

Total plasma thyroxine and cortisol levels are elevated because of estrogen-induced increases in their respective binding globulins. The level of plasma free cortisol also rises modestly and may contribute to maternal adipose tissue gain and to mammary gland development.

Parathyroid hormone (PTH) secretion also increases. PTH augments maternal plasma levels of 1, 25-$(OH)_2$ vitamin D, which in turn increases dietary calcium absorption. This enhances the supply of calcium for the growing fetal skeleton.

MATERNAL-FETAL METABOLISM

During pregnancy, the average gain in maternal weight is 11 kg. Approximately half of this can be attributed to changes in maternal tissues and half to the fetus and placenta. The mother must ingest approximately 300 extra kcal and 30 extra g of protein daily to support fetal development, enlarge maternal energy stores, and sustain growth of certain tissues.

For the first 4½ months of pregnancy, the mother is in an anabolic state, and the conceptus represents an insignificant nutritional drain. This phase is characterized by normal or even increased maternal sensitivity to insulin. Maternal plasma levels of glucose, free fatty acids, glycerol, and amino acids are normal or slightly decreased. Dietary carbohydrate and protein loads are rapidly used. Maternal lipogenesis is favored, glycogen stores are expanded, and protein synthesis is enhanced. This supports early growth of

the breasts and uterus and prepares the mother to withstand the later metabolic demands of the enlarging fetus.

During the second half of pregnancy the mother shifts into a catabolic state aptly described as "accelerated starvation." Insulin sensitivity is replaced by insulin resistance. This resistance causes elevation of postprandial plasma levels of glucose and amino acids as the uptake of dietary carbohydrate, protein, and fat by maternal tissues is reduced. Consequently, the diffusion of glucose and the facilitated transport of amino acids across the placenta into the fetus are accelerated. During maternal fasting intervals, plasma glucose and amino acid levels fall more rapidly than in non-pregnant women, because the fetus continues to siphon off these substances. Maternal lipolysis is excessively stimulated, ensuring alternate oxidative fuels for the mother and even for the fetus, to whom ketoacids and free fatty acids can be transferred across the placenta. HCS is the key hormone responsible for maternal insulin resistance and for lipid mobilization during fasting in this later stage of pregnancy. Elevated estrogen, progesterone, and cortisol levels also contribute to antagonizing insulin action.

PARTURITION

The exact mechanisms for normal termination of human pregnancy are still unclear. Studies in numerous animal models suggest possible roles for cortisol, estrogens, progesterone, relaxin, oxytocin, prostaglandins, and catecholamines in the initiation, maintenance, and termination of labor. Because much species variation exists, it is uncertain what can be applied to humans. Figure 45-8 shows some current notions of the endocrine regulation of parturition.

Once the contents of the uterus reach some critical size, stretching of the uterine muscle fibers increases their contractility. Thus uncoordinated uterine contractions begin at least 1 month before the end of gestation. However, some signal from the fetus indicating readiness may initiate the process of labor. In sheep a fetal adrenal product, probably cortisol, has been strongly implicated. Fetal cortisol production in humans does rise sharply during the last few weeks of gestation. However, the evidence for a rapid surge of cortisol immediately preceding the onset of labor in humans is contradictory. Nonetheless, the late gesta-

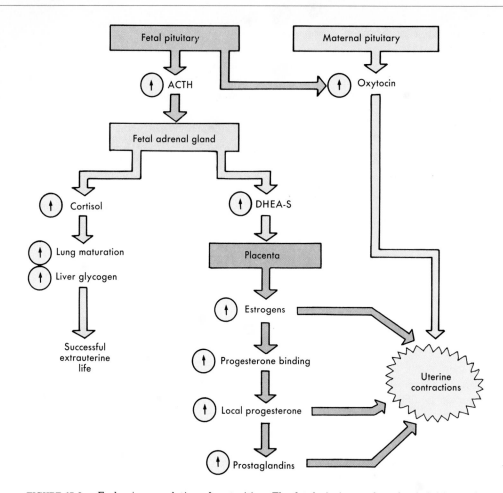

FIGURE 45-8   Endocrine regulation of parturition. The fetal pituitary-adrenal axis initiates signals that decrease the ratio of effective progesterone to estrogen in the myometrium. This leads to uterine contractions, which are mediated by prostaglandins. Oxytocin may contribute to labor but is not essential. However, oxytocin sustains uterine contractions after expulsion of the fetus so as to minimize maternal loss of blood. Cortisol prepares the fetus to maintain its own supply of oxygen and glucose after birth. *ACTH,* Adrenocorticotropic hormone; *DHEA-S,* dehydroepiandrosterone sulfate.

tional increase in cortisol secretion is important for preparing the human fetus for the abrupt transition to extrauterine life by stimulating lung maturation and by increasing stores of glycogen in the liver.

The appropriate intrauterine ratio of progesterone to estrogen content helps suppress contractions throughout pregnancy by keeping uterine concentrations of prostaglandins low. In some species, but not in humans, a sharp drop in maternal plasma progesterone levels precedes parturition. In humans, however, rising estrogen levels near term may increase placental protein binding of progesterone, and this may reduce the **effective** progesterone/estrogen ratio in

the myometrium. Progesterone is thought to inhibit uterine contractions by preventing prostaglandin synthesis. A decrease in this progesterone effect and an increase in the estrogen effect would augment the synthesis of prostaglandins. Prostaglandins increase free intracellular calcium concentrations in the myometrium and trigger uterine contractions. A key role for prostaglandins in termination of pregnancy is supported by the high levels found in maternal plasma and in the amniotic fluid during labor. Also, these compounds are effective for inducing abortion.

Once labor is initiated, both maternal and fetal oxytocin may help sustain and increase the strength of uterine contractions, although labor can proceed in the absence of oxytocin. After delivery, myometrial contractions, stimulated by maternal oxytocin, act to constrict the uterine vessels and prevent excessive bleeding.   •

BIBLIOGRAPHY
Journal Articles

Belchetz PE et al: Hypophysial responses to continuous and intermittent delivery of hypothalamic gonadrotropin-releasing hormone, Science 202:631, 1978.

Bryant-Greenwood GD: Relaxin as a new hormone, Endocr Rev 3:62, 1982.

Goebelsmann U: Protein and steroid hormones in pregnancy, J Reprod Med 23:166, 1979.

Hoff JD et al: Hormonal dynamics at midcycle: a reevaluation, J Clin Endocrinol Metab 57:797, 1983.

Keyes PL and Wiltbank MD: Endocrine regulation of the corpus luteum, Annu Rev Physiol 50:465, 1988.

Liu JH and Yen SS: Induction of midcycle gonadotropin surge by ovarian steroids in women: a critical evaluation, J Clin Endocrinol Metab 57:797, 1983.

McCarty KS Jr et al: Oestrogen and progesterone receptors: physiological and pathological considerations, J Clin Endocrinol Metab 12:133, 1983.

McLachlan RI et al: Circulating immunoreactive inhibin levels during the normal human menstrual cycle, J Clin Endocrinol Metab 65:954, 1987.

McNatty KP et al: The production of progesterone, androgens, and estrogens by granulosa cells, thecal tissue, and stromal tissue from human ovaries in vitro, J Clin Endocrinol Metab 49:687, 1979.

McNatty KP et al: The microenvironment of the human antral follicle: interrelationships among the steroid levels in antral fluid, the population of the granulosa cells, and the status of the oocyte in vivo and in vitro, J Clin Endocrinol Metab 49:851, 1979.

Nathanielsz PW: Endocrine mechanisms of parturition, Annu Rev Physiol 40:411, 1978.

Pohl CR and Knobil E: The role of the central nervous system in the control of ovarian function in higher primates, Annu Rev Physiol 44:583, 1982.

Richards JS and Hedin L: Molecular aspects of hormone action in ovarian follicular development, ovulation, and luteinization, Annu Rev Physiol 50:441, 1988.

Richelson LS et al: Relative contributions of aging and estrogen deficiency to postmenopausal bone loss, N Engl J Med 311:1273, 1984.

Simpson ER and McDonald PC: Endocrine physiology of the placenta, Annu Rev Physiol 43:163, 1981.

Thoburn GD et al: Endocrine control of parturition, Physiol Rev 59:863, 1979.

Tsang BK et al: Androgen biosynthesis in human ovarian follicles, cellular source, gonadotropic control, and adenosine 3',5'-monophosphate mediation, J Clin Endocrinol Metab 48:153, 1979.

Books and Monographs

Fisher DA: Fetal endocrinology: endocrine disease and pregnancy. In Degroot LJ et al, editors: Endocrinology, vol 3, New York, 1989, Grune & Stratton, Inc.

Kenigsberg D et al: The ovary: development and control of follicular maturation and ovulation. In Degroot LJ et al, editors: Endocrinology, vol 3, New York, 1989, Grune & Stratton, Inc.

Marshall JL and Odell WD: The menstrual cycle—hormonal regulation, mechanisms of anovulation, and responses of the reproductive tract to steroid hormones. In Degroot LJ et al, editors: Endocrinology, vol 3, New York, 1989, Grune & Stratton, Inc.

Ross GT et al: The ovary. In Yen SSC and Jaffe RB: Reproductive endocrinology, Philadelphia, 1986, WB Saunders Co.

Yen SSC: The human menstrual cycle. In Yen SSC and Jaffe RB: Reproductive endocrinology, Philadelphia, 1986, WB Saunders Co.

EXERCISE PHYSIOLOGY

LORING B. ROWELL

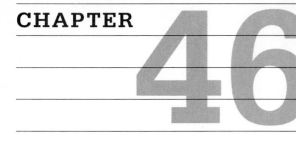

PART IX

CHAPTER 46

Integration of Body Systems in Exercise

The greatest demands ever placed on cardiovascular, respiratory, and metabolic control systems occur during exercise. The transport of oxygen to the tissues can increase by as much as 14-fold in normally active young people and can exceed 25-fold in athletes engaged in endurance competition. Although this chapter emphasizes the cardiovascular, respiratory, and metabolic adjustments to dynamic, upright exercise, the functions of many other systems are discussed as they affect this complex, highly integrated adjustment. This chapter also explores the steps and possible limiting factors in the transport of oxygen from the air to the lungs, blood, capillaries, and finally the tissues and their mitochondria.

Figure 46-1 illustrates the potential barriers that can limit the transport of oxygen, any one of which can be greatly enhanced by disease. An understanding of the potential weak links in this chain of oxygen transport is fundamental to both physiology and medicine. To examine the functional limits of the cardiovascular and respiratory systems, one must also consider the regulatory limitations imposed by metabolic, neural, and humoral systems.

Exercise can maximally activate local and reflex vasodilator mechanisms in the active skeletal muscles. This vasodilation must be opposed by powerful vasoconstrictor mechanisms in other tissues to preserve blood pressure or to maintain perfusion of one region or organ over another. The potential mismatch between the capacity of the heart to pump blood and the ability of certain organs to vasodilate and receive the blood must be prevented by precise autonomic nervous regulation. The major homeostatic reflexes are activated as various organs compete for a **limited** cardiac output. In the end the heart and brain have priority and their blood flows must be preserved; that is, arterial blood pressure must also be maintained. Exercise engages most systems of the body to some degree, and all must be regulated.

SCALING RESPONSES

Maximal Oxygen Uptake

The first consideration is how the various systems respond to exercise. A way to compare the responses of different individuals is also necessary, as well as an understanding of why the responses differ. The simplest starting point is the *maximal oxygen consumption*($\dot{V}O_{2max}$) because it provides an objective and reproducible measure of a functional upper limit; that is, it is a **full-scale** measurement.

The $\dot{V}O_{2max}$ is characterized by an O_2 uptake that cannot be exceeded by a greater increase in work intensity or the mass of active muscle (Figure 46-2).

The Fick Principle and Physiologic Meaning of Maximal Oxygen Uptake

The maximal rate of oxygen utilization ($\dot{V}O_{2max}$) can be derived from the Fick equation (see Chapter 17). $\dot{V}O_{2max}$ equals the product of maximal heart rate (HR_{max}) times maximal stroke volume (SV_{max}) times maximal arteriovenous (AV) oxygen difference, or $([O_2]_{pv} - [O_2]_{pa})_{max}$:

$$\dot{V}O_{2max} = HR_{max} \times SV_{max} \times ([O_2]_{pv} - [O_2]_{pa})_{max} \quad \textbf{(1)}$$

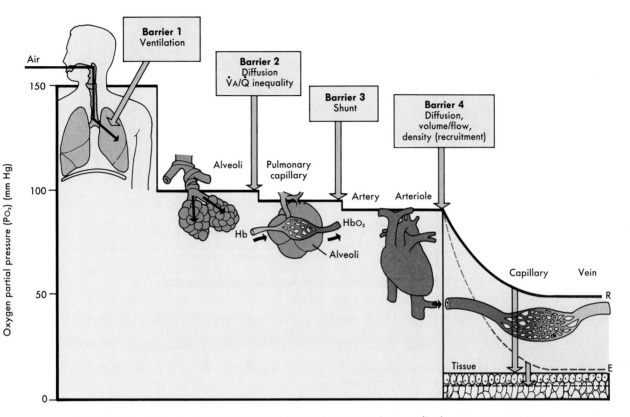

FIGURE 46-1   The cascade of changes in the partial pressure of oxygen (Po_2) as it passes into the airways and then into the alveoli, where it is diluted with CO_2 and water vapor. Oxygen then passes from the alveoli into the blood, but its passage can be hindered by diffusion across the alveolar membrane or by poor matching of ventilation $(\dot{V}a)$ and blood flow, or perfusion $(\dot{Q})$. Small shunts may reduce the concentration of O_2 in pulmonary venous and arterial blood (shunts are normally negligible). Oxygen is provided to the tissues by cardiac pumping. In the tissue capillaries O_2 must diffuse across to the cells and into the mitochondria. Effectiveness of O_2 delivery to tissues depends on the number of open capillaries, capillary blood volume, and blood flow. Note the fall in venous O_2 tension from rest (R) to exercise (E) when more capillaries open with exercise. HbO_2, oxyhemoglobin. (Redrawn from West JB: Respiratory physiology—the essentials, Baltimore, 1974, Williams & Wilkins.)

where pv is pulmonary venous, and pa is pulmonary arterial. Therefore, $\dot{V}o_{2max}$ measures the capacity of the combined cardiovascular and respiratory systems to **deliver** blood and O_2 to the tissues. As shown later, it is **not** a measure of the capacity of the tissues to **utilize** O_2. In subsequent sections of this chapter the control and functional limits of these variables in the Fick equation are discussed.

Figure 46-3 shows the relationships of the variables of the Fick equation and total O_2 consumption up to its maximal value in three groups of individuals: endurance athletes, normally active individuals, and patients with mitral stenosis. These groups were

selected because they exemplify those having high, normal, and low values for $\dot{V}o_{2max}$. The patients had a partly obstructed mitral valve that limited left ventricular filling and stroke volume, but other complications had not yet developed (no heart failure or pulmonary dysfunction). The average values for each determinant of $\dot{V}o_{2max}$ shown in Table 46-1 emphasize an important point: the differences in the $\dot{V}o_{2max}$ of the three groups can be traced mainly to their different stroke volumes inasmuch as maximal heart rates and values for $([O_2]_{pv} - [O_2]_{pa})_{max}$ were similar. Thus the magnitude of stroke volume usually accounts for the differences in maximal cardiac output and thus for

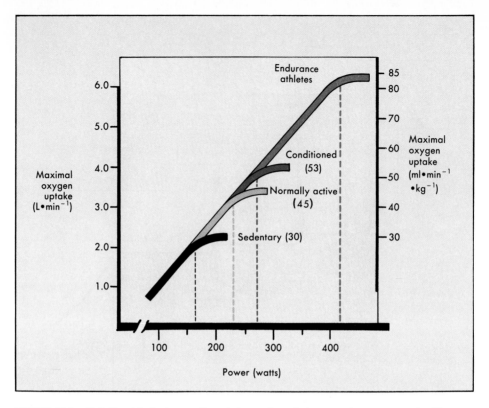

FIGURE 46-2   Relationship between O_2 uptake and work in normal young people. The true maximal oxygen uptake $\dot{V}O_{2max}$ is reached when O_2 uptake no longer rises with increasing workloads that employ 40% to 50% of the total muscle mass. Values for $\dot{V}O_{2max}$ range from 30 ml·min^{-1} per Kg of body weight for sedentary people up to 85 ml·min^{-1} per Kg in elite endurance athletes. Note the small rise from 45 to 53 ml·min^{-1} per Kg in normally active people after conditioning. (Redrawn from Rowell LB: Human circulation: regulation during physical stress, New York, 1986, Oxford University Press.)

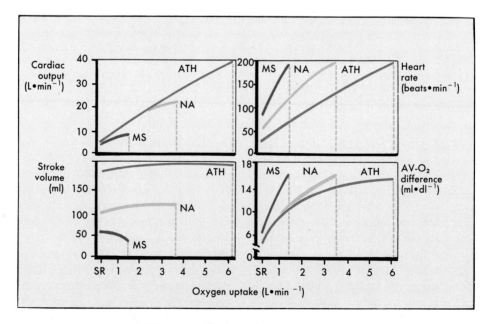

FIGURE 46-3   For legend see opposite page.

Table 46-1 Physiologic Basis for Differences in Maximal Oxygen Uptake ($\dot{V}O_{2max}$) in Three Groups*

	$\dot{V}O_{2\,max}$ (ml·min^{-1})	=	Heart Rate (beats·min^{-1})	×	Stroke Volume (ml)	×	Arteriovenous Oxygen Difference (ml·dl^{-1})
ATH	6250	=	190	×	205	×	16
NA	3500	=	195	×	112	×	16
MS	1400	=	190	×	43	×	17

Adapted from Rowell LB: Human circulation: regulation during physical stress, New York, 1986, Oxford University Press.

$\dot{V}O_{2max}$ in normal individuals. However, anything that reduces heart rate and arterial O_2 content, or raises venous O_2 content (reduced AV-O_2 difference), will reduce the $\dot{V}O_{2max}$ and thus lower the functional capacity of the system. The sensitivity of $\dot{V}O_{2max}$ to small changes in its determinants has made exercise a powerful diagnostic tool in clinical medicine, allowing one to identify a weak link in the chain of O_2 transport.

Relative Oxygen Uptake: A Basis for Comparison

The lack of significant differences in total oxygen extraction among the three groups illustrated in Figure 46-3 suggests that they share a similar distribution of cardiac output among regions extracting little O_2 (e.g., splanchnic region and kidneys) and regions extracting the most O_2 (active muscle). If a substantial fraction of the cardiac output was distributed to splanchnic organs and kidneys, total O_2 extraction or systemic AV-O_2 difference would be reduced. Figure 46-4 shows that blood flow to these major regions decreases in inverse proportion to $\dot{V}O_2$ and is ultimately reduced by 70% to 80% at $\dot{V}O_{2max}$ in each group. Scaling these regional blood flow responses against the percentage of $\dot{V}O_{2max}$ (the "relative" $\dot{V}O_2$) in Figure 46-4 reveals some underlying similarities among the three groups. When comparison is based on **relative** rather than **absolute** demands on the cardiovascular system, their responses become indistinguishable. This suggests that the O_2 delivery system is somehow regulated in relation to its capacity.

The reductions in splanchnic and renal blood flows as well as blood flows to inactive muscle and other inactive regions are caused mainly by sympathetic neural vasoconstriction (see later section). Release of norepinephrine from sympathetic axon terminals and ultimate diffusion into the plasma causes the increase in plasma norepinephrine concentration; the adrenal medulla releases epinephrine but very little norepinephrine in humans. This rise in plasma norepinephrine concentration parallels the rise in heart rate (for heart rates greater than approximately 100 beats·min^{-1}) and is inversely proportional to the decline in splanchnic and renal blood flows (Figure 46-5). In general, plasma norepinephrine concentration in humans provides a rough index of sympathetic nervous activity. It is a rough index because as it is released, norepinephrine is quickly taken up by sympathetic nerve terminals and by the liver and is also broken down by enzymes in the nerves, tissues, and blood.

FIGURE 46-3 Overall cardiovascular responses to graded dynamic exercise (starting from supine rest [SR]) in three groups of people whose values for $\dot{V}O_{2max}$ (marked by *vertical dashed lines*) are very low (patients with "pure" mitral stenosis [MS]), normal (normally active [NA]), or very high (endurance athletes [ATH]). Note extreme differences in maximal cardiac outputs and stroke volumes and similarities in maximal heart rates and maximal arteriovenous oxygen (AV-O_2) differences. (Redrawn from Rowell LB: Human circulation: regulation during physical stress, New York, 1986, Oxford University Press.)

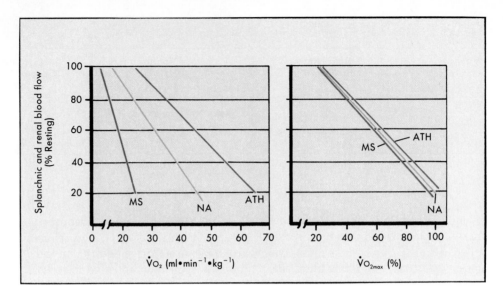

FIGURE 46-4   Splanchnic and renal blood flows as percentage of resting values (1500 and 1200 ml·min^{-1}, respectively) versus absolute O_2 uptake $(\dot{V}O_2)$ per Kg of body weight *(left panel)* and versus relative O_2 uptake, or percentage of $\dot{V}O_{2max}$ *(right panel)*. Same groups as in Figure 46-3. (Redrawn from Rowell LB: Human circulation: regulation during physical stress, New York, 1986, Oxford University Press.)

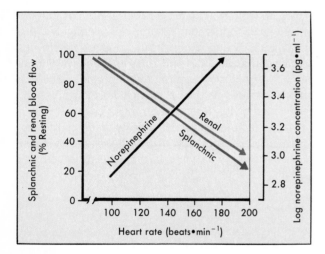

FIGURE 46-5   Splanchnic and renal blood flows (percentage of resting values) and log plasma norepinephrine concentration versus heart rate. Norepinephrine, derived mainly from neuronal leakage, reflects the progressive rise in sympathetic outflow to the heart to raise its rate and to visceral organs to reduce their blood flow after heart rate exceeds 100 beats·min^{-1}. (Redrawn from Rowell LB: Human circulation: regulation during physical stress, New York, 1986, Oxford University Press.)

TRANSITION FROM REST TO EXERCISE: INITIAL ADJUSTMENTS

Passive Effects of Standing Up: Orthostasis

The sudden changes in the circulation caused by moving from a supine to an upright posture stem from a gravitationally induced shift of about 600 ml of blood into veins situated below the heart (see Chapter 23). The human heart is typically about 1.2 to 1.5 m above the feet. Approximately 75% of the total blood volume is below the level of the heart and mostly in veins, which are 20 to 30 times more distensible than arteries. Initially the column of blood between the heart and feet is broken up by venous valves: thus most blood does not surge into the legs and cause syncope. This may happen rapidly, however, in patients with defective venous valves. Normally, continued flow of blood into dependent veins gradually fills them with blood. The rising transmural pressure distends the veins, opens their valves, and ultimately creates a continuous hydrostatic column of blood between the heart and feet. Blood that was previously in the thorax

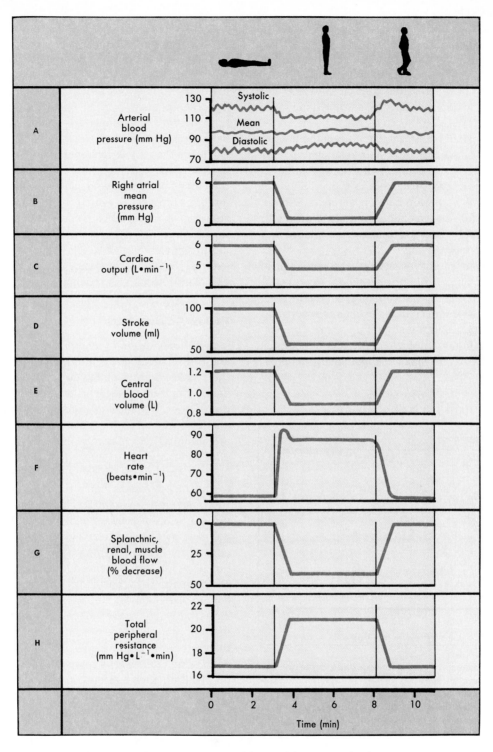

FIGURE 46-6 Cardiovascular responses to sudden upright posture (at 3 minutes, *middle column*) and then activating muscle pump by gently contracting leg muscles without movement (at 8 minutes, *right column*). The time course of the changes is approximate. Note the rise in heart rate does not compensate for the fall in stroke volume, but the rise in total peripheral resistance is sufficient to maintain arterial mean pressure. Note the fall in aortic pulse pressure. (Redrawn from Rowell LB: Human circulation: regulation during physical stress, New York, 1986, Oxford University Press.)

and provided the ventricles with a filling pressure of about 5 mm Hg shifts into pelvic and leg veins, causing ventricular filling pressure to fall close to zero and stroke volume to decrease by 40%, as shown in Figure 46-6.

This decrease in stroke volume reveals the unique importance of the Frank-Starling relationship (see Chapter 17) as a determinant of stroke volume in humans whose hearts are subject to great changes in filling pressure whenever body position is altered. Again, these are passive effects associated only with gravitational shifts of blood volume into a large and distensible system of veins.

If veins were as stiff as arteries, the problem of hydrostatic pooling would be eliminated. However, it would also eliminate the capacitance function of veins, that is, the large changes in volume associated with small changes in transmural pressure. This volume *reservoir* function of veins permits substantial loss of blood with only small changes in arterial pressure (Figure 46-7).

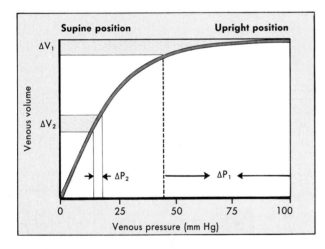

FIGURE 46-7 Schematic of volume-pressure curve for the human venous system showing its stiff characteristics in the legs during upright posture *(right portion of curve)* and its distensible nature during supine posture *(left portion of curve)*. The advantage of the stiff portion is that change in volume (Δv_1) is small despite a large rise in pressure (ΔP_1) on standing up. (In foot veins, pressure would rise from 15 to 100 mm Hg during standing). The advantage of the distensible characteristics in supine posture is that during hemorrhage, the loss of volume (note $\Delta v_2 = \Delta v_1$) would cause only about a 5 mm Hg fall in venous pressure. Thus the venous system can buffer volume losses. (Redrawn from Rowell LB: Human circulation: regulation during physical stress, New York, 1986, Oxford University Press.)

The shape of a venous volume-pressure curve is of fundamental importance. It represents a compromise in design to meet two requirements. The initial steep (high-compliance) portion used during supine posture allows adjustment to hemorrhage and also causes pooling problems on standing. Large volumes can leave the venous system with only small changes in distending pressure. In contrast, the flat region of the curve shows that veins become stiffer at high distending pressures, which helps limit pooling on standing. This venous volume–venous pressure relationship involves not only adjustments occurring during upright posture, but also those occurring during exercise, when maintenance of an adequate ventricular filling pressure is crucial.

Reflex Adjustments

Activity of stretch receptors in the carotid sinuses and the aortic arch (the arterial baroreceptors) and in the atrial and great pulmonary vessels (cardiopulmonary receptors) is reduced on standing. Because of the sudden declines in cardiopulmonary pressures and in aortic and carotid pulse pressures, baroreflexes are initiated (see Chapter 22), and mean pressure often does not fall. Parasympathetic (vagus) nerve fibers to the heart and sympathetic adrenergic fibers to the blood vessels constitute the efferent arm of these powerful reflexes. The inhibition of vagal activity starts immediately and causes heart rate to increase within one or two beats, whereas the vasoconstrictor effect of sympathetic activation on resistance vessels requires several seconds (see Figure 46-6). Sympathetic vasoconstriction reduces the blood flow to skin, skeletal muscle, splanchnic organs, and kidneys by 30% to 40%. Glomerular filtration rate falls in proportion to the decrease in renal blood flow, and therefore urine production is reduced. Regional vasoconstriction raises total peripheral resistance enough to maintain mean arterial pressure and may even increase it slightly despite the 20% decrease in cardiac output. The low central venous pressure and the diminished arterial pulse pressure increase sympathetic neural activity via the baroreflexes. This is not a stable situation, however; even though blood flow to the legs is reduced by vasoconstriction, venous pooling continues but much more slowly because of the vasoconstriction. Eventually the dependent veins fill with blood, causing stroke volume and cardiac output to fall to a point at which blood pressure can no longer be maintained and fainting occurs. For a time vaso-constriction maintains capillary hydrostatic pressures low enough to minimize outward filtration. When venous pressure rises in the dependent regions, however, the accompanying rise in capillary pressure causes net filtration and edema.

The veins could have an active, reflex role in limiting the volume of blood that can be pooled below the heart. By contracting their smooth muscle, veins are thought to decrease their diameter and actively expel blood, and their contracted walls become stiffer and less able to expand. However, most evidence shows that reflex constriction of veins in the arms and legs (where measurements can be made) does not accompany upright posture. Veins of the splanchnic region are richly innervated and can constrict when their sympathetic nerve supply is activated. Whether they constrict during orthostasis is not known.

Although splanchnic veins hold about 20% of the total blood volume, most of the blood that leaves these veins does so because of the purely passive effects of **vasoconstriction** on their transmural pressure; vasoconstriction refers to active constriction of arterioles and all resistance vessels, whereas venoconstriction refers to active constriction of veins. The constriction of arterioles decreases pressures downstream in the veins, and the elastic recoil of venous walls forces blood out and toward the right atrium, which has essentially zero pressure in upright individuals (see Figures 46-6 and 46-7). The important point is that vasoconstriction often mimics venoconstriction by reducing venous transmural pressure and thus decreasing venous volume, as shown by the steep portion of the curve in Figure 46-7. These same principles apply during exercise, hemorrhage, and other stresses during which vasoconstriction of large inactive regions passively reduces their venous volume. The displaced blood flows back to the heart and helps to maintain ventricular filling pressure.

Onset of Exercise

Mechanical Adjustments: Muscle and Respiratory Pumps Humans cannot remain upright without syncope unless they activate the muscle pump, which acts as a second heart on the venous, or return, side of the circulation. For this reason upright posture has been referred to as "exercise on a stationary base."

Figure 46-8 illustrates the operation of the muscle pump. Contraction of leg muscles compresses the veins, and because of their valves, blood is forced in only one direction—toward the heart. This pump can generate pressures as high as 90 mm Hg. Figure 46-6 shows that as soon as muscles contract, the central circulatory effects of orthostasis on thoracic blood volume and ventricular filling pressures are immediately reversed. Stroke volume, along with filling pressure or preload (Frank-Starling, or length-tension, effects), are rapidly restored to levels present during supine rest. Thereafter, cardiac output simply increases in proportion to heart rate with little or no further increase in stroke volume (see Figure 46-3). Activation of the muscle pump rapidly lowers the pressures in leg veins. As a result, capillary pressure is reduced to a level where reabsorption can proceed, and edema is reversed.

The respiratory pump is not nearly as effective as the muscle pump. Breathing changes the transmural pressures in the great veins that enter the thorax. A deep **inspiration** lowers intrathoracic pressure and thereby increases the pressure gradient from the extrathoracic inferior vena cava to the right atrium. This transiently increases blood flow to the heart.

Metabolic Adjustments Muscle contraction initiates extreme, rapid vasodilation in the active skeletal muscles resulting from the release of vasoactive metabolites from the active muscle cells. For example, the glycogen stored in muscle is a high-molecular-weight polymer that lacks osmotic effects. With exercise, glycogen breaks down into osmotically active glucose molecules. These molecules in turn are metabolized, and in the process they generate various osmotically active and vasoactive compounds. The rise in tissue osmolarity causes water to move into skeletal muscle cells, where osmotic pressure is increasing. Also, capillary hydrostatic pressure increases with rising blood pressure during exercise, causing water to move from the blood to the interstitial spaces.

Metabolic Vasodilators Mildly vasoactive substances in muscle include potassium and hydrogen ions (K^+, H^+), high CO_2, low O_2, adenosine, and most of the intermediates of glycolysis and the tricarboxylic-acid cycle. The combination of high K^+ and high osmolality with low O_2 tension is particularly effective in causing limited vasodilation. Most vasoactive substances (e.g., K^+ and osmotically active particles) appear only transiently in the venous blood that

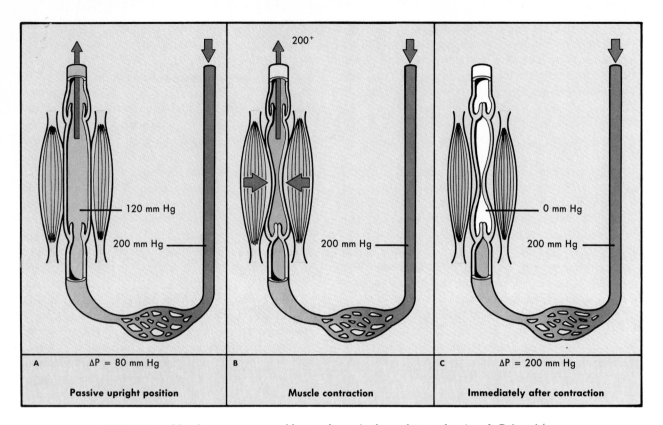

A ΔP = 80 mm Hg	B	C ΔP = 200 mm Hg
Passive upright position	**Muscle contraction**	**Immediately after contraction**

FIGURE 46-8   Muscle pump, or second heart, shown in three phases of action. **A,** Relaxed (upright posture). Summed dynamic and hydrostatic pressures are 200 mm Hg on the arterial and 120 mm Hg on the venous side of the capillary bed. **B,** Muscle contraction compresses the veins, blocking arterial inflow and creating high local venous pressures with venous valves causing unidirectional flow toward the heart. **C,** Immediately after contraction, pressure in emptied veins momentarily falls to zero (valves prevent back flow and back pressure), and net arterial driving pressure is momentarily raised to 200 mm Hg. ΔP, change in pressure. (Redrawn from Rowell LB: Human circulation: regulation during physical stress, New York, 1986, Oxford University Press.)

drains muscle; that is, they disappear in minutes even though muscle contraction and vasodilation are maintained. However, adenosine release from an active muscle persists. Vasodilation appears to be triggered by one mechanism but maintained by another.

No substances have been found that, when infused into the muscle, will simulate the enormous vasodilation observed with strong, voluntary muscular contractions. Normal patterns of motor unit recruitment plus an active muscle pump appear to be essential parts of the normal vasodilator process. Possibly the myogenic effects of compression and relaxation of the vessel walls during the contraction of the surrounding muscles contribute to the increased blood flow.

Reflex Adjustments   Despite the sudden, extreme vasodilation in active skeletal muscle just described, mean arterial pressure normally does not fall at the onset of exercise. A fall in mean arterial pressure is prevented by a reflex reduction in the activity of the rapidly responding parasympathetic nerves, which causes heart rate to increase. The decrease in mean arterial pressure also elicits a reflex increase in activity in the slower-acting sympathetic nerves, which not only raises heart rate but also increases vascular resistance and reduces blood flow in various organs. The magnitude of these effects increases with the intensity of exercise.

The rapid rise in cardiac output is aided by the inotropic action of the norepinephrine released from sympathetic nerve terminals in the myocardium (see

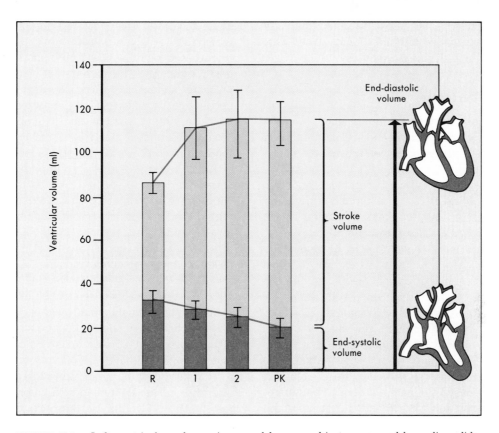

FIGURE 46-9   Left ventricular volumes in normal human subjects measured by radionuclide scintography during upright seated rest *(R)* and three loads of upright exercise classified as mild *(1)*, moderate *(2)*, and exhausting, or peak *(PK)*. Note the rise in end-diastolic volume with exercise and the progressive fall in end-systolic volume. Ejection fraction (end-systolic volume/end-diastolic volume) rose from 60% to 85%. (Redrawn from Poliner LR et al: Circulation 62:528, 1980.)

Chapter 18). (As already noted, in humans the adrenal medulla releases very little norepinephrine.) Norepinephrine, as with the epinephrine subsequently released from the adrenal medulla, affects the availability of calcium to the myocardial cells by raising calcium influx across the sarcolemma. The effect is an increase in stroke volume, a reduced end-systolic volume, or greater ejection fraction, resulting from more vigorous contraction (Figure 46-9).

Two important consequences of regional vasoconstriction are (1) it lessens the fall in total peripheral resistance when active muscle vessels dilate, and (2) it causes the passive displacement of blood volume away from the veins downstream from the constricted arterioles and displaces that blood back toward the heart. This is especially important in vascular regions such as the large splanchnic venous bed, which contains 20% of the total blood volume. Thus vasoconstriction assists the muscle pump to raise the ventricular preload. As mentioned, arteriolar vasoconstriction can resemble venoconstriction through its purely passive effects of lowering transmural downstream pressure and thus mobilizing blood volume in the downstream veins.

OVERALL CARDIOVASCULAR CONTROL DURING EXERCISE

After the initial 1 or 2 minutes of exercise, cardiovascular responses tend to stabilize or at least change more slowly with time. This section deals with the regulation of heart rate, stroke volume, total and regional oxygen extraction, and regional blood flow during maintained dynamic exercise.

Control of Heart Rate

Heart rate and cardiac output somehow increase in a highly predictable manner and in close proportion to the relative demands for oxygen transport made on the circulation. Little is known about the nature of the signals that govern these cardiovascular responses, that is, not only heart rate and cardiac output, but also blood pressure and peripheral resistance. These last two responses are discussed later.

Most of the rise in heart rate up to about 100 beats·min^{-1} is achieved by a withdrawal of vagal activity. Interestingly, at or near this heart rate the norepinephrine released from sympathetic nerve terminals begins to appear in the bloodstream. The concentration of norepinephrine rises sharply in the blood as the heart rate increases above 100 beats·min^{-1} (see Figure 46-5). Above about 100 beats·min^{-1} the increase in heart rate is achieved mainly by increased sympathetic drive. The norepinephrine released from the sympathetic nerves in the heart acts on β-adrenergic receptors on the automatic cells in the sinoatrial (SA) node.

Stroke Volume and Ventricular Performance

Figure 46-3 shows that during mild to maximal exercise, stroke volume is maintained at the high levels present during supine rest. Stroke volume may even increase slightly as heart rate approaches maximal values of 190 to 195 beats·min^{-1}. This means that the volume ejected with each heartbeat remains nearly constant despite a decrease in ventricular filling time from 0.55 second in an individual at rest with heart rate at 70 beats·min^{-1} to 0.12 second at a heart rate of 195 beats·min^{-1} when the individual is exercising. Examination of the events of the cardiac cycle (see Chapter 17), reveals the effect of an 80% reduction in cycle duration on ventricular volume. Clearly the speed of ventricular emptying and ventricular filling must increase dramatically to maintain stroke volume.

Extrinsic Factors Figure 46-9 shows how both extrinsic and intrinsic myocardial factors can affect ventricular volumes. The rise in end-diastolic volume when a person shifts from upright rest to exercise again illustrates the Frank-Starling relationship, in which the rise in ventricular filling pressure (*preload*) increases myocardial fiber length and causes a stronger contraction. The extent that ventricular filling pressure increases during exercise is not known because of the difficulty in separating the changes in pressure within the cardiac chambers from the breathing-induced changes in pressure of the intrathoracic regions surrounding the heart.

Aortic blood pressure is an important component of the ventricular *afterload*. This pressure opposes the left ventricular contraction and must be exceeded within the left ventricle before blood can be ejected into the aorta (see Chapter 17). Effects of afterload are often described in terms of velocity of contraction, called the force-velocity relationship. A large afterload decreases the velocity of contraction, so that stroke volume is reduced because ejection time is no longer sufficient for optimal cardiac ejection. A reduction in afterload has the opposite effect on stroke volume. A fitting analogy is that one can move a 5-pound weight faster than a 50-pound weight.

Normally, aortic mean pressure (or ventricular afterload) increases from approximately 90 mm Hg at rest to about 115 mm Hg during maximal exercise (Figure 46-10).

Intrinsic Factors Three alterations that occur during exercise can increase the intrinsic contractile state of the myocardium: (1) increased heart rate, (2) greater activity of cardiac sympathetic nerves, and (3) high circulating concentrations of epinephrine.

The effect of changing heart rate, or the interval between beats, on contractility is called the *interval-strength relationship* (see Chapter 18). The rate of contraction affects the quantity of calcium available to the myocardial cells. Contractility increases with an increase in heart rate because Ca^{++} enters the myocytes during the action potential plateau, and when heart rate increases, there are more plateaus per minute.

The norepinephrine released from cardiac sympathetic nerves and the epinephrine released from the adrenal medulla and carried to the heart by the circulation both have powerful positive inotropic effects through their activation of cardiac β$_1$-adrenergic receptors. The concentration of epinephrine rises abruptly when exercise is vigorous enough to require 75% or more of the $\dot{V}O_{2max}$ (Figure 46-11). Activation of the β$_1$-adrenergic receptors increases the Ca^{++} conductance of the myocytes during the action

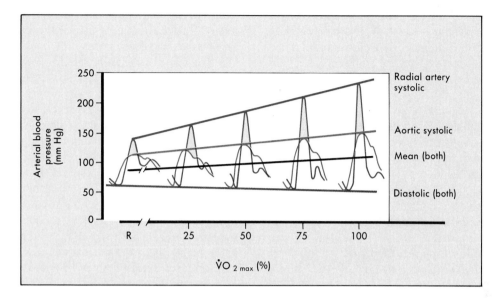

FIGURE 46-10 Simultaneously measured pressure waves from the aorta and the radial artery during rest *(R)* and upright exercise at four fractions of $\dot{V}O_{2max}$. Peripheral wave amplification caused large increases in radial arterial pressure. Note the small rise in mean arterial pressure. (Redrawn from Rowell LB: Human circulation: regulation during physical stress, New York, 1986, Oxford University Press.)

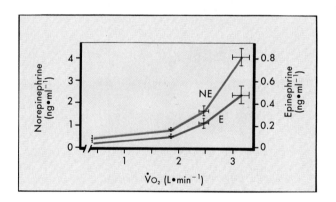

FIGURE 46-11 Central venous concentrations of norepinephrine and epinephrine at rest and at submaximal and maximal $\dot{V}O_2$.

potential plateau and thus tends to raise the intracellular concentration of Ca^{++}.

The importance of these intrinsic factors is revealed in Figure 46-9, which shows the progressive reduction in end-systolic volume as the intensity of exercise is increased. Because of a more vigorous contraction induced by the catecholamines, the left ventricle ejects up to 85% of its end-diastolic volume.

Oxygen Extraction and Arteriovenous Oxygen Difference

Arterial Oxygen Content Arterial O_2 content is determined by the concentration of hemoglobin and its O_2-binding capacity, by alveolar ventilation and alveolar PO_2, by pulmonary diffusion capacity, and finally, by the degree of any venous-arterial shunting (see Figure 46-1). The hemoglobin concentration rises by about 10% to 15% during intense exercise; this is attributable to plasma water loss into muscle cells. Nevertheless, the content of O_2 in the arterial blood remains quite constant over a wide range of O_2 uptakes, as shown in Figure 46-12. In subjects with a sharp rise in hemoglobin concentration and thus in the arterial O_2 capacity, the arterial saturation (O_2 content/O_2 capacity) may decrease slightly. This desaturation is sometimes great (e.g., less than 85% saturation) in endurance athletes with very high maximal cardiac outputs (and thus with very high pulmonary blood flows). The limitations that cause this desaturation are described later. At high cardiac output a small change in arterial O_2 content will greatly affect total O_2 transport.

Mixed Venous Oxygen Content As shown in Figure 46-12, the O_2 content in the pulmonary artery (or in the mixed venous blood draining all organs) falls progressively as the metabolic rate increases. O_2 content reaches values as low as 2 or 3 $ml \cdot dl^{-1}$ and approaches the low values observed in the femoral venous blood draining the exercising legs at levels of exercise that approach $\dot{V}O_{2max}$ (2 $ml \cdot dl^{-1}$ or lower). In fact, the O_2 content of the venous blood draining inactive regions (e.g., splanchnic, inactive limbs) approaches the low content in the femoral venous blood; the level falls within the shaded area in Figure 46-12. Total extraction of O_2 approaches 85% at exercise levels near $\dot{V}O_{2max}$. If only the legs are exercised, how does one explain the widened AV-O_2 difference across so many organs and the extraction of 85% of all the O_2 in blood? Two factors explain this: (1) O_2 extraction by active muscle increases greatly, and (2) the regional vasoconstriction that occurs reduces blood flow to other organs so that their O_2 uptake is maintained by increased O_2 extraction.

Many factors determine how much O_2 is extracted by active muscle. The rise in blood temperature and fall in blood pH in the muscle will shift the O_2 dissociation curve toward the right (see Chapter 27), so that blood holds less O_2 at a given PO_2 (Bohr shift). However, the most important factor is the number of open capillaries per muscle fiber. This capillary density determines the diffusion distances for O_2 between capillaries and muscle cells. Also, the mean transit time for red blood cells through the capillaries

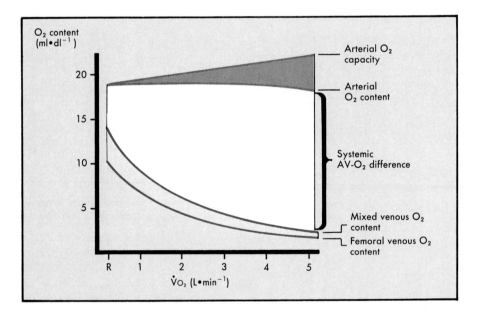

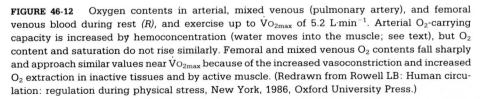

FIGURE 46-12   Oxygen contents in arterial, mixed venous (pulmonary artery), and femoral venous blood during rest *(R)*, and exercise up to $\dot{V}O_{2max}$ of 5.2 $L \cdot min^{-1}$. Arterial O_2-carrying capacity is increased by hemoconcentration (water moves into the muscle; see text), but O_2 content and saturation do not rise similarly. Femoral and mixed venous O_2 contents fall sharply and approach similar values near $\dot{V}O_{2max}$ because of the increased vasoconstriction and increased O_2 extraction in inactive tissues and by active muscle. (Redrawn from Rowell LB: Human circulation: regulation during physical stress, New York, 1986, Oxford University Press.)

must be long enough to permit the unloading of O_2. The mean transit time varies with the ratio of the capillary blood volume to the capillary blood flow.

In resting skeletal muscle, few capillaries are perfused and diffusion distances are great. Therefore O_2 extraction is not very efficient, and AV-O_2 difference is low. The mean transit time of the red cells through so few capillaries must be short because of the low capillary volume relative to the high resting flow (Figure 46-13). When muscle contracts, the number of perfused capillaries increases enormously and diffusion distances become shorter. Despite the high blood flow, the total capillary blood volume increases because of the greater number of open capillaries. Thus the mean transit time is sufficiently long to permit efficient O_2 transfer between red cells and muscle cells. A close positive linear relationship exists between the capillary/muscle fiber ratio and the magnitude of $\dot{V}O_{2max}$.

The second factor that explains the high total O_2 extraction is the regional vasoconstriction. This shifts blood flow away from those organs with high flow and

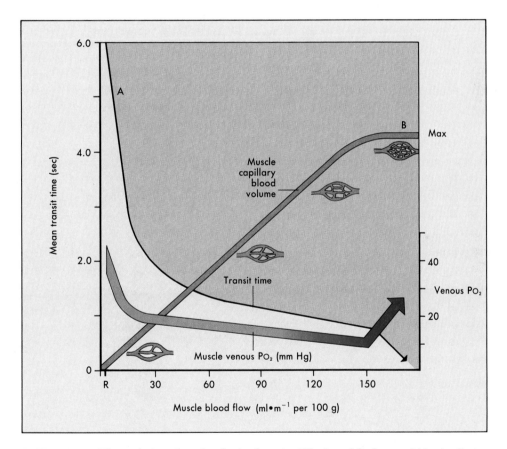

FIGURE 46-13   Theoretical explanation for inadequate diffusion of O_2 from red blood cells to active muscle during intense exercise. Line *A* shows estimated mean transit times for red cells through muscle capillaries with increasing blood flow. Line *B* shows the relative rise in muscle capillary blood volume caused by recruitment of more capillaries with increasing metabolism and blood flow. As long as capillary volume increases with blood flow, mean transit time is sufficiently long to permit equilibration of O_2 between capillaries and muscle fibers. When all capillaries are open and capillary volume can no longer increase, further increases in blood flow reduce mean transit time so that time to **unload** O_2 is too brief. O_2 extraction falls, and venous O_2 content **rises.** The same can occur in the lung. The effect there, however, is a **fall** in pulmonary venous and arterial O_2 content when mean transit time falls, since time to **load** O_2 is too brief. (Redrawn from Saltin B: Am J Cardiol 55:420, 1985.)

low O_2 extraction (e.g., splanchnic, renal, resting muscle, skin) and redirects that flow to regions that require high blood flow and high O_2 extraction during exercise (i.e., to active muscle). In this way most O_2 can be extracted from the blood.

Regional Blood Flow During Exercise: Cardiac Output Distribution

Visceral Organs Figures 46-4 and 46-5 describe the negative linear relationship between splanchnic and renal blood flows and the rise in O_2 uptake—up to maximal values. Note how these events change in close inverse relation to the rise in plasma norepinephrine concentration. Plasma renin activity also increases along with norepinephrine, presumably because of greater sympathetic outflow to the kidney (via a β-adrenergic mechanism). The increase in plasma renin activity is not associated with a fall in sodi-

um load to the kidney or with a fall in renal perfusion pressure, which are other events that can trigger renin release.

The amount of O_2 that can be "transferred" to active skeletal muscle simply by redirecting blood flow away from visceral organs through regional vasoconstriction is illustrated in Table 46-2. This example shows that a 77% reduction in splanchnic blood flow alone would make 1150 ml of blood and 230 ml of O_2 available to active muscle each minute. Splanchnic O_2 uptake is maintained at 60 $ml \cdot min^{-1}$ by an increase in the splanchnic AV-O_2 difference from 4 to 17 $ml \cdot dl^{-1}$. The same calculations apply to the kidney, if the renal blood flow is assumed to be 1200 $ml \cdot min^{-1}$. Such calculations show that 950 ml of blood flow or 190 ml of O_2 could be transferred from the kidneys to the skeletal muscles each minute. When the estimated changes in the skin and inactive muscle are added (these regions also

Table 46-2 Additional Oxygen and Blood Flow Provided by Regional Vasoconstriction During "Maximal" Exercise

SPLANCHNIC

Oxygen uptake ($ml \cdot min^{-1}$)	=	Blood flow ($ml \cdot min^{-1}$)	×	AV-O_2 difference ($ml \cdot dl^{-1}$)
Rest 60	=	1500	×	4
Exercise 60	=	350 $\Delta1150$	×	17
ΔOxygen available 230 $ml \cdot min^{-1}$	=	ΔBlood flow	×	Arterial O_2 content
	=	1150 $ml \cdot min^{-1}$	×	20 $ml \cdot dl^{-1}$

RENAL

ΔOxygen available 190 $ml \cdot min^{-1}$	=	950 $ml \cdot min^{-1}$	×	20 $ml \cdot dl^{-1}$

OTHER (SKIN, INACTIVE MUSCLE, ETC.)

ΔOxygen available 180 $ml \cdot min^{-1}$	=	900 $ml \cdot min^{-1}$	×	20 $ml \cdot dl^{-1}$

Totals
ΔOxygen available
600 $ml \cdot min^{-1}$

ΔBlood flow
3000 $ml \cdot min^{-1}$

Adapted from Rowell LB: Human circulation: regulation during physical stress, New York, 1986, Oxford University Press.

show vasoconstriction), then the total blood flow made available by vasoconstriction is approximately 3 $L \cdot min^{-1}$, which provides 600 ml of O_2 for use by the active muscles each minute. This explains how persons can extract 85% of all oxygen from the blood.

Without regional vasoconstriction, the systemic AV-O_2 difference would be less. $\dot{V}O_{2max}$ would also be less by an amount that would depend on the cardiac output. Loss of this regional vasoconstriction in patients with low cardiac outputs (see Figure 46-3) would reduce $\dot{V}O_{2max}$ by approximately 40%. In contrast, in endurance athletes $\dot{V}O_{2max}$ would be only about 10% lower because of their high maximal cardiac outputs. Thus the importance of regional vasoconstriction increases as maximal cardiac output decreases. About half the ability to increase O_2 uptake in patients with low cardiac output is provided by this sympathetically mediated redistribution of cardiac output.

Regional vasoconstriction also counteracts some of the decrease in total peripheral resistance during exercise; the effects of this also depend on the magnitude of maximal cardiac output. In patients with mitral stenosis, loss of this adjustment would cause a 33% reduction in arterial blood pressure. Conversely, in athletes with high maximal cardiac outputs, the decline in pressure would be negligible.

These extreme reductions in gastrointestinal and hepatic blood flow have no obvious consequences during heavy exercise done over minutes. Many metabolic functions of the liver and absorption of water from the gut remain normal. Digestion of food in the gut may be slowed because of the low blood flow. Despite its low blood flow, the liver releases large quantities of glucose into the blood during exercise. Increased concentration of catecholamines in the liver stimulate glycogenolysis. Since renal blood flow and glomerular filtration rate parallel one another, the effect of decreased renal blood flow is reduced urine formation and thus greater water conservation. However, when exercise is intense and prolonged, body temperature rises, and splanchnic and renal function can be severely impaired. This impairment stems from the cycle of rising organ O_2 uptake caused by increasing temperature in the face of falling blood flow, which in turn raises organ temperature even more. The consequences of this dangerous situation occur frequently in long-distance runners whose high motivation (and high endorphin levels?) enables

extreme exertion in hot environments. Prolonged splanchnic and renal ischemia causes injury to these organs and explains the subsequent gastrointestinal and renal malfunction.

Another important feature of splanchnic vasoconstriction is its effect on the distribution of blood volume. Vasomotor adjustments change vascular resistance, but they also regulate the distribution of blood volume and thereby help to maintain adequate ventricular filling pressure and thus cardiac performance. One serious consequence of preventing splanchnic vasoconstriction in exercise is that splanchnic blood flow and blood volume rise in proportion to increased arterial pressure. The splanchnic region is so compliant and sequesters so much blood that the volume of blood available for filling the heart is reduced; thus stroke volume and cardiac output can no longer be maintained. **The peripheral circulation affects cardiac performance by controlling the volume of blood available to fill the ventricles.**

Skin and Skeletal Muscle The circulation to skin and inactive skeletal muscle is reduced by the increase in sympathetic vasoconstrictor activity during the first minutes of exercise. Thereafter, blood flow to the skin gradually increases along with the rise in body temperature. The increased temperature is sensed by thermosensitive neurons in the hypothalamus and the spinal cord. The thermoregulatory reflexes activate a vasodilatory system that supplies cutaneous arterioles and causes skin blood flow to rise in proportion to the higher body temperature. The increased skin blood flow per degree rise in body core temperature is much less during exercise than at rest (Figure 46-14), presumably because the sympathetic vasoconstrictor outflow associated with exercise partly overrides sympathetic vasodilator outflow. This competition between vasodilation and vasoconstriction of skin diminishes heat loss. This accounts for part of the increase in body temperature during exercise. In short, skin is under the competing control of the thermoregulatory system and other reflexes that cause vasoconstriction. Thus the body temperature maintained during exercise is a result of competing homeostatic controls (regulation of temperature, pressure, flow, and volume distribution).

Sympathetic vasoconstrictor outflow to skeletal muscle also increases. This was once thought not to affect blood flow to active muscle, because the local metabolites released during activity either prevent the release of norepinephrine from the sympathetic

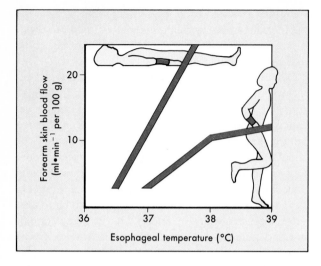

FIGURE 46-14   Skin blood flow per 100 g of human forearm during whole-body heating of supine resting subjects and during moderate upright exercise in hot conditions. The rate of increasing skin blood flow and the level reached is much greater in supine resting subjects in relation to body core temperature. In addition to the low rate of increasing skin blood flow during exercise, flow also approaches an upper limit as exercise is prolonged and body temperature rises to higher levels. This suggests that skin vasoconstriction is overpowering vasodilation in reducing blood flow. (Redrawn from Rowell LB: Human circulation: regulation during physical stress, New York, 1986, Oxford University Press.)

nerve endings or somehow blunt effects of norepinephrine on the α-receptors of the vascular smooth muscle cells. We now know that activation of sympathetic nerves to metabolically active regions can greatly reduce their blood flow. This response has been seen in both the coronary and the skeletal muscle circulations. Why is sympathetic nervous outflow to active skeletal muscle increased? Why is blood flow reduced to an organ already demanding most of the oxygen? These questions are discussed later. It suffices to say here that vasoconstriction must occur in active muscle to maintain blood pressure whenever muscle requires more blood flow than the heart can provide.

Heart and Brain Myocardial O_2 uptake and coronary blood flow increase sharply during exercise. Coronary extraction of O_2 is always so high that flow must increase in direct proportion to cardiac O_2 uptake. Myocardial O_2 requirements and coronary blood flow are most closely correlated with the product (the *pressure-rate product*) of heart rate and aortic

systolic pressure. That is, cardiac O_2 consumption is mainly determined by the number of times each minute the left ventricle must generate a given blood pressure. Surprisingly the evidence suggests that even the heart receives some sympathetic vasoconstrictor outflow during exercise, and that neural activity can significantly reduce total coronary blood flow. The current hypothesis is that this vasoconstriction is discretely directed to subepicardial vessels so as to redirect some of their blood flow to subendocardial vessels, which are more compressed by ventricular contractions and less well perfused during exercise.

Cerebral vessels are not subject to the constrictor effects of increased sympathetic nervous activity during exercise. Cerebral blood flow tends to increase only slightly, even when the exercise level approaches maximal exertion, despite the rise in blood pressure. This relative constancy of cerebral blood flow provides a good example of *autoregulation* (see Chapter 22).

Mechanisms of Regional Vasoconstriction

Sympathetic Nervous Activity Most evidence shows that the sympathetic nervous system is the primary cause of regional vasoconstriction. Increases in sympathetic nerve activity have been directly measured, even in humans, and blockade of these nerves or of their vascular α-adrenergic receptors blunts or abolishes regional vasoconstriction. Figure 46-5 shows that splanchnic and renal vasoconstriction are closely paralleled by increasing overflow of norepinephrine from the sympathetic nerve endings into the plasma. Nevertheless, other factors can contribute to the vasomotor adjustments. Such factors may serve to maintain the responses or to amplify the sympathetic neural effects.

Hormonal Factors Hormonal effects are particularly important under conditions of prolonged intense exercise, during which blood volume, body hydration, osmolality, metabolism, and temperature are changing.

1. *Renin-angiotensin system and aldosterone.* In general, plasma renin activity increases in parallel with plasma norepinephrine concentration. This reflects the increase in sympathetic nervous outflow to the kidneys and other organs when the exercise is strenuous. Renin is released from the kidney into blood, where it reacts with renin substrate (an α_2-

globulin) to form the inactive polypeptide, angiotensin I (see Chapters 33 and 39). In its passage through various organs, angiotensin I is converted to angiotensin II by an enzyme located in the vascular endothelium. Angiotensin II is a potent vasoconstrictor that acts directly on vascular smooth muscle. It also potentiates the effects of sympathetic vasoconstrictor activity by increasing norepinephrine release and by delaying its neuronal reuptake, thereby increasing norepinephrine availability at the adrenergic receptors. In addition to its vasomotor effects, angiotensin II stimulates the release of the mineralocorticoid hormone, aldosterone. This hormone increases the reabsorption of sodium by the kidneys, which in turn causes renal retention of water and expansion of plasma volume. This also diminishes water loss if the subject is sweating profusely. Although blockade of the renin-angiotensin system during brief, moderate exercise has no effect on cardiac output, blood pressure, and heart rate, this system's actions become important when fluid and electrolyte balance is disturbed, as in prolonged, intense exercise, which is usually accompanied by some dehydration.

2. *Vasopressin.* Vasopressin (antidiuretic hormone) is released from the posterior lobe of the neurohypophysis of the pituitary gland (see Chapter 39). Vasopressin is a potent vasoconstrictor; per mole, it is more potent than angiotensin II, which far exceeds the potency of norepinephrine. Vasopressin increases the reabsorption of water at the collecting tubules of the kidney and is thus an antidiuretic (or water-retaining) hormone.

Any release of vasopressin during brief periods of even maximal exercise is probably unimportant. Release of vasopressin, as with renin, becomes important whenever dehydration accompanies prolonged, intense exertion because of vasopressin's effects on salt and water retention. This would be especially true when warm temperatures cause profuse sweating and severe dehydration. The power of renin and vasopressin as vasoconstrictors becomes potentially important as the dehydrated runner develops hypotension because of the shrinkage of blood volume and because the sympathetic vasomotor system has already exerted its maximal effects. In prolonged, strenuous exercise, as in severe hemorrhage, these hormones can serve as essential backup mechanisms to sympathetic vasoconstrictor activity. Their additional vasoconstrictor effects could delay a disabling fall in blood pressure.

3. *Epinephrine and norepinephrine.* Figure 46-11 shows the increases in central venous blood concentrations of epinephrine and norepinephrine during mild to maximal exercise. The major source of circulating epinephrine is the chromaffin tissue of the adrenal medulla. Epinephrine concentrations become sufficiently high to contribute to the metabolic and hemodynamic responses to exercise. When epinephrine is infused into resting subjects, the metabolic rate, the rate of lipolysis, and free fatty acid concentrations increase significantly, yet the concentration in the blood increases only two- to threefold. At about 300 pg·ml^{-1} (10 times the resting level) epinephrine increases glycogenolysis in both liver and skeletal muscle. The liver releases glucose into the bloodstream. In muscle, glycolysis and lactate release are augmented (muscles do not release glucose). These levels of epinephrine are often observed during moderate exercise.

Hemodynamic responses also increase in proportion to epinephrine concentration. At the highest concentration seen during brief periods of maximal exercise (400 to 500 pg·ml^{-1}):

Heart rate increases by 24%.
Systolic blood pressure increases by 26%.
Mean arterial pressure decreases by 10%.
End-systolic volume decreases by 30%.
Stroke volume increases by 40%.
Cardiac output increases by 74%.
Total peripheral resistance decreases by 48%.
Left ventricular ejection fraction increases by 75%.

These changes reflect the inotropic (contractility) and chronotropic (rate) effects of epinephrine on the heart and also its β_2-adrenergic (vasodilator) effects on splanchnic and skeletal muscle vasculature (i.e., lowered mean arterial pressure and total peripheral resistance). One should not assume, however, that epinephrine is the principal cause of the similar changes that accompany exercise. Cardiac contractility and heart rate could just as easily be increased by the norepinephrine released through its sympathetic nerves. The vasodilation in skeletal muscle is presumably caused by local release of vasoactive metabolites. Nevertheless, when the ventricles have been denervated (experimentally in animals or after heart transplants in humans), excellent cardiac performance can be maintained during exercise. The loss of ability to raise heart rate after denervation is compensated for by the Frank-Starling mechanism (filling pressure and ventricular fiber length are increased because of the lower heart rate) and by the inotropic effects of epinephrine (more blood is ejected with each heartbeat).

Norepinephrine is a neurotransmitter, and increases in its plasma concentration during exercise and other stresses stem primarily from its spillover into blood after release from axon terminals of postganglionic sympathetic neurons. Skeletal muscle is the major source. Thus, unlike a hormone, transport to some target cells by the circulation is not required to explain norepinephrine's physiologic effects. In humans the adrenal medulla releases little norepinephrine. However, when sympathetic activity and neuronal spillover of norepinephrine are high, as during intense exercise, its concentrations reach plasma levels at which it can act as a hormone. As such, norepinephrine can contribute, along with epinephrine, to the β_1-adrenergic effects on the heart and can slightly oppose epinephrine's β_2-adrenergic (vasodilator) effects on the peripheral vasculature. When norepinephrine is infused into resting subjects, its "hormonal" effects appear at plasma concentrations that exceed $1000 \, pg \cdot ml^{-1}$. During maximal exercise, plasma norepinephrine concentrations typically reach $4000 \, pg \cdot ml^{-1}$, or about 12 to 15 times more than resting levels. Circulating norepinephrine probably does not exert any additional effect on regions already powerfully vasoconstricted by sympathetic nerves. Concentrations of norepinephrine in the synaptic junctions between nerve endings and blood vessels exceed (by orders of magnitude) the highest plasma concentrations of the transmitter after it has leaked into the bloodstream. Metabolic vasodilation in active muscle will oppose the weak effects of circulating norepinephrine. One possible target, however, could be skin vessels, which contain only α-adrenergic receptors. Skin vessels are vasoconstricted during maximal exercise, which explains the pallor as the subject becomes exhausted. However, this vasoconstriction could also originate from the neurogenic vasoconstriction so widespread during intense exercise.

LIMITS ON OXYGEN TRANSPORT

Figure 46-1 traces the cascade of partial pressures of oxygen progressing from the mouth down to the mitochondria of active skeletal muscle and points to four potential barriers that can limit oxygen transport.

Figure 46-15 summarizes the possible limitations to total oxygen consumption: restrictions in (1) pulmonary gas exchange, (2) maximal cardiac output, (3) peripheral vascular adjustments, and (4) the metabolic capacity of active muscle.

To discover the limitations in any system, one must stress it to the point at which some function is clearly impaired. Exercise has become a valuable diagnostic tool because of the way defects are amplified when demands are increased. Different limitations can exist in different individuals, especially when function is compromised by disease.

One can rule out the fourth potential barrier, metabolic limitations. If O_2 consumption was limited by the metabolic capacity of skeletal muscle, the addition of extra O_2 to blood (by having the subject breathe O_2-rich gas mixtures) would not be able to increase O_2 uptake; however, it does. Thus it is not the capacity of muscle to consume O_2 that limits the ability to raise total O_2 consumption. Furthermore, if the $\dot{V}O_{2max}$ was limited by the metabolic capacity of the muscle, simply bringing more muscles into play during maximal exercise would raise total O_2 uptake. That is, $\dot{V}O_{2max}$ would be a function of the active muscle mass; however, it is not. Once 40% to 50% of total muscle mass is maximally active, a true $\dot{V}O_{2max}$ is reached. If 20% to 30% more muscle becomes maximally active, this does not raise O_2 uptake further, despite the total demand for 20% to 30% more O_2. **Therefore the circulation cannot supply the oxygen required by the additional working muscles; demand exceeds supply.**

Also, direct biochemical measurements in human skeletal muscle biopsy specimens reveal that its oxidative capacity per unit mass of tissue far exceeds the capacity of the circulation to supply it with O_2 when more than a few kilograms of muscle are active. Exceptions do occur, however, in patients with metabolic diseases that reduce the capacity of tissues to consume O_2.

Limitations in Pulmonary Gas Exchange

In normal young people, ventilation rates are always adequate, and the gradients in PO_2 from the alveolus to the pulmonary capillaries and to arterial blood are only a few mm Hg. Thus arterial PO_2 and O_2 content are normally well maintained even at exercise levels that require $\dot{V}O_{2max}$. However, in some endurance athletes with a very high maximal cardiac out-

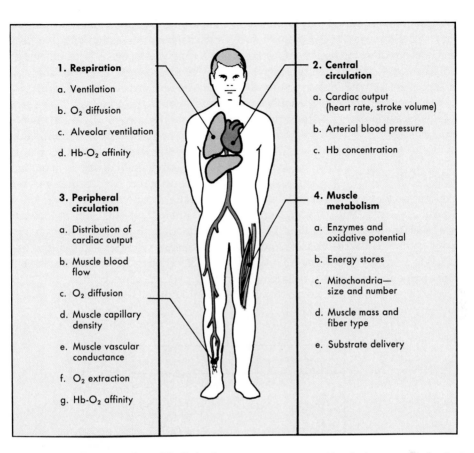

FIGURE 46-15   Summary of possible limitations to oxygen consumption during exercise. See text for discussion. (Redrawn from Rowell LB: Human Circulation: regulation during physical stress, New York, 1986, Oxford University Press.)

put and $\dot{V}O_{2max}$, arterial PO_2 declines sharply when the exercise level approaches $\dot{V}O_{2max}$, because pulmonary function does not keep pace with cardiovascular function. This condition occurs much more frequently in patients with pulmonary disease. One cause of this arterial desaturation in athletes is a relative hypoventilation, possibly stemming from mechanical limits to such high rates of ventilation. That is, the athletes can raise ventilation to such high levels that resistance to airflow in the airways of the lung becomes great enough to raise the cost of breathing and to limit further increases in ventilation. The major cause, however, and the reason for discussing it here, is the failure to attain diffusion equilibrium for O_2 in the pulmonary capillaries because of the very high blood flow and the brief mean transit time for the red cells through these capillaries. This limitation in the O_2 transport system is fundamental and can restrict

the unloading of O_2 in skeletal muscle as well (i.e., the same problem but in reverse).

Figure 46-13 illustrates the probable cause of the short mean transit time. In Figure 46-13 the unloading of O_2 in muscle is used as an example; the same points apply to O_2 loading in the lung. As blood flow increases, mean transit time (volume/flow) decreases only gradually because capillary blood volume rises with blood flow. However, when the capillaries are fully distended so that their volume can no longer increase, the mean transit time must fall sharply with further increases in blood flow. Consequently, blood does not remain in the pulmonary capillaries long enough for O_2 to diffuse from the alveolus to the red blood cells, or in skeletal muscle long enough for O_2 to diffuse from the red cells to the muscle cells. In exercising individuals the gradient for O_2 from the alveolus to the pulmonary capillaries increases from

5 mm Hg at rest to about 50 mm Hg at maximal exercise.

In some patients, severe diffusion limitations arise from pulmonary diseases that affect alveolar and pulmonary vascular structures. Well-ventilated regions sometimes are poorly perfused and poorly ventilated regions are overperfused, leading to ventilation/perfusion ($\dot{V}A/\dot{Q}$) inequalities.

Central Circulatory Limitations

Is the cardiac output truly maximal when the exercise is near the level requiring $\dot{V}o_{2max}$? If an upper limit on O_2 uptake is reached, and if that limit is far below the metabolic potential of active muscle, muscle blood flow (i.e., O_2 delivery) clearly imposes the limit on the O_2 consumption. Furthermore, addition of more muscle to that task raises neither cardiac output

nor O_2 uptake, as shown in Figure 46-16. This diagram shows the effect of adding heavy arm exercise at a time when severe leg exercise had raised cardiac output and oxygen uptake **almost** to maximal values. The added increments in cardiac output and O_2 uptake when the arms began exercising are far below those required by both limbs when exercised separately. The most striking feature is the sudden fall in leg blood flow after adding arm work; arterial blood pressure remained constant. If cardiac output is truly at its maximum, any further vasodilation or rise in vascular conductance (e.g., in the arms) must be counteracted by sufficient vasoconstriction to cause an equal and opposite change in conductance (e.g., in the legs). Otherwise, arterial blood pressure would fall; blood pressure equals cardiac output divided by total vascular conductance.

Another important example of a limitation in car-

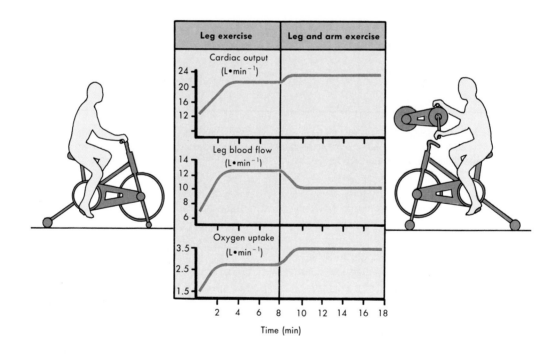

FIGURE 46-16   An experimental demonstration of the limited ability of the heart to provide blood flow to a large mass of active muscle. A large mass of muscle is heavily engaged (leg exercise, *left column*) so that cardiac output and O_2 uptake are close to maximal values. Additional engagement of another muscle group (arm plus leg exercise) raises cardiac output and O_2 uptake to maximal values, but this is not enough to supply the extra O_2 demanded by the arms. That is, the combined demands for blood flow exceed the pumping capacity of the heart. Active leg muscle (and arm muscle) has to vasoconstrict and decrease muscle blood flow (and O_2 uptake) in order to keep blood pressure from falling. (Redrawn from Rowell LB: Human circulation: regulation during physical stress, New York, 1986, Oxford University Press.)

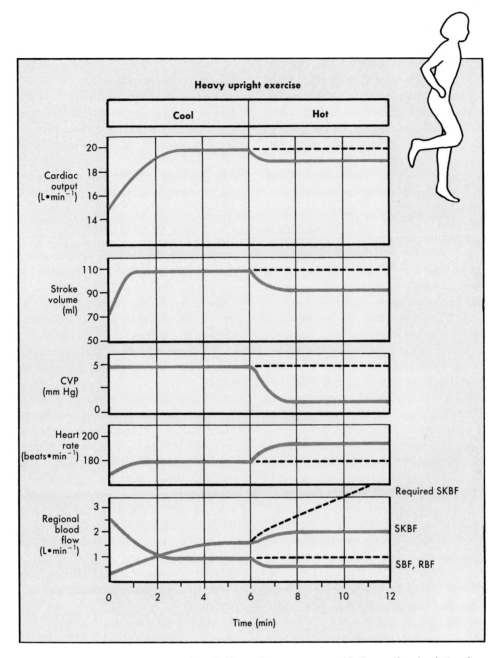

FIGURE 46-17   Schematic illustration of effects of cutaneous vasodilation on the circulation during heavy upright exercise. When skin arterioles dilate, a large cutaneous venous plexus fills with blood, lowering thoracic blood volume, central venous pressure *(CVP)*, and ventricular filling pressure. The fall in stroke volume is so large that the rise in heart rate is insufficient to maintain cardiac output. Additional vasoconstriction of splanchnic and renal beds *(bottom panel)* further reduce splanchnic blood flow *(SBF)* and renal blood flow *(RBF)*. This can only provide a few hundred additional milliliters per minute to skin, but total skin blood flow *(SKBF)* is still far below the level required for sufficient heat exchange. Body temperature rises. Cardiac output is not able to increase and meet combined needs of skin and working muscle.

diac output occurs when severe heat stress is imposed during maximal exercise. When cardiac output is at or near maximum, vasodilation of skin must be compensated for by vasoconstriction elsewhere. Figures 46-4 and 46-5 show that when the exercise level approaches Vo_{2max} under normothermic conditions, the splanchnic organs and kidneys are already maximally vasoconstricted. Thus they could not provide the additional blood needed to perfuse cutaneous vessels, which, when maximally vasodilated, can receive 7 $L \cdot min^{-1}$. Figure 46-17 shows that blood pressure must be maintained by vasoconstriction in active muscle. Thus Vo_{2max} would decrease in proportion to the blood flow and O_2 that are redistributed to the skin. Skin extracts little O_2, so the additional O_2 it receives is wasted in exchange for a greater heat loss.

The second adjustment is in the skin itself, which never vasodilates as much as it would if the same heat stress were applied when the subject was at rest (see Figure 46-14). That is, cutaneous vasodilation is counteracted by an increase in sympathetic vasoconstrictor activity, which is directed to cutaneous arterioles and causes enough vasoconstriction to reduce skin blood flow and lessen any fall in blood pressure. This is a dangerous situation because the cutaneous vasoconstriction reduces the rate of heat loss. This explains why hyperthermia and heat injury occur so often in endurance athletes who exercise strenuously in hot weather. When hyperthermia becomes severe enough to threaten the central nervous system, cutaneous vessels may dilate, which causes blood pressure to fall and precipitates a sudden collapse. Either way, regulation fails.

Thus far, all the available evidence indicates that cardiac pumping capacity is the major limitation in O_2 transport from the lungs to the tissues. Also, the heart cannot provide the blood flow needed by both the skin and muscle circulations during heavy exercise in hot environments.

Peripheral Circulatory Limitations

In addition to the limitations in cardiac pumping capacity just described, peripheral circulatory function can also limit O_2 uptake by the tissues, especially when diseases of the blood vessels or the autonomic nervous system are present. Two peripheral circulatory limitations are proposed most often:

1. O_2 delivery to the tissues, particularly skeletal muscle, is limited by the capacity of blood vessels to dilate.
2. O_2 delivery is limited by diffusion of O_2 from blood to the tissues, that is, capillary blood flow and blood volume, mean transit time, and diffusion distances (see Figures 46-13 and 46-15).

Figure 46-16 reveals that total demands for blood flow to muscle can exceed the pumping capacity of the heart. This clearly shows that large masses of muscle can vasodilate to a point at which the heart can no longer provide them with adequate blood flow. The only way to protect blood pressure is to increase sympathetic vasoconstrictor outflow to the active muscles to reduce their vasodilation. The potential disparity between maximal cardiac output and the highest blood flow requirements of active muscle can be seen by determining how much blood flow can increase in a small mass of muscle. This blood flow, expressed per 100 g of muscle multiplied by the total mass of active muscle (e.g., 15 kg or 50% of total muscle mass), reveals how much cardiac output would have to rise to meet the total need. For example, during maximal exertion of the human quadriceps muscle, which weighs 2 to 3 kg, its blood flow reaches 250 $ml \cdot min^{-1}$ per 100 g of muscle. This peak value for muscle blood flow is similar to that found in horses, dogs, and other mammals during intense, whole-body exercise. These animals, however, have maximal cardiac outputs that are two to three times greater per kilogram of body weight than those of humans. Humans could never perfuse muscle at these levels during severe, whole-body exercise such as running, in which 50% of the total muscle mass (or 15 kg in a 75 kg man) is active. To maintain blood pressure, the heart would have to pump 36 L of blood to active muscle each minute, plus an additional 3 to 4 L per minute to other organs. Maximal cardiac output for normal young people is usually about 25 L per minute (Figure 46-3). Since 40% to 50% of total muscle mass is normally activated during severe exercise to reach Vo_{2max}, one can conclude that the blood vessels in muscle cannot vasodilate up to their potential because the capacity of human hearts to pump blood is too small. Thus a mismatch exists between cardiac pumping capacity and the demands of large masses of active muscle for blood flow. Therefore to maintain blood pressure, this vasodilation must be counteracted by vasoconstriction in active muscle. This mismatch must be the critical limitation of whole-body exercise.

A second limitation in total O_2 transport involves

the diffusion of O_2 from the capillaries to the active muscle cells. With increased muscle activity, tissue O_2 uptake increases relatively more than does blood flow. Thus O_2 extraction by the muscle increases from 30% to 90%. This high extraction is made possible by an enormous increase in the number of perfused capillaries *(capillary recruitment),* which greatly increases the surface area and decreases the diffusion distances between the cells and capillaries. This capillary recruitment increases total capillary blood volume in parallel with the rising blood flow, as in Figure 46-13. Therefore the mean transit time decreases little, which allows adequate time for O_2 unloading. Since physical conditioning (see following discussion) increases the extraction of O_2 by muscle and also increases its capillary density, capillary density may have been one of the factors that limited Vo_{2max} before conditioning.

The distribution of cardiac output to the various organs does not usually limit O_2 transport because the progressive reduction in blood flow to nonexercising regions maximizes the extraction of O_2 all over the body. Loss of this regional vasoconstriction, as in autonomic nervous system disorders, would reduce the efficiency of O_2 extraction and thereby reduce Vo_{2max} proportionally.

Cardiac performance is greatly influenced by the peripheral circulation (see Chapter 23). Therefore the vasomotor system can affect cardiac function by modulating cardiac preload and afterload. As mentioned earlier, failure to vasoconstrict the highly compliant or distensible splanchnic regions during exercise would increase the splanchnic blood volume. This would decrease stroke volume and cardiac output by decreasing the thoracic blood volume and the ventricular filling pressure. Vasodilation of the skin circulation results in the same consequences. The cutaneous venous plexus, as with the splanchnic venous system, is large and highly compliant. As body temperature rises during exercise, cutaneous blood vessels are reflexly vasodilated and cutaneous venous volume increases. The result is an immediate decrease in ventricular filling pressure and stroke volume. The tachycardia associated with moderate exercise in a hot environment compensates for the fall in stroke volume, and therefore cardiac output does not fall. However, the tachycardia does not provide enough additional cardiac output to supply the skin. Intense vasoconstriction of visceral organs is required to transfer additional blood flow to the skin

and also to compensate for some of the blood volume that has moved into the cutaneous veins.

The larger the volume of blood close to the skin surface, the greater are the rates of heat loss and the lower is the ventricular filling pressure. Gradually a competition develops between skin and muscle for cardiac output. Since the heart cannot provide enough blood flow for both regions, body temperature rises. This illustrates the sensitivity of the human cardiovascular system to the interorgan distribution of blood volume and flow. The relative distribution of cardiac output between highly compliant regions, which sequester volume when their flow rises (i.e., splanchnic and cutaneous), and noncompliant or essentially "stiff" regions (e.g., skeletal muscle) profoundly affects the performance of the heart. This interplay between cutaneous vasodilation and ventricular performance is the most serious problem in human temperature regulation. Often the heart cannot keep pace with the demands during exercise.

In summary, the limitations in O_2 transport and Vo_{2max} are primarily in the cardiac pumping capacity and secondarily in the diffusion of O_2 from the alveoli to the blood and from the blood to the tissue. Since O_2 uptake is limited by O_2 transport and not by the metabolic capacity of the tissues, the Vo_{2max} is a true measure of the functional capacity of the cardiovascular system.

Physical Conditioning

Cardiovascular Adjustments Physical conditioning increases the O_2 transport capacity and the Vo_{2max}. The major factors that limit these variables have been identified, and these are clearly the ones that must be altered by conditioning. First, a higher maximal cardiac output is achieved by an increase in stroke volume; maximal heart rate is not increased. This typically accounts for 50% of the increase in Vo_{2max}. The mechanism for the increase in stroke volume is unknown. Since ejection fraction normally reaches 85% in maximal exercise, a further improvement in contractility is unlikely to occur with conditioning. Some claim an increase in blood volume augments ventricular filling pressure, but direct evidence is lacking. Clearly, however, if mechanical constraints of the pericardium on cardiac filling are lessened, stroke volume could increase greatly. After the pericardium has been cut, maximal stroke volume increases substantially in dogs, and indirect evidence indicates this may occur in certain patients after cardiac surgery.

The other half of the adjustment in $\dot{V}o_{2max}$ results from the increase in the systemic AV-O_2 difference. This increase is not ascribable to a more efficient distribution of cardiac output during exercise; inactive regions are similarly constricted in conditioned and unconditioned people at the same relative intensity of exercise. Improved O_2 extraction is attributable to the increase in muscle capillary density and the associated decrease in diffusion distances. The number of capillaries per muscle fiber increases. Despite the rise in muscle blood flow, adequate mean transit time is apparently preserved by the rise in capillary blood volume. Diffusion of O_2 is also facilitated by an increase in muscle myoglobin concentration. Also, more O_2 is unloaded at the muscle at a given Po_2. This results from the greater rise in temperature and fall in pH (Bohr effect) associated with the higher intensities of exercise required to achieve $\dot{V}o_{2max}$.

Metabolic Adjustments

Physical conditioning increases the activity of oxidative enzymes in skeletal muscle by 40%, but this does not ordinarily contribute to the increase in $\dot{V}o_{2max}$. However, this adaptation can explain the 300% increase in the capacity for prolonged heavy exercise. This great metabolic adjustment occurs despite a rise in $\dot{V}o_{2max}$ of only 15% and despite the absence of significant changes in muscle blood flow at submaximal levels of O_2 uptake. The increased activity of oxidative enzymes means that a much greater fraction of the pyruvate formed during glycolysis can be oxidized to CO_2 and water in the mitochondria. Also, more of the active acetate needed for oxidative reactions of the tricarboxylic-acid cycle can be provided by the oxidation of free fatty acids. This spares muscle glycogen and enables the muscle to oxidize a greater quantity of lipids, the major form of energy storage.

The cardiovascular and metabolic adjustments just described are reversed quickly within a few days to a few weeks by deconditioning.

MATCHING CIRCULATORY AND METABOLIC FUNCTIONS

What signals initiate the close matching between blood flow and metabolism during exercise? What makes heart rate, blood pressure, and ventilation rise? Surprisingly, these problems are still unsolved, but three important theories have emerged.

The first theory states that centrally generated neural activity, called motor commands (*central command*), simultaneously activates cardiovascular and skeletal muscle motor systems. These command signals may originate from motor neurons in the motor cortex, the basal ganglia, the cerebellum, or the spinal cord. The activation of the cardiovascular system appears to be proportional to the number of motor units recruited to maintain a given force. For example, if muscle force generation is impaired by neuromuscular blockade, more motor units must be recruited to maintain a given force. The cardiovascular responses are augmented together with the increased motor command and the greater perception of effort. Several other tests also support the idea that central command signals help to establish the cardiovascular responses to exercise.

According to a second theory, the signals that match the cardiovascular and metabolic responses originate from the chemosensitive group III and IV afferent nerve fibers, which are densely distributed within skeletal muscle. The central idea here is that any accumulation of metabolites within the muscle because of a mismatch between blood flow and metabolism, would activate these fibers and trigger reflexes that act to restore muscle blood flow. This scheme applies as well to conditions in which blood flow is too high and excessive washout of metabolites would reduce feedback from muscle. The theory is especially attractive, because most of the cardiac output is distributed to active skeletal muscle during exercise. It is now clear that powerful, pressure-raising reflexes originate from these group III and IV afferent fibers when muscle blood flow falls below a critical level, which causes lactate and other metabolites to accumulate. The reflex occurs in patients with atherosclerotic arterial disease; a restriction of blood flow to active muscle causes ischemia and a rise in blood pressure followed by severe pain called *intermittent claudication*. The reflex also seems to be important during intense exercise, when the ratio of metabolism to blood flow is very high. However, the reflex apparently is not active during mild exercise, when this ratio is so low that substantial changes in flow cause little accumulation of metabolites. One must remember that these chemoreflexes, as with the baroreflexes, act as pressure-raising reflexes, but they respond to reduced blood flow instead of reduced blood pressure.

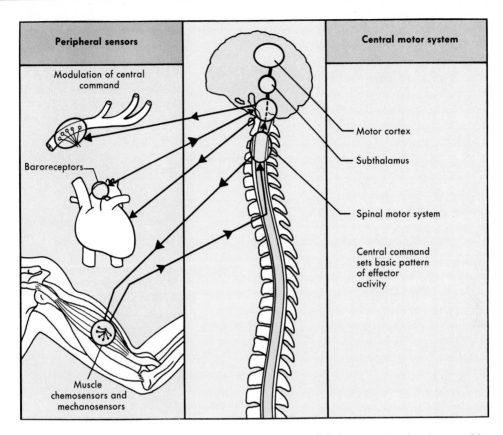

FIGURE 46-18   Proposed scheme for cardiovascular control during exercise showing possible interaction between centrally generated motor signals (central command) and cardiovascular signals presumed to set basic patterns of effector activity. These signals are in turn modified by reflexes from baroreceptors and chemosensitive (chemosensors) and mechanosensitive (mechanosensors) nerves in skeletal muscle as appropriate error signals develop. (Adapted from Rowell LB: Human circulation: regulation during physical stress, New York, 1986, Oxford University Press.)

The third theory states that arterial baroreflexes originating from mechanoreceptors in the carotid sinus and aortic arch would correct any imbalance between the fall in muscle vascular resistance and the rise in blood flow and thus would maintain blood pressure. Baroreflexes were once thought to be unimportant, because blood pressure does not fall when muscle blood vessels dilate at the onset of exercise and because blood pressure rises **above** resting levels during exercise. It is now clear that the arterial pressure at which the baroreflex is most sensitive to changes in pressure (called the operating point, or *set point*) is somehow shifted upward toward a higher pressure during exercise. Thus the arterial pressure is closely regulated at this new level (this is often called "resetting" the set point). Baroreflexes do not, as once thought, become insensitive to perturbations during exercise. This means that during severe whole-body exercise, when muscle vasodilation starts to overwhelm the pumping capacity of the heart, a decline in blood pressure below its regulated level will trigger a baroreflex. A powerful, pressure-raising reflex clearly is needed to constrict arterioles in active muscle and keep blood pressure from falling.

However, the maintenance of blood pressure in the situation just described could also originate from a muscle chemoreflex triggered by a fall in muscle blood flow. Also, if blood flow were insufficient to meet the metabolic needs of the muscle, the muscle would rapidly fatigue, and more motor units would need to be recruited to maintain force. This means an increased central motor command. Thus all three theories are probably valid, and each mechanism most likely contributes to overall cardiovascular regulation during intense exercise. Muscle chemoreflexes may not be operating during mild exercise.

Figure 46-18 proposes a scheme for cardiovascular control during exercise. The central command or centrally generated somatomotor and cardiovascular motor signals set the basic patterns of effector activity, that is, cardiac, vagal, and sympathetic activity and sympathetic vasomotor activity. This activity is then modulated (1) by muscle chemoreflexes, which provide a flow-sensitive, blood pressure–raising reflex, and (2) by arterial baroreceptors, which provide pressure-sensitive, blood pressure–modulating reflexes. Feedback from the mechanosensitive group III fibers in the muscle may also provide important feedback about the rate and magnitude of muscle force development.

BIBLIOGRAPHY
Journal Articles

Amberson WR: Physiologic adjustments to the standing posture, Maryland Univ Sch Med Bull 27:127, 1943.

Blackmon JR et al: Physiological significance of maximal oxygen intake in pure mitral stenosis, Circulation 36:497, 1967.

Brengelmann GL: Circulatory adjustments to exercise and heat stress, Annu Rev Physiol 45:191, 1983.

Dempsey JA: Is the lung built for exercise? Med Sci Sports Exerc 18:143, 1986.

Dempsey JA et al: Exercise-induced arterial hypoxemia in healthy persons at sea-level, J Physiol (Lond) 355:161, 1984.

Eldridge FL et al: Stimulation by central command of locomotion, respiration, and circulation during exercise, Respir Physiol 59:313, 1985.

Goldstein DS et al: Relationship between plasma norepinephrine and sympathetic neural activity, Hypertension 5:552, 1983.

Hainsworth R: Vascular capacitance: its control and importance, Rev Physiol Biochem Pharmacol, 105:101, 1986.

Milvy P, editor: The marathon: physiological, medical, epidemiological, and psychological studies, Ann NY Acad Sci 301, 1977.

Moreno AH and Burchell AR: Respiratory regulation of splanchnic and systemic venous return in normal subjects and in patients with hepatic cirrhosis, Surg Gynecol Obstet 154:257, 1982.

Poliner LR et al: Left ventricular performance in normal subjects: a comparison of the responses to exercise in the upright and supine positions, Circulation 62:528, 1980.

Pollack AA and Wood EH: Venous pressure in the saphenous vein at the ankle in man during exercise and changes in posture, J Appl Physiol 1:649, 1949.

Robinson BF et al: Control of heart rate by the autonomic nervous system: studies in man on the interrelation between baroreceptor mechanisms and exercise, Circ Res 19:400, 1966.

Rowell LB: Human cardiovascular adjustments to exercise and thermal stress, Physiol Rev 54:75, 1974.

Rowell LB: Reflex control of the cutaneous vasculature, J Invest Dermatol 69: 154, 1977.

Rowell LB et al: Splanchnic blood flow and metabolism in heat-stressed man, J Appl Physiol 24:475, 1968.

Saltin B: Hemodynamic adaptations to exercise, Am J Cardiol 55:42D, 1985.

Secher NH et al: Central and regional circulatory effects of adding arm exercise to leg exercise, Acta Physiol Scand 100:288, 1977.

Shepherd JT et al: Static (isometric) exercise: retrospection and introspection, Circ Res 48(suppl 1):179, 1981.

Stratton JR et al: Hemodynamic effects of epinephrine: concentration-effect study in humans, J Appl Physiol 58:1199, 1985.

Wagner PD et al: Pulmonary gas exchange in humans exercising at sea level and simulated altitude, J Appl Physiol 61:260, 1986.

Books and Monographs

Blomqvist CG and Stone HL: Cardiovascular adjustments to gravitational stress. In Handbook of physiology, section 2: The cardiovascular system—peripheral circulation and organ blood flow, vol III, Bethesda, Md, 1983, American Physiological Society.

Feigl EO: Coronary circulation. In Patton HD et al, editors: Textbook of physiology, Philadelphia, 1989, WB Saunders Co.

Gauer OH and Thron HL: Postural changes in the circulation. In Handbook of physiology, section 2: Circulation, vol III, Washington, DC, 1965, American Physiological Society.

Johnson RH et al: Neurocardiology: the interrelationships between dysfunction in the nervous and cardiovascular systems, London, 1984, WB Saunders Co.

Katz AM: Physiology of the heart, New York, 1977, Raven Press.

Rothe CF: Venous system: physiology of the capacitance vessels. In Handbook of physiology, section 2: The cardiovascular system—peripheral circulation and organ blood flow, vol III, Bethesda, Md, 1983, American Physiological Society.

Rowell LB: Cardiovascular adjustments to thermal stress. In Handbook of physiology, section 2: The cardiovascular system—peripheral circulation and organ blood flow, vol III, Bethesda, Md, 1983, American Physiological Society.

Rowell LB: Human circulation: regulation during physical stress, New York, 1986, Oxford University Press.

Rowell LB and Johnson JM: Role of the splanchnic circulation in reflex control of the cardiovascular system. In Sheperd AP and Granger DN, editors: Physiology of intestinal circulation, New York, 1984, Raven Press.

Saltin B and Gollnick PG: Skeletal muscle adaptability: significance for metabolism and performance. In Handbook of physiology, section 10: Skeletal muscle, Bethesda, Md, 1980, American Physiological Society.

Shepherd JT: Circulation to skeletal muscle. In Handbook of physiology, section 2: The cardiovascular system—peripheral circulation and organ blood flow, vol II, Bethesda, Md, 1983, American Physiological Society.

Sparks HV Jr: Effect of local metabolic factors on vascular smooth muscle. In Handbook of physiology, section 2: The cardiovascular system—vascular smooth muscle, vol II, Bethesda, Md, 1980, American Physiological Society.

West JB: Respiratory physiology—the essentials, Baltimore, 1974, Williams & Wilkins.

Index